T0297422

Studies in Computational Intelligence

Volume 622

Series editor

About this Series

The series "Studies in Computational Intelligence" (SCI) publishes new developments and advances in the various areas of computational intelligence—quickly and with a high quality. The intent is to cover the theory, applications, and design methods of computational intelligence, as embedded in the fields of engineering, computer science, physics and life sciences, as well as the methodologies behind them. The series contains monographs, lecture notes and edited volumes in computational intelligence spanning the areas of neural networks, connectionist systems, genetic algorithms, evolutionary computation, artificial intelligence, cellular automata, self-organizing systems, soft computing, fuzzy systems, and hybrid intelligent systems. Of particular value to both the contributors and the readership are the short publication timeframe and the worldwide distribution, which enable both wide and rapid dissemination of research output.

More information about this series at http://www.springer.com/series/7092

Van-Nam Huynh · Vladik Kreinovich
Songsak Sriboonchitta
Editors

Causal Inference in Econometrics

Editors
Van-Nam Huynh
School of Knowledge Science
Japan Advanced Institute of Science and Technology
Ishikawa
Japan

Songsak Sriboonchitta
Faculty of Economics
Chiang Mai University
Chiang Mai
Thailand

Vladik Kreinovich
Department of Computer Science
University of Texas at El Paso
El Paso, TX
USA

ISSN 1860-949X ISSN 1860-9503 (electronic)
Studies in Computational Intelligence
ISBN 978-3-319-27283-2 ISBN 978-3-319-27284-9 (eBook)
DOI 10.1007/978-3-319-27284-9

Library of Congress Control Number: 2015956358

Printed on acid-free paper

This Springer imprint is published by SpringerNature
The registered company is Springer International Publishing AG Switzerland

Preface

Econometrics is the application of mathematical, statistical, and computational methods to economic data. Econometrics adds empirical content to economic theory, allowing theories to be tested and used for forecasting and policy evaluation.

The ultimate goal of economics in general—and of econometrics in particular—is not only to describe the economic phenomena, but also to improve them, i.e., to make production, distribution, and consumption of goods and services more efficient and more fair. To be able to effectively control economic phenomena, it is important to understand the causal relation between them. In view of this importance, the main emphasis of this volume is on casual inference in econometrics.

Analysis of causal inference is one of the most difficult tasks in data analysis in general and in analyzing economic data in particular: when we observe that two phenomena are related, it is often difficult to decide whether one of these phenomena causally influences the other, or whether these two phenomena have a common cause.

To get a good understanding of causal inference, it is important to have models of economic phenomena which are as accurate as possible. It is therefore important not only to further improve traditional econometric models, but also to consider nontraditional economic models, such as Computable General Equilibrium (CGE) models (that properly take into account non-economic factors such as government regulations and tax policy), fuzzy models (that take into account expert knowledge formulated in imprecise natural-language terms), and models obtained by using nonparametric techniques of machine learning (in particular, neural networks) and data mining, techniques that uncover the general dependencies from the data itself—instead of the usual assumption that the model belongs to a certain predefined parametric family of models.

This volume contains several state-of-the-art papers which are directly or indirectly related to causal inference in econometrics. Some of these papers directly deal with causal inference. Others deal with models that seem promising for the future analysis of causal inference. These papers provide theoretical analysis of the corresponding mathematical, statistical, computational, and economical models.

Several other papers describe applications of the related econometric models and techniques to real-life economic situations.

We hope that this versatile volume will help practitioners to learn how to apply new econometric techniques, and help researchers to further improve the existing models and to come up with new ideas on how to best detect and analyze causality in economics.

We want to thank all the authors for their contributions and all anonymous referees for their thorough analysis and helpful comments.

The publication of this volume is partly supported by the Chiang Mai School of Economics (CMSE), Thailand. Our thanks to Dean Pisit Leeahtam and CMSE for providing crucial support. Our special thanks to Prof. Hung T. Nguyen for his valuable advice and constant support.

We would also like to thank Prof. Janusz Kacprzyk (Series Editor) and Dr. Thomas Ditzinger (Senior Editor, Engineering/Applied Sciences) for their support and cooperation in this publication.

January 2016

Van-Nam Huynh
Vladik Kreinovich
Songsak Sriboonchitta

Contents

Part I
Fundamental Theory

Validating Markov Switching VAR Through Spectral Representations

Monica Billio and Maddalena Cavicchioli

Abstract We develop a method to validate the use of Markov Switching models in modelling time series subject to structural changes. Particularly, we consider multivariate autoregressive models subject to Markov Switching and derive close-form formulae for the spectral density of such models, based on their autocovariance functions and stable representations. Within this framework, we check the capability of the model to capture the relative importance of high- and low-frequency variability of the series. Applications to U.S. macroeconomic and financial data illustrate the behaviour at different frequencies.

1 Introduction

Issues about persistence and frequency variability of time series are often raised in macroeconomics and finance. In particular, a large literature in econometrics has developed parametric tools to capture the low-frequency behaviour of some time series, which exhibit strong persistence in time. According to these recent findings, the persistency characterizing some time series should be taken into account when modelling the series as a non-linear model. In this paper we propose a method to validate the use of Markov Switching (MS) models through the use of their spectral density functions. We first apply some new tools recently proposed in Cavicchioli [1] to detect the presence of structural changes in the data. Then we derive close-form formulae for the spectral representations of Markov Switching VAR processes which are necessary to evaluate high- and low-frequency variability of time series. The aim is twofold: from one side, we investigate non-linear features of the data to correctly specify the parametric model, from the other, we check the correct specification

M. Billio
Department of Economics, Università Cà Foscari, Cannaregio 873, 30121 Venezia, Italy
e-mail: billio@unive.it

M. Cavicchioli (✉)
Department of Economics, University of Modena and Reggio Emilia,
Viale Berengario 51, 41121 Modena, Italy
e-mail: maddalena.cavicchioli@unimore.it

V.-N. Huynh et al. (eds.), *Causal Inference in Econometrics*,
Studies in Computational Intelligence 622, DOI 10.1007/978-3-319-27284-9_1

analyzing their frequency contents through the spectral density function. If the empirically detected persistency is captured by the chosen parametrization, then we can be more confident in the application of our parametric model. Our results are related to the work of Krolzig [6] in terms of state space representation and stable representation and to the paper of Pataracchia [7] where a different Markovian representation has been considered. However, note that in the latter paper, it is assumed that the constant term (which is also governed by Markov chain) is zero. Here we find more general and useful expressions. Thus our primary interest is to test non-linearity in the data, study their behaviour at different frequencies through spectral functions and validate the chosen model in relation with its empirical counterpart. The plan of the paper is the following. In Sect. 2 we study the spectral density functions of MS VAR(0) and VAR(p) processes in close-form, both from their switching state-space representations and from stable VARMA representations. Section 3 investigates the presence of structural changes in real data. Then we check the ability of the chosen models to capture high- and low-frequency variability, using arguments from Sect. 2. Section 4 concludes. Derivation of some formulae can be found in the Appendix.

2 Spectra of Markov Switching VAR

2.1 The Case of Hidden Markov Process

Let us consider the model

$$\mathbf{y}_t = \boldsymbol{\nu}_{s_t} + \boldsymbol{\Sigma}_{s_t}\mathbf{u_t} \tag{2.1}$$

where $\mathbf{u}_t \sim IID(\mathbf{0}, \mathbf{I}_K)$, that is, $E(\mathbf{u}_t) = \mathbf{0}$, $E(\mathbf{u}_t\,\mathbf{u}_t^{'}) = \mathbf{I}_K$ and $E(\mathbf{u}_t\,\mathbf{u}_\tau^{'}) = \mathbf{0}$ if $t \neq \tau$. Furthermore, $\mathbf{y}_t$, $\boldsymbol{\nu}_{s_t}$ and $\mathbf{u}_t$ are $K \times 1$, $\boldsymbol{\Sigma}_{s_t}$ is $K \times K$ and (s_t) follows an irreducible and ergodic M-state Markov chain. Let $\mathbf{P} = (p_{ij})_{i,j=1,\ldots,M}$ be the transition matrix of the chain, where $p_{ij} = Pr(s_t = j|s_{t-1} = i)$. Ergodicity implies the existence of a stationary vector of probabilities $\boldsymbol{\pi} = (\pi_1 \ldots \pi_M)^{'}$ satisfying $\boldsymbol{\pi} = \mathbf{P}^{'}\boldsymbol{\pi}$ and $\mathbf{i}_M^{'}\boldsymbol{\pi} = 1$, where $\mathbf{i}_M$ denotes the $(M \times 1)$ vector of ones. Irreducibility implies that $\pi_m > 0$ for $m = 1, \ldots, M$, meaning that all unobservable states are possibile. An useful representation for (s_t) is obtained by letting $\boldsymbol{\xi}_t$ denote a random $(M \times 1)$ vector whose mth element is equal to unity if $s_t = m$ and zero otherwise. Then the Markov chain follows a VAR(1) process

$$\boldsymbol{\xi}_t = \mathbf{P}^{'}\boldsymbol{\xi}_{t-1} + \mathbf{v}_t$$

where $\mathbf{v}_t = \boldsymbol{\xi}_t - E(\boldsymbol{\xi}_t|\boldsymbol{\xi}_{t-1})$ is a zero mean martingale difference sequence.

Consequently, we have the following standard properties ($h > 0$):

$$E(\boldsymbol{\xi}_t) = \boldsymbol{\pi} \qquad E(\boldsymbol{\xi}_t\boldsymbol{\xi}_t^{'}) = \mathbf{D} = \mathrm{diag}(\pi_1 \ldots \pi_M)$$
$$E(\boldsymbol{\xi}_t\boldsymbol{\xi}_{t+h}^{'}) = \mathbf{DP^h} \qquad \mathbf{v}_t \sim IID(\mathbf{0}, \mathbf{D} - \mathbf{P}^{'}\mathbf{DP})$$

Define $\boldsymbol{\Lambda} = (\boldsymbol{\nu}_1 \dots \boldsymbol{\nu}_M)$ and $\boldsymbol{\Sigma} = (\boldsymbol{\Sigma}_1 \dots \boldsymbol{\Sigma}_M)$. We get a first state space representation of (2.1)

$$\begin{cases} \mathbf{y}_t = \boldsymbol{\Lambda}\boldsymbol{\xi}_t + \boldsymbol{\Sigma}(\boldsymbol{\xi}_t \otimes \mathbf{I}_K)\mathbf{u}_t \\ \boldsymbol{\xi}_t = \mathbf{P}'\boldsymbol{\xi}_{t-1} + \mathbf{v}_t \end{cases} \tag{2.2}$$

In fact, for $s_t = m$, $\boldsymbol{\xi}_t = \mathbf{e}_m$ the mth column of the identity matrix $\mathbf{I}_M$. So we get

$$\begin{aligned} \mathbf{y}_t &= \begin{pmatrix} \boldsymbol{\nu}_1 & \dots & \boldsymbol{\nu}_M \end{pmatrix} \begin{pmatrix} 0 \\ \vdots \\ 1 \\ \vdots \\ 0 \end{pmatrix} + \begin{pmatrix} \boldsymbol{\Sigma}_1 & \dots & \boldsymbol{\Sigma}_M \end{pmatrix} \begin{pmatrix} \mathbf{0} \\ \vdots \\ \mathbf{I}_K \\ \vdots \\ \mathbf{0} \end{pmatrix} \mathbf{u}_t \\ &= \boldsymbol{\nu}_m + \boldsymbol{\Sigma}_m \mathbf{I}_K \mathbf{u}_t = \boldsymbol{\nu}_m + \boldsymbol{\Sigma}_m \mathbf{u}_t. \end{aligned}$$

The transition equation in (2.2) differs from a stable linear VAR(1) process by the fact that one eigenvalue of $\mathbf{P}'$ is equal to one, and the covariance matrix is singular due to the adding-up restriction. For analytical purposes, a slightly different formulation of the transition equation in (2.2) is more useful, where the identity $\mathbf{i}_M'\boldsymbol{\xi}_t = 1$ is eliminated; see Krolzig [6], Chap. 3. This procedure alters the state-space representation by using a new $(M-1)$-dimensional state vector

$$\boldsymbol{\delta}_t = \begin{pmatrix} \xi_{1,t} - \pi_1 \\ \vdots \\ \xi_{M-1,t} - \pi_{M-1} \end{pmatrix}.$$

The transition matrix $\mathbf{F}$ associated with $\boldsymbol{\delta}_t$ is given by

$$\mathbf{F} = \begin{pmatrix} p_{1,1} - p_{M,1} & \dots & p_{M-1,1} - p_{M,1} \\ \vdots & & \vdots \\ p_{1,M-1} - p_{M,M-1} & \dots & p_{M-1,M-1} - p_{M,M-1} \end{pmatrix}.$$

The eigenvalues of $\mathbf{F}$ are less than 1 in absolute value. Here the relations

$$\xi_{M,t} = 1 - \sum_{m=1}^{M-1} \xi_{mt} \qquad \pi_M = 1 - \sum_{m=1}^{M-1} \pi_m$$

have been used. Then we have

$$\boldsymbol{\xi}_t - \boldsymbol{\pi} = \mathbf{P}'(\boldsymbol{\xi}_{t-1} - \boldsymbol{\pi}) + \mathbf{v}_t$$

hence

$$\boldsymbol{\delta}_t = \mathbf{F}\,\boldsymbol{\delta}_{t-1} + \mathbf{w}_t$$

where

$$\mathbf{w}_t = (\mathbf{I}_{M-1} \qquad -\mathbf{i}_{M-1})\mathbf{v}_t.$$

This gives a second (unrestricted) state-space representation

$$\mathbf{y}_t = \boldsymbol{\Lambda}\boldsymbol{\pi} + \boldsymbol{\Lambda}(\boldsymbol{\xi}_t - \boldsymbol{\pi}) + \boldsymbol{\Sigma}((\boldsymbol{\xi}_t - \boldsymbol{\pi}) \otimes \mathbf{I}_K)\mathbf{u}_t + \boldsymbol{\Sigma}(\boldsymbol{\pi} \otimes \mathbf{I}_K)\mathbf{u}_t$$

hence

$$\begin{cases} \mathbf{y}_t = \boldsymbol{\Lambda}\boldsymbol{\pi} + \widetilde{\boldsymbol{\Lambda}}\boldsymbol{\delta}_t + \widetilde{\boldsymbol{\Sigma}}(\boldsymbol{\delta}_t \otimes \mathbf{I}_K)\mathbf{u}_t + \boldsymbol{\Sigma}(\boldsymbol{\pi} \otimes \mathbf{I}_K)\mathbf{u}_t \\ \boldsymbol{\delta}_t = \mathbf{F}\,\boldsymbol{\delta}_{t-1} + \mathbf{w}_t \end{cases} \tag{2.3}$$

where

$$\widetilde{\boldsymbol{\Lambda}} = (\boldsymbol{\nu}_1 - \boldsymbol{\nu}_M \ldots \boldsymbol{\nu}_{M-1} - \boldsymbol{\nu}_M) \qquad\qquad \widetilde{\boldsymbol{\Sigma}} = (\boldsymbol{\Sigma}_1 - \boldsymbol{\Sigma}_M \ldots \boldsymbol{\Sigma}_{M-1} - \boldsymbol{\Sigma}_M).$$

We then have the following standard properties:

$$\begin{aligned} &E(\boldsymbol{\delta}_t) = 0 && E(\boldsymbol{\delta}_t\boldsymbol{\delta}_t^{'}) = \widetilde{\mathbf{D}} \\ &E(\boldsymbol{\delta}_t\boldsymbol{\delta}_{t+h}^{'}) = \widetilde{\mathbf{D}}(\mathbf{F}^{'})^h, \quad h > 0 && \mathbf{w}_t \sim IID(\mathbf{0}, \widetilde{\mathbf{D}} - \mathbf{F}\widetilde{\mathbf{D}}\mathbf{F}^{'}) \end{aligned}$$

where

$$\widetilde{\mathbf{D}} = \begin{pmatrix} \pi_1(1-\pi_1) \ldots & -\pi_1\pi_{M-1} \\ \vdots & \vdots \\ -\pi_{M-1}\pi_1 \ \ldots & \pi_{M-1}(1-\pi_{M-1}) \end{pmatrix}.$$

The autocovariance function of the process $(\mathbf{y}_t)$ in (2.3) is given by

$$\begin{aligned} &\Gamma_{\mathbf{y}}(0) = \widetilde{\boldsymbol{\Lambda}}\widetilde{\mathbf{D}}\widetilde{\boldsymbol{\Lambda}}^{'} + \widetilde{\boldsymbol{\Sigma}}(\widetilde{\mathbf{D}} \otimes \mathbf{I}_K)\widetilde{\boldsymbol{\Sigma}}^{'} + \boldsymbol{\Sigma}((\mathbf{D}\mathbf{P}_\infty) \otimes \mathbf{I}_K)\boldsymbol{\Sigma}^{'} \\ &\Gamma_{\mathbf{y}}(h) = \widetilde{\boldsymbol{\Lambda}}\mathbf{F}^h\widetilde{\mathbf{D}}\widetilde{\boldsymbol{\Lambda}}^{'}, \qquad h > 0 \end{aligned}$$

where $\mathbf{D}\mathbf{P}_\infty = \boldsymbol{\pi}\boldsymbol{\pi}^{'}$ and $\mathbf{P}_\infty = \lim_n \mathbf{P}^n = \mathbf{i}_M\boldsymbol{\pi}^{'}$. The multivariate spectral matrix describes the spectral density functions of each element of the state vector in the diagonal terms. The off-diagonal terms are defined cross spectral density functions and they are typically complex numbers. Here we are only interested in the diagonal terms. Therefore, we can compute them, without loss of generality, considering the summation

$$F_{\mathbf{y}}(\omega) = \sum_{h=-\infty}^{+\infty} \Gamma_{\mathbf{y}}(|h|)e^{-i\omega h}$$

where the frequency ω belongs to $[-\pi, \pi]$; see also Pataracchia [7] where a different spectral representation was obtained. Since the spectral radius $\rho(\mathbf{F})$ of $\mathbf{F}$ is less than 1, the spectral density matrix of the process $(\mathbf{y}_t)$ in (2.3) is given by

$$F_{\mathbf{y}}(\omega) = Q + 2\widetilde{\boldsymbol{\Lambda}}\mathbf{F}\mathcal{R}e\{(\mathbf{I}_{M-1}e^{i\omega} - \mathbf{F})^{-1}\}\widetilde{\mathbf{D}}\widetilde{\boldsymbol{\Lambda}}' \tag{2.4}$$

where $\mathcal{R}e$ denotes the real part of the complex matrix $(\mathbf{I}_{M-1}e^{i\omega} - \mathbf{F})^{-1}$, and

$$Q = \widetilde{\boldsymbol{\Lambda}}\widetilde{\mathbf{D}}\widetilde{\boldsymbol{\Lambda}}' + \widetilde{\boldsymbol{\Sigma}}(\widetilde{\mathbf{D}} \otimes \mathbf{I}_K)\widetilde{\boldsymbol{\Sigma}}' + \boldsymbol{\Sigma}((\mathbf{D}\mathbf{P}_\infty) \otimes \mathbf{I}_K)\boldsymbol{\Sigma}'.$$

Complete derivation of Formula (2.4) is given in the Appendix. An alternative approach to the same problem is based on a stable representation of (2.3). Set $\mu_{\mathbf{y}} = \boldsymbol{\Lambda}\boldsymbol{\pi}$. From (2.3) we get

$$\boldsymbol{\delta}_t = F(L)^{-1}\mathbf{w}_t$$

where $F(L) = \mathbf{I}_{M-1} - \mathbf{F}\,L$ (here L is the lag operator). Substituting this relation into the measurement equation in (2.3) yields

$$|F(L)|(\mathbf{y}_t - \mu_{\mathbf{y}}) = \widetilde{\boldsymbol{\Lambda}}F(L)^*\mathbf{w}_t + \widetilde{\boldsymbol{\Sigma}}(F(L)^*\mathbf{w}_t \otimes \mathbf{I}_K)\mathbf{u}_t + |F(L)|\boldsymbol{\Sigma}(\boldsymbol{\pi} \otimes \mathbf{I}_K)\mathbf{u}_t$$

where $F(L)^*$denotes the adjoint matrix of $F(L)$ and $|F(L)|$ is the determinant of $F(L)$. Thus we get a stable VARMA(p^*, q^*) representation of the process $(\mathbf{y}_t)$ in (2.3)

$$\phi(L)(\mathbf{y}_t - \mu_{\mathbf{y}}) = \theta(L)\boldsymbol{\epsilon}_t \tag{2.5}$$

where $p^* = q^* \leq M - 1$, $\phi(L) = |F(L)|$ is scalar and

$$\theta(L) = (\widetilde{\boldsymbol{\Lambda}}F(L)^* \quad \widetilde{\boldsymbol{\Sigma}}(F(L)^* \otimes \mathbf{I}_K) \quad |F(L)|\mathbf{I}_K).$$

See also Cavicchioli [1], Theorem 6. The error term is also given by

$$\boldsymbol{\epsilon}_t = (\mathbf{w}_t' \quad \mathbf{u}_t'(\mathbf{w}_t' \otimes \mathbf{I}_K) \quad \mathbf{u}_t'(\boldsymbol{\pi}' \otimes \mathbf{I}_K)\boldsymbol{\Sigma}')'$$

with variance matrix

$$\Xi = Var(\boldsymbol{\epsilon}_t) = \mathrm{diag}(\widetilde{\mathbf{D}} - \mathbf{F}\widetilde{\mathbf{D}}\mathbf{F}', (\widetilde{\mathbf{D}} - \mathbf{F}\widetilde{\mathbf{D}}\mathbf{F}') \otimes \mathbf{I}_K, \boldsymbol{\Sigma}((\mathbf{D}\mathbf{P}_\infty) \otimes \mathbf{I}_K)\boldsymbol{\Sigma}').$$

Using (2.5) the spectral density matrix of the process $(\mathbf{y}_t)$ in (2.3) is also given by

$$F_{\mathbf{y}}(\omega) = \frac{\theta(e^{i\omega})\Xi\theta'(e^{-i\omega})}{|\phi(e^{i\omega})|^2}.$$

In fact, we can apply a well-known result (see, for example, Gourieroux and Monfort [4], Chap. 8, Formula 8.3, p. 257). The spectral density of a VARMA process

$$\Phi(L)\mathbf{y}_t = \Theta(L)\boldsymbol{\epsilon}_t,$$

with $Var(\boldsymbol{\epsilon}) = \boldsymbol{\Omega}$, is given by

$$F_{\mathbf{y}}(\omega) = \frac{1}{2\pi}\Phi^{-1}(exp(i\omega))\Theta(exp(i\omega))\boldsymbol{\Omega}\overline{\Theta(exp(i\omega))'}\ \overline{\Phi^{-1}(exp(i\omega))'} \tag{2.6}$$

This formula can be applied when $\det \Phi(z)$ has all its roots outside the unit circle. Moreover, we can also write $F_{\mathbf{y}}(\omega)$ as

$$F_{\mathbf{y}}(\omega) = \frac{1}{2\pi}\frac{\Phi^*(exp(i\omega))\Theta(exp(i\omega))\boldsymbol{\Omega}\overline{\Theta(exp(i\omega))'}\ \overline{\Phi^*(exp(i\omega))'}}{|\det \Phi(exp(i\omega))|^2}$$

where Φ^* denotes the adjoint matrix of Φ. Here, we apply these formulae ignoring the coefficient. Written in this form $F_{\mathbf{y}}(\omega)$ is a matrix whose elements are rational functions of $exp(i\omega)$. This property is a characteristic of the VARMA process.

2.2 *The Case of MS VAR(p)*

Let us consider the MS VAR(p), $p > 0$, process

$$A(L)\mathbf{y}_t = \boldsymbol{\nu}_{s_t} + \boldsymbol{\Sigma}_{s_t}\mathbf{u}_t \tag{2.7}$$

where $A(L) = \mathbf{I}_K - \mathbf{A}_1 L - \cdots - \mathbf{A}_p L^p$ is a $(K \times K)$-dimensional lag polynomial. Assume that there are no roots on or inside the unit circle of the complex plane, i.e., $|A(z)| \neq 0$ for $|z| \leq 1$. Reasoning as above, the process $(\mathbf{y}_t)$ in (2.7) admits a stable VARMA(p^*, q^*) with $p^* \leq M + p - 1$ and $q^* \leq M - 1$:

$$\Psi(L)(\mathbf{y}_t - \mu_{\mathbf{y}}) = \theta(L)\boldsymbol{\epsilon}_t \tag{2.8}$$

where $\Psi(L) = |F(L)|A(L) = \phi(L)A(L)$ and $\theta(L)\boldsymbol{\epsilon}_t$ is as in (2.4). If we want the autoregressive part of the stable VARMA in (2.8) to be scalar, we have to multiply (2.8) on the left with the adjoint $A(L)^*$ to give a stable VARMA(p', q') representation, where the bounds satisfy $p' \leq M + Kp - 1$ and $q' \leq M + (K-1)p - 1$. Thus the spectral density matrix of the process $(\mathbf{y}_t)$ in (2.8) is given by

$$\begin{aligned} F_{\mathbf{y}}(\omega) &= \frac{A^{-1}(e^{i\omega})\theta(e^{i\omega})\Xi\theta'(e^{-i\omega})[A'(e^{-i\omega})]^{-1}}{|\phi(e^{i\omega})|^2} \\ &= \frac{A^*(e^{i\omega})\theta(e^{i\omega})\Xi\theta'(e^{-i\omega})A^{*'}(e^{-i\omega})}{|\phi(e^{i\omega})|^2|\det A(e^{i\omega})|^2}. \end{aligned}$$

From the above section we can also obtain the matrix expression

$$F_{\mathbf{y}}(\omega) = A^{-1}(e^{i\omega})Q[A'(e^{-i\omega})]^{-1} + 2A^{-1}(e^{i\omega})\widetilde{\boldsymbol{\Lambda}}\mathbf{F} \\ \times \mathcal{R}e\{(\mathbf{I}_{M-1}e^{i\omega} - \mathbf{F})^{-1}\}\widetilde{\mathbf{D}}\widetilde{\boldsymbol{\Lambda}}'[A'(e^{-i\omega})]^{-1}. \tag{2.9}$$

A similar result can be obtained for a Markov Switching VAR(p), $p > 0$, process

$$A_{s_t}(L)\mathbf{y}_t = \boldsymbol{\nu}_{s_t} + \boldsymbol{\Sigma}_{s_t}\mathbf{u}_t \tag{2.10}$$

where we assume that the state variable is independent of the observables.

Define

$$\mathbf{A}(L) = (A_1(L) \dots A_M(L))$$

where

$$A_m(L) = \mathbf{I}_K - \mathbf{A}_{1,m}L - \cdots - \mathbf{A}_{p,m}L^p$$

for $m = 1, \dots, M$. Recall that $s_t \in \{1, \dots, M\}$. Then (2.10) can be written in the form

$$\mathbf{A}(L)(\boldsymbol{\xi}_t \otimes \mathbf{I}_K)\mathbf{y}_t = \boldsymbol{\Lambda}\boldsymbol{\xi}_t + \boldsymbol{\Sigma}(\boldsymbol{\xi}_t \otimes \mathbf{I}_K)\mathbf{u}_t.$$

Assume that $B(L) = \mathbf{A}(L)(\boldsymbol{\pi} \otimes \mathbf{I}_K)$ is invertible. Then the spectral density matrix of the process $(\mathbf{y}_t)$ in (2.10) is given by

$$F_{\mathbf{y}}(\omega) = \frac{B^{-1}(e^{i\omega})\theta(e^{i\omega})\Xi\theta'(e^{-i\omega})[B'(e^{-i\omega})]^{-1}}{|\phi(e^{i\omega})|^2}. \tag{2.11}$$

Finally, we can also obtain the matrix expression

$$F_{\mathbf{y}}(\omega) = B^{-1}(e^{i\omega})Q[B'(e^{-i\omega})]^{-1} + 2B^{-1}(e^{i\omega})\widetilde{\boldsymbol{\Lambda}}\mathbf{F} \\ \times \mathcal{R}e\{(\mathbf{I}_{M-1}e^{i\omega} - \mathbf{F})^{-1}\}\widetilde{\mathbf{D}}\widetilde{\boldsymbol{\Lambda}}'[B'(e^{-i\omega})]^{-1}. \tag{2.12}$$

3 Frequency Variability in Real Data

A recent paper by Müller and Watson [5] has proposed a framework to study how successful are time series models in explaining low-frequency variability. In fact, some econometric models (local-to-unity or fractional) were specifically designed to capture low-frequency variability of the data. However, they provide reliable guidance for empirical analysis only if they are able to accurately describe not only low-frequency behaviour of the time series, but also high-frequency. In particular, the authors focus on lower frequencies than the business cycle, that is a period greater than eight years, and some inference is proposed on the low-frequency component of the series of interest by computing weighted averages of the data, where the weights are low-frequency trigonometric series. We propose to look at the relative importance of low- and high-frequencies in a time series from a different prospective.

We assume that a suitable parametric model should be able to capture the relative importance of the different frequencies which characterize the behaviour of the series. Our aim is to study some empirical questions of interest from Müller and Watson [5] with a different approach. In particular, we firstly use recent test from Cavicchioli [1] to correctly parametrize the process we are considering. Then, by using simple Maximum Likelihood Estimation (MLE) expressions from Cavicchioli [2] we proceed to estimate the model. Finally, using spectral density results presented in Sect. 2, we check if the chosen model is able to extract frequency variability of the initial process. Following Müller and Watson [5], we investigate the following questions: (1) after accounting for a deterministic linear trend, is real gross domestic product (GDP) consistent with a I(1) model? (2) is the term spread consistent with the I(0) model, that is are long term and short term interest rates cointegrated? (3) are absolute daily returns, which are characterized by "slow decay of autocorrelations" consistent with an I(1) or I(0) model? We study those questions allowing the possibility that those series may be affected by structural changes and, if it is the case, they should be taken into account when fitting a model on the data. We take postwar quarterly U.S. data and focus on a period greater than 32 quarters, that is frequencies lower than the business cycle, as in Müller and Watson [5]. In particular, we consider quarterly values (1952:Q1–2005:Q3) of the logarithm of de-trended real GDP and de-meaned term spread—difference between interest rates for 10 years and 1 year U.S Treasury bonds. Moreover, we observe daily absolute returns (January 2nd, 1957–September 30th, 2013) computed as the logarithm of the ratio between consecutive closing prices from S&P500. Data are taken from the FRED database. Before proceeding with our analysis, we plot sample periodograms of the data in order to have a preliminary idea on the different behaviour of the series. In Fig. 1 we recognize a mixed pattern of low- and high-frequency cycles for real GDP and bond spread which produces uncertainty on the relative importance of the two components. On the contrary, we recognize the explosion at the low-frequency in the periodogram of absolute returns, as we expected from "long-memory" considerations. To correctly estimate the process, the first step is to test linearity or non-linearity of the model and, if it is case, the number of regimes which characterizes the time series. For the determination of regime number, we use results from Cavicchioli [1].

With regard to the real GDP, we select a linear model, that is one regime is sufficient to describe the data. We include one lag for the autoregressive model (as suggested by standard information criteria for the AR model) and plot its spectral density in Fig. 2 (upper panel). The spectra of this model is typical of an autoregressive model; here low-frequencies are the most important, giving credit to an I(1) model. However, if we take the first differences of the series, we somehow depurate the process from the stochastic trend (not only from the linear deterministic one, as before). Here the test suggests a MS(2) AR(1) model and the spectra in Fig. 2 (lower panel) retains only high-frequency movements. It suggests that the long-run pattern characterized by two phases of the economy is captured by the switching model.

When considering the Treasury bond spread, a 2-state switching model is selected. Thus, we estimate a MS(2) AR(1) model which turns to be as follows

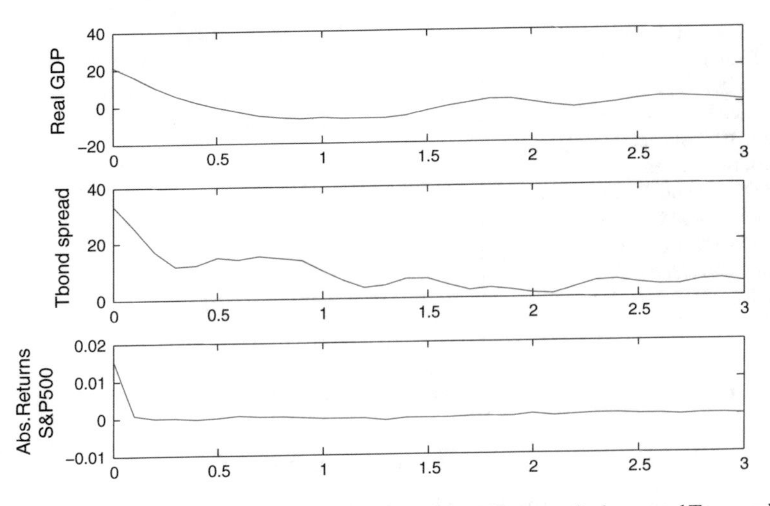

Fig. 1 Periodograms of logarithm of detrended U.S. real GDP (*top panel*), demeaned Treasury bond spread as the difference between interest rates for 10 years and 1 year U.S Treasury bonds (*middle panel*); those two series have quarterly frequency and the period is from 1952:Q1 to 2005:Q3. Absolute daily returns (*bottom panel*) as the logarithm of the ratio between consecutive closing prices from S&P500 (January 2nd, 1957–September 30th, 2013). Data are taken from FRED database

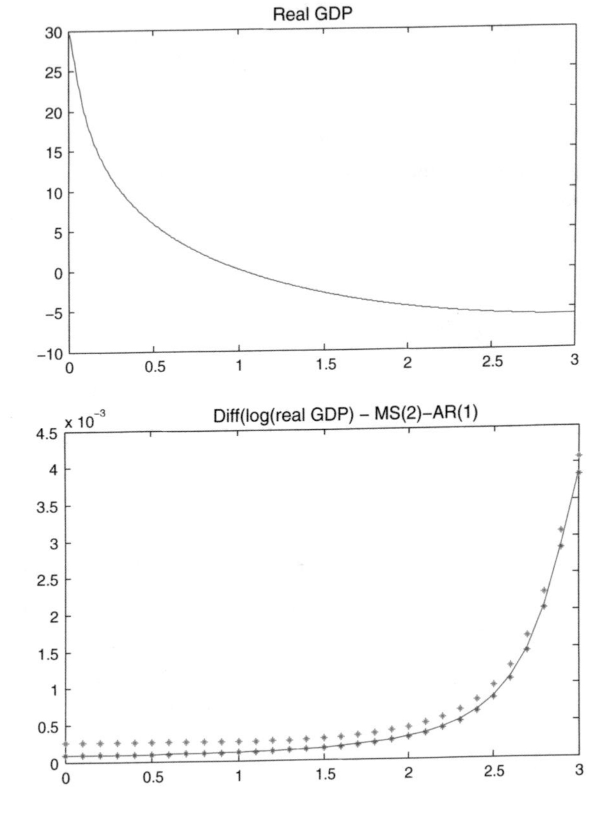

Fig. 2 Spectral density functions (*solid lines*) for the logarithm of detrended U.S. real GDP modelled as a linear AR(1) (*upper panel*) and for the logarithm of differenced U.S. real GDP modelled as MS(2) AR(1) (*lower panel*) along with 95 % confidence interval bands (*starred lines*). Both series have quarterly frequency for the period 1952:Q1–2005:Q3. Data are taken from FRED database

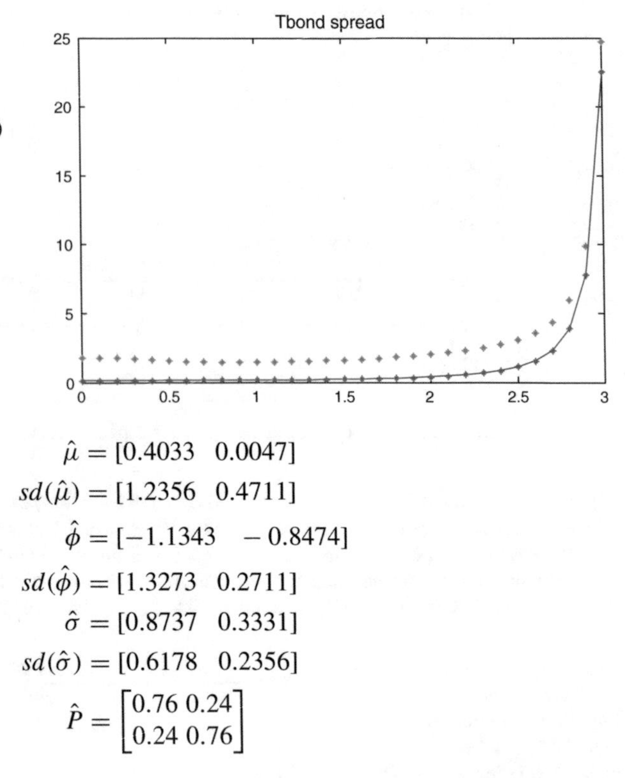

Fig. 3 Spectral density (*solid line*) of demeaned Treasury bond spread (difference between interest rates for 10 years and 1 year) at quarterly frequency (1952:Q1–2005:Q3) modelled as a MS(2) AR(1), along with 95 % confidence interval bands (*starred lines*). Data are taken from FRED database

$$\hat{\mu} = [0.4033 \quad 0.0047]$$
$$sd(\hat{\mu}) = [1.2356 \quad 0.4711]$$
$$\hat{\phi} = [-1.1343 \quad -0.8474]$$
$$sd(\hat{\phi}) = [1.3273 \quad 0.2711]$$
$$\hat{\sigma} = [0.8737 \quad 0.3331]$$
$$sd(\hat{\sigma}) = [0.6178 \quad 0.2356]$$
$$\hat{P} = \begin{bmatrix} 0.76 & 0.24 \\ 0.24 & 0.76 \end{bmatrix}$$

Then we use the estimated values in the spectral formulae of Sect. 2 to depict the spectral representation of the data, which is in Fig. 3. The spectrum suggests that only high-frequencies of the series matter. This seems to be consistent with an I(0) model, where high-frequency variability dominates the process.

Finally, we evaluate the behaviour of absolute returns from S&P500, which typically suffer of "long memory". Here a 3-state switching model suits the data and we estimate a MS(3) AR(1). The estimated parameters are the following

$$\hat{\mu} = [-0.0408 \quad 0.0004 \quad 0.0049]$$
$$sd(\hat{\mu}) = [0.1276 \quad 0.0009 \quad 0.0084]$$
$$\hat{\phi} = [0.2013 \quad -0.0001 \quad -0.0819]$$
$$sd(\hat{\phi}) = [4.8271 \quad 0.2609 \quad 0.5717]$$
$$\hat{\sigma} = [0.0737 \quad 0.0005 \quad 0.0048]$$
$$sd(\hat{\sigma}) = [0.0471 \quad 0.0003 \quad 0.0031]$$
$$\hat{P} = \begin{bmatrix} 0.98 & 0.00 & 0.02 \\ 0.02 & 0.98 & 0.00 \\ 0.00 & 0.04 & 0.96 \end{bmatrix}$$

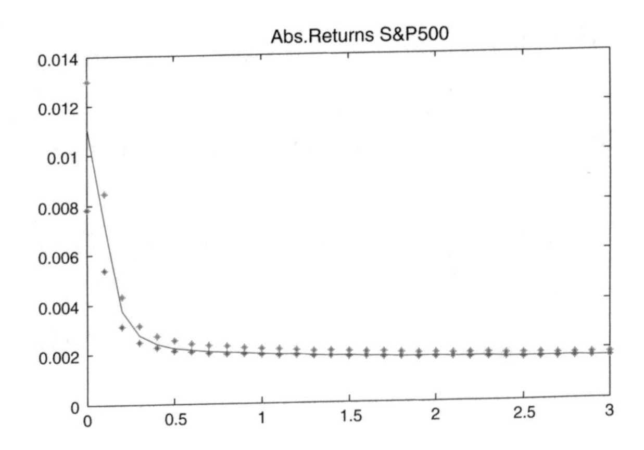

Fig. 4 Spectral density (*solid line*) of absolute daily returns (logarithm of the ratio between consecutive closing prices from S&P500) for the period January 2nd, 1957–September 30th, 2013, modelled as a MS(3) AR(1), along with 95 % confidence interval bands (*starred lines*). Data are taken from FRED database

Using Formula (2.12), we construct the spectral density of the process having the above estimated parameters. This is plotted in Fig. 4 and it is very close to the sample periodogram, with very tight confidence bands, opening up a room for considering structural change rather than long memory attributes of the process. Moreover, from the estimated values, we recognize a first regime of high-volatility and negative returns, a second regime of low volatility and high returns and a third state of moderate volatility and average returns. In particular, estimated transition probabilities show that regimes are very persistent, which is also in line with the conclusion given by Diebold and Inoue [3].

4 Conclusion

In this work we study multivariate AR models subject to Markov Switching in the most general form and derive close-form formulae for the spectral density functions of such processes. The spectral densities of these models can be very useful as a tool to infer information on the persistency characterizing the series and to check the correct parametrization of the process. In particular, after having assessed linearity or non-linearity of the series, spectral analysis gives some insights on the relative importance of high- and low- frequency variability and help to validate the assumed model. We applied the introduced methods to some macroeconomic and financial data to evaluate their frequency variability via spectral analysis.

Appendix

Derivation of Formula (2.4):
The spectral density of the process $(\mathbf{y_t})$ in (2.3) is given by

$$F_{\mathbf{y}}(\omega) = \sum_{h=-\infty}^{+\infty} \Gamma_{\mathbf{y}}(|h|)e^{-i\omega h} = \Gamma_{\mathbf{y}}(0) + \sum_{h=1}^{+\infty} \Gamma_{\mathbf{y}}(h)e^{-i\omega h} + \sum_{k=-\infty}^{-1} \Gamma_{\mathbf{y}}(k)e^{-i\omega k}$$
$$= \Gamma_{\mathbf{y}}(0) + \sum_{h=1}^{+\infty} \Gamma_{\mathbf{y}}(h)e^{-i\omega h} + \sum_{h=1}^{+\infty} \Gamma_{\mathbf{y}}(k)e^{i\omega h}.$$

Note that

$$\left(\sum_{h=1}^{n} A^h\right)(I - A) = (A + A^2 + \cdots + A^n)(I - A) = A - A^{n+1}$$

which is equal to A when n goes to infinity with the spectral radius of A less than 1. Hence

$$\left(\lim_{n\to+\infty} \sum_{h=1}^{n} A^h\right)(I - A) = A \quad \text{and} \quad \sum_{h=1}^{+\infty} A^h = A(I - A)^{-1}.$$

It turns out that spectral density of the process in (2.3) is given by

$$F_{\mathbf{y}}(\omega) = \widetilde{\boldsymbol{\Lambda}}\widetilde{\mathbf{D}}\widetilde{\boldsymbol{\Lambda}}' + \widetilde{\boldsymbol{\Sigma}}(\widetilde{\mathbf{D}} \otimes \mathbf{I}_K)\widetilde{\boldsymbol{\Sigma}}' + \boldsymbol{\Sigma}((\mathbf{D}\mathbf{P}_\infty) \otimes \mathbf{I}_K)\boldsymbol{\Sigma}'$$
$$+ \sum_{h=1}^{+\infty} \widetilde{\boldsymbol{\Lambda}}\mathbf{F}^h\widetilde{\mathbf{D}}\widetilde{\boldsymbol{\Lambda}}' e^{-i\omega h} + \sum_{h=1}^{+\infty} \widetilde{\boldsymbol{\Lambda}}\mathbf{F}^h\widetilde{\mathbf{D}}\widetilde{\boldsymbol{\Lambda}}' e^{i\omega h}$$
$$= Q + \widetilde{\boldsymbol{\Lambda}} \sum_{h=1}^{+\infty} (\mathbf{F}e^{-i\omega})^h\widetilde{\mathbf{D}}\widetilde{\boldsymbol{\Lambda}}' + \widetilde{\boldsymbol{\Lambda}} \sum_{h=1}^{+\infty} (\mathbf{F}e^{i\omega})^h\widetilde{\mathbf{D}}\widetilde{\boldsymbol{\Lambda}}'$$
$$= Q + \widetilde{\boldsymbol{\Lambda}}(\mathbf{F}e^{-i\omega})(\mathbf{I} - \mathbf{F}e^{-i\omega})^{-1}\widetilde{\mathbf{D}}\widetilde{\boldsymbol{\Lambda}}' + \widetilde{\boldsymbol{\Lambda}}(\mathbf{F}e^{i\omega})(\mathbf{I} - \mathbf{F}e^{i\omega})^{-1}\widetilde{\mathbf{D}}\widetilde{\boldsymbol{\Lambda}}'$$
$$= Q + 2\widetilde{\boldsymbol{\Lambda}}\mathbf{F}\mathcal{R}e\{(\mathbf{I}_{M-1}e^{i\omega} - \mathbf{F})^{-1}\}\widetilde{\mathbf{D}}\widetilde{\boldsymbol{\Lambda}}'$$

where $\mathcal{R}e$ denotes the real part of the complex matrix $(\mathbf{I}_{M-1}e^{i\omega} - \mathbf{F})^{-1}$, and

$$Q = \widetilde{\boldsymbol{\Lambda}}\widetilde{\mathbf{D}}\widetilde{\boldsymbol{\Lambda}}' + \widetilde{\boldsymbol{\Sigma}}(\widetilde{\mathbf{D}} \otimes \mathbf{I}_K)\widetilde{\boldsymbol{\Sigma}}' + \boldsymbol{\Sigma}((\mathbf{D}\mathbf{P}_\infty) \otimes \mathbf{I}_K)\boldsymbol{\Sigma}'. \qquad \square$$

References

1. Cavicchioli, M.: Determining the number of regimes in Markov-switching VAR and VMA models. J. Time Ser. Anal. **35**(2), 173–186 (2014)
2. Cavicchioli, M.: Analysis of the likelihood function for Markov switching VAR(CH) models. J. Time Ser. Anal. **35**(6), 624–639 (2014)
3. Diebold, F.X., Inoue, A.: Long memory and regime switching. J. Econom. **105**, 131–159 (2001)
4. Gourieroux, C., Monfort, A.: Time Series and Dynamic Models. Cambridge University Press, Cambridge (1997)
5. Müller, U.K., Watson, M.W.: Testing models of low frequency variability. Econometrica **76**(5), 979–1016 (2008)
6. Krolzig, H.M.: Markov-Switching Vector Autoregressions: Modelling, Statistical Inference and Application to Business Cycle Analysis. Springer, Berlin (1997)
7. Pataracchia, B.: The spectral representation of Markov switching ARMA models. Econ. Lett. **112**, 11–15 (2011)

Rapid Optimal Lag Order Detection and Parameter Estimation of Standard Long Memory Time Series

G.S. Dissanayake

Abstract Objective of this paper is to highlight the rapid assessment (in a few minutes) of fractionally differenced standard long memory time series in terms of parameter estimation and optimal lag order assessment. Initially, theoretical aspects of standard fractionally differenced processes with long memory and related state space modelling will be discussed. An efficient mechanism based on theory to estimate parameters and detect optimal lag order in minimizing processing speed and turnaround time is introduced subsequently. The methodology is extended using an available result in literature to present rapid results of an optimal fractionally differenced standard long memory model. Finally, the technique is applied to a couple of real data applications illustrating it's feasibility and importance.

Keywords ARFIMA process · Long memory · Fractional difference · Spectral density · Stationarity

1 Introduction

Background Information and Literature Review:

The stochastic analysis of time series began with the introduction of Autoregressive Moving Average (ARMA) model by [34], it's popularization by [6] and subsequent developments of a number of path breaking research endeavours. In particular, in the early 1980s the introduction of long memory processes became an extensive practice among time series specialists and econometricians. In their papers [18, 23] proposed the class of fractionally integrated autoregressive moving average (ARFIMA or FARIMA) processes, extending the traditional autoregressive integrated moving average (ARIMA) series with a fractional degree of differencing. A hyperbolic decay of the autocorrelation function (acf), partial autocorrelation function (pacf) and an unbounded spectral density peak at or near the origin are two special characteristics of the ARFIMA family in contrast to exponential decay of the acf and a

G.S. Dissanayake (✉)
School of Mathematics and Statistics, University of Sydney, Sydney, NSW, Australia
e-mail: gnanadarshad@gmail.com

V.-N. Huynh et al. (eds.), *Causal Inference in Econometrics*,
Studies in Computational Intelligence 622, DOI 10.1007/978-3-319-27284-9_2

bounded spectrum at the origin in the traditional ARMA family. In addition to the mle (maximum likelihood estimation) approach, [16] have considered the estimation of parameters of ARFIMA using the frequency domain approach. References [11, 30, 31] have considered the estimation of ARFIMA parameters using the smoothed periodogram technique. Additional expositions presented in [1, 4, 5, 7, 10, 17, 29, 32] and references their in provide a comprehensive discussion about long memory series estimation. In yet another development, fractionally differenced long memory model parameters were estimated using maximum likelihood and least squares with their convergence rates, limiting distribution and strong consistency by [35]. Another interesting parameter assessment study employing state space modelling of ARFIMA series could be found in [19, 27].

An optimal lag order for the parent model of the ARFIMA series known as the Gegenbauer autoregressive moving average (GARMA) process driven by Gaussian white noise was established using state space modelling in [13] through the validation of the model mean square error (MSE) by the predictive accuracy. The technique was extended to a different dimension in [14] through the replacement of Gaussian white noise by heteroskedastic errors employing a Gegenbauer process highlighted in [12].

Unfortunately the fractionally differenced long memory model parameter assessment often has a high turnaround time or a very slow processing speed depending on the series length, lag order and the number of replications in delivering Monte Carlo evidence. In such a context a rapid mechanism to estimate model parameters and optimal lag order of even the simplest fractionally differenced standard long memory model in the form of an ARFIMA series is seemingly absent in the current literature. In lieu of it this paper addresses the void by presenting a mechanism that revolves around recent advancements in information technology in slashing the turnaround time and processing speed as an original contribution.

In summary, consideration of this paper will be given to a certain class of long memory ARFIMA processes generated by Gaussian white noise. In following it the paper will comprise of Sect. 2 providing preliminaries of fractionally differenced ARFIMA processes with long memory. ARFIMA processes and the use of truncated state space representations and Kalman Filter (KF) in estimating it's long memory version will be considered in Sects. 3 and 4. The state space methodology will follow the work of [2, 3, 8, 9, 13, 15, 19, 21, 22, 27, 28]. This will be followed by Sect. 4 illustrating simulation results. Corroborating real data applications will be presented in Sects. 5 and 6 will comprise of concluding remarks.

2 Preliminaries

Certain preliminary definitions and concepts that are useful in comprehending the material in the subsequent sections of this paper are introduced next for clarity and completeness.

Definition 1.1
A stochastic process is a family of random variables $\{X_t\}$, indexed by a parameter t, where t belongs to some index set $\mathcal{T}$.

In terms of stochastic processes the concept of *stationarity* plays an important role in many applications.

Definition 1.2
A stochastic process $\{X_t; t \in \mathcal{T}\}$ is said to be strictly stationary if the probability distribution of the process is invariant under translation of the index, i.e., the joint probability distributions of $(X_{t_1}, \ldots, X_{t_n})$ is identical to that of $(X_{t_1+k}, \ldots, X_{t_n+k})$, for all $n \in \mathcal{Z}^+$(Set of positive integers), $(t_1, \ldots, t_n) \in \mathcal{T}, k \in \mathcal{Z}(set \quad of \quad integers)$. i.e.,

$$F(x_1, \ldots, x_n; t_1, \ldots, t_n) = F(x_1, \ldots, x_n; t_{1+k}, \ldots, t_{n+k}), \tag{1}$$

Definition 1.3
A stochastic process $\{X_t\}$ is said to be a Gaussian process if and only if the probability distribution associated with any set of time points is multivariate normal.

In particular, if the multivariate moments $E(X_{t_1}^{s_1} \ldots X_{t_n}^{s_n})$ depend only on the time differences, the process is called stationary up to order s, when $s \leq s_1 + \cdots + s_n$.

Note that, the second order stationarity is obtained by setting $s = 2$ and this weak stationarity asserts that the mean μ is a constant (i.e., independent of t) and the covariance function $\gamma_{t\tau}$ is dependent only on the time difference. That is,
$E(X_t) = \mu$, for all t
and
$Cov(X_t, X_\tau) = E[(X_t - \mu)(X_\tau - \mu)] = \gamma_{|t-\tau|}$, for all t, τ.
Time difference $k = |t - \tau|$ is called the *lag*. The corresponding autocovariance function is denoted by γ_k.

Definition 1.4
The acf (autocorrelation function) of a stationary process $\{X_t\}$ is a function whose value at lag k is

$$\rho_k = \gamma_k / \gamma_0 = Corr(X_t, X_{t+k}), \quad for \quad all \quad t, k \in \mathcal{Z}, \tag{2}$$

Definition 1.5
The pacf (partial autocorrelation function) at lag k of a stationary process $\{X_t\}$ is the additional correlation between X_t and X_{t+k} when the linear dependence of X_{t+1} through to X_{t+k-1} is removed.

Definition 1.6
A time series is a set of observations on X_t, each being recorded at a specific time t, where $t \in (0, \infty)$.

Let $\{X_t\}$ be a stationary time series with autocovariance at lag k, $\gamma_k = Cov(X_t, X_{t+k})$, acf (autocorrelation at lag k) $\rho_k = \mathrm{Corr}(X_t, X_{t+k})$ and the normalized spectrum or spectral density function (sdf), $f(\omega) = \frac{1}{2\pi} \sum_{k=-\infty}^{\infty} \rho_k e^{-i\omega k}; -\pi < \omega < \pi$, where ω is the Fourier frequency. There are two main types of time series uniquely identified

by the behaviour of ρ_k and $f(\omega)$. They are classified as $memory$ types and their basic definitions in advance time series analysis are given as follows:
Short Memory: A stationary time series $\{X_t\}$ is *short memory* if $\sum |\rho_k| < \infty$. Then ρ_k decays exponentially depicting properties of $\rho_k \sim a^k$ for some $|a| < 1$ resulting in a finite spectrum at $\omega = 0$ or $lim_{\omega->0} f(\omega)$ existing and being bounded.
A stationary ARMA process is, therefore, short memory.
Long Memory: A stationary time series $\{X_t\}$ is a standard *long memory* case if $\rho_k \sim k^{-d}, d > 0$ for large k and $\sum |\rho_k| = \infty$. Then ρ_k decays hyperbolically resulting in $lim_{\omega->0} f(\omega)$ being unbounded at $\omega = 0$.

In her paper [20] provides a number of alternative characteristics of long memory processes. Interestingly, [33] introduced a characteristic-based clustering method to capture the characteristic of long-range dependence (self-similarity).
Note: Processes in which the decay of ρ_k takes a shape in between exponential and hyperbolic arcs are known as *intermediate memory*. It implies the acf plot of such a process will be neither exponential nor hyperbolic but corresponding to a curve in between the shapes.
Remark: Partial autocorrelation function (pacf) of each memory type will provide corresponding shapes related to that of the acf.

2.1 *Fractionally Differenced Long Memory Processes*

When a long memory process is subject to the technique of fractional differencing it becomes a fractionally differenced long memory series. Due to it's importance in time series econometrics as a method, fractional differencing becomes the next topic of interest.

2.1.1 Fractional Differencing

Suppose that $\{Y_t\}$ is a long memory stationary time series with an unbounded spectrum at the origin. It can be shown that the time series Y_t can be transformed to a short memory series X_t through a fractional filter of the form

$$X_t = (1 - B)^d Y_t, \ d \in (0, 0.5),$$

where d could take any real fractional value within the open interval $(0, 0.5)$ such that the full ARFIMA(p,d,q) model becomes

$$\phi(B)(1 - B)^d Y_t = \theta(B)\varepsilon_t, \tag{3}$$

where $\phi(B)$ and $\theta(B)$ are stationary AR(p) and invertible MA(q) operators, B the backshift operator such that $\phi(B) = 1 - \phi_1 B - \cdots - \phi_p B^p$, $\theta(B) = 1 + \theta_1 B + \cdots + \theta_q B^q$ have zeros outside the unit circle and $\{\varepsilon_t\} \sim WN(0, \sigma^2)$.

See for example [18, 23] for details.

The ARFIMA process is a special case of the GARMA(p,d,q) model given by

$$\phi(B)(1-2uB+B^2)^d Y_t = \theta(B)\varepsilon_t, \quad (4)$$

in which the polynomial index $u = 1$ reduces the Gegenbauer expression $(1-2uB+B^2)$ to $(1-B)^{2d}$ resulting in a standard long memory model with a polynomial power term $2d$.

The above fractional differencing is used in long memory time series modelling and analysis.

Note: A fractionally differenced stationary ARFIMA process is considered long memory if the memory parameter $d \in (0, 0.5)$.

Remark: For convenience a fractionally differenced ARFIMA(0,d,0) series is considered hereafter in the analysis of the paper.

3 State Space Representation of an ARFIMA Time Series

Consider the Wold representation of a Gaussian ARFIMA(0,d,0) process with $\varepsilon_t \sim$ NID$(0, \sigma^2)$ given by

$$X_t = \psi(B)\varepsilon_t = \sum_{j=0}^{\infty} \psi_j \varepsilon_{t-j}, \quad (5)$$

where $\psi_0 = 1, \sum_{j=0}^{\infty} \psi_j^2 = \infty$ and each of the coefficients are defined by the equation $\psi_j = \frac{\Gamma(j+d)}{\Gamma(d)\Gamma(j+1)}$.

Now the mth order moving average approximation to (5) is obtained by truncating the right hand side at lag m, such that

$$X_{t,m} = \sum_{j=0}^{m} \psi_j \varepsilon_{t-j}, \quad (6)$$

where $\{X_{t,m}\}$ is a truncated ARFIMA process that will vary with the chosen truncation lag order m, which is fixed and finite.

3.1 State Space Representation of ARFIMA Model

In this approach, a dynamic time series is transformed into a suitable equivalent system comprising of two fundamental (*Measurement/Observation* and *State/Transient*) equations. As has been shown in the literature, this equivalent state space representation is not unique for time series. It will be similar to the state space representation of the ARFIMA series of [9]. Following this approach, (6) is equivalent to:

$$\begin{aligned} X_{t,m} &= Z\alpha_t + \varepsilon_t, \\ \alpha_{t+1} &= T\alpha_t + H\varepsilon_t, \end{aligned} \tag{7}$$

where $\alpha_{t+1} = [X(t+1|t), X(t+2|t), X(t+3|t), \ldots, X(t+m|t)]'$ is the $m \times 1$ state vector with elements
$\alpha_{j,t+1} = E(X_{t+j,m}|\mathscr{F}_t)$, $\mathscr{F}_t = \{X_{t,m}, X_{t-1,m}, \ldots,\}$, and

$$Z = [1, 0, \ldots, 0], \quad T = \begin{bmatrix} 0 & 1 & 0 & \cdots & 0 \\ 0 & 0 & 1 & \ddots & 0 \\ \vdots & \vdots & \ddots & \ddots & 0 \\ \vdots & \cdots & \cdots & 0 & 1 \\ 0 & 0 & \cdots & \cdots & 0 \end{bmatrix}, \quad H = \begin{bmatrix} \psi_1 \\ \psi_2 \\ \vdots \\ \vdots \\ \psi_m \end{bmatrix},$$

Z, T, H are suitably chosen matrices with dimensions $1 \times m$, $m \times m$, and $m \times 1$.

Vector α_t consists of stochastic elements that have evolved from the process. H consists of an m number of $\psi_\bullet$ coefficients. The specification comprises of the initial state distribution, $\alpha_1 \sim \mathrm{N}(a_1, P_1)$, where $a_1 = 0$ and P_1 is the Toeplitz matrix with elements $p_{hk} = \sum_j \psi_j \psi_{j+|h-k|}$, such that the state space configuration will be based on the Wold representation of (6).

Note: Similar results could be obtained using the corresponding $AR(m')$ approximation by truncating (5) such that $\pi(B) = (1-B)^d \approx \sum_{j=0}^{m'} d_j \varepsilon_{t-j}$. See [9, 19] for a comparison of the two approximations in the fractionally integrated case.

The most popular algorithm utilized by state space modelling specialists for estimation and prediction is known as the KF and becomes the focal point of Sect. 3.2.

3.2 KF and Estimation Process

KF was introduced by [25, 26] to provide estimates of parameters in a state space model of a time series or a linear dynamic system disturbed by Gaussian white noise. Approximate maximum likelihood estimation and prediction of time series parameters can be executed by adopting a state space approach coupled with a set of recursions called the KF.

Due to the presence of stochastic elements in the system it uses a series of measurements observed over time containing random variations (noise) to return pseudo-innovations of the model in creating the pseudo log-likelihood and quasi profile likelihood functions. This gives the optimal quasi maximum likelihood estimates (QMLE's) of the model parameters as shown in Tables 1 and 2 in Sect. 4. Let $\{x_t, t = 1, \ldots, n\}$ be a time series. The likelihood function of an approximating MA(m) model could be evaluated using the KF set of recursions for $t = 1, \ldots, n$:

Table 1 QMLE results due to the MA approximation

m	29	30	31	32	33	34	35
$\hat{d}$	0.2285	0.2288	0.2304	0.2317	0.2320	0.2376	0.2362
$\hat{\sigma}$	0.9664	0.9625	0.9622	0.9650	0.9649	0.9571	0.9660
Model-bias	−0.0051	−0.0087	−0.0074	−0.0033	−0.0031	−0.0053	0.0022

Table 2 QMLE results due to AR approximation

m	9	10	11	12	13
$\hat{d}$	0.2238	0.2223	0.2215	0.2222	0.2243
$\hat{\sigma}$	0.9638	0.9633	0.9643	0.9657	0.9662
Model-bias	−0.0124	−0.0144	−0.0142	−0.0121	−0.0095

$$\begin{aligned} \nu_t &= x_t - Za_t, & f_t &= ZP_tZ', \\ & & K_t &= (TP_tZ')/f_t, \\ a_{t+1} &= Ta_t + K_t\nu_t, & P_{t+1} &= TP_tT' + HH' - K_tK_t'/f_t, \end{aligned} \tag{8}$$

where K_t is the *Kalman gain* (which shows the effect of estimate of the previous state to the estimate of the current state), and f_t the prediction error variance

The KF returns the pseudo-innovations ν_t, such that if the MA(m) approximation were the true model, $\nu_t \sim \text{NID}(0, \sigma^2 f_t)$, so that the quasi log-likelihood of (d, σ^2) is (apart from a constant term)

$$\ell(d, \sigma^2) = -\frac{1}{2}\left(n \ln \sigma^2 + \sum_{t=1}^{n} \ln f_t + \frac{1}{\sigma^2}\sum_{t=1}^{n}\frac{\nu_t^2}{f_t}\right). \tag{9}$$

The scale parameter σ^2 can be concentrated out of the likelihood function, so that

$$\hat{\sigma}^2 = \sum_t \frac{\nu_t^2}{f_t},$$

and the quasi profile log likelihood is

$$\ell_{\sigma^2}(d) = -\frac{1}{2}\left[n(\ln \hat{\sigma}^2 + 1) + \sum_{t=1}^{n} \ln f_t\right]. \tag{10}$$

The maximisation of (10) can be performed by a quasi-Newton algorithm, after a reparameterization which constrains d in the subset of $\mathscr{R}(0, 0.5)$. For convenience we use the following reparameterization: $\theta_1 = \exp(2d)/(1 + \exp(2d))$. Furthermore, a discussion about the KF formulation of the likelihood function can be found in [24]. The simulation results given next provides an interesting assessment.

4 Simulation Results

The Monte Carlo simulation experiment was based on the state space model of the previous section and was executed through the KF recursive algorithm, which is already established in the literature. To propel the speed of it Fast Fourier transforms (FFT) were utilized and to a certain degree it could also be classified as a hybrid model. It enabled to slash the processor speed and turnaround time illustrating the rapidity of optimal lag order detection and estimation in proposing a creative component. The programs were parallelized and run on high speed multiple servers with random access memory (RAM) capacities ranging from 24–1024 gigabytes using the MATLAB R2011b software version. QMLE's of d and σ^2 due to an approximate likelihood through a state space approach using MA and AR approximations for an ARFIMA model driven by Gaussian white noise are shown in Tables 1 and 2 utilizing fast FFT.

By using FFT to convert functions of time to functions of frequency, the above state space approach is illustrated by fitting a Gaussian ARFIMA process with model initial values of $d = 0.2$, $\sigma = 1$ and the results are as follows in terms of both the MA and AR approximations:

For this particular simulation the likelihood is monotonically increasing with m due to the use of a log likelihood function. Figure 1 displays the implied spectral density of $X_{t,m}$ corresponding to the above parameter estimates. For $m > 1$ they are characterised by a spectral peak around the frequency $cos^{-1}0$ and are side lobes due to the truncation of the MA filter (A Fourier series oscillation overshoot near a discontinuity that does not die out with increasing frequency but approaches a finite limit known as *Gibbs phenomenon*). The autoregressive estimates do not suffer from the Gibbs phenomenon. It is illustrated in Fig. 2.

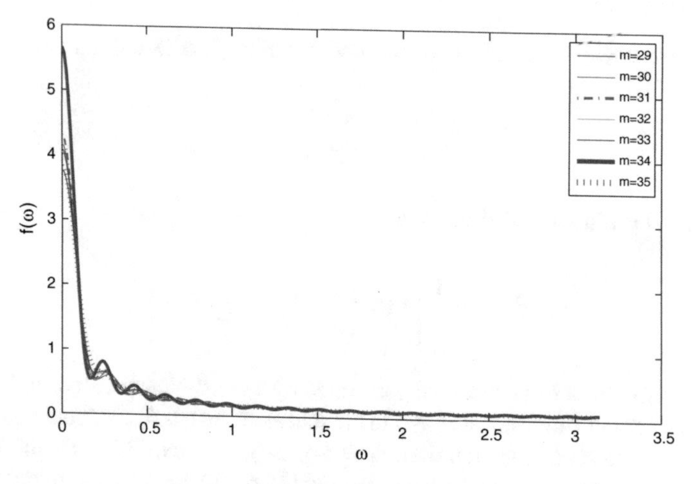

Fig. 1 Spectral density of an ARFIMA(0,d,0) series using MA approximation

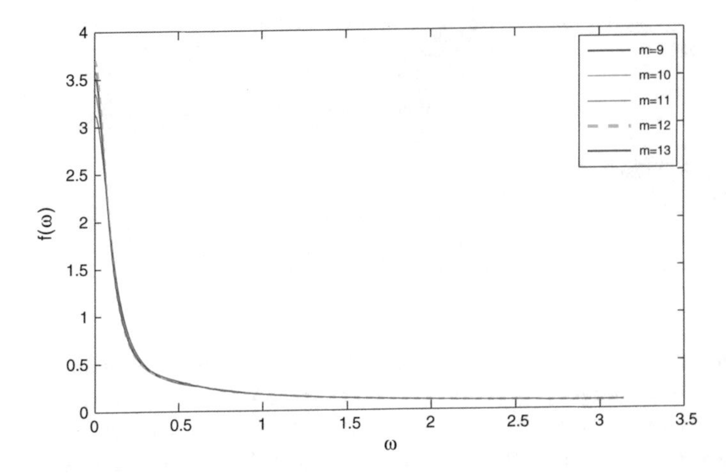

Fig. 2 Spectral density of an ARFIMA(0,d,0) series using AR approximation

By recalling from Sect. 1 that the state space methodology presented in Sect. 3 was utilized by [13] for a GARMA(0,d,0) series in concluding that the optimal lag order falls within [29, 35] for an MA and [9, 13] for an AR approximation. The result is applicable to the ARFIMA(0,d,0) model considered in this paper, since it is a special case of the GARMA(0,d,0) series (Refer: the explanation related to Eq. (2) in Sect. 2). By using the result and the minimum model-bias of estimators in Tables 1 and 2 it is clear that for the series generated through the MA approximation the **optimal lag order** is **30**, and for the series created through the AR approximation it is **10**. More importantly the processing or turnaround times taken to deliver the results were: **approximately** 3 **minutes** for the MA representation and **approximately** 5 **minutes** for the AR representation. It clearly proves the cost effectiveness of the process, since if run without the FFT's the processing time will vary between 45 min to 17 h depending on the utilized server, number of iterations and incorporated lags. The difference in processing turnaround times of the two estimation techniques (MA and AR) is due to the varying Monte Carlo error as explained in [13].

The simulation study of the MA approximation also revealed that the asymptotic variance of long memory parameter d was $\frac{\pi^2}{24n}$ independent of the series length (n) corroborating the result of [13]. These developments motivated the author to apply the methodology of this paper towards real applications involving Nile river outflow and Australia Consumer Price Index (ACPI) data, which depict characteristics of standard long memory (based on literature) and becomes the topic of the next section.

5 Empirical Evidence

Nile River outflow and ACPI data have been premier real data sets utilized by time series econometricians and statisticians over the years due to their close relationship with standard long memory. The chosen data sets have a significant impact in

Table 3 QMLE results for Nile River Data

Method	$\hat{d}$
MA-Approximation	0.291 (0.012)
AR-Approximation	0.278 (0.019)

Table 4 QMLE results for ACPI data

Method	$\hat{d}$
MA-Approximation	0.331 (0.009)
AR-Approximation	0.319 (0.015)

econometrics, since an assessment of the Nile river outflow is important to irrigation and agricultural production yields that affect the economies of many third world African nations, while an ACPI benchmark affects the economic stability of a developed first world country. The long memory feature becomes evident from the *sdf* plots of the datasets with infinite peaks close to the origin. Furthermore the *acf* and *pacf* plots depict long memory through hyperbolically decaying arcs. In lieu of it Nile River data from 1870 to 2011 and ACPI data from 3rd quarter, 1948 to 2nd quarter 2015 were considered and fitted to the hybrid ARFIMA(0,d,0) state space model discussed in this paper. The datasets were downloaded from https://datamarket.com and http://www.abs.gov.au websites and the corresponding results are provided in Tables 3 and 4.
Note: The values in brackets right adjacent to the parameter estimates of d are the standard errors. From the estimate values it is evident that the long memory property is preserved since with both the approximations of the applications $0 < \hat{d} < 0.5$.

Concluding remarks are provided next based on the details and results of the preceding sections.

6 Concluding Remarks

The facts provided in Sects. 1–5 of this paper highlight and address various issues that are prevalent in the current literature. In terms of such issues one major void is the lack of a rapid estimation process for standard long memory models. In lieu of it a hybrid state space modelling technique based on the combined utilization of the KF and FFT is introduced as a novel contribution to the existing body of knowledge. Thereafter by employing the method and an existing result in the literature an optimal lag order is established for a standard long memory process by way of two distinct estimation techniques as a secondary contribution. It is established by ascertaining the smallest bias of each estimator within the optimal lag order interval. Finally, the developed methodology is applied to a real data sets in hydrology and economics in symbolizing the long memory attribute.

References

1. Andel, J.: Long memory time series models. Kybernetika **22**(2), 105–123 (1986)
2. Anderson, B.D.O., Moore, J.B.: Optimal Filtering. Prentice-Hall, New York (1979)
3. Aoki, M.: State Space Modeling of Time Series. Springer, Berlin (1990)
4. Beran, J.: Statistics for Long-Memory Processes. Chapman and Hall, New York (1994)
5. Beran, J., Feng, Y., Ghosh, S., Kulik, R.: Long-Memory Processes Probabilistic Properties and Statistical Methods. Springer, Berlin (2013)
6. Box, P., Jenkins, G.M.: Time Series Analysis: Forecasting and Control. Holden-Day, San Francisco (1970)
7. Brockwell, P.J., Davis, R.A.: Time Series: Theory and Methods. Springer, New York (1991)
8. Brockwell, P.J., Davis, R.A.: Introduction to Time Series and Forecasting. Springer, New York (1996)
9. Chan, N.H., Palma, W.: State space modeling of long-memory processes. Ann. Stat. **26**(2), 719–740 (1998)
10. Chan, N.H., Palma, W.: Estimation of long-memory time series models: a Survey of different likelihood-based methods. Adv. Econom. **20**(2), 89–121 (2006)
11. Chen, G., Abraham, B., Peiris, M.S.: Lag Window Estimation of the degree of differencing in fractionally integrated time series models. J. Time Ser. Anal. **15**(5), 473–487 (1994)
12. Dissanayake, G.S., Peiris, M.S.: Generalized fractional processes with conditional heteroskedasticity. Sri Lankan J. Appl. Stat. **12**, 1–12 (2011)
13. Dissanayake, G.S., Peiris, M.S., Proietti, T.: State space modeling of gegenbauer processes with long memory. computational statistics and data analysis. Ann. Comput. Financ. Econom. (2014) http://dx.doi.org/10.1016/j.csda.2014.09.014
14. Dissanayake, G.S., Peiris, M.S., Proietti, T.: Estimation of generalized fractionally differenced processes with conditionally heteroskedastic errors. In: Rojas Ruiz, I., Ruiz Garcia, G. (eds.) International Work Conference on Time Series. Proceedings ITISE 2014, pp. 871–890. Copicentro Granada S L (2014). ISBN 978-84-15814-97-9
15. Durbin, J.: Time Series Analysis by State Space Methods. Number 24 in Oxford Statistical Science Series. Oxford University Press, Oxford (2001)
16. Geweke, J., Porter-Hudak, S.: The estimation and application of long memory time series models. J. Time Ser. Anal. **4**(4), 221–238 (1983)
17. Giraitis, L., Koul, H.L., Surgailis, D.: Large Sample Inference for Long Memory Processes. Imperial College Press, London (2012)
18. Granger, C.W.J., Joyeux, R.: An introduction to long-memory time series models and fractional differencing. J. Time Ser. Anal. **1**, 15–29 (1980)
19. Grassi, S., Santucci de Magistris, P.: When long memory meets the Kalman filter: a comparative study. Comput. Stat. Data Anal. **76**, 301–319 (2014)
20. Guegan, D.: How can we define the concept of long memory? An Econometric survey. Econom. Rev. **24**(2), 113–149 (2005)
21. Harvey, A.C.: Forecasting, Structural Time Series Models and the Kalman Filter. Cambridge University Press, Cambridge (1989)
22. Harvey, A.C., Proietti, T.: Readings in Unobserved Components Models. Oxford University Press, Oxford (2005)
23. Hosking, J.R.M.: Fractional Differencing. Biometrika **68**, 165–176 (1981)
24. Jones, R.H.: Maximum likelihood fitting of ARMA models to time series with missing observations. Technometrics **22**, 389–395 (1980)
25. Kalman, R.E.: A new approach to linear filtering and prediction problems. Trans. Am. Soc. Mech. Eng. **83D**, 35–45 (1961)
26. Kalman, R.E., Bucy, R.S.: New results in linear filtering and prediction theory. Trans. Am. Soc. Mech. Eng. **83**, 95–108 (1961)
27. Palma, W.: Long Memory Time Series Theory and Methods. Wiley, New Jersey (2007)
28. Pearlman, J.G.: An algorithm for the exact likelihood of a high-order autoregressive-moving average process. Biometrika **67**(1), 232–233 (1980)

29. Rangarajan, G., Ding, M. (eds.): Processes with Long-Range Correlations. Springer, Berlin (2003)
30. Reisen, V.: Estimation of the fractional difference parameter in the ARIMA(p, d, q) model using the smoothed periodogram. J. Time Ser. Anal. **15**, 315–335 (1994)
31. Reisen, V., Abraham, B., Lopes, S.: Estimation of parameters in ARFIMA processes: a simulation study. Commun. Stat. Simul. Comput. **30**(4), 787–803 (2001)
32. Teyssiere, G., Kirman, A. (eds.): Long Memory in Economics. Springer, Berlin (2007)
33. Wang, X., Smith, K., Hyndman, R.J.: Characteristic-based clustering for time series data. Data Min. Knowl. Discov. **13**, 335–364 (2006)
34. Whittle, P.: Hypothesis Testing in Time Series. Almqvist and Wiksells, Uppsala (1951)
35. Yajima, Y.: On estimation of long-memory time series models. Aust. N. Z. J. Stat. **27**(3), 303–320 (1985)

Spatial Econometric Analysis: Potential Contribution to the Economic Analysis of Smallholder Development

Renato Villano, Euan Fleming and Jonathan Moss

Abstract The stars appear to be aligned for a sustained effort to improve information to rural development policy makers about the impact space has on the opportunities for development of the ubiquitous smallholder households in rural areas of Southeast Asian countries. The influences of spatially heterogeneous resource constraints on farming activities, distance to markets and institutions, and spatial interaction among smallholders can now be better accounted for in modelling work as a result of improvements in analytical methodologies, the growing availability of so-called 'big data' and access to spatially defined information in panel data sets. The scope for taking advantage of these advances is demonstrated with two examples from a Southeast Asian country, the Philippines: spillovers and neighbourhood effects in impact studies and the development of sophisticated spatial stochastic frontier models to measure and decompose productivity growth on smallholdings.

Keywords Autoregressive · Philippines · Rural development · Space · Smallholder

1 Introduction

Space has long been recognised as a major factor influencing the welfare of smallholders producing food, fibre and beverages in Southeast Asian and other developing countries (where a smallholding can be considered as a plot of land, typically less than 2 ha in developing countries, that is sufficient for a single farming family unit with a limited resource base to sustain itself). These small production units are typically spatially dispersed, subject to variations in soils, topography, vegetation and climate, and frequently are located long distances from agricultural markets and support services such as financial institutions and research and extension agencies. The adverse effects of distance from markets are exacerbated by inadequate or absent infrastructure and other public goods, and lack of competition among buyers of their outputs in more remote areas.

R. Villano · E. Fleming (✉) · J. Moss
UNE Business School, University of New England, Armidale, NSW, Australia
e-mail: efleming@une.edu.au

V.-N. Huynh et al. (eds.), *Causal Inference in Econometrics*,
Studies in Computational Intelligence 622, DOI 10.1007/978-3-319-27284-9_3

Marketing margins (the difference between what it costs an individual or firm to produce a good or service and what revenue it receives from its sale) tend to comprise a substantial proportion of the retail prices of agricultural products. This is especially so where agricultural industries rely on export markets as a destination for a significant part of their output. Furthermore, smallholders are not uniformly distributed in rural areas; there is often a high degree of clustering, which may vary according to the product being produced. Clustering is especially strong with many horticultural products that are produced in a land-intensive manner. Hence, space has had an enduring influence on the economics of smallholder development. Given this importance of space, it is surprising that spatial analysis has not been more prominent in agricultural development studies, in terms of spatial econometrics, spatial equilibrium analysis (analysis of the volumes and trade direction across spatial entities) or methods of exploratory data analysis such as cluster analysis. We are concerned specifically with the potential role of spatial econometric analysis in this sphere. The overriding factor that has retarded the use of spatial econometric analyses to date has been in assembling the necessary data to undertake them effectively. This drawback has been reduced in recent years and, together with advances in econometric methods, there is now greater opportunity for a renewed emphasis on accumulating knowledge of spatial impacts to inform agricultural policy making, particularly how it affects smallholders.

In the next section of the paper, we outline these advances in data availability, focusing on geographic information system (GIS) and global positioning system (GPS) mapping, so-called 'big data' and panel data collection to reflect the presence of spatial and temporal heterogeneity (the uneven distribution of phenomena of interest across space and time). In Sect. 3, we review the existing literature on spatial econometric analysis, concentrating on recent progress in methodology and analyses of smallholder development. This review leads us to examine potential areas for analysis using spatial econometric methods in a Philippines context. We assess opportunities to use spatial econometric analysis for assessing the impact of rural development projects on smallholders, and for measuring and decomposing productivity of smallholder farming and explaining factors influencing it. In the final section, we consider the prospects for more fruitful use of spatial econometric analysis for smallholder policy formulation in the future and draw some conclusions.

2 Advances in Data to Capture Spatial Heterogeneity

2.1 GIS and GPS Mapping

GIS is a term used to define the set of tools available for collecting, storing, manipulating, analysing and displaying spatial data [73]. GIS tools operate on standard geographical primitives such as points, lines, areas and continuously varying surfaces. While remote sensing, scanning of the earth by satellite or high-flying aircraft, is a common technique to obtain conventional cartography and thematic (land-use) maps, GPS have been the main factor behind the advancement of spatial data in agriculture [21]. GPS is a worldwide radio-navigation network formed by a constellation of satellites and their ground stations which allows users anywhere in the world with a compatible receiver to accurately locate geographical positions through triangulation. This technology allows data to be geo-referenced by linking spatial dimensions. Coupled with low-cost personal computers and an increasing prevalence of GPS-capable smart phones and tablets [34, 84], this technology is resulting in rapid collection and mapping of agricultural inputs and outputs.

A fundamental principle of spatial analyses is the calculation of distance between geo-referenced data (that is, data defined according to location). Progress has been made recently in calculating the shortest distance between two locations using one of three different travel-distance estimation methods. The first method calculates the shortest distance 'as the crow flies' from each cell to a specific location using the orthodromic algorithm, which calculates the shortest distance between two locations on the surface of a sphere. The second method, called the Dijkstra algorithm [15], calculates the shortest distance between locations along a road network. The third method is called the least-cost algorithm and is based on the Dijkstra algorithm. It accounts for differences in travel speed and fuel consumption along different road classes, formations and gradients [14, 64]). This algorithm determines the shortest route while simultaneously accounting for travel time and fuel consumption. These three methods are now discussed in further detail.

First, the orthodromic algorithm estimates the shortest distance between geographical coordinates. For each cell in the area of interest, the distance to each location is determined using the Haversine formula, an equation for calculating the distance between two locations on a sphere using their latitudes and longitudes. For smallholders, access to markets and institutions that support and facilitate their activities is not a simple matter of a direct distance, and so this algorithm provides an inaccurate estimation of the likely distance between locations. However, it has the advantage of requiring no information on the road network.

Second, the calculation of minimum travel distance using the Dijkstra algorithm allows the shortest route from each cell to each location to be estimated along a road network. The algorithm proposed by [15] has been applied and assessed in minimum distance calculation studies by numerous authors (e.g. [17, 55, 83]). To use the Dijkstra algorithm, a road network is divided into nodes (intersections) and branches (road sections). The shortest path between two locations is solved by an iterative process described by [15]. As the distance of the origin node is known, this node is labelled with a distance of zero. The algorithm requires data on the road network. For example, [53] obtained the data for his study in Australia from several shapefiles (geospatial vector data formats) in the GeoScience Australia database [29]. The distance of each connected road section was determined using the Euclidean distance.

Third, the least-cost travel algorithm to estimate distance builds on the Dijkstra algorithm by including several physical characteristics of the road that may be critical for determining the optimal route. Several authors have shown that road class and formation are important factors when calculating shortest feasible routes. For example, [18, 35] applied weights or penalties to the Dijkstra algorithm to account for different road characteristics. This method allows for optimal routes to be determined based on a cost penalty as well as distance. It is desirable to use cost penalties that reflect the impact of road class and formation on travel. Moss [53], for example, used GIS shapefiles containing road network data including class and formation from [29]. Considering topography when assessing the time and cost of travel is also desirable but uncommon. Road gradient can be included in the least-cost algorithm as a cost penalty, using topographical data.

For the latter two algorithms, not all start locations (smallholdings) are adjacent to a road—a common situation for farmers in rural areas of developing countries. Therefore, an additional 'starting distance' algorithm needs to be used to find the nearest adjacent road for each source location. This algorithm is initialised by subdividing all cells into three sets: cells assessed and for which no roads intersect; cells from which the next iteration will assess whether any road intersects; and a set containing all cells not yet assessed. The first step is to assess whether the starting location cell is intersected by the road network. If so, the minimum start distance to the road network is zero. Otherwise, all cells adjacent to the starting location cell are placed in the second set and assessed to determine whether any are intersected by part of the road network. Once a cell with an intersecting road is found, the distance from the starting location to this cell is determined. This location becomes the starting node for the Dijkstra algorithm and the calculated distance is added to the optimal distance. If no cell is found in the adjacent bundle, all adjacent cells to the cells currently in the second set are found and moved to it while simultaneously moving the cells currently in the second set to the first set.

2.2 *Big Data*

We follow [63] who, from 12 options, favoured defining big data as either 'The belief that the more data you have the more insights and answers will rise automatically from the pool of ones and zeros' or 'A new attitude by businesses, non-profits, government agencies, and individuals that combining data from multiple sources could lead to better decisions'. Specifically in relation to smallholder agriculture, the more comprehensive the data sets on the spatially heterogeneous environments in which smallholders operate, the more likely it is that estimated models will capture this spatial heterogeneity and accurately represent smallholder performance and opportunities for progress.

Comprehensive spatial data sets of interest include topography, climate, hydrological systems, soils and geology, current land-use cover and land-use suitability, livestock intensity levels, harvested areas, crop yields and cropping frequency, population density, transportation networks and access to nearest markets. They are becoming increasingly available through resources such as the [25, 52, 76] Geodatabase, often at a resolution sufficient to capture spatial heterogeneity at a smallholder level. It is not just a question of big data now being more readily available; the time and cost involved in its capture and use have decreased substantially.

2.3 *Increased Availability of Panel Data Sets*

Over time, data-collection programs have accumulated more panels that have improved econometric analyses generally and spatial econometric analyses specifically. The advantages of panel data sets for econometric modelling have been well-researched in the literature. Hsiao [38] outlined these advantages as: increasing the degrees of freedom and reducing collinearity among explanatory variables, thereby increasing estimation efficiency; enabling research questions to be explored that would not be possible with either time-series or cross-sectional data (especially important for spatial econometric analyses); better control for missing or unobserved variables; simplification of computation and inference in certain circumstances; and more accurate predictions for individual outcomes (such as an individual's behaviour, which can be crucial in ascertaining how smallholders are likely to respond to a particular intervention). The use of panel data allows us to control for individual heterogeneity, which could result in biased results if not taken into account. More importantly, the use of panel data enables study of the dynamics of adjustment that are particularly crucial in studying smallholder agriculture. For example, it enables the analyst to shed light on the patterns and sources of productivity growth, changes in the dynamics of farming systems (defined by [16]) as 'a population of individual farm systems that have broadly similar resource bases, enterprise patterns, household livelihoods and constraints, and for which similar development strategies and interventions would be appropriate'), and differential impacts of government policies and other interventions.

3 Review of Existing Literature on Spatial Econometric Analysis of Smallholder Development

3.1 *Recent Progress in Spatial Econometric Modelling*

Lesschen et al. [45] is a good place to start to review statistical methods for analysing the spatial dimension—in their case, specifically in respect of changes in land use and farming systems. They began with a summary of geographic ways to represent data such as point, polygon and raster data (a matrix of cells in a grid formation) and how to link different data representations. The methods covered include techniques for exploratory data analysis (factor analysis, principal component analysis, canonical correlation analysis and cluster analysis), various techniques for regression analysis, Bayesian analysis and artificial neural networks. They then proceeded to discuss modelling issues relevant to the spatial analysis of land use and farming systems, among them spatial autocorrelation.

More recently, [22] reviewed progress and prospects in the application of spatial econometric modelling in the wake of the publication of an introductory book on spatial econometrics by [44]. In particular, they considered 'the argument in favour of the spatial Durbin model, the use of indirect effects as a more valid basis for testing whether spatial spillovers are significant, the use of Bayesian posterior model probabilities to determine which spatial weights matrix best describes the data, and [LeSage and Pace's] contribution to the literature on spatiotemporal models' [22].

Arbia et al. [7] edited an issue of the journal, Economic Modelling, that featured 10 papers reflecting recent advances in the application of spatial econometric methods. While its context is at the international level, the study by [24] in which they estimated a theoretical growth model (a model for forecasting growth in national income per head) accounting for technological interdependence among economies has relevance to analysing growth at more disaggregated levels. This is especially so in the way the authors examine the impact of spillover effects:

> Technological interdependence is assumed to operate through spatial externalities. The magnitude of the physical capital externalities at steady state, which is not usually identified in the literature, is estimated using a spatial econometric specification. Spatial externalities are found to be significant. This spatially augmented Solow model yields a conditional convergence equation which is characterized by parameter heterogeneity. A locally linear spatial autoregressive specification is then estimated providing a convergence speed estimate for each country of the sample [24].

This statement by [24] introduces a key concept for spatial econometric analysts: spatial externality. A negative spatial externality occurs where the action of one neighbour adversely affects the welfare of another neighbour adjacent to or near it, but the neighbour causing the impact does not have to recompense the adversely affected neighbour. A positive spatial externality occurs where the action of one neighbour increases the welfare of another neighbour adjacent to or near it, but the neighbour generating the beneficial impact is unable to claim recompense from the beneficiary. Spatial interdependence is an essential element for a spatial externality to occur. Technological interdependence, the presence of interdependence between production methods, is one particular kind of spatial interdependence that is important when analysing a production system such as farming.

Schmidtner [68, p. 2] provided a good overall account of the development of spatial econometric analysis. She began by citing [56] who posited its five key attributes as the role of spatial interdependence, asymmetry of spatial relations, importance of explanatory factors located in other spaces, differentiation between ex post and ex ante interaction and explicit modelling of space. Schmidtner [68] then recounted how [5] had developed these ideas within a formal framework of econometric model estimation and specification tests. Use of spatial econometric methods enables unbiased and consistent results to be obtained [43].

Baylis et al. [11] discussed recent contributions made in the theoretical and empirical literature on spatial econometric methods for panel data. Of particular interest in this paper is the focus they place on applications in agricultural economics. These applications cover finance and risk management, production and land economics, development economics and environmental economics, with special emphasis on climate change and agricultural applications. They opined that the spatial econometric methods they describe 'hold great potential for applied researchers' [11].

Finally, [9, 42] provide arguably the most authoritative and up-to-date reviews of spatial econometric models. The chapter by [42] is particularly valuable. The authors rigorously discuss applicable spatial econometric techniques using panel data. They begin with a discussion of static panel data models in which the spatial effects are specified in the disturbance or regression part of the model, or both. Table 12.1 provides a good summary of different static spatial panel models. They then consider dynamic panel data models that account for state dependence. Table 12.2 summarises the various spatial dynamic panel data models. Next, other models with cross-section dependence are examined. Finally, there is a valuable section on testing for spatial effect.

3.2 *Use of Spatial Econometric Analysis to Study Spillovers and Spatial Interaction*

Spatial interdependence takes two related forms: a spatial externality called the 'neighbourhood effect'—where neighbours interact with each other or take collective action (frequently with concomitant learning processes)—and a spillover—where actions by one neighbour have an impact on the performance of other neighbours. The latter requires neighbours to be interdependent but not to interact with each other. It is often difficult to disentangle the two actions, which can be integral parts of the same process.

3.2.1 Neighbourhood Effect

Holloway and Lapar [37] published a seminal paper on neighbourhood effects, employing spatial econometric analysis. They estimated two quantities defining such a spatial externality: the propensity for neighbours to make the same decision and the magnitude of the neighbourhood. Employing recent advances in Bayesian estimation and model comparison, they applied this method to a sample of smallholders (producing mainly rice and onions but also raising livestock) in northern Philippines.

The difficulty in teasing out a neighbourhood effect from a spillover is exemplified in the paper by [72] who discussed the neighbourhood effect between farms in England. They were concerned with 'coordinated environmental action' [72] and its effect on performance which entailed spillover effects in the form of greater environmental benefits.

3.2.2 Spillover Effect

Spatial spillovers at the micro level in the context of smallholder production refer to impacts of activity undertaken by one unit involved in the production or provision of a good or service on the performance of smallholders in proximity to that unit. They are commonly mentioned in relation to the introduction of new production technologies in which case they tend to be positive effects that raise the productivity and ability to innovate of smallholders. The context can change: spillovers may be analysed across production units, districts, provinces or even countries.

Min and Jiaying [50] applied spatial autocorrelation analysis to study the determinants of spatial disparities in agricultural mechanisation in China. They found evidence of spillovers in mechanisation across provinces.

Spillovers affecting smallholders need not be between farms and often follow from the provision of public goods. For example, [13] undertook a spatial econometric analysis of factors influencing smallholder income growth in Kenya, focusing particularly on the spatial spillovers in public good impacts. His spatial impact model comprised three modes of endogenous and exogenous interaction effects, and correlated effects (shared unobserved factors that result in similar behavioural patterns). He found strong support for the use of spatial models.

3.2.3 Organic Farming with Neighbourhood Effects and Spillovers

Organic farms (those applying technologies without the use of chemicals or synthetic fertilizers during production or processing) are particularly sensitive to spatial externalities, and hence to spillovers and neighbourhood effects. Schmidtner [68] estimated spatial econometric models to study conversion to organic farming in Germany, using county-level data. Her results revealed that 'a high share of organically managed land in a region seems to be an ideal precondition for the decision of a farmer to convert to organic production [and] available technical and juridical knowledge in a region 'as well as positive external effects at the plot level...might increase the diffusion of organic farming as an innovation' [68]. She concluded that 'incentives to stimulate clusters of organic farming could support the exploitation of economies of scale external to the farm'.

Parker and Munroe [59] were also concerned with organic farming, investigating what they termed 'edge-effect externalities' (comparable to neighbourhood effects) and the location and production patterns of organic farming in the Central Valley of California in USA. They defined edge-effect externalities as 'spatial externalities whose marginal impacts decrease as distance from the border generating the negative impact increases' [59]. Conflicts arise, according to [59], because certain production processes in farming systems are incompatible, leading to production losses between neighbouring farms. The authors posited that the magnitude of these losses depends on both the scale of each activity and patterns of land use. They used a generalised method of moments spatially autoregressive model to show how parcel geometry and surrounding land uses affect the probability that a specified land area may be certified as organic.

The above definition of edge-effect externalities implies they are negative externalities. But, as indicated above, positive spatial externalities also exist in smallholder farming, as reflected above by neighbourhood effects. For example, [46] showed how they can be brought about by the demonstration effect and information and knowledge flow between neighbouring smallholders. Addressing the question whether spatial spillovers exist in the adoption of organic dairy farming, they tested the hypothesis that 'neighboring farmers can help to reduce the uncertainty of organic conversion by lowering the fixed costs of learning about the organic system' [46]. Their estimated reduced-form econometric model of the decision to convert to organic dairying using a 'spatially explicit 10-year panel dataset' of dairy farms in southwestern Wisconsin provided evidence that the presence of neighbouring organic dairy farms positively influences a farmer's decision to convert to organic production.

In a similar manner, [82] investigated factors influencing farmers' decision to convert to organic farming in Honduras by estimating a 'spatially explicit adoption model' [82]. They noted that previous studies had focused on the influence of 'spatial patterns in the diffusion and adoption of agricultural technologies in general and organic agriculture in particular'. Factors influencing these spatial patterns for which they tested included 'the availability of information in the farmer's neighbourhood, social conformity concerns and perceived positive external effects of the adoption decision'. One of their key findings was that positive productivity spillovers to neighbouring farm plots reduce the probability of adoption of organic farming methods.

Sutherland et al. [72] used organic farming as a proxy for coordination in their multi-disciplinary analysis of the potential impacts of undertaking similar environmental actions on many farms in a confined area. They concluded that 'encouraging local farmer co-ordination can have clear environmental benefits without high economic cost, but must be undertaken with caution—specifically regarding the trade-offs between benefits, local geophysical and social characteristics, and assumptions made about inter-farmer trust' [72].

3.2.4 Impact of Productivity on Rural Welfare

Minten and Barrett [51] collated spatially explicit data for Madagascar on soil conditions, temperature, altitude and rainfall patterns in GIS format and then overlayed them with a map of communes (local government entities). In model estimation they corrected for spatial autocorrelation when conducting regression analyses to demonstrate that communes adopting improved agricultural technologies have higher levels of productivity and 'enjoy lower food prices, higher real wages for unskilled workers, and better welfare indicators' [51, p. 797].

3.2.5 Analysing Convergence

Paraguas and Dey [58] examined convergence in total factor productivity (TFP) (a productivity measure involving all factors of production) among participants in the Indian aquaculture industry by estimating a spatial econometric model. They tested for the presence of spatial autocorrelation and spatial heterogeneity, and concluded that spillover effects on TFP growth existed and that convergence was taking place.

Pede et al. [61] used province-level data in the Philippines to estimate a conditional convergence growth model (a model for estimating how long it would take to achieve convergence of income per head across countries or regions subject to the structural constraints each faces). They used the estimated model to analyse the relationship between growth and inequality and reported results showing inequality to have a positive impact on growth in income per head. But the magnitude of this effect was found to be unstable across regions. The literature on spatial convergence of productivity is growing. A recent inter-regional example is the paper by [54].

3.2.6 Using Spatial Econometric Analysis in Multi-level Models

Herrero et al. [36] reported on their attempt to project future changes in smallholder farming systems in Kenya by linking socio-economic scenarios with regional and household models. The method they chose was multi-scale models and scenarios that they found useful for linking global change and local-scale processes.

3.3 Environmental and Land Use Applications

3.3.1 Capturing the Effects of Spatial Heterogeneity on the Environment

Bockstael [12] wrote a landmark paper on the important spatial perspective of the environmental implications of agricultural production and capturing the economic behaviour causing land use change in a spatially disaggregated manner. She made the salient observations that:

> The spatial pattern and distribution of land use at one scale or another has important environmental consequences that range from local water quality to global biodiversity. In addition, the pattern of land use affects costs of production and exchange [12].
>
> [Economists'] treatment of space, in any manner, has been largely superficial. We often use cross-sectional data that are inherently spatial but we rarely exploit the underlying spatial relationships or acknowledge them in our econometrics. Or we aggregate spatially disperse data causing artificially sharp intraregional distinctions and unrealistic interregional uniformity [12].

Allaire [3] studied the determinants of participation by farmers in agriculture-based environmental programs in France by estimating a spatial Durbin probit model.

They implicitly modelled the spatial interactions between farmers' decisions, paying particular attention to the quality of the prediction of participation rates by using successive geographical scales to improve the representation of spatial diffusion of these programs.

3.3.2 Valuing Farmland

Bockstael [12] outlined econometric and other modelling problems induced by the spatial heterogeneity associated with attempting to value land. One major problem she identified was omitted variables that are spatially correlated:

> In both the hedonic model and the land use conversion model, we can expect that the omitted variables will be spatially correlated. Almost any variable of importance that affects the value of a parcel of land or its likelihood of conversion will be highly correlated with its neighbor's. There are as yet no satisfactory solutions to the problem of spatial autocorrelation in discrete choice models [12].

Methodological advances have since provided solutions to a significant degree, particularly where the advent of big data has reduced the problem of omitted variables. In a much-cited paper, [66] estimated a hedonic equation for farmland in areas of USA where non-irrigated farming is possible. Among the contributions they claimed for the paper is the development of a data set that integrates the spatial distribution of soil and climatic variables within counties. They also let the error terms be spatially correlated to obtain more accurate results. These results are presented to show the potential impacts on farmland values of a range of warming scenarios.

Baylis et al. [11] constructed and estimated a hedonic model of farmland values in USA using panel data. Their estimates included spatial error and lag models with fixed and random effects.

3.3.3 Representing the Forest-Agriculture Spatial Interface

A few studies have examined spatial elements of land use involving forestry, often focusing on the spatial interface with agriculture. Moss [53] investigated the potential of different land-use options to produce carbon offsets in selected regions in Australia. Options were assessed using spatio-temporal simulation models developed at differing levels of resolution, ranging from 1.1 km^2 cells in a spatial grid to the farm scale.

Wheeler et al. [81] used a panel data set to investigate the determinants of forest clearing in Indonesia. They observed that, until recently, empirical research into the economic dynamics of forest clearing has been hindered by a lack of spatially disaggregated time-series data. However, they claimed that the emergence of more effective ways of translating satellite images into credible estimates of forest clearing have made it possible to conduct econometric analyses of these dynamics at a high level of spatial and temporal disaggregation.

Upton et al. [77] employed a panel data set of afforestation over small-area boundaries from vector and raster data sources in Ireland. They used these data to estimate a spatial econometric model of the impacts of various physical, economic and policy factors on afforestation rates.

Vera-Diaz et al. [78] assessed the environmental impacts of soybean expansion and infrastructure projects in the Amazon Basin in Brazil. Subsequently, [79] combined ecological and economic data with satellite imaging and GIS analysis to model the impacts of infrastructure projects on land use in the same area using spatial econometric techniques. This model has been used to estimate changes in the returns to land for different infrastructure projects [28].

Roy Chowdhury [65] used remote sensing and spatial modelling to quantify land use change in Mexico, and to analyse factors inducing this change in the largest protected area in the country. Rates of land change were calculated for 'prevailing tenure regimes and for reserve core versus buffer zones by employing GIS layers delineating the respective boundaries' [65].

Entwisle et al. [23] estimated village-level models of the impacts of population on land planted to upland crops in a district in Thailand. They partitioned land into spatial units corresponding to villages, constructed radial buffers of 3 km around the nuclear village centres, and accounted for spatial autocorrelation by estimating a spatial error model with the proportion of land planted to upland crops as the dependent variable. Their main findings were that changes in land use associated with population change appear to radiate outwards from village centres a and that growth in the population of households better predicts the proportion of land planted upland crops than growth in village population as a whole.

3.4 Accounting for Space in Analyses of Technology Adoption and Productivity

3.4.1 Analysing Technology Adoption

Staal et al. [71] undertook a pioneering study in spatial econometric analysis of smallholder development in using GIS to study the location and adoption of technology by dairy smallholders in Kenya. They concluded that although their study required 'large geo-referenced data sets and high resolution GIS layers, the methodology demonstrates the potential to better unravel the multiple effects of location on farmer decisions on technology and land use' [71].

Edirisinghe [20] examined the spatial interdependence of production choice by smallholder rubber producers in Sri Lanka. Drawing heavily on the modelling work by [37], he applied a Bayesian spatial autoregressive probit model to measure the impact of this production choice on the choice made by neighbouring smallholders.

3.4.2 Estimating Yield Functions

The increased use of GPS and GIS for managing crop production is exemplified by [6]. These authors applied spatial econometric methodology to estimate crop yield functions to optimise fertiliser application in Argentina.

Florax [26] undertook a spatial econometric analysis of millet yield in the West African Sahel. The method enabled them to capture local differences in soil variation, ecological characteristics and yield.

Ward [80] used spatially specific data to estimate a cereal yield response function using an estimator for spatial error models to control for endogenous sample selection. While they found that their estimated elasticities did not differ greatly from those of simpler models that ignore spatial features in the data, it cannot be assumed that elasticities will be universally similar across other regions and farming systems.

Traditional methods of controlling for spatial heterogeneity in measuring and decomposing TFP To date, space has mostly been incorporated in studies concerned with measuring and explaining TFP using crude devices such as district or regional categorical variables. A recent advance, but still quite limited approach, has been to estimate individual regional production frontiers and a meta-frontier for all regions. It is commonly used for studies of the productivity of smallholders to capture differences in agronomic, climatic and other physical conditions that they face. There are numerous examples of this approach, with a recent one of Philippine rice producers being [49].

Recent innovations in accounting for spatial heterogeneity and spillovers in TFP studies [67] used Bayesian procedures to estimate a stochastic frontier model with a latent spatial structure to control for spatial variations in outputs of Brazilian farmers. They specified independent normal or conditional autoregressive priors for such spatial effects. Two model comparison criteria were applied to lend support their contention that latent spatial effects should be taken into account.

Areal et al. [8] considered the effect of spatial factors that affected production and efficiency levels on 215 dairy farms in England between 2000 and 2005. Using Bayesian methods, they included a spatial lag to account for the spatial component and firm-specific effect in model estimations to predict technical inefficiencies. They found that results for the conditional posterior of the spatial dependence parameter were sensitive to the specification of the spatial weight matrix, which they concluded may be due to whether a connectivity matrix or a distance-based spatial matrix is used, and also to the cut-off size chosen.

Barrios and Lavado [10] demonstrated how a standard stochastic frontier model could be augmented by a sparse spatial autoregressive component for a single cross-section of data and a spatial-temporal component for a panel data set. They provided two illustrations using Philippine rural data to explain how technical efficiency is explained by exogenous factors when estimating such a model.

Hughes et al. [39] measured productivity for individual farms in broadacre farming in Australia. In addition to the normal input-output data set used in productivity analysis, they matched it with spatial climate data that enabled them to estimate a farm-level stochastic frontier model. Productivity was then decomposed into its efficiency components.

Pavlyuk [60] specified a spatial stochastic frontier model that included spatial lags, spatial autoregressive disturbances and spatial autoregressive inefficiencies. He applied the model to a data set of 122 European airports to obtain inefficiency estimates. A key conclusion drawn was that spatial relationships and spatial heterogeneity were present in the airport industry.

Eberhardt and Teal [19] modelled the nature of the cross-section dependence for 128 countries using panel data employing a standard empirical model that was extended to include common correlated effects estimators. These estimators enabled the authors to 'account for the presence of unobserved common factors with heterogeneous factor loadings by introducing cross-section averages for the dependent and independent variables into the regression model, each with a country-specific parameter' [19]. They concluded that the 'agro-climatic environment drives similarity in TFP evolution across countries with heterogeneous production technology' [19].

Fusco and Vidoli [27] recommended an expansion of the composed error term of stochastic frontier models into three rather than the standard two components, with the additional term linked to a spatial lag. They tested this method for its ability to control for spatial, global and local heterogeneity using simulated data on production in the Italian wine industry. We further discuss the material in this paper below.

Moura e Sa Cardoso and Ravishankar [54] applied standard stochastic frontier analysis to estimate production inefficiencies to assess the degree to which a given region's observed output falls short of the most efficient output. They then modelled region-specific efficiencies as outcomes of the level of human capital and tested for convergence among regional efficiency levels.

Pede et al. [62] investigated spatial dependency among technical efficiencies in rice production separately for rainfed and irrigated ecosystems in the Philippines using panel data. Their results divulged evidence of spatial correlation among these technical efficiency estimates, with stronger spatial dependency among farmers in the rainfed ecosystem. They also used spatial econometric analysis to find evidence of spatial dependence in both household and farm plot neighbourhoods.

The most exciting recent methodological advances in measuring and decomposing TFP have been the studies by [1, 30–32, 75]. These authors have merged methods common in the efficiency and productivity literature with spatial econometric methods in an innovative fashion. Glass et al. [31] decomposed TFP growth in 40 European countries using a spatial autoregressive frontier model. Adetutu et al. [1] used a multistage spatial method to analyse the effects of efficiency and TFP growth on pollution in the same 40 European countries. Glass et al. [32] blended traditional non-spatial stochastic frontier models with key contributions to spatial econometrics to develop a spatial autoregressive stochastic frontier model for panel data for 41 European countries. They showed how to specify the spatial autoregressive frontier to allow efficiency to vary over time and across regions.

In an important recent breakthrough, [75] developed a stochastic frontier model based on Bayesian methods to decompose inefficiency 'into an idiosyncratic and a spatial, spillover component' [75]. They derived exact posterior distributions of parameters and proposed computational schemes based on Gibbs sampling with data augmentation ‘to conduct simulation-based inference and efficiency measurement’ of regions in Italy for the period from 1970 to 1993.

3.4.3 Combining Impact Analysis with Measuring and Decomposing TFP

Studies have begun to emerge that combine the analysis of impacts from interventions in farming with measuring and decomposing TFP. For example, [2] estimated a spatial autoregressive stochastic production model to predict technical efficiencies of matched subsamples of farmers participating in an agricultural extension program in Tanzania. Samples of treated and control agents of farmers were obtained by using propensity score matching. Travnikar and Juvancic [74] used a spatial econometric approach to examine the impact of investment support on agricultural holdings in Slovenia. They confirmed a positive relationship between farm investment support and agricultural labour productivity, and spatial spillovers in agricultural labour productivity.

We discuss the opportunities that these methods provide for impact analysis and analysing smallholder productivity in the Philippines in the next section.

4 Potential Areas for Analysis Using Spatial Econometric Methods: Examples from the Philippines

The literature reviewed above exhibits four prominent features of interest in the context of this paper. First, there is a wide range of methods now available to conduct spatial econometric analyses suited to meeting a variety of research goals. Second, the volume of such analyses of smallholder production and rural development that have been undertaken using these methods remains small, notably in Southeast Asian countries with the prominent exceptions of [37, 62] in the Philippines and [23] in Thailand. Third, panel data sets are crucial for the effective use of many of these methods. Finally, the research papers vary in their coverage of all the elements needed to conduct analyses to achieve their research aims: knowledge of the spatially heterogeneous production environment; formulation and estimation of the most suitable spatial econometric model; use of panel data sets; use of big data to reduce omitted variable problems; and application of a suitable distance measure.

It would take too much space covering the many potentially rewarding smallholder and rural development research possibilities for Southeast Asian countries. We confine ourselves to two key areas where recent methodological advances provide promising research avenues in one country, the Philippines: analysing smallholder response to interventions in the assessment of the impact of rural development interventions on smallholders; and measuring, decomposing and explaining TFP growth in smallholder farming. First, we briefly describe the nature of spatial heterogeneity of rural Philippines, especially in relation to the dominant rice farming systems.

4.1 Spatial Heterogeneity in the Rural Sector of the Philippines: Example of Rice Ecosystems

Rural areas in the Philippines exhibit a range of environments that influence the ways in which production technologies are applied by smallholders and contribute to spatial inequalities in income and welfare. For the predominant agricultural activity, rice, various ecosystems can be observed that experience temporal variations in yields within a cropping year where more than one rice crop is produced and/or various cropping rotations exist [48, 49]. Inter-year variations in rice and rotational crop yields are also a feature of rice-based ecosystems.

Smallholders operate in a wide range of physical environments of differing altitude, climate and soil types that are largely beyond their control. According to the Philippine Atmospheric, Geophysical and Astronomical Services Administration (PAGASA), climatic conditions in the country are classified into four different types based on rainfall distribution. Type 1 climate has two pronounced seasons–dry from November to April and wet during the rest of the year. For Type 2 climate, there is no dry season but minimum monthly rainfall occurs from March to May and maximum rainfall is pronounced from November to January. In contrast, Type 3 climate has no pronounced maximum rain period with a short dry season lasting only from one to three months. Lastly, Type 4 climate has rainfall evenly distributed throughout the year and has no dry season. Attempts have been made to classify the environments in which rice is grown in these rice-based farming systems. Most use the water regime as a basis for their classification system but it may also relate to topography and the ability of the soil to retain water. The general classes of rice production are the irrigated, rainfed lowland, upland and flood-prone environments [33, 40] depicted in Fig. 1. All ecosystems are characterised by the natural resources of water and land, and by the adaptation of rice plants to them. Irrigated rice may be found at any point in the toposequence (defined by Allaby (2012) as a 'sequence of soils in which distinctive soil characteristics are related to topographic situation if water delivery is available').

The classification described here is that being used by the International Rice Research Institute [40]. An irrigated environment is where rice is grown in bunded, puddled fields with assured irrigation for one or more crops in one year. The irrigated ecosystem is divided into the irrigated wet season and the irrigated dry season with rainfall variability and supplementary irrigation as the basis for classification.

In an upland environment, rice is grown in areas where no effort is made to impound water and where there is no natural flooding of the land. Upland rice is planted in areas at higher slopes of an undulating landscape where the groundwater table is at least 50 cm below the surface. People who practise shifting cultivation

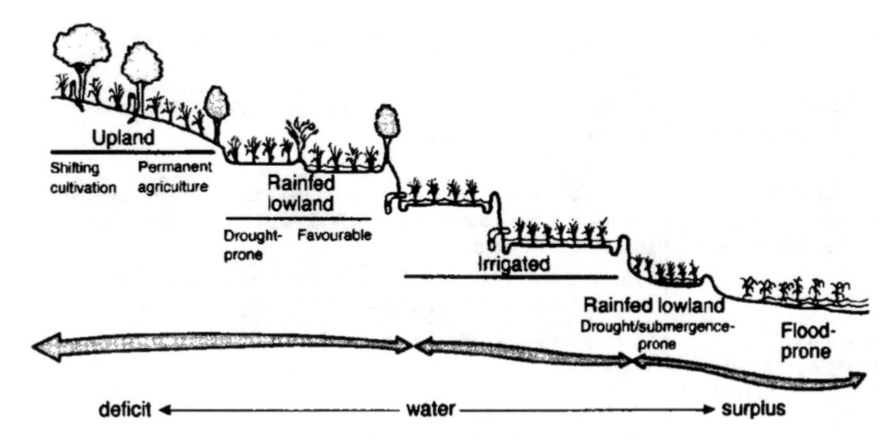

Fig. 1 Rice ecosystem characteristics. *Source* [40], adapted from [33]

allocate a significant part of their land to upland rice, which is usually inter-cropped with maize, cassava and other crops.

Another important rice ecosystem is the flood-prone environment. The fields are slightly sloping or depressed. Rice may be submerged for 10 consecutive days during crop growth. It is usually seeded immediately before the arrival of the floodwaters, and little more is done until the time arrives for the crop to be harvested. In some areas, the rice may be transplanted once or twice as the floodwaters advance, in an attempt to save the young rice seedlings from drowning if the floods rise too rapidly for the seedlings to survive.

The major characteristics of farming systems of the major rice production environments are presented in Table 1, which demonstrates the environmental variations within ecosystems. This can be demonstrated by discussing the important rainfed lowland ecosystem. The literature reveals that rainfed rice environments have been characterised for various purposes at different scales using a large range of techniques [70]. The research ranges from broad regional-scale characterisation to detailed farm-level studies.

The rainfed lowland environments are where the water supply for rice plants is principally provided by rainfall, run-off water and underground water. The rainfed lowland rice fields are usually bunded. The bunds serve to retain floodwaters, as well as rainwater, which fall during the growing season. Rice fields, in general, are submerged or flooded with water that may exceed 50 cm for no more than 10 consecutive days [41].

Depending on environmental conditions, rainfed lowlands may be classified into favourable and unfavourable ecosystems (drought-prone, submergence-prone, drought and submergence-prone and medium-deep water). In favourable areas, which are intermediate between rainfed and irrigated ecosystems, field water cannot be completely controlled but rainfall is usually adequate and well distributed. The favourable rainfed areas account for 20 % of the total rainfed lowlands [47] and the remaining

Table 1 Characteristics of rice-farming systems in major rice production environments

Rice environment	Characteristics
Irrigated rice, with assured year-round water	Continuous rice, with one or two crops per year, and occasionally a third rice crop or an upland crop
Diversion irrigation and (favourable) rainfed lowland rice	Where the monsoon season is six months or more: Dry-seeded rice, followed by transplanted rice (irrigated when water is available), followed by an upland crop
	Where the monsoon season is less than six months: Transplanted rice (irrigated when water is available), followed by upland rice
In rainfed, drought-prone areas (mostly on alluvial terraces)	One dry- or wet-seeded rice crop. Upland crops may follow in good seasons
In flood-prone (deepwater and floating rice) areas	One transplanted or wet-seeded rice crop. (In some deepwaterareas, double transplanting is practised.)
In upland rice areas	Under shifting cultivation: Dry-seeded rice, often interplanted, e.g., with maize, and followed by another upland crop, e.g., cassava. Under mechanised cultivation: Sole crop dry-seeded rice, grown annually for several years, after which a grass pasture may be established

Source [33]

80 % of the rainfed lowland area is less favourable. Rice in the latter area suffers from varying degrees of drought, submergence and both drought and submergence [57].

Rainfed lowland rice crops may suffer from both drought and flood. At higher terraces, these environments may be drought-prone. In the back swamps, poor soil conditions due to poor drainage are common. Farmers often cultivate rainfed lowland rice at several toposequence levels such that on one farm some fields may be drought-prone while others may be flood-prone in the same season.

The predominant cropping system in the rainfed environments is a single crop of rice although, in some areas, farmers are able to grow rice and a post-rice crop in the following season. For example, in the rainfed areas of the northern provinces of the Philippines, farmers plant legumes, wheat, maize or vegetables as a second crop but usually on a smaller area.

Understanding why variations in productivity exist between smallholders in different biophysical environments helps crop policymakers, scientists and extension staff to develop and extend new technologies to smallholders and improve existing ones that will significantly increase productivity. Furthermore, variations in productivity among groups of smallholders in different locations means that improved ways of doing things have to be appropriate to their resource endowments, socioeconomic conditions and biophysical and climatic environments. It requires the measurement of the impacts of innovations over time to gain an appreciation whether it has been possible to construct a new production frontier to increase production and incomes. Rainfed lowland farmers have started adopting the new varieties, but the yields have remained lower in their environments than in more favourable environments. The major concerns are low soil fertility and fertiliser use, drought and flood problems, the lack of location-specific varieties and production technologies, poor weed management, inadequate availability and quality of inputs, inadequate and ineffective and uneven extension support to farmers, slow adoption of recommended technologies, and poor rural infrastructure.

4.2 Assessment of Smallholder Response to Rural Development Interventions

We consider the most promising avenue in estimating spatio-temporal models of smallholder response to project, program and policy interventions in rural areas in the Philippines to be that pursued by [20, 37]. Holloway and Lapar [37, p. 41] posited that 'in targeting public resources to particular regions it is important to have some understanding of the extent to which proactive agents influence neighbours, the extent to which this "ripple-effect" is passed on and the range of its geographical dispersion'. As mentioned above, [37] examined the propensity for neighbouring smallholders in the Philippines to make the same decision employing Bayesian model estimation methods. A key feature of their approach is the way in which they modelled the interaction between farmers and how it affects an individual farmer's decisions. This interaction clearly has a spatial dimension.

Holloway and Lapar [37] stressed the need to get right the specification of the 'spatial-weight' or 'spatial-contiguity' matrix, which can be done by comparing alternative specifications of the spatial weights to arrive at the 'correct neighbourhood size'. They do this by using a Bayesian model comparison procedure. Their model of market participation was undertaken in three steps: specifying the relationship that a household has between its propensity to enter the market and its observed characteristics; modelling the impact of neighbouring households' decisions to affect the decisions of the household; and formulating and estimating a spatial autoregressive probit model (SAPM) and spatial error probit model (SEPM) of the discrete choice of the household. They spell out five features of their probability model that direct their estimation approach.

Data were obtained from a household survey conducted in the study site of the Crop-Animal Systems Research Project using structured questionnaires [37]. Among other variables, it includes a measure of distance between the household and the market. It was found that a positive neighbourhood effect exists for market participation. Among five SAPM and five SEPM models that were estimated, the largest marginal and maximised likelihood values were found for a single model, the SAPM spanning a three-purok radius (where a purok comprises 10–20 households).

Holloway and Lapar [37, p. 40] hinted at the wide range of application of SAPMs and SEPMs to study neighbourhood effects and spillovers:

> Neighbourhood effects may arise from information provided by participant neighbours about markets, prices, product quality or quantity. This is, in one sense, the impact of regional-specific neighbourhood social capital provided by participating neighbours.

It shows that there is considerable potential for this form of analysis in studies of the impacts of interventions in rural areas of the Philippines.

Adoption of improved production practices including new technologies is a key area of concern for smallholder development in that agricultural research, development and extension outcomes are a major avenue for improving smallholder farming practices, and thereby the incomes and living standards of the families of smallholders. Edirisinghe [20], for instance, investigated the spatial correlation in production choice by smallholders in Sri Lanka by analysing the choice of production of sheet rubber. Like [37], he modelled spatial relationships by estimating a SAPM to discover that a strong spatial relationship (neighbourhood effect) existed. He inferred from this result that it has important implications for designing extension services by demonstrating that extending advice to all smallholders is unnecessary for reaching and benefiting all farmers.

4.3 *Measuring, Decomposing and Explaining TFP Growth in Smallholder Farming*

Several studies have attempted to account for firm-specific heterogeneity in the measurement of TFP and decomposition into its efficiency components. In initial empirical examples, heterogeneity has been taken into account by adding contextual variables in the model to correct for differences between firms. But accounting for the individual effects on estimates means that the spatial dimension is ignored. We summarise just three of a number of useful approaches. First, we follow [27] by defining a spatial stochastic frontier model as:

$$\log(y_i) = \log(f(x_i; \beta)) + v_i - u_i \tag{1}$$

$$\log(y_i) = \log(f(x_i; \beta)) + v_{it} - \left(1 - \rho \sum_i w_i\right)^{-1} \hat{u}_i \tag{2}$$

where u_i is the inefficiency term that is dependent on w is a matrix that includes relative spatial information and ρ is spatial lag parameter $(0 < \rho < 1)$; $v_i \sim iidN(0, \sigma_v^2)$; $u_i \sim iidN^+(0, 1 - \rho \sum_i w_i)^{-2} \sigma_{\hat{u}}^2$; and v_i and u_i are independently distributed of each other and $\hat{u}_i \sim N(0, \sigma_{\hat{u}}^2)$. Details of the density function and log-likelihood function are provided by [27, pp. 682–683].

Second, [32] demonstrated how to formulate and estimate a spatial autoregressive stochastic frontier model using annual data for 41 European countries over 22 panels. Their specification of the frontier allowed efficiency to vary over time and across the cross-sections, a highly desirable attribute. The calculation of efficiency followed a standard composed error structure whereas the spatial frontier was estimated innovatively by accounting for the endogenous spatial autoregressive variable with a multivariate iterative search procedure combining algorithms used in spatial econometrics with established stochastic frontier analysis. A key element of their results was that of asymmetry between efficiency spillovers between countries.

Third, given both spatial and temporal variations in production conditions in rice-based ecosystems, it is desirable to formulate models that capture both of these dimensions. This has been achieved in the study by [31] who measured and decomposed TFP growth using a spatial autoregressive frontier model. Glass et al. [31] explained how the marginal effect of an explanatory variable in the model is a function of the spatial autoregressive variable. The panel data set is for 40 European countries over the period from 1995 to 2008. The usual components of TFP change were measured technical efficiency change, scale efficiency change and technological change and distance between countries was captured as either economic distance or a composite variable of geographic and economic distance. Results revealed that the largest spillover effect of TFP growth was the positive change in scale, which was much larger across economic distance than across the composite measure of geographic and economic distance.

Following [31] a spatial autoregressive production frontier model may be used to evaluate the TFP growth in rice production. A general model is specified as:

$$y_{it} = \kappa + \alpha_i + \tau_t + f(x,t)_{it} + \lambda \sum_{j=1}^{N} w_{it} y_{it} + z_{it}\phi + \varepsilon_{it} \qquad (3)$$
$$i = 1, \ldots, N; t = 1, \ldots, T$$

where N is a cross-section of farms; T is the fixed time dimension; y_it is the output of the ith farm, α is a farm-specific fixed effect; τ_t is a time period effect; $f(x,t)_{it}$ represents a functional form of the production technology; x is a vector of inputs; λ is the spatial autoregressive parameter; w_it is an element of the spatial matrix W; z_{it} is a vector of farm or location-specific characteristics and ϕ is the associated vector of parameters; and ϵ_{it} is an *iid* disturbance for i and t with zero mean and variance σ^2. The specification of the above spatial autoregressive frontier allows efficiency to vary over time and between farms. The formulation of marginal effects and TFP decomposition can be obtained following [44] and [31, p. 292].

Combining impact analysis with the measurement of TFP by estimating a spatial stochastic frontier, as done by [2, 74], could be exploited to understand the effects on smallholder productivity and income of a multitude of government policies and development plans implemented in the Philippines.

In the words of [69], the spatial stochastic frontier shows great promise, and of [75], 'Clearly, further theoretical and empirical research on the subject would be of great interest'.

5 Prospects and Conclusions

The timing is auspicious for rural development policy makers in Southeast Asian countries to consider the potential to apply spatial econometric methods when analysing options for smallholder development. The examples outlined above in terms of options to study smallholders in the Philippines support this contention. Panel data sets are becoming more common and individual smallholder data now regularly come with GIS coordinates to identify their location. This enables the easy collation of data on distances between smallholders and other points of interest in a rural area.

Methods of analysis are improving and an increasingly broad set of powerful techniques are becoming available to tackle specific research questions. In particular, substantial methodological progress has been made in the estimation of spatial panel data models. It is incumbent on national researchers to apprise policy makers of the analytical potential. Two essential elements in research processes will be: the training of researchers in accounting for spatial heterogeneity in production conditions, application of suitable spatial econometric techniques and interpretation of model results; and use of spatial distance measures that truly capture the time and costs of travel between two points.

Advances in big data have been beneficial in enabling analysts to account better for spatial heterogeneity in income-earning conditions and potential in rural areas. It has helped reduce model misspecification. But there are issues arising from the potential for private ownership of some of the data, symbolised by the entry of a multinational agribusiness firm, Monsanto, among others, into the big-data collection and prescriptive-planting arenas:

> Monsanto's prescriptive-planting system, FieldScripts, had its first trials last year and is now on sale in four American states. Its story begins in 2006 with a Silicon Valley startup, the Climate Corporation. Set up by two former Google employees, it used remote sensing and other cartographic techniques to map every field in America (all 25 m of them) and superimpose on that all the climate information that it could find. By 2010 its database contained 150 billion soil observations and 10 trillion weather-simulation points.
>
> ...farmers distrust the companies peddling this new method. They fear that the stream of detailed data they are providing on their harvests might be misused. Their commercial secrets could be sold, or leak to rival farmers; the prescriptive-planting firms might even use the data to buy underperforming farms and run them in competition with the farmers; or the companies could use the highly sensitive data on harvests to trade on the commodity markets, to the detriment of farmers who sell into those markets [4].

Researchers too might find they are restricted from using such big data sources on individual farms in the future as the trend towards big data collection on farms extends globally. Another risk is a recent trend towards diminishing agricultural research budgets that endanger the continued collection of rich panel data sets of farming activities. The capability of national research centres to maintain and collect extensive longitudinal panel and spatially explicit data rest on the funding allocated to these institutions. Even if current data-gathering processes are continued, there is a risk that the quality of the data will diminish without the skilled research personnel needed to identify influential elements of spatial heterogeneity and to manage these processes.

Our concluding observation is that the systems perspective should prevail: the whole is greater than the sum of the parts. *Skilled analysts with detailed understanding of the production environment and its spatial heterogeneity* are needed to formulate and estimate panel data models that integrate methodological advances in spatial econometric analysis with use of big data and an appropriate measure of distance. Integrating all these components will provide much more insightful and useful results than modelling without even one of these elements when analysing smallholder development.

References

1. Adetutu, M., Glass, A.J., Kenjegalieva, K., Sickles, R.C.: The effects of efficiency and TFP growth on pollution in Europe: a multistage spatial analysis. SSRN Electron. J. **10** (2014)
2. Affuso, A.: Spatial autoregressive stochastic frontier analysis: an application to an impact evaluation study. http://ssrn.com/abstract=1740382 (2010). Accessed 7 July 2015
3. Allaire, G., Cahuzac, E., Simioni, M.: Spatial diffusion and adoption determinants of European agri-environmental supports related to extensive grazing in France. Paper presented to 5emes Journees, INRA-SFER-CIRAD, Dijon, December 2011. www.sfer.asso.fr/content/download/.../D2. Accessed 29 June 2015
4. Anon. Digital disruption on the farm: managers in the most traditional of industries distrust a promising new technology. The Economist, 24 May 2014. http://www.economist.com/news/business/21602757-managers-most-traditional-industries-distrust-promising-new-technology-digital. Accessed 18 July 2015
5. Anselin, L.: Spatial Econometric: Methods and Models. Kluwer Academic Publishers, Dordrecht (1988)
6. Anselin, L., Bongiovanni, R., Lowenberg-DeBoer, J.: A spatial econometric approach to the economics of site-specific nitrogen. Am. J. Agric. Econ. **86**(3), 675–687 (2004)
7. Arbia, G., Lopez Bazo, E., Moscone, F.: Frontiers in spatial econometrics modelling. Econ. Model. **29**(1), 1–2 (2012)
8. Areal, F.J.K., Balcombe, K., Tiffin, R.: Integrating spatial dependence into stochastic frontier analysis. MPRA Paper, 24961 (2010)
9. Baltagi, B.H.: Spatial panels. In: Ullah, A., Giles, D.E.A. (eds.) Handbook of Empirical Economics and Finance. Chapman and Hall, Taylor and Francis Group, Boca Raton, Florida (2011)
10. Barrios, E.B., Lavado, R.F.: Spatial stochastic frontier models. Philippine Institute of Development Studies Discussion Paper, (2010-08) (2010)
11. Baylis, K., Paulson, N.D., Piras, G.: Spatial approaches to panel data in agricultural economics: a climate change application. J. Agric. Appl. Econ. **43**(3), 325–338 (2011)
12. Bockstael, N.E.: Modeling economics and ecology: the importance of a spatial perspective. Am. J. Agric. Econ. **78**(5), 1168–1180 (1996)
13. Chamberlin, J.: Infrastructure, services, and smallholder income growth: evidence from Kenyan panel data. In: Paper Presented at the 4th International Conference of the African Association of Agricultural Economists, 22–25 Sept, Hammamet, Tunisia (2013)
14. Choi, Y., Nieto, A.: Optimal haulage routing of off-road dump trucks in construction and mining sites using Google earth and a modified least-cost path algorithm. Autom. Constr. **20**(7), 982–997 (2011)
15. Dijkstra, E.W.: A note on two problems in connection with graphs. Numerische Math. **1**(1), 269–271 (1959)

16. Dixon, J., Gulliver, A., Gibbon, D.: Farming systems and poverty: improving farmers livelihoods in a changing world (2001). ftp://ftp.fao.org/docrep/fao/004/ac349e/ac349e00.pdf. Accessed 17 Aug 2015
17. Dreyfus, S.E.: An appraisal of some shortest-path algorithms. Oper. Res. **17**(3), 395–412 (1969)
18. Dubuc, S.: Gis-based accessibility analysis for network optimal location model: an application for bio-energy projects of the mineral industry in the UK. Eur. J. Geogr. **407**, 1–24 (2007)
19. Eberhardt, M., Teal, F.: No mangoes in the Tundra: spatial heterogeneity in agricultural productivity analysis. Oxf. Bull. Econ. Stat. **75**(6), 914–939 (2013)
20. Edirisinghe, J.: Spatial dependence of production choice: application of Bayesian spatial autoregressive probit model on smallholder rubber producers. J. Agric. Environ. Int. Dev. **107**(2), 213–227 (2013)
21. Ehsani, R., Durairaj, C.D.: Spatial food and agricultural data. Systems analysis and modelling in food and agriculture, encyclopedia of life support systems EOLSS, Developed under the Auspices of the UNESCO, EOLSS Publishers, Paris, France (2002)
22. Elhorst, J.P.: Applied spatial econometrics: raising the bar. Spat. Econ. Anal. **5**(1), 9–28 (2010)
23. Entwisle, B., Walsh, S.J., Rindfuss, R.R., VanWey, L.K.: Population and upland crop production in Nang Rong, Thailand. Popul. Environ. **26**(6), 449–470 (2005)
24. Ertur, C., Koch, W.: Growth, technological interdependence and spatial externalities: theory and evidence. J. Appl. Econom. **22**, 1033–1062 (2007)
25. FAOGeoNetwork. Geonetwork User Manual, Release 2.6.4. http://www.fao.org/geonetwork/srv/en/main.home (2012). Accessed 29 July 2015
26. Florax, R.J.G.M., Voortman, R.L.: Spatial dimensions of precision agriculture: a spatial econometric analysis of millet yield on Sahelian coversands. Agric. Econ. **27**(2), 426–443 (2002)
27. Fusco, E., Vidoli, F.: Spatial stochastic frontier models: controlling spatial global and local heterogeneity. Int. Rev. Appl. Econ. **27**(5), 679–694 (2013)
28. GDAE. Trade, agricultural expansion, and climate change in the Amazon basin: a research program of Tufts Global Development and Environment Institute. http://www.ase.tufts.edu/gdae/policy_research/amazon_project.html (2015). Accessed 14 July 2015
29. GeoScienceAustralia. National scale: free data downloads. https://www.ga.gov.au/products/servlet/controller? (2012). Accessed 23 Dec 2012
30. Glass, A., Kenjegalieva, K., Sickles, R.C.: A spatial autoregressive production frontier model for panel data: with an application to European countries. http://ssrn.com/abstract=2227720 (2013). Accessed 7 July 2015
31. Glass, A., Kenjegalieva, K., Paez-Farrell, J.: Productivity growth decomposition using a spatial autoregressive frontier model. Econ. Lett. **119**, 291–295 (2013)
32. Glass, A.J., Kenjegalieva, K., Sickles, R.C.: Estimating efficiency spillovers with state level evidence for manufacturing in the US. Econ. Lett. **123**, 154–159 (2014)
33. Greenland, D.J.: The sustainability of rice farming. International Rice Research Institute and CAB International, Wallingford (1997)
34. Guan, S., Shikanai, T., Minami, T., Nakamura, M., Ueno, M., Setouchi, H.: Development of a system for recording farming data by using a cellular phone equipped with GPS. Agric. Inf. Res. **15**(3), 214–254 (2006)
35. Healey, S.P., Blackard, J.A., Morgan, T.A., Loeffler, D., Jones, G., Songster, J., Brandt, J.P., Moisen, G.G., DeBlander, L.T.: Changes in timber haul emissions in the context of shifting forest management and infrastructure. Carbon Balance Manag. **4**(9), 1–11 (2009)
36. Herrero, M., Thornton, P.K., Bernues, A., Baltenweck, I., Vervoort, J., van de Steeg, J., Makokha, S., van Wijk, M.T., Karanja, S., Rufino, M.C., Staal, S.J.: Exploring future changes in smallholder farming systems by linking socio-economic scenarios with regional and household models. Glob. Environ. Change **24**, 165–182 (2014)
37. Holloway, G., Lapar, M.L.A.: How big is your neighbourhood? spatial implications of market participation among filipino smallholders. J. Agric. Econ. **58**(1), 37–60 (2007)
38. Hsiao, C.: Analysis of Panel Data. Cambridge University Press, Cambridge (2003)
39. Hughes, N., Davidson, A., Lawson, K., Jackson, T., Sheng, Y.: Productivity pathways: climate-adjusted production frontiers for the Australian broadacre cropping industry, ABARES

Research Report 11.5. Technical report, Australian Bureau of Agricultural and Resource Economics and Science, Canberra (2011)
40. IRRI. 1993–1995 IRRI Rice Almanac. International Rice Research Institute, Los Banos, Laguna (1993)
41. IRRI. RiceWeb: a compendium of facts and figures from the world of rice. International Rice Research Institute, Los Banos, Laguna (2001)
42. Lee, L.F., Yu, J.: Spatial panel data models. In: Baltagi, B.H. (ed.) The Oxford Handbook of Panel Data. Oxford University Press, Oxford (2015)
43. LeSage, J.: The theory and practice of spatial econometrics. http://www.spatial--econometrics.com (1999)
44. LeSage, J.P., Pace, R.K.: Introduction to Spatial Econometrics. Taylor and Francis, Boca Raton (2009)
45. Lesschen, J.B., Berburg, P.H., Staal, S.J.: Statistical methods for analysing the spatial dimension of changes in land use and farming. LUCC Report Series 7, The International Livestock Research Institute, Nairobi, Kenya and Land Use Cover Change Focus 3 Office, Wageningen University, Netherlands (2005)
46. Lewis, D.J., Braham, B.L., Robinson, B.: Are there spatial spillovers in adoption of clean technology? the case of organic dairy farming. Land Econ. **87**(2), 250–267 (2011)
47. Mackill, D.J., Coffman, W.R., Garrity, D.P.: Rainfed Lowland Rice Improvement. International Rice Research Institute, Los Banos, Laguna (1996)
48. Mariano, M.J., Villano, R., Fleming, E.: Are irrigated farming ecosystems more productive than rainfed farming systems in rice production in the Philippines? Agric. Ecosyst. Environ. **139**(4), 603–610 (2010)
49. Mariano, M.J., Villano, R., Fleming, E.: Technical efficiency of rice farms in different agroclimatic zones in the Philippines: an application of a stochastic metafrontier model. Asian Econ. J. **25**(3), 245–269 (2011)
50. Min, M., Jiaying, C.: A spatial econometrics analysis on regional disparities of agricultural mechanization in China. In: Contributed Paper to the 1st International Conference on Agro-Geoinformatics, Beijing, 2–4 August 2012
51. Minten, B., Barrett, C.B.: Agricultural technology, productivity, and poverty in Madagascar. World Dev. **36**(5), 797–822 (2008)
52. MODIS. Moderate resolution imaging spectroradiometer: data, Online database, National Aeronautics and Space Administration Online Database. http://modis.gsfc.nasa.gov/data/ (2015). Accessed 29 July 2015
53. Moss, J.F.: Potential contribution of land-use change to climate policy: a spatio-temporal analysis in an Australian catchment. Ph.d. thesis, University of New England, Armidale, NSW (2014)
54. Moura e Sa Cardoso, C., Ravishankar, G.: Productivity growth and convergence: a stochastic frontier analysis. J. Econ. Stud. **42**, 224–236 (2014)
55. Nannicini, G., Baptiste, P., Barbier, G., Krob, D., Liberti, L.: Fast paths in large-scale dynamic road networks. Comput. Optim. Appl. **45**(1), 143–158 (2010)
56. Paelinck, J., Klaassen, L.: Spatial Econometrics. Saxon House, Farnborough (1979)
57. Pandey, S.: Rainfed lowland rice research: challenges and priorities for the 21st century. In: Cooper, M., Fukai, S., Salisbury, J. (eds) Breeding Strategies for Rainfed Lowland Rice in Drought-Prone Environments, ACIAR Proceedings No. 77. Australian Centre for International Agricultural Research, Canberra (1997)
58. Paraguas, F.J., Dey, M.: Aquaculture productivity convergence in India: a spatial econometric perspective. Agric. Econ. Res. Rev. **19**, 121–134 (2006)
59. Parker, D.C., Munroe, D.K.: The geography of market failure: edge-effect externalities and location and production patterns of organic farming. Ecol. Econ. **60**, 821–833 (2007)
60. Pavlyuk, D.: Maximum likelihood estimator for spatial stochastic frontier models. Munich Personal RePEc Archive Paper, (43390). http://mpra.ub.uni--muenchen.de/43390/ (2012). Accessed 28 July 2015
61. Pede, V.O., Sparks, A.H., McKinley, J.D.: Regional income inequality and economic growth: a spatial econometrics analysis for provinces in the Philippines. Contributed Paper Presented at the 56th AARES Annual Conference, Fremantle, Western Australia, 7–10 February 2012

62. Pede, V.O., McKinley, J., Singbo, A., Kajisa, K.: Spatial dependency of technical efficiency in rice farming: the case of Bohol, Philippines. Paper Presented at the 2015 Agricultural and Applied Economics Association and Western Agricultural Economics Association Annual Meeting, San Francisco, CA, 26–28 July 2015
63. Press, G.: Big data definitions: what's yours? http://www.forbes.com/sites/gilpress/2014/09/03/12--big--data--definitions--whats--yours/ (2015). Accessed 7 July 2015
64. Rees, W.G.: Least-cost paths in mountainous terrain. Comput. Geosci. **30**(3), 203–209 (2004)
65. Roy Chowdhury, R.R.: Landscape change in the Calakmul Biosphere Reserve, Mexico: modeling the driving forces of smallholder deforestation in land parcels. Appl. Geogr. **26**, 129–152 (2006)
66. Schlenker, W., Hanemann, W.M., Fisher, A.C.: The impact of global warming on U.S. agriculture: an econometric analysis of optimal growing conditions. Rev. Econ. Stat. **88**(1), 113–115 (2006)
67. Schmidt, A.M., Moreira, A.R.B., Helfand, S.M., Fonseca, T.C.O.: Spatial stochastic frontier models: accounting for unobserved local determinants of inefficiency. J. Product. Anal. **31**, 101–112 (2009)
68. Schmidtner, E.B.: Spatial econometric methods in agricultural economics: selected case studies in German agriculture. Ph.d. dissertation, University of Hohenheim (2013)
69. Sickles, R.C., Hao, J., Shang, C.: Panel data and productivity measurement. In: Baltagi, B.H. (ed.) The Oxford Handbook of Panel Data. Oxford University Press, Oxford (2015)
70. Singh, V.P., Toung, T.P., Kam, S.P.: Characterizing rainfed rice environments: An overview of the biophysical aspects. In: Tuong, T.P., Kam, S.P., Wade, L., Pandey, S., Bouman, B.A.M., Hardy, B. (eds.) Characterizing and Understanding Rainfed Environments, pp. 3–32. International Rice Research Institute, Los Banos, Laguna (2000)
71. Staal, S.J., Baltenweck, I., Waithaka, M.M., deWolff, T., Njoroge, L.: Location and uptake: integrated household and GIS analysis of technology adoption and land use, with application to smallholder dairy farms in Kenya. Agric. Econ. **27**, 295–315 (2002)
72. Sutherland, L., Gabriel, D., Hathaway-Jenkins, L., Pascual, U., Schmutz, U., Rigby, D., Godwin, R., Sait, S.M., Sakrabani, R., Kunin, W.E., Benton, T.G., Stagl, S.: The neighbourhood effect: a multidisciplinary assessment of the case for farmer co-ordination in agri-environmental programmes. Land Use Policy **29**, 502–512 (2012)
73. Tomlin, C.D.: GIS and Cartographic Modelling. ESRI Press, Redland (2012)
74. Travnikar, T., Juvancic, L.: Application of spatial econometric approach in the evaluation of rural development policy: the case of measure modernisation of agricultural holdings. Stud. Agric. Econ. **115**, 98–103 (2013)
75. Tsionas, E.G., Michaelides, P.G.: A spatial stochastic frontier model with spillovers: evidence for Italian regions. Scott. J. Polit. Econ. (in press), pp. 1–14. http://onlinelibrary.wiley.com/doi/10.1111/sjpe.12081/full (2015). Accessed 30 July 2015
76. UNEP. Environmental data explorer. Online database, United Nations Environment Programme. http://geodata.grid.unep.ch/ (2014). Accessed 29 July 2015
77. Upton, V., ODonoghue, C., Ryan, M.: The physical, economic and policy drivers of land conversion to forestry in Ireland. J. Environ. Manag. **132**, 79–86 (2014)
78. Vera-Diaz, M.D.C., Kaufmann, R.K., Nepstad, D.C.: The environmental impacts of soybean expansion and infrastructure development in Brazils Amazon basin. GDAE Working Paper, 2009. http://www.ase.tufts.edu/gdae/Pubs/wp/09-05TransportAmazon.pdf. Accessed 12 July 2015
79. Vera-Diaz, M.D.C., Kaufmann, R.K., Nepstad, D.C.: Transportation infrastructure development and soybean expansion in the Amazon basin: a spatial approach. Unpublished manuscript (2009)
80. Ward, P.S., Florax, R.J.G.M., Flores-Lagunes, A.: Climate change and agricultural productivity in Sub-Saharan Africa: a spatial sample selection model. Eur. Rev. Agric. Econ. **41**(2), 199–216 (2014)
81. Wheeler, D., Hammer, D., Kraft, R., Dasgupta, S., Blankespoor, B.: Economic dynamics and forest clearing: a spatial econometric analysis for Indonesia (2011)

82. Wollni, M., Andersson, C.: Spatial patterns of organic agriculture adoption: evidence from Honduras. Paper Presented at the AEL (Research Committee on Development Economics) Conference, Munich, 21–22 June 2013
83. Zhan, F.B., Noon, C.E.: Shortest path algorithms: an evaluation using real road networks. Transp. Sci. **32**(1), 65–73 (1998)
84. Zhong, D., Zhu, X., Zhong, L., Xu, J.: Research and design of cross-platform farmland data acquisition system based on PhoneGap. Guangdong Agric. Sci. **19**, 41 (2014)

Consistent Re-Calibration in Yield Curve Modeling: An Example

Mario V. Wüthrich

Abstract Popular yield curve models include affine term structure models. These models are usually based on a fixed set of parameters which is calibrated to the actual financial market conditions. Under changing market conditions also parametrization changes. We discuss how parameters need to be updated with changing market conditions so that the re-calibration meets the premise of being free of arbitrage. We demonstrate this (consistent) re-calibration on the example of the Hull–White extended discrete-time Vasiček model, but this concept applies to a wide range of related term structure models.

1 Introduction

Popular stochastic models for interest rate modeling include affine term structure models such as the Vasiček [10] and the Cox-Ingersoll-Ross (CIR) [1] models. These models are based on spot rate processes $(r_t)_{t\geq 0}$. No-arbitrage arguments are then applied to obtain zero-coupon bond prices, yield curves, forward rates and option prices from these spot rate processes. These term structure models have in common that prices are calculated under a risk neutral measure for known *static* model parameters. In practice, one often makes an explicit model choice and then calibrates the model parameters to the actual financial market conditions. Since financial market conditions may change over time, models are permanently re-calibrated to new financial market situations. These re-calibrations imply that model parameters change over time. Therefore, these model parameters should be understood *dynamically* as stochastic processes.

If we work under the premise of no-arbitrage, then re-calibration is severely restricted by side constraints. The study of these side constraints has led to an

M.V. Wüthrich (✉)
RiskLab, Department of Mathematics, ETH Zurich, Rämistrasse 101,
8092 Zurich, Switzerland
e-mail: mario.wuethrich@math.ethz.ch

M.V. Wüthrich
Swiss Finance Institute, Zurich, Switzerland

V.-N. Huynh et al. (eds.), *Causal Inference in Econometrics*,
Studies in Computational Intelligence 622, DOI 10.1007/978-3-319-27284-9_4

interesting and tractable class of so-called *consistent re-calibration* (CRC) models, see Harms et al. [3] and Richter–Teichmann [9]. The aim of this paper is to revisit and discuss CRC models with the "simplest" tractable term structure model at hand. Already this simple model leads to an interesting family of term structure models which turns out to be rather powerful and which naturally combines spot rate models with the Heath-Jarrow-Morton (HJM) [5] framework. To highlight this relationship we choose the Hull–White extended discrete-time (one-factor) Vasiček model. We admit that in many cases this model is not appropriate for real world yield curve modeling but it is suitable for this presentation because it allows to show all essential features of CRC. More appropriate models for real world modeling are considered in Harms et al. [3, 4]. Interestingly, these CRC models also allow to identify the market-price of risk which describes the differences between the risk neutral (pricing) measure and the real world measure under which we collect (historical) observations. The market-price of risk identification is a notoriously difficult problem but it seems rather straightforward in the present set-up.

Organization of this manuscript. In Sect. 2 we revisit the discrete-time one-factor Vasiček model, we discuss its Hull–White extension, and in Theorem 1 we provide the calibration of the Hull–White extension to the actually observed market yield curve. In Sect. 3 we describe the CRC algorithm which allows for stochastic modeling of model parameters. This section is complemented by the HJM representation of the CRC algorithm. In Sect. 4 we introduce the market-price of risk which describes the differences between the risk neutral pricing measure and the real world probability measure. In Sect. 5 we give interpretations to the parameter processes and we provide calibration and estimation of all model parameters. Finally, in Sects. 6 and 7 we summarize our results and we support our conclusions with an example based on the Swiss currency CHF. The proofs of the statements are provided in the appendix.

2 Hull–White Extended Discrete-Time Vasiček Model

In this section we revisit the Hull–White extended discrete-time (one-factor) Vasiček model. Fix a grid size $\Delta > 0$ and consider the discrete time grid $\{0, \Delta, 2\Delta, 3\Delta, \ldots\} = \Delta\mathbb{N}_0$. Choose a (sufficiently rich) filtered probability space $(\Omega, \mathcal{F}, \mathbb{P}^*, \mathbb{F})$ with discrete-time filtration $\mathbb{F} = (\mathcal{F}_t)_{t\in\mathbb{N}_0}$, where $t \in \mathbb{N}_0$ refers to time point $t\Delta$. Assume that $\mathbb{P}^*$ denotes a risk neutral measure (equivalent martingale measure) for the bank account numeraire (money market account). A stochastic process $\mathbf{X} = (X_t)_{t\in\mathbb{N}_0}$ will be called $\mathbb{F}$-adapted if every component X_t of $\mathbf{X}$ is $\mathcal{F}_t$-measurable.

2.1 Discrete-Time (One-Factor) Vasiček Model

Choose parameters $b, \beta, \sigma > 0$ and define the spot rate process $(r_t)_{t\in\mathbb{N}_0}$ as follows: choose initial value $r_0 \in \mathbb{R}$ and define recursively for $t > 0$

$$r_t = b + \beta r_{t-1} + \sigma \varepsilon_t^*, \tag{1}$$

with $(\varepsilon_t^*)_{t \in \mathbb{N}_0}$ being $\mathbb{F}$-adapted and ε_t^* being independent of $\mathscr{F}_{t-1}$ having a (one-dimensional) standard Gaussian distribution under $\mathbb{P}^*$ for $t > 0$. Under the risk neutral measure $\mathbb{P}^*$, the no-arbitrage price at time $t\Delta$ of the zero-coupon bond (ZCB) with maturity date $m\Delta > t\Delta$ is given by

$$P(t, m) = P_\Delta(t, m) = \mathbb{E}^* \left[\exp \left\{ -\Delta \sum_{k=t}^{m-1} r_k \right\} \middle| \mathscr{F}_t \right].$$

For notational convenience we drop the lower index Δ in $P_\Delta(t, m)$ (Δ is kept fixed throughout the paper). Assume $\beta \neq 1$. The price of the ZCB in this discrete-time (one-factor) Vasiček model has the following affine term structure, see Theorem 3.6 in Wüthrich–Merz [11],

$$P(t, m) = \exp \{A(t, m) - r_t \; B(t, m)\},$$

with $A(m-1, m) = 0$ and $B(m-1, m) = \Delta$, and recursively for $0 \leq t < m-1$

$$\begin{aligned} A(t, m) &= A(t+1, m) - bB(t+1, m) + \frac{\sigma^2}{2} B(t+1, m)^2, \\ B(t, m) &= \frac{\Delta}{1-\beta} \left(1 - \beta^{m-t}\right). \end{aligned} \tag{2}$$

In the case $\beta = 1$ we have $B(t, m) = (m-t)\Delta$ and $A(t, m)$ is given by (2).

The yield curve takes at time $t\Delta$ the following affine form in the discrete-time Vasiček model: for maturity dates $m\Delta > t\Delta$ we have

$$Y^{\text{Vasi}}(t, m) = -\frac{1}{(m-t)\Delta} \log P(t, m) = -\frac{A(t, m)}{(m-t)\Delta} + r_t \, \frac{B(t, m)}{(m-t)\Delta},$$

with spot rate at time $t\Delta$ given by $Y^{\text{Vasi}}(t, t+1) = r_t$.

The possible shapes of these Vasiček yield curves $(Y^{\text{Vasi}}(t, m))_{m>t}$ are rather limited because of the restrictive functional form and because we only have three parameters $b, \beta, \sigma > 0$ for model calibration. Typically, this does not allow to match the observed market yield curve. Therefore, we consider the Hull–White [7] extended version of the discrete-time Vasiček model.

2.2 Hull–White Extended Version of the Vasiček Model

For the Hull–White [7] extended version of the discrete-time Vasiček model we replace spot rate process (1) as follows: set initial value $r_0 \in \mathbb{R}$ and define recursively for $t > 0$

$$r_t = b_t^{(k)} + \beta r_{t-1} + \sigma \varepsilon_t^*, \tag{3}$$

with constant $b > 0$ in (1) being replaced by t-dependent constants $b_t^{(k)} > 0$ and all other terms being the same (ceteris paribus). The upper index $^{(k)}$ will be playing an important role in the following text. This is going to be explained in detail below.

Assume $\beta \neq 1$. The no-arbitrage price of the ZCB under $\mathbb{P}^*$ in the Hull–White extended discrete-time Vasiček model has affine term structure

$$P^{(k)}(t,m) = \exp\left\{A^{(k)}(t,m) - r_t\, B(t,m)\right\}, \tag{4}$$

with $A^{(k)}(m-1,m) = 0$ and $B(m-1,m) = \Delta$, and recursively for $0 \le t < m-1$

$$A^{(k)}(t,m) = A^{(k)}(t+1,m) - b_{t+1}^{(k)} B(t+1,m) + \frac{\sigma^2}{2} B(t+1,m)^2, \tag{5}$$

$$B(t,m) = \frac{\Delta}{1-\beta}\left(1 - \beta^{m-t}\right). \tag{6}$$

The yield curve then takes at time $t\Delta$ the following affine form in the Hull–White extended discrete-time Vasiček model: for maturity dates $\Delta m > \Delta t$ we have

$$Y^{(k)}(t,m) = -\frac{A^{(k)}(t,m)}{(m-t)\Delta} + r_t\,\frac{B(t,m)}{(m-t)\Delta}. \tag{7}$$

Observe that the Hull–White extended Vasiček yield curves $(Y^{(k)}(t,m))_{m>t}$ have a flexible form because the Hull–White extension $(b_t^{(k)})_t$ provides (infinitely) many degrees of freedom for model calibration. This we are going to explain next.

2.3 *Calibration of Hull–White Extension*

The spot rate process (3) has the Markov property and initialization at time zero is rather arbitrary. In other words, we could consider any other initial time point $k\Delta \in \Delta\mathbb{N}_0$, and study the (time-shifted) spot rate process $(r_{k+s})_{s\in\mathbb{N}_0}$ starting in $r_k \in \mathbb{R}$ at time $k\Delta$ and for $t > k$

$$r_t = b_t^{(k)} + \beta r_{t-1} + \sigma \varepsilon_t^*.$$

The choice of the upper index in $b_t^{(k)}$ now gains an explicit meaning: if we start the spot rate process $(r_{k+s})_{s\in\mathbb{N}_0}$ at time $k\Delta$ we can calibrate the Hull–White extension $(b_t^{(k)})_{t>k}$ to the actual market conditions at that time point $k\Delta$. More specifically, the Hull–White extension is calibrated as follows. Assume that there is a fixed final time-to-maturity date $M\Delta$ and assume that there is an observable market yield curve at time $k\Delta$ given by the vector

$$\mathbf{y}_k^{\mathrm{mkt}} = \left(y^{\mathrm{mkt}}(k, k+1), \ldots, y^{\mathrm{mkt}}(k, k+M)\right)' \in \mathbb{R}^M,$$

where $y^{\mathrm{mkt}}(k,t)$ is the yield rate at time $k\Delta$ for maturity date $t\Delta > k\Delta$. For the same time-to-maturity dates we have Hull–White extended discrete-time Vasiček yield curve based on Hull–White extension $(b_t^{(k)})_{t>k}$, see (7),

$$\mathbf{y}_k^{(k)} = \left(-\frac{A^{(k)}(k,k+1)}{\Delta} + r_k \frac{B(k,k+1)}{\Delta}, \ldots, -\frac{A^{(k)}(k,k+M)}{M\Delta} + r_k \frac{B(k,k+M)}{M\Delta}\right)'.$$

The upper index in $\mathbf{y}_k^{(k)}$ refers to the chosen Hull–White extension $(b_t^{(k)})_{t>k}$ and the lower index to time point $k\Delta$ when the yield curve $\mathbf{y}_k^{(k)}$ is observed. Assume $r_k = y^{\mathrm{mkt}}(k, k+1)$, $\beta \neq 1$ and $\sigma > 0$ are given. The aim is to calibrate Hull–White extension $(b_t^{(k)})_{t>k}$ at time $k\Delta$ such that we obtain identity

$$\mathbf{y}_k^{(k)} = \mathbf{y}_k^{\mathrm{mkt}}. \tag{8}$$

That is, the Hull–White extension $(b_t^{(k)})_{t>k}$ is calibrated such that the model perfectly fits the observed market yield curve $\mathbf{y}_k^{\mathrm{mkt}}$ at time $k\Delta$. This calibration problem can be solved explicitly by induction. We define for $\beta \neq 1$ the left-lower-triangular matrix $\mathsf{C}(\beta) = (c_{i,j})_{1 \le i,j \le M-1} \in \mathbb{R}^{(M-1)\times(M-1)}$ by

$$c_{i,j} = \frac{\Delta}{1-\beta}\left(1 - \beta^{i+1-j}\right) 1_{\{j \le i\}} = B(j, i+1)\, 1_{\{j \le i\}}.$$

For $\mathbf{y} = (y_1, \ldots, y_M)' \in \mathbb{R}^M$ we define vector $\mathbf{z}(\beta, \sigma, \mathbf{y}) = (z_1, \ldots, z_{M-1})' \in \mathbb{R}^{M-1}$ by

$$z_s = \frac{\sigma^2}{2} \sum_{j=k+1}^{k+s} B(j, k+s+1)^2 - y_1 B(k, k+s+1) + (s+1)\Delta\, y_{s+1}, \tag{9}$$

for $s = 1, \ldots, M-1$.

Theorem 1 *Assume $\beta \neq 1$ and $\sigma > 0$ are given parameters. Denote by $\mathbf{y}_k^{(k)}$ the yield curve at time $k\Delta$ obtained from the Hull–White extended discrete-time Vasiček model* (3)–(6) *with Hull–White extension $\mathbf{b}^{(k)} = (b_{k+1}^{(k)}, \ldots, b_{k+M-1}^{(k)})' \in \mathbb{R}^{M-1}$. Identity $\mathbf{y}_k^{(k)} = \mathbf{y}_k^{\mathrm{mkt}}$ holds if and only if*

$$\mathbf{b}^{(k)} = \mathbf{b}^{(k)}(\beta, \sigma, \mathbf{y}_k^{\mathrm{mkt}}) = \mathsf{C}(\beta)^{-1} \mathbf{z}(\beta, \sigma, \mathbf{y}_k^{\mathrm{mkt}}).$$

Theorem 1 explains how the Hull–White extension $\mathbf{b}^{(k)}$ needs to be calibrated so that the resulting yield curve in the Hull–White extended discrete-time Vasiček model exactly matches the observed market yield curve $\mathbf{y}_k^{\mathrm{mkt}}$ for given parameters β and σ. Theorem 1 motivates the basic concept of consistent re-calibration (CRC) models.

Let in a first step the parameters β and σ reflect *actual* financial market conditions. In a second step we then calibrate the Hull–White extension $\mathbf{b}^{(k)}$ using Theorem 1 so that the resulting model does not allow for arbitrage under a change to actual market parameters β and σ. This algorithm termed CRC algorithm is going to be explained next.

3 Consistent Re-Calibration Models

The crucial extension works as follows: assume that also model parameters β and σ (may) vary over time and we aim at constantly re-calibrating the Hull–White extension in a consistent way. We will show that this allows for (stochastic) modeling of model parameters (which could be interpreted as a state space model) and naturally leads to a Heath-Jarrow-Morton (HJM) [5] point of view for term structure modeling.

3.1 Consistent Re-Calibration Algorithm

Assume that $(\beta_k)_{k\in\mathbb{N}_0}$ and $(\sigma_k)_{k\in\mathbb{N}_0}$ are $\mathbb{F}$-adapted processes with $\beta_k \neq 1$ and $\sigma_k > 0$ for all $k \in \mathbb{N}_0$, a.s. The CRC algorithm reads as follows.

(i) Initialization $k = 0$. Assume that the market yield curve observation at time 0 is given by $\mathbf{y}_0^{\mathrm{mkt}}$. The $\mathcal{F}_0$-measurable Hull–White extension at time 0 is then obtained by, see Theorem 1,

$$\mathbf{b}^{(0)} = \mathsf{C}(\beta_0)^{-1}\mathbf{z}(\beta_0, \sigma_0, \mathbf{y}_0^{\mathrm{mkt}}).$$

This makes the Hull–White extended discrete-time Vasiček yield curve (for given parameters β_0 and σ_0) identically equal to $\mathbf{y}_0^{\mathrm{mkt}}$.

(ii) Spot rate dynamics from $k \to k+1$. Assume $(r_k, \beta_k, \sigma_k, \mathbf{b}^{(k)})$ are given. The spot rate process $(r_{k+s})_{s\in\mathbb{N}_0}$ is then for $t > k$ given by

$$r_t = b_t^{(k)} + \beta_k r_{t-1} + \sigma_k \varepsilon_t^*. \tag{10}$$

Note that parameters β_k and σ_k are now $\mathcal{F}_k$-measurable. This provides $\mathcal{F}_{k+1}$-measurable yield rates at time $(k+1)\Delta$ for $m > k+1$

$$Y^{(k)}(k+1, m) = -\frac{A^{(k)}(k+1, m)}{(m-(k+1))\Delta} + r_{k+1}\frac{B^{(k)}(k+1, m)}{(m-(k+1))\Delta},$$

with $A^{(k)}(m-1, m) = 0$ and $B^{(k)}(m-1, m) = \Delta$, and recursively for $k+1 \le t < m-1$

$$A^{(k)}(t,m) = A^{(k)}(t+1,m) - b_{t+1}^{(k)} B^{(k)}(t+1,m) + \frac{\sigma_k^2}{2} B^{(k)}(t+1,m)^2,$$

$$B^{(k)}(t,m) = \frac{\Delta}{1-\beta_k}\left(1 - \beta_k^{m-t}\right).$$

This is exactly the no-arbitrage price under $\mathbb{P}^*$ if parameters β_k and σ_k remain constant for all $t > k$. Note that also $B^{(k)}$ now depends on k.

(iii) Parameter update and re-calibration at $k+1$. Assume that at time $(k+1)\Delta$ the parameters (β_k, σ_k) are updated to $(\beta_{k+1}, \sigma_{k+1})$. This parameter update may be initiated through a new $\mathscr{F}_{k+1}$-measurable yield curve observation

$$\mathbf{y}_{k+1}^{(k)} = \left(Y^{(k)}(k+1,k+2), \ldots, Y^{(k)}(k+1,k+1+M)\right)' \in \mathbb{R}^M,$$

or through a change of financial market conditions. This parameter update $(\beta_k, \sigma_k) \mapsto (\beta_{k+1}, \sigma_{k+1})$ requires *re-calibration* of the Hull–White extension, otherwise there is arbitrage. This re-calibration provides $\mathscr{F}_{k+1}$-measurable Hull–White extension at time $(k+1)\Delta$ given by

$$\mathbf{b}^{(k+1)} = \mathsf{C}(\beta_{k+1})^{-1}\mathbf{z}(\beta_{k+1}, \sigma_{k+1}, \mathbf{y}_{k+1}^{(k)}). \tag{11}$$

The resulting yield curve $\mathbf{y}_{k+1}^{(k+1)}$ under these updated parameters is identically equal to $\mathbf{y}_{k+1}^{(k)}$. Note that this CRC makes the upper index in $\mathbf{y}_k^{(k)}$ superfluous because the Hull–White extension is re-calibrated to the new parameters such that the resulting yield curve remains unchanged. Therefore, we will just write $\mathbf{y}_k$ in the remaining text.

(iv) Iteration. Iterate items (ii)–(iii) for $k \geq 0$. □

Remark 1 There is a well-hidden issue in the above CRC algorithm, namely, if we want to calibrate $\mathbf{b}^{(k)} \in \mathbb{R}^M$, $k > 0$, and we start this algorithm at time 0 then the maturities in $\mathbf{y}_0^{\text{mkt}}$ need to go well beyond $M\Delta$. Either these maturities are observable and the length of $\mathbf{b}^{(k)}$ is reduced in every step of the CRC algorithm or an appropriate extrapolation method beyond the latest available maturity date is applied in every step.

3.2 Heath-Jarrow-Morton Representation of the CRC Algorithm

We analyze the yield curve dynamics $(\mathbf{y}_k)_{k\geq 0}$ obtained from the CRC algorithm of Sect. 3.1. Due to re-calibration (11) it fulfills the following identity (we drop the upper index $^{(k)}$ in $Y(\cdot,\cdot)$)

$$Y(k+1,m) = -\frac{A^{(k)}(k+1,m)}{(m-(k+1))\Delta} + r_{k+1}\frac{B^{(k)}(k+1,m)}{(m-(k+1))\Delta}$$
$$= -\frac{A^{(k+1)}(k+1,m)}{(m-(k+1))\Delta} + r_{k+1}\frac{B^{(k+1)}(k+1,m)}{(m-(k+1))\Delta}. \tag{12}$$

The first line describes the spot rate dynamics of step (ii) of the CRC algorithm. This is based on the $\mathscr{F}_k$-measurable parameters $(\beta_k, \sigma_k, \mathbf{b}^{(k)})$. The second line describes the parameter update of step (iii) of the CRC algorithm and is based on the $\mathscr{F}_{k+1}$-measurable parameters $(\beta_{k+1}, \sigma_{k+1}, \mathbf{b}^{(k+1)})$, after re-calibration. Note that the CRC algorithm has two degrees of freedom because the Hull–White extension $\mathbf{b}^{(k+1)}$ is used for consistency property (11). Our aim is to express $Y(k+1,m)$ as a function of r_k and $Y(k,m)$. This provides the following theorem.

Theorem 2 (HJM representation) *Assume the Hull–White extended discrete-time Vasiček model satisfies the CRC algorithm (i)–(iv) of Sect. 3.1. The yield curve dynamics $(\mathbf{y}_k)_{k\geq 0}$ has the following HJM representation under $\mathbb{P}^*$ for $m > k+1$*

$$(m-(k+1))\Delta\, Y(k+1,m) = (m-k)\Delta\, Y(k,m) - \Delta\, Y(k,k+1)$$
$$+ \frac{\sigma_k^2}{2}\, B^{(k)}(k+1,m)^2 + \sigma_k\, B^{(k)}(k+1,m)\, \varepsilon_{k+1}^*,$$

with $B^{(k)}(k+1,m) = \frac{\Delta}{1-\beta_k}(1-\beta_k^{m-(k+1)})$.

Remark 2

- Observe that in Theorem 2 a magic simplification happens. Simulating the CRC algorithm (i)-(iv) to future time points $k\Delta > 0$ does not require calculation of the Hull–White extensions $(\mathbf{b}^{(k)})_{k\in\mathbb{N}_0}$ according to (11), the knowledge of parameter process $(\beta_k, \sigma_k)_{k\geq 0}$ is sufficient. The Hull–White extensions are fully encoded in the yield curves $\mathbf{y}_k$, $k \in \mathbb{N}_0$, and we can avoid inversion of (potentially) high dimensional matrices $\mathsf{C}(\beta_k)$.
- Theorem 2 provides a one-factor HJM model for the yield curve. This can easily be extended to a multi-factor model by starting with a multi-factor Vasiček model with i.i.d. multivariate Gaussian innovations ε_t^*, $t \in \mathbb{N}$, under $\mathbb{P}^*$, see Harms et al. [4].
- One could also directly start with the HJM framework. However, then we would lose the interpretation of the terms in the HJM equation. Moreover, this approach combines classical spot rate models with HJM representations in such away that well-known term structure models are combined with CRC of parameters.
- We can easily replace the Gaussian innovations ε_t^*, $t \in \mathbb{N}$, by other distributions, for instance, gamma distributions. These replacements may also allow for closed form solutions in the CRC algorithm of Sect. 3.1 and in the HJM representation of Theorem 2. However, in general, the disappearance of the Hull–White extension $\mathbf{b}^{(k)}$ in the HJM representation will not (completely) happen but it will occur in a weakened form in that only the first term $b_{k+1}^{(k)}$ of the Hull–White extension needs to be determined.

There are still two missing pieces in this modeling approach, namely, the description of the parameter process $(\beta_k, \sigma_k)_{k\geq 0}$ and the modeling of the real world measure $\mathbb{P}$. So far we have described the process under a risk neutral pricing measure $\mathbb{P}^*$, and next we are going to model the process under the real world measure, which will allow for yield curve prediction.

4 Real World Dynamics and Market-Price of Risk

All previous derivations were done under a risk neutral measure $\mathbb{P}^*$ for the bank account numeraire. In order to calibrate the model to real data we also need to be able to track the processes under the real world measure $\mathbb{P}$. We present a specific change of measure which provides tractable solutions for the spot rate process dynamics also under the real world measure. For simplicity we choose a sufficiently large finite time horizon $T \in \mathbb{N}$ and we assume that $\mathscr{F}_T = \mathscr{F}$. Choose a two-dimensional $\mathbb{F}$-adapted process $(\lambda_k)_{k\in\mathbb{N}_0} = (\lambda_k^{(0)}, \lambda_k^{(1)})_{k\in\mathbb{N}_0}$ that describes the market-price of risk parameter process. We define the following $\mathbb{P}^*$-density process $(\xi_{k+s})_{s\in\mathbb{N}_0}$

$$\xi_{k+s} = \exp\left\{-\frac{1}{2}\sum_{v=k+1}^{k+s}\left(\lambda_{v-1}^{(0)} + \lambda_{v-1}^{(1)} r_{v-1}\right)^2 - \sum_{v=k+1}^{k+s}\left(\lambda_{v-1}^{(0)} + \lambda_{v-1}^{(1)} r_{v-1}\right)\varepsilon_v^*\right\}, \tag{13}$$

with spot rate process $(r_{k+s})_{s\in\mathbb{N}_0}$ given by (10). The real world probability measure $\mathbb{P}$ is then obtained by the Radon-Nikodym derivative

$$\frac{d\mathbb{P}}{d\mathbb{P}^*} = \xi_T. \tag{14}$$

An immediate consequence is (see also Lemma 3.19 in Wüthrich–Merz [11]) that

$$\varepsilon_{k+s+1} = \varepsilon_{k+s+1}^* + \lambda_{k+s}^{(0)} + \lambda_{k+s}^{(1)} r_{k+s}$$

is standard Gaussian-distributed under $\mathbb{P}$, conditionally given $\mathscr{F}_{k+s}$. The spot rate process $(r_{k+s})_{s\in\mathbb{N}_0}$ given by (10) is described under the real world measure $\mathbb{P}$ for $t > k$ by

$$\begin{aligned} r_t &= b_t^{(k)} - \sigma_k \lambda_{t-1}^{(0)} + \left(\beta_k - \sigma_k \lambda_{t-1}^{(1)}\right) r_{t-1} + \sigma_k \varepsilon_t \\ &= a_{t-1}^{(k)} + \alpha_{t-1}^{(k)} r_{t-1} + \sigma_k \varepsilon_t, \end{aligned} \tag{15}$$

where we define for $t \geq k$ the parameter processes

$$a_t^{(k)} = b_{t+1}^{(k)} - \sigma_k \lambda_t^{(0)} \quad \text{and} \quad \alpha_t^{(k)} = \beta_k - \sigma_k \lambda_t^{(1)}, \tag{16}$$

with $(a_t^{(k)}, \alpha_t^{(k)})$ being $\mathscr{F}_t$-measurable. Formula (15) describes the spot rate process under the real world measure $\mathbb{P}$ and formula (10) under the risk neutral measure $\mathbb{P}^*$ using change of measure (14). We are going to calibrate this process to real data below.

Theorem 2 has provided the HJM view of the re-calibrated Hull–White extended discrete-time Vasiček model under the risk neutral measure $\mathbb{P}^*$. In order to predict yield curves we need to study the dynamics under the real world measure $\mathbb{P}$. This is done in the following corollary.

Corollary 1 *Assume the Hull–White extended discrete-time Vasiček model satisfies the CRC algorithm (i)–(iv) of Sect. 3.1. The yield curve dynamics* $(\mathbf{y}_k)_{k\geq 0}$ *has the following HJM representation for* $m > k+1$ *under the real world measure* $\mathbb{P}$ *satisfying* (14)

$$\begin{aligned}(m-(k+1))\Delta\, Y(k+1,m) = {} & (m-k)\Delta\, Y(k,m) \\ & - \left[\Delta + \sigma_k \lambda_k^{(1)} B^{(k)}(k+1,m)\right] Y(k,k+1) \\ & + \left[\frac{\sigma_k^2}{2} B^{(k)}(k+1,m)^2 - \sigma_k \lambda_k^{(0)} B^{(k)}(k+1,m)\right] \\ & + \sigma_k B^{(k)}(k+1,m)\varepsilon_{k+1},\end{aligned}$$

with $B^{(k)}(k+1,m) = \frac{\Delta}{1-\beta_k}(1-\beta_k^{m-(k+1)})$.

We obtain two additional drift terms and $-\sigma_k \lambda_k^{(1)} B^{(k)}(k+1,m) Y(k,k+1)$ and $-\sigma_k \lambda_k^{(0)} B^{(k)}(k+1,m)$ which are characterized by the market-price of risk parameter $\lambda_k = (\lambda_k^{(0)}, \lambda_k^{(1)})$.

5 Choice of Parameter Process

The last missing piece is the modeling of the parameter process $(\beta_t, \sigma_t)_{t\in\mathbb{N}_0}$. In this section we give different interpretations to this parameter process which will lead to different modeling approaches.

From (10) we get the one-step ahead development $k \to k+1$

$$r_{k+1} = b_{k+1}^{(k)} + \beta_k r_k + \sigma_k \varepsilon_{k+1}^*, \tag{17}$$

with $\mathscr{F}_k$-measurable parameters $b_{k+1}^{(k)}$, β_k and σ_k. Thus, on the one hand $(r_t)_{t\in\mathbb{N}_0}$ evolves according to (17) and on the other hand parameters $(\beta_t, \sigma_t)_{t\in\mathbb{N}_0}$ evolve according to the financial market environment. Note that the Hull–White extension $(\mathbf{b}^{(t)})_{t\in\mathbb{N}_0}$ is fully determined through (11). In order to distinguish the two evolvements of $(r_t)_{t\in\mathbb{N}_0}$ and $(\beta_t, \sigma_t)_{t\in\mathbb{N}_0}$, respectively, we need to assume that the evolvement of the financial market conditions is more viscous than the changes in the spot rate process.

This implies that the evolvement of the parameter process $(\beta_t, \sigma_t)_{t\in\mathbb{N}_0}$ is taking place at a lower pace or a wider time scale compared to $(r_t)_{t\in\mathbb{N}_0}$. We now give different interpretations and calibration procedures for the parameter process.

5.1 Pricing Model Approach Interpretation

$\triangleright$ *Mean reversion rate.* Parameter β is the mean reversion rate of an AR(1) process. Set for illustrative purposes $b_{k+1}^{(k)} \equiv b$, $\beta_k \equiv \beta \in (0,1)$ and $\sigma_k \equiv \sigma$ in (17), then we obtain for $t \in \mathbb{N}$

$$r_t|\mathscr{F}_0 \sim \mathscr{N}\left(\beta^t r_0 + (1-\beta^t)\frac{b}{1-\beta}, \ \sigma^2\frac{1-\beta^{2t}}{1-\beta^2}\right), \qquad \text{under } \mathbb{P}^*.$$

The mean reversion rate $\beta \in (0,1)$ determines at which speed the spot rate process $(r_t)_{t\in\mathbb{N}_0}$ returns to its long-term mean $b^* = b/(1-\beta)$, because $\beta^t \to 0$ for $t \to \infty$. A sensible choice of β_k at time $k\Delta$ will adapt this mean reversion rate to the actual financial market conditions.

$\triangleright$ *Instantaneous variance approach I.* Parameter σ_k plays the role of the instantaneous volatility parameter of the spot rate process $(r_t)_{t\in\mathbb{N}_0}$. The choice of σ_k should reflect actual financial market conditions. This instantaneous volatility parameter can, for instance, be modeled by a Heston [6] like approach, i.e. where $(\sigma_t^2)_{t\geq 0}$ has a Cox-Ingersoll-Ross (CIR) [1] process

$$d\sigma_t^2 = \kappa\left(\theta - \sigma_t^2\right)dt + g\sqrt{\sigma_t^2}dW_t, \tag{18}$$

in which $(W_t)_{t\geq 0}$ is a standard Brownian motion that maybe correlated with $(\varepsilon_k^*)_{k\in\mathbb{N}}$ under $\mathbb{P}^*$. In a more discrete time fashion we could choose a GARCH model for the modeling of the instantaneous variance $(\sigma_k^2)_{k\in\mathbb{N}_0}$. In applications, process (18) should be chosen more viscous than the one of the spot rate process, in order to distinguish and separate the two.

$\triangleright$ *Instantaneous variance approach II.* A second approach links σ_k in (17) directly to the spot rate level r_k at time $k\Delta$. If we replace $(\sigma_t^2)_{t\geq 0}$ in (18) by $(r_t)_{t\geq 0}$ (assuming continuous time $t \in \mathbb{R}_+$ for the moment) we obtain the CIR [1] interest model. In a similar spirit we could adapt the discrete-time spot rate version to

$$r_{k+1} = b_{k+1}^{(k)} + \beta_k r_k + g\sqrt{|r_k|}\varepsilon_{k+1}^*, \tag{19}$$

for a given parameter $g > 0$. However, there is a fundamental difference between our CRC model using (19) and the CIR interest model. In our model the volatility level $\sqrt{|r_k|}$ is only used for the spot rate dynamics in the next period and the yield curve is calculated using the HJM representation of Theorem 2 which is based on the Hull–White extended discrete-time Vasiček model. Whereas the CIR model provides a different affine term structure model which results in non-central χ^2-distributed yields.

This idea can be carried forward by replacing (19) by the spot rate model

$$r_{k+1} = b_{k+1}^{(k)} + \beta_k r_k + g(r_k)\varepsilon_{k+1}^*, \tag{20}$$

for an appropriate function $g\colon \mathbb{R} \to \mathbb{R}_+$, i.e. the instantaneous volatility parameter σ_k is a function $g(r_k)$ of the actual spot rate level r_k at time $k\Delta$. This idea reflects the empirical study of Deguillaume et al. [2] and a possible choice of $g(\cdot)$ could be of the form of Fig. 7 in [2].

The approaches presented in this section are fully based on the risk neutral measure $\mathbb{P}^*$ and calibration often uses actual option prices in order to choose the parameters according to actual market conditions.

5.2 *Historical Calibration of the Prediction Model*

Assume the real world measure $\mathbb{P}$ on $(\Omega, \mathscr{F})$ is defined according to Sect. 4. Moreover, assume for the moment that $a_k^{(k)}$ and $\alpha_k^{(k)}$, given in (15), and the instantaneous volatility parameter σ_k are constant in $k \in \{s-K, \dots, s\}$ for a small window length K, and denote them by a, α and σ. This is motivated by the assumption that the parameter process is more viscous than the spot rate process and, therefore, is almost constant over small window lengths. Under these assumptions we can try to estimate parameters (a, α, σ) at time $k\Delta$ from historical observations. We choose a fixed (small) window length $K \in \mathbb{N}$ and we assume that we have observations $r_{(k-K):k} = (r_t)_{t=k-K,\dots,k}$ for this window length. The log-likelihood function of $r_{(k-K):k}$ under (15) for constant parameters in $\{k-K, \dots, k\}$ is then given by

$$\ell_{r_{(k-K):k}}(a, \alpha, \sigma) \propto \sum_{t=k-K+1}^{k} \left(-\log\sigma - \frac{1}{2\sigma^2}(r_t - a - \alpha r_{t-1})^2 \right).$$

To maximize the log-likelihood function over the parameters a, α and σ we need to solve the following system of equations, which provides the maximum likelihood estimators (MLEs),

$$\frac{\partial \ell_{r_{(k-K):k}}}{\partial a}(a, \alpha, \sigma) = 0, \quad \frac{\partial \ell_{r_{(k-K):k}}}{\partial \alpha}(a, \alpha, \sigma) = 0 \quad \text{and} \quad \frac{\partial \ell_{r_{(k-K):k}}}{\partial \sigma}(a, \alpha, \sigma) = 0.$$

The MLEs at time $k\Delta$ for window length K are then given by, see Proposition 3.7 in Wüthrich–Merz [11],

$$\widehat{\alpha}_k^{\text{MLE}} = \frac{K\sum_{t=k-K+1}^{k} r_t r_{t-1} - \sum_{t=k-K+1}^{k} r_t \sum_{t=k-K+1}^{k} r_{t-1}}{K\sum_{t=k-K+1}^{k} r_{t-1}^2 - (\sum_{t=k-K+1}^{k} r_{t-1})^2}, \tag{21}$$

$$\widehat{a}_k^{\text{MLE}} = \frac{1}{K}\sum_{t=k-K+1}^{k} \left(r_t - \widehat{\alpha}_k^{\text{MLE}} r_{t-1}\right), \tag{22}$$

$$\widehat{\sigma_k^2}^{\text{MLE}} = \frac{1}{K}\sum_{t=k-K+1}^{k} \left(r_t - \widehat{\alpha}_k^{\text{MLE}} r_{t-1} - \widehat{a}_k^{\text{MLE}}\right)^2. \tag{23}$$

Note that $\widehat{\alpha}_k^{\text{MLE}}$, $\widehat{a}_k^{\text{MLE}}$ and $\widehat{\sigma_k^2}^{\text{MLE}}$ are the MLEs for the parameters α, a and σ^2 at time $k\Delta$ (for window length K). This motivates the "best prediction model" at time $k\Delta$ by setting

$$a_k^{(k)} = \widehat{a}_k^{\text{MLE}}, \qquad \alpha_k^{(k)} = \widehat{\alpha}_k^{\text{MLE}} \qquad \text{and} \qquad \sigma_k^2 = \widehat{\sigma_k^2}^{\text{MLE}} \tag{24}$$

for the one-step ahead prediction from time $k\Delta$ to time $(k+1)\Delta$ in (15). This calibration was purely done under the (historical) real world measure $\mathbb{P}$ and we have not used any no-arbitrage arguments. The choice of the window length K crucially determines the viscosity of the parameter process.

Note that (24) describes the spot rate process under the real world measure $\mathbb{P}$. This is not sufficient for price predictions because we also need to understand the risk neutral measure $\mathbb{P}^*$ through the market-price of risk parameter $(\lambda_k)_{k\in\mathbb{N}_0}$, see Corollary 1. This is exactly what we are going to consider in the next section.

5.3 *Continuous-Time Modeling Motivated Inference*

In this section we elaborate the differences between the risk neutral measure and the real world measure. The basic idea is that volatility is the same under both measures which in turn allows to identify the residual parts under both measures. These considerations are inspired by continuous time modeling which will be reflected by choosing a small grid size Δ in our context, we also refer to Harms et al. [3].

5.3.1 Historical Inference on Spot Rates

Assume spot rate process $(r_t)_{t\in\mathbb{N}_0}$ is given under $\mathbb{P}$ by $r_0 \in \mathbb{R}$ and for $t > 0$

$$r_t = a + \alpha r_{t-1} + \sigma\varepsilon_t.$$

The n-step ahead prediction of $(r_t)_{t\in\mathbb{N}_0}$ for $n > 0$ has distribution

$$r_n|_{\mathscr{F}_0} \sim \mathcal{N}\left(\alpha^n r_0 + (1-\alpha^n)a^*\,,\; \sigma^2\frac{1-\alpha^{2n}}{1-\alpha^2}\right), \qquad \text{under } \mathbb{P},$$

with long-term mean $a^* = a/(1-\alpha)$ for $\alpha \neq 1$. We define

$$\gamma = \alpha^n \qquad \text{and} \qquad \tau^2 = \sigma^2\frac{1-\gamma^2}{1-\gamma^{2/n}}.$$

This implies for the n-step ahead prediction

$$r_n|_{\mathscr{F}_0} \sim \mathcal{N}\left(\gamma r_0 + (1-\gamma)a^*\,,\; \tau^2\right), \qquad \text{under } \mathbb{P}. \tag{25}$$

Observe that we remain in the same family of models. Assume we have observations $r_0, r_n, r_{2n}, \ldots$ for a fixed time grid $(\Delta n)\mathbb{N}_0 \stackrel{\text{def.}}{=} \{0, \Delta n, 2\Delta n, \ldots\}$. This allows to estimate γ, a^* and τ^2 on that time grid $(\Delta n)\mathbb{N}_0$, for instance, using MLE as explained in (21)–(23). From these estimates we can reconstruct α and σ^2 on time grid $\Delta\mathbb{N}_0$ as follows:

$$\alpha = \gamma^{1/n} \;=\; \exp\left\{\frac{1}{n}\log\gamma\right\} \;=\; 1 + \frac{1}{n}\log\gamma + o(1/n) \qquad \text{as } n \to \infty,$$

$$\sigma^2 = \tau^2\frac{1-\gamma^{2/n}}{1-\gamma^2} \;=\; \frac{1}{n}\frac{2\tau^2\log\gamma^{-1}}{1-\gamma^2} + o(1/n) \qquad \text{as } n \to \infty.$$

For the asymptotic statements we revert the interpretation of n, that is, we assume that a fixed time grid $\kappa\mathbb{N}_0$ is given for parameter estimation. Then we refine this time grid according to $\Delta = \Delta(n) = \kappa/n$ for large n. Therefore, the spot rate process $(r_t)_{t\in\mathbb{N}_0}$ has on $\Delta = \Delta(n)$ approximation $\widetilde{r}_0 = r_0$ and for $t > 0$

$$\widetilde{r}_t = \frac{1}{n}a^*\log\gamma^{-1} + \left(1 - \frac{1}{n}\log\gamma^{-1}\right)\widetilde{r}_{t-1} + \sqrt{\frac{1}{n}}\sqrt{\frac{2\tau^2\log\gamma^{-1}}{1-\gamma^2}}\,\varepsilon_t,$$

for γ, a^* and τ^2 describing the dynamics on time grid $\kappa\mathbb{N}_0$. The innovation of this approximation is given by

$$\mathsf{D}(\widetilde{r}_t) \stackrel{\text{def.}}{=} \widetilde{r}_t - \widetilde{r}_{t-1} = \frac{1}{n}\left(a^* - \widetilde{r}_{t-1}\right)\log\gamma^{-1} + \sqrt{\frac{1}{n}}\sqrt{\frac{2\tau^2\log\gamma^{-1}}{1-\gamma^2}}\,\varepsilon_t. \tag{26}$$

The crucial observation is that the drift term and the random term live asymptotically for $n \to \infty$ on different scales, namely, we have

$$\mathbb{E}\left[(\mathsf{D}(\widetilde{r}_t))^2 \middle| \mathscr{F}_{t-1} \right] = \frac{1}{n^2} \left[(a^* - \widetilde{r}_{t-1}) \log \gamma^{-1} \right]^2 + \frac{1}{n} \frac{2\tau^2 \log \gamma^{-1}}{1-\gamma^2}.$$

Therefore, if the grid size $\Delta = \Delta(n)$ is sufficiently small, i.e. n sufficiently large, the volatility part dominates the drift part and we may approximate

$$\mathbb{E}\left[(\mathsf{D}(\widetilde{r}_t))^2 \middle| \mathscr{F}_{t-1} \right] \approx \frac{1}{n} \frac{2\tau^2 \log \gamma^{-1}}{1-\gamma^2} \approx \sigma^2. \tag{27}$$

This approximation suggests to estimate σ with the realized volatility, i.e. for window length K and observation $r_{(k-K):k}$ we set realized variance estimator for observations $r_{(k-K):k}$

$$\widehat{\sigma_k^2}^{\text{RVar}} = \frac{1}{K} \sum_{t=k-K+1}^{k} (r_t - r_{t-1})^2 = \frac{1}{K} \sum_{t=k-K+1}^{k} (\mathsf{D}(r_t))^2. \tag{28}$$

Note that the realized variance estimator differs from the MLE $\widehat{\sigma_k^2}^{\text{MLE}}$ given in (23). We are going to analyze this difference numerically. Approximation (27) motivates to carry forward this idea to yield curve observations for general times-to-maturity. This is done in the next section.

5.3.2 Historical Realized Co-Variation

We can carry forward the idea of the last section to calculate realized co-variations within the discrete-time Vasiček model of Sect. 2.1. The innovation of the yield rate for fixed time-to-maturity $m\Delta$ is in the discrete-time Vasiček model given by

$$Y^{\text{Vasi}}(t, t+m) - Y^{\text{Vasi}}(t-1, t-1+m) = \frac{B(t, t+m)}{m\Delta} (r_t - r_{t-1}). \tag{29}$$

Note that this assumes viscosity of parameters over short periods. From formula (26) we immediately obtain approximation (for small $\Delta = \Delta(n)$)

$$\begin{aligned} \mathsf{D}(Y_{t,t+m}^{\text{Vasi}}) &\overset{\text{def.}}{=} Y^{\text{Vasi}}(t, t+m) - Y^{\text{Vasi}}(t-1, t-1+m) \\ &\approx \frac{B(t, t+m)}{m\Delta} \mathsf{D}(\widetilde{r}_t) \\ &= \frac{B(t, t+m)}{m\Delta} \left(\frac{1}{n} \left(a^* - \widetilde{r}_{t-1} \right) \log \gamma^{-1} + \sqrt{\frac{1}{n}} \sqrt{\frac{2\tau^2 \log \gamma^{-1}}{1-\gamma^2}} \varepsilon_t \right). \end{aligned}$$

If we assume that times-to-maturity $m_1\Delta$ and $m_2\Delta$ are fixed and for $\Delta = \Delta(n)$ sufficiently small we get approximation for the co-variation, see also (27),

$$\Sigma_t^{\mathrm{CoV}}(m_1, m_2) \stackrel{\text{def.}}{=} \mathbb{E}\left[\mathsf{D}(Y_{t,t+m_1}^{\mathrm{Vasi}})\mathsf{D}(Y_{t,t+m_2}^{\mathrm{Vasi}})\middle|\mathscr{F}_{t-1}\right] \approx \frac{(1-\beta^{m_1})(1-\beta^{m_2})}{m_1 m_2 (1-\beta)^2}\,\sigma^2. \tag{30}$$

Note that the latter is interesting for several reasons:

- Formula (30) allows for direct cross-sectional calibration of β (which is the mean reversion parameter under the risk neutral measure $\mathbb{P}^*$). That is, the mean reversion parameter β for pricing can directly be estimated from real world observations (without knowing the market-price of risk). This is always based on the assumption that the volatility dominates the drift term.
- We have restricted ourselves to a one-factor Vasiček model which provides comonotonic ZCB prices for different times-to-maturity. Of course, this comonotonicity property needs to be relaxed and co-variations (30) are helpful for determining the number of factors needed, for instance, using a principal component analysis (applied to time series of co-variations).

Assume we have market yield curve observations $\mathbf{y}_t^{\mathrm{mkt}}$ for a time window $t \in \{k-K, \ldots, k\}$. The co-variations can be estimated at time $k\Delta$ by the realized co-variation estimators

$$\widehat{\Sigma}_k^{\mathrm{CoV}}(m_1, m_2) = \frac{1}{K} \sum_{t=k-K+1}^{k} \mathsf{D}(y_{t,t+m_1}^{\mathrm{mkt}})\mathsf{D}(y_{t,t+m_2}^{\mathrm{mkt}}),$$

for

$$\mathsf{D}(y_{t,t+m}^{\mathrm{mkt}}) = y^{\mathrm{mkt}}(t, t+m) - y^{\mathrm{mkt}}(t-1, t-1+m).$$

5.3.3 Cross-Sectional Mean Reversion Parameter Estimate

The realized variance estimator $\widehat{\sigma_k^2}^{\mathrm{RVar}}$ and the realized co-variation estimators $\widehat{\Sigma}_k^{\mathrm{CoV}}(m_1, m_2)$ for given times-to-maturity $m_1\Delta$ and $m_2\Delta$ allow for direct estimation of the mean reversion parameter β at time $k\Delta$. We define the following estimator, see (30),

$$\widehat{\beta_k}^{\mathrm{RCoV}} = \arg\min_{\beta} \left\{ \sum_{m_1, m_2} w_{m_1,m_2} \left(\frac{\widehat{\Sigma}_k^{\mathrm{CoV}}(m_1, m_2)}{\widehat{\sigma_k^2}^{\mathrm{RVar}}} - \frac{(1-\beta^{m_1})(1-\beta^{m_2})}{m_1 m_2 (1-\beta)^2} \right)^2 \right\}, \tag{31}$$

for given symmetric weights $w_{m_1,m_2} = w_{m_2,m_1} \geq 0$. This optimization needs care because it is quite sensitive and it should be modified to the more robust version

$$\frac{\widehat{\Sigma}_k^{\text{CoV}}(m_1, m_2)}{\widehat{\sigma_k^2}^{\text{RVar}}} - \frac{(1-\beta^{m_1})(1-\beta^{m_2})}{m_1 m_2 (1-\beta)^2}$$

$$= \frac{\widehat{\Sigma}_k^{\text{CoV}}(m_1, m_2)}{\widehat{\sigma_k^2}^{\text{RVar}}} - \frac{\left(\sum_{l_1=0}^{m_1-1} \beta^{l_1}\right)\left(\sum_{l_2=0}^{m_2-1} \beta^{l_2}\right)}{m_1 m_2}.$$

We require $(\widehat{\sigma_k^2}^{\text{RVar}})^{-1} \widehat{\Sigma}_k^{\text{CoV}}(m_1, m_2) \in [1/(m_1 m_2), 1]$ for obtaining mean reversion rate $\beta \in [0, 1]$. Therefore, we define its truncated normalized version by

$$\widehat{\Xi}_k(m_1, m_2) = \max\left\{(m_1 m_2)^{-1}, \min\left\{\left(\widehat{\sigma_k^2}^{\text{RVar}}\right)^{-1} \widehat{\Sigma}_k^{\text{CoV}}(m_1, m_2)\ , \ 1\right\}\right\}. \quad (32)$$

Below we use the following estimator for weights $w_m \geq 0$

$$\widehat{\widehat{\beta}}_k^{\text{RCoV}} = \arg\min_{\beta} \left\{ \sum_m w_m \left(\widehat{\Xi}_k(m, m) - \frac{\left(\sum_{l=0}^{m-1} \beta^l\right)^2}{m^2} \right)^2 \right\}. \quad (33)$$

Remark 3 We would like to emphasize that there is a subtle difference between calibration of spot rate volatility in Sect. 5.3.1 and the corresponding method for co-variations of Sect. 5.3.2. For the spot rate process we know that the (one-dimensional) observation r_n always lies within the range of possible observations in the AR(1) model described by (25). For the multi-dimensional observation $\mathbf{y}_t^{\text{mkt}}$ this is not necessarily the case because (29) prescribes a specific form of the range based on the previous observation $\mathbf{y}_{t-1}^{\text{mkt}}$ and the functional form of $B(t, t+m)$. If the observation is too different from the range, then calibration (33) will fail. In the example below we will observe exactly this failure after the financial crisis 2008, because (33) possesses monotonicity properties that are not present in the observed data after 2008.

5.3.4 Inference on Market-Price of Risk Parameters

Finally, we aim at determining the market-price of risk parameter process $(\lambda_k)_{k\in\mathbb{N}_0}$ under the given density process $(\xi_{k+s})_{s\in\mathbb{N}_0}$ assumption (13). From (16) we get $\lambda_k^{(1)} = \sigma_k^{-1}(\beta_k - \alpha_k^{(k)})$. This immediately motivates inference of $\lambda_k^{(1)}$ by

$$\widehat{\lambda}_k^{(1)} = \left(\widehat{\sigma_k^2}^{\text{RVar}}\right)^{-1/2} \left(\widehat{\widehat{\beta}}_k^{\text{RCoV}} - \widehat{\alpha}_k^{\text{MLE}}\right). \quad (34)$$

For inference of $\lambda_k^{(0)}$ we use the first term $b_{k+1}^{(k)}$ of the Hull–White extension which is explicitly given in the proof of Theorem 1 and reads as

$$\frac{\sigma_k^2}{2}\Delta - r_k(1+\beta_k) + 2y^{\mathrm{mkt}}(k,k+2) = b_{k+1}^{(k)} = a_k^{(k)} + \sigma_k \lambda_k^{(0)}. \tag{35}$$

This motivates the following inference of $\lambda_k^{(0)}$

$$\widehat{\lambda}_k^{(0)} = \left(\widehat{\sigma_k^2}^{\mathrm{RVar}}\right)^{-1/2} \left(\frac{\widehat{\sigma_k^2}^{\mathrm{RVar}}}{2}\Delta - r_k(1+\widehat{\beta_k}^{\mathrm{RCoV}}) + 2y^{\mathrm{mkt}}(k,k+2) - \widehat{a}_k^{\mathrm{MLE}} \right). \tag{36}$$

6 Conclusions

We have discussed an extension of classical spot rate models to allow for time dependent model parameters. This extension is done such that the model remains consistent in the sense that parameter updates and re-calibration respect the premise of no-arbitrage. This leads to a natural separation of the drift term into two parts: one correcting for the no-arbitrage drift and a second one describing the market-price of risk. We have exemplified this with the discrete-time one-factor Vasiček model at hand, but this concept of CRC applies to a wide range of continuous and discrete-time models, for more examples we refer to Harms et al. [3, 4].

Within the discrete-time one-factor Vasiček model, formulas (11), (21)–(23), (28), (31), (34) and (36) fully specify the inference of past parameters and the market-prices of risk. The only choices of freedom there still remain in this inference analysis are the choice of the window length K (which determines the viscosity of the parameters) and the choices of the weights w_m attached to the different times-to-maturity m in (33). The latter will require a good balance between short and long times-to-maturity in order to make sure that not one end of the yield curve dominates the inference process. For the example below we only choose short times-to-maturity, accounting for the fact that the one-factor Vasiček model is not able to model the whole range of possible times-to-maturity. For the latter one should switch to multi-factor models which are, for instance, considered in Harms et al. [4].

Going forward we will need to model stochastically the parameter process and the market-price of risk process in order to predict future yield curves. An alternative way that avoids explicit stochastic modeling of these processes is to push the yield curve period by period through Corollary 1 and then make path-wise inference of the parameters as described above. This reflects an empirical (or bootstrap) modeling approach where parameters are fully encoded in the (new) yield curve observations.

In the next section we provide an example that is based on the Swiss currency CHF. Since the discrete-time one-factor Vasiček model cannot capture the entire range of possible times-to-maturity, we only consider the short end of the CHF yield curve. Typically, one needs at least a three-factor model in order to reflect short and long times-to-maturity simultaneously in an appropriate way.

7 Swiss Currency CHF Example

We choose a business daily grid size $\Delta = 1/252$. As historical observations, we choose the Swiss Average Rates (SARs) for the Swiss currency CHF. The SARs are available for times-to-maturity up to 3 months with the following explicit times-to-maturity: the SAR Over-Night (SARON) corresponds to a time-to-maturity of Δ and the SAR Tomorrow-Next (SARTN) to a time-to-maturity of 2Δ. The latter is not completely correct because SARON is a collateral over-night rate and tomorrow-next is a call money rate for receiving money tomorrow which has to be paid back the next business day. Moreover, we have SARs for times-to-maturity of 1 week (SAR1W), 2 weeks (SAR2W), 1 month (SAR1M) and 3 months (SAR3M), see also Jordan [8] for background information on SARs. The data is available from December 8, 1999, until September 15, 2014. We illustrate the data in Fig. 1 (lhs) and the rooted realized variations $\sqrt{\widehat{\Sigma}_k^{\text{CoV}}(m,m)}$, $m \in \{1, 2, 5, 10, 21, 63\}$, for a window length of $K = 126$ business days is given in Fig. 1 (rhs). The vertical dotted lines indicate the financial crisis in 2003 and the beginning of the financial crisis 2008.

Swiss currency interest data is known to be difficult for modeling. This is true in particular after the financial crisis 2008, see Fig. 1. One reason for this modeling challenge is that the Swiss National Bank (SNB) has strongly intervened at the financial market after 2008, both to protect the Swiss banking sector which has two very large international banks (compared to the Swiss GDP), and also to support the Swiss export industry and the Swiss tourism which heavily depend on exchange rates. One instrument applied by the SNB was a floor on the exchange rate EUR/CHF. Of course, these interventions heavily influence the SARs and the yield curves and these kind of actions cannot be captured by a simple one-factor CRC Vasiček model. However, we would still like to perform the analysis. We therefore only consider short

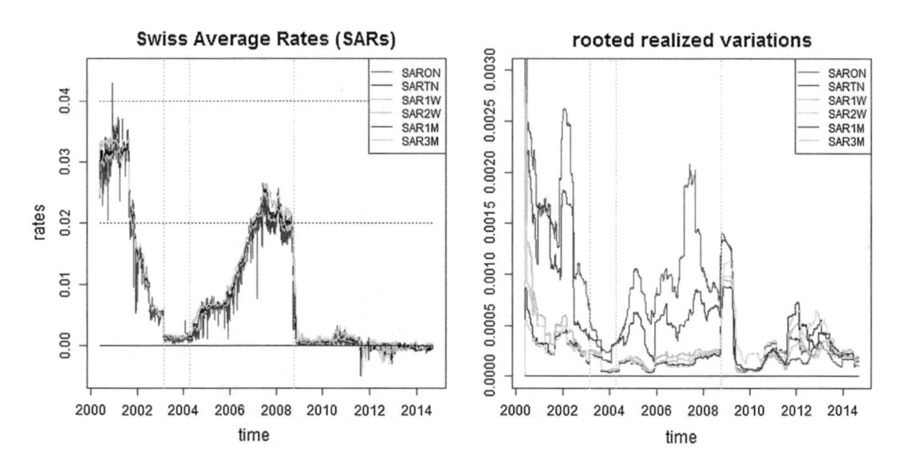

Fig. 1 (lhs) Swiss average rates (SARs) and (rhs) rooted realized variations $\sqrt{\widehat{\Sigma}_k^{\text{CoV}}(m,m)}$ for window length $K = 126$ from December 8, 1999, until September 15, 2014

times-to-maturity, because if we would include longer times-to-maturity we should (at least) choose a multi-factor spot rate model which allows for more flexibility between the short and the long end of the yield curve.

In Fig. 1 we see that the interest rate level is roughly 0 % after 2008. Also interestingly, the order of the levels of realized variations $\widehat{\Sigma}_k^{\mathrm{CoV}}(m,m)$ changes in 2008 in the times-to-maturity $m \in \{1, 2, 5, 10, 21, 63\}$. We will see that this change in order causes severe problems in the calibration of the one-factor Vasiček model (see also Remark 3).

In Fig. 2 we provide the MLEs of the mean reversion rate $\widehat{\alpha}_k^{\mathrm{MLE}}$, see (21), and the intercept $\widehat{a}_k^{\mathrm{MLE}}$, see (22). For window length $K = 126$ (which provides good smoothing through viscosity but also a time lag) we obtain quite reasonable values, with mean reversion rates in the interval $(0.5, 1)$. Mostly they are well bounded away from 1 which usually means that we cannot reject an AR(1) model. Intercepts $\widehat{a}_k^{\mathrm{MLE}}$ also look reasonable, after the financial crisis 2008 intercepts collapse to 0, see Fig. 2 (rhs) and Fig. 1 (lhs). This means that the long-term rate of the spot rate process is around 0 after 2008. Finally, in Fig. 3 (lhs) we provide the MLEs $\sqrt{\widehat{\sigma_k^2}^{\mathrm{MLE}}}$ for the estimated volatility parameter, see (23). We observe that the volatility adapts to the level of the spot rate which is in the spirit of Deguillaume et al. [2], and we could now investigate whether there is an explicit functional form. This provides all MLEs (24) under the real world measure $\mathbb{P}$.

Next we investigate the rooted realized variance estimator $\sqrt{\widehat{\sigma_k^2}^{\mathrm{RVar}}}$, see (28). This estimator is based on the assumption that for sufficiently small grid size Δ we can neglect the drift term because the volatility term is the dominant one. The results are presented in Fig. 3 (rhs). This should be compared to estimator $\sqrt{\widehat{\sigma_k^2}^{\mathrm{MLE}}}$ in Fig. 3 (lhs). We see that both estimators provide very similar results which, of course, supports the methods applied. In Fig. 4 (lhs) we plot the ratio between the two estimates.

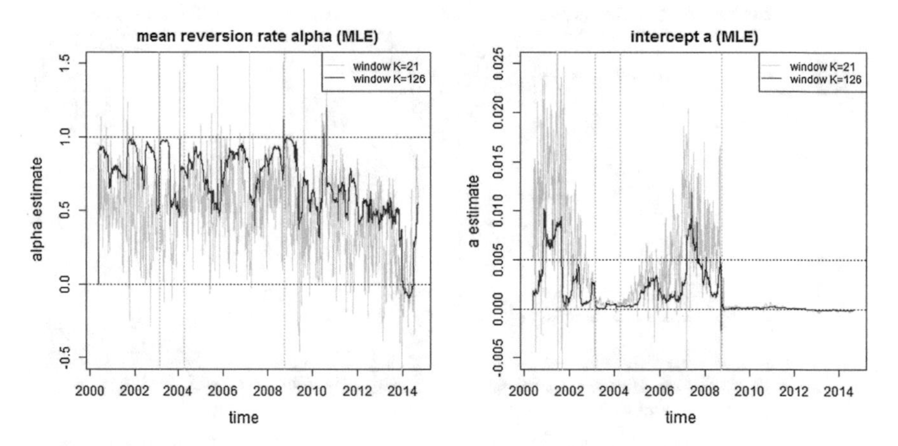

Fig. 2 MLEs (lhs) mean reversion rate $\widehat{\alpha}_k^{\mathrm{MLE}}$ and (rhs) intercept $\widehat{a}_k^{\mathrm{MLE}}$ for window lengths $K = 21, 126$ under the real world measure $\mathbb{P}$ from December 8, 1999, until September 15, 2014

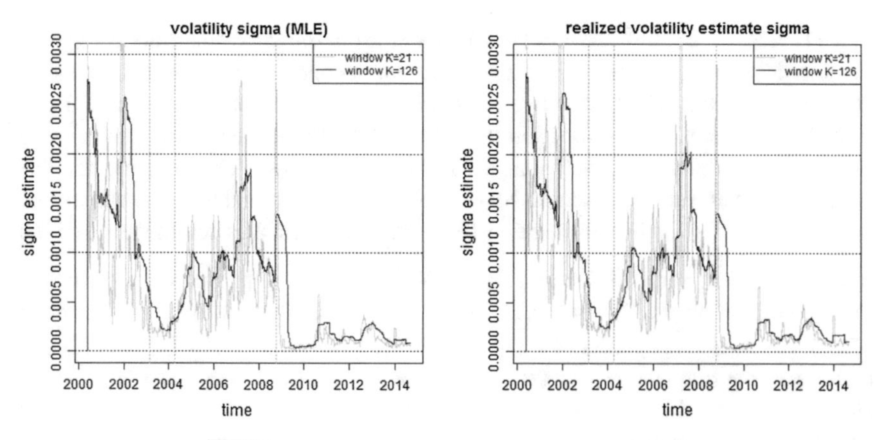

Fig. 3 (lhs) MLEs $\sqrt{\widehat{\sigma_k^2}^{\text{MLE}}}$ and (rhs) realized volatility estimator $\sqrt{\widehat{\sigma_k^2}^{\text{RVar}}}$ for window lengths $K = 21, 126$ from December 8, 1999, until September 15, 2014

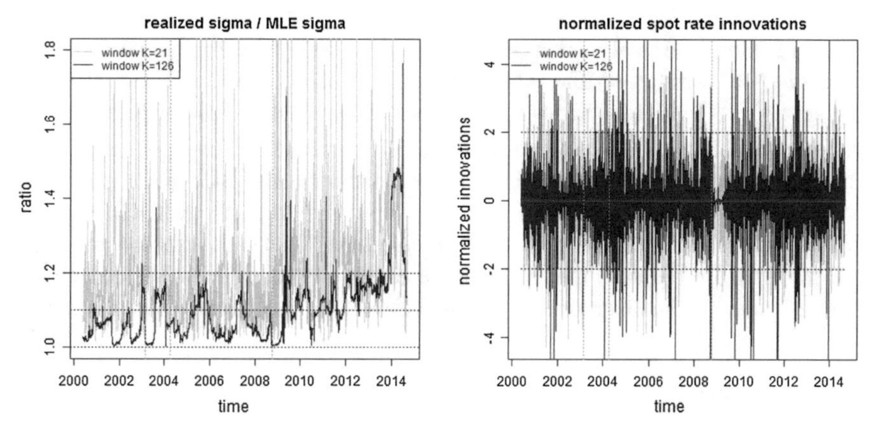

Fig. 4 (lhs) ratio $\sqrt{\widehat{\sigma_k^2}^{\text{RVar}}}/\sqrt{\widehat{\sigma_k^2}^{\text{MLE}}}$ and (rhs) normalized spot rate innovations $\mathsf{D}(r_k)/\sqrt{\widehat{\sigma_k^2}^{\text{RVar}}}$ for window lengths $K = 21, 126$ from December 8, 1999, until September 15, 2014

Before 2008 this ratio is close to 1, however, it is not always close to one which says that for a thorough estimation we need to carefully consider the drift term, or choose a smaller grid size Δ if possible. Figure 4 (rhs) gives the normalized spot rate innovations $\mathsf{D}(r_k)/\sqrt{\widehat{\sigma_k^2}^{\text{RVar}}}$ which are roughly centered but look too heavy-tailed to support the Gaussian innovation assumption for $(\varepsilon_t)_{t\in\mathbb{N}_0}$.

In Fig. 1 (rhs) we provide the rooted realized variations $\sqrt{\widehat{\Sigma}_k^{\text{CoV}}(m,m)}$ for times-to-maturity $m \in \{1, 2, 5, 10, 21, 63\}$, see (30), and in Fig. 5 (lhs) their truncated normalized counterparts $\widehat{\Xi}_k(m,m)$, see (32). We see that after the financial crisis 2008 the truncated normalized values $\widehat{\Xi}_k(m,m)$ look strange. This is caused by the changed order of realized variations $\widehat{\Sigma}_k^{\text{CoV}}(m,m)$ for times-to-maturity

$m \in \{1, 2, 5, 10, 21, 63\}$, see Fig. 1 (rhs). Since the one-factor Vasiček model requires monotonicity in these times-to-maturity m it cannot cope with the new situation after 2008 (see also Remark 3). Moreover, also the SARTN observations provide difficulties. We have already mentioned in the introduction to this section that the quality of SARTN is slightly different compared to the other SARs. This may explain this picture and therefore we exclude SARTN for the further analysis, i.e. we choose weights $w_m = 1_{\{m \neq 2\}}$ in optimization (33). This then provides the mean reversion rate estimates $\widehat{\widehat{\beta}}_k^{\text{RCoV}}$ given in Fig. 5 (rhs). We can now compare it to the estimated mean reversion rates $\widehat{\alpha}_k^{\text{MLE}}$ of Fig. 2 (rhs). From this we then estimate market-price of risk parameter $\widehat{\lambda}_k^{(1)}$, see (34). The results are presented in Fig. 6 (lhs). We see that before the financial crisis 2008 this market-price of risk parameter is negative. However, we should not look at this parameter individually, but always at the total market-price of risk drift that is determined by the following sum, see (13),

$$\widehat{\lambda}_k(r_k) = \widehat{\lambda}_k^{(0)} + \widehat{\lambda}_k^{(1)} r_k. \tag{37}$$

In order to study the total market-price of risk drift we first determine the first component $b_{k+1}^{(k)}$ of the Hull–White extension, see (35), and compare it to the estimated intercept $\widehat{a}_k^{\text{MLE}}$. These time series are presented in Fig. 6 (rhs). From their difference we then estimate the second market-price of risk parameter $\widehat{\lambda}_k^{(0)}$ using formula (36). This provides the graphs in Fig. 7 (lhs). We see that this market-price of risk parameter estimates have opposite signs compared to $\widehat{\lambda}_k^{(1)}$ of Fig. 6 (lhs). We can then estimate the total market-price of risk drift $\widehat{\lambda}_k(r_k)$ defined in (37). The results are presented in Fig. 7 (rhs). In the time frame where we can use the one-factor Vasiček model (before 2008) we obtain a positive total market-price of risk drift $\widehat{\lambda}_k(r_k)$ (maybe this

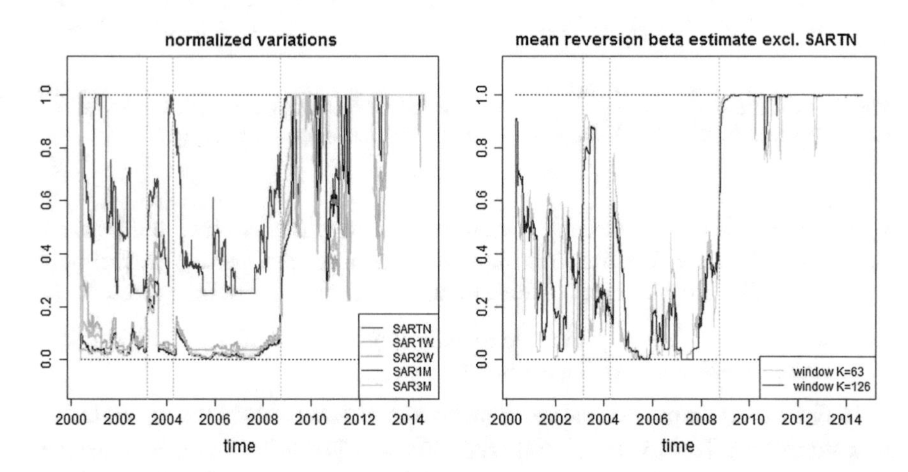

Fig. 5 (lhs) truncated normalized variations $\widehat{\Xi}_k(m, m)$ for $m \in \{2, 5, 10, 21, 63\}$ for window length $K = 126$ and (rhs) estimated mean reversion rate $\widehat{\widehat{\beta}}_k^{\text{RCoV}}$ for window lengths $K = 21, 126$ from December 8, 1999, until September 15, 2014

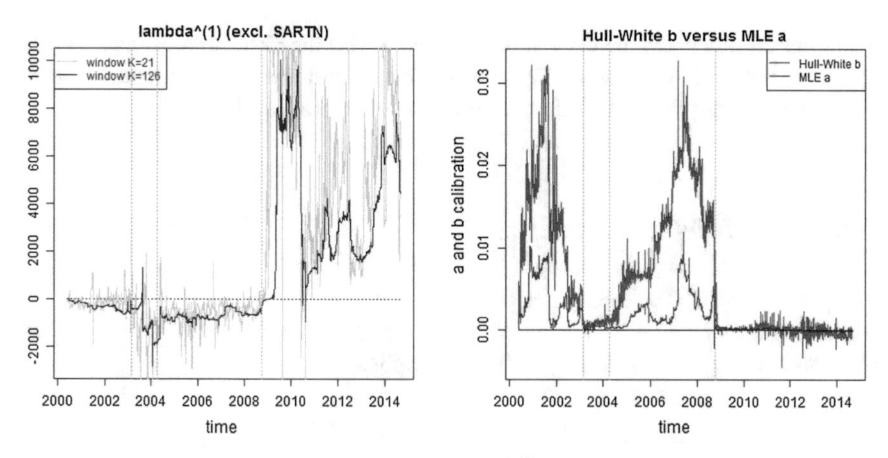

Fig. 6 (lhs) estimated market-price of risk parameter $\widehat{\lambda}_k^{(1)}$ for window lengths $K = 21, 126$ and (rhs) comparison Hull–White extension $b_{k+1}^{(k)}$ and intercept $\widehat{a}_k^{\text{MLE}}$ for window length $K = 126$ from December 8, 1999, until September 15, 2014

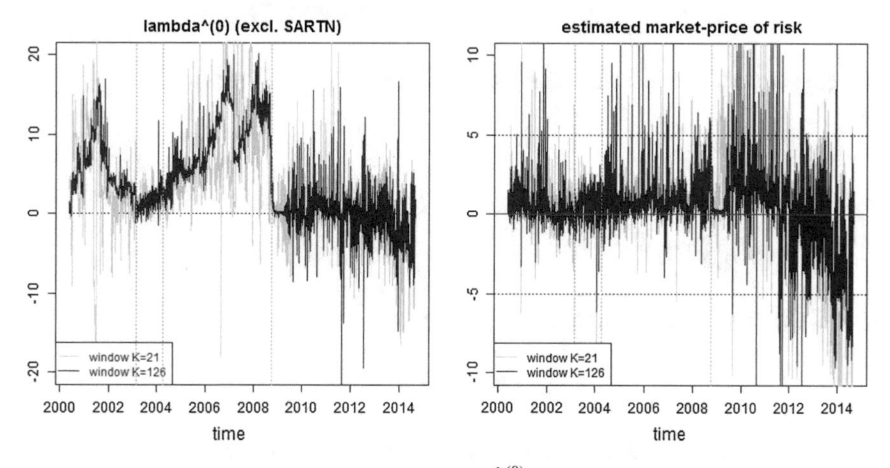

Fig. 7 (lhs) estimated market-price of risk parameter $\widehat{\lambda}_k^{(0)}$ and (rhs) total market-price of risk drift $\widehat{\lambda}_k(r_k) = \widehat{\lambda}_k^{(0)} + \widehat{\lambda}_k^{(1)} r_k$ for window lengths $K = 21, 126$ from December 8, 1999, until September 15, 2014

still needs some smoothing). Of course, this makes perfect sense because market risk aversion requires a positive drift of prices which results in an overall positive market-price of risk term.

For yield curve prediction we now need to model the processes $(\alpha_k^{(k)}, \sigma_k)_k$ and $(\lambda_k)_k$ for future time periods, see Figs. 2 (lhs), 3, 6 (lhs) and 7 (lhs). Based on these quantities all other parameters can be calculated with the CRC algorithm and yield curve prediction is then obtain by considering conditional expectations of future yield curves under the real world measure $\mathbb{P}$, using Corollary 1.

Appendix: proofs

Proof (Theorem 1*)* The theorem is proved by induction.
(i) Initialization $t = k + 1$. We initialize by calculating the first term $b_t^{(k)} = b_{k+1}^{(k)}$ of $\mathbf{b}^{(k)}$. We have $A^{(k)}(k+1, k+2) = 0$. This implies, see (5),

$$A^{(k)}(k, k+2) = -b_{k+1}^{(k)} B(k+1, k+2) + \frac{\sigma^2}{2} B(k+1, k+2)^2.$$

From (8) we have

$$A^{(k)}(k, k+2) = r_k B(k, k+2) - 2\Delta y^{\mathrm{mkt}}(k, k+2).$$

Merging the last two identities and using $r_k = y^{\mathrm{mkt}}(k, k+1)$ provides

$$\begin{aligned}
&b_{k+1}^{(k)} B(k+1, k+2) \\
&\quad = \frac{\sigma^2}{2} B(k+1, k+2)^2 - y^{\mathrm{mkt}}(k, k+1) B(k, k+2) + 2\Delta y^{\mathrm{mkt}}(k, k+2) \\
&\quad = z_1(\beta, \sigma, \mathbf{y}_k^{\mathrm{mkt}}).
\end{aligned}$$

This is exactly the first component of the identity

$$\mathsf{C}(\beta)\mathbf{b}^{(k)} = \mathbf{z}(\beta, \sigma, \mathbf{y}_k^{\mathrm{mkt}}). \tag{38}$$

(ii) Induction step $t \to t + 1 < M$. Assume we have calibrated $b_{k+1}^{(k)}, \ldots, b_t^{(k)}$ and these correspond to the first $t - k$ components of (38). The aim is to determine $b_{t+1}^{(k)}$. We have $A^{(k)}(t+1, t+2) = 0$ and iteration implies

$$A^{(k)}(k, t+2) = -\sum_{j=k+1}^{t+1} b_j^{(k)} B(j, t+2) + \sum_{j=k+1}^{t+1} \frac{\sigma^2}{2} B(j, t+2)^2.$$

From (8) we obtain

$$A^{(k)}(k, t+2) = r_k B(k, t+2) - (t+2-k)\Delta y^{\mathrm{mkt}}(k, t+2).$$

Merging the last two identities and using $r_k = y^{\mathrm{mkt}}(k, k+1)$ provides

$$\begin{aligned}\sum_{j=k+1}^{t+1} & b_j^{(k)} B(j, t+2) \\ &= \sum_{j=k+1}^{t+1} \frac{\sigma^2}{2} B(j, t+2)^2 - y^{\mathrm{mkt}}(k, k+1) B(k, t+2) \\ &\quad + (t+2-k) \Delta y^{\mathrm{mkt}}(k, t+2) \\ &= z_{t+1-k}(\beta, \sigma, \mathbf{y}_k^{\mathrm{mkt}}).\end{aligned}$$

Observe that this exactly corresponds to the $(t+1-k)$th component of (38). This proves the claim. □

Proof (Theorem 2*)* Using (12) and (10) for $t = k+1$ we have

$$\begin{aligned}&(m-(k+1))\, \Delta\, Y(k+1, m) \\ &\quad = -A^{(k)}(k+1, m) + \left(b_{k+1}^{(k)} + \beta_k r_k + \sigma_k \varepsilon_{k+1}^*\right) B^{(k)}(k+1, m).\end{aligned}$$

We add and subtract $-A^{(k)}(k, m) + r_k B^{(k)}(k, m)$,

$$\begin{aligned}(m-(k+1)) \Delta\, Y(k+1, m) = &- A^{(k)}(k, m) + r_k B^{(k)}(k, m) \\ &+ A^{(k)}(k, m) - A^{(k)}(k+1, m) - r_k B^{(k)}(k, m) \\ &+ \left(b_{k+1}^{(k)} + \beta_k r_k + \sigma_k \varepsilon_{k+1}^*\right) B^{(k)}(k+1, m).\end{aligned}$$

We have the following two identities, the second simply follows from the definition of $A^{(k)}(k, m)$,

$$\begin{aligned}-A^{(k)}(k, m) + r_k B^{(k)}(k, m) &= (m-k) \Delta\, Y(k, m), \\ A^{(k)}(k, m) - A^{(k)}(k+1, m) &= -b_{k+1}^{(k)} B^{(k)}(k+1, m) + \frac{\sigma_k^2}{2} B^{(k)}(k+1, m)^2.\end{aligned}$$

Therefore, the right-hand side of the previous equality can be rewritten and provides

$$\begin{aligned}(m-(k+1)) \Delta\, Y(k+1, m) = (m-k) \Delta\, Y(k, m) &+ \frac{\sigma_k^2}{2} B^{(k)}(k+1, m)^2 \\ &+ \sigma_k B^{(k)}(k+1, m) \varepsilon_{k+1}^* \\ &- r_k \left(B^{(k)}(k, m) - \beta_k B^{(k)}(k+1, m)\right).\end{aligned}$$

Observe that the bracket on the third line is equal to Δ and that $r_k = Y(k, k+1)$. This proves the claim. □

References

1. Cox, J.C., Ingersoll, J.E., Ross, S.A.: A theory of the term structure of interest rates. Econometrica **53**(2), 385–407 (1985)
2. Deguillaume, N., Rebonato, R., Pogudin, A.: The nature of the dependence of the magnitude of rate moves on the rates levels: a universal relationship. Quant. Financ. **13**(3), 351–367 (2013)
3. Harms, P., Stefanovits, D., Teichmann, J., Wüthrich, M.V.: Consistent Recalibration of Yield Curve Models (2015). arXiv:1502.02926
4. Harms, P., Stefanovits, D., Teichmann, J., Wüthrich, M.V.: Consistent Re-calibration of the Discrete Time Multifactor Vasiček Model. Working paper (2015)
5. Heath, D., Jarrow, R., Morton, A.: Bond pricing and the term structure of interest rates: a new methodology for contingent claim valuation. Econometrica **60**(1), 77–105 (1992)
6. Heston, S.L.: A closed-form solution for options with stochastic volatility with applications to bond and currency options. Rev. Financ. Stud. **6**(2), 327–343 (1993)
7. Hull, J., White, A.: Branching out. Risk **7**, 34–37 (1994)
8. Jordan, T.J.: SARON—an innovation for the financial markets. In: Launch event for Swiss Reference Rates, Zurich, 25 August 2009
9. Richter, A., Teichmann, J.: Discrete Time Term Structure Theory and Consistent Recalibration Models (2014). arXiv:1409.1830
10. Vasiček, O.: An equilibrium characterization of the term structure. J. Financ. Econ. **5**(2), 177–188 (1977)
11. Wüthrich, M.V., Merz, M.: Financial Modeling. Actuarial Valuation and Solvency in Insurance, Springer, Heidelberg (2013)

Autoregressive Conditional Duration Model with an Extended Weibull Error Distribution

Rasika P. Yatigammana, S.T. Boris Choy and Jennifer S.K. Chan

Abstract Trade duration and daily range data often exhibit asymmetric shape with long right tail. In analysing the dynamics of these positively valued time series under autoregressive conditional duration (ACD) models, the choice of the conditional distribution for innovations has posed challenges. A suitably chosen distribution, which is capable of capturing unique characteristics inherent in these data, particularly the heavy tailedness, is proved to be very useful. This paper introduces a new extension to the class of Weibull distributions, which is shown to perform better than the existing Weibull distribution in ACD and CARR modelling. By incorporating an additional shape parameter, the Weibull distribution is extended to the extended Weibull (EW) distribution to enhance its flexibility in the tails. An MCMC based sampling scheme under a Bayesian framework is employed for statistical inference and its performance is demonstrated in a simulation experiment. Empirical application is based on trade duration and daily range data from the Australian Securities Exchange (ASX). The performance of EW distribution, in terms of model fit, is assessed in comparison to two other frequently used error distributions, the exponential and Weibull distributions.

Keywords ACD models · CARR models · Extended Weibull · Bayesian inference

R.P. Yatigammana · S.T.B. Choy (✉)
Discipline of Business Analytics, The University of Sydney, Sydney, NSW 2006, Australia
e-mail: boris.choy@sydney.edu.au

R.P. Yatigammana
e-mail: tyat9428@uni.sydney.edu.au

J.S.K. Chan
School of Mathematics and Statistics, The University of Sydney, Sydney, NSW 2006, Australia
e-mail: jennifer.chan@sydney.edu.au

V.-N. Huynh et al. (eds.), *Causal Inference in Econometrics*,
Studies in Computational Intelligence 622, DOI 10.1007/978-3-319-27284-9_5

1 Introduction

With the recent advancements in computational technology, data capturing and storage capabilities, the use of high frequency data (HFD) has gathered considerable momentum in recent years. Consequently, there has been a surge in the interest of research in many business related areas such as economics and finance. This was propelled by the enhanced availability and easy access to HFD linked with financial market transactions. As an important economic variable, the irregularly spaced trade durations convey useful information about market dynamics. The literature on market micro structure bears evidence to the economic significance of this variable [1]. Most often this is modelled by autoregressive conditional duration (ACD) models introduced by [2].

Empirical evidence suggests that duration data generally has a unimodal distribution with high density around zero as the majority of transactions have durations close to zero. The data display an asymmetric shape with a long right tail [3, 4]. Such conditional distribution for the data has proven to be challenging to model parametrically, in spite of there being significant methodological developments in analysing the dynamics of ACD models. Frequently, the conditional distribution has been modelled by the exponential and Weibull distributions while the log-normal, generalised gamma and Burr have also been considered. However, most of these distributions have shown limitations in the specification of conditional duration distribution [5–7]. They are unable to capture some of the characteristics of duration distributions precisely, especially the long right tail. This may have a negative impact on forecasting and hence trading strategy formulation. Consequently, the choice of standard parametric error distributions in the application to a dynamic duration model is still of much interest in the literature.

The increased prevalence of rich data sources, particularly in financial markets across the globe motivates the development of flexible financial time series models to capture the subtle movements and intricacies of HFD distributions. In this context, we propose a variation of the Weibull distribution with an extra parameter to add flexibility in the tail behaviour. This will be referred to as the extended Weibull (EW) distribution. The proposed distribution will prove useful to develop financial risk management strategies and evaluating properties such as the Value-at-risk (VaR) and Time-at-risk (TaR) for optimum capital allocation.

The modelling structure for an ACD model is not confined to modelling duration data alone, but has been extended to other positive valued time series. One example is the daily range of an asset price, which is defined as the difference between the highest and lowest log asset price within a trading day. The daily range could be used as an efficient measure of the local volatility of an asset price [8]. The conditional autoregressive range (CARR) model [9] analyses the daily range data and showed improved performance in out-of-sample volatility forecasts over other frequently used volatility models. The CARR model shares the same model structure as the ACD model and both models belong to the family of multiplicative error models [10]. In this paper, we assess the performance of the ACD and CARR models with the EW distribution by simulation and demonstrate their applicability through two real data sets from the Australian Securities Exchange (ASX), one of trade durations for a stock and the other of daily range data for a market index.

In summary, the main objective of this paper is to introduce the EW distribution to model the conditional distribution of positively valued time series in an ACD or CARR framework. Secondly, we derive the salient features and moments that characterise this distribution. Thirdly we assess the effectiveness of the estimation procedure based on Bayesian methodology through an extensive simulation study. The fourth objective is to produce empirical applications involving two different data sets to ascertain the applicability and comparative performance of EW distribution.

The remainder of this paper is organised as follows. Section 2 introduces the EW distribution with illustrations on the shapes of its probability density function (pdf) and some properties. Section 3 discusses the model formulation under the ACD model framework. Then the estimation of parameters based on the Bayesian approach is described in Sect. 4. A simulation experiment is performed in Sect. 5 while Sect. 6 reports the outcomes of the empirical application for two stock market data on trade durations and daily range. Further, this section compares the proposed model with two models using exponential and Weibull distributions, respectively. Section 7 summarises the results and concludes the paper.

2 Extended Weibull Distribution

Compared to the Weibull distribution, the EW distribution has an extra shape parameter, which allows for more flexibility in skewness and kurtosis. Suppose that X is a random variable following the EW distribution, denoted by $EW(\lambda, k, \gamma)$, with scale parameter $\lambda > 0$, shape parameter $k > 0$ and the additional shape parameter $\gamma > 0$. Then X has the following pdf

$$f(x) = \left(1 + \frac{1}{\gamma^k}\right) \frac{k}{\lambda^k} x^{k-1} e^{-\left(\frac{x}{\lambda}\right)^k} \left[1 - e^{-\left(\frac{\gamma x}{\lambda}\right)^k}\right], x > 0. \tag{1}$$

From (1), it is clear that EW distribution provides more flexibility than the Weibull distribution. When $\gamma \to \infty$, it becomes the Weibull (λ, k) distribution. Figure 1 illustrates the varying shapes of the EW pdf for selected parameter values. Further, it is evident that for the shape parameter $k < 1$, the distribution is more skewed and the asymmetry is accentuated by the additional γ parameter as well. However, the impact of γ is low when both $k < 1$ and $\lambda < 1$.

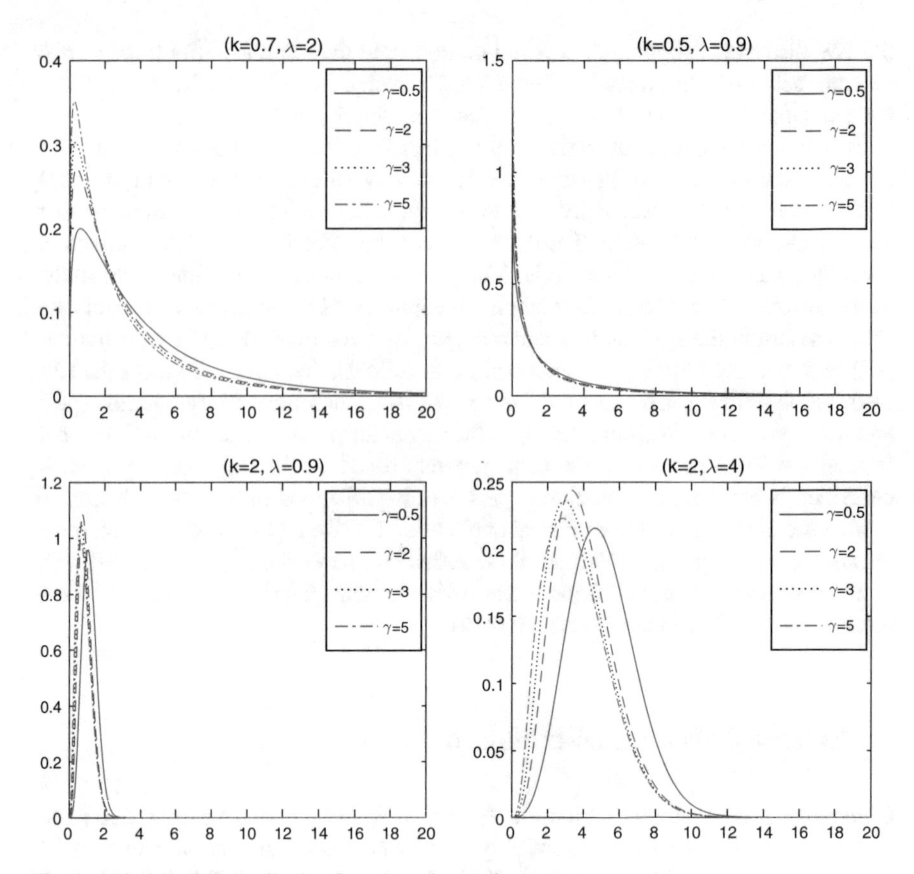

Fig. 1 Pdf of the EW distribution for selected sets of parameters

2.1 Properties of EW Distribution

The general expression for the rth moment of X is given by

$$E(X^r) = \int_0^\infty \left(1 + \frac{1}{\gamma^k}\right) \frac{k}{\lambda^k} x^{r+k-1} e^{-\left(\frac{x}{\lambda}\right)^k} \left[1 - e^{-\left(\frac{\gamma x}{\lambda}\right)^k}\right] dx$$

$$= \left(1 + \frac{1}{\gamma^k}\right) \lambda^r \;\; \Gamma\left(1 + \frac{r}{k}\right) \left[1 - \frac{1}{(1+\gamma^k)^{1+\frac{r}{k}}}\right].$$

The mean (μ), variance (σ^2), skewness (ρ) and kurtosis (ζ), which describes the characteristics of the EW distribution can be obtained accordingly from $E(X^r)$. See Appendix for derivations. The cumulative distribution function (cdf) $F(x)$, survivor function $S(x)$ and hazard function $h(x)$ of the EW distribution are given by

$$F(x) = 1 + \frac{1}{\gamma^k} e^{-\left(\frac{x}{\lambda}\right)^k} \left[e^{-\left(\frac{\gamma x}{\lambda}\right)^k} - 1 - \gamma^k \right], \tag{2}$$

$$S(x) = \left[\gamma^k + 1 - e^{-\left(\frac{\gamma x}{\lambda}\right)^k} \right] \frac{1}{\gamma^k} e^{-\left(\frac{x}{\lambda}\right)^k}, \tag{3}$$

$$h(x) = \frac{\left(\gamma^k + 1\right) k x^{k-1} \left(1 - e^{-\left(\frac{\gamma x}{\lambda}\right)^k}\right)}{\left[\gamma^k + 1 - e^{-\left(\frac{\gamma x}{\lambda}\right)^k} \right] \lambda^k}, \tag{4}$$

respectively. For the derivations, see Appendix.

The intensity of transaction arrivals has important implications in analysing durations. Figure 2 displays the various shapes of hazard function with different sets of parameter values as in Fig. 1 for comparison. For $k > 1$, the hazard function is

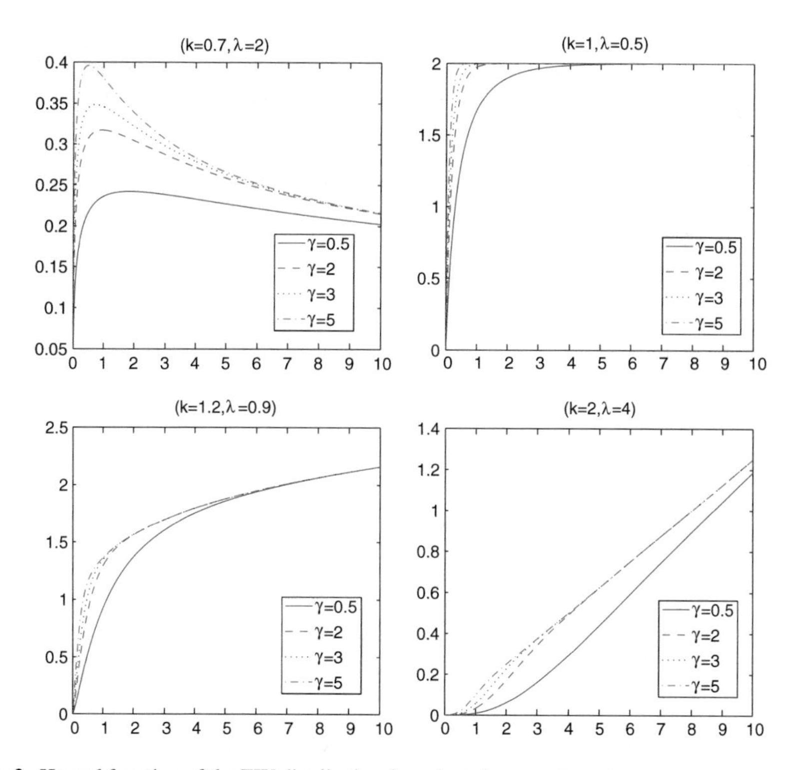

Fig. 2 Hazard function of the EW distribution for selected parameter values

monotonically increasing at different rates. However, for $0 < k < 1$, it is unimodal, implying non-monotonocity. Further, this characteristic is more prominent for larger values of γ. In the special case of $k = 1$, the hazard function converges to a constant rate given by $1/\lambda$, as x increases. Grammig and Maurer [3] assert that a distribution with non-monotonic hazard function can better capture the behaviour of durations.

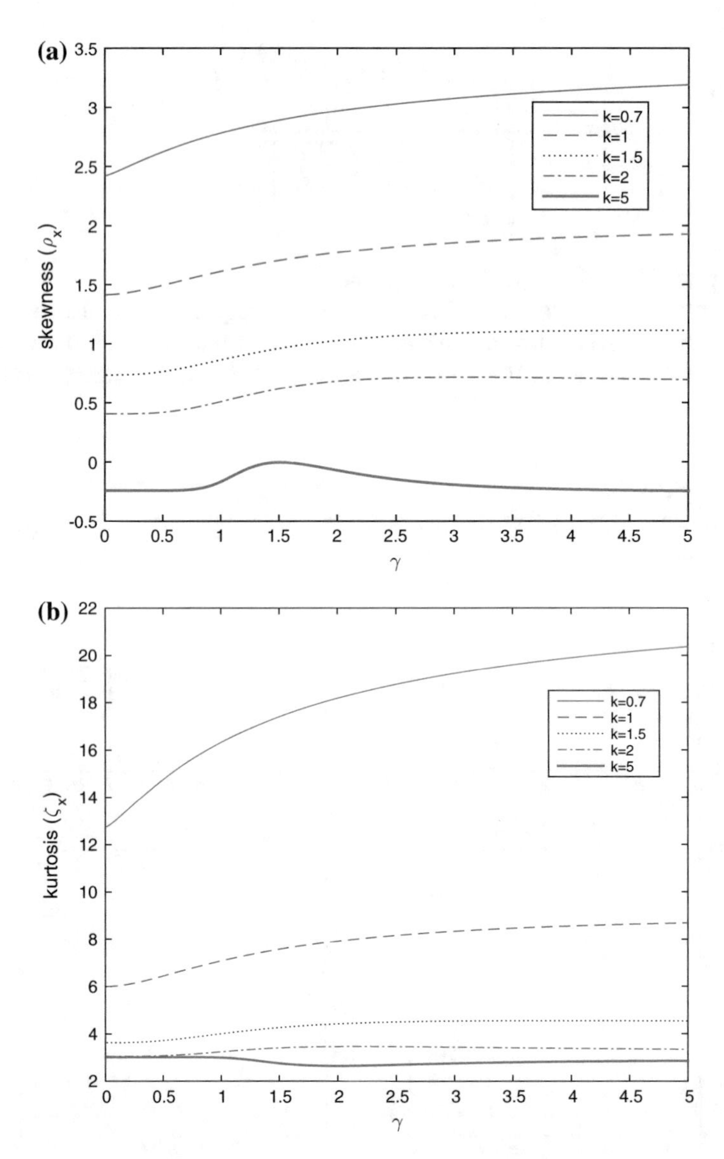

Fig. 3 **a** Variation of skewness and **b** variation of kurtosis over γ for selected values of k for the EW distribution

The relative impact of γ on the skewness and kurtosis of the distribution, across various values of k, is presented in Fig. 3. For smaller k (i.e. $k < 2$), both the skewness and kurtosis tend to increase with γ. However, for larger k, there exists a non-monotonic relationship between these characteristics and γ. For a given value of γ, skewness is inversely related to k, although there is no such a distinct relationship in the case of kurtosis, particularly when $k > 2$.

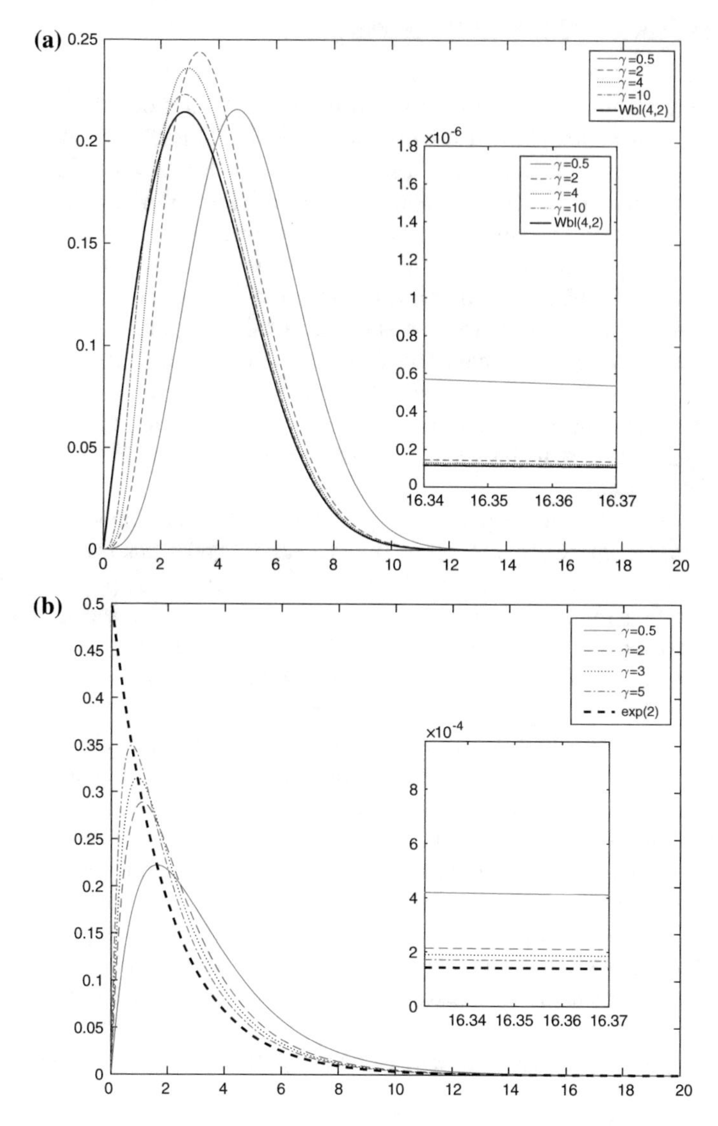

Fig. 4 Comparison of the EW distribution with **a** the Weibull and **b** the exponential with similar shape and scale parameters

In comparison to the Weibull distribution, the EW distribution tends to capture more of the right tail indicating more flexibility, irrespective of the value of γ. Figure 4a compares the two distributions for similar values of k and λ with varying γ. The same phenomenon is observed in the case of the exponential distribution where the EW density is plotted for $k = 1$ in Fig. 4b.

3 ACD Model with EW Distribution

In modelling positively valued time series under an ACD model framework, the exponential distribution is too restrictive and the Weibull distribution is superior to the exponential distribution as it has a shape parameter that increases its modelling flexibility considerably. Then our proposed EW distribution is motivated by an additional shape parameter γ, which entails better capturing of the heavier tail than the Weibull distribution. Hence, this distribution can provide a better fit.

Under the ACD model, the unit mean restriction for the innovations has to be imposed. One option is to consider a standardised distribution, incorporating the mean as a standardised parameter. The alternative is to directly restrict the mean to be unity so that one parameter (usually the scale parameter) can be expressed as a function of others, under the unit mean condition. In either case there will be one less (free) parameter, reducing the number of parameters for the EW distribution to two instead of three.

ACD model is generally applied to irregularly spaced financial market transaction data. The most primary economic time series analysed under this framework is the time interval between two consecutive trades of a given stock, popularly known as trade durations. In such a scenario, let $\{t_0, t_1, \ldots, t_i, \ldots\}$ be a sequence of time points the stock under consideration is traded in the stock exchange, such that $t_0 \leq t_1 \leq \cdots \leq t_i \leq \cdots$. Here t_0 denotes the starting time point and t_T is the last time point of the observed sequence, where T is the length of the series. Modelling a financial point process in a duration framework, as waiting times characterises the point process as a discrete time series. In a generic sense, $x_i = t_i - t_{i-1}$ denotes the ith duration between two transactions that occur at times t_i and t_{i-1}. Therefore, ignoring simultaneous transactions, which is the preferred practice as evidenced in literature, the sequence $\{x_1, x_2, \ldots, x_T\}$ will generate a time series of positive measurements. ACD models are concerned with modelling such positive valued time series. On the other hand, positively valued time series also arise in the study of price volatility using price range such as the intraday high low prices. CARR models analyse range data applying an identical model structure and hence the following description is equally valid for CARR models.

Let $x_i, i = 1, \ldots, T$ be a time series of trade durations under the ACD modelling framework,

$$x_i = \psi_i \epsilon_i \tag{5}$$

in a multiplicative error structure, where the deterministic component, $\psi_i = E(X_i)$ is the conditional expectation of the ith duration, given the past information, that is,

$$\psi_i = E(x_i | x_{i-1}, \ldots, x_1).$$

The main assumption here is that the standardized durations

$$\epsilon_i = \frac{x_i}{\psi_i}$$

are independent and identically distributed (i.i.d), having a positive support and a unit mean, and ψ_i is formulated under the ACD(p, q) model as

$$\psi_i = \alpha_0 + \sum_{j=1}^{p} \alpha_j x_{i-j} + \sum_{l=1}^{q} \beta_l \psi_{i-l},$$

where p and q are non-negative integers. The following restrictions

$$\alpha_0 > 0, \alpha_j \geq 0, \beta_l \geq 0, \sum_{j=1}^{p} \alpha_j + \sum_{l=1}^{q} \beta_l < 1 \tag{6}$$

ensure positivity and stationarity of durations, respectively.

For most practical purposes, a basic version of ACD model suffices and hence an ACD (1, 1) model is considered in this paper. The conditional expectation ψ_i is estimated via the following recursive formula

$$\psi_i = \alpha_0 + \alpha_1 x_{i-1} + \beta \psi_{i-1} \tag{7}$$

The long term mean is $\dfrac{\alpha_0}{1 - \alpha_1 - \beta}$. The random disturbance ϵ_i is assumed to follow the $EW(\lambda, k, \gamma)$ distribution; imposing the unit mean restriction using (11) yields

$$\lambda = \frac{k\gamma^k \left(1 + \gamma^k\right)^{1/k}}{\Gamma(1/k)\left[\left(1 + \gamma^k\right)^{1+1/k} - 1\right]} \tag{8}$$

Then the distributional form of duration X_i is as follows:

$$X_i \sim EW(\lambda \psi_i, k, \gamma)$$

Substituting for λ in (8), the parameters to be estimated are $\boldsymbol{\theta} = (\alpha_0, \alpha_1, \beta, k, \gamma)$.

Under this framework, the conditional likelihood has the following expression

$$L(\boldsymbol{x}|\boldsymbol{\theta}) = \prod_{i=1}^{T} \frac{1}{\psi_i} \frac{k}{\lambda^k} \left(1 + \frac{1}{\gamma^k}\right) \left(\frac{x_i}{\psi_i}\right)^{k-1} e^{-\left(\frac{x_i}{\lambda\psi_i}\right)^k} \left[1 - e^{-\left(\gamma \frac{x_i}{\lambda\psi_i}\right)^k}\right] \quad (9)$$

The ACD model incorporating EW errors is denoted as an EW-ACD model. The exponential and Weibull distributions are also considered for comparison and their ACD models are denoted as EACD and WACD models respectively.

4 Bayesian Estimation Methodology

The statistical inference is carried out using the Bayesian simulation approach, which allows simultaneous finite sample inference. The main advantage of using Bayesian techniques is the ability to incorporate prior knowledge into the estimation process. Further, positivity and stationarity constraints given in (6) can be directly incorporated into the prior distribution. For fast convergence of the Markov chain, the parameter generation is executed in blocks [11].

Without prior knowledge, most non-informative priors we adopted are flat over the feasible region, with the likelihood function dominating the inference. Mostly a uniform prior is adopted for $\boldsymbol{\alpha} = (\alpha_0, \alpha_1, \beta)$, over the constraint region in (6), ensuring the enforcement of these restrictions. We also choose the uniform prior for the shape parameter of k, while ensuring $k > 0$. On the other hand an inverse prior is assumed for γ, with positivity constraint, $f(\gamma) \propto \frac{1}{\gamma}$. Under the assumption of independence of the individual blocks of parameters, the prior distribution can be specified as follows

$$f(\boldsymbol{\theta}) = f_\alpha(\boldsymbol{\alpha}) f_k(k) f_\gamma(\gamma)$$

The joint posterior pdf of $\boldsymbol{\theta}$ is proportional to the product of the prior density $f(\boldsymbol{\theta})$ and the likelihood function of (9). If $\boldsymbol{\theta}_J$, $J = 1, 2, 3$ represents a parameter block, $\boldsymbol{\alpha}$, k or γ at a given simulation step, then $\boldsymbol{\theta}_{-J}$ represents the vector of paramters excluding $\boldsymbol{\theta}_J$. For each updating step, the posterior distributions for the chosen element $\boldsymbol{\theta}_J$ conditional on the data and other parameters, in the MCMC setup is

$$f(\boldsymbol{\theta}_J|\boldsymbol{x}, \boldsymbol{\theta}_{-J}) \propto L(\boldsymbol{x}|\boldsymbol{\theta}) f(\boldsymbol{\theta}_J) \quad (10)$$

where $L(\boldsymbol{x}|\boldsymbol{\theta})$ is given in (9). The posterior distributions for each choice of $\boldsymbol{\theta}_J$ do not have a standard distributional form due to lack of conjugacy between the likelihood function and the prior distributions. Therefore, the Metropolis-Hastings (MH) algorithm is used to generate samples for each block of parameters from (10). This method was introduced by [12] involving a symmetric transition density, which was later generalised by [13].

We adopted a two stage MH method to draw parameters in an adaptive MCMC sampling scheme, similar to [14]. In the first stage, random walk (RW) Metropolis algorithm is employed, to sample parameters from the posterior distribution. The algorithm uses a multivariate normal (MVN) proposal density $q(.|.)$ with its mean at the current value $\boldsymbol{\theta}_J^{(k-1)}$ at iteration k and the covariance matrix being a multiple of that of a certain matrix, that is, $a\boldsymbol{\Sigma}_{\theta_J}$, where $\boldsymbol{\Sigma}_{\theta_J}$ is set to be an identity matrix for convenience. The RW metropolis proposal density has proven to be quite useful in Bayesian inference, where the acceptance rate of the draws could be tuned by adjusting the scalar a attached to the covariance matrix to ensure optimal acceptance probability, around 0.44 for one dimension and could be as low as 0.23 as the dimensionality increases [15]. The sample mean $\bar{\boldsymbol{\theta}}_J$ and sample covariance $\boldsymbol{S}_{\theta_J}$ are formed using M iterates of $\boldsymbol{\theta}_J$ after burn-in, from the first stage. In the second stage, an independent kernel MH algorithm is applied for parameter sampling, using the Gaussian proposal distribution with mean $\bar{\boldsymbol{\theta}}_J$ and covariance $\boldsymbol{S}_{\theta_J}$.

In the simulation and empirical analyses, the burn-in sample is 15,000 iterates from the MCMC to ensure proper convergence and 10,000 iterates are sampled thereafter for estimation purposes, during the first stage. The burn-in value is chosen based on the trace plots of the iterates. In the second stage, we use an independent MVN density $q(.)$, with the mean and the covariance being estimated using the 10,000 values generated from the first stage, after the burn-in. In this stage, a sample of 10,000 iterates is generated from the thus formulated independent proposal after a burn-in of 5000 and the sample average form the parameter estimates.

5 Simulation Study

In order to assess the performance of the estimation methodology, a simulation study is performed prior to its empirical applications.

5.1 Random Variates Generation

Simulation of the innovation ϵ_i of the ACD model from $f(\epsilon)$ is not straight forward as the cdf is not invertible. Therefore, the rejection sampling method is used to draw ϵ_i from the EW distribution. Weibull (λ, k) is used as the envelope distribution $g(\epsilon)$, as the EW pdf encompasses the Weibull pdf.

The algorithm for generating a sample from $f(\epsilon)$ as developed by [16] is given as follows:

1. Sample ϵ from $g(\epsilon)$.
2. Generate a uniform (0,1) random number u.
3. Compute $q = f(\epsilon)/Mg(\epsilon)$. where M is a quantity such that $M \geq \sup_\epsilon f(\epsilon)/g(\epsilon)$.
4. If $u < q$, accept ϵ as a realisation of $f(\epsilon)$.
5. Otherwise reject the value and repeat from step 1 onwards.

The supremum of $f(\cdot)/g(\cdot)$ happens to be $(1 + 1/\gamma^k)$; Therefore M could be fixed at this value as it is > 1. Accordingly $q = 1 - e^{-(\frac{\gamma\epsilon}{\lambda})^k}$.

5.2 Simulation

This simulation study validates the effectiveness of the proposed estimation methodology and is conducted under four sets of values for (k, γ), where the series length is $T = 10{,}000$ and the number of replications is $R = 200$. As the empirical evidence suggested relatively large values for γ, the simulated examples were generated by fixing relatively high values for γ, except for one case. The parameters of the mean equation are kept fixed as its estimation is fairly straight forward but the estimation of shape parameters (k, γ) are more problematic.

The simulation results presented in Table 1 display the true values, average of posterior mean estimates, standard errors and 95 % credible intervals of the 200 replications, for the model parameters. Coverage percentages indicate the number

Table 1 Simulation study of 200 replications from ACD (1, 1) with EW innovations and T = 10,000 for four sets of (k, γ)

Parameter	True value	Estimate	Std. Error	95 % CI	Coverage (%)
α_0	0.05	0.0535	0.0065	(0.0417, 0.0671)	91.4
α_1	0.10	0.1031	0.0078	(0.0886, 0.1191)	95.9
β	0.85	0.8436	0.0118	(0.8193, 0.8654)	93.4
k	0.50	0.4994	0.0088	(0.4846, 0.5188)	89.8
γ	0.5	0.6539	0.5694	(0.1126, 1.2137)	85.8
α_0	0.05	0.0574	0.0067	(0.0456, 0.0716)	80.6
α_1	0.10	0.1069	0.0063	(0.0951, 0.1196)	82.1
β	0.85	0.8360	0.0109	(0.8133, 0.8559)	75.0
k	1.40	1.3789	0.0299	(1.3177, 1.4343)	89.3
γ	5.0	4.7003	0.5707	(3.6176, 5.8500)	92.4
α_0	0.05	0.0514	0.0064	(0.0398, 0.0648)	96.0
α_1	0.10	0.1008	0.0072	(0.0872, 0.1155)	96.0
β	0.85	0.8479	0.0113	(0.8245, 0.8691)	97.0
k	0.70	0.6930	0.0180	(0.6583, 0.7264)	89.0
γ	10.0	9.6000	2.7240	(5.0399, 15.4392)	90.0
α_0	0.05	0.0514	0.0065	(0.0395, 0.0651)	94.5
α_1	0.10	0.1015	0.0076	(0.0873, 0.1170)	92.5
β	0.85	0.8474	0.0117	(0.8229, 0.8690)	94.0
k	0.70	0.6991	0.0092	(0.6802, 0.7168)	94.5
γ	100.0	104.78	20.898	(69.738, 153.04)	90.0

of times the credible interval contains the true parameter value, as a percentage of R. True values of all parameters are contained within the average 95 % credible intervals of the posterior samples and the estimates are very close to their true values. The precision of estimates is fairly good except for γ, particularly when it is less than one. Coverage percentages are quite satisfactory, except for β of the second panel. Overall, the estimation of model parameters appears to perform well under the proposed methodology.

6 Empirical Analysis

This section demonstrates the application of the ACD (1, 1) model with the EW distribution fitted to two real life financial time series of trade duration and daily range respectively.

6.1 Trade Duration Data

Adjusted duration data of Telstra (TLS) stock traded in the Australian Securities Exchange (ASX) during the one week period from 1 to 7 October, 2014 is considered in the analysis. The relevant tick by tick trade data can be obtained from the Securities Industry Research Centre of Asia-Pacific (SIRCA) in Australia. The original durations were based on trades occurred during normal trading hours, ignoring overnight intervals and zero durations. Thereafter, data was adjusted for its daily seasonality. The observed trade durations are generally subjected to intraday seasonality or 'diurnal' effect. Engle and Russell [2] and several other authors have recognised this to be a deterministic component. This factor should be accounted for, prior to carrying out any empirical analysis on the stochastic properties of duration processes. The estimation of the deterministic diurnal factor, was done via a cubic spline with knots at each half hour interval [17]. Thus adjusted duration, x_i, which is referred to as duration hereafter, is modelled under this framework.

The total length of the series is 19473. The time series plot of Fig. 5a reveals the clustering effect generally observed in trade durations. On the other hand, Fig. 5b shows the excessive amount of values close to zero and the long right tail which are common characteristics of such data.

The summary statistics reported in Table 2 indicate overdispersion, generally prevalent in trade durations. Moreover, the series is positively skewed with a heavy tail according to the values of skewness and kurtosis. Sample autocorrelation function (ACF) of the adjusted durations is given in Fig. 6. The ACF clearly shows longterm serial dependence in the data, although the values appear to be small in magnitude.

Three competing models EACD (1, 1), WACD (1, 1) and EW-ACD (1, 1) are fitted to TLS trade durations. The parameter estimates are given in Table 3 together with the standard errors in parentheses. All the parameter estimates are significant at

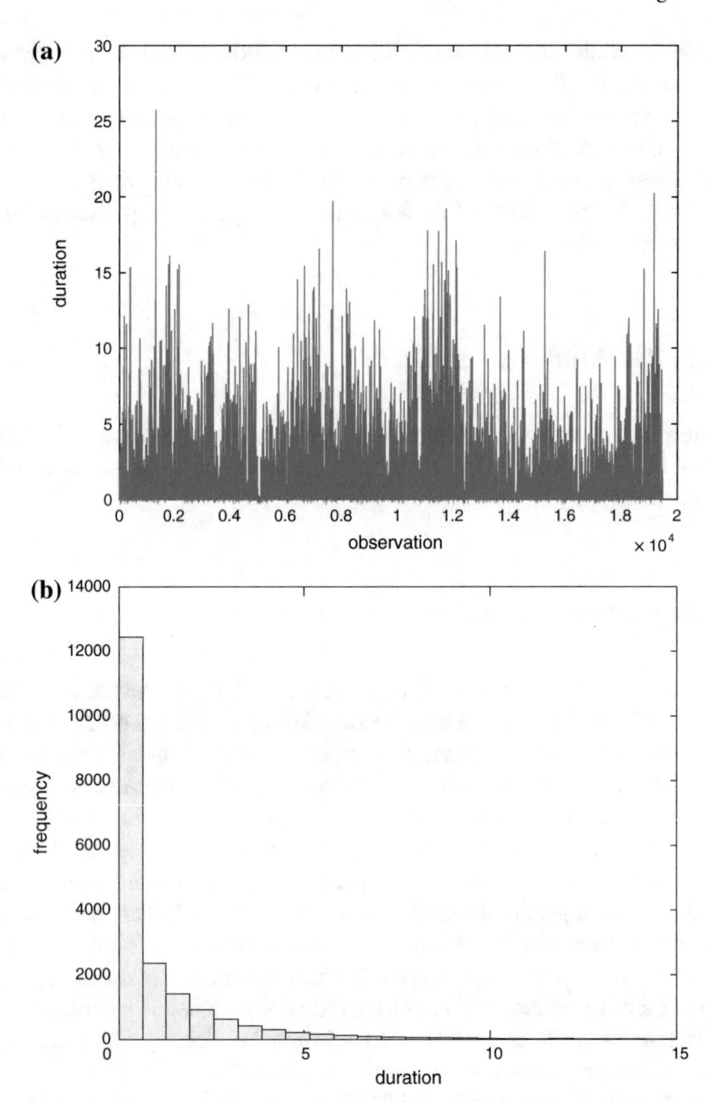

Fig. 5 **a** Time series plot and **b** histogram of adjusted trade durations of TLS stock during the period 1 to 7 October, 2014

Table 2 Summary statistics of TLS trade durations during the period 1 to 7 October, 2014

Obs.	Min	Max	Mean	Median	Std	Skewness	Kurtosis
19473	0.0013	25.7623	1.0000	0.2236	1.7809	3.4867	20.7528

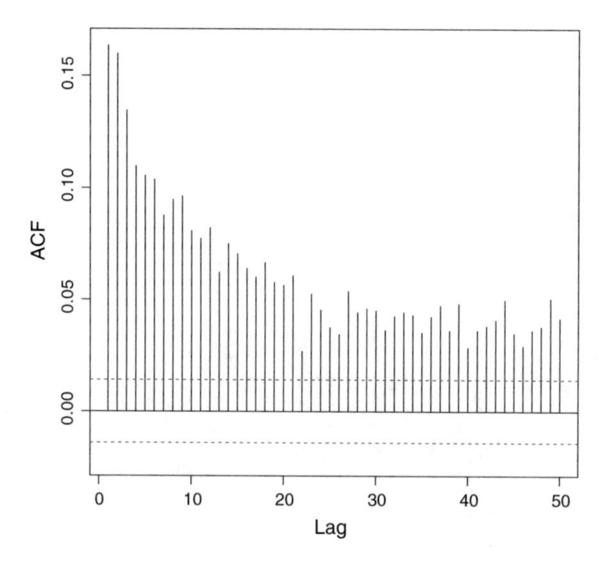

Fig. 6 Sample ACF for adjusted durations of TLS stock during the period 1 to 7 October, 2014

Table 3 Parameter estimates from ACD (1, 1) with errors from exponential, Weibull and EW distributions fitted to TLS duration data

Parameter	EACD	WACD	EW-ACD
α_0	0.0503 (0.0025)	0.0472 (0.0040)	0.0531 (0.0044)
α_1	0.1759 (0.0056)	0.2726 (0.0120)	0.2714 (0.0119)
β	0.7868 (0.0062)	0.7257 (0.0119)	0.7268 (0.0119)
k		0.4988 (0.0026)	0.4734 (0.0031)
γ			2088.44 (213.62)
BIC	33547.54	12195.25	11662.60
DIC	33523.86	12161.14	11641.02

the 5 % level. The EW-ACD model seems to be the best performing model, in terms of model fit based on both BIC and DIC. For WACD as well as EW-ACD models, the estimated shape parameter, k is less than one. This indicates a monotonously decreasing hazard function for adjusted TLS trade durations. It is reasonable for the high liquid asset. Although the extra shape parameter γ of EW distribution is considerably large, showing heavy skewness and approaching Weibull distribution it still possesses a heavier tail than Weibull distribution and hence a better model fit. That is the main reason for its superior performance.

The fitted conditional expected mean of trade durations $\hat{\psi}_i$ from the EW-ACD model is plotted against durations of TLS in Fig. 7a. The model seems to adequately capture the mean durations. On the other hand, the P-P plot of the residuals from the same model confirms adequate fit of the distribution as depicted in Fig. 7b. In regions of high probability mass for near zero durations, the difference between

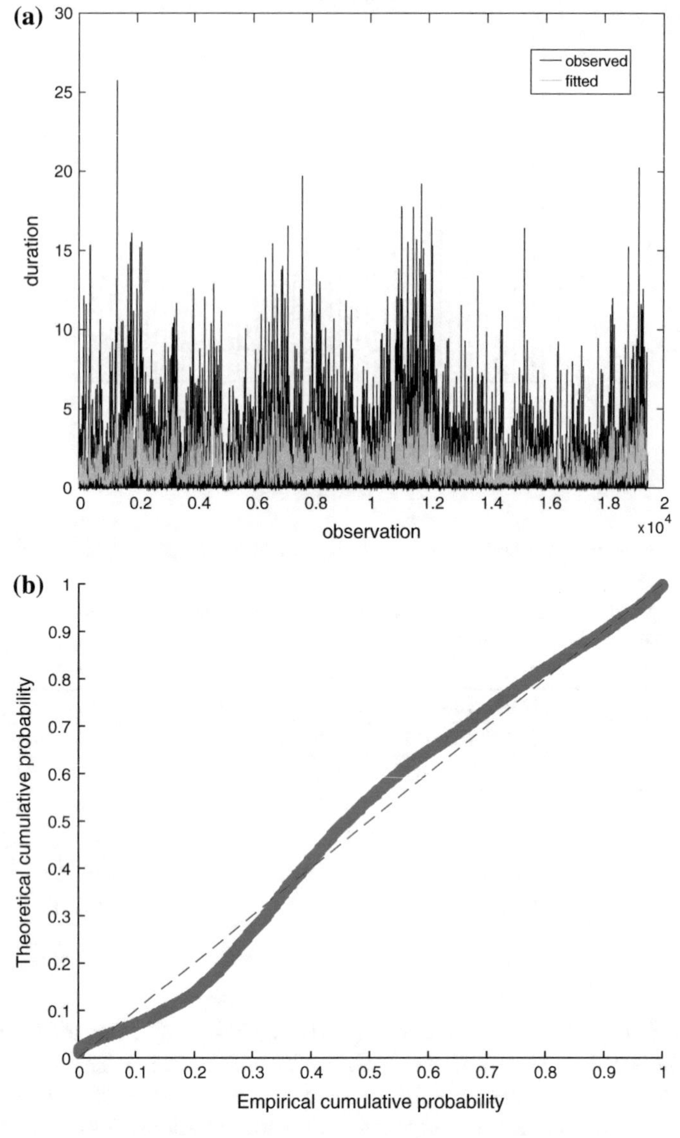

Fig. 7 **a** Fitted conditional mean of adjusted trade durations and **b** P-P plot of residuals from the EW-ACD model fitted to TLS duration data during the period 1 to 7 October, 2014

theoretical and empirical cdfs is more apparent and negative than those in the low density regions for higher level durations, where the difference is less and positive. The pattern of the plot is consistent with heavier tail than observed and represent a uni-modal distribution. For further details on P-P plots, refer [18].

6.2 Daily Range Data

As mentioned earlier, the ACD model can be fitted to a time series with non-negative observations, such as the daily range and trade duration. Hence, the second real life example is on stock volatility modelling based on the daily range. We consider a time series of daily range of the log of All Ordinaries index (AOI) of ASX for the period from 1 May 2009 to 26 April 2013, consisting of 1008 observations. The data can be downloaded from Yahoo Finance. The daily percentage log-range R_i is given by

$$R_i = 100 \times [\ln(\max_z P_{i_z}) - \ln(\min_z P_{i_z})],$$

where P_{i_z} is the AOI measured at discrete time z in day i and max (min) is the maximum (minimum) of P_{i_z} over all time z in day i.

Time series plot, histogram and sample ACF are given in Figs. 8a, b and 9, respectively. Volatility clustering, heavy tailedness and long term serial dependence appear to be common features of the AOI daily range series. On the other hand, the histogram shows that it has a uni-modal distribution having the mode shifted away from zero and has a relatively lower skewness and kurtosis, in comparison to the duration data, as indicated by the descriptive statistics in Table 4. Hence, the two series considered differ in their distributional shapes.

Again, three models EACD (1, 1), WACD (1, 1) and EW-ACD (1, 1) are fitted to the range data. The parameter estimates are reported in Table 5 together with the standard errors in parentheses. All the estimates are significant, at the usual 5 % level, except α_0 of the EACD model. The best performing model is again the EW-ACD model, while the worst performer is the EACD model, in terms of both BIC and DIC. The estimated shape parameters, k for both WACD and EW-ACD models are greater than one, contrary to those of duration data. This indicates a monotonously increasing hazard function for the range data series of AOI. This is consistent with the phenomenon of volatility clustering where large volatility tends to be followed by large volatility and vice versa. Furthermore, the extra shape parameter γ of the EW distribution has a small value, which is less than 2, catering to the relatively low skewness prevalent in range data.

The fitted conditional expected mean of daily range from the EW-ACD model is plotted against the observed daily range of AOI in Fig. 10a. The model seems to adequately capture the average volatility. On the other hand, the pdf plot fitted to residuals from the same model displayed in Fig. 10b indicates a good fit of the EW distribution and hence a suitable distribution for the residuals. The mean, standard deviation, skewness and kurtosis of the standardised residuals are 1.000, 0.4364, 0.7248 and 3.6147 respectively, indicating positive skewness and higher kurtosis than normal. Obviously, the skewness and kurtosis are lower than the original data after modelling the mean structure. The hazard and the cdf are plotted in Fig. 11a and b respectively, to get an idea about their behaviour.

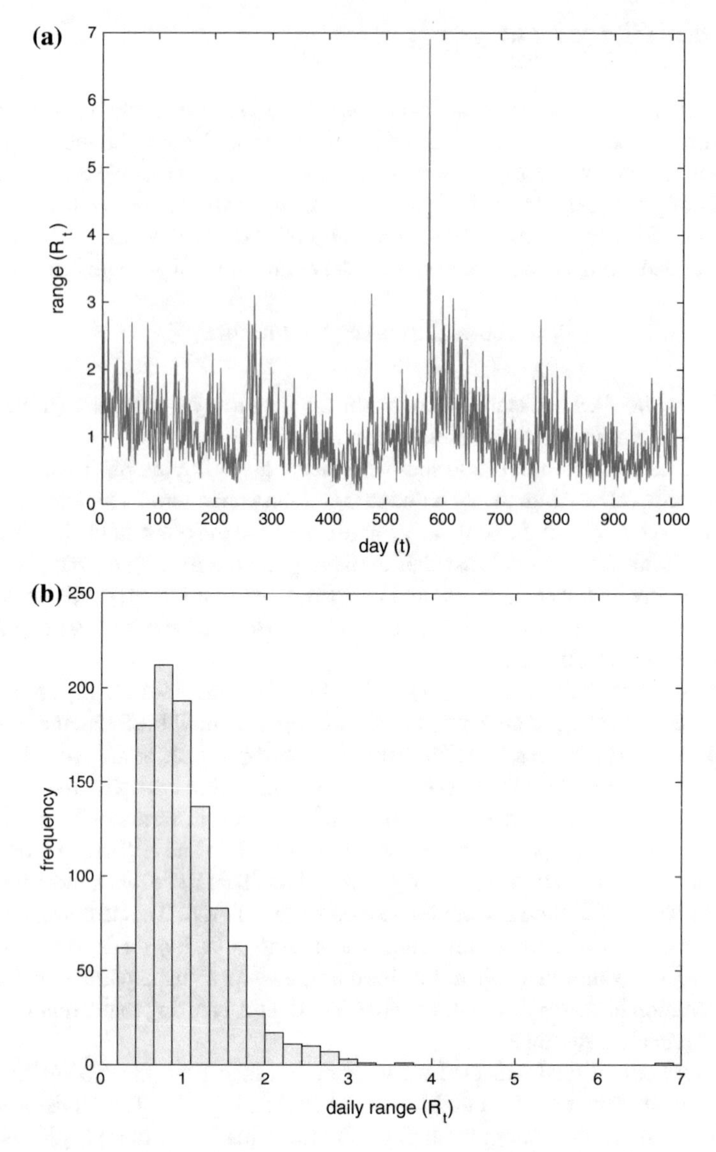

Fig. 8 **a** Time series plot and **b** histogram of daily range of AOI during the period 1 May 2009 to 26 April 2013

7 Conclusion

A new distribution named as Extended Weibull (EW) is developed to allow a more flexible error distribution in ACD models. An additional shape parameter included in the variant form of the existing Weibull distribution provides this added flexibility.

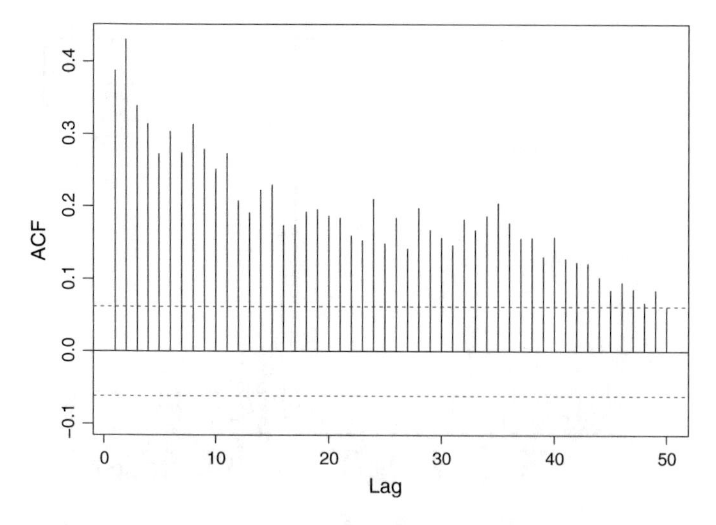

Fig. 9 Sample ACF of daily range of AOI for the period from 1 May 2009 to 26 April 2013

Table 4 Summary statistics of AOI daily range during the period 1 May, 2009 to 26 April, 2013

Obs.	Min	Max	Mean	Median	Std	Skewness	Kurtosis
1008	0.2021	6.9083	1.0291	0.9199	0.5393	2.2998	17.9846

Table 5 Parameter estimates of the ACD (1, 1) model with errors from the exponential, Weibull and EW distributions fitted to AOI range data

Parameter	EACD	WACD	EW-ACD
α_0	0.0975 (0.0603)	0.1988 (0.0142)	0.1142 (0.0157)
α_1	0.2659 (0.0655)	0.3215 (0.0136)	0.2788 (0.0213)
β	0.6418 (0.1047)	0.4861 (0.0107)	0.6112 (0.0269)
k		2.3493 (0.0400)	1.8521 (0.0661)
γ			1.6577 (0.2496)
BIC	2033.56	1153.86	1056.98
DIC	2015.98	1042.03	984.08

This parameter tends to capture heavier tails better than the Weibull distribution, which is a commonly used error distribution, due to its simplicity. In the presence of high skewness and kurtosis, this parameter tends to be large, in general, especially for small values of k.

The main attributes of the EW distribution are investigated, including the derivation of first four moments, cdf, survivor and hazard functions. The flexibility of the distribution is envisaged not only in terms of different shapes of the density function but also the hazard function. Interestingly, unlike the exponential or the Weibull

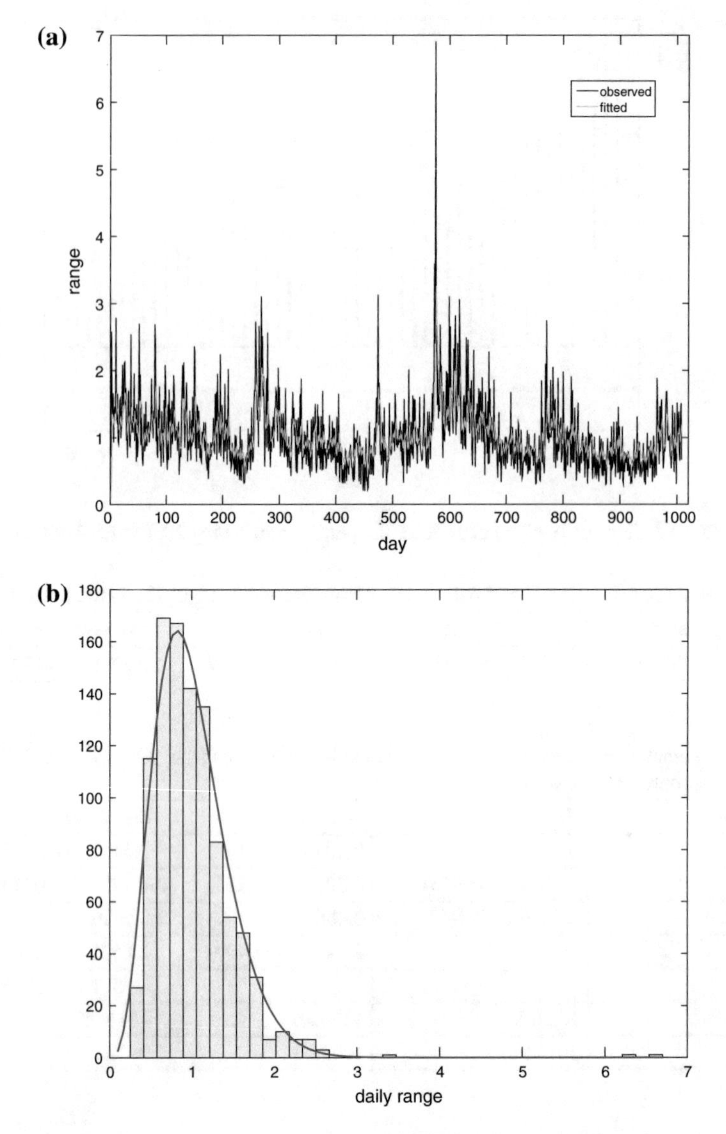

Fig. 10 **a** Fitted conditional mean of AOI daily range and **b** fitted pdf of residuals from the EW-ACD model fitted to AOI daily range during the period 1 May, 2009 to 26 April, 2013

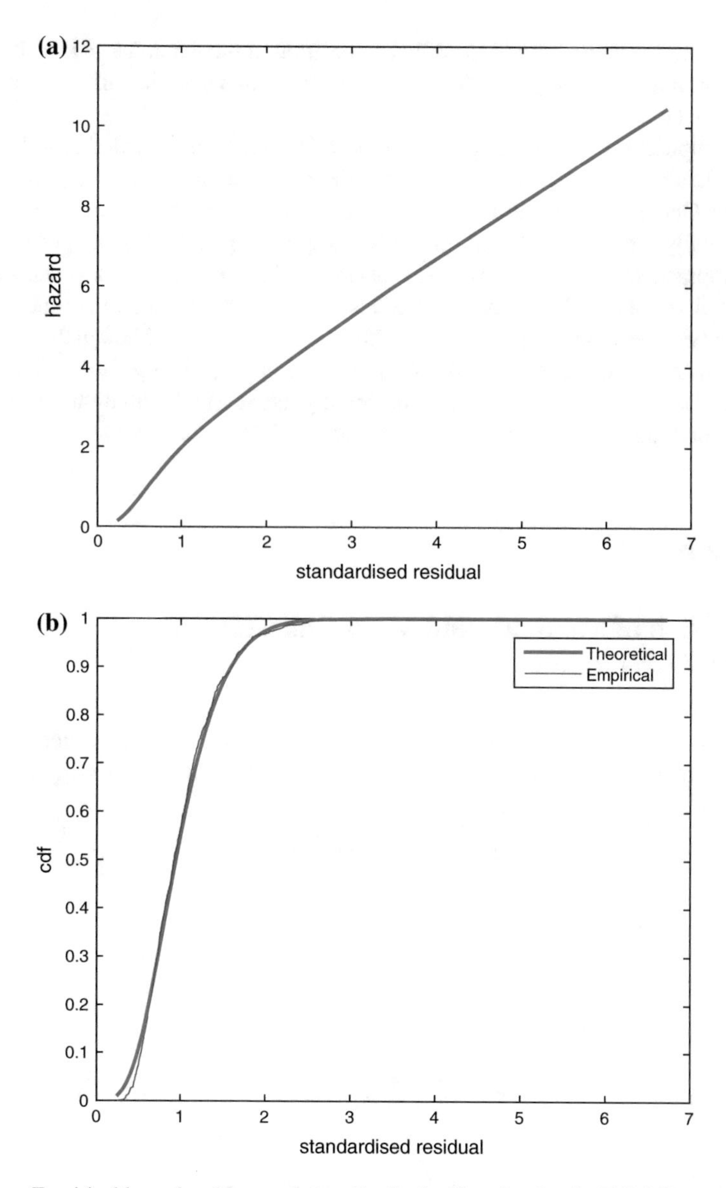

Fig. 11 **a** Empirical hazard and **b** cumulative distribution function for the EW-ACD residuals of AOI daily range

distributions, the hazard function could be non-monotonic when $0.5 < k < 1$, which is more prominent when $\gamma > 0.5$. This is a useful feature, particularly in modelling duration data.

The empirical performance of the EW distribution is investigated based on two real life data sets from the ASX, which share some common features but yet different in nature. One is the trade durations of TLS, characterising the mean duration and the other is daily range data of AOI, characterising average volatility. Its performance was compared with two other widely considered distributions, the exponential and Weibull distributions. In terms of both data sets, the EW distribution outperformed the other two distributions, irrespective of the magnitude of γ. Although theoretically the EW distribution converges to the Weibull distribution, when γ tends to infinity, it showed an improvement in model fit. This highlights the EW distribution's usefulness as a potential contender for the error distribution of ACD models.

Appendix

Calculation of Moments and Main Characteristics for EW Distribution

Let X be a random variable following the EW distribution with parameters λ, k and γ. The distribution of X will be denoted as EW(λ,k,γ) with the following pdf

$$f(x) = \left(1 + \frac{1}{\gamma^k}\right) \frac{k}{\lambda^k} x^{k-1} e^{-\left(\frac{x}{\lambda}\right)^k} \left[1 - e^{-\left(\frac{\gamma x}{\lambda}\right)^k}\right].$$

1. Derivation of mean, $E(X)$

$$\begin{aligned} E(X) = \mu &= \int_0^\infty x f(x) dx \\ &= \frac{k}{\lambda^k}\left(1 + \frac{1}{\gamma^k}\right)\left[\int_0^\infty x^k e^{-\left(\frac{x}{\lambda}\right)^k} dx - \int_0^\infty x^k e^{-\left(\frac{1+\gamma^k}{\lambda^k}\right)(X)^k} dx\right] \\ &= \lambda\left(1 + \frac{1}{\gamma^k}\right) \Gamma\left(1 + \frac{1}{k}\right)\left[1 - \frac{1}{(1+\gamma^k)^{1+\frac{1}{k}}}\right] \\ &= \frac{\lambda}{k\gamma^k} \Gamma\left(\frac{1}{k}\right)\left[\frac{(1+\gamma^k)^{1+\frac{1}{k}} - 1}{(1+\gamma^k)^{\frac{1}{k}}}\right] \end{aligned} \tag{11}$$

2. Derivation of variance, $Var(X)$

$$
\begin{aligned}
Var(X) &= \sigma^2 = E(X^2) - [E(X)]^2 \\
E(X^2) &= \frac{k}{\lambda^k}\left(1+\frac{1}{\gamma^k}\right)\left[\int_0^\infty x^{k+1} e^{-\left(\frac{x}{\lambda}\right)^\alpha} dx - \int_0^\infty x^{k+1} e^{-\left(\frac{1+\gamma^k}{\lambda^k}\right)(x)^k} dx\right] \\
&= \lambda^2\left(1+\frac{1}{\gamma^k}\right)\Gamma\left(1+\frac{2}{k}\right)\left[1-\frac{1}{(1+\gamma^k)^{1+\frac{2}{k}}}\right] \\
\sigma^2 &= \lambda^2\left(1+\frac{1}{\gamma^k}\right)\left\{\Gamma\left(1+\frac{2}{k}\right)\left[1-\frac{1}{(1+\gamma^k)^{1+\frac{2}{k}}}\right]\right. \\
&\left.\quad - \left(1+\frac{1}{\gamma^k}\right)\left[\Gamma\left(1+\frac{1}{k}\right)\right]^2\left[1-\frac{1}{(1+\gamma^k)^{1+\frac{1}{k}}}\right]^2\right\} \qquad (12)
\end{aligned}
$$

3. Derivation of skewness, $Skew(X)$

$$
Skew(X) = \rho = \frac{E\left[X-E(X)\right]^3}{\sigma^3} = \frac{S_1}{S_2} \qquad (13)
$$

where
$$
\begin{aligned}
E\left[X-E(X)\right]^3 &= \frac{k}{\lambda^k}\left(1+\frac{1}{\gamma^k}\right)\left(\int_0^\infty (x-\mu)^3 x^{k-1} e^{-\left(\frac{x}{\lambda}\right)^k} dx\right. \\
&\left.\quad - \int_0^\infty (x-\mu)^3 x^{k-1} e^{-\left(\frac{1+\gamma^k}{\lambda^k}\right)x^k} dx\right], \\
S_1 &= \Gamma\left(1+\frac{3}{k}\right)\left[1-\frac{1}{(1+\gamma^k)^{1+\frac{3}{k}}}\right] \\
&\quad - \frac{3\mu}{\lambda}\Gamma\left(1+\frac{2}{k}\right)\left[1-\frac{1}{(1+\gamma^k)^{1+\frac{2}{k}}}\right] \\
&\quad + \frac{3\mu^2}{\lambda^2}\Gamma\left(1+\frac{1}{k}\right)\left[1-\frac{1}{(1+\gamma^k)^{1+\frac{1}{k}}}\right] - \frac{\mu^3}{\lambda^3}\left(1-\frac{1}{1+\gamma^k}\right), \\
S_2 &= \left[1+\frac{1}{\gamma^k}\right]^{1/2}\left\{\Gamma\left(1+\frac{2}{k}\right)\left[1-\frac{1}{(1+\gamma^k)^{1+\frac{2}{k}}}\right]\right. \\
&\left.\quad - \left(1+\frac{1}{\gamma^k}\right)\left[\Gamma\left(1+\frac{1}{k}\right)\right]^2\left[1-\frac{1}{(1+\gamma^k)^{1+\frac{1}{k}}}\right]^2\right\}^{\frac{3}{2}}
\end{aligned}
$$

4. Derivation of kurtosis, $Kurt(X)$

$$Kurt(X) = \zeta = \frac{E\left[X - E(X)\right]^4}{\sigma^4} = \frac{K_1}{K_2} \tag{14}$$

$$\text{where} \quad E\left[X - E(X)\right]^4 = \frac{k}{\lambda^k}\left(1 + \frac{1}{\gamma^k}\right)\left[\int_0^\infty (x-\mu)^4 x^{k-1} e^{-\left(\frac{x}{\lambda}\right)^k} dx - \int_0^\infty [x-\mu)^4 x^{k-1} e^{-\left(\frac{1+\gamma^k}{\lambda^k}\right)x^k} dx\right],$$

$$\begin{aligned} K_1 = {} & \Gamma\left(1+\frac{4}{k}\right)\left[1 - \frac{1}{(1+\gamma^k)^{1+\frac{4}{k}}}\right] - \frac{4\mu}{\lambda}\Gamma\left(1+\frac{3}{k}\right)\left[1 - \frac{1}{(1+\gamma^k)^{1+\frac{3}{k}}}\right] \\ & + \frac{6\mu^2}{\lambda^2}\Gamma\left(1+\frac{2}{k}\right)\left[1 - \frac{1}{(1+\gamma^k)^{1+\frac{2}{k}}}\right] - \frac{4\mu^3}{\lambda^3}\Gamma\left(1+\frac{1}{k}\right)\left[1 - \frac{1}{(1+\gamma^k)^{1+\frac{1}{k}}}\right] \\ & + \frac{\mu^4}{\lambda^4}\left(1 - \frac{1}{1+\gamma^k}\right), \end{aligned}$$

$$\begin{aligned} K_2 = {} & \left(1+\frac{1}{\gamma^k}\right)\left\{\Gamma\left(1+\frac{2}{k}\right)\left[1 - \frac{1}{(1+\gamma^k)^{1+\frac{2}{k}}}\right]\right. \\ & \left. - \left(1+\frac{1}{\gamma^k}\right)\left[\Gamma\left(1+\frac{1}{k}\right)\right]^2\left[1 - \frac{1}{(1+\gamma^k)^{1+\frac{1}{\alpha}}}\right]^2\right\}^2 \end{aligned}$$

5. Derivation of the cumulative distribution function, cdf, $F(x)$

$$\begin{aligned} F(x) &= \int_0^x f(y)dy \\ &= \frac{k}{\lambda^k}\left(1+\frac{1}{\gamma^k}\right)\left[\int_0^x y^{k-1} e^{-\left(\frac{y}{\lambda}\right)^k} dy - \int_0^x y^{k-1} e^{-\left(\frac{1+\gamma^k}{\lambda^k}\right)y^k} dy\right] \\ &= 1 + \frac{1}{\gamma^k} e^{-\left(\frac{x}{\lambda}\right)^k}\left[e^{-\left(\frac{\gamma x}{\lambda}\right)^k} - 1 - \gamma^k\right] \end{aligned} \tag{15}$$

References

1. Pacurar, M.: Autoregressive conditional duration models in finance: a survey of the theoretical and empirical literature. J. Econ. Surv. **22**(4), 711–751 (2008)
2. Engle, R.F., Russell, J.R.: Autoregressive conditional duration: a new model for irregularly spaced transaction data. Econometrica, pp. 1127–1162 (1998)

3. Grammig, J., Maurer, K.-O.: Non-monotonic hazard functions and the autoregressive conditional duration model. Econom. J. **3**(1), 16–38 (2000)
4. Bhatti, C.R.: The birnbaum-saunders autoregressive conditional duration model. Math. Comput. Simul. **80**(10), 2062–2078 (2010)
5. Bauwens, L., Giot, P., Grammig, J., Veredas, D.: A comparison of financial duration models via density forecasts. Int. J. Forecast. **20**(4), 589–609 (2004)
6. Allen, D., Chan, F., McAleer, M., Peiris, S.: Finite sample properties of the qmle for the log-acd model: application to australian stocks. J. Econom. **147**(1), 163–185 (2008)
7. Allen, D., Lazarov, Z., McAleer, M., Peiris, S.: Comparison of alternative acd models via density and interval forecasts: evidence from the australian stock market. Math. Comput. Simul. **79**(8), 2535–2555 (2009)
8. Parkinson, M.: The extreme value method for estimating the variance of the rate of return. J. Bus. 61–65 (1980)
9. Chou, R.Y.: Forecasting financial volatilities with extreme values: the conditional autoregressive range (carr) model. J. Money, Credit Bank. 561–582 (2005)
10. Engle, R.: New frontiers for arch models. J. Appl. Econom. **17**(5), 425–446 (2002)
11. Carter, C.K., Kohn, R.: On gibbs sampling for state space models. Biometrika **81**(3), 541–553 (1994)
12. Metropolis, N., Rosenbluth, A.W., Rosenbluth, M.N., Teller, A.H., Teller, E.: Equation of state calculations by fast computing machines. J. Chem. Phys. **21**(6), 1087–1092 (1953)
13. Hastings, W.K.: Monte carlo sampling methods using markov chains and their applications. Biometrika **57**(1), 97–109 (1970)
14. Chen, C.W., So, M.K., Gerlach, R.H.: Assessing and testing for threshold nonlinearity in stock returns. Aust. N. Z. J. Stat. **47**(4), 473–488 (2005)
15. Gelman, A., Roberts, G., Gilks, W.: Efficient metropolis jumping rules. Bayesian Stat. **5**(42), 599–608 (1996)
16. Von Neumann, J.: 13. Various Techniques used in Connection with Random Digits (1951)
17. Smith, M., Kohn, R.: Nonparametric regression using bayesian variable selection. J. Econom. **75**(2), 317–343 (1996)
18. Wilk, M.B., Gnanadesikan, R.: Probability plotting methods for the analysis for the analysis of data. Biometrika **55**(1), 1–17 (1968)

Across-the-Board Spending Cuts Are Very Inefficient: A Proof

Vladik Kreinovich, Olga Kosheleva, Hung T. Nguyen and Songsak Sriboonchitta

Abstract In many real-life situations, when there is a need for a spending cut, this cut is performed in an across-the-board way, so that each budget item is decreased by the same percentage. Such cuts are ubiquitous, they happen on all levels, from the US budget to the university budget cuts on the college and departmental levels. The main reason for the ubiquity of such cuts is that they are perceived as fair and, at the same time, economically reasonable. In this paper, we perform a quantitative analysis of this problem and show that, contrary to the widely spread positive opinion about across-the-board cuts, these cuts are, on average, very inefficient.

1 Formulation of the Problem: Are Across-the-Board Spending Cuts Economically Reasonable

Across-the-board spending cuts are ubiquitous. When a department or even a country faces an unexpected decrease in funding, it is necessary to balance the budget by making some spending cuts.

V. Kreinovich (✉)
Department of Computer Science, University of Texas at El Paso,
500 W. University, El Paso, TX 79968, USA
e-mail: vladik@utep.edu

O. Kosheleva
University of Texas at El Paso, 500 W. University, El Paso, TX 79968, USA
e-mail: olgak@utep.edu

H.T. Nguyen
Department of Mathematical Sciences, New Mexico State University,
Las Cruces, NM 88003, USA
e-mail: hunguyen@nmsu.edu

H.T. Nguyen · S. Sriboonchitta
Faculty of Economics, Chiang Mai University, Chiang Mai, Thailand
e-mail: songsakecon@gmail.com

V.-N. Huynh et al. (eds.), *Causal Inference in Econometrics*,
Studies in Computational Intelligence 622, DOI 10.1007/978-3-319-27284-9_6

In many such situations, what is implemented is an across-the board cut, when all the spending items are decreased by the same percentage. For example, all the salaries are decreased by the same percentage.

The ubiquity of such cuts is motivated largely by the fact that since they apply to everyone on the same basis, they are fair.

Across-the-board cuts may sound fair, but are they economically efficient? The fact that such cuts are fair do not necessarily mean that they are economically efficient. For example, if we consistently take all the wealth of a country and divide it equally between all its citizens, this may be a very fair division, but, because of its lack of motivations to work harder, this clearly will not be a very economically efficient idea.

Current impression. The current impression that across-the-board cuts may not be economically optimal, but they are economically reasonable; see, e.g., [1, 3–6, 8–11].

What we show in this paper. In this paper, we perform a quantitative analysis of the effect of across-the-board cuts, and our conclusion is that their economic effect is much worse than it is usually perceived.

Comment. To make our argument as convincing as possible, we tried our best to make this paper—and its mathematical arguments—as mathematically simple and easy-to-read as we could.

2 Let Us Formulate the Problem in Precise Terms

Formulation of the problem in precise terms. Let us start by formulating this problem in precise terms.

What is given. First, we need to describe what we had before the need appeared for budget cuts. Let us denote the overall spending amount by x, and the amount originally allocated to different spending categories by $x_1, x_2, \ldots, x_n$, so that

$$\sum_{i=1}^{n} x_i = x.$$

Sometimes, it turns out that the original estimate x for the spending amount was too optimistic, and instead we have a smaller amount $y < x$.

What we need to decide. Based on the decrease amount $y < x$, we need to select new allocations, i.e., select the values $y_1 \le x_1, \ldots, y_n \le x_n$ for which $\sum\limits_{i=1}^{n} x_i = x$.

What is an across-the-board spending cut. An across-the-board spending cut means that for each i, we take $y_i = (1 - \delta) \cdot x_i$, where the common value $\delta > 0$ is determined by the condition that $(1 - \delta) \cdot x = y$. Thus, this value δ is equal to

$$\delta = 1 - \frac{y}{x}.$$

What we plan to analyze. We want to check whether the across-the-board spending cut $y_i = (1 - \delta) \cdot x_i$ is economically reasonable, e.g., to analyze how it compares with the optimal budget cut.

We need to describe, in precise terms, what is better and what is worse for the economy. To make a meaningful comparison between different alternative versions of budget cuts, we need to have a clear understanding of which economical situations are preferable. In other words, we need to be able to consistently compare any two different situations.

It is known that such a linear (total) order on the set of all possible alternatives can be, under reasonable conditions, described by a real-valued functions $f(y_1, \ldots, y_n)$ defined on the set of such alternatives: for every two alternatives $(y_1, \ldots, y_n)$ and $(y'_1, \ldots, y'_n)$, the one with the larger value of this function is preferable (see, e.g., [12]):

- if $f(y_1, \ldots, y_n) > f(y'_1, \ldots, y'_n)$, then the alternative $(y_1, \ldots, y_n)$ is preferable;
- on the other hand, if $f(y'_1, \ldots, y'_n) > f(y_1, \ldots, y_n)$, then the alternative $(y'_1, \ldots, y'_n)$ is preferable.

The objective function should be monotonic. The more money we allocate to each item i, the better. Thus, the objective function should be increasing in each of its variables: if $y_i < y'_i$ for some i and $y_i \le y'_i$ for all i, then we should have $f(y_1, \ldots, y_n) < f(y'_1, \ldots, y'_n)$.

We consider the generic case. In this paper, we do not assume any specific form of the objective function $f(y_1, \ldots, y_n)$. Instead, we will show that the same result—that across-the-board cuts are not efficient—holds for all possible objective functions (of course, as long as they satisfy the above monotonicity condition). So, whether our main objective is:

- to increase the overall GDP,
- or to raise the average income of all the poor people,
- or, alternatively, to raise the average income of all the rich people,

no matter what is our goal, across-the-board cuts are a far-from-optimal optimal way to achieve this goal.

Resulting formulation of the problem. We assume that the objective function $f(y_1, \ldots, y_n)$ is given.

We have the initial amount x. Based on this amount, we selected the values $x_1, \ldots, x_n$ for which $f(x_1, \ldots, x_n)$ attains the largest possible value under the constraint that $\sum\limits_{i=1}^{n} x_i = x$. Let us denote the value of the objective function corresponding to this original budget allocation by f_x.

Now, we are given a different amount $y < x$. Ideally, we should now select the values $y_1, \ldots, y_n$ for which $f(y_1, \ldots, y_n)$ attains the largest possible value under the constraint that $\sum_{i=1}^{n} y_i = y$. Due to monotonicity, the resulting best-possible value f_y of the objective function $f(y_1, \ldots, y_n)$ is smaller than the original value f_x.

In the across-the-board arrangement, instead of selecting the optimal values y_i, we select the across-the-board values $y_i = (1 - \delta) \cdot x_i$, where $\delta = 1 - \frac{y}{x}$. The resulting allocation of funds is, in general, not as good as the optimal one. Thus, the resulting value of the objective function f_δ is, in general, smaller than f_y.

To decide how economically reasonable are across-the-board cuts, we need to compare:

- the *optimal* decrease $f_x - f_y$ in the value of the objective function, with
- the decrease $f_x - f_\delta$ caused by using across-the-board spending cuts.

3 Analysis of the Problem

Possibility of linearization. Usually, the relative size of the overall cut does not exceed 10 %; usually it is much smaller. By economic standards, a 10 % cut is huge, but from the mathematical viewpoint, it is *small*—in the sense that terms which are quadratic in this cut can be safely ignored. Indeed, the square of $0.1 = 10\,\%$ is $0.01 = 1\,\% \ll 10\,\%$.

Thus, if we expand the dependence of the objective function $f(y_1, \ldots, y_n)$ in Taylor series around the point $(x_1, \ldots, x_n)$, i.e., if we consider the dependence

$$f(y_1, \ldots, y_n) = f(x_1, \ldots, x_n) - \sum_{i=1}^{n} c_i \cdot (x_i - y_i) + \cdots$$

$$= f_x - \sum_{i=1}^{n} c_i \cdot (x_i - y_i) + \cdots, \quad (1)$$

where $c_i \stackrel{\text{def}}{=} \frac{\partial f}{\partial y_i}$, then we can safely ignore terms which are quadratic in terms of the differences and conclude that

$$f(y_1, \ldots, y_n) = f_x - \sum_{i=1}^{n} c_i \cdot \Delta y_i,$$

where we denoted $\Delta y_i \stackrel{\text{def}}{=} x_i - y_i \geq 0$, and thus, that:

$$f_x - f(y_1, \ldots, y_n) = \sum_{i=1}^{n} c_i \cdot \Delta y_i. \quad (2)$$

Comment. Since the objective function $f(x_1, \ldots, x_n)$ is monotonic in each of the variables, all the partial derivatives c_i are non-negative: $c_i \geq 0$.

Linearization simplifies the problem: general idea. Let us describe how the use of linearization simplifies the computation of the two differences $f_x - f_y$ and $f_x - f_\delta$.

Linearization simplifies the problem: case of optimal spending cuts. Let us start with the computation of the difference $f_x - f_y$ corresponding to the optimal spending cuts. The optimal arrangement $(y_1, \ldots, y_n)$ is the one that maximizes the value of the objective function $f(y_1, \ldots, y_n)$ under the constraint

$$\sum_{i=1}^{n} y_i = y.$$

Maximizing the value of the objective function $f(y_1, \ldots, y_n)$ is equivalent to minimizing the difference $f_x - f(y_1, \ldots, y_n)$, which, according to the formula (2), is equivalent to minimizing the sum $\sum\limits_{i=1}^{n} c_i \cdot \Delta y_i$.

To make the problem easier to solver, let us also describe the constraint $\sum\limits_{i=1}^{n} y_i = y$ in terms of the new variables Δy_i. This can be achieved if we subtract this constraint from the formula $\sum\limits_{i=1}^{n} x_i = x$. As a result, we get an equality $\sum\limits_{i=1}^{n} \Delta y_i = \Delta y$, where we denoted $\Delta y \stackrel{\text{def}}{=} x - y$.

Thus, due to the possibility of linearization, the corresponding optimization problem takes the following form: minimize the sum $\sum\limits_{i=1}^{n} c_i \cdot \Delta y_i$ under the constraint

$$\sum_{i=1}^{n} \Delta y_i = \Delta y.$$

Let us prove that this minimum is attained when $\Delta y_{i_0} = \Delta y$ for the index i_0 corresponding to the smallest possible value of the derivative c_i, and $\Delta y_i = 0$ for all other indices $i \neq i_0$.

Indeed, for the arrangement when $\Delta y_{i_0} = \Delta y$ and $\Delta y_i = 0$ for all $i \neq i_0$, the minimized sum attains the value

$$\sum_{i=1}^{n} \Delta y_i = c_{i_0} \cdot \Delta y = \left(\min_i c_i\right) \cdot \Delta y.$$

Let us prove that for every other arrangement, we have a larger (or equal) value of the difference $f_x - f(y_1, \ldots, y_n)$. Indeed, by our choice of i_0, we have $c_i \geq c_{i_0}$ for all i. Thus, due to $\Delta y_i \geq 0$, we have $c_i \cdot \Delta y_i \geq c_{i_0} \cdot \Delta y_i$, and therefore,

$$\sum_{i=1}^{n} c_i \cdot \Delta y_i \geq \sum_{i=1}^{n} c_{i_0} \cdot \Delta y_i = c_{i_0} \cdot \left(\sum_{i=1}^{n} \Delta y_i\right) = c_{i_0} \cdot \Delta y.$$

Thus, the difference $f_x - f_y$ corresponding to the optimal spending cuts is equal to

$$f_x - f_y = \left(\min_i c_i\right) \cdot \Delta y. \tag{3}$$

Linearization simplifies the problem: case of across-the-board spending cuts. For across-the-board spending cuts, we have $y_i = (1 - \delta) \cdot x_i$ and hence,

$$\Delta y_i = x_i - y_i = \delta \cdot x_i.$$

The coefficient δ can be obtained from the condition that $(1 - \delta) \cdot x = y$, i.e., that $\Delta y = x - y = \delta \cdot x$, thus $\delta = \dfrac{\Delta y}{x}$.

Substituting the corresponding values Δy_i into the linearized expression for the objective function, we conclude that

$$\sum_{i=1}^{n} c_i \cdot \Delta y_i = \sum_{i=1}^{n} c_i \cdot \delta \cdot x_i = \delta \cdot \sum_{i=1}^{n} c_i \cdot x_i = \frac{\Delta y}{x} \cdot \sum_{i=1}^{n} c_i \cdot x_i = \Delta y \cdot \sum_{i=1}^{n} c_i \cdot \delta x_i,$$

where we denoted $\delta x_i \stackrel{\text{def}}{=} \dfrac{x_i}{x}$. From the constraint $\sum\limits_{i=1}^{n} x_i = x$, one can conclude that $\sum\limits_{i=1}^{n} \delta x_i = 1$. Thus, the resulting decrease $f_x - f_\delta$ is equal to:

$$f_x - f_\delta = \Delta y \cdot \sum_{i=1}^{n} c_i \cdot \delta x_i. \tag{4}$$

What we need to compare. To compare the decreases in the value of the objective function corresponding to the optimal cuts and to the across-the-board cuts, we therefore need to compare the expressions (3) and (4).

Let us treat the values c_i and δx_i as random variables. The values of c_i and Δx_i depend on many factors which we do not know beforehand, so it makes sense to treat them as random variables. In this case, both expressions (3) and (4) become random variables.

How we compare the random variables. Because of the related uncertainty, sometimes, the difference $f_x - f_\delta$ may be almost optimal, and sometimes, it may be much larger than the optimal difference $f_x - f_y$.

A reasonable way to compare two random variables is to compare their mean values. This is what we mean, e.g., when we say that Swedes are, on average taller than Americans: that the average height of a Swede is larger than the average height of an American.

It is reasonable to assume that the variables c_i and δx_i are all independent. Since we have no reason to believe that the variables c_i corresponding to different budget items and/or the variables δx_j are correlated, it makes sense to assume that these variables are independent. This conclusion is in line with the general Maximum Entropy approach to dealing with probabilistic knowledge: if there are several possible probability distributions consistent with our knowledge, it makes sense to select the one which has the largest uncertainty (entropy; see, e.g., [2, 7]), i.e., to select a distribution for which the entropy

$$S = -\int \rho(x) \cdot \ln(\rho(x))\, dx$$

attains the largest possible value, where $\rho(x)$ is the probability density function (pdf).

In particular, for the case when for two random variables, we only know their marginal distributions, with probability densities $\rho_1(x_1)$ and $\rho_2(x_2)$, the Maximum Entropy approach selects the joint probability distribution with the probability density $\rho(x_1, x_2) = \rho_1(x_1) \cdot \rho_2(x_2)$ that corresponds exactly to the case when these two random variables are independent.

Consequence of independence. In general, the mean $E[X + Y]$ of the sum is equal to the sum $E[X] + E[Y]$ of the means $E[X]$ and $E[Y]$. So, from the formula (4), we conclude that

$$E[f_x - f_\delta] = \Delta y \cdot \sum_{i=1}^{n} E[c_i \cdot \delta x_i].$$

Since we assume that for each i, the variables c_i and δx_i are independent, we conclude that

$$E[f_x - f_\delta] = \Delta y \cdot \sum_{i=1}^{n} E[c_i] \cdot E[\delta x_i]. \tag{5}$$

Here, we have no reason to believe that some values δx_i are larger, so it makes sense to assume that they have the same value of $E[\delta x_i]$. From the fact that $\sum_{i=1}^{n} \delta x_i = 1$, we conclude that $\sum_{i=1}^{n} E[\delta x_i] = 1$, i.e., that $n \cdot E[\delta x_i] = 1$. Thus, $E[\delta x_i] = \frac{1}{n}$, and the formula (5) takes the form

$$E[f_x - f_\delta] = \Delta y \cdot \frac{1}{n} \cdot \sum_{i=1}^{n} E[c_i]. \tag{6}$$

Let us select distributions for c_i. Now, we need to compare:

- the value (6) corresponding to across-the-board cuts with
- the expected value of the optimal difference (3):

$$E[f_x - f_y] = \Delta y \cdot E\left[\min_i c_i\right]. \quad (7)$$

In both cases, the only remaining random variables are c_i, so to estimate these expressions, we need to select appropriate probability distributions for these variables.

We do not have much information about the values c_i. We know that $c_i \geq 0$. We also know that these values cannot be too large. Thus, we usually know an upper bound c on these values. Thus, for each i, the only information that we have about the corresponding random variable c_i is that it is located on the interval $[0, c]$.

Under this information, the Maximum Entropy approach recommends that we select the uniform distribution on this interval. This recommendation is in perfect accordance with common sense: if we have no reason to believe that some values from this interval are more probable or less probable then others, then it is reasonable to assume that all these values have the exact same probability, i.e., that the distribution is indeed uniform.

Let us use the selected distributions to estimate the desired mean decreases (6) **and** (7). For the uniform distribution on the interval $[0, c]$, the mean value is known to be equal to the midpoint $\frac{c}{2}$ of this interval. Substituting $E[c_i] = \frac{c}{2}$ into the formula (6), we conclude that

$$E[f_x - f_\delta] = \frac{1}{2} \cdot \Delta y \cdot c. \quad (8)$$

To compute the estimate (7), let us first find the probability distribution for the minimum $m \stackrel{\text{def}}{=} \min_i c_i$. This distribution can be deduced from the fact that for each value v, the minimum m is greater than v if and only if each of the coefficients c_i is greater than v:

$$m \geq v \Leftrightarrow (c_1 > v) \,\&\, \ldots \,\&\, (c_n > v).$$

Thus,

$$\text{Prob}(m > v) = \text{Prob}((c_1 > v) \,\&\, \ldots \,\&\, (c_n > v)).$$

Since the variables $c_1, \ldots, c_n$ are all independent, we have

$$\text{Prob}(m > v) = \text{Prob}(c_1 > v) \cdot \ldots \cdot \text{Prob}(c_n > v).$$

For each i, the random variable c_i is uniformly distributed on the interval $[0, c]$, so $\text{Prob}(c_i > v) = \frac{c - v}{c}$, and thus,

$$\text{Prob}(m > v) = \left(\frac{c - v}{c}\right)^n.$$

So, the cumulative distribution function (cdf)

$$F_m(v) = \text{Prob}(m \le v) = 1 - \text{Prob}(m > v)$$

is equal to:

$$F_m(v) = 1 - \left(\frac{c - v}{c}\right)^n.$$

By differentiating the cdf, we can get the formula for the corresponding probability density function (pdf)

$$\rho_m(v) = \frac{dF_m(v)}{dv} = \frac{n}{c^n} \cdot (c - v)^{n-1}.$$

Based on this pdf, we can compute the desired mean value:

$$E[m] = \int_0^c v \cdot \rho_m(v)\, dv = \int_0^c v \cdot \frac{n}{c^n} \cdot (c - v)^{n-1}\, dv.$$

By moving the constant factor outside the integral and by introducing a new auxiliary variable $w = c - v$ for which $v = c - w$ and $dv = -dw$, we can reduce this integral expression to a simpler-to-integrate form

$$E[m] = \frac{n}{c^n} \cdot \int_0^c (c - w) \cdot w^{n-1}\, dw = \frac{n}{c^n} \cdot \left(c \cdot \int_0^c w^{n-1}\, dw - \int_0^c w^n\, dw\right)$$
$$= \frac{n}{c^n} \cdot \left(c \cdot \frac{c^n}{n} - \frac{c^{n+1}}{n+1}\right) = \frac{n}{c^n} \cdot c^{n+1} \cdot \left(\frac{1}{n} - \frac{1}{n+1}\right)$$
$$= c \cdot n \cdot \frac{1}{n \cdot (n+1)} = \frac{c}{n+1}.$$

Substituting the resulting expression

$$E\left[\min_i c_i\right] = \frac{c}{n+1}$$

into the formula (7), we conclude that

$$E[f_x - f_\delta] = \frac{1}{n+1} \cdot \Delta y \cdot c, \tag{9}$$

which is indeed much smaller than the expression (8).

Conclusion: across-the-board spending cuts are indeed very inefficient. In this paper, we compared the decreases in the value of the objective function for two possible ways of distributing the spending cuts:

- the optimal spending cuts, and
- the across-the-board spending cuts.

The resulting mean decreases are provided by the expressions (8) and (9). By comparing these expressions, we can conclude that the average decrease caused by the across-the-board cuts is $\frac{n+1}{2}$ larger than what is optimally possible, where n is the overall number of different budget items.

This result shows that on average, across-the-board cuts are indeed very inefficient.

Acknowledgments We acknowledge the partial support of the Center of Excellence in Econometrics, Faculty of Economics, Chiang Mai University, Thailand.

This work was also supported in part by the National Science Foundation grants HRD-0734825 and HRD-1242122 (Cyber-ShARE Center of Excellence) and DUE- 0926721.

References

1. Bergner, J.: The Case for Across-the-Board Spending Cuts, Mercatus Center, George Mason University (2013)
2. Chokr, B., Kreinovich, V.: How far are we from the complete knowledge: complexity of knowledge acquisition in Dempster–Shafer approach. In: Yager, R.R., Kacprzyk, J., Pedrizzi, M. (eds.) Advances in the Dempster–Shafer Theory of Evidence, pp. 555–576. Wiley, New York (1994)
3. Congressional Budget Office, CBO Estimated Impact of Automatic Budget Enforcement Procedures Specified in the Budget Control Act, Washington, DC, September 2011
4. Congressional Budget Office, Economic Effects of Policies Contributing to Fiscal Tightening in 2013, Washington, DC, November 2012
5. Congressional Budget Office, The Budget and Economic Outlook: Fiscal Years 2013–2023, Washington, DC, February 2013
6. Edwards, C.: We Can Cut Government: Canada Did, Cato Policy Report, vol. 34(3), pp. 1–4, May/June (2012)
7. Jaynes, E.T., Bretthorst, G.L.: Probability Theory: The Logic of Science. Cambridge University Press, Cambridge (2003)
8. Kogan, R.: How the Potential 2013 Across-the-Board Cuts in the Debt-Limit Deal Would Occur, Center on Budget and Policy Priorities, Washington, DC, USA, November 2011
9. Kogan, R.: How the Across-the-Board Cuts in the Budget Control Act Will Work, Center on Budget and Policy Priorities, Washington, DC, USA, April 2012
10. Kogan, R.: Timing of the 2014 Sequestration, Center on Budget and Policy Priorities, Washington, DC, USA, August 2013
11. Schotter, A.: Microeconomics: A Modern Approach. Cengage Learning, Boston (2009)
12. Suppes, P., Krantz, D.M., Luce, R.D., Tversky, A.: Foundations of Measurement, vol. I–III. Academic Press, San Diego (1989)

Invariance Explains Multiplicative and Exponential Skedactic Functions

Vladik Kreinovich, Olga Kosheleva, Hung T. Nguyen and Songsak Sriboonchitta

Abstract In many situations, we have an (approximately) linear dependence between several quantities: $y \approx c + \sum_{i=1}^{n} a_i \cdot x_i$. The variance $v = \sigma^2$ of the corresponding approximation error $\varepsilon = y - \left(c + \sum_{i=1}^{n} a_i \cdot x_i\right)$ often depends on the values of the quantities $x_1, \ldots, x_n$: $v = v(x_1, \ldots, x_n)$; the function describing this dependence is known as the *skedactic function*. Empirically, two classes of skedactic functions are most successful: multiplicative functions $v = c \cdot \prod_{i=1}^{n} |x_i|^{\gamma_i}$ and exponential functions $v = \exp\left(\alpha + \sum_{i=1}^{n} \gamma_i \cdot x_i\right)$. In this paper, we use natural invariance ideas to provide a possible theoretical explanation for this empirical success; we explain why in some situations multiplicative skedactic functions work better and in some exponential ones. We also come up with a general class of invariant skedactic function that includes both multiplicative and exponential functions as particular cases.

V. Kreinovich (✉)
Department of Computer Science, University of Texas at El Paso,
500 W. University, El Paso, TX 79968, USA
e-mail: vladik@utep.edu

O. Kosheleva
University of Texas at El Paso, 500 W. University, El Paso, TX 79968, USA
e-mail: olgak@utep.edu

H.T. Nguyen
Department of Mathematical Sciences, New Mexico State University,
Las Cruces, NM 88003, USA
e-mail: hunguyen@nmsu.edu

H.T. Nguyen · S. Sriboonchitta
Faculty of Economics, Chiang Mai University, Chiang Mai, Thailand
e-mail: songsakecon@gmail.com

V.-N. Huynh et al. (eds.), *Causal Inference in Econometrics*,
Studies in Computational Intelligence 622, DOI 10.1007/978-3-319-27284-9_7

1 Why Are Multiplicative and Exponential Skedactic Functions Empirically Successful: Formulation of the Problem

Linear dependencies are ubiquitous. In many practical situations, a quantity y depends on several other quantities $x_1, \ldots, x_n$: $y = f(x_1, \ldots, x_n)$. Often, the ranges of x_i are narrow: $x_i \approx x_i^{(0)}$ for some $x_i^{(0)}$, so the differences $\Delta x_i \stackrel{\text{def}}{=} x_i - x_i^{(0)}$ are relatively small. In such situations, we can expand the dependence of y on $x_i = x_i^{(0)} + \Delta x_i$ in Taylor series and keep only linear terms in the resulting expansion:

$$y = f(x_1, \ldots, x_n) = f(x_1^{(0)} + \Delta x_1, \ldots, x_n^{(0)} + \Delta x_n) \approx a_0 + \sum_{i=1}^{n} a_i \cdot \Delta x_i,$$

where $a_0 \stackrel{\text{def}}{=} f\left(x_1^{(0)}, \ldots, x_n^{(0)}\right)$ and $a_i \stackrel{\text{def}}{=} \dfrac{\partial f}{\partial x_i}$. Substituting $\Delta x_i = x_i - x_i^{(0)}$ into this formula, we get

$$y \approx c + \sum_{i=1}^{n} a_i \cdot x_i,$$

where $c \stackrel{\text{def}}{=} a_0 - \sum_{i=1}^{n} a_i \cdot x_i^{(0)}$.

Linear dependencies are approximate. Usually, in addition to the quantities $x_1, \ldots, x_n$ that provide the most influence on y, there are also many other quantities that (slightly) influence y, so many that it is not possible to take all of them into account. Since we do not take these auxiliary quantities into account, the above linear dependence is only approximate.

The corresponding approximation errors $\varepsilon \stackrel{\text{def}}{=} y - \left(c + \sum_{i=1}^{n} a_i \cdot x_i\right)$ depend on un-observed quantities and thus, cannot be predicted based only on the values of the observed quantities $x_1, \ldots, x_n$. It is therefore reasonable to view these errors as random variables.

Skedactic functions. A natural way to describe a random variable is by its moments, starting with the mean—the first moment—and the variance—which enables us to compute the second moment. If the first moment is not 0, i.e., if the linear approximation is biased, we can always correct this bias by appropriately updating the constant c.

Next, we need to know the second moment which, since the mean is 0, coincides with the variance v. In general, for different values of x_i, we may have different values of the variance. For example, in econometrics, if we are trying to predict how investment x_1 in an industry affects its output y, clearly larger investments result not only in larger output, but also in larger output variations.

The function $v(x_1, \ldots, x_n)$ that describes how the variance depends on the values of the quantities $x_1, \ldots, x_n$ is known as the *skedactic function.*

Which skedactic functions are empirically successful. In econometric applications, two major classes of skedactic functions have been empirically successful: multiplicative functions (see, e.g., [2], [3, Sect. 9.3], and [4])

$$v(x_1, \ldots, x_n) = c \cdot \prod_{i=1}^{n} |x_i|^{\gamma_i}$$

and exponential functions ([5], Chap. 8)

$$v(x_1, \ldots, x_n) = \exp\left(\alpha + \sum_{i=1}^{n} \gamma_i \cdot x_i\right).$$

According to the latest review [4]:

- neither of this functions has a theoretical justification, and
- in most situations, the multiplication function results in more accurate estimates.

What we do in this paper. In this paper, we use reasonable invariance ideas to provide a possible theoretical explanation for the empirical success of multiplicative and exponential skedactic functions.

We also use invariance to come up with a more general class of skedactic functions to use when neither multiplicative nor exponential functions provide a sufficiently accurate description of the desired dependence.

2 Natural Invariances

Scaling. Many economics quantities correspond to prices, wages, etc. and are therefore expressed in terms of money. The numerical value of such a quantity depends on the choice of a monetary unit. For example, when a European country switches to Euro from its original currency, the actual incomes do not change (at least not immediately), but all the prices and wages get multiplied by the corresponding exchange rate k: $x_i \to x_i' = k \cdot x_i$.

Similarly, quantities that describe the goods, such as amount of oil or amount of sugar, also change their numerical values when we use different units: for example, for the oil production, we get different numerical values when we use barrels and when we use metric tons.

When the numerical value of a quantity gets thus re-scaled (multiplied by a constant), the value of its variance gets multiplied by the square of this constant.

Scale-invariance. Since changing the measuring units for measuring $x_1, \ldots, x_n$ does not change the corresponding economic situations, it makes sense to require that the skedactic function also does not change under such re-scaling: namely, for each combination of re-scalings on inputs, there should be an appropriate re-scaling of the output after which the dependence remains the same.

In precise terms, this means that for every combination of numbers $k_1, \ldots, k_n$, there should exist a value $k = k(k_1, \ldots, k_n)$ with the following property:

$$v = v(x_1, \ldots, x_n) \text{ if and only if } v' = v(x_1', \ldots, x_n'), \text{ where } v' = k \cdot v \text{ and } x_i' = k_i \cdot x_i.$$

Shift and shift-invariance. While most economic quantities are scale-invariant, some are not: e.g., the unemployment rate is measured in percents, there is a fixed unit. Many such quantities, however, can have different numerical values depending on how we define a starting point.

For example, we can measure unemployment in absolute units, or we can measure it by considering the difference $x_i - k_i$ between the actual unemployment and the ideal level $k_i > 0$ which, in the opinion of the economists, corresponds to full employment.

In general, for such quantities, we have a *shift* transformation $x_i \to x_i' = x_i + k_i$. To consider dependence on such quantities, it is therefore reasonable to consider skedactic functions which are shift-invariant, i.e., for which for every combinations of numbers $(k_1, \ldots, k_n)$, there exists a number k for which

$$v = v(x_1, \ldots, x_n) \text{ if and only if } v' = f(x_1', \ldots, x_n'), \text{ where } v' = k \cdot v \text{ and } x_i' = x_i + k_i.$$

3 Case of Scale Invariance: Definitions and the Main Result

Definition 1 We say that a non-negative measurable function $v(x_1, \ldots, x_n)$ is *scale-invariant* if for every n-tuple of real numbers $(k_1, \ldots, k_n)$, there exists a real number $k = k(k_1, \ldots, k_n)$ for which, for every $x_1, \ldots, x_n$ and v, the following two conditions are equivalent to each other:

- $v = v(x_1, \ldots, x_n)$;
- $v' = v(x_1', \ldots, x_n')$, where $v' = k \cdot v$ and $x_i' = k_i \cdot x_i$.

Proposition 1 *A skedactic function is scale-invariant if and only it has the form* $v(x_1, \ldots, x_n) = c \cdot \prod_{i=1}^{n} |x_i|^{\gamma_i}$ *for some values* c *and* γ_i.

Comment. For reader's convenience, all the proofs are placed in the last Proofs section.

Discussion. Thus, scale-invariance explains the use of multiplicative skedactic functions.

4 Case of Shift-Invariance: Definitions and the Main Result

Definition 2 We say that a non-zero non-negative measurable function $v(x_1, \ldots, x_n)$ is *shift-invariant* if for every n-tuple of real numbers $(k_1, \ldots, k_n)$, there exists a real number $k = k(k_1, \ldots, k_n)$ for which, for every $x_1, \ldots, x_n$ and v, the following two conditions are equivalent to each other:

- $v = v(x_1, \ldots, x_n)$;
- $v' = v(x'_1, \ldots, x'_n)$, where $v' = k \cdot v$ and $x'_i = x_i + k_i$.

Proposition 2 *A skedactic function is scale-invariant if and only it has the form* $v(x_1, \ldots, x_n) = \exp\left(\alpha + \prod_{i=1}^{n} \gamma_i \cdot x_i\right)$ *for some values* α *and* γ_i.

Discussion. Thus, shift-invariance explains the use of exponential skedactic functions. The fact that most economic quantities are scale-invariant explains why, in general, multiplicative skedactic functions are more empirically successful.

5 General Case

General case: discussion. A general case is when some of the inputs are scale-invariant and some are shift-invariant. Without losing generality, let us assume that the first m variables $x_1, \ldots, x_m$ are scale-invariant, while the remaining variables $x_{m+1}, \ldots, x_n$ are shift-invariant.

Definition 3 Let $m \le n$ be an integer. We say that a non-zero non-negative measurable function $v(x_1, \ldots, x_n)$ is *m-invariant* if for every n-tuple of real numbers $(k_1, \ldots, k_n)$, there exists a real number $k = k(k_1, \ldots, k_n)$ for which, for every $x_1, \ldots, x_n$ and v, the following two conditions are equivalent to each other:

- $v = v(x_1, \ldots, x_n)$;
- $v' = v(x'_1, \ldots, x'_n)$, where $v' = k \cdot v$, $x'_i = k_i \cdot x_i$ for $i \le m$, and $x'_i = x_i + k_i$ for $i > m$.

Proposition 3 *A skedactic function is m-invariant if and only it has the form*

$$v(x_1, \ldots, x_n) = \exp\left(\alpha + \sum_{i=1}^{m} \gamma_i \cdot \ln(|x_i|) + \sum_{i=m+1}^{n} \gamma_i \cdot x_i\right) \tag{1}$$

for some values μ *and* γ_i.

Discussion. For $m = n$, this formula leads to a multiplicative skedactic function, with $c = \exp(\alpha)$. For $m = 0$, this formula leads to the exponential skedactie function. For intermediate values $m = 1, 2, \ldots, n - 1$, we get new expressions that may be useful when neither multiplicative not exponential skedactic functions work well.

6 Proofs

Proof of Proposition 1. It is easy to check that the multiplicative skedactic function is indeed scale-invariant: we can take $k = \prod_{i=1}^{n} |k_i|^{\gamma_i}$.

Let us prove that, vice versa, if a skedactic function is scale-invariant, then it is multiplicative. Indeed, the above equivalence condition means that for every $k_1, \ldots, k_n$, $v = v(x_1, \ldots, x_n)$ implies that $v' = v(x'_1, \ldots, x'_n)$, where $v' = k \cdot v$ and $x'_i = k_i \cdot x_i$. Substituting the expressions for v' and k'_i into the equality $v' = v(x'_1, \ldots, x'_n)$, we conclude that $k \cdot v = v(k_1 \cdot x_1, \ldots, k_n \cdot x_n)$.

We know that $k = k(k_1, \ldots, k_n)$ and $v = v(x_1, \ldots, x_n)$. Thus, we conclude that

$$k(k_1, \ldots, k_n) \cdot v(x_1, \ldots, x_n) = v(k_1 \cdot x_1, \ldots, k_n \cdot x_n). \tag{2}$$

From this equation, we infer that

$$k(k_1, \ldots, k_n) = \frac{v(k_1 \cdot x_1, \ldots, k_n \cdot x_n)}{v(x_1, \ldots, x_n)}. \tag{3}$$

The right-hand side of this formula is a non-negative measurable function, so we can conclude that the ratio $k(k_1, \ldots, k_n)$ is also non-negative and measurable.

Let us now consider two different tuples $(k_1, \ldots, k_n)$ and $(k'_1, \ldots, k'_n)$. If we first use the first re-scaling, i.e., go from x_i to $x'_i = k_i \cdot x_i$, we get

$$v(x'_1, \ldots, x'_n) = v(k_1 \cdot x_1, \ldots, k_n \cdot x_n) = k(k_1, \ldots, k_n) \cdot v(x_1, \ldots, x_n). \tag{4}$$

If we then apply, to the new values x'_i, an additional re-scaling $x'_i \to x''_i = k'_i \cdot x'_i$, we similarly conclude that

$$v(x''_1, \ldots, x''_n) = v(k'_1 \cdot x'_1, \ldots, k'_n \cdot x'_n) = k(k'_1, \ldots, k'_n) \cdot v(x'_1, \ldots, x'_n). \tag{5}$$

Substituting the expression (4) for $v(x'_1, \ldots, x'_n)$ into this formula, we conclude that

$$v(x''_1, \ldots, x''_n) = k(k'_1, \ldots, k'_n) \cdot k(k_1, \ldots, k_n) \cdot v(x_1, \ldots, x_n). \tag{6}$$

On the other hand, we could get the values x''_i if we directly multiply each value x_i by the product $k''_i \stackrel{\text{def}}{=} k'_i \cdot k_i$:

$$x''_i = k'_i \cdot x'_i = k'_i \cdot (k_i \cdot x_i) = (k'_i \cdot k_i) \cdot x_i = k''_i \cdot x_i.$$

For the new values k''_i, the formula (4) takes the form

$$\begin{aligned} v(x''_1, \ldots, x''_n) &= k(k''_1, \ldots, k''_n) \cdot v(x_1, \ldots, x_n) \\ &= k(k'_1 \cdot k_1, \ldots, k'_n \cdot k_n) \cdot v(x_1, \ldots, x_n). \end{aligned} \tag{7}$$

The left-hand sides of the formulas (6) and (7) are the same, hence the right-hand sides are also equal, i.e.,

$$\begin{aligned} & k(k'_1 \cdot k_1, \ldots, k'_n \cdot k_n) \cdot v(x_1, \ldots, x_n) \\ & \quad = k(k'_1, \ldots, k'_n) \cdot k(k_1, \ldots, k_n) \cdot v(x_1, \ldots, x_n). \end{aligned} \tag{8}$$

If the skedactic function is always equal to 0, then it is multiplicative, with $c = 0$. If it is not everywhere 0, this means that its value is different from 0 for some combination of values $x_1, \ldots, x_n$. Substituting these values into the formula (8) and dividing both sides by $v(x_1, \ldots, x_n) \neq 0$, we conclude that

$$k(k'_1 \cdot k_1, \ldots, k'_n \cdot k_n) = k(k'_1, \ldots, k'_n) \cdot k(k_1, \ldots, k_n). \tag{9}$$

When $k_i = k'_i = -1$ for some i and $k'_i = k_i = 1$ for all other i, we get

$$1 = k(1, \ldots, 1) = k(k_1, \ldots, k_n) \cdot k(k_1, \ldots, k_n) = k^2(k_1, \ldots, k_n). \tag{10}$$

Since the function k is non-negative, this means that $k(k_1, \ldots, k_n) = 1$. Thus, from the formula (9), we can conclude that the value $k(k_1, \ldots, k_n)$ does not change if we change the signs of k_i, i.e., that

$$k(k_1, \ldots, k_n) = k(|k_1|, \ldots, |k_n|).$$

Taking logarithms of both sides of the formula (9), and taking into account that $\ln(a \cdot a') = \ln(a) + \ln(a')$, we conclude that

$$\ln(k(k'_1 \cdot k_1, \ldots, k'_n \cdot k_n)) = \ln(k(k'_1, \ldots, k'_n)) + \ln(k(k_1, \ldots, k_n)). \tag{11}$$

Let us now define an auxiliary function

$$K(K_1, \ldots, K_n) \stackrel{\text{def}}{=} \ln(k(\exp(K_1), \ldots, \exp(K_n)).$$

Since the function $k(k_1, \ldots, k_n)$ is measurable, the function $K(K_1, \ldots, K_n)$ is also measurable.

Since $\exp(a + a') = \exp(a) \cdot \exp(a')$, we conclude that when $k_i = \exp(K_i)$ and $k'_i = \exp(K'_i)$, then $k_i \cdot k'_i = \exp(K_i) \cdot \exp(K'_i) = \exp(K_i + K'_i)$. Thus, from (11), we conclude that for the new function $K(K_1, \ldots, K_n)$, we get

$$K(K'_1 + K_1, \ldots, K'_n + K_n) = K(K'_1, \ldots, K'_n) + K(K_1, \ldots, K_n). \tag{12}$$

Functions that satisfy the property (12) are known as *additive*. It is known (see, e.g., [1]) that every measurable additive function is linear, i.e., has the form

$$K(K_1, \ldots, K_n) = \sum_{i=1} \gamma_i \cdot X_i \tag{13}$$

for some values γ_i.

From $K(K_1, \ldots, K_n) = \ln(k(\exp(K_1), \ldots, \exp(K_n)))$, it follows that

$$k(\exp(K_1), \ldots, \exp(K_n)) = \exp(K(K_1, \ldots, K_n)) = \exp\left(\sum_{i=1}^{n} \gamma_i \cdot K_i\right).$$

For each $k_1, \ldots, k_n$, we have

$$k(k_1, \ldots, k_n) = k(|k_1|, \ldots, |k_n|).$$

For $K_i = \ln(|k_i|)$, we have $\exp(K_i) = |k_i|$, hence

$$k(k_1, \ldots, k_n) = \exp\left(\sum_{i=1}^{n} \gamma_i \cdot \ln(|k_i|)\right) = \prod_{i=1}^{n} |k_i|^{\gamma_i}. \tag{14}$$

From (2), we can now conclude that

$$v(x_1, \ldots, x_n) = k(x_1, \ldots, x_n) \cdot v(1, \ldots, 1).$$

Substituting expression (14) for $k(x_1, \ldots, x_n)$ into this formula and denoting $c \stackrel{\text{def}}{=} v(1, \ldots, 1)$, we get the desired formula for the multiplicative skedastic function $v(x_1, \ldots, x_n) = c \cdot \prod_{i=1}^{n} |x_i|^{\gamma_i}$. The proposition is proven.

Proof of Proposition 2. It is easy to check that the exponential skedactic function is indeed shift-invariant: we can take $k = \exp\left(\sum_{i=1}^{n} \gamma_i \cdot k_i\right)$.

Let us prove that, vice versa, if a skedactic function is shift-invariant, then it is exponential. Indeed, the above equivalence condition means that for every $k_1, \ldots, k_n$, $v = v(x_1, \ldots, x_n)$ implies that $v' = v(x'_1, \ldots, x'_n)$, where $v' = k \cdot v$ and $x'_i = x_i + k_i$. Substituting the expressions for v' and k'_i into the equality $v' = v(x'_1, \ldots, x'_n)$, we conclude that $k \cdot v = v(x_1 + k_1, \ldots, x_n + k_n)$.

We know that $k = k(k_1, \ldots, k_n)$ and $v = v(x_1, \ldots, x_n)$. Thus, we conclude that

$$k(k_1, \ldots, k_n) \cdot v(x_1, \ldots, x_n) = v(x_1 + k_1, \ldots, x_n + k_n). \tag{15}$$

From this equation, we infer that

$$k(k_1, \ldots, k_n) = \frac{v(x_1 + k_1, \ldots, x_n + k_n)}{v(x_1, \ldots, x_n)}. \tag{16}$$

The right-hand side of this formula is a non-negative measurable function, so we can conclude that the ratio $k(k_1, \ldots, k_n)$ is also non-negative and measurable.

Let us now consider two different tuples $(k_1, \ldots, k_n)$ and $(k'_1, \ldots, k'_n)$. If we first use the first shift, i.e., go from x_i to $x'_i = x_i + k_i$, we get

$$v(x'_1, \ldots, x'_n) = v(x_1 + k_1, \ldots, x_n + k_n) = k(k_1, \ldots, k_n) \cdot v(x_1, \ldots, x_n). \quad (17)$$

If we then apply, to the new values x'_i, an additional shift $x'_i \to x''_i = x'_i + k'_i$, we similarly conclude that

$$v(x''_1, \ldots, x''_n) = v(x'_1 + k'_1, \ldots, x'_n + k'_n) = k(k'_1, \ldots, k'_n) \cdot v(x'_1, \ldots, x'_n). \quad (18)$$

Substituting the expression (17) for $v(x'_1, \ldots, x'_n)$ into this formula, we conclude that

$$v(x''_1, \ldots, x''_n) = k(k'_1, \ldots, k'_n) \cdot k(k_1, \ldots, k_n) \cdot v(x_1, \ldots, x_n). \quad (19)$$

On the other hand, we could get the values x''_i if we directly shift each value x_i by the sum $k''_i \stackrel{\text{def}}{=} k'_i + k_i$:

$$x''_i = x'_i + k'_i = (x_i + k_i) + k'_i = x_i + (k_i + k'_i) = x_i + k''_i$$

For the new values k''_i, the formula (17) takes the form

$$\begin{aligned} v(x''_1, \ldots, x''_n) &= k(k''_1, \ldots, k''_n) \cdot v(x_1, \ldots, x_n) \\ &= k(k_1 + k'_1, \ldots, k_n + k'_n) \cdot v(x_1, \ldots, x_n). \end{aligned} \quad (20)$$

The left-hand sides of the formulas (19) and (20) are the same, hence the right-hand sides are also equal, i.e.,

$$\begin{aligned} &k(k_1 + k'_1, \ldots, k_n + k'_n) \cdot v(x_1, \ldots, x_n) \\ &= k(k'_1, \ldots, k'_n) \cdot k(k_1, \ldots, k_n) \cdot v(x_1, \ldots, x_n). \end{aligned} \quad (21)$$

Since the skedactic function is assumed to be non-zero, its value is different from 0 for some combination of values $x_1, \ldots, x_n$. Substituting these values into the formula (21) and dividing both sides by $v(x_1, \ldots, x_n) \neq 0$, we conclude that

$$k(k_1 + k'_1, \ldots, k_n + k'_n) = k(k'_1, \ldots, k'_n) \cdot k(k_1, \ldots, k_n). \quad (22)$$

Taking logarithms of both sides of the formula (22), and taking into account that $\ln(a \cdot a') = \ln(a) + \ln(a')$, we conclude that

$$\ln(k(k_1 + k'_1, \ldots, k_n + k'_n)) = \ln(k(k'_1, \ldots, k'_n)) + \ln(k(k_1, \ldots, k_n)). \quad (23)$$

Thus, the function $\ln(k(k_1, \ldots, k_n))$ is measurable and additive, and hence [1] has the form

$$\ln(k(k_1, \ldots, k_n)) = \sum_{i=1}^{n} \gamma_i \cdot k_i.$$

Hence, by taking exp of both sides, we conclude that

$$k(k_1, \ldots, k_n) = \exp\left(\sum_{i=1}^{n} \gamma_i \cdot k_i\right). \tag{24}$$

From (15), we can now conclude that

$$v(x_1, \ldots, x_n) = k(x_1, \ldots, x_n) \cdot v(0, \ldots, 0).$$

Substituting expression (24) for $k(x_1, \ldots, x_n)$ into this formula and denoting $\alpha \stackrel{\text{def}}{=} \ln(v(0, \ldots, 0))$, so that $v(0, \ldots, 0) = \exp(\alpha)$, we get the desired formula for the exponential skedastic function $v(x_1, \ldots, x_n) = \exp\left(\alpha + \sum_{i=1}^{n} \gamma_i \cdot x_i\right)$. The proposition is proven.

Proof of Proposition 3. It is easy to check that the skedactic function described in the formulation of Proposition 3 is indeed m-invariant: we can take

$$k = \prod_{i=1}^{m} |k_i|^{\gamma_i} \cdot \exp\left(\sum_{i=m+1}^{n} \gamma_i \cdot k_i\right).$$

Let us prove that, vice versa, if a skedactic function is m-invariant, then it has the desired form. Indeed, the above equivalence condition means that for every $k_1, \ldots, k_n$, $v = v(x_1, \ldots, x_n)$ implies that $v' = v(x'_1, \ldots, x'_n)$, where $v' = k \cdot v$, $x'_i = k_i \cdot x_i$ for $i \le m$, and $x'_i = x_i + k_i$ for $i > m$. Substituting the expressions for v' and k'_i into the equality $v' = v(x'_1, \ldots, x'_n)$, we conclude that

$$k \cdot v = v(k_1 \cdot x_1, \ldots, k_m \cdot x_m, x_{m+1} + k_{m+1}, \ldots, x_n + k_n).$$

We know that $k = k(k_1, \ldots, k_n)$ and $v = v(x_1, \ldots, x_n)$. Thus, we conclude that

$$k(k_1, \ldots, k_n) \cdot v(x_1, \ldots, x_n)$$
$$= v(k_1 \cdot x_1, \ldots, k_m \cdot x_m, x_{m+1} + k_{m+1}, \ldots, x_n + k_n). \tag{25}$$

From this equation, we infer that

$$k(k_1, \ldots, k_n) = \frac{v(k_1 \cdot x_1, \ldots, k_m \cdot x_m, x_{m+1} + k_{m+1}, \ldots, x_n + k_n)}{v(x_1, \ldots, x_n)}. \tag{26}$$

The right-hand side of this formula is a non-negative measurable function, so we can conclude that the ratio $k(k_1, \dots, k_n)$ is also non-negative and measurable.

Let us now consider two different tuples $(k_1, \dots, k_n)$ and $(k'_1, \dots, k'_n)$. If we first use the transformation corresponding to the first tuple, i.e., go from x_i to $x'_i = k_i \cdot x_i$ for $i \le m$ and to $x'_i = x_i + k_i$ for $i > m$, we get

$$v(x'_1, \dots, x'_n) = v(k_1 \cdot x_1, \dots, k_m \cdot x_m, x_{m+1} + k_{m+1}, \dots, x_n + k_n)$$
$$= k(k_1, \dots, k_n) \cdot v(x_1, \dots, x_n). \quad (27)$$

If we then apply, to the new values x'_i, an additional transformation $x'_i \to x''_i = k'_i \cdot x'_i$ for $i \le m$ and $x'_i \to x''_i = x'_i + k'_i$ for $i > m$, we similarly conclude that

$$v(x''_1, \dots, x''_n) = v(k'_1 \cdot x'_1, \dots, k'_m \cdot x'_m, x'_{m+1} + k'_{m+1}, \dots, x'_n + k'_n)$$
$$= k(k'_1, \dots, k'_n) \cdot v(x'_1, \dots, x'_n). \quad (28)$$

Substituting the expression (27) for $v(x'_1, \dots, x'_n)$ into this formula, we conclude that

$$v(x''_1, \dots, x''_n) = k(k'_1, \dots, k'_n) \cdot k(k_1, \dots, k_n) \cdot v(x_1, \dots, x_n). \quad (29)$$

On the other hand, we could get the values x''_i if we directly apply to the tuple x_i the transformation corresponding to the product $k''_i = k'_i \cdot k_i$ for $i \le m$ and to the sum $k''_i = k_i + k'_i$ for $i > m$:

$$x''_i = k'_i \cdot x'_i = k'_i \cdot (k_i \cdot x_i) = (k'_i \cdot k_i) \cdot x_i = k''_i \cdot x_i$$

for $i \le m$ and

$$x''_i = x'_i + k'_i = (x_i + k_i) + k'_i = x_i + (k_i + k'_i) = x_i + k''_i$$

for $i > m$.

For the new values k''_i, the formula (27) takes the form

$$v(x''_1, \dots, x''_n) = k(k''_1, \dots, k''_n) \cdot v(x_1, \dots, x_n)$$
$$= k(k'_1 \cdot k_1, \dots, k'_m \cdot k_m, k_{m+1} + k'_{m+1}, \dots, k_n + k'_n) \cdot v(x_1, \dots, x_n). \quad (30)$$

The left-hand sides of the formulas (29) and (30) are the same, hence the right-hand sides are also equal, i.e.,

$$k(k'_1 \cdot k_1, \dots, k'_m \cdot k_m, k_{m+1} + k'_{m+1}, \dots, k_n + k'_n) \cdot v(x_1, \dots, x_n)$$
$$= k(k'_1, \dots, k'_n) \cdot k(k_1, \dots, k_n) \cdot v(x_1, \dots, x_n). \quad (31)$$

Since we assume that the skedactic function is non-zero, its value is different from 0 for some combination of values $x_1, \dots, x_n$. Substituting these values into the formula (31) and dividing both sides by $v(x_1, \dots, x_n) \ne 0$, we conclude that

$$k(k'_1 \cdot k_1, \ldots, k'_m \cdot k_m, k_{m+1} + k'_{m+1}, \ldots, k_n + k'_n)$$
$$= k(k'_1, \ldots, k'_n) \cdot k(k_1, \ldots, k_n). \quad (32)$$

When $k_i = k'_i = -1$ for some $i \le m$ and $k'_i = k_i = 1$ for all other i, we get

$$1 = k(1, \ldots, 1) = k(k_1, \ldots, k_n) \cdot k(k_1, \ldots, k_n) = k^2(k_1, \ldots, k_n). \quad (33)$$

Since the function k_i is non-negative, this means that $k(k_1, \ldots, k_n) = 1$. Thus, from the formula (32), we can conclude that the value $k(k_1, \ldots, k_n)$ does not change if we change the signs of k_i for $i \le m$, i.e., that

$$k(k_1, \ldots, k_m, k_{m+1}, \ldots, k_n) = k(|k_1|, \ldots, |k_m|, k_{m+1}, \ldots, k_n).$$

Taking logarithms of both sides of the formula (32), and taking into account that $\ln(a \cdot a') = \ln(a) + \ln(a')$, we conclude that

$$\ln(k(k'_1 \cdot k_1, \ldots, k'_m \cdot k_m, k_{m+1} + k'_{m+1}, \ldots, k_n + k'_n))$$
$$= \ln(k(k'_1, \ldots, k'_n)) + \ln(k(k_1, \ldots, k_n)). \quad (34)$$

Let us now define an auxiliary function

$$K(K_1, \ldots, K_n) \stackrel{\text{def}}{=} \ln(k(\exp(K_1), \ldots, \exp(K_m), K_{m+1}, \ldots, K_n)).$$

Since the function $k(k_1, \ldots, k_n)$ is measurable, the function $K(K_1, \ldots, K_n)$ is also measurable.

Since $\exp(a + a') = \exp(a) \cdot \exp(a')$, for $i \le m$, we conclude that when $k_i = \exp(K_i)$ and $k'_i = \exp(K'_i)$, then $k_i \cdot k'_i = \exp(K_i) \cdot \exp(K'_i) = \exp(K_i + K'_i)$. Thus, from (34), we conclude that for the new function $K(K_1, \ldots, K_n)$, we get

$$K(K'_1 + K_1, \ldots, K'_n + K_n) = K(K'_1, \ldots, K'_n) + K(K_1, \ldots, K_n). \quad (35)$$

The function $K(K_1, \ldots, K_n)$ is measurable and additive and hence [1] has the form

$$K(K_1, \ldots, K_n) = \sum_{i=1}^{n} \gamma_i \cdot X_i \quad (36)$$

for some values γ_i.

From

$$K(K_1, \ldots, K_n) = \ln(k(\exp(K_1), \ldots, \exp(K_m), K_{m+1}, \ldots, K_n)),$$

it follows that

$$k(\exp(K_1), \ldots, \exp(K_m), K_{m+1}, \ldots, K_n) = \exp(K(K_1, \ldots, K_n)) = \exp\left(\sum_{i=1}^{n} \gamma_i \cdot K_i\right).$$

For each $k_1, \ldots, k_n$, we have

$$k(k_1, \ldots, k_m, k_{m+1}, \ldots, k_n) = k(|k_1|, \ldots, |k_m|, k_{m+1}, \ldots, k_n).$$

Let us take $K_i = \ln(|k_i|)$ for $i \le m$ and $K_i = k_i$ for $i > m$, then we have $\exp(K_i) = |k_i|$ for $i \le m$ and $K_i = k_i$ for $i > m$. Hence,

$$k(k_1, \ldots, k_n) = \exp\left(\sum_{i=1}^{m} \gamma_i \cdot \ln(|k_i|) + \sum_{i=m+1}^{n} \gamma_i \cdot k_i\right). \quad (37)$$

From (25), we can now conclude that

$$v(x_1, \ldots, x_n) = k(x_1, \ldots, x_n) \cdot v(1, \ldots, 1, 0, \ldots, 0).$$

Substituting expression (37) for $k(x_1, \ldots, x_n)$ into this formula, we get the desired formula for the skedastic function

$$v(x_1, \ldots, x_n) = \exp\left(\alpha + \sum_{i=1}^{m} \gamma_i \cdot \ln(|x_i|) + \sum_{i=m+1}^{n} \gamma \cdot x_i\right),$$

with $\alpha = \ln(v(1, \ldots, 1, 0, \ldots, 0))$. The proposition is proven.

Acknowledgments We acknowledge the partial support of the Center of Excellence in Econometrics, Faculty of Economics, Chiang Mai University, Thailand.
This work was also supported in part by the National Science Foundation grants HRD-0734825 and HRD-1242122 (Cyber-ShARE Center of Excellence) and DUE- 0926721.

References

1. Aczél, J., Dhombres, J.: Functional Equations in Several Variables. Cambridge University Press, Cambridge (2008)
2. Harvey, A.C.: Estimating regression models with multiplicative heteroscedasticity. Econometrica **44**, 461–465 (1976)
3. Judge, G.G., Hill, R.C., Griffiths, W.E., Lütkepohl, H., Lee, T.-C.: Introduction to the Theory and Practice of Econometrics. Wiley, New York (1988)
4. Romano, J.P., Wolf, M.: Resurrecting Weighted Least Squares, University of Zurich, Department of Economics, Working paper no. 172 (2015) SSRN http://wwrn.com/abstract=2491081
5. Wooldridge, J.M.: Introductory Econometrics. South-Western, Mason (2012)

Why Some Families of Probability Distributions Are Practically Efficient: A Symmetry-Based Explanation

Vladik Kreinovich, Olga Kosheleva, Hung T. Nguyen and Songsak Sriboonchitta

Abstract Out of many possible families of probability distributions, some families turned out to be most efficient in practical situations. Why these particular families and not others? To explain this empirical success, we formulate the general problem of selecting a distribution with the largest possible utility under appropriate constraints. We then show that if we select the utility functional and the constraints which are invariant under natural symmetries—shift and scaling corresponding to changing the starting point and the measuring unit for describing the corresponding quantity x—then the resulting optimal families of probability distributions indeed include most of the empirically successful families. Thus, we get a symmetry-based explanation for their empirical success.

1 Formulation of the Problem

Some families of probability distributions are empirically successful. Theoretically, we can have infinite many different families of probability distributions, but in practice, only a few families have been empirically successful; see, e.g., [20].

V. Kreinovich (✉)
Department of Computer Science, University of Texas at El Paso, 500 W. University, El Paso, TX 79968, USA
e-mail: vladik@utep.edu

O. Kosheleva
University of Texas at El Paso, 500 W. University, El Paso, TX 79968, USA
e-mail: olgak@utep.edu

H.T. Nguyen
Department of Mathematical Sciences, New Mexico State University, Las Cruces, NM 88003, USA
e-mail: hunguyen@nmsu.edu

H.T. Nguyen · S. Sriboonchitta
Faculty of Economics, Chiang Mai University, Chiang Mai, Thailand
e-mail: songsakecon@gmail.com

V.-N. Huynh et al. (eds.), *Causal Inference in Econometrics*,
Studies in Computational Intelligence 622, DOI 10.1007/978-3-319-27284-9_8

In some cases, there is a clear theoretical explanation for the families' success, but not always. For some of these families, there is a good theoretical explanation for their success. For example, the Central Limit theorem explains the ubiquity of normal distributions in situations where we have a joint effect of numerous small factors.

However, for many other empirically successful families of distributions, there is no clear theoretical explanation for their empirical success.

What we do in this paper. In this paper, for many empirically successful families of 1-D distributions, we provide a theoretical explanation of their success.

2 Our Main Idea

What we want: reminder. In general, we are looking for a family which is the best among all the families that satisfy appropriate constraints.

Natural symmetries. In selecting appropriate objective functions (which describe what is the best) and appropriate constraints, we use the fact that in practical applications, the numerical value of the corresponding quantity x depends:

- on the choice of the starting point for its measurement and
- on the choice of the measuring unit.

If we change the starting point to the one which is x_0 units smaller, then all the values shift by x_0: $x \to x + x_0$. Similarly, if we change the original measuring unit to a one which is λ times smaller, then all the values are scaled by λ: $x \to \lambda \cdot x$.

For example, if we replace a meter with a centimeter, a 100 times smaller measuring unit, then all numerical values multiply by 100: 2 m becomes 200 cm. Another example: shift and scaling describe the transition between Celsius to Fahrenheit.

Invariance. Since these shifts and scaling do not change the corresponding quantities—just change the numbers that represent their values—it is therefore reasonable to require that the appropriate objective functions and constraints do not change (= are invariant) under these transformations.

What we do in this paper: an idea. Since it is reasonable to restrict ourselves to invariant objective functions and invariant constraints, we describe all such objective functions and constraints. We then describe the distributions which are optimal for thus selected objective functions and constraints.

It turns out that the resulting optimal families indeed include many empirically successful families of distributions. Thus, our approach explains the empirical success of many such families.

Comments. The fact that natural symmetries explain the empirical success of families of probability distributions is in good accordance with modern physics, where symmetries are one of the main ways to generate new physical theories; see, e.g., [11].

This fact is also in good accordance with the fact that many empirically successful formulas from neural networks, expert systems, etc., can also be explained by the corresponding symmetries [31].

It should be noted that in this paper, we only consider 1-D families. Our preliminary results show that a symmetry-based explanation of empirically successful families can be extended to the multi-D case as well. Indeed, one of the ways to describe a multi-D distribution $F(x_1, \ldots, x_n) = \text{Prob}(X_1 \le x_1 \,\&\, \ldots \,\&X_n \le x_n)$ is to describe the corresponding marginal distributions $F_i(x_i) = \text{Prob}(X_i \le x_i)$ and the corresponding *copula*, i.e., a function $C(u_1, \ldots, u_n)$ for which $F(x_1, \ldots, x_n) = C(F_1(x_1), \ldots, F_n(x_n))$ [25, 26, 30]. In [24], we have shown that symmetries can explain the empirically successful families of copulas.

3 Which Objective Functions Are Invariant?

We should maximize utility. According to decision theory, decisions of a rational agent are equivalent to maximizing a certain objective function known as *utility*; see, e.g., [12, 27, 29, 40].

Localness property. Based on partial information about the probability distribution, we want to reconstruct the values $\rho(x)$ corresponding to all possible x. It is reasonable to require that if have two distribution which differ only in some local region, and the first distribution is better, then if we replace a common distribution outside this region by another common distribution, the first distribution will still be better.

It is known (see, e.g., [13]) that each utility function with this property is either a sum or a product of functions depending only on a local value $\rho(x)$. Since maximizing the product is equivalent to maximizing its logarithm, and logarithm of the product is equal to the sum of logarithms, we can thus conclude, without losing generality, that the utility function is a sum of functions of $\rho(x)$. In the continuous case, with infinitely many variables $\rho(x)$, we have the limit of the sums, i.e., an integral. Thus, the general expression of an objective function with the localness property is $\int A(\rho(x), x)\, dx$.

Shift-invariance. We want the resulting criterion not to change if we simply shift x, i.e., replace each numerical value x with the shifted value $x + x_0$. Thus, the above integral expression should not change—which means that there should be no explicit dependence on x, i.e., that we should have $\int A(\rho(x))\, dx$.

Scale-invariance: formulation of the requirement. We also want the resulting comparison not to change if we simply re-scale x, i.e., replace each numerical value x with the re-scaled value $y = \lambda \cdot x$. In terms of the re-scaled values, the pdf changes to $\lambda^{-1} \cdot \rho(\lambda^{-1} \cdot y)$, so the new objective function has the form

$$\int A(\lambda^{-1} \cdot \rho(\lambda^{-1} \cdot y))\, dy.$$

By changing the variable to $x = \lambda^{-1} \cdot y$, we get $\lambda \cdot \int A(\mu \cdot \rho(x))\,dx$, where we denoted $\mu \stackrel{\text{def}}{=} \lambda^{-1}$.

Scale-invariance: analysis of the requirement. Scale-invariance means, in particular, that if we add a small deviation $\delta\rho(x)$ to the original distribution in such a way that the value of the objective function does not change, then the value of the re-scaled objective function should not change either. The fact that we still get a pdf means that $\int \delta\rho(x)\,dx = 0$. For small deviations, $A(\rho(x) + \delta\rho) = A(\rho(x)) + A'(\rho(x)) \cdot \delta\rho(x)$. Thus, the fact that the value of the re-scaled objective function does not change means that $\int A'(\rho(x)) \cdot \delta\rho(x)\,dx = 0$. Similarly, the fact that the value of the original objective function does not change means that

$$\int A'(\mu \cdot \rho(x)) \cdot \delta\rho(x)\,dx = 0.$$

So, we arrive at the following requirement: for every function $\delta\rho(x)$ for which $\int \delta\rho(x)\,dx = 0$ and $\int A'(\rho(x)) \cdot \delta\rho(x)\,dx = 0$, we should have

$$\int A'(\mu \cdot \rho(x)) \cdot \delta\rho(x)\,dx = 0.$$

Functions form a Hilbert space—an infinite-dimensional analog of the Euclidean space, with the scalar (dot) product $\langle a, b\rangle \stackrel{\text{def}}{=} \int a(x) \cdot b(x)\,dx$. In these terms, the condition $\int A'(\rho(x)) \cdot \delta\rho(x)\,dx = 0$ can be written as $\langle A'(\rho), \delta\rho\rangle = 0$, i.e., as the condition that the functions $A'(\rho(x))$ and $\delta\rho(x)$ are orthogonal: $\delta\rho \perp A'(\rho)$. In these terms, the invariance requirement means that any function $\delta\rho$ which is orthogonal to 1 and to $A'(\rho(x))$ should also be orthogonal to the function $A'(\mu \cdot \rho(x))$. From this geometric reformulation, one can see that the function $A'(\mu \cdot \rho(x))$ should belong to the linear space spanned by 1 and $A'(\rho(x))$, i.e., that we should have

$$A'(\mu \cdot \rho(x)) = a(\mu, \rho) + b(\mu, \rho) \cdot A'(\rho(x)) \tag{1}$$

for some constants $a(\mu, \rho)$ and $b(\mu, \rho)$.

Let us show that the values $a(\mu, \rho)$ and $b(\mu, \rho)$ do not depend on the pdf $\rho(x)$. Indeed, if we plug in two different values x_1 and x_2 into the formula (1), we get a system of two linear equations for two unknowns $a(\mu, \rho)$ and $b(\mu, \rho)$:

$$A'(\mu \cdot \rho(x_1)) = a(\mu, \rho) + b(\mu, \rho) \cdot A'(\rho(x_1));$$

$$A'(\mu \cdot \rho(x_2)) = a(\mu, \rho) + b(\mu, \rho) \cdot A'(\rho(x_2)).$$

From this system, we conclude that

$$b(\mu, \rho) = \frac{A'(\mu \cdot \rho(x_2)) - A'(\mu \cdot \rho(x_1))}{A'(\rho(x_2)) - A'(\rho(x_1))}$$

and $a(\mu, \rho) = A'(\mu \cdot \rho(x_1)) - b(\mu, rho) \cdot A'(\rho(x_1))$. From these formulas, we see that the values of $a(\mu, \rho)$ and $b(\mu, \rho)$ depend only on the values $\rho(x_1)$ and $\rho(x_2)$ and thus, do not depend on any other value $\rho(x)$ for $x \neq x_1, x_2$.

If we start with some other values x_1', x_2' which are different from x_1 and x_2, we conclude that $a(\mu, \rho)$ and $b(\mu, \rho)$ do not depend on the values $\rho(x_1)$ and $\rho(x_2)$ either. Thus, a and b do not depend on $\rho(x)$ at all: $a(\mu, \rho) = a(\mu)$, $b(\mu, \rho) = b(\mu)$, and the Eq. (1) takes the form

$$A'(\mu \cdot \rho(x)) = a(\mu) + b(\mu) \cdot A'(\rho(x)). \tag{2}$$

It is reasonable to consider the case when the function $A(\rho)$ is twice differentiable—we can do that since any continuous function can be approximated, with any given accuracy, by twice differentiable functions. In this case, the derivative A' is differentiable. From the above expression for a and b in terms of A', we conclude that the functions $a(\mu)$ and $b(\mu)$ are also differentiable. Differentiating both side of the equality (2) with respect to μ and taking $\mu = 1$, we get

$$\rho \cdot \frac{dA'}{d\rho} = a'(1) + b'(1) \cdot A'(\rho).$$

We can separate A and ρ if we multiply both sides by $d\rho$ and divide both sides by ρ and by the right-hand side; we then get

$$\frac{dA'}{a'(1) + b'(1) \cdot A'} = \frac{d\rho}{\rho}.$$

Let us consider two possible cases: $b'(1) = 0$ and $b'(1) \neq 0$.

When $b'(1) = 0$, then integrating this equation leads to the following expression for the derivative A' of the desired function $A(\rho)$: $A' = a'(1) \cdot \ln(\rho) + \text{const}$. Now, the second integration leads to $A(\rho) = a'(1) \cdot \rho \cdot \ln(\rho) + c_1 \cdot \rho + c_2$. Since for the term $c_1 \cdot \rho$, the integral is always constant $\int (c_1 \cdot \rho(x))\,dx = c_1 \cdot \int \rho(x)\,dx = c_1$, optimizing the expression $\int A(\rho(x))\,dx$ is equivalent to optimizing the entropy $-\int \rho(x) \cdot \ln(\rho(x))\,dx$.

When $b'(1) \neq 0$, then for $B \stackrel{\text{def}}{=} A' + \dfrac{a'(1)}{b'(1)}$, we get $\dfrac{dB}{b'(1) \cdot B} = \dfrac{d\rho}{\rho}$, so integration leads to $\ln(B) = b'(1) \cdot \ln(\rho) + \text{const}$ and $B = C \cdot \rho^{\beta}$ for $\beta \stackrel{\text{def}}{=} b'(1)$. Hence, $A'(\rho) = B - \text{const} = C \cdot \rho^{\beta} + \text{const}$. Integrating one more time, when $\beta \neq -1$, we get $A(\rho) = \text{const} \cdot \rho^{\beta+1} + c_1 \cdot \rho + c_2$. Similarly to the above case, optimizing the expression $\int A(\rho(x))\,dx$ is equivalent to optimizing *generalized entropy* $\int (\rho(x))^{\alpha}\,dx$, for $\alpha = \beta + 1$.

When $\beta = -1$, integration leads to $A(\rho) = \text{const} \cdot \ln(\rho) + c_1 \cdot \rho + c_2$, so optimizing $\int A(\rho(x))\,dx$ is equivalent to optimizing another case of *generalized entropy* $\int \ln(\rho(x))\,dx$.

Conclusion: which objective functions are invariant. There are exactly three shift- and scale-invariant objective functions: entropy $-\int \rho(x) \cdot \ln(\rho(x))\,dx$ and generalized entropy $\int \ln(\rho(x))\,dx$ and $\int (\rho(x))^\alpha\,dx$, for $\alpha \neq 1$.

Comment. From the purely mathematical viewpoint, this result is similar to a classification of invariant objective functions for selecting the best image [21].

4 Which Constraints Are Invariant?

Shift-invariance: formulation of the problem. We say that the constraints

$$\int f_i(x) \cdot \rho(x)\,dx = c_i$$

corresponding to the functions $f_i(x)$ ($1 \leq i \leq n$) are *shift-invariant* if the values of the corresponding quantities $\int f_i(x) \cdot \rho(x)\,dx$ uniquely determine the values of these quantities for a scaled distribution.

To be more precise, after the re-scaling, for the same quantity, the original numerical value x is replaced by the new value $y = x + x_0$. In the new scale, the probability density $\rho_y(y)$ has the form $\rho_y(y) = \rho(y - x_0)$. Thus, when we compute the constraint-related quantities based on shifted values, we get the integrals $\int f_i(y) \cdot \rho(y - x_0)\,dy$. Our requirements is that the values of all these new integrals should be uniquely determined based on the values of n original integrals.

Analysis of the problem. In the new integral, we can consider $x = y - x_0$ as the new variable; in this case, each new integral takes the form

$$\int f_i(x + x_0) \cdot \rho(x)\,dx.$$

Thus, our requirement is that if for every pdf, we know the values $\int f_i(x) \cdot \rho(x)\,dx$, then we can uniquely determine the values $\int f_i(x + x_0) \cdot \rho(x)\,dx$. So, for every small change $\delta\rho(x)$ for which:

- $\rho(x) + \delta\rho(x)$ remains a pdf, i.e., $\int \delta\rho(x) = 0$ and
- the values of the original integrals do not change, i.e., $\int f_i(x) \cdot \delta\rho(x) = 0$,

the values of the new integrals shall also remain unchanged, i.e., we should have $\int f_i(x + x_0) \cdot \delta\rho(x)\,dx = 0$.

In geometric terms, this means that any function $\delta\rho$ which is orthogonal to all the functions $1, f_1, \ldots, f_n$, should also be orthogonal to the function $f_i(x + x_0)$. Thus, similarly to the case of invariant objective functions, we conclude that the function $f_i(x + x_0)$ should belong to the linear space spanned by 1 and f_i, i.e., that we should have

$$f_i(x + x_0) = c_i(x_0) + \sum_{j=1}^{n} c_{ij}(x_0) \cdot f_j(x) \tag{3}$$

for some constants $c_i(x_0)$ and $c_{ij}(x_0)$.

It is reasonable to consider the case of differentiable functions $f_i(x)$. In this case, by selecting sufficiently many values x_k, we can get a system of linear equations from which we can uniquely determine $c_i(x_0)$ and $c_{ij}(x_0)$:

$$f_i(x_k + x_0) = c_i(x_0) + \sum_{j=1}^{n} c_{ij}(x_0) \cdot f_j(x_k).$$

By the Cramer's rule [35], a solution to a system of linear equations is a differentiable function of its parameters $f_i(x_k + x_0)$. Since the functions f_i are differentiable, we conclude that the functions $c_i(x_0)$ and $c_{ij}(x_0)$ are differentiable as well—as compositions of differentiable functions.

Differentiating both sides of the formula (3) with respect to x_0 and taking $x_0 = 0$, we conclude that

$$\frac{df_i}{dx} = c_i'(0) + \sum_{j=1}^{n} c_{ij}'(0) \cdot f_j(x).$$

Thus, the functions $f_i(x)$ together with a constant function 1 are solutions to a system of linear differential equations with constant coefficients. Solutions to such systems are known (see, e.g., [38]): they are linear combinations of functions of the type $x^k \cdot \exp(a \cdot x) \cdot \sin(\omega \cdot x + \varphi)$, where $k \geq 0$ is a natural number and $a + \omega \cdot \mathrm{i}$ is an eigenvalue of the corresponding matrix. So, we arrive at the following conclusion:

Conclusion: shift-invariant constraints. The functions $f_i(x)$ corresponding to shift-invariant constraints are linear combinations of the functions of the type
$x^k \cdot \exp(a \cdot x) \cdot \sin(\omega \cdot x + \varphi)$.

Scale-invariance: formulation of the problem. We say that the constraints

$$\int f_i(x) \cdot \rho(x)\, dx = c_i$$

corresponding to the functions $f_i(x)$ $(1 \leq i \leq n)$ are *scale-invariant* if the values of the corresponding quantities $\int f_i(x) \cdot \rho(x)\, dx$ uniquely determine the values of these quantities for a shifted distribution.

After the re-scaling, for the same quantity, the original numerical value x is replaced by the new value $y = \lambda \cdot x$. In the new scale, the probability density $\rho_y(y)$ has the form $\rho_y(y) = \lambda^{-1} \cdot \rho(\lambda^{-1} \cdot y)$. Thus, when we compute the constraint-related quantities based on scaled values, we get the integrals $\int \lambda^{-1} \cdot f_i(y) \cdot \rho(\lambda^{-1} \cdot y)\, dy$. Our requirement is that the values of all these new integrals should be uniquely determined based on the values of n original integrals.

Analysis of the problem. In the new integral, we can consider $x = \lambda^{-1} \cdot y$ as the new variable; in this case, each new integral takes the form

$$\int f_i(\lambda \cdot x) \cdot \rho(x)\, dx.$$

Thus, our requirement is that if for every pdf, we know the values $\int f_i(x) \cdot \rho(x)\, dx$, then we can uniquely determine the values $\int f_i(\lambda \cdot x) \cdot \rho(x)\, dx$. Similar to the shift-invariant case, we can conclude that

$$f_i(\lambda \cdot x) = c_i(\lambda) + \sum_{j=1}^{n} c_{ij}(\lambda) \cdot f_j(x)$$

for some differentiable function $c_i(\lambda)$ and $c_{ij}(\lambda)$. Differentiating both sides of this equality relative to λ and taking $\lambda = 1$, we conclude that

$$x \cdot \frac{df_i(x)}{dx} = c_i'(1) + \sum_{j=1}^{n} c_{ij}'(1) \cdot f_j(x).$$

Here, $\frac{dx}{x} = dz$ for $z = \ln(x)$. Thus, if we express all the functions $f_i(x)$ in terms of z, i.e., consider $f_i(x) = F_i(\ln(x))$, with $F_i(z) \stackrel{\text{def}}{=} f_i(\exp(z))$, then for the new functions $F_i(z)$, we get a system of linear differential equations with constant coefficients:

$$\frac{dF_i(z)}{dz} = c_i'(1) + \sum_{j=1}^{n} c_{ij}'(1) \cdot F_j(z).$$

We already know that each solution is a linear combination of functions of the type $z^k \cdot \exp(a \cdot z) \cdot \sin(\omega \cdot z + \varphi)$. Substituting $z = \ln(x)$ into this formula, we conclude that each function $f_i(x)$ is a linear combination of functions the type

$$(\ln(x))^k \cdot x^a \cdot \sin(\omega \cdot \ln(x) + \varphi).$$

Comment. Note that scaling only related values of the same sign, so we may have two different expressions for $x < 0$ and for $x > 0$.

If instead of scaling relative to 0, we have scaling relative to some other value x_0, i.e., transformations $x - x_0 \to \lambda \cdot (x - x_0)$, then we get expressions

$$(\ln(x - x_0))^k \cdot (x - x_0)^a \cdot \sin(\omega \cdot \ln(x - x_0) + \varphi).$$

In this case, we may have different expressions for $x \le x_0$ and for $x \ge x_0$.

Conclusion: scale-invariant constraints. Each function $f_i(x)$ corresponding to scale-invariant constraints is a linear combinations of functions

$$(\ln(x - x_0))^k \cdot (x - x_0)^a \cdot \sin(\omega \cdot \ln(x - x_0) + \varphi),$$

where k is a natural number. Note that we may have different expressions for $x < x_0$ and for $x > x_0$.

Which constraints are both shift- and scale-invariant? To answer this question, let us check when shift-invariant constraints are also scale-invariant. We cannot have $a \neq 0$, since then the corresponding function grows too fast for scale constraints. We similarly cannot have $b \neq 0$, so the only remaining terms are monomials x^k. Thus, each function corresponding to shift- and scale-invariant constraints is a linear combination of monomials, i.e., a polynomial:

Conclusion: shift- and scale-invariant constraints. Each function $f_i(x)$ corresponding to shift- and scale-invariant constraints is a polynomial.

5 Invariant Objective Functions and Constraints: Summary

Let us summarize our results.

Symmetry-based criteria. There are three possible symmetry-based criteria: entropy $-\int \rho(x) \cdot \ln(\rho(x))\,dx$ and generalized entropy $\int \ln(\rho(x))\,dx$ and $\int (\rho(x))^\alpha\,dx$, for $\alpha \neq 1$.

Constraints which are both shift- and scale-invariant. The only such constraints correspond to polynomials $P(x)$.

Shift-invariant constraints. A function $f(x)$ corresponding to each such constraint is a linear combination of functions of the type $x^k \cdot \exp(a \cdot x) \cdot \sin(\omega \cdot x + \varphi)$, where $k \geq 0$ is a natural number.

Scale-invariant constraints. For scaling around a point x_0, a function $f(x)$ corresponding to each such constraint is a linear combinations of functions

$$(\ln(x - x_0))^k \cdot (x - x_0)^a \cdot \sin(\omega \cdot \ln(x - x_0) + \varphi),$$

where k is a natural number. Note that we may have different expressions for $x < x_0$ and for $x > x_0$.

In particular, we can have scale-invariant constraint $\int f(x) \cdot \rho(x) = 0$ with $f(x) = 1$ for $x \geq x_0$ and $f(x) = 0$ for $x < x_0$. Since $\rho(x) \geq 0$, this constraint implies that $\rho(x) = 0$ for all $x \geq x_0$. Similarly, we can have a constraint implying that $\rho(x) = 0$ for all $x \leq x_0$.

By combining two such constraints, we get a restriction of a distribution to an interval.

Optimizing an invariant objective function under invariant constraints: general formulas. In general, we optimize an invariant objective function $J(\rho)$ under the constraints $\int \rho(x)\,dx = 1$ and $\int f_i(x)\cdot\rho(x)\,dx = c_i$ for several invariant constraints $f_i(x)$. For this constraint optimization problem, the Lagrange multiplier methods results in an unconditional optimization of a functional

$$J(\rho) + \lambda \cdot \left(\int \rho(x)\,dx - 1\right) + \sum_i \left(\int f_i(x)\cdot\rho(x)\,dx - c\right),$$

where λ and λ_i are the corresponding Lagrange multipliers. Differentiating this expression with respect to $\rho(x)$ and equating the resulting derivative to 0, we get the following equations:

$$\ln(\rho(x)) = -1 + \lambda + \sum_i \lambda_i \cdot f_i(x) \text{ for the usual entropy;}$$

$$-(\rho(x))^{-1} = \lambda + \sum_i \lambda_i \cdot f_i(x) \text{ for } J(\rho) = \int \ln(\rho(x))\,dx; \text{ and}$$

$$(-\alpha)\cdot(\rho(x))^{\alpha-1} = \lambda + \sum_i \lambda_i \cdot f_i(x) \text{ for } J(\rho) = \int (\rho(x))^{\alpha}\,dx.$$

6 Resulting Distributions

Let us now list effective families of distributions which result from our approach, i.e., which are optimal with respect to some symmetry-based criterion under some invariant constraints. In our listing:

- we start with the case when all constraints are both shift- and scale-invariant,
- then we consider the case when all constraints are scale-invariant, with respect to the same value x_0,
- then we consider the case when all constraints are shift-invariant,
- finally, we will consider the case when different constraints are invariant relative to different transformations.

For each of these cases, we consider first situations when the objective function is the usual entropy, and then situations when a generalized entropy is used.

Comments.

- Some distributions have several different symmetry-based justifications. For example, among all distributions located on an interval, the uniform distribution has the largest possible entropy and also the largest possible generalized entropy.

- For most distributions, we mention one or two practical situations in which this particular distribution has been effectively used. Of course, our list of examples does not exhaust all efficient applications of the distribution: for many distributions, the number of practical applications can fill a book (and such books have been published; see, e.g., [22]).
- While our symmetry-based approach explains many empirically successful probability distributions, not all such distributions can be thus described. For example, many infinitely divisible distributions do not have an analytical representations and thus, cannot be represented in this form. It should be mentioned that this omissions is not so bad, since most of such distributions already have a theoretical explanation for their success.

6.1 All Constraints Are Both Shift- and Scale-Invariant, Objective Function is Entropy

Towards a general formula. In this case, we optimize entropy $-\int \rho(x) \cdot \ln(\rho(x))\, dx$ under constraints $\int \rho(x)\, dx$ and constraints of the type $\int P_i(x) \cdot \rho(x)\, dx = c_i$ for some polynomials $P_i(x)$. For this constraint optimization problem, the Lagrange multiplier method leads to optimizing the expression

$$-\int \rho(x) \cdot \ln(\rho(x))\, dx + \lambda \cdot \left(\int \rho(x)\, dx - 1\right) + \sum_i \lambda_i \cdot \left(\int P_i(x) \cdot \rho(x)\, dx - c_i\right),$$

where λ and λ_i are Lagrange multipliers.

Differentiating this expression with respect to $\rho(x)$ and equating the derivative to 0, we conclude that

$$-\ln(\rho(x)) - 1 + \lambda + \sum_i \lambda_i \cdot P_i(x) = 0,$$

hence

$$\ln(\rho(x)) = -1 + \lambda + \sum_i \lambda_i \cdot P_i(x).$$

The right-hand side of this formula is a linear combination of polynomials and is, thus, also a polynomial. We will denote this polynomial by $P(x)$. From $\ln(\rho(x)) = P(x)$, we conclude that the corresponding pdf has the form $\rho(x) = \exp(P(x))$ for some polynomial $P(x)$. This is a general formula for such optimal probability distributions.

Example of a successful probability distribution of this type. The most widely used distribution, the normal distribution with probability density

$$\rho(x) = \frac{1}{\sqrt{2\pi}} \cdot \exp\left(-\frac{(x-\mu)^2}{\sigma^2}\right),$$

is exactly of this type. It is a well-known fact that of all the distributions with given mean and given variance, the normal distribution has the largest possible entropy.

6.2 All Constraints Are Both Shift- and Scale-Invariant, Objective Function is Generalized Entropy

Towards a general formula. In this case, we optimize generalized entropy $\int \ln(\rho(x))\,dx$ or $(\rho(x))^\alpha\,dx$ under constraints $\int \rho(x)\,dx$ and constraints of the type $\int P_i(x) \cdot \rho(x)\,dx = c_i$ for some polynomials $P_i(x)$. For this constraint optimization problem, we optimize

$$\int \ln(\rho(x))\,dx + \lambda \cdot \left(\int \rho(x)\,dx - 1\right) + \sum_i \lambda_i \cdot \left(\int P_i(x) \cdot \rho(x)\,dx - c_i\right) \text{ or}$$

$$\int (\rho(x))^\alpha\,dx + \lambda \cdot \left(\int \rho(x)\,dx - 1\right) + \sum_i \lambda_i \cdot \left(\int P_i(x) \cdot \rho(x)\,dx - c_i\right).$$

Differentiating this expression with respect to $\rho(x)$ and equating the derivative to 0, we conclude that

$$-\alpha \cdot (\rho(x))^{\alpha-1} = \lambda + \sum_i \lambda_i \cdot P_i(x) = 0,$$

hence

$$(\rho(x))^{\alpha-1} = -\frac{1}{\alpha} \cdot \left(\lambda + \sum_i \lambda_i \cdot P_i(x)\right).$$

The right-hand side $P(x)$ of this formula is a polynomial, so we get the following general formula for such optimal probability distributions: $\rho(x) = (P(x))^\beta$, where $\beta = \frac{1}{\alpha - 1}$.

Example of a successful probability distribution of this type. An example of such a distribution is *Cauchy distribution*, with probability density

$$\rho(x) = \frac{\Delta}{\pi} \cdot \frac{1}{1 + \frac{(x-\mu)^2}{\Delta^2}}.$$

This distribution is actively used in physics, to describe resonance energy distribution and the corresponding widening of spectral lines; see, e.g. [19]. It is also used to estimate the uncertainty of the results of data processing [23].

Comment. It should be mentioned that while formally, we get the Cauchy distribution, the above derivation is not fully straightforward: it includes a constraint of the type $\int x^2 \cdot \rho(x)\,dx = \text{const}$, but for the Cauchy distribution, the corresponding integral is infinite. So, to make the above derivation mathematically correct, we should first consider the problem limited to distributions located on an interval $[-T, T]$ and then tend T to infinity.

6.3 All Constraints Are Scale-Invariant Relative to the Same Value x_0, Objective Function is Entropy

Half-normal distribution. By combining the constrains on mean and second moment with the constraint implying that $\rho(x) = 0$ for $x \le 0$, we get a *half-normal* distribution, i.e., a distribution whose pdf for $x \le 0$ is 0, and for $x > 0$ is twice that of the normal distribution.

Generalized Gamma distribution. For $f_1(x) = \ln(x)$, $f_2(x) = x^\alpha$, and for a scale-invariant constraint corresponding to $x \ge 0$, optimization leads to $\ln(\rho(x)) = \lambda + \lambda_1 \cdot \ln(x) + \lambda_2 \cdot x^\alpha$, i.e., to the *Generalized Gamma distribution*

$$\rho(x) = \text{const} \cdot x^{\lambda_1} \cdot \exp(\lambda_2 \cdot x^\alpha)$$

which is efficiently used in survival analysis in social sciences [4].

Several probability distributions are particular cases of this general formula. Let us list some of them in alphabetic order.

Particular cases of Generalized Gamma: chi-square distribution. When λ_1 is a natural number and $\alpha = 2$, we get the *chi-square distribution* used to check how well the model fits the data. Under the name of *Nakagami distribution*, this distribution is also used to model attenuation of wireless signals traversing multiple paths [34].

Particular cases of Generalized Gamma: inverse Gamma distribution. When $\alpha = -1$, then we get the *inverse Gamma distribution* which is often used as a prior distribution in Bayesian analysis [3, 14], e.g., to describe the prior distribution of the variance. In particular, when $2\lambda_1$ is a negative integer, we get the *scaled-inverse chi-square distribution*, and for specific values of α_2, we get the *inverse chi-square distribution*.

Particular cases of Generalized Gamma: exponential distribution. When $\lambda_1 = 0$ and $\alpha = 1$, we get the *exponential distribution* $\rho(x) = \text{const} \cdot \exp(-k \cdot x)$. This distribution describe the time between consecutive events, e.g., in queuing theory, in radioactive decay, etc. (It should be noted that the exponential distribution can also be obtained by using generalized entropy.)

Particular cases of Generalized Gamma: Gamma distribution. When $\alpha = 1$, we get the *Gamma distribution* which is often used as a prior distribution in Bayesian analysis. In particular, when $\lambda_1 = k$ is a natural number, we get the *Erlang distribution* that describe the time during which k consecutive events occur [6].

Particular cases of Generalized Gamma: Fréchet distribution. When $\lambda_1 = 0$, we get the *Fréchet distribution* which describes the frequency of extreme events, such as the yearly maximum and minimum stock prices in economics [10] and yearly maximum rainfalls in hydrology [5].

Particular cases of Generalized Gamma: half-normal distribution. When $\lambda_1 = 0$ and $\alpha = 2$, we get the above-described *half-normal distribution.*

Particular cases of Generalized Gamma: inverse Gamma distribution. When $\alpha = -1$, we get the *inverse Gamma distribution* which is used to describe a prior distribution for variance in Bayesian analysis.

Particular cases of Generalized Gamma: Rayleigh distribution. When $\lambda_1 = 1$ and $\alpha = 2$, we get the *Rayleigh distribution* which is used to describe the length of random vectors—e.g., the distribution of wind speed in meteorology.

Particular cases of Generalized Gamma: type-2 Gumbel (Weibull) distribution. When $\lambda_1 = \alpha - 1$, we get the *type-2 Gumbel (Weibull) distribution* which is used to describe the frequency of extreme events and time to failure.

Further generalization of Generalized Gamma: 3-parametric Gamma distribution. For $f_1(x) = \ln(x - \mu)$, $f_2(x) = (x - \mu)^\alpha$, and for a scale-invariant constraint corresponding to $x \geq \mu$, optimization leads to

$$\ln(\rho(x)) = \lambda + \lambda_1 \cdot \ln(x - \mu) + \lambda_2 \cdot (x - \mu)^\alpha,$$

i.e., to the *3-parametric Gamma distribution*

$$\rho(x) = \text{const} \cdot (x - \mu)^{\lambda_1} \cdot \exp(\lambda_2 \cdot (x - \mu)^\alpha)$$

which is efficiently used in hydrology [41, 42].

Inverse Gaussian (Wald) distribution. For scale-invariant constraints $f_1(x) = \ln(x)$, $f_2(x) = x$, $f_3(x) = x^{-1}$, and for a scale-invariant constraint leading to $x > 0$, optimization leads to $\ln(\rho(x)) = \lambda + \lambda_1 \cdot \ln(x) + \lambda_2 \cdot x + \lambda_4 \cdot x^{-1}$, i.e., to $\rho(x) = \text{const} \cdot x^{\lambda_1} \cdot \exp(\lambda_2 \cdot x + \lambda_3 \cdot x^{-1})$. In particular, for $\lambda_1 = -1.5$, we get the *inverse Gaussian (Wald) distribution*. This distribution describes the time a Brownian Motion with positive drift takes to reach a fixed positive level.

Laplace distribution. For a scale-invariant constraint $f_1(x) = |x - \mu|$, optimization leads to $\ln(\rho(x)) = \lambda + \lambda_1 \cdot |x - \mu|$, so we get *Laplace distribution*

$$\rho(x) = \text{const} \cdot \exp(\lambda_1 \cdot |x - \mu|).$$

This distribution has many applications [22]. For example, it is used:

- in *speech recognition*, as a prior distribution for the Fourier coefficients [9];
- in *databases*, where, to preserve privacy, each record is modified by adding a Laplace-generated noise [8].

Lévy (van der Waals) distribution. For scale-invariant constraints

$$f_1(x) = \ln(x - \mu),$$

$f_2(x) = (x - \mu)^{-1}$, and for a scale-invariant constraint equivalent to $x - \mu > 0$, optimization leads to $\ln(\rho(x)) = \lambda + \lambda_1 \cdot \ln(x - \mu) + \lambda_2 \cdot (x - \mu)^{-1}$, i.e., to $\rho(x) = \text{const} \cdot (x - \mu)^{\lambda_1} \cdot \exp(\lambda_2 \cdot (x - \mu)^{-1})$. In particular, for $\lambda_1 = -1.5$, we get the *Lévy (van der Waals) distribution*. This distribution is used in spectroscopy, to describe different spectra [7].

Log-normal distribution. For scale-invariant constraints $f_1(x) = \ln(x)$ and $f_2(x) = (\ln(x))^2$ and for a scale-invariant constraint equivalent to $x > 0$, optimization leads to $\ln(\rho(x)) = \lambda + \lambda_1 \cdot \ln(x) + \lambda_2 \cdot (\ln(x))^2$, i.e., to the *log-normal distribution* $\rho(x) = \text{const} \cdot x^{\lambda_1} \cdot \exp(\lambda_2 \cdot (\ln(x))^2)$. This distribution describes the product of several independent random factors. It has many applications. In particular, it is used in econometrics to describe:

- the compound return of a sequence of multiple trades,
- a long-term discount factor, etc.

6.4 All Constraints Are Shift-Invariant, Objective Function Is Entropy

Gumbel distribution. For a shift-invariant constraint $f_1(x) = \exp(k \cdot x)$, optimization leads to $\ln(\rho(x)) = \lambda + \lambda_1 \cdot \exp(k \cdot x)$, i.e., to the *Gumbel distribution* $\rho(x) = \text{const} \cdot \exp(\lambda_1 \cdot \exp(k \cdot x))$ which is used to describe the frequency of extreme events.

Type I Gumbel distribution. For shift-invariant constraints $f_1(x) = x$ and $f_2(x) = \exp(k \cdot x)$, optimization leads to $\ln(\rho(x)) = \lambda + \lambda_1 \cdot x + \lambda_2 \cdot \exp(k \cdot x)$ and thus, to $\rho(x) = \text{const} \cdot \exp(\lambda_1 \cdot x + \lambda_2 \cdot \exp(k \cdot x))$. In particular, for $\lambda_1 = k$, we get *type I Gumbel distribution* which is used to decrease frequencies of extreme values.

6.5 All Constraints Are Shift-Invariant, Objective Function Is Generalized Entropy

Hyperbolic secant distribution. When we use the objective function $\int \ln(\rho(x))\,dx$ and a shift-invariant constraint $f_1(x) = \exp(k \cdot x) + \exp(-k \cdot x)$, optimization leads to $(\rho(x))^{-1} = -\lambda - \lambda_1 \cdot (\exp(k \cdot x) + \exp(-k \cdot x)) = -\lambda + c \cdot \cosh(k \cdot x)$. Thus, we get $\rho(x) = \text{const} \cdot \dfrac{1}{-\lambda + c \cdot \cosh(k \cdot x)}$. The requirement that $\int \rho(x)\,dx = 1$ leads to $\lambda = 0$, so we get a *hyperbolic secant distribution*. This distribution is similar to the normal one, but it has a more acute peak and heavier tails, so it is used when we have a distribution which is close to normal but has heavier tails.

6.6 Different Constraints Have Different Symmetries, Objective Function Is Entropy

In some cases, to get the desired distribution, we need to combine constraints with symmetries of different type. Let us give examples of the resulting distributions.

Uniform distribution. If we impose constraints leading to $x \geq a$ and $x \leq b$, then the largest values of the entropy is attained on the uniform distribution on the interval $[a, b]$.

Comment. It should be noted that the same result holds if we use generalized entropy.

Beta and arcsine distribution. Constraints $\int \ln(x) \cdot \rho(x)\,dx$ and $\int \ln(a - x) \cdot \rho(x)\,dx$ are both scale-invariant, but the first one is scale-invariant relative to $x_0 = 0$, while the second one is scale-invariant relative to $x_0 = a$. Optimizing entropy under these constraints – and under similarly scale-invariant constraints implying that $0 \leq x \leq a$, results in $\rho(x) = A \cdot x^{\alpha} \cdot (a - x)^{\beta}$ for some A, α, and β.

This formula describes a *Beta distribution* on the interval $[0, a]$. This distribution has numerous practical applications in many areas including agriculture [18], epidemiology [43], geosciences [16], meteorology [39], population genetics [1], and project management [28].

In particular, for $a = 1$ and $\alpha = \beta = 0.5$, we get the *arcsine distribtion*, with probability density $\rho(x) = \dfrac{1}{\pi \cdot \sqrt{x \cdot (1 - x)}}$. This distribution describes, for example, the measurement error caused by an external sinusoidal signal coming at a random moment of time [37].

Beta prime (F-) distribution. For scale-invariant constraints $f_1(x) = \ln(x)$ and $f_2(x) = \ln(x + a)$ and for a constraint leading to $x > 0$, optimizing entropy leads to $\ln(\rho(x)) = \lambda + \lambda_1 \cdot \ln(x) + \lambda_2 \cdot \ln(x + a)$, i.e., to the *Beta prime (F-) distribution* $\rho(x) = \text{const} \cdot x^{\lambda_1} \cdot (x + a)^{\lambda_2}$.

Log distribution. Let us impose scale-invariant constraints $f_1(x) = x$ and $f_2(x) = \ln(x)$, and constraints leading to $x \geq a$ and $x \leq b$. Then the largest entropy occurs when for $x \in [a, b]$, we have $\ln(\rho(x)) = \lambda + \lambda_1 \cdot x + \lambda_2 \cdot \ln(x)$, hence $\rho(x) = \text{const} \cdot \exp(\lambda_1 \cdot x) \cdot x^{\lambda_2}$. For $\lambda_1 = -1$, we get the *log distribution*.

Generalized Pareto distribution. For scale-invariant constraint $f_1(x) = \ln(x + x_0)$ and a scale-invariant constraint leading to $x > x_m$, optimization leads to $\ln(\rho(x)) = \lambda + \lambda_1 \cdot \ln(x + x_0)$, hence to the *Generalized Pareto distribution*

$$\rho(x) = \text{const} \cdot (x + x_0)^{\lambda_1}.$$

This distribution describes the frequency of large deviations in economics, in geophysics, and in other applications areas [10]. The case $x_0 = 0$ is known as the *Pareto distribution.*

Comment. The Generalized Pareto distribution can also be derived by using generalized entropy.

Gompertz distribution. For shift-invariant constraints $f_1(x) = \exp(b \cdot x)$ and $f_2(x) = x$, and a scale-invariant constraint leading to $x > 0$, optimization leads to $\ln(\rho(x)) = \lambda + \lambda_1 \cdot x + \lambda_2 \cdot \exp(b \cdot x)$, hence to *Gompertz distribution*

$$\rho(x) = \text{const} \cdot \exp(\lambda_1 \cdot x) \cdot \exp(\lambda_2 \cdot \exp(b \cdot x)).$$

This distribution describes aging and life expectancy [2, 36]; it is also used to in software engineering, to describe the "life expectancy" of software [33].

Reciprocal and U-quadratic distribution. For a scale-invariant constraint $f_1(x) = \ln(x - \beta)$ and scale-invariant constraints corresponding to $x \geq a$ and $a \leq b$, optimization leads to $\rho(x) = A \cdot x^{\alpha}$ for $x \in [a, b]$. In particular:

- for $\alpha = -1$ and $\beta = 0$, we get the *reciprocal distribution* $\rho(x) = \text{const} \cdot x^{-1}$. This distribution is used in *computer arithmetic*, to describe the frequency with which different numbers occur [17, 32];
- for $\alpha = 2$, we get the *U-quadratic distribution* $\rho(x) = \text{const} \cdot (x - \beta)^2$; this distribution is often effectively used to describe random quantities with a bimodal distribution.

Comment. Please note that, as we show later in this paper, both distributions can also be obtained by using generalized entropy.

Truncated normal distribution. By combining the constrains on mean and second moment with the constraint implying that $\rho(x) = 0$ for $x \leq a$ and for $x \geq b$, we get the *truncated normal distribution*, i.e., a normal distribution limited to the interval $[a, b]$. This distribution is actively used in econometrics, to model quantities about which we only know lower and upper bounds [15].

von Mises distribution. For a shift-invariant constraint $f_1(x) = \cos(x - \mu)$ and for scale-invariant criteria corresponding to $x \geq -\pi$ and $x \leq \pi$, optimization leads to $\ln(\rho(x)) = \lambda + \lambda_1 \cdot \cos(x - \mu)$, i.e., to the *von Mises distribution*

$$\rho(x) = \text{const} \cdot \exp(\lambda_1 \cdot \cos(x - \mu))$$

which is frequently used to describe random angles $x \in [-\pi, \pi]$.

6.7 Different Constraints Have Different Symmetries, Objective Function is Generalized Entropy

Uniform distribution. If we impose constraints leading that $x \geq a$ and $x \leq b$, then the largest values of the generalized entropy is attained on the uniform distribution on the interval $[a, b]$.

Comment. It should be noted that the same result holds if we use the usual entropy.

Exponential and Erlang distribution. For the objective function $\int (\rho(x))^2 \, dx$, for the shift-invariant constraint $f_1(x) = x^k \cdot \exp(-a \cdot x)$, and a scale-invariant constraint corresponding to $x > 0$, optimization leads to the *Erlang distribution*

$$\rho(x) = \text{const} \cdot x^k \cdot \exp(-a \cdot x);$$

in particular, for $k = 0$, we get an *exponential distribution* $\rho(x) = \text{const} \cdot \exp(-a \cdot x)$.

Comment. Exponential distribution can also be obtained by using the usual entropy.

Generalized Pareto distribution. For the objective function $\int (\rho(x))^\alpha \, dx$ and for the scale- and shift-invariant constraint $f_1(x) = x$ and a scale-invariant constraint leading to $x > 0$, optimization leads to $-\alpha \cdot (\rho(x))^{\alpha-1} = \lambda + \lambda_1 \cdot x$, hence to the *Generalized Pareto distribution* $\rho(x) = \text{const} \cdot (x + x_0)^{-\gamma}$, where $\gamma = -1/(\alpha - 1)$.

Comment. The Generalized Pareto distribution can also be derived by using the usual entropy.

Raised cosine distribution. For the objective function $\int (\rho(x))^2 \, dx$, for a shift-invariant constraint $f_1(x) = \cos(\omega \cdot x + \varphi)$ and scale-invariant constraints corresponding to $x \geq a$ and $x \leq b$, optimization leads to the *raised cosine distribution* $\rho(x) = c_1 + c_2 \cdot \cos(\omega \cdot x + \varphi)$.

Reciprocal distribution. For the generalized entropy $\int \ln(\rho(x)) \, dx$, a scale- and shift-invariant constraint $f_1(x) = x$ and scale-invariant constraints corresponding to $x \geq a$ and $a \leq b$, optimization leads to the reciprocal distribution $\rho(x) = A \cdot x^{-1}$ for $x \in [a, b]$.

Comment. Please note that this distribution can also be obtained by using the usual entropy.

U-quadratic distribution. For the generalized entropy $\int (\rho(x))^2 \, dx$, a scale- and shift-invariant constraint $f_1(x) = (x - \beta)^2$ and scale-invariant constraints corresponding to $x \geq a$ and $a \leq b$, optimization leads to the U-quadratic distribution $\rho(x) = \text{const} \cdot (x - \beta)^2$ for $x \in [a, b]$.

Comment. Please note that this distribution can also be obtained by using the usual entropy.

7 Conclusion

In the previous section, we listed numerous families of distributions which are optimal if we optimize symmetry-based utility functions under symmetry-based constraints. One can see that this list includes many empirically successful families of distributions—and that most empirically successful families of distributions are on this list. Thus, we indeed provide a symmetry-based explanation for the empirical success of these families.

Acknowledgments We acknowledge the partial support of the Center of Excellence in Econometrics, Faculty of Economics, Chiang Mai University, Thailand.
This work was also supported in part by the National Science Foundation grants HRD-0734825 and HRD-1242122 (Cyber-ShARE Center of Excellence) and DUE- 0926721.

References

1. Balding, D.J., Nichols, R.A.: A method for quantifying differentiation between populations at multi-allelic loci and its implications for investigating identity and paternity. Genetica **96**(1–2), 3–12 (1995)
2. Benjamin, B., Haycocks, H.W., Pollard, J.: The Analysis of Mortality and Other Actuarial Statistics. Heinemann, London (1980)
3. Bernardo, J.M., Smith, A.F.M.: Bayesian Theory. Wiley, New York (1993)
4. Box-Steffensmeier, J.M., Jones, B.S.: Event History Modeling: A Guide for Social Scientists. Cambridge University Press, New York (2004)
5. Coles, S.: An Introduction to Statistical Modeling of Extreme Values. Springer, Berlin (2001)
6. Cox, D.R.: Renewal Theory. Wiley, New York (1967)
7. Croxton, C.A.: Statistical Mechanics of the Liquid Surface. Wiley, New York (1980)
8. Dwork, C., McSherry, F., Nissim, K., Smith, A.: Calibrating noise to sensitivity in private data analysis. In: Proceedings of the Theory of Cryptography Conference TCC'2006, Springer, Heidelberg(2006)
9. Eltoft, T., Taesu, K., Lee, T.-W.: On the multivariate Laplace distribution. IEEE Signal Process. Lett. **13**(5), 300–303 (2006)
10. Embrechts, P., Klüppelberg, C., Mikosch, T.: Modelling Extremal Events for Insurance and Finance. Springer, Heidelberg (2013)
11. Feynman, R.P., Leighton, R.B., Sands, M.: Feynman Lectures on Physics. Basic Books, New York (2011)
12. Fishburn, P.C.: Utility Theory for Decision Making. Wiley, New York (1969)
13. Fishburn, P.C.: Nonlinear Preference and Utility Theory. The John Hopkins Press, Baltimore (1988)
14. Gelman, A., Carlin, J.B., Stern, H.S., Vehtari, A., Rubin, D.B.: Bayesian Data Analysis. Chapman and Hall/CRC, Boca Raton (2013)
15. Greene, W.H.: Econometric Analysis. Prentice Hall, Upper Saddle River (2011)
16. Gullco, R.S., Anderson, M.: Use of the Beta distribution to determine well-log shale parameters. SPE Reserv. Eval. Eng. **12**(6), 929–942 (2009)
17. Hamming, R.W.: On the distribution of numbers. Bell Syst. Tech. J. **49**(8), 1609–1625 (1970)
18. Haskett, J.D., Pachepsky, Y.A., Acock, B.: Use of the beta distribution for parameterizing variability of soil properties at the regional level for crop yield estimation. Agric. Syst. **48**(1), 73–86 (1995)
19. Hecht, E.: Optics. Addison-Wesley, New York (2001)

20. Johnson, N.L., Kotz, S., Balakrishnan, N.: Continuous Univariate Distributions, vol. 1. Wiley, New York (1994). vol. 2 (1995)
21. Kosheleva, O.: Symmetry-group justification of maximum entropy method and generalized maximum entropy methods in image processing. In: Erickson, G.J., Rychert, J.T., Smith, C.R. (eds.) Maximum Entropy and Bayesian Methods, pp. 101–113. Kluwer, Dordrecht (1998)
22. Kotz, S., Kozubowski, T.J., Podgórski, K.: The Laplace Distribution and Generalizations: A Revisit with Applications to Communications. Economics, Engineering and Finance. Birkhauser, Boston (2001)
23. Kreinovich, V., Ferson, S.: A new Cauchy-based black-box technique for uncertainty in risk analysis. Reliab. Eng. Syst. Saf. **85**(1–3), 267–279 (2004)
24. Kreinovich, V., Nguyen, H.T., Sriboonchitta, S.: Why Clayton and Gumbel copulas: a symmetry-based explanation. In: Huynh, V.-N., Kreinovich, V., Sriboonchitta, S., Suriya, K. (eds.) Uncertainty Analysis in Econometrics, with Applications, pp. 79–90. Springer Verlag, Berlin, Heidelberg (2013)
25. Kurowicka, D., Cooke, R.M.: Uncertainty Analysis with High Dimensional Dependence Modelling. Wiley, New York (2006)
26. Kurowicka, D., Joe, H. (eds.): Dependence Modeling: Vine Copula Handbook. World Scientific, Singapore (2010)
27. Luce, D.R., Raiffa, H.: Games and Decisions. Introduction and Critical Survey, Wiley, New York (1957)
28. Malcolm, D.G., Roseboom, J.H., Clark, C.E., Fazar, W.: Application of a technique for research and development program evaluation. Oper. Res. **7**(5), 646–669 (1958)
29. Myerson, R.B.: Game theory: Analysis of Conflict. Harvard University Press, Cambridge (1991)
30. Nelsen, R.B.: An Introduction to Copulas. Springer, New York (2007)
31. Nguyen, H.T., Kreinovich, V.: Applications of Continuous Mathematics to Computer Science. Kluwer, Dordrecht (1997)
32. Nguyen, H.T., Kreinovich, V., Longpré, L.: Dirty pages of logarithm tables, lifetime of the Universe, and (subjective) probabilities on finite and infinite intervals. Reliab. Comput. **10**(2), 83–106 (2004)
33. Ohishi, K., Okamura, H., Dohi, T.: Gompertz software reliability model: estimation algorithm and empirical validation. J. Syst. Softw. **82**(3), 535–543 (2009)
34. Parsons, J.D.: The Mobile Radio Propagation Channel. Wiley, New York (1992)
35. Poole, D.: Linear Algebra: A Modern Introduction. Cengage Learning, Boston (2014)
36. Preston, S.H., Heuveline, P., Guillot, M.: Demography: Measuring and Modeling Population Processes. Blackwell, Oxford (2001)
37. Rabinovich, S.G.: Measurement Errors and Uncertainties: Theory and Practice. Springer, Heidelberg (2005)
38. Robinson, J.C.: An Introduction to Ordinary Differential Equations. Cambridge University Press, Cambridge (2004)
39. Sulaiman, M.Y., Oo, W.M.H., Wahab, M.A., Zakaria, A.: Application of beta distribution model to Malaysian sunshine data. Renew. Energy **18**(4), 573–579 (1999)
40. Suppes, P., Krantz, D.M., Luce, R.D., Tversky, A.: Foundations of Measurement. Geometrical, Threshold, and Probabilistic Representations, vol. II. Academic Press, California (1989)
41. van Nooijen, R., Gubareva, T., Kolechkina, A., Gartsman, B.: Interval analysis and the search for local maxima of the log likelihood for the Pearson III distribution. Geophys. Res. Abstr. **10**, EGU2008–A05006 (2008)
42. van Nooijen, R., Kolechkina, A.: In: Nehmeier, M. (ed.) Two Applications of Interval Analysis to Parameter Estimation in Hydrology. Abstracts of the 16th GAMM-IMACS International Symposium on Scientific Computing, Computer Arithmetic and Validated Numerics SCAN'2014. Würzburg, Germany, 21–26 Sept 2014, p. 161
43. Wiley, J.A., Herschkorn, S.J., Padian, N.S.: Heterogeneity in the probability of HIV transmission per sexual contact: the case of male-to-female transmission in penile-vaginal intercourse. Stat. Med. **8**(1), 93–102 (1989)

The Multivariate Extended Skew Normal Distribution and Its Quadratic Forms

Weizhong Tian, Cong Wang, Mixia Wu and Tonghui Wang

Abstract In this paper, the class of multivariate extended skew normal distributions is introduced. The properties of this class of distributions, such as, the moment generating function, probability density function, and independence are discussed. Based on this class of distributions, the extended noncentral skew chi-square distribution is defined and its properties are investigated. Also the necessary and sufficient conditions, under which a quadratic form of the model has an extended noncentral skew chi-square distribution, are obtained. For illustration of our main results, several examples are given.

1 Introduction

In many real-world problems, assumptions of normality are violated as data possess some level of skewness. The class of skew normal distributions is an extension of the normal distribution, allowing for the presence of skewness. Azzalini [1, 2] defined the random vector $\mathbf{V}$ to be a *multivariate skew-normal distribution*, denoted by $\mathbf{V} \sim SN_n(\boldsymbol{\mu}, \Sigma, \boldsymbol{\alpha})$, if its density function is

$$f_{\mathbf{V}}(\mathbf{x}; \boldsymbol{\mu}, \Sigma, \boldsymbol{\alpha}) = 2\phi_n(\mathbf{x}; \boldsymbol{\mu}, \Sigma)\Phi\left(\boldsymbol{\alpha}'\Sigma^{-1/2}(\mathbf{x} - \boldsymbol{\mu})\right), \qquad \mathbf{x} \in \Re^n, \tag{1}$$

W. Tian · C. Wang · T. Wang (✉)
Department Mathematical Sciences, New Mexico State University,
Las Cruces, NM, USA
e-mail: xjlaojiu@nmsu.edu

C. Wang
e-mail: cong960@nmsu.edu

T. Wang
e-mail: twang@nmsu.edu

M. Wu
College of Applied Sciences, Beijing University of Technology,
Beijing 100022, China
e-mail: wumixia@bjut.edu.cn

V.-N. Huynh et al. (eds.), *Causal Inference in Econometrics*,
Studies in Computational Intelligence 622, DOI 10.1007/978-3-319-27284-9_9

where $\phi_n(\mathbf{x}; \boldsymbol{\mu}, \Sigma)$ is the n-dimensional normal density function with mean vector $\boldsymbol{\mu}$ and covariance matrix Σ, and $\Phi(\cdot)$ is the standard normal distribution function. Since then, the class of multivariate skew normal distributions have been studied by many authors, see Azzalini [3], Arellano–Valle [4], Gupta [10], and Szkely [17].

In financial applications, specialized models for time series are used, notably the ARCH model and its variants. Since the presence of skewness in financial time series is a feature not easily accounted for by classical formulations, the tools discussed here become natural candidates for consideration. De Luca and Loperfido [8] and De Luca et al. [9] have constructed a autoregressive conditional heteroskedasticity (GARCH) formulation for multivariate financial time series where asymmetric relationships exist among a group of stock markets, with one market playing a leading role over the others. Their construction links naturally with the concepts implied by the multivariate skew normal distribution, when one considers the different effect on the secondary markets induced by 'good news' and 'bad news' from the leading market. Corns and Satchell [7] have proposed a GARCH-style model where the random terms have skew normal distribution, regulated by two equations, one as in usual GARCH models which pertains to the scale factor, conditionally on the past, and an additional equation of analogous structure which regulates the skewness parameter. Li et al. [5] proposed two extended versions of Type I Wang transform using two versions of skew-normal distributions on risk measure. Tian et al. [19] provided a new skew normal risk measure and has been shown that the new skew normal risk measure satisfied the classic capital asset pricing model.

Quadratic forms are important in testing the second order conditions that distinguish maxima from minima in economic optimization problems, in checking the concavity of functions that are twice continuously dierentiable and in the theory of variance in statistics. For the case where the location parameter was zero, the quadratic forms under skew normal settings were studied by Gupta and Huang [10], Genton et al. [11], Huang and Chen [13] and Loperfido [15]. For the general cases, the noncentral skew chi-square distribution was defined and new versions of Cochran's theorem were obtained in Wang et al. [18]. The matrix version of Cochran's theorem under multivariate skew normal settings was discussed in Ye et al. [20]. Also applications of using noncentral skew chi-square distribution in linear mixed model and variance components models were obtained in Ye et al. [21, 22].

The class of distributions used in this paper is the modified version of Kumar [14] defined as follows. A random variable Z is said to have an *extended skew generalized normal distribution* with parameters $\gamma \geq -1$, $\alpha_1 \in \Re$, and $\alpha_2 \geq 0$, if its probability density function is given by

$$f(z; \alpha_1, \alpha_2, \gamma) = \frac{2}{2+\gamma}\phi(z)\left[1 + \gamma\Phi\left(\frac{\alpha_1 z}{\sqrt{1+\alpha_2 z^2}}\right)\right], \tag{2}$$

where $\phi(\cdot)$ and $\Phi(\cdot)$ are the standard normal density function and distribution function, respectively. For the case where $\alpha_1 = \alpha$ and $\alpha_2 = 0$, the distribution with the density f given in (2) is called the *extended skew normal distribution*, denoted by $Z \sim ESN(\alpha, \gamma)$.

In this paper, the class of multivariate extended skew normal distributions is defined and its properties such as density functions, moment generating functions, and dependence are discussed. The usefulness of this extra parameter γ is emphasized based on its density graphs. In order to study distributions of quadratic forms of this model, the extended noncentral skew chi-square distribution is defined. Also the necessary and sufficient conditions, under which a quadratic form is extended noncentral skew chi-square distributed, are obtained. Our results are extensions of those given in Gupta and Huang [10], Genton et al. [11], Huang and Chen [13] and Loperfido [15], Wang et al. [18], Ye and Wang [21], and Ye et al. [22].

The organization of this paper is given as follows. The definition of the class of multivariate extended skew normal distributions and its properties are given in Sect. 2. Extended noncentral skew chi-square distribution with properties is discussed in Sect. 3. Necessary and sufficient conditions, under which a quadratic form is an extended noncentral skew chi-square distributed, are obtained in Sect. 4. Several examples are given for the illustration of our main results.

2 The Multivariate Extended Skew Normal Distribution

Let $\mathcal{M}_{n\times k}$ be the set of all $n \times k$ matrices over the real field $\Re$ and $\Re^n = \mathcal{M}_{n\times 1}$. For any B, $B \in \mathcal{M}_{n\times k}$, B' and $r(B)$ denote the transpose and rank of B, respectively.

Definition 1 The random vector $\mathbf{Z} \in \Re^k$ follows a *multivariate extended skew-normal distribution*, denoted by $\mathbf{Z} \sim ESN_k(\Sigma, \boldsymbol{\alpha}, \gamma)$, if its density function is given by

$$f_Z(\mathbf{z}; \Sigma, \boldsymbol{\alpha}) = \frac{2}{2+\gamma}\phi_k(\mathbf{z}; \Sigma)\left[1 + \gamma\Phi\left(\boldsymbol{\alpha}'\mathbf{z}\right)\right], \qquad (3)$$

where $\phi_k(\mathbf{z}; \Sigma)$ is the k-dimensional normal density function with mean vector $\mathbf{0}$ and covariance matrix Σ, and $\Phi(\cdot)$ is the standard normal distribution function.

Remark 1 several special cases of $ESN_k(\Sigma, \boldsymbol{\alpha}, \gamma)$ are listed as follows.
(i) When $\boldsymbol{\alpha} = 0$ or $\gamma = 0$, $ESN_k(\Sigma, \boldsymbol{\alpha}, \gamma)$ is reduced to $N_k(\mathbf{0}, \Sigma)$, the multivariate normal distribution $N_k(\mathbf{0}, \Sigma)$.
(ii) As both γ and $\boldsymbol{\alpha}$ tend to $+\infty$, $ESN_k(\Sigma, \boldsymbol{\alpha}, \gamma)$ is reduced to the multivariate truncated normal distribution of $N_k(\mathbf{0}, \Sigma)$ for $\mathbf{x} \geq 0$.
(iii) When $\gamma = -1$, $ESN_k(\Sigma, \boldsymbol{\alpha}, \gamma)$ is reduced to $SN_k(\Sigma, -\boldsymbol{\alpha})$. Also it is reduced to $SN_k(\Sigma, \boldsymbol{\alpha})$ as γ tends to ∞.
(iv) For cases where $\gamma > 0$, $ESN_k(\Sigma, \boldsymbol{\alpha}, \gamma)$ is the mixture of a normal and a skew normal distributions.

For emphasis the usefulness of this extra parameter γ, the plots of density functions of the bivariate extended skew normal distribution with $\boldsymbol{\alpha} = \begin{pmatrix} 5 \\ -2 \end{pmatrix}$ and

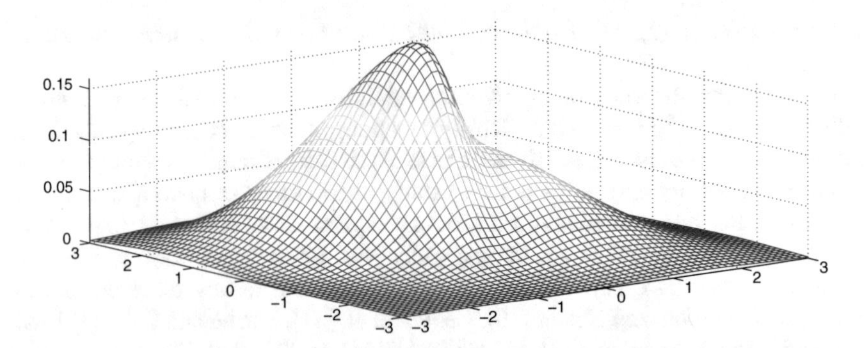

Fig. 1 Density curves of $ESN_2(\Sigma, (5, -2)', -0.5)$

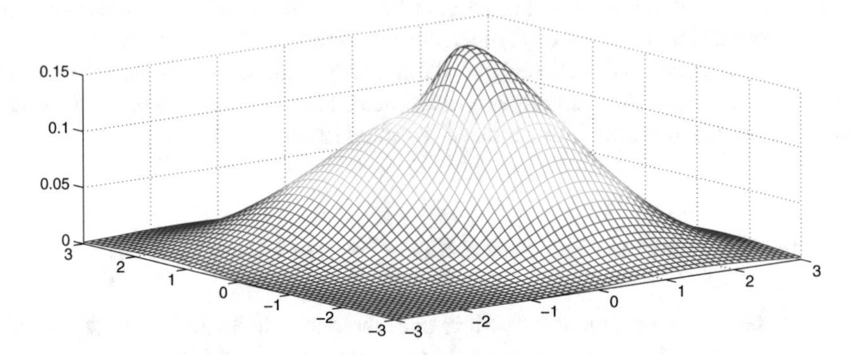

Fig. 2 Density curves of $ESN_2(\Sigma, (5, -2)', 0.5)$

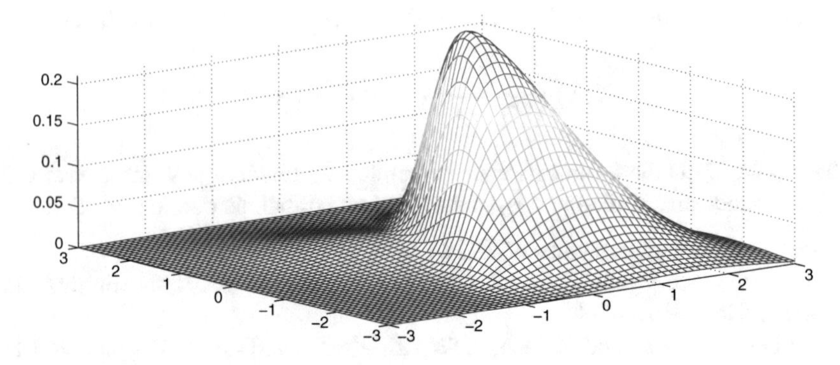

Fig. 3 Density curves of $ESN_2(\Sigma, (5, -2)', 10)$

$\Sigma = \begin{pmatrix} 1 & 0.5 \\ 0.5 & 1 \end{pmatrix}$ for different values of γ. Figure 1 is the density function corresponding to $\gamma = -0.5$. From Fig. 1, we can see that the negative value of γ changes the directions of skewness as it changes the sign of $\boldsymbol{\alpha}$. From Figs. 2, 3 and 4, which

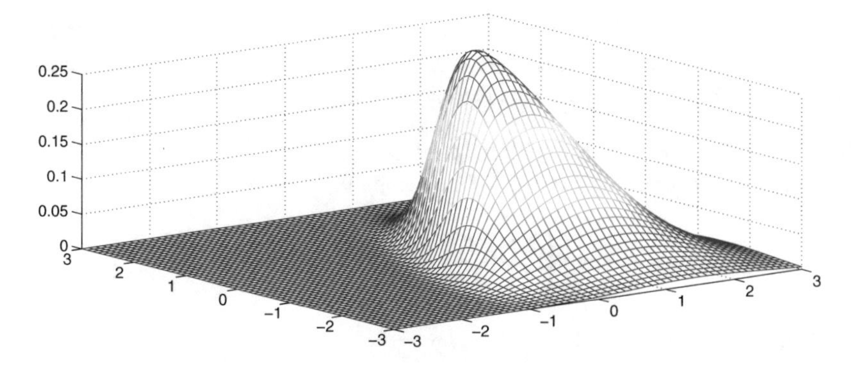

Fig. 4 Density curves of $ESN_2(\Sigma, (5, -2)', 1000)$

corresponding to $\gamma = 0.5$, $\gamma = 10$, and $\gamma = 1000$, we can see both peaks and shapes of density curves change as values of γ varies.

The following Lemmas will be used to prove the properties of our multivariate extended skew normal distribution.

Lemma 1 (Azzalini and Capitanio [3]) *If* $\mathbf{V} \sim SN_n(\Sigma, \boldsymbol{\alpha})$, *and* A *is a non-singular* $n \times n$ *matrix, then*

$$A'\mathbf{V} \sim SN_n(A'\Sigma A, A^{-1}\boldsymbol{\alpha}).$$

Lemma 2 (Zacks [23] and Chen [6]) *Let* $\mathbf{U} \sim N_n(\boldsymbol{0}, \Sigma)$. *For any scalar* $p \in \Re$ *and* $\mathbf{q} \in \Re^n$, *we have*

$$E[\Phi(p + \mathbf{q}'\mathbf{U})] = \Phi\left\{\frac{p}{(1 + \mathbf{q}'\Sigma\mathbf{q})^{1/2}}\right\}. \tag{4}$$

Proposition 1 *If* $\mathbf{Z} \sim ESN_k(\Sigma, \boldsymbol{\alpha}, \gamma)$, *and* A *is a non-singular* $n \times n$ *matrix, then*

$$A'\mathbf{Z} \sim ESN_n(A'\Sigma A, A^{-1}\boldsymbol{\alpha}, \gamma).$$

The proof of Proposition 1 is similar to that of Lemma 1.

Now we will extend about definition of multivariate extended skew-normal distribution to a more general case.

Definition 2 Let $\mathbf{Z} \sim ESN_k(I_k, \boldsymbol{\alpha}, \gamma)$. The random vector $\mathbf{Y} = \boldsymbol{\mu} + B'\mathbf{Z}$ is said to have a *generalized multivariate extended skew normal distribution* with location parameter $\boldsymbol{\mu} \in \Re^n$, scale parameter $B \in M_{k\times n}$, skewness parameter $\boldsymbol{\alpha} \in \Re^k$, and extended parameter $\gamma \geq -1$, denoted by $\mathbf{Y} \sim E\mathbb{S}N_n(\boldsymbol{\mu}, B, \boldsymbol{\alpha}, \gamma)$.

Remark 2 Note that the Definition 1 is the special case of Definition 2 where

$$ESN_n(\Sigma, \boldsymbol{\alpha}, \gamma) = E\mathbb{S}N_n(\mathbf{0}, \Sigma^{\frac{1}{2}}, \Sigma^{\frac{1}{2}}\boldsymbol{\alpha}, \gamma).$$

Indeed, we know that if $\mathbf{Y} = \Sigma^{\frac{1}{2}}\mathbf{Z}$ with $\mathbf{Z} \sim ESN(I_n, \Sigma^{\frac{1}{2}}\boldsymbol{\alpha}, \gamma)$, then we obtain the result by the Proposition 1. Also the multivariate skew normal distribution given in Wang et al. [18] is a special case of Definition 2 where $\gamma = -1$.

Theorem 1 *Suppose that* $\mathbf{Y} \sim E\mathbb{S}N_n(\boldsymbol{\mu}, B, \boldsymbol{\alpha}, \gamma)$.
(i) The moment generating function of $\mathbf{Y}$ *is given by*

$$M_{\mathbf{Y}}(\mathbf{t}) = \frac{2}{2+\gamma} \exp\left\{\mathbf{t}'\boldsymbol{\mu} + \frac{\mathbf{t}'B'B\mathbf{t}}{2}\right\}\left[1 + \gamma\Phi\left(\frac{\boldsymbol{\alpha}'B\mathbf{t}}{(1+\boldsymbol{\alpha}'\boldsymbol{\alpha})^{\frac{1}{2}}}\right)\right]. \tag{5}$$

(ii) The density of $\mathbf{Y}$*, if it exits, is given by*

$$f(\mathbf{y}; \boldsymbol{\mu}, B, \boldsymbol{\alpha}, \gamma) = \frac{2}{2+\gamma}\phi_k(\mathbf{y}, \boldsymbol{\mu}, \Sigma)\left\{1 + \gamma\Phi\left(\frac{\boldsymbol{\alpha}'B\Sigma^{-1}(\mathbf{y}-\boldsymbol{\mu})}{[1+\boldsymbol{\alpha}'(I_k - B\Sigma^{-1}B')\boldsymbol{\alpha}]^{\frac{1}{2}}}\right)\right\}, \tag{6}$$

where $\Sigma = B'B$.
(iii) The mean vector and covariance matrix of $\mathbf{Y}$ *are*

$$E[\mathbf{Y}] = \boldsymbol{\mu} + \frac{\gamma}{2+\gamma}\sqrt{\frac{2}{\pi}}\frac{B'\boldsymbol{\alpha}}{(1+\boldsymbol{\alpha}'\boldsymbol{\alpha})^{\frac{1}{2}}}$$

and

$$Cov(\mathbf{Y}) = B'\left(I_k - \frac{\gamma^2}{(2+\gamma)^2}\frac{2}{\pi}\frac{\boldsymbol{\alpha}\boldsymbol{\alpha}'}{1+\boldsymbol{\alpha}'\boldsymbol{\alpha}}\right)B.$$

Proof For (i), the moment generating function of $\mathbf{Y}$ is,

$$\begin{aligned}
M_{\mathbf{Y}}(\mathbf{t}) &= \frac{2}{\gamma+2}\int_{\Re^k}\frac{1}{(2\pi)^{k/2}}\exp\left\{\mathbf{t}'\boldsymbol{\mu} + \mathbf{t}'B'\mathbf{z} - \frac{1}{2}\mathbf{z}'\mathbf{z}\right\}\left[1+\gamma\Phi(\boldsymbol{\alpha}'\mathbf{z})\right]d\mathbf{z} \\
&= \frac{2}{\gamma+2}\int_{\Re^k}\frac{1}{(2\pi)^{k/2}}\exp\left\{\mathbf{t}'\boldsymbol{\mu} + \mathbf{t}'B'\mathbf{Z} - \frac{1}{2}\mathbf{Z}'\mathbf{Z}\right\}d\mathbf{Z} \\
&\quad + \frac{2}{\gamma+2}\int_{\Re^k}\frac{1}{(2\pi)^{k/2}}\exp\left\{\mathbf{t}'\boldsymbol{\mu} + \mathbf{t}'B'\mathbf{Z} - \frac{1}{2}\mathbf{Z}'\mathbf{Z}\right\}\Phi(\boldsymbol{\alpha}'\mathbf{z})d\mathbf{z} \\
&= \frac{2}{\gamma+2}\int_{\Re^k}\frac{1}{(2\pi)^{k/2}}\exp\left\{-\frac{1}{2}(\mathbf{z}-B\mathbf{t})'(\mathbf{z}-B\mathbf{t})\right\}\exp\left\{\mathbf{t}'\boldsymbol{\mu} + \frac{\mathbf{t}'B'B\mathbf{t}}{2}\right\}d\mathbf{z} \\
&\quad + \frac{2}{\gamma+2}\int_{\Re^k}\frac{1}{(2\pi)^{k/2}}\exp\left\{-\frac{1}{2}(\mathbf{z}-B\mathbf{t})'(\mathbf{z}-B\mathbf{t})\right\}\exp\left\{\mathbf{t}'\boldsymbol{\mu} + \frac{\mathbf{t}'B'B\mathbf{t}}{2}\right\}\Phi(\boldsymbol{\alpha}'\mathbf{z})d\mathbf{z} \\
&= \frac{2}{\gamma+2}\exp\left\{\mathbf{t}'\boldsymbol{\mu} + \frac{\mathbf{t}'B'B\mathbf{t}}{2}\right\}\left[1+\gamma\int_{\Re^k}\frac{1}{(2\pi)^{k/2}}\exp\left\{-\frac{1}{2}(\mathbf{z}-B\mathbf{t})'(\mathbf{z}-B\mathbf{t})\right\}\Phi(\boldsymbol{\alpha}'\mathbf{z})d\mathbf{z}\right] \\
&= \frac{2}{\gamma+2}\exp\left\{\mathbf{t}'\mu + \frac{\mathbf{t}'B'B\mathbf{t}}{2}\right\}\left[1+\gamma\Phi\left(\frac{\boldsymbol{\alpha}'B\mathbf{t}}{(1+\boldsymbol{\alpha}'\boldsymbol{\alpha})^{\frac{1}{2}}}\right)\right],
\end{aligned}$$

where, by Lemma 2,

$$\int_{\Re^k} \frac{1}{(2\pi)^{k/2}} \exp\left\{-\frac{1}{2}(\mathbf{z} - B\mathbf{t})'(\mathbf{z} - B\mathbf{t})\right\} \Phi(\boldsymbol{\alpha}'\mathbf{z})d\mathbf{z} = \Phi\left(\frac{\boldsymbol{\alpha}'B\mathbf{t}}{(1+\boldsymbol{\alpha}'\boldsymbol{\alpha})^{\frac{1}{2}}}\right).$$

(ii) Let $\mathbf{Z}_1 \sim ESN_n(I_n, \boldsymbol{\alpha}_1, \gamma)$ and $\mathbf{Y}_1 = \boldsymbol{\mu} + \Sigma^{\frac{1}{2}}\mathbf{Z}_1$ with $\Sigma = B'B$. By Proposition 1 and (3), the density of $\mathbf{Y}_1$ is

$$f_{\mathbf{Y}_1}(\mathbf{y}_1) = \frac{2}{2+\gamma}\phi_n(\mathbf{y}_1 - \boldsymbol{\mu}; \Sigma)\left[1 + \gamma\Phi(\boldsymbol{\alpha}_1'\Sigma^{-\frac{1}{2}}(\mathbf{y}_1 - \boldsymbol{\mu}))\right].$$

By part (i),

$$M_{\mathbf{Y}_1}(\mathbf{t}) = \frac{2}{\gamma+2}\exp\left\{\mathbf{t}'\boldsymbol{\mu} + \frac{\mathbf{t}'\Sigma\mathbf{t}}{2}\right\}\left[1 + \gamma\Phi\left(\frac{\boldsymbol{\alpha}_1'\Sigma^{\frac{1}{2}}\mathbf{t}}{(1+\boldsymbol{\alpha}_1'\boldsymbol{\alpha}_1)^{\frac{1}{2}}}\right)\right].$$

Let $\boldsymbol{\alpha}_1 = \dfrac{\Sigma^{-\frac{1}{2}}B'\boldsymbol{\alpha}}{[1+\boldsymbol{\alpha}'(I_k - B\Sigma^{-1}B')\boldsymbol{\alpha}]^{\frac{1}{2}}}$, we obtain

$$1 + \boldsymbol{\alpha}_1'\boldsymbol{\alpha}_1 = \frac{1+\boldsymbol{\alpha}'\boldsymbol{\alpha}}{1+\boldsymbol{\alpha}'(I_k - B\Sigma^{-1}B')\boldsymbol{\alpha}}$$

and

$$\frac{\boldsymbol{\alpha}_1'\Sigma^{\frac{1}{2}}\mathbf{t}}{(1+\boldsymbol{\alpha}_1'\boldsymbol{\alpha}_1)^{\frac{1}{2}}} = \frac{\boldsymbol{\alpha}'B\mathbf{t}}{(1+\boldsymbol{\alpha}'\boldsymbol{\alpha})^{\frac{1}{2}}}.$$

Therefore, we obtain $M_{\mathbf{Y}}(\mathbf{t}) = M_{\mathbf{Y}_1}(\mathbf{t})$, so that $f_{\mathbf{Y}}(\mathbf{y}; \boldsymbol{\mu}, B, \boldsymbol{\alpha}, \gamma) = f_{\mathbf{Y}_1}(\mathbf{y}_1)$, the desired result follows.

(iii) The mean vector $E(\mathbf{Y})$ and the covariance $Cov(\mathbf{Y})$ of $\mathbf{Y}$ can be obtained using the moment generating function of $\mathbf{Y}$. □

Theorem 2 *Suppose that* $\mathbf{Z} \sim ESN_k(I_k, \boldsymbol{\alpha}, \gamma)$. *Let* $\mathbf{Y}_i = \boldsymbol{\mu}_i + B_i'\mathbf{Z}$ *with* $B_i \in M_{k\times n_i}$. *Then* $\mathbf{Y}_1$ *and* $\mathbf{Y}_2$ *are independent if and only if*
(i) $B_1'B_2 = 0$ *and*
(ii) either $B_1'\boldsymbol{\alpha} = \mathbf{0}$ *or* $B_2'\boldsymbol{\alpha} = \mathbf{0}$.

Proof The "if" part is trivial. We only prove "only if" part. Let $\mathbf{Y} = \boldsymbol{\mu} + B'\mathbf{Z}$ with $\boldsymbol{\mu}' = (\boldsymbol{\mu}_1', \boldsymbol{\mu}_2')$ and $B = (B_1, B_2)$.
Suppose $\mathbf{Y}_1$ and $\mathbf{Y}_2$ are independent. Then we have $M_{\mathbf{Y}}(\mathbf{t}) = M_{\mathbf{Y}_1}(\mathbf{t}_1)M_{\mathbf{Y}_2}(\mathbf{t}_2)$ with $\mathbf{t}' = (\mathbf{t}_1', \mathbf{t}_2')$, where

$$M_{\mathbf{Y}}(\mathbf{t}) = \frac{2}{2+\gamma}\exp\left\{\mathbf{t}'\boldsymbol{\mu} + \frac{\mathbf{t}'B'B\mathbf{t}}{2}\right\}\left[1 + \gamma\Phi\left(\frac{\alpha'B\mathbf{t}}{(1+\boldsymbol{\alpha}'\boldsymbol{\alpha})^{\frac{1}{2}}}\right)\right],$$

$$M_{\mathbf{Y}_1}(\mathbf{t_1}) = \frac{2}{2+\gamma}\exp\left\{\mathbf{t_1}'\boldsymbol{\mu}_1 + \frac{\mathbf{t_1}'B_1'B_1\mathbf{t_1}}{2}\right\}\left[1+\gamma\Phi\left(\frac{\alpha' B_1\mathbf{t_1}}{(1+\boldsymbol{\alpha}'\boldsymbol{\alpha})^{\frac{1}{2}}}\right)\right],$$

and

$$M_{\mathbf{Y}_2}(\mathbf{t_2}) = \frac{2}{2+\gamma}\exp\left\{\mathbf{t_2}'\boldsymbol{\mu}_2 + \frac{\mathbf{t_2}'B_2'B_2\mathbf{t_2}}{2}\right\}\left[1+\gamma\Phi\left(\frac{\alpha' B_2\mathbf{t_2}}{(1+\boldsymbol{\alpha}'\boldsymbol{\alpha})^{\frac{1}{2}}}\right)\right].$$

After simplification, we obtain,

$$\frac{\frac{2}{2+\gamma}\left[1+\gamma\Phi\left(\frac{\alpha' B_1\mathbf{t_1}}{(1+\boldsymbol{\alpha}'\boldsymbol{\alpha})^{\frac{1}{2}}}\right)\right]\left[1+\gamma\Phi\left(\frac{\alpha' B_2\mathbf{t_2}}{(1+\boldsymbol{\alpha}'\boldsymbol{\alpha})^{\frac{1}{2}}}\right)\right]}{1+\gamma\Phi\left(\frac{\alpha' B_1\mathbf{t_1}+\alpha' B_2\mathbf{t_2}}{(1+\boldsymbol{\alpha}'\boldsymbol{\alpha})^{\frac{1}{2}}}\right)} = \exp\{\mathbf{t}_1' B_1' B_2\mathbf{t_2}\}, \tag{7}$$

for all $\mathbf{t}_1 \in \Re^{n_1}$ and $\mathbf{t}_2 \in \Re^{n_2}$.
Assume that $B_1'B_2 \neq 0$, then we choose $\mathbf{s}_1 \in \Re^{n_1}$ and $\mathbf{s}_2 \in \Re^{n_2}$ such that $\mathbf{s}_1'B_1'$ $B_2\mathbf{s}_2 > 0$. Let $\mathbf{t}_1 = c\mathbf{s}_1, \mathbf{t}_2 = c\mathbf{s}_2, c \in \Re$. Then the right hand side of (7) approaches to $\pm\infty$ as $c \to \infty$. Note that, the left hand side of (7) is bounded for all $\mathbf{t} \in \Re^n$ when $\gamma \neq -1$, and it approach 0 or ∞ when $\gamma = -1$. Therefore, $B_1'B_2 = 0$ as (7) holds for all $\mathbf{t} \in \Re^n$.
Now from (7), we have

$$\frac{2}{2+\gamma}\left[1+\gamma\Phi\left(\frac{\alpha' B_1\mathbf{t_1}}{(1+\boldsymbol{\alpha}'\boldsymbol{\alpha})^{\frac{1}{2}}}\right)\right]\left[1+\gamma\Phi\left(\frac{\alpha' B_2\mathbf{t_2}}{(1+\boldsymbol{\alpha}'\boldsymbol{\alpha})^{\frac{1}{2}}}\right)\right] = 1+\gamma\Phi\left(\frac{\alpha' B_1\mathbf{t_1}+\alpha' B_2\mathbf{t_2}}{(1+\boldsymbol{\alpha}'\boldsymbol{\alpha})^{\frac{1}{2}}}\right),$$

for all $\mathbf{t}_1 \in \Re^{n_1}$ and $\mathbf{t}_2 \in \Re^{n_2}$.

Note that we do not need the condition (ii) if $\gamma = 0$. Now, assume that both $\boldsymbol{\alpha}'B_1 \neq 0$ and $\boldsymbol{\alpha}'B_2 \neq 0$. There exist $\mathbf{s}_1 \in \Re^{n_1}$ and $\mathbf{s}_2 \in \Re^{n_2}$ such that $\boldsymbol{\alpha}'B_1\mathbf{s}_1 > 0$ and $\boldsymbol{\alpha}'B_2\mathbf{s}_2 > 0$. Let $\mathbf{t}_1 = c_1\mathbf{s}_1$ and $\mathbf{t}_2 = c_2\mathbf{s}_2, c_1, c_2 \in \Re$. Then as $c_1 \to \infty$ we have,

$$\frac{1+\gamma}{2+\gamma}\left[1+\gamma\Phi\left(\frac{\alpha' B_2\mathbf{s_2}}{(1+\boldsymbol{\alpha}'\boldsymbol{\alpha})^{\frac{1}{2}}}\right)\right] = 1+\gamma,$$

for all $\mathbf{s_2} \in \Re^{n_2}$, so that $\boldsymbol{\alpha}'B_2\mathbf{s_2}$ must be zero. Similarly, as $c_2 \to \infty$, we obtain that $\boldsymbol{\alpha}'B_2\mathbf{s_1}$ must be zero. Therefore the desired result follows. □

From Theorem 2, it is easy to prove the following result.

Corollary 1 *Let $\mathbf{Y} \sim E\mathbb{S}N_n(\boldsymbol{\mu}, B, \boldsymbol{\alpha}, \gamma)$. Then $A_1\mathbf{Y}$ and $A_2\mathbf{Y}$, with $A_i \in M_{m_i \times n}$, are independent if and only if*
(i) $A_1B'BA_2' = 0$ and
(ii) either $A_1B'\boldsymbol{\alpha} = \mathbf{0}$ or $A_2B'\boldsymbol{\alpha} = \mathbf{0}$.

Example 1 Let $\mathbf{Y} \sim E\mathbb{S}N_n(\mathbf{0}, B, \boldsymbol{\alpha}, \gamma)$, $\bar{Y} = \mathbf{1}_n'\mathbf{Y}/n$ and $\mathbf{W} = (I_n - \bar{J}_n)\mathbf{Y}$, with $\bar{J}_n = \mathbf{1}_n\mathbf{1}_n'/n$ and $\mathbf{1}_n$ is the column vector of 1's in $\Re^n$. From Corollary 1, we know that $\bar{Y}$ and $\mathbf{W}$ are independent if and only if $\mathbf{1}_n' B'B(I_n - \bar{J}_n) = 0$ and $\mathbf{1}_n' B\boldsymbol{\alpha} = 0$ or $(I_n - \bar{J}_n)B'\boldsymbol{\alpha} = \mathbf{0}$.

In particular, if $B'B = I_n$, we have $\mathbf{1}_n' B'B(I_n - \bar{J}_n) = 0$ and $\bar{Y}$ and $\mathbf{W}$ are not independent if both $\mathbf{1}_n B\boldsymbol{\alpha} \neq 0$ and $(I_n - \bar{J}_n)B'\boldsymbol{\alpha} \neq \mathbf{0}$. For example, even if $\boldsymbol{\alpha} = (1, 0, \ldots, 0)'$ and $B = I_n$, then $\mathbf{1}_n B\boldsymbol{\alpha} = 1$ and $(I_n - \bar{J}_n)B'\boldsymbol{\alpha} = (1 - \frac{1}{n}, -\frac{1}{n}, \ldots, -\frac{1}{n})'$. □

3 Extended Noncentral Skew Chi-Square Distributions

The noncentral skew chi-square distribution, with k degrees of freedom, noncentrality λ and skewness parameters δ_1, δ_2, was defined by Wang et al. [18] ant Ye et al. [20]. Here we use a similar method to define the extended noncentral skew chi-square distribution under the multivariate extended skew normal settings.

Definition 3 Let $\mathbf{X} \sim E\mathbb{S}N_k(\boldsymbol{\nu}, I_k, \boldsymbol{\alpha}, \gamma)$. The distribution of $\mathbf{X}'\mathbf{X}$ is defined as the *extended noncentral skew chi-square distribution* with k degrees of freedom, noncentrality parameter $\lambda = \boldsymbol{\nu}'\boldsymbol{\nu}$, skewness parameters $\delta_1 = \boldsymbol{\alpha}'\boldsymbol{\nu}$, $\delta_2 = \boldsymbol{\alpha}'\boldsymbol{\alpha}$ and extended parameter γ, denoted by $\mathbf{X}'\mathbf{X} \sim ES\chi_k^2(\lambda, \delta_1, \delta_2, \gamma)$.

In the following, we discuss the properties of the extended noncentral skew chi-square distribution.

Theorem 3 *Suppose that* $\mathbf{Y} \sim E\mathbb{S}N_n(\boldsymbol{\mu}, B, \boldsymbol{\alpha}, \gamma)$. *Let* $Q = \mathbf{Y}'W\mathbf{Y}$ *with symmetric* $W \in M_{n\times n}$. *Then the M.G.F of* Q *is given by*

$$M_Q(t) = \frac{2}{2+\gamma} \frac{\exp\left\{t\boldsymbol{\mu}'W\boldsymbol{\mu} + 2t^2\boldsymbol{\mu}'WB'(I_k - 2tBWB')^{-1}BW\boldsymbol{\mu}\right\}}{|I_k - 2tBWB'|^{\frac{1}{2}}} \times \left[1 + \gamma\Phi\left(\frac{2t\boldsymbol{\alpha}'(I_k - 2tBWB')^{-1}BW\boldsymbol{\mu}}{\left[1 + \boldsymbol{\alpha}'(I_k - 2tBWB')^{-1}\boldsymbol{\alpha}\right]^{\frac{1}{2}}}\right)\right], \tag{8}$$

for $t \in \Re$ *such that* $2|t| < \sigma(BWB')$, *where* $\sigma(A)$ *is the spectral of* A.

Proof The moment generating function of Q is

$$\begin{aligned} M_Q(t) &= E\left(\exp\{t\mathbf{Y}'W\mathbf{Y}\}\right) \\ &= E\left(\exp\left\{t\mu'W\mu + t(\mathbf{Z}'BWB'\mathbf{Z} + 2\mu'WB'\mathbf{Z})\right\}\right) \\ &= \frac{2\exp\{t\mu'W\mu\}}{\gamma+2}\int_{\Re^k}\frac{1}{2\pi^{k/2}}\exp\left\{t\left(\mathbf{z}'BWB'\mathbf{z} + 2\mu'WB'\mathbf{z} - \frac{1}{2}\mathbf{z}'\mathbf{z}\right)\right\}d\mathbf{z} \\ &\quad + \frac{2\gamma\exp\{t\mu'W\mu\}}{\gamma+2}\int_{\Re^k}\frac{1}{2\pi^{k/2}}\exp\left\{t\left(\mathbf{z}'BWB'\mathbf{z} + 2\mu'WB'\mathbf{z} - \frac{1}{2}\mathbf{z}'\mathbf{z}\right)\right\}\Phi(\alpha'\mathbf{z})d\mathbf{z}, \end{aligned}$$

by a similar argument given in proof of (i) of Theorem 1,

$$M_Q(t) = \frac{2}{2+\gamma} \frac{\exp\left\{t\boldsymbol{\mu}'W\boldsymbol{\mu} + 2t^2\boldsymbol{\mu}'WB'(I_k - 2tBWB')^{-1}BW\boldsymbol{\mu}\right\}}{|I_k - 2tBWB'|^{\frac{1}{2}}}$$
$$\times \left[1 + \gamma\Phi\left(\frac{2t\boldsymbol{\alpha}'(I_k - 2tBWB')^{-1}BW\boldsymbol{\mu}}{\left[1 + \boldsymbol{\alpha}'(I_k - 2tBWB')^{-1}\boldsymbol{\alpha}\right]^{\frac{1}{2}}}\right)\right]. \qquad \square$$

Corollary 2 *Let* $\mathbf{X} \sim E\mathbb{S}N_k(\boldsymbol{\nu}, I_k, \boldsymbol{\alpha}, \gamma)$*, we have* $U = \mathbf{X}'\mathbf{X} \sim ES\chi_k^2(\lambda, \delta_1, \delta_2, \gamma)$*, then*
(i) the moment generating function of U is

$$M_U(t) = \frac{2}{2+\gamma} \frac{\exp\left\{t(1-2t)^{-1}\lambda\right\}}{(1-2t)^{k/2}} \left[1 + \gamma\Phi\left(\frac{2t(1-2t)^{-1}\delta_1}{\left[1 + (1-2t)^{-1}\delta_2\right]}\right)\right],$$

for $t \in \Re$ *such that* $2|t| < 1$*,*
(ii) the mean of U is

$$E(U) = \lambda + k + \frac{4\gamma\delta_1}{(\gamma+2)\sqrt{2\pi(1+\delta_2)}},$$

(iii) the variance of U is

$$Var(U) = 4\gamma + 2k + \frac{4\delta_1\left[\sqrt{2\pi}\gamma(\gamma+2)(2+\delta_2) - 2\gamma^2\delta_1(1+\delta_2)^{\frac{1}{2}}\right]}{\pi(\gamma+2)^2(1+\delta_2)^{\frac{3}{2}}},$$

where λ*,* δ_1 *and* δ_2 *are given in definition 3.*

Remark 3 Note that Theorem 3 and Corollary 2 extend the results of Wang et al. [18] from the case where $\gamma = -1$ to cases where $\gamma > -1$. Also, when $\gamma = 0$ or $\delta_1 = 0$, $ES\chi_k^2(\lambda, \delta_1, \delta_2, \gamma)$ is reduced to noncentral χ^2 distribution with degrees of freedom k.

Theorem 4 *The density function of* $U \sim ES\chi_k^2(\lambda, \delta_1, \delta_2, \gamma)$ *with* $\lambda \neq 0$ *and* $k > 1$*, is*

$$f_U(u; \lambda, \delta_1, \delta_2, \gamma) = \frac{2\exp\left\{-\frac{1}{2}(\lambda+u)\right\}}{(\gamma+2)\Gamma\left(\frac{1}{2}\right)\Gamma\left(\frac{k}{2}\right)2^{k/2}} u^{\frac{k}{2}-1} {}_0F_1\left(\frac{k}{2}, \frac{\lambda u}{4}\right)$$
$$+ \frac{\gamma\exp\left\{-\frac{1}{2}(\lambda+u)\right\}}{(\gamma+2)\Gamma\left(\frac{1}{2}\right)\Gamma\left(\frac{k-1}{2}\right)2^{k/2-1}} h(u; \lambda, \delta_1, \delta_2) \quad u > 0, \quad (9)$$

where

$$h(x; \lambda, \delta_1, \delta_2) = \int_{-\sqrt{x}}^{\sqrt{x}} \exp\left(\lambda^{1/2}s_1\right)\left(x - s_1^2\right)^{\frac{k-3}{2}} \Phi\left\{\alpha_0\left(s_1 - \lambda^{1/2}\right)\right\} ds_1,$$

$\alpha_0 = \frac{\lambda^{-1/2}\delta_1}{(1+\delta_2-\delta_1^2/\lambda)^{1/2}}$, *and* ${}_0F_1(b, z)$ *denotes the confluent hypergeometric function (Muirhead [16]).*

Proof Let $\mathbf{X} \sim E\mathbb{S}N_k(\boldsymbol{\nu}, I_k, \boldsymbol{\alpha}, \gamma)$, and $\mathbf{Y} = K\mathbf{X}$, where K is an $k \times k$ orthogonal matrix whose elements in the first row are $\nu_i/\sqrt{\lambda}$, $i = 1, 2, \ldots, k$. W have $\mathbf{Y} \sim E\mathbb{S}N_k(K\boldsymbol{\nu}, I_k, K\boldsymbol{\alpha}, \gamma)$ with $K\boldsymbol{\nu} = (\sqrt{\lambda}, 0, \ldots, 0)$, and for $k > 1$,

$$U = \mathbf{X}'\mathbf{X} = \mathbf{Y}'\mathbf{Y} = y_1^2 + \sum_{i=2}^{k} y_i^2.$$

Note that $Y_1 \sim E\mathbb{S}N(\sqrt{\lambda}, 1, \alpha_0, \gamma)$ and $(Y_2, \ldots, Y_k)' \sim E\mathbb{S}N_{k-1}(0, I_{k-1}, \bar{\boldsymbol{\alpha}}_2, \gamma)$. By the Definition 3, we have $\sum_{i=2}^{k} Y_i^2 \sim \chi^2_{k-1}$. Then the conditional density function of U given $Y_1 = y_1$ is

$$f_{U|Y_1=y_1}(u) = \frac{\left(u-y_1^2\right)^{\frac{k-3}{2}}}{\Gamma\left(\frac{k-1}{2}\right)2^{\frac{k-1}{2}}} \exp\left\{-\frac{u-y_1^2}{2}\right\}, \quad u \geq y_1^2.$$

Therefore, the density function of U is

$$\begin{aligned}
f_U(u) &= \int_{-\sqrt{u}}^{\sqrt{u}} f_{U|Y_1=y_1}(u) f_{Y_1}(y_1) dy_1 \\
&= \frac{1}{(\gamma+2)\Gamma\left(\frac{1}{2}\right)\Gamma\left(\frac{k-1}{2}\right)2^{\frac{k-2}{2}}} \int_{-\sqrt{u}}^{\sqrt{u}} \exp\left\{-\frac{u-y_1^2+(y_1-\sqrt{\lambda})^2}{2}\right\}\left(u-y_1^2\right)^{\frac{k-3}{2}} dy_1 \\
&\quad + \frac{\lambda}{(\gamma+2)\Gamma\left(\frac{1}{2}\right)\Gamma\left(\frac{k-1}{2}\right)2^{\frac{k-2}{2}}} \int_{-\sqrt{u}}^{\sqrt{u}} \exp\left\{-\frac{u-y_1^2+\left(y_1-\sqrt{\lambda}\right)^2}{2}\right\} \\
&\quad \times \left(u-y_1^2\right)^{\frac{k-3}{2}} \Phi\left(\alpha_0(y_1-\sqrt{\lambda})\right) dy_1
\end{aligned}$$

Note that

$$\begin{aligned}
&\frac{1}{(\gamma+2)\Gamma\left(\frac{1}{2}\right)\Gamma\left(\frac{k-1}{2}\right)2^{\frac{k-2}{2}}} \int_{-\sqrt{u}}^{\sqrt{u}} \exp\left\{-\frac{u-y_1^2+\left(y_1-\sqrt{\lambda}\right)^2}{2}\right\}\left(u-y_1^2\right)^{\frac{k-3}{2}} dy_1 \\
&= \frac{\exp\left\{-\frac{1}{2}(u+\lambda)\right\}}{(\gamma+2)\Gamma\left(\frac{1}{2}\right)\Gamma\left(\frac{k-1}{2}\right)2^{\frac{k-2}{2}}} \int_{-\sqrt{u}}^{\sqrt{u}} \exp\left\{\sqrt{\lambda}y_1\right\}\left(u-y_1^2\right)^{\frac{k-3}{2}} dy_1.
\end{aligned}$$

Let $y_1 = \sqrt{u}\cos\theta$, we obtain

$$= \frac{\exp\left\{-\frac{1}{2}(u+\lambda)\right\}}{(\gamma+2)\Gamma\left(\frac{1}{2}\right)\Gamma\left(\frac{k-1}{2}\right)2^{\frac{k-2}{2}}}\int_0^{\pi}\exp\left\{\sqrt{\lambda}\sqrt{u}\cos\theta\right\}\left(u\sin^2\theta\right)^{\frac{k-3}{2}}\sqrt{u}\sin\theta d\theta$$

$$= \frac{\exp\left\{-\frac{1}{2}(u+\lambda)\right\}}{(\gamma+2)\Gamma\left(\frac{1}{2}\right)\Gamma\left(\frac{k-1}{2}\right)2^{\frac{k-2}{2}}}\int_0^{\pi}\exp\left\{\sqrt{\lambda}\sqrt{u}\cos\theta\right\}(\sin\theta)^{k-2}u^{\frac{k-2}{2}}d\theta$$

$$= \frac{2\exp\left\{-\frac{1}{2}(\lambda)\right\}}{(\gamma+2)}\frac{\exp\left\{-\frac{1}{2}u\right\}}{\Gamma\left(\frac{k-1}{2}\right)2^{\frac{k}{2}}}u^{\frac{m}{2}-1}\frac{\Gamma\left(\frac{k}{2}\right)}{\Gamma\left(\frac{1}{2}\right)\Gamma\left(\frac{k-1}{2}\right)}\int_0^{\pi}\exp\left\{\sqrt{\lambda}\sqrt{u}\cos\theta\right\}(\sin\theta)^{k-2}d\theta$$

$$= \frac{2\exp\left\{-\frac{1}{2}(\lambda)\right\}}{(\gamma+2)}\frac{\exp\{-\frac{1}{2}u\}}{\Gamma\left(\frac{k-1}{2}\right)2^{\frac{k}{2}}}u^{\frac{m}{2}-1}{}_0F_1\left(\frac{k}{2};\frac{\lambda u}{4}\right).$$

Similarly,

$$\frac{\gamma}{(\gamma+2)\Gamma\left(\frac{1}{2}\right)\Gamma\left(\frac{k-1}{2}\right)2^{\frac{k-2}{2}}}\int_{-\sqrt{u}}^{\sqrt{u}}\exp\left\{-\frac{u-y_1^2+\left(y_1-\sqrt{\lambda}\right)^2}{2}\right\}\left(u-y_1^2\right)^{\frac{k-3}{2}}\Phi(\alpha_0(y_1-\sqrt{\lambda}))dy_1$$

$$= \frac{\gamma\exp\left\{-\frac{1}{2}(u+\lambda)\right\}}{(\gamma+2)\Gamma\left(\frac{1}{2}\right)\Gamma\left(\frac{k-1}{2}\right)2^{\frac{k-2}{2}}}\int_{-\sqrt{u}}^{\sqrt{u}}\exp\left(\lambda^{1/2}s_1\right)\left(u-y_1^2\right)^{\frac{m-3}{2}}\Phi\left\{\alpha_0\left(y_1-\lambda^{1/2}\right)\right\}dy_1.\square$$

Remark 4 From the density function given in (9), we know that the extended non-central skew chi-square distribution is uniquely described by the parameters k, λ, δ_1, δ_2 and γ. Similarly, in order to emphasize the usefulness of the extended parameter γ in our noncentral extended skew chi-square distributions for $k = 5, \delta_1 = -5, \delta_2 = 10$ with different values of γ are plotted in Fig. 5. In Fig. 5, we can see that the peaks and shapes of $ES\chi_5(5, -5, 10, \gamma)$ are changed and are reduced to $S\chi_5(5, -5, 10)$, as γ goes to ∞, which is given in Ye et al. [21].

4 The Distribution of Quadratic Form of Y

In this section, we talk about the distribution of quadratic forms of $\mathbf{Y} \sim E\mathbb{S}N_n(\boldsymbol{\mu}, B, \boldsymbol{\alpha}, \gamma)$. We will say the matrix M is idempotent if and only if $MM = M$.

Lemma 3 (Wang et al. [18]) *Let us have $B \in M_{k\times n}$, $\boldsymbol{\mu} \in \Re^n$, $\lambda > 0$, $m \leq k$ be a positive integer, and $W \in M_{n\times n}$ be nonnegative definite. If*

$$\frac{\exp\left[t\boldsymbol{\mu}'W\boldsymbol{\mu}+2t^2\boldsymbol{\mu}'WB'(I_k-2tBWB')^{-1}BW\boldsymbol{\mu}\right]}{|I_k-2tBWB'|^{\frac{1}{2}}} = \frac{\exp\left\{\frac{t}{1-2t}\lambda\right\}}{(1-2t)^{m/2}},$$

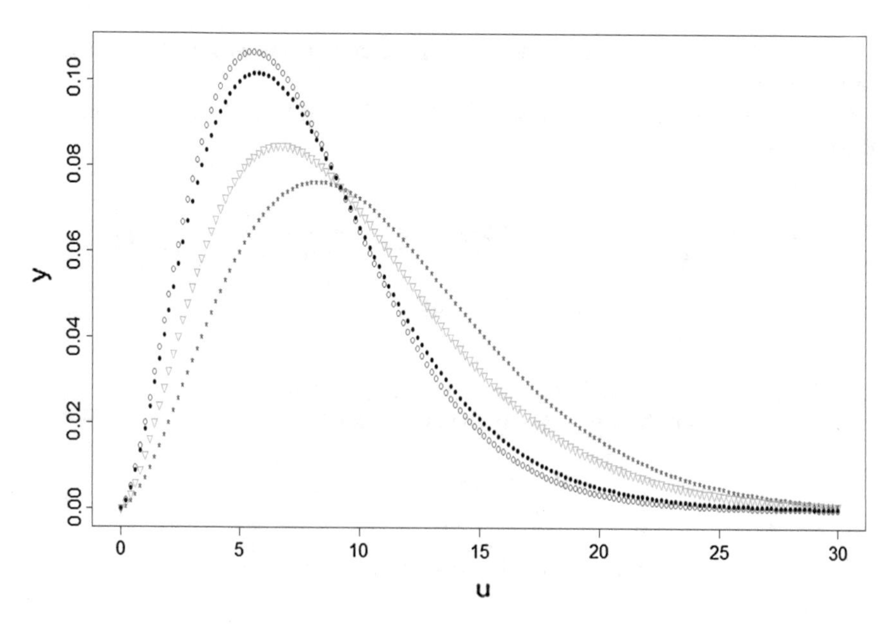

Fig. 5 Density curves of $ES\chi_5(5, -5, 10, -0.5)$ (*Blue* and *stars*), $ES\chi_5(5, -5, 10, 0.5)$ (*Green* and *triangles*), $ES\chi_5(5, -5, 10, 10)$ (*Red* and *circles*) and $ES\chi_5(5, -5, 10, 1000)$(*Black* and *dots*)

for all t with $2|t| < min\{1, \sigma(BW'B)\}$, *then*
(i) BWB' *is idempotent of rank m, and*
(ii) $\lambda = \boldsymbol{\mu}'W\boldsymbol{\mu} = \boldsymbol{\mu}'WB'BW\boldsymbol{\mu}$.

Theorem 5 *Let* $\mathbf{Y} \sim E\mathbb{S}N_n(\boldsymbol{\mu}, B, \boldsymbol{\alpha}, \gamma)$ *and* $Q = \mathbf{Y}'W\mathbf{Y}$ *with nonnegative definite* $W \in M_{n\times n}$. *Then the necessary and sufficient conditions under which* $Q \sim ES\chi_k^2(\lambda, \delta_1, \delta_2, \gamma)$ *for some* $\boldsymbol{\alpha}$ *or* γ *including 0, are*
(i) BWB' *is idempotent of rank m,*
(ii) $\lambda = \boldsymbol{\mu}'W\boldsymbol{\mu} = \boldsymbol{\mu}'WB'BW\boldsymbol{\mu}$,
(iii) $\delta_1 = \boldsymbol{\mu}'WB'\boldsymbol{\alpha}/c$,
(iv) $\delta_2 = \boldsymbol{\alpha}'BWB'\boldsymbol{\alpha}/c^2$,
where $c = \sqrt{1 + \boldsymbol{\alpha}'P_2P_2'\boldsymbol{\alpha}}$, *and* $P = (P_1 \quad P_2)$ *is an orthogonal matrix in* $M_{k\times k}$ *such that*

$$BWB' = P\begin{pmatrix} I_m & \mathbf{0} \\ \mathbf{0} & \mathbf{0} \end{pmatrix}P' = P_1P_1'.$$

Proof For "if" part, assume that conditions (i)-(iv) hold. By (i), there exists an orthogonal matrix $P \in M_{k\times k}$ such that

$$BWB' = P\begin{pmatrix} I_m & \mathbf{0} \\ \mathbf{0} & \mathbf{0} \end{pmatrix}P' = P_1P_1',$$

where $P = (P_1 \quad P_2)$. It is easy to see that (i) and $BWB'BW = BW$ are equivalent for nonnegative definite W.
Let $\mathbf{Z} \sim ESN_k(I_k, \boldsymbol{\alpha}, \gamma)$ and $\mathbf{Z}_* = P_1'\mathbf{Z}$. Then $\mathbf{Z}_* \sim ESN_m(I_m, P_1\boldsymbol{\alpha}, \gamma)$ and the density of $\mathbf{Z}_*$ is

$$f_{\mathbf{Z}_*}(\mathbf{z}_*) = \frac{2}{2+\gamma}\phi_m(\mathbf{z}_*; I_m)\left[1 + \gamma\Phi\left(\frac{\boldsymbol{\alpha}'P_1\mathbf{z}_*}{[1 + \boldsymbol{\alpha}'(I_k - P_1P_1')\alpha]^{\frac{1}{2}}}\right)\right]$$
$$= \frac{2}{2+\gamma}\phi_m(\mathbf{z}_*; I_m)\left[1 + \gamma\Phi\left(\boldsymbol{\alpha}_*'\mathbf{z}_*\right)\right].$$

Therefore, we have $\mathbf{Z}_* \sim ESN(I_m, \boldsymbol{\alpha}_*, \gamma)$ with $\boldsymbol{\alpha}_* = P_1'\boldsymbol{\alpha}/c$.
Let $\boldsymbol{\nu} = P_1BW\boldsymbol{\mu}$ and $\mathbf{X} = \nu + \mathbf{Z}_*$. By the Definition 3,

$$\mathbf{X}'\mathbf{X} \sim ES\chi^2_m(\lambda, \delta_1, \delta_2, \gamma),$$

with $\lambda = \boldsymbol{\nu}'\boldsymbol{\nu} = \boldsymbol{\mu}'WB'BWB'BW\boldsymbol{\mu}$, $\delta_1 = \boldsymbol{\nu}'\boldsymbol{\alpha}_* = \dfrac{\boldsymbol{\mu}'WB'\boldsymbol{\alpha}}{c}$ and $\delta_2 = \boldsymbol{\alpha}_*'\boldsymbol{\alpha}_* = \dfrac{\boldsymbol{\alpha}'BWB'\boldsymbol{\alpha}}{c^2}$.
Next, we need to show $Q \overset{d}{=} \mathbf{X}'\mathbf{X}$.

$$Q = \mathbf{Y}'W\mathbf{Y} = (\boldsymbol{\mu} + B'\mathbf{Z})'W(\boldsymbol{\mu} + B'\mathbf{Z})$$
$$= \boldsymbol{\mu}'W\boldsymbol{\mu} + 2\mathbf{Z}'BW\boldsymbol{\mu} + \mathbf{Z}'BWB'\mathbf{Z}$$
$$= (\boldsymbol{\nu} + P_1'\mathbf{Z})'(\boldsymbol{\nu} + P_1'\mathbf{Z}) = \mathbf{X}'\mathbf{X},$$

so that the desired result follows.
For "only if" part, suppose $Q = \mathbf{Y}'W\mathbf{Y} \sim ES\chi^2_m(\lambda, \delta_1, \delta_2, \gamma)$, we have

$$\begin{aligned} M_Q(t) &= \frac{2}{2+\gamma}\frac{\exp\left\{t\boldsymbol{\mu}'W\boldsymbol{\mu} + 2t^2\boldsymbol{\mu}'WB'(I_k - 2tBWB')^{-1}BW\boldsymbol{\mu}\right\}}{|I_k - 2tBWB'|^{\frac{1}{2}}} \\ &\quad \times\left[1 + \gamma\Phi\left(\frac{2t\boldsymbol{\alpha}'(I_k - 2tBWB')^{-1}BW\boldsymbol{\mu}}{\left[1 + \boldsymbol{\alpha}'(I_k - 2tBWB')^{-1}\boldsymbol{\alpha}\right]^{\frac{1}{2}}}\right)\right] \\ &= \frac{2}{2+\gamma}\frac{\exp\left\{t(1-2t)^{-1}\lambda\right\}}{(1-2t)^{k/2}}\left[1 + \gamma\Phi\left(\frac{2t(1-2t)^{-1}\delta_1}{\left[1 + (1-2t)^{-1}\delta_2\right]}\right)\right], \end{aligned} \tag{10}$$

for all t such that $2|t| < min(1, \sigma(BWB'))$.
Note that let $\boldsymbol{\alpha} = 0$ or $\gamma = 0$,

$$\frac{\exp\left\{t\boldsymbol{\mu}'W\boldsymbol{\mu} + 2t^2\boldsymbol{\mu}'WB'(I_k - 2tBWB')^{-1}BW\boldsymbol{\mu}\right\}}{|I_k - 2tBWB'|^{\frac{1}{2}}} = \frac{\exp\left\{t(1-2t)^{-1}\lambda\right\}}{(1-2t)^{k/2}}.$$

By Lemma 3, we get (i) and (iii).
By (10), we obtain

$$\frac{2t\boldsymbol{\alpha}'(I_k - 2tBWB')^{-1}BW\boldsymbol{\mu}}{\left[1+\boldsymbol{\alpha}'(I_k - 2tBWB')^{-1}\boldsymbol{\alpha}\right]^{\frac{1}{2}}} = \frac{2t(1-2t)^{-1}\delta_1}{\left[1+(1-2t)^{-1}\delta_2\right]},$$

for all $t \in \Re$.
For $t = 0$,

$$\frac{\boldsymbol{\alpha}'BW\boldsymbol{\mu}}{(1+\boldsymbol{\alpha}'\boldsymbol{\alpha})^{1/2}} = \frac{\delta_1}{(1+\delta_2)^{1/2}}. \tag{11}$$

For $t \neq 0$, by Taylor expansion, the coefficient of t term should be equal to each other,

$$\boldsymbol{\alpha}'BW\boldsymbol{\mu}\delta_2^{\frac{1}{2}} = \left[\boldsymbol{\alpha}'BWB'(BWB')\boldsymbol{\alpha}\right]^{\frac{1}{2}} = \left(\boldsymbol{\alpha}'P_1P_1'\boldsymbol{\alpha}\right)^{\frac{1}{2}}\delta_1. \tag{12}$$

Compare (11) with (12), we have $\delta_1 = \boldsymbol{\mu}'WB'\boldsymbol{\alpha}/c$ and $\delta_2 = \boldsymbol{\alpha}'BWB'\boldsymbol{\alpha}/c^2$, as desired. □

Example 2 Consider the one-way balance random effect model with extended skew-normal random errors given by

$$\mathbf{Y} = \mu\mathbf{1}_n + (I_a \otimes \mathbf{1}_m)\mathscr{E}_1 + \mathscr{E}_0, \tag{13}$$

where $n = am$, Y is a n-dimensional random variable, $\mu \in \Re$, $\mathscr{E}_1 \sim N_a(0, \sigma_a^2 I_a)$, $\mathscr{E}_0 \sim E\mathbb{S}N_n(0, \sigma_0^2 I_n, \mathbf{1}_n\alpha, \gamma)$ with $\alpha \in \Re$, $\gamma \geq -1$, and $\mathscr{E}_1$ and $\mathscr{E}_0$ are independent.
Consider quadratic forms of $\mathbf{Y}$:

$$q_i = \mathbf{Y}'W_i\mathbf{Y}, \quad i = 1, 2, \quad W_1 = I_a \otimes \bar{J}_m, \quad \text{and} \quad W_2 = I_a \otimes (I_m - \bar{J}_m).$$

For testing $H_0 : \sigma_a^2 = 0$ vs $H_1 : \sigma_a^2 \neq 0$, we can rewrite it into an equivalent form

$$H_0 : \sigma^2 = \sigma_0^2 \quad vs \quad H_1 : \sigma^2 \neq \sigma_0^2.$$

Then under H_0, we know that

$$\mathbf{Y} \sim E\mathbb{S}N_n(\mathbf{1}_n\mu, \sigma^2 I_n, \mathbf{1}_n\alpha, \gamma).$$

(i) For the distribution of q_1, note that $W_1^2 = W_1$, $r(W_1) = a$ and $P_2P_2' = I_n - W_1$, we obtain $c = 1$, $\lambda = \mu\mathbf{1}_n'W_1\mathbf{1}_n\mu = n\mu^2$, $\delta_1 = \mu\mathbf{1}_n'W_1\mathbf{1}_n\alpha = n\alpha\mu$, and $\delta_2 = \alpha^2\mathbf{1}_n'W_1\mathbf{1}_n\mathbf{1}_n = n\alpha^2$.
Therefore, by Theorem 5,

$$q_1 \sim ES\chi_a^2(n\mu^2, n\alpha\mu, n\alpha^2, \gamma).$$

(ii) Similarly, $W_2^2 = W_2$, $r(W_2) = n - a$ and $P_2 P_2' = I_n - W_2$, we obtain $c = \sqrt{1 + n\alpha^2}$, $\lambda = \mu \mathbf{1}_n' W_2 \mathbf{1}_n \mu = 0$, $\delta_1 = \mu \mathbf{1}_n' W_2 \mathbf{1}_n \alpha / c = 0$ and $\delta_2 = \alpha^2 \mathbf{1}_n' W_2 \mathbf{1}_n \mathbf{1}_n / c^2 = 0$.
i.e,

$$q_2 \sim ES\chi^2_{n-a}(0, 0, 0, \gamma),$$

which is χ^2_{n-a}, chi-square distribution of $n - a$ degrees of freedom.
(iii) Since $B = I_n$, we obtain $W_1 B' B W_2 = 0$ and $W_2 B' \mathbf{1}_n \alpha = 0$. By Corollary 1, we know, $W_1 \mathbf{Y}$ and $W_2 \mathbf{Y}$ are independent and hence q_1 and q_2 are independent. □

Example 3 Let $\mathbf{Y} \sim E\mathbb{S}N_4(\mathbf{1}_4\mu, B, \mathbf{1}_4\alpha, \gamma)$ and $q = \mathbf{Y}' W \mathbf{Y}$, where $\mu \in \Re$, $W = I_2 \otimes \bar{J}_2$ and

$$B = \frac{1}{3}\begin{pmatrix} 1 & 2 & 0 & 0 \\ 2 & 1 & 0 & 0 \\ 0 & 0 & 1 & 2 \\ 0 & 0 & 2 & 1 \end{pmatrix}.$$

Note that

$$BWB' = \frac{1}{2}\begin{pmatrix} 1 & 1 & 0 & 0 \\ 1 & 1 & 0 & 0 \\ 0 & 0 & 1 & 1 \\ 0 & 0 & 1 & 1 \end{pmatrix},$$

which is idempotent of rank 2, so that the condition (i) of Theorem 5 holds. Also the orthogonal matrix P in Theorem 5 is

$$P = \begin{pmatrix} \frac{\sqrt{2}}{2} & 0 & \frac{\sqrt{2}}{2} & 0 \\ \frac{\sqrt{2}}{2} & 0 & -\frac{\sqrt{2}}{2} & 0 \\ 0 & \frac{\sqrt{2}}{2} & 0 & \frac{\sqrt{2}}{2} \\ 0 & \frac{\sqrt{2}}{2} & 0 & -\frac{\sqrt{2}}{2} \end{pmatrix} \quad with \quad P_2 = \begin{pmatrix} \frac{\sqrt{2}}{2} & 0 \\ -\frac{\sqrt{2}}{2} & 0 \\ 0 & \frac{\sqrt{2}}{2} \\ 0 & -\frac{\sqrt{2}}{2} \end{pmatrix},$$

we obtain $c = \sqrt{1 + \alpha \mathbf{1}_4' P_2 P_2' \mathbf{1}_4 \alpha} = 1$. For the calculation of parameters we have

$$\mathbf{1}_4' W \mathbf{1}_4 = 4, \quad \mathbf{1}_4' W B' B W \mathbf{1}_4 = 4, \quad \mathbf{1}_4' W B' \mathbf{1}_4 = 4, \quad \text{and} \quad \mathbf{1}_4' B W B' \mathbf{1}_4 = 4$$

so that

$$\lambda = 4\mu^2, \qquad \delta_1 = 4\alpha\mu, \quad \text{and} \quad \delta_2 = 4\alpha^2.$$

Therefore, by Theorem 5,

$$q \sim ES\chi_2^2(4\mu^2, 4\alpha\mu, 4\alpha^2, \gamma). \quad \square$$

References

1. Azzalini, A.: A class of distributions which includes the normal ones. Scand. J. Stat. **12**, 171–178 (1985)
2. Azzalini, A., Dalla, A.: The multivariate skew-normal distribution. Biometrika **83**(4), 715–726 (1996)
3. Azzalini, A., Capitanio, A.: Statistical applications of the multivariate skew normal distribution. J. R. Stat. Soc.: Ser. B (Stat. Methodol.) **61**(3), 579–602 (1999)
4. Arellano-Valle, R., Ozan, S., Bolfarine, H., Lachos, V.: Skew normal measurement error models. J. Multivar. Anal. **96**(2), 265–281 (2005)
5. Li, B., Wang, T., Tian, W.: In: Huynh, V., Krennovich, V., Sriboonchitta, S., Suriya, K. (eds.) Uncertainty Analysis in Econometrics with Applications. Risk measures and asset pricing models with new versions of Wang transform, pp. 155–166. Springer, Berlin (2013)
6. Chen, J., Gupta, A.: Matrix variate skew normal distributions. Statistics **39**(3), 247–253 (2005)
7. Corns, T.R.A., Satchell, S.E.: Modelling conditional heteroskedasticity and skewness using the skew-normal distribution one-sided coverage intervals with survey data. Metron **LXVIII**, 251C263 (2010). [224]
8. De Luca, G., Loperfido, N.M.R.: A skew-in-mean GARCH model Chap. 12, pages 205C222. In: Genton, M.G. (ed.) Skew-elliptical Distributions and their Applications: A Journey Beyond Normality. Chapman and Hall/CRC, Boca Raton (2004). [224]
9. De Luca, G., Genton, M.G., Loperfido, N.: A multivariate skew-GARCH model. Adv. Economet. **20**, 33C57 (2005). [224]
10. Gupta, A., Huang, W.: Quadratic forms in skew normal variates. J. Math. Anal. Appl. **273**(2), 558–564 (2002)
11. Genton, M., He, L., Liu, X.: Moments of skew-normal random vectors and their quadratic forms. Stat. Probab. Lett. **51**(4), 319–325 (2001)
12. Huang, W., Chen, Y.: Generalized skew-Cauchy distribution. Stat. Probab. Lett. **77**(11), 1137–1147 (2007)
13. Huang, W., Chen, Y.: Quadratic forms of multivariate skew normal-symmetric distributions. Stat. Probab. Lett. **76**(9), 871–879 (2006)
14. Kumar, C., Anusree, M.: On an extended version of skew generalized normal distribution and some of its properties. Commun. Stat.-Theory Methods **44**(3), 573–586 (2015)
15. Loperfido, N.: Quadratic forms of skew-normal random vectors. Stat. Probab. Lett. **54**(4), 381–387 (2001)
16. Muirhead, R.: Aspects of Multivariate Statistical Theory. Wiley, New York (2009)
17. Szkely, G., Rizzo, M.: A new test for multivariate normality. J. Multivar. Anal. **93**(1), 58–80 (2005)
18. Wang, T., Li, B., Gupta, A.: Distribution of quadratic forms under skew normal settings. J. Multivar. Anal. **100**(3), 533–545 (2009)
19. Tian, W., Wang, T., Li, B.: Risk Measures with Wang Transforms under flexible Skew-generalized Settings. J. Intell. Technol. Appl. Stat. **7**(2), 107–145 (2014)
20. Ye, R., Wang, T., Gupta, A.: Distribution of matrix quadratic forms under skew-normal settings. J. Multivar. Anal. **131**, 229–239 (2014)
21. Ye, R., Wang, T.: Inferences in linear mixed models with skew-normal random effects. Acta Mathematica Sinica, English Series **31**(4), 576–594 (2015)
22. Ye, R., Wang, T., Sukparungsee, S., Gupta, A.: Tests in variance components models under skew-normal settings. Metrika, 1–20 (2015)
23. Zacks, S.: Parametric Statistical Inference: Basic Theory and Modern Approaches. Elsevier, Philadelphia (2014)

Multiple Copula Regression Function and Directional Dependence Under Multivariate Non-exchangeable Copulas

Zheng Wei, Tonghui Wang and Daeyoung Kim

Abstract In this paper, the multiple directional dependence between response variable and covariates using non-exchangeable copulas based regression is introduced. The general measure for the multiple directional dependence in the joint behavior is provided. Several multivariate non-exchangeable copula families including skew normal copula, and the generalized Farlie-Gumbel-Morgenstern copula models are investigated. For the illustration of main results, several examples are given.

1 Introduction

The copulas have been increasingly popular for modeling multivariate data as they account for the dependence structure and provide flexible representations of the multivariate distribution. And the copula based modeling has been extensively applied to many areas including actuarial sciences, finance, neuroscience, and weather research (e.g. Sriboonchitta et al. [14, 15]). The notion of copula has been introduced by Sklar [13]. It has become one of the most significant new tool in statistical modeling. Copulas, multivariate distributions with standard uniform marginals, contain the most of the multivariate dependence structure properties and do not depend on the marginals of the variables. Standard references for a detailed overview of copula theory, see Joe [6], Nelsen [12], Wei et al. [22], and Wei et al. [21].

The analysis of directional dependence in nonexperimental study has been applied in various research contexts such as exchange rates (e.g. Dodge and Rousson [4]),

Z. Wei · D. Kim
Department of Mathematics and Statistics, University of Massachusetts Amherst, Amherst, MA, USA
e-mail: wei@math.umass.edu

D. Kim
e-mail: daeyoung@math.umass.edu

Z. Wei · T. Wang (✉)
Department of Mathematical Sciences, New Mexico State University, Las Cruces, NM, USA
e-mail: twang@nmsu.edu

V.-N. Huynh et al. (eds.), *Causal Inference in Econometrics*,
Studies in Computational Intelligence 622, DOI 10.1007/978-3-319-27284-9_10

deficit hyperactivity disorder (e.g. Kim and Kim [9]), gene networks from gene expression data (Kim et al. [8]), and the development of aggression in adolescence (Von Eye and Wiedermann [19]; Kim and Kim [7]). The directional dependence is an asymmetric dependence/interaction between the variables. Sungur [16, 17] argued that the best way of understanding the directional dependence between variables is to study it through the copula regression model. As noted in Sungur [16], in order to study the directional dependence stemming from the joint behavior of the variables of interest, the non-exchangeable copulas should be utilized. This is because the exchangeable copulas describes only directional dependence originating from the marginals of the variables.

However, it is challenging to construct the multivariate non-exchangeable copulas that allows delineation of the asymmetric dependence among multivariate variables. In this paper we propose two new classes of multivariate non-exchangeable copula families. The first is the skew normal copula based on the class of skew normal distributions (Azzalini [1–3]). The second is the generalized Farlie-Gumbel-Morgenstern (FGM) copulas extended from the works of Úbeda-Flores [18]. We illustrate the application of the proposed multivariate non-exchangeable copulas to examples of multiple regression-based directional dependence.

This paper is organized as follows. The multivariate copula based directional dependence and the multiple directional dependence measures, and their properties are given in Sect. 2. Multivariate skew normal copulas are provided in Sect. 3.1. Multivariate non-exchangeable copulas, which generalize the FGM copulas, are proposed in Sect. 3.2.

2 Multivariate Copula Based Directional Dependence

2.1 *Multivariate Non-exchangeable Copulas*

In order to investigate the directional dependence for multivariate data through copulas, we will first define subcopulas (Nelsen [12]) and then define copulas as subcopulas with the unit hypercube domain. Then we introduce an exchangeability property commonly assumed in the copula and show its limitation in terms of the conditional distribution of the copula.

Definition 1 A *k-dimensional* **copula** is a function $C : [0,1]^k \mapsto [0,1]$, satisfying following properties:

(a) C is grounded, i.e., if at least one $u_i = 0$, $C(u_1, \ldots, u_k) = 0$;

(b) For every $u_i \in [0,1]$, $C_i(u_i) \equiv C(1, \ldots, 1, u_i, 1, \ldots, 1) = u_i$;

(c) C is k-increasing in the sense that, for any $J = \prod_{i=1}^{k}[u_i, v_i] \subseteq [0,1]^k$ with $u_i, v_i \in [0,1]$,

$$volC(J) = \sum_{\boldsymbol{a}} sgn(\boldsymbol{a})C(\boldsymbol{a}) \geq 0,$$

where the summation is over all vertices $\boldsymbol{a}$ of J, and for $\boldsymbol{a} = (a_1, \ldots, a_k)^T$, with $(a_1, \ldots, a_k)^T$ is the transpose of $(a_1, \ldots, a_k)^T$, and $a_i = u_i$ or v_i,

$$sgn(\boldsymbol{a}) = \begin{cases} 1, & \text{if} \quad a_i = v_i \quad \text{for an even number of } i's, \\ -1, & \text{if} \quad a_i = v_i \quad \text{for an odd number of } i's. \end{cases}$$

Let $\boldsymbol{X} = (X_0, X_1, \ldots, X_k)^T$ be a $(k+1)$-dimensional random vector having the joint cumulative distribution function (CDF) $H(x_0, x_1, \ldots, x_k)$, and marginal CDF's $F_0(x_0)$, $F_1(x_1)$, ..., $F_k(x_k)$. Sklar's theorem (Sklar [13]) states that if the marginals of $\boldsymbol{X}$ are continuous, then there exist a unique copula C such that $H(x_0, x_1, \ldots, x_k) = C(F_0(x_0), F_1(x_1), \ldots, F_k(x_k))$. One can easily see that the copula C of $\boldsymbol{X}$ represents information on dependence structures of H on a quantile scale: for the random vector $\boldsymbol{X}$ with continuous margins, $C(u_0, u_1, \ldots, u_k) = H(F_0^{-1}(u_0), F_1^{-1}(u_1), \ldots, F_k^{-1}(u_k))$ where F^{-1} denotes the generalized inverse function of F such that $F^{-1}(u) = \inf\{x : F(x) \geq u\}$.

For an absolutely continuous copula C, the copula density is defined as

$$c(u_0, u_1, \ldots, u_k) = \frac{\partial^{k+1} C(u_0, u_1, \ldots, u_k)}{\partial u_0 \partial u_1 \cdots \partial u_k}.$$

Given a random vector $\boldsymbol{X}$, with an absolute continuous H with strictly increasing continuous marginals $F_0(x_0), F_1(x_1), \ldots, F_k(x_k)$, the copula density defined above is given by

$$c(u_0, u_1, \ldots, u_k) = \frac{h(F_0^{-1}(u_0), F_1^{-1}(u_1), \ldots, F_k^{-1}(u_k))}{f_0(F_0^{-1}(u_0)) f_1(F_1^{-1}(u_1)) \ldots f_k(F_k^{-1}(u_k))}, \tag{1}$$

where $h(x_0, x_1, \ldots, x_k)$ is the joint density of $H(x_0, x_1, \ldots, x_k)$ and $f_0, f_1, \ldots, f_k$ are the marginal densities.

The conditional distribution of copulas can often provide the insightful information on dependence structures between the variables of interest.

Remark 1 For a $(k+1)$-dimensional random vector $\boldsymbol{X} = (X_0, X_1, \ldots, X_k)^T$ with the corresponding copula $C(u_0, u_1, \ldots, u_k)$ where $U_i = F_i(x_i)$ and $i = 0, 1, \ldots, k$, the conditional distribution of X_0 given $X_1, \ldots, X_k$ is given by,

$$\begin{aligned} F_{X_0|X_1,\ldots,X_k}(x_0 \mid x_1, \ldots, x_k) &\equiv P(X_0 \leq x_0 \mid X_1 = x_1, \ldots, X_k = x_k) \\ &= \left. \frac{\partial^k C(u_0, u_1, \ldots, u_k)}{\partial u_1 \cdots \partial u_k} \right|_{u_0 = F_0(x_0), u_1 = F_1(x_1), \ldots, u_k = F_k(x_k)}. \end{aligned}$$

For a $(k+1)$ standard uniform variables U_0 and $\boldsymbol{U} = (U_1, \ldots, U_k)^T$ with a $(k+1)$-copula C, the conditional distribution of U_0 given $\boldsymbol{U} = \boldsymbol{u}$ is given by,

$$C_{U_0|U}(u_0) \equiv P(U_0 \leq u_0|\boldsymbol{U} = \boldsymbol{u}) = \frac{\partial^k C(u_0, u_1, \ldots, u_k)}{\partial u_1 \cdots \partial u_k},$$

where $\boldsymbol{u} = (u_1, \ldots, u_k)^T$.

For the bivariate continuous random vector $(X_0, X_1)^T$ with $(U_0, U_1)^T \equiv (F_0(X_0), F_1(X_1))^T$, the conditional distribution of the corresponding copula $C_{U_0|U_1}(u_0)$ implies the probability that X_0 is less than or equal to its u_0th quantile given that X_1 reaches its u_1th quantile.

An assumption commonly made in the copula for applied research is the exchangeable (or symmetric) dependence structure between the variables of interest, and several families of exchangeable copulas including Archimedean copulas and all meta-elliptical copulas are widely used in the literatures (Nelsen [12]).

Definition 2 A $(k+1)$-dimensional random vector $\boldsymbol{X} = (X_0, X_1, \ldots, X_k)^T$ is **exchangeable** if the joint distribution of $\boldsymbol{X}$ is the same as the joint distribution of $\boldsymbol{X}_\sigma = (X_{\sigma(0)}, X_{\sigma(1)}, \ldots, X_{\sigma(k)})^T$ for any permutation $\sigma \in \Gamma$ where Γ denotes the set of all permutations on the set $\{0, 1, \ldots, k\}$. A $(k+1)$-copula C is **exchangeable** if it is the distribution function of a $(k+1)$-dimensional exchangeable uniform random vector $\boldsymbol{U} = (U_0, U_1, \ldots, U_k)^T$ satisfying $C(u_0, u_1, \ldots, u_k) = C(u_{\sigma(0)}, u_{\sigma(1)}, \ldots, u_{\sigma(k)})$.

The exchangeability assumption imposes a strong symmetry property on the copula describing the dependence structure of the data. For example, if the copula of the bivariate random vector (X_0, X_1) is exchangeable, then $C_{U_0|U_1}(u_0) \equiv P(U_0 \leq u_0|U_1 = u_1)$ is equal to $C_{U_1|U_0}(u_1) \equiv P(U_1 \leq u_1|U_0 = u_0)$. That is, the probability that X_1 is larger than its u_1th quantile given that X_0 reaches its u_0th quantile is equal to the probability computed the other way around.

2.2 Directional Dependence Using Copula-Based Multiple Regression

The approach to the directional dependence using non-exchangeable copula developed in Sungur [16, 17] is limited to the bivariate regression case. However, in practice, the problem of directional dependence often occur in the setting of multivariate data. In the following, we will extend the method of the directional dependence using copulas to multiple regression problems.

Definition 3 For a $(k+1)$-copula $C(u_0, u_1, \ldots, u_k)$ of U_0 and $\boldsymbol{U}$ with $\boldsymbol{U} = (U_1, \ldots, U_k)^T$, the **copula-based multiple regression function** of U_0 on $\boldsymbol{U} = \boldsymbol{u}$ is defined by

$$r^C_{U_0|U}(\boldsymbol{u}) \equiv E(U_0|\boldsymbol{U}=\boldsymbol{u}) = \int_0^1 u_0 \frac{c(u_0, u_1, \ldots, u_k)}{c_U(\boldsymbol{u})} du_0$$
$$= 1 - \frac{1}{c_U(\boldsymbol{u})} \int_0^1 C_{U_0|U}(u_0) du_0, \tag{2}$$

where $C_U(\boldsymbol{u}) = C(1, u_1, \ldots, u_k)$ is the marginal distribution of $\boldsymbol{U}$, and

$$c(u_0, u_1, \ldots, u_k) = \frac{\partial^{k+1} C(u_0, u_1, \ldots, u_k)}{\partial u_0 \partial u_1 \cdots \partial u_k}, \qquad c_U(\boldsymbol{u}) = \frac{\partial^k C_U(\boldsymbol{u})}{\partial u_1 \cdots \partial u_k},$$
$$C_{U_0|U}(u_0) = \frac{\partial^k C(u_0, u_1, \ldots, u_k)}{\partial u_1 \cdots \partial u_k}$$

are the joint copula density of U_0 and $\boldsymbol{U}$, the marginal copula density of $\boldsymbol{U}$ and the conditional distribution of U_0 given $\boldsymbol{U}$, respectively.

In multiple regression problems for the multivariate data, there are two types of setting where one can consider the directional dependence (Wiedermann and von Eye [23]). The first setting is where all variables of interest can be conceptually the outcome variable due to no a priori distinction between outcome variables and predictors. The second setting is where some variables are theoretically known to be predictors of the other variables, and so there is need to adjust the effects of the predictors on the directional dependence between the other variables of interest.

Accordingly, we define the directional dependence in joint behavior for the $(k+1)$ uniform random variates U_0 and $\boldsymbol{U} = (U_1, \ldots, U_d, \ldots, U_k)^T$ with the corresponding copula $C(u_0, u_1, \ldots, u_d, \ldots, u_k)$.

Definition 4 The $(k + 1)$ uniform random variates U_0 and $\boldsymbol{U}$ have no multiple directional dependence in joint behavior if the $(k + 1)$ copula-based multiple regression functions $r^C_{U_{\sigma(0)}|U_\sigma}(\boldsymbol{u}_\sigma)$ for all permutations $\sigma \in \Gamma$ are all equal to each other where Γ denotes the set of all permutations on the set $\{0, 1, \ldots, k\}$ and $U_\sigma = (U_{\sigma(1)}, \ldots, U_{\sigma(k)})^T$.

From the definitions above, we can say that (conditional) directional dependence in joint behavior exists if there are at least two copula-based multiple regression functions which differ. As an example of the Definition 4, for the three uniform variates (U_0, U_1, U_2),

$$r^C_{U_0|(U_1,U_2)=(\omega_1,\omega_2)}(\omega_1, \omega_2) = r^C_{U_1|(U_0,U_2)=(\omega_1,\omega_2)}(\omega_1, \omega_2) = r^C_{U_2|(U_0,U_1)=(\omega_1,\omega_2)}(\omega_1, \omega_2)$$

implies no directional dependence in joint behavior where

$$r^C_{U_0|U_1,U_2} = 1 - \frac{1}{\frac{\partial^2 C(u_0=1,u_1,u_2)}{\partial u_1 \partial u_2}} \int_0^1 \frac{\partial^2 C(u_0,u_1,u_2)}{\partial u_1 \partial u_2} du_0,$$

$$r^C_{U_1|U_0,U_2} = 1 - \frac{1}{\frac{\partial^2 C(u_0,u_1=1,u_2)}{\partial u_0 \partial u_2}} \int_0^1 \frac{\partial^2 C(u_0,u_1,u_2)}{\partial u_0 \partial u_2} du_1,$$

$$r^C_{U_2|U_0,U_1} = 1 - \frac{1}{\frac{\partial^2 C(u_0,u_1,u_2=1)}{\partial u_0 \partial u_1}} \int_0^1 \frac{\partial^2 C(u_0,u_1,u_2)}{\partial u_0 \partial u_1} du_2.$$

Remark 2 For the multivariate exchangeable copulas satisfying $C(u_0, u_1, \ldots, u_k) = C(u_{\sigma(0)}, u_{\sigma(1)}, \ldots, u_{\sigma(k)})$ for all permutations σ, there is no directional dependence because all copula-based multiple regression functions of interest are the same. In order to detect the directional dependence in joint behavior for the data, therefore, one needs to consider using the multivariate non-exchangeable copulas. We will focus on the development of multivariate non-exchangeable copulas in Sect. 3.

The following proposition gives a basic property of the copula-based multiple regression function for uniform random variates, which will be useful for further development.

Proposition 1 *Let $r^C_{U_0|\boldsymbol{U}}(\boldsymbol{u})$ be the copula-based multiple regression function of U_0 on $\boldsymbol{U}$. Then*

$$E[r^C_{U_0|\boldsymbol{U}}(\boldsymbol{U})] = \frac{1}{2}.$$

Proof From the definition of copula-based multiple regression function, we have

$$\begin{aligned}
E[r^C_{U_0|\boldsymbol{U}}(\boldsymbol{U})] &= \int_{[0,1]^k} \left(1 - \frac{1}{c_{\boldsymbol{U}}(\boldsymbol{u})} \int_0^1 C_{U_0|\boldsymbol{U}}(u_0)\right) c_{\boldsymbol{U}}(\boldsymbol{u}) du_0 d\boldsymbol{u} \\
&= 1 - \int_{[0,1]^k} \left(\int_0^1 C_{U_0|\boldsymbol{U}}(u_0)\right) du_0 d\boldsymbol{u} \\
&= 1 - \int_0^1 \int_{[0,1]^k} \frac{\partial^k C(u_0, u_1, \ldots, u_k)}{\partial u_1 \cdots \partial u_k} d\boldsymbol{u} du_0 \\
&= 1 - \int_0^1 [C(u_0, 1, \ldots, 1) - C(u_0, 0, \ldots, 0)] du_0 \\
&= 1 - \int_0^1 u_0 du_0 = \frac{1}{2}.
\end{aligned}$$

□

In order to quantitatively measure the directional dependence given the copula-based multiple regression, say $r^C_{U_0|U}(\boldsymbol{\omega})$, we propose the following measure,

$$\rho^2_{(U\to U_0)} = \frac{Var(r^C_{U_0|U}(\boldsymbol{U}))}{Var(U_0)} = \frac{E[(r^C_{U_0|U}(\boldsymbol{U})-1/2)^2]}{1/12} = 12E[(r^C_{U_0|U}(\boldsymbol{U}))^2]-3. \quad (3)$$

Note that the form of the copula-based multiple regression used to compute the measure in Eq. (3) depends on Definition 4. If $U_0, U_1, \ldots, U_k$ are independent, then all copula-based multiple regression functions are equal to 1/2 (e.g., $r^C_{U_0|U}(\boldsymbol{u}) = 1/2$) and so the corresponding directional dependence measures in Eq. (3) are all zero.

Remark 2.2. The idea of the measure in Eq. (3) is similar to the idea of the multiple coefficient of correlation $R^2 = \frac{SSR}{SST}$ under the multiple regression model designed to measure the proportion of regression sum of squares (SSR) and total sum of squares (SST), of a dependent variable given independent variables. Thus, the proposed measure in Eq. (3) can also be interpreted as the proportion of variance of a tentative outcome variable explained by the other variables and so they can be used to compare the predictive powers of given copula-based regression functions.

3 Multivariate Non-exchangeable Copulas and Their Application to Directional Dependence

In this section we propose two new multivariate non-exchangeable copulas, skew normal copula and new generalization of Farlie-Gumbel-Morgenstern (FGM) copulas. We then show their application to the directional dependence in the framework of copula-based multiple regression.

3.1 Skew Normal Copulas

The class of skew normal distributions together with its extensions and applications have been an important topic in past three decades, see Azzalini [2, 3], Gupta et al. [5], Wang et al. [20], and Ye et al. [24]. A random variable X is said to have the skew normal distribution with the parameter $\lambda \in \mathbb{R}$, denoted by $X \sim SN(\lambda)$, if its probability density function is

$$f(x;\lambda) = 2\phi(x)\Phi(\lambda x), \qquad x \in \mathbb{R}, \quad (4)$$

where $\phi(\cdot)$ and $\Phi(\cdot)$ are the probability density function and the CDF of a standard normal variable. The following result of Azzalini [3] will be useful in building the multivariate skew normal distribution,

Lemma 1 *If Z_0 and Z_1 are independent standard normal random variables with mean 0 and variance 1, denoted by $N(0, 1)$, and $\delta \in (-1, 1)$, then*

$$X = \delta|Z_0| + \sqrt{1-\delta^2}Z_1, \tag{5}$$

is $SN(\lambda)$ where $\lambda = \delta/\sqrt{1-\delta^2}$.

Let $\mathbf{Z} = (Z_0, \ldots, Z_k)^T$ be the multivariate normal distributed random vector with standard normal marginals, independent of $Z \sim N(0, 1)$. The joint distribution of Z and $\mathbf{Z}$ is given by

$$\begin{pmatrix} Z \\ \mathbf{Z} \end{pmatrix} = N_{k+2}\left(\mathbf{0}, \begin{pmatrix} 1 & \mathbf{0}^T \\ \mathbf{0} & R \end{pmatrix}\right), \tag{6}$$

where $N_k(\boldsymbol{\mu}, \Sigma)$ is the k-dimensional multivariate normal distribution with mean vector $\boldsymbol{\mu}$ and covariance matrix Σ, and R is a $(k+1)\times(k+1)$-dimensional correlation matrix of $\mathbf{Z}$. Consider random variables $X_0, X_1, \ldots, X_k$ given by

$$X_j = \delta_j|Z| + \sqrt{1-\delta_j^2}Z_j, \tag{7}$$

where $j = 0\ldots, k$ and $\delta_j \in (0, 1)$. By Lemma 1, $X_j \sim SN(\lambda_j)$ where $\lambda_j = \delta_j/\sqrt{1-\delta_j^2}$. Also, the joint density function of $\boldsymbol{X} = (X_0, X_1, \ldots, X_k)^T$ (see Azzalini [3]) is

$$h(\boldsymbol{x}) = 2\phi_{k+1}(\boldsymbol{x}; \Sigma)\Phi(\boldsymbol{\alpha}^T\boldsymbol{x}), \tag{8}$$

where

$$\boldsymbol{\alpha}^T = \frac{\boldsymbol{\lambda}^T R^{-1}\Delta^{-1}}{(1+\boldsymbol{\lambda}^T R^{-1}\boldsymbol{\lambda})^{1/2}}, \qquad \Sigma = \Delta(R + \boldsymbol{\lambda}\boldsymbol{\lambda}^T)\Delta,$$

$$\Delta = diag((1-\delta_0^2)^{1/2}, \ldots, (1-\delta_k^2)^{1/2}), \qquad \boldsymbol{\lambda} = (\lambda_0, \ldots, \lambda_k)^T,$$

and $\phi_{k+1}(\boldsymbol{x}; \Sigma)$ denotes the density function of $N_{k+1}(\mathbf{0}, \Sigma)$. The random vector $\boldsymbol{X}$ with the joint probability density function in Eq (8) is said to be the multivariate skew normal distribution with the location parameter $\mathbf{0}$, scale parameter Σ, and the skewness parameter $\boldsymbol{\alpha}$, denoted by $\boldsymbol{X} \sim SN_{k+1}(\Sigma, \boldsymbol{\alpha})$.

We now define the multivariate non-exchangeable skew normal copula using the multivariate skew normal distribution defined above.

Definition 5 A $(k+1)$-copula C is said to be a **skew normal copula** if

$$C(u_0, u_1, \ldots, u_k) = H\left(F_0^{-1}(u_0; \lambda_0), F_1^{-1}(u_1; \lambda_1), \ldots, F_k^{-1}(u_k; \lambda_k); \Sigma, \boldsymbol{\alpha}\right), \tag{9}$$

where $F_j^{-1}(u_j; \lambda_j)$ denotes the inverse of $F_j(x_j, \lambda_j)$ with $X_j \sim SN(\lambda_j)$, Σ and $\boldsymbol{\alpha}$ are given in (8), and H is the CDF of $SN_{k+1}(\Sigma, \boldsymbol{\alpha})$.

Using the definition of the copula density in Eq (1), the multivariate non-exchangeable skew normal copula density is given by

$$c(u_0, u_1, \ldots, u_k) = \frac{h(F_0^{-1}(u_0; \lambda_0), F_1^{-1}(u_1; \lambda_1), \ldots, F_k^{-1}(u_k; \lambda_k); \Sigma, \boldsymbol{\alpha})}{f_0(F_0^{-1}(u_0; \lambda_0)) f_1(F_1^{-1}(u_1; \lambda_1)) \ldots f_k(F_k^{-1}(u_k; \lambda_k))}. \quad (10)$$

where $h(x_0, x_1, \ldots, x_k; \Sigma, \boldsymbol{\alpha})$ is the joint skew normal density in Eq. (8) and $f_j(x_j; \lambda_j)$ is the marginal density of skew normal variables, $X_j \sim SN(\lambda_j)$.

Remark 3 It is easy to see that the skew normal copula (9) is exchangeable if and only if $\delta_0 = \delta_1 = \ldots = \delta_k$ and all off diagonal elements of R equal each other. Therefore, the skew normal copula can be used for the analysis of directional dependence unless the condition of exchangeability above described holds.

Remark 4 The regression function based on the non-exchangeable skew normal copula in Eq. (9). For the skew normal copula given in Definition 5, the copula regression function of U_0 given $\boldsymbol{U} = \boldsymbol{u}$ is

$$r^C_{U_0|U}(\boldsymbol{u}) = 1 - \frac{K}{c_U(\boldsymbol{u})} \int_0^1 \int_{-\infty}^{F_0^{-1}(u_0;\lambda_0)} \phi_{k+1}(x_0, \boldsymbol{w}; (\Sigma^{-1} - \tilde{I}_{k+1})^{-1}) \quad (11)$$

$$\times \frac{\Phi(\alpha_0 x_0 + \sum_{i=1}^{k} \alpha_i w_i)}{\prod_{i=1}^{k} \Phi(\lambda_i w_i)} dx_0 du_0,$$

where

$$\boldsymbol{w} = (w_1, \ldots, w_k)^T = (F_1^{-1}(u_1; \lambda_1), \ldots, F_k^{-1}(u_k; \lambda_k))^T,$$

$K = \frac{(2\pi)^{k/2}\sqrt{|(\Sigma^{-1} - \tilde{I}_{k+1})^{-1}|}}{2^{k-1}\sqrt{|\Sigma|}}$, $\tilde{I}_{k+1} = diag(0, I_k)$, and I_k is the $k \times k$ identity matrix.

Example 1 The density function of $\boldsymbol{X} = (X_0, X_1, X_2)^T$ in (8) is

$$h(x_0, x_1, x_2) = 2\phi_3(x_0, x_1, x_2; \Sigma)\Phi(\alpha_0 x_0 + \alpha_1 x_1 + \alpha_2 x_2), \quad (12)$$

where Σ and $\boldsymbol{\alpha} = (\alpha_0, \alpha_1, \alpha_2)$ are given in (8). To be specific,

$$\Sigma = \begin{pmatrix} 1 & \sigma_{01} & \sigma_{02} \\ \sigma_{01} & 1 & \sigma_{12} \\ \sigma_{02} & \sigma_{12} & 1 \end{pmatrix},$$

and $\sigma_{ij} = \delta_i \delta_j + \rho_{ij}\sqrt{1 - \delta_i^2}\sqrt{1 - \delta_j^2}$. Note that the marginal density of X_0, X_1 and X_2,

$$f_i(x_i) = 2\phi(x_i)\Phi(\lambda_i x_i), \qquad i = 0, 1, 2. \tag{13}$$

where $\lambda_i = \delta_i/\sqrt{1-\delta_i^2}$. The condition density of X_0 given $X_1 = x_1, X_2 = x_2$ is

$$h(x_0|x_1, x_2) = \phi_c(x_0|x_1, x_2)\frac{\Phi(\alpha_0 x_0 + \alpha_1 x_1 + \alpha_2 x_2)}{\Phi(\bar{\alpha}_1 x_1 + \bar{\alpha}_2 x_2)}, \tag{14}$$

where $\phi_c(x_0|x_1, x_2)$ is the conditional density associated with the normal variable with standardised marginals and correlation Σ, $(\bar{\alpha}_1, \bar{\alpha}_2)^T$ is the shape parameter of the marginal distribution of $(X_1, X_2)^T$, to be specific,

$$\bar{\alpha}_1 = \frac{\alpha_1 + \frac{\sigma_{01}-\sigma_{12}\sigma_{02}}{1-\sigma_{12}^2}\alpha_0}{\sqrt{1+\alpha_0^2\rho_c^2}}, \quad \bar{\alpha}_2 = \frac{\alpha_2 + \frac{\sigma_{02}-\sigma_{12}\sigma_{01}}{1-\sigma_{12}^2}\alpha_0}{\sqrt{1+\alpha_0^2\rho_c^2}},$$

$$\rho_c^2 = 1 - \frac{\sigma_{01}^2 - 2\sigma_{01}\sigma_{02}\sigma_{12} + \sigma_{02}^2}{1-\sigma_{12}^2}.$$

The moment generating function of X_0 given $X_1 = x_1, X_2 = x_2$ is obtained as,

$$M_{x_0|x_1,x_2}(t) = \exp\left\{\mu_c t + \rho_c^2 t^2/2\right\} \frac{\Phi\left(\frac{\alpha_0\rho_c^2 t + \alpha_0\mu_c + \alpha_1 x_1 + \alpha_2 x_2}{\sqrt{1+\alpha_0^2\rho_c^2}}\right)}{\Phi(\bar{\alpha}_1 x_1 + \bar{\alpha}_2 x_2)}, \tag{15}$$

where $\mu_c = \frac{(\sigma_{01}-\sigma_{02}\sigma_{12})x_1 + (\sigma_{02}-\sigma_{01}\sigma_{12})x_2}{1-\sigma_{12}^2}$.

The mean regression of X_0 given $X_1 = x_1, X_2 = x_2$,

$$E\left(X_0|X_1 = x_1, X_2 = x_2\right) = \frac{1}{\Phi(\bar{\alpha}_1 x_1 + \bar{\alpha}_2 x_2)}\left\{\mu_c \Phi\left(\frac{\alpha_0\mu_c + \alpha_1 x_1 + \alpha_2 x_2}{\sqrt{1+\alpha_0^2\rho_c^2}}\right)\right.$$
$$\left. + \left(\frac{\alpha_0\rho_c^2}{\sqrt{1+\alpha_0^2\rho_c^2}}\right)\phi\left(\frac{\alpha_0\mu_c + \alpha_1 x_1 + \alpha_2 x_2}{\sqrt{1+\alpha_0^2\rho_c^2}}\right)\right\}. \tag{16}$$

For $\delta_0 = 0.3, \delta_1 = 0.5, \delta_2 = 0.8$ and $\rho_{ij} = 0.5$, for all $i, j = 0, 1, 2$, the plot of the mean regression function of X_0 given $X_1 = x_1, X_2 = x_2$ in Eq. (16) is given in Fig. 1.

For the case where $k = 2$, the skew normal copula model in (9) is,

$$C_{\boldsymbol{\theta}}^{SN}(u_0, u_1, u_2) = H(F_0^{-1}(u_0, \lambda_0), F_1^{-1}(u_1, \lambda_1), F_2^{-1}(u_2, \lambda_2); \Sigma, \boldsymbol{\alpha})$$

where H is the CDF of (12). We obtain the skew normal copula regression function of U_0 given $U_1 = u_1, U_2 = u_2$,

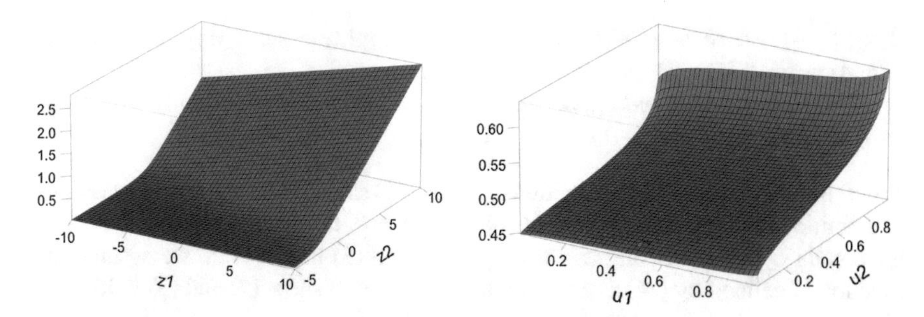

Fig. 1 The plot of the regression function $E[X_0|X_1 = x_1, X_2 = x_2]$ in (16) (*left*) and the plot of the skew normal copula regression function $r^C_{U_0|U_1,U_2}(u_1, u_2)$ in (17) (*right*) for $\delta_0 = 0.3$, $\delta_1 = 0.5$, $\delta_2 = 0.8$, and $\rho_{ij} = 0.5$, for all $i, j = 0, 1, 2$

$$r^C_{U_0|U_1,U_2}(u_1, u_2) = 1 - \frac{K}{c_{12}(u_1, u_2)} \int_0^1 \int_{-\infty}^{F_0^{-1}(u_0;\lambda_0)} \phi_3(x_0, w_1, w_2; (\Sigma^{-1} - \tilde{I}_3)^{-1}) \times \frac{\Phi(\alpha_0 x_0 + \alpha_1 w_1 + \alpha_2 w_2)}{\Phi(\lambda_1 w_1)\Phi(\lambda_2 w_2)} dx_0 du_0, \quad (17)$$

where $K = \frac{\pi\sqrt{|(\Sigma^{-1}-\tilde{I}_3)^{-1}|}}{\sqrt{|\Sigma|}}$, $(w_1, w_2)^T = (F_1^{-1}(u_1; \lambda_1), F_2^{-1}(u_2; \lambda_2))^T$, $\tilde{I}_3 = diag(0, I_2)$, and I_2 is the 2×2 identity matrix. The plot of the copula regression function $r^C_{U_0|U_1,U_2}(u_1, u_2)$ for $\delta_0 = 0.3, \delta_1 = 0.5, \delta_2 = 0.8$ and $\rho_{ij} = 0.5$, for all $i, j = 0, 1, 2$ in Eq. (17) is given in Fig. 1.

3.2 Multivariate Non-exchangeable Generalized FGM Copula

Úbeda-Flores [18] proposed one class of bivariate non-exchangeable FGM copulas. We provide a trivariate extension as follows.

Proposition 2 *Let f_0, f_1 and f_2 be non-zero real functions defined on* $[0, 1]$ *and C be the function on* $[0, 1]^3$ *given by*

$$C(u_0, u_0, u_1) = u_0u_1u_2 + f_0(u_0)f_1(u_1)u_2 + f_0(u_0)f_2(u_2)u_1 + f_1(u_1)f_2(u_2)u_0. \quad (18)$$

Then C is a copula if and only if
(1) $f_i(0) = f_i(1) = 0$, $i = 0, 1, 2$;
(2) f_i *absolutely continuous,* $i = 0, 1, 2$; *and*
(3) $\min\{\alpha_0\beta_1+\alpha_0\beta_2+\alpha_1\beta_2, \beta_0\alpha_1+\alpha_0\beta_2+\alpha_1\beta_2, \alpha_0\beta_1+\beta_0\alpha_2+\alpha_1\beta_2, \alpha_0\beta_1+\alpha_0\beta_2+\beta_1\alpha_2, \beta_0\alpha_1+\beta_0\alpha_2+\alpha_1\beta_2, \beta_0\alpha_1+\alpha_0\beta_2+\beta_1\alpha_2, \alpha_0\beta_1+\beta_0\alpha_2+\beta_1\alpha_2, \beta_0\alpha_1+\beta_0\alpha_2+$

$\beta_1\alpha_2\} \geq -1$, *where* $\alpha_i = \inf\{f_i'(u_i); u_i \in A_i\} < 0$, *and* $\beta_i = \sup\{f_i'(u_i); u_i \in A_i\} > 0$, *with* $A_i = \{u_i \in I; f_i'(u_i)$ *exists* $\}$, *for* $i = 0, 1, 2$.

Furthermore, the copula given in (18) is exchangeable if and only if $f_0(u) \equiv f_1(u) \equiv f_2(u)$ *for all* $u \in [0, 1]$.

Proof It is immediate that the function given by (18) satisfies the boundary conditions in definition of copula if and only if $f_i(0) = f_i(1) = 0$, $i = 0, 1, 2$. We will show that the copula C is 3-increasing if and only if (2) and (3) holds. First, we assume the function C defined by (18) is an 3-copula, and we will show (2) and (3) holds. Let $F_i(x, y)$, $i = 0, 1, 2$, denote the functions defined on the set $T = \{(x, y) \in [0, 1]^2 | x < y\}$ by

$$F_i(x, y) = \frac{f_i(y) - f_i(x)}{y - x}, \qquad \text{for} \quad i = 0, 1, 2.$$

Then, C is 3-increasing if and only if

$$-1 \leq F_0(u_0, v_0)F_1(u_1, v_1) + F_0(u_0, v_0)F_2(u_2, v_2) + F_1(u_1, v_1)F_2(u_2, v_2). \quad (19)$$

Hence, C is n-increasing if and only if the following inequalities holds:

$$\begin{aligned} -1 &\leq \gamma_0\delta_1 + \gamma_0\delta_2 + \gamma_1\delta_2, & -1 &\leq \delta_0\gamma_1 + \gamma_0\delta_2 + \gamma_1\delta_2, \\ -1 &\leq \gamma_0\delta_1 + \delta_0\gamma_2 + \gamma_1\delta_2, & -1 &\leq \gamma_0\delta_1 + \gamma_0\delta_2 + \delta_1\gamma_2, \\ -1 &\leq \delta_0\gamma_1 + \delta_0\gamma_2 + \gamma_1\delta_2, & -1 &\leq \delta_0\gamma_1 + \gamma_0\delta_2 + \delta_1\gamma_2, \\ -1 &\leq \gamma_0\delta_1 + \delta_0\gamma_2 + \delta_1\gamma_2, & -1 &\leq \delta_0\gamma_1 + \delta_0\gamma_2 + \delta_1\gamma_2, \end{aligned}$$

where $\gamma_i = \inf\{F_i(u_i, v_i) : u_i < v_i, f_i(u_i) > f_i(u_i)\}$ and $\delta_i = \sup\{F_i(u_i, v_i) : u_i < v_i, f_i(u_i) < f_i(u_i)\}$, $i = 0, 1, 2$. Since $f_i(0) = f_i(1) = 0$ and f_i's are non-zero, the sets above are non-empty. Also, since (19) holds for all $(u_i, v_i) \in T$, we know that $F_i(u_i, v_i)$ is bounded for $i = 0, 1, 2$. Therefore, we have

$$\begin{aligned} \gamma_i = \inf\{F_i(u_i, v_i) : u_i < v_i, f_i(u_i) > f_i(u_i)\} &= \inf\{F_i(u_i, v_i) : u_i < v_i\} \\ &= \inf\{f_i'(u), u \in A\} = \alpha_i < 0, \\ \delta_i = \sup\{F_i(u_i, v_i) : u_i < v_i, f_i(u_i) < f_i(u_i)\} &= \sup\{F_i(u_i, v_i) : u_i < v_i\} \\ &= \sup\{f_i'(u), u \in A\} = \beta_i < 0. \end{aligned}$$

In summary, we have shown that if C is a copula, then (1), (2), and (3) are true. Conversely, the proof follows the same steps backwards, which completes the proof. □

Using the generalized FGM copulas given in (18), the theorem below gives the copula-based regression functions.

Theorem 1 *For the generalized FGM copula given in (18), the copula regression function is*

$$r^C_{U_0|U_1,U_2}(u_1,u_2) = 1 - \frac{\frac{1}{2}(1+f_1'(u_1)f_2'(u_2)) + (f_1'(u_1)+f_2'(u_2))\int_0^1 f_0(u_0)du_0}{1+f_1'(u_1)f_2'(u_2)}.$$

Proof The proof is straight forward. □

To illustrate the application of the two proposed multivariate non-exchangeable FGM copulas and the corresponding copula regression functions to the directional dependence, two examples are given below.

Example 2 Let $f_i(u_i) = \rho_i u_i(1-u_i)^{p_i}$, $i = 0, 1, 2$ in (18). The copula C is given by

$$\begin{aligned}C(u_0,u_1,u_2) &= u_0u_1u_2 + \rho_0\rho_1 u_0(1-u_0)^{p_0}u_1(1-u_1)^{p_1}u_2 + \rho_0\rho_2(1-u_0)^{p_0}\\ &\quad\times u_2(1-u_2)^{p_2}u_1 + \rho_1\rho_2 u_1(1-u_1)^{p_1}u_2(1-u_2)^{p_2}u_0.\end{aligned}$$

After simplifications, we obtain $\alpha_i = \min\{-\rho_i\left(\frac{p_i-1}{p_i+1}\right)^{p_i-1}, \rho_i\}$ and $\beta_i = \max\{\rho_i, -\rho_i\left(\frac{p_i-1}{p_i+1}\right)^{p_i-1}\}$. Therefore, if $\rho_i^2\left(\frac{p_i-1}{p_i+1}\right)^{p_i-1} \le \frac{1}{3}$, (e.g. if $-1 \le \rho_i \le 1$ and $p_i \ge 2$,) for all $i = 0, 1, 2$, C is a copula by Proposition 2. From Theorem 1, the copula regression function is

$$\begin{aligned}r^C_{U_0|U_1,U_2}(u_1,u_2) &= 1 - \frac{1}{c_{12}(u_1,u_2)}\int_0^1 C_{U_0|U_1,U_2}(u_0)du_0\\ &= 1 - \frac{1}{c_{12}(u_1,u_2)}[1/2 + Beta(2,p_0+1)[f_1'(u_2)+f_2'(u_2)]\\ &\quad + 1/2 f_1'(u_1)f_2'(u_2)],\end{aligned}$$

where $c_{12}(u_1,u_2) = 1+\rho_1\rho_2(1-u_1)^{p_1-1}[1-(1+p_1)u_1](1-u_2)^{p_2-1}[1-(1+p_2)u_2]$, $f_i'(u_i) = \rho_i(1-u_i)^{p_i-1}[1-(1+p_i)u_i]$, $i = 1, 2$, and $Beta(a,b) = \int_0^1 t^{a-1}(1-t)^{b-1}dt$.

Furthermore, for $p_0 = 2, p_1 = 3, p_2 = 4$, $\rho_0 = 0.9$, $\rho_1 = 0.8$, and $\rho_2 = 0.9$, the measures of the directional dependence in joint behavior for three copula-based regression functions are $\rho^2_{(U_1,U_2\to U_0)} = 0.007$, $\rho^2_{(U_0,U_1\to U_2)} = 0.002$, and $\rho^2_{(U_0,U_2\to U_1)} = 0.003$.

References

1. Azzalini, A.: The Skew-Normal Probability Distribution and Related Distributions, such as the Skew-t, http://azzalini.stat.unipd.it/SN/
2. Azzalini, A., Capitanio, A.: Statistical applications of the multivariate skew normal distribution. J. R. Stat. Soc. Ser. B (Stat. Methodol.) **61**(3), 579–602 (1999)
3. Azzalini, A., Dalla Valle, A.: The multivariate skew-normal distribution. Biometrika **83**(4), 715–726 (1996)
4. Dodge, Y., Rousson, V.: On asymmetric properties of the correlation coeffcient in the regression setting. Am. Stat. **55**(1), 51–54 (2001)
5. Gupta, A.K., González-Farías, G., Domínguez-Molina, J.A.: A multivariate skew normal distribution. J. Multivar. Anal. **89**(1), 181–190 (2004)
6. Joe, H.: Multivariate Models and Multivariate Dependence Concepts. CRC Press (1997)
7. Kim, S., Kim, D.: Directional dependence analysis using skew-normal copula-based regression. Book Chapter In: Statistics and Causality: Methods for Applied Empirical Research, eds. by Wiedermann, W., von Eye, A. Wiley (2015)
8. Kim, J.M., Jung, Y.S., Sungur, E.A., Han, K.H., Park, C., Sohn, I.: A copula method for modeling directional dependence of genes. BMC Bioinform. **9**(1), 225 (2008)
9. Kim, D., Kim, J.M.: Analysis of directional dependence using asymmetric copula-based regression models. J. Stat. Comput. Simul. **84**(9), 1990–2010 (2014)
10. Kotz, S., Balakrishnan, N., Johnson, N.L.: Continuous Multivariate Distributions, vol. 1, Models and Applications (vol. 59). Wiley (2002)
11. McNeil, A.J., Frey, R., Embrechts, P.: Quantitative Risk Management: Concepts. Princeton University Press, Techniques and Tools (2005)
12. Nelsen, R.B.: An Introduction to Copulas, 2nd edn, Springer (2006)
13. Sklar, A.: Fonctions de répartition á n dimensions et leurs marges. Publ. Inst. Statist. Univ. Paris **8**, 229–231 (1959)
14. Sriboonchitta, S., Liu, J., Kreinovich, V., Nguyen, H.T.: A vine copula approach for analyzing financial risk and co-movement of the Indonesian, Philippine and Thailand stock markets. In: Modeling Dependence in Econometrics, pp. 245–257 (2014)
15. Sriboonchitta, S., Nguyen, H.T., Wiboonpongse, A., Liu, J.: Modeling volatility and dependency of agricultural price and production indices of Thailand: static versus time-varying copulas. Int. J. Approx. Reas. **54**(6), 793–808 (2013)
16. Sungur, E.A.: A note on directional dependence in regression setting. Commun. Stat. Theory Methods **34**, 1957–1965 (2005)
17. Sungur, E.A.: Some observations on copula regression functions. Dempster Commun. Stat. Theory Methods **34**, 1967–1978 (2005)
18. Úbeda-Flores, M.: A new class of bivariate copulas. Stat. Prob. Lett. **66**(3), 315–325 (2004)
19. VonEye, A., Wiedermann, W.: On direction of dependence in latent variable contexts. Educ. Psychol. Measur. **74**(1), 5–30 (2014)
20. Wang, T., Li, B., Gupta, A.K.: Distribution of quadratic forms under skew normal settings. J. Multivar. Anal. **100**(3), 533–545 (2009)
21. Wei, Z., Wang, T., Nguyen, P.A.: Multivariate dependence concepts through copulas. Int. J. Approx. Reason. (2015) **65**, 24–33
22. Wei, Z., Wang, T., Panichkitkosolkul, W.: Dependence and association concepts through copulas. In: Modeling Dependence in Econometrics, 113–126
23. Wiedermann, W., VonEye, A.: Direction of effects in multiple linear regression models. Multivar. Behav. Res. **50**(1), 23–40 (2015)
24. Ye, R., Wang, T., Gupta, A.K.: Distribution of matrix quadratic forms under skew-normal settings. J. Multivar. Anal. **131**, 229–239 (2014)

On Consistency of Estimators Based on Random Set Vector Observations

Zheng Wei, Tonghui Wang and Baokun Li

Abstract In this paper, the characterization of the joint distribution of random set vector by the belief function is investigated. A routine of calculating the bivariate coarsening at random model of finite random sets is obtained. In the context of reliable computations with imprecise data, we show that the maximum likelihood estimators of parameters in CAR model are consistent. Several examples are given to illustrate our results.

1 Introduction

Random sets can be used to model imprecise observations of random variables where the outcomes are set valued instead of real valued. The theory of random sets is viewed as a natural generalization of multivariate statistical analysis. Random set data can also be viewed as imprecise or incomplete observations which are frequent in today's technological societies. The distribution of the univariate random set and its properties can be found in [1, 8, 11]. Recently, the characterization of joint distributions of random sets on co-product spaces was discussed by Schmelzer [9, 10], and Wei et al. [14–16]. In this paper, this characterization is extended to the distributions of bivariate discrete random set vector.

In the univariate discrete application, we usually partition the set E into finitely subsets $A_i \in \mathscr{A}$, the σ-field of subsets of E, $i = 1, 2, \ldots, N$. Consider the

Z. Wei
Department of Mathematics and Statistics, University of Massachusetts,
Amherst, MA, USA
e-mail: wei@math.umass.edu

Z. Wei · T. Wang (✉)
Department of Mathematical Sciences, New Mexico State University,
Las Cruces, NM, USA
e-mail: twang@nmsu.edu

B. Li
School of Statistics, Southwestern University of Finance and Economy,
Qingyang, China
e-mail: bali@swufe.edu.cn

V.-N. Huynh et al. (eds.), *Causal Inference in Econometrics*,
Studies in Computational Intelligence 622, DOI 10.1007/978-3-319-27284-9_11

random set $\mathscr{S}$ that takes values from the class $\{A_1, A_2, \ldots, A_N\}$ with probability $f_{\mathscr{S}}(A_i) = P(\mathscr{S} = A_i)$, and the random variable X with probability $P_X(A_i) = P(X \in A_i), i = 1, 2, \ldots, N$. Heitjan and Rubin [5] provide a definition of *coarsening at random* (CAR) model:

$$P(\mathscr{S} = A | X = x) = \text{constant}, \qquad \text{for any } A \in \mathscr{A}, \quad x \in A.$$

The importance of the univariate CAR model has been discussed by Li and Wang [7], Grünwald and Halpern [4] and Jaeger [6]. The examples and applications of coarsening at random are discussed in Heitjan and Rubin [5], Gill et al. [2], Tsiatis [12], Nguyen [8], Tsiatis et al. [13]. In this paper, the bivariate CAR model is introduced based on the joint distribution of random set vector. Also, a routine for calculating the bivariate CAR model solutions is provided.

For coarsened data (e.g. data grouping), suppose a bivariate random vector (X, Y) with the sample space $E_1 \times E_2$ is distributed according to the density $f(x, y|\boldsymbol{\theta})$ with respect to some measure. Furthermore, assume that instead of observing the random vector (X, Y) directly, only the set valued observations are obtained. One typical such type of data coarsening is grouping. Under simple grouping, the random set vector $(\mathscr{S}_1, \mathscr{S}_2)$ is a function of random vector (X, Y). One motivation example is by considering the problem of heaping in epidemiologic studies of populations of smokers. The distributions of cigarettes smoked per day tend to have large heaps at integral multiples of twenty (a pack of cigarettes), particularly for a heavy smoker. This is because the person who smoke only a few cigarettes are more likely to report the exact number they smoke, but that a heavy smoker would tend to report the number of cigarettes in complete packs. The statistician's goal being to draw inferences about the parameter $\boldsymbol{\theta}$. The general theory of the likelihood function for the univariate random set has been developed by Heitjan and Rubin [5]. In this paper, the consistency of the likelihood estimators based on random set vector observations is obtained.

This paper is organized as follows. The characterization of the joint distribution of random set vector by its joint belief functions is obtained in Sect. 2. As an application of random set vector, the bivariate CAR model and its properties are investigated in Sect. 3. Also, the computational aspects of CAR model are provided. The consistency of the likelihood estimators based on random set vector observations is developed in Sect. 4. To illustrate our main results, several examples are given.

2 Characterization of the Joint Belief Function of Discrete Random Set Vector

Throughout this paper, let $(\Omega, \mathscr{A}, P)$ be a probability space and let E_1 and E_2 be finite sets, where Ω is sample space, $\mathscr{A}$ is a σ-algebra on subsets of Ω and P is a probability measure. Recall that a finite random set $\mathscr{S}$ with values in power set of a finite E is a map $\mathscr{S} : \Omega \to 2^E$ such that $\mathscr{S}^{-1}(\{A\}) = \{\omega \in \Omega : \mathscr{S}(\omega) = A\} \in \mathscr{A}$ for any $A \subseteq E$. Let $f : 2^E \to [0, 1]$ be $f(A) = P(\mathscr{S} = A)$, then f is a probability density function of $\mathscr{S}$ on 2^E. In the following, we will extend this definition to the cases of the random set vector.

Definition 1 A **random set vector** $(\mathscr{S}_1, \mathscr{S}_2)$ with values in $2^{E_1} \times 2^{E_2}$ is a map $(\mathscr{S}_1, \mathscr{S}_2) : \Omega \to 2^{E_1} \times 2^{E_2}$ such that $\{\omega \in \Omega : \mathscr{S}_1(\omega) = A, \mathscr{S}_2(\omega) = B\} \in \mathscr{A}$, for any $A \subseteq E_1$ and $B \subseteq E_2$. The function $h : 2^{E_1} \times 2^{E_2} \to [0, 1]$, $h(A, B) = P(\mathscr{S}_1(\omega) = A, \mathscr{S}_2(\omega) = B)$, $A \subseteq E_1$ and $B \subseteq E_2$, is called a **joint probability density function** of $(\mathscr{S}_1, \mathscr{S}_2)$.

Inspired by the distribution of univariate random sets, we are going to define axiomatically the concept of joint distribution functions of the random set vector $(\mathscr{S}_1, \mathscr{S}_2)$.

Let $(\mathscr{S}_1, \mathscr{S}_2)$ be a (nonempty) random set vector on $2^{E_1} \times 2^{E_2}$, and $H : 2^{E_1} \times 2^{E_2} \to [0, 1]$ be

$$H(A, B) = P(\mathscr{S}_1 \subseteq A, \mathscr{S}_2 \subseteq B) = \sum_{C \subseteq A} \sum_{D \subseteq B} h(C, D), \quad A \in 2^{E_1}, \quad B \in 2^{E_2}. \tag{1}$$

It can be shown that H satisfies the following properties:

(i) $H(\emptyset, \emptyset) = H(\emptyset, B) = H(A, \emptyset) = 0$, and $H(E_1, E_2) = 1$;

(ii) H is **monotone of infinite order on each component**, i.e., for any B in 2^{E_2} and any distinct sets $A_1, A_2, \ldots, A_k$ in 2^{E_1}, $k \geq 1$,

$$H\left(\bigcup_{i=1}^{k} A_i, B\right) \geq \sum_{\emptyset \neq I \subseteq \{1,2,\ldots,k\}} (-1)^{|I|+1} H\left(\bigcap_{i \in I} A_i, B\right), \tag{2}$$

and for any $A \in 2^{E_1}$ and any distinct sets $B_1, B_2, \ldots, B_\ell$ in 2^{E_2}, $\ell \geq 1$,

$$H\left(A, \bigcup_{j=1}^{\ell} B_j\right) \geq \sum_{\emptyset \neq J \subseteq \{1,2,\ldots,\ell\}} (-1)^{|J|+1} H\left(A, \bigcap_{j \in J} B_j\right); \tag{3}$$

and

(iii) $H(.,.)$ is **jointly monotone of infinite order**, i.e., for distinct sets $A_1, A_2, \ldots, A_k$ in 2^{E_1} and distinct $B_1, B_2, \ldots, B_\ell$ in 2^{E_2}, where k, ℓ are positive integers,

$$\begin{aligned} H\left(\bigcup_{i=1}^{k} A_i, \bigcup_{j=1}^{\ell} B_j\right) \geq & - \sum_{\emptyset \neq I \subseteq \{1,2,\ldots,k\}} \sum_{\emptyset \neq J \subseteq \{1,2,\ldots,\ell\}} (-1)^{|I|+|J|} H\left(\bigcap_{i \in I} A_i, \bigcap_{j \in J} B_j\right) \\ & + \sum_{\emptyset \neq I \subseteq \{1,2,\ldots,k\}} (-1)^{|I|+1} H\left(\bigcap_{i \in I} A_i, \bigcup_{j=1}^{\ell} B_j\right) \\ & + \sum_{\emptyset \neq J \subseteq \{1,2,\ldots,\ell\}} (-1)^{|J|+1} H\left(\bigcup_{i=1}^{k} A_i, \bigcap_{j \in J} B_j\right). \end{aligned} \tag{4}$$

It turns out that the properties (i), (ii) and (iii) of H above characterize the joint distribution function of a (nonempty) random set vector.

Definition 2 A set function $H : 2^{E_1} \times 2^{E_2} \to [0, 1]$ satisfying the properties (i), (ii) and (iii) is said to be the **joint belief function** of random set vector $(\mathscr{S}_1, \mathscr{S}_2)$.

Given any given joint belief function H of $(\mathscr{S}_1, \mathscr{S}_2)$, there exists a probability density function $h : 2^{E_1} \times 2^{E_2} \to [0, 1]$ corresponding to H. In fact, let $H : 2^{E_1} \times 2^{E_2} \to [0, 1]$ be such that

(i) $H(\emptyset, \emptyset) = H(\emptyset, B) = H(A, \emptyset) = 0$, and $H(E_1, E_2) = 1$,

(ii) H is monotone of infinite order on each component, and

(iii) H is joint monotone of infinite order. Then for any $(A, B) \in 2^{E_1} \times 2^{E_2}$, there exists a nonnegative set function $h : 2^{E_1} \times 2^{E_2} \to [0, 1]$, called the **Möbius inverse** of H, such that

$$H(A, B) = \sum_{C \subseteq A} \sum_{D \subseteq B} h(C, D) \tag{5}$$

and

$$\sum_{C \subseteq E_1} \sum_{D \subseteq E_2} h(C, D) = 1. \tag{6}$$

The function $h : 2^{E_1} \times 2^{E_2} \to [0, 1]$ defined by

$$h(A, B) = \sum_{C \subseteq A} \sum_{D \subseteq B} (-1)^{|A \setminus C| + |B \setminus D|} H(C, D), \tag{7}$$

where $A \setminus C = A \cap C^c$ and C^c is the complement of C.

Given a set function $H : 2^{E_1} \times 2^{E_2} \to [0, 1]$, it is natural to ask whether if it is a well-defined joint belief function. By the conditions in (i), (ii) and (iii) of H, we only need to check all distinct sets $A_1, \ldots, A_k$ and $B_1, \ldots, B_\ell$.

Similar to conditions (i), (ii) and (iii) of H, there is a property called completely monotone in each component, given by Schmelzer [9, 10] as follows.

A set function $H_1 : 2^{E_1} \times 2^{E_2} \to [0, 1]$ is said to be **completely monotone in each component**, if for any $k \geq 2$ and $(A_i, B_i) \in 2^{E_1} \times 2^{E_2}$, $i = 1, 2 \ldots, k$,

$$H_1\left(\bigcup_{i=1}^{k} A_i, \bigcup_{i=1}^{k} B_i\right) \geq \sum_{\emptyset \neq I \subseteq \{1,2,\ldots,k\}} (-1)^{|I|+1} H_1\left(\bigcap_{i \in I} A_i, \bigcap_{i \in I} B_i\right). \tag{8}$$

It can be shown that (8) is equivalent to (ii) and (iii).

3 Bivariate CAR Models

As an application of the joint belief function of random set vector, bivariate CAR models are introduced and its properties are discussed in this section. Also, the computational aspects of CAR model are provided.

Let $(\Omega, \mathscr{A}, P)$ be a probability space, E_1 and E_2 be finite sets. Consider bivariate random vector $(X, Y) : \Omega \to E_1 \times E_2$, and random set vector $(\mathscr{S}_1, \mathscr{S}_2) : \Omega \to 2^{E_1} \times 2^{E_2}$. A **coarsening of** (X, Y) is a non-empty random set vector $(\mathscr{S}_1, \mathscr{S}_2)$ on $E_1 \times E_2$ such that $P(X \in \mathscr{S}_1, Y \in \mathscr{S}_2) = 1$.

Definition 3 A coarsening $(\mathscr{S}_1, \mathscr{S}_2)$ is said to be a **bivariate coarsening at random (CAR)** of the bivariate random vector (X, Y), if for all $(A, B) \in (2^{E_1} \setminus \{\phi\}) \times (2^{E_2} \setminus \{\phi\})$, there exists a number $\pi(A, B)$ such that

$$P(\mathscr{S}_1 = A, \mathscr{S}_2 = B | X = x, Y = y) = \pi(A, B) \tag{9}$$

for any $x \in A$, $y \in B$. The constant $\pi(A, B)$ is called a **bivariate CAR probability**.

Proposition 1 *The condition of bivariate CAR of (X, Y) given in (9) is equivalent to*

$$P(\mathscr{S}_1 = A, \mathscr{S}_2 = B | X = x, Y = y) = P(\mathscr{S}_1 = A, \mathscr{S}_2 = B | X \in A, Y \in B) \tag{10}$$

for any $(A, B) \in (2^{E_1} \setminus \{\phi\}) \times (2^{E_2} \setminus \{\phi\})$ and $x \in A$, $y \in B$.

Proof Assume that the condition (9) holds. Since $(\mathscr{S}_1, \mathscr{S}_2)$ is a coarsening of (X, Y), we have $(\mathscr{S}_1 = A, \mathscr{S}_2 = B) \subseteq (X \in A, Y \in B)$ so that $P(X \in A, Y \in B | \mathscr{S}_1 = A, \mathscr{S}_2 = B) = 1$. Also,

$$\begin{aligned}
P(X \in A, Y \in B | \mathscr{S}_1 = A, \mathscr{S}_2 = B) &= \sum_{x \in A, y \in B} P(X = x, Y = y | \mathscr{S}_1 = A, \mathscr{S}_2 = B) \\
&= \sum_{x \in A, y \in B} \frac{P(\mathscr{S}_1 = A, \mathscr{S}_2 = B | X = x, Y = y) P(X = x, Y = y)}{P(\mathscr{S}_1 = A, \mathscr{S}_2 = B)} \\
&= \frac{\pi(A, B)}{P(\mathscr{S}_1 = A, \mathscr{S}_2 = B)} \sum_{x \in A, y \in B} P(X = x, Y = y) \\
&= \frac{\pi(A, B)}{P(\mathscr{S}_1 = A, \mathscr{S}_2 = B)} P(X \in A, Y \in B) = 1.
\end{aligned}$$

Therefore, which indicate that (10) holds,

$$\pi(A, B) = \frac{P(\mathscr{S}_1 = A, \mathscr{S}_2 = B)}{P(X \in A, Y \in B)} = P(\mathscr{S}_1 = A, \mathscr{S}_2 = B | X \in A, Y \in B).$$

It is easy to show the condition (10) implies the condition (9). □

Remark 3.1 The characterization of CAR mechanism for univariate random set by utilizing the concept of uniform multi-cover was given in Gill and Grunwald [3]. Their result can be extended to the bivariate CAR mechanism. Note that

$$\sum_{A\ni x, B\ni y} \pi(A, B) = 1, \text{ for any } x \in E_1, y \in E_2. \tag{11}$$

Indeed, we know that $\pi(A, B) = 0$, for any set A that does not contain x or set B that does not contain y. For any $x \in E_1$ and $y \in E_2$, we have

$$\begin{aligned} 1 &= \sum_{A\subseteq E_1, B\subseteq E_2} P(\mathscr{S}_1 = A, \mathscr{S}_2 = B | X = x, Y = y) \\ &= \sum_{A\not\ni x \text{ or } B\not\ni y} P(\mathscr{S}_1 = A, \mathscr{S}_2 = B | X = x, Y = y) \\ &\quad + \sum_{A\ni x, B\ni y} P(\mathscr{S}_1 = A, \mathscr{S}_2 = B | X = x, Y = y) \\ &= \sum_{A\ni x, B\ni y} P(\mathscr{S}_1 = A, \mathscr{S}_2 = B | X = x, Y = y) = \sum_{A\ni x, B\ni y} \pi(A, B). \end{aligned}$$

Note that the following Theorem is an extension of the result given in Gill et al. [2] from univariate case to bivariate case.

Theorem 1 *Let $(\mathscr{S}_1, \mathscr{S}_2)$ be a non-empty bivariate random set vector with the joint density $h(A, B)$, $(A, B) \in (2^{E_1} \setminus \{\phi\}) \times (2^{E_2} \setminus \{\phi\})$. Then there exist bivariate CAR probabilities, $\pi(A, B)$, and a joint mass function $p(x, y)$ of a bivariate random vector (X, Y) on $E_1 \times E_2$ such that*

$$h(A, B) = \pi(A, B) p(A, B) \quad \textit{for all } (A, B) \in 2^{E_1 \times E_2} \setminus \{\emptyset\},$$

where $p(A, B) = \sum_{y\in B} \sum_{x\in A} p(x, y)$. Furthermore, for each A, B with $h(A, B) > 0$, $\pi(A, B)$ and $p(A, B)$ are uniquely determined by the joint distribution of bivariate random set vector $(\mathscr{S}_1, \mathscr{S}_2)$.

Remark 3.2 Note that the Theorem 1 proves the existence of the values $\pi(A, B)$ and $p(A, B)$, it does not explain how to compute $\pi(A, B)$ and $p(x, y)$ based on the values $h(A, B)$. In the following, a computational method for computing values $p(x, y)$ and $\pi(A, B)$ is given, based on the joint density $h(A, B)$ of $(\mathscr{S}_1, \mathscr{S}_2)$.

Assume $E_1 = \{x_1, x_2, \ldots, x_{N_1}\}$, $E_2 = \{y_1, y_2, \ldots, y_{N_2}\}$. From Theorem 1, we know that

$$\pi(A, B) = \frac{h(A, B)}{p(A, B)} = \frac{h(A, B)}{\sum_{x\in A} \sum_{y\in B} p(x, y)} \quad \text{for all } (A, B) \in 2^{E_1 \times E_2} \setminus \{\phi\}.$$

Table 1 Joint density of $(\mathscr{S}_1, \mathscr{S}_2)$

$h_{\mathscr{C}}$	{3}	{4}	{5}	{3,4}	{3,5}	{4,5}	{3,4,5}
{1}	5/48	13/144	11/144	0	0	0	1/16
{2}	1/12	1/12	1/12	0	0	0	1/12
{1,2}	1/16	11/144	13/144	0	0	0	5/48

Table 2 Bivariate CAR probabilities

$\pi(\cdot,\cdot)$	{3}	{4}	{5}	{3,4}	{3,5}	{4,5}	{3,4,5}
{1}	0.5825	0.5412	0.5001	0	0	0	0.1254
{2}	0.5418	0.5005	0.4595	0	0	0	0.1661
{1,2}	0.1879	0.2292	0.2702	0	0	0	0.1042

Note that (11) can be rewritten as:

$$1 = \sum_{A \ni x} \sum_{B \ni y} \pi(A, B) = \sum_{A \ni x} \sum_{B \ni y} \frac{h(A, B)}{\sum_{s \in A} \sum_{t \in B} p(s, t)} \quad \text{for any } x \in E_1, y \in E_2. \quad (12)$$

The expression (12) is a nonlinear system of $N_1 N_2$ equations with unknowns $p(x_i, y_j)$'s $i = 1, 2, \ldots, N_1$, $j = 1, 2, \ldots, N_2$.

We can now use, e.g., BB-package in CRAN-R for solving this system of $N_1 N_2$ nonlinear equations with $N_1 N_2$ unknowns. Once we have determined the $N_1 N_2$ values $p(x_i, y_j)$, we can then compute, for every pair of sets $A \subseteq E_1$, $B \subseteq E_2$, the value $\pi(A, B) = h(A, B)/p(A, B)$.

Example 1 Suppose the joint density $h_C(A, B)$ is given in Table 1.
By using the above method, we obtain the joint mass function, $p(x, y)$, bivariate random vector (X, Y) on $E_1 \times E_2$,

$$p(1,3) = 0.1788, \quad p(1,4) = 0.1668, \quad p(1,5) = 0.1527,$$
$$p(2,3) = 0.1538, \quad p(2,4) = 0.1665, \quad p(2,5) = 0.1814.$$

The corresponding bivariate CAR probabilities are listed in Table 2.

4 The Likelihood Function of Random Set Vector Observations

Suppose a random vector (X, Y) is distributed according to the joint density $f(x, y|\boldsymbol{\theta})$ with respect to some measure ν, where $\boldsymbol{\theta} \subseteq \Theta$ is the vector of parameter. We would like to make statistical inferences about $\boldsymbol{\theta}$. Furthermore, assume that instead of observing (X, Y) directly, one only observes $(\mathscr{S}_1, \mathscr{S}_2) = (\mathscr{S}_1(X), \mathscr{S}_2(Y))$, a vector of subsets within which (X, Y) has fallen. The following definition of likelihood function generalizes Heitjan and Rubin's definition into bivariate case.

One particularly straightforward type of data coarsening is grouping. Under simple grouping, the random set vector $(\mathscr{S}_1, \mathscr{S}_2)$ is a function of random vector (X, Y). The conditional distribution of $(\mathscr{S}_1, \mathscr{S}_2)$ given $(X = x, Y = y)$ is

$$r(A, B|x, y, \boldsymbol{\theta}) = \begin{cases} 1, & \text{if } A = \mathscr{S}_1(x), \text{ and } B = \mathscr{S}_2(y) \\ 0, & \text{otherwise.} \end{cases}$$

Hence the likelihood function arising from (A, B), the observed value of $(\mathscr{S}_1, \mathscr{S}_2)$, is

$$\begin{aligned} L_G(\boldsymbol{\theta}|A, B) &= \int_{E_2}\int_{E_1} r(A, B|x, y, \boldsymbol{\theta}) f(x, y|\boldsymbol{\theta}) dxdy \\ &= \int_{B}\int_{A} f(x, y|\boldsymbol{\theta}) dxdy \end{aligned} \tag{13}$$

Now consider a more general form of grouping in which the precision of reporting is a function of a random variables G_1 and G_2 with sample space Γ_1 and Γ_2, respectively. Conditional on $X = x, Y = y$, the joint distribution of G_1 and G_2 is given by $g(s, t; x, y, \boldsymbol{\gamma})$. The random variables G_1 and G_2 determines the precision of reporting in the sense that the value of G_1 and G_2 determines which of a collection of possible mappings $X \to \mathscr{S}_1(X)$ and $Y \to \mathscr{S}_2(Y)$ to use in coarsening X and Y. In this case we only be able to observe the coarsened data $(\mathscr{S}_1(X, G_1), \mathscr{S}_2(Y, G_2))$. Therefore, the conditional distribution of $(\mathscr{S}_1, \mathscr{S}_2)$ given $(X = x, Y = y, G_1 = g_1, G_2 = g_2)$ is

$$r(A, B|x, y, g_1, g_2, \boldsymbol{\theta}, \boldsymbol{\gamma}) = \begin{cases} 1, & \text{if } A = \mathscr{S}_1(x, g_1), \text{ and } B = \mathscr{S}_2(y, g_2), \\ 0, & \text{otherwise.} \end{cases}$$

In lots of cases, random variables G_1 and G_2 are not directly observed, but can at best be inferred from the observed coarse data. One motivation example is by considering the problem of heaping in epidemiologic studies of populations of smokers. The distributions of cigarettes smoked per day tend to have large heaps at integral multiples of twenty (a pack of cigarettes), particularly for a heavy smoker. This is because the person who smoke only a few cigarettes are more likely to report the exact number they smoke, but that a heavy smoker would tend to report the number of cigarettes in complete packs.

Assume that G_1 and G_2 are not directly observed, but can at best be inferred from the observed coarse value (A, B) of $(\mathscr{S}_1(x, g_1), \mathscr{S}_2(y, g_2))$. That is, if $x \in A, y \in B$, and (A, B) is consistent with g_1 and g_2, then $(A, B) = (\mathscr{S}_1(x, g_1), \mathscr{S}_2(y, g_2))$ will be observed. In this case, the distribution of $(\mathscr{S}_1, \mathscr{S}_2)$ given x, y and $\boldsymbol{\gamma}$ is

$$k(A, B|x, y, \boldsymbol{\gamma}) = \int_{\Gamma_1}\int_{\Gamma_2} r(A, B|x, y, g_1, g_2, \boldsymbol{\theta}, \boldsymbol{\gamma}) g(g_1, g_2|x, y, \boldsymbol{\gamma}) dg_2 dg_1.$$

The following likelihood function for coarsened data of X and Y generalized the likelihood function given by Heitjan and Rubin [5],

$$L_G(\boldsymbol{\theta}, \boldsymbol{\gamma}|A, B) = \int_B\int_A f(x, y|\boldsymbol{\theta}) k(A, B|x, y, \boldsymbol{\gamma}) dx dy. \tag{14}$$

Denote

$$P_{\boldsymbol{\theta}}(A, B) = P_{\boldsymbol{\theta}}(X \in A, Y \in B) = \int_B\int_A f(x, y|\boldsymbol{\theta}) dx dy.$$

Let $E_1 \times E_2$ be the finite sample space of random vector (X, Y), assume that $(\mathscr{S}_1, \mathscr{S}_2)$ is a bivariate CAR of (X, Y), i.e., $k(A, B|x, y, \boldsymbol{\gamma})$ takes the same value for all $x \in A$, $y \in B$, by the notation introduced in last section, we have $k(A, B|x, y, \boldsymbol{\gamma}) = \pi(A, B)$. Furthermore, by Theorem 1, the bivariate random set vector $(\mathscr{S}_1, \mathscr{S}_2)$ has the following joint density:

$$h_{\boldsymbol{\theta}}(\mathscr{S}_1 = A, \mathscr{S}_2 = B) = \pi(A, B) P_{\boldsymbol{\theta}}(A, B) = \pi(A, B) \int_B\int_A f(x, y|\boldsymbol{\theta}) dx dy.$$

Let $(A_1, B_1), \ldots, (A_n, B_n)$ be a random sample of the random set vector. The likelihood function (14) for the parameter $\boldsymbol{\theta}$ is:

$$L(\boldsymbol{\theta}|(A_1, B_1), \ldots, (A_n, B_n)) = C((A_1, B_1), \ldots, (A_n, B_n)) \prod_{i=1}^{n} P_{\boldsymbol{\theta}}(A_i, B_i), \tag{15}$$

where

$$C((A_1, B_1), \ldots, (A_n, B_n)) = \prod_{i=1}^{n} \pi(A_i, B_i),$$

which is independent of parameter $\boldsymbol{\theta}$. Let

$$L_1(\boldsymbol{\theta}|(A_1, B_1), \ldots, (A_n, B_n)) = \prod_{i=1}^{n} P_{\boldsymbol{\theta}}(A_i, B_i), \tag{16}$$

then maximizing the likelihood function $L(\boldsymbol{\theta}|(A_1, B_1), \ldots, (A_n, B_n))$ is equivalent to maximizing the function $L_1(\boldsymbol{\theta}|(A_1, B_1), \ldots, (A_n, B_n))$ over all $\boldsymbol{\theta} \in \Theta$.

For the proof of our main results, an extended version of Weak Law of Large Numbers (WLLN) is needed and given as follows.

Lemma 1 *Let Z be a random vector as a real valued function of $(\mathscr{S}_1, \mathscr{S}_2)$, which is a random set vector over the finite range $\mathscr{A}_1 \times \mathscr{A}_2$ and the density $h_{\boldsymbol{\theta}}(A, B)$. Let Z_i be a sequence of i.i.d. random vectors as Z, then*

$$\frac{1}{n}\sum_{i=1}^{n} Z_i \rightarrow_P E_{\boldsymbol{\theta}}(Z), \text{ as } n \rightarrow \infty.$$

Proof Since the support for the random variable Z is finite, then

$$E_{\boldsymbol{\theta}}(Z) = \sum_{A \in \mathscr{A}_1} \sum_{B \in \mathscr{A}_2} h_{\boldsymbol{\theta}}(A, B) Z(A, B)$$

exists. Applying the WLLN for usual random variables, the lemma is proved. □

The following lemma is a modification of Wald's argument and will be used in the proof of our main results.

Lemma 2 *Let $(\mathscr{S}_1, \mathscr{S}_2)$ be the bivariate CAR of the random vector (X, Y), and $\{(\mathscr{S}_1^n, \mathscr{S}_2^n), n \geq 1\}$ be a sequence of i.i.d. random sets distributed as $(\mathscr{S}_1, \mathscr{S}_2)$ on a finite support $\mathscr{A}_1 \times \mathscr{A}_2$. Then, for any fixed $\boldsymbol{\theta} \neq \boldsymbol{\theta}_0$,*

$$\lim_{n\rightarrow\infty} \boldsymbol{P}_{\boldsymbol{\theta}_0}[L_1(\boldsymbol{\theta}_0|(A_1, B_1), \ldots, (A_n, B_n)) > L_1(\boldsymbol{\theta}|(A_1, B_1), \ldots, (A_n, B_n))] = 1,$$

where $\boldsymbol{P}_{\boldsymbol{\theta}_0}$ is the joint probability of $\{(\mathscr{S}_1^i, \mathscr{S}_2^i)\}_{i=1}^n$, and $L_1(\boldsymbol{\theta}|(A_1, B_1), \ldots, (A_n, B_n))$ is given in (16).

Proof The event $L_1(\boldsymbol{\theta}_0|(A_1, B_1), \ldots, (A_n, B_n)) > L_1(\boldsymbol{\theta}|(A_1, B_1), \ldots, (A_n, B_n))$ is equivalent to the event that

$$\frac{1}{n}\sum_{i=1}^{n} \log\left[\frac{P_{\boldsymbol{\theta}}(A_i, B_i)}{P_{\boldsymbol{\theta}_0}(A_i, B_i)}\right] < 0.$$

By Lemma 1, the left hand-side converges to $E_{\boldsymbol{\theta}_0}\left[\log \frac{P_{\boldsymbol{\theta}}(\mathscr{S}_1, \mathscr{S}_2)}{P_{\boldsymbol{\theta}_0}(\mathscr{S}_1, \mathscr{S}_2)}\right]$, the expected value of a function of a random set. Jensen's Inequality applied with the strictly convex function $\log(x)$ yields

$$E_{\boldsymbol{\theta}_0}\left[\log\left(\frac{P_{\boldsymbol{\theta}}(\mathscr{S}_1,\mathscr{S}_2)}{P_{\boldsymbol{\theta}_0}(\mathscr{S}_1,\mathscr{S}_2)}\right)\right] \le \log\left[E_{\boldsymbol{\theta}_0}\left(\frac{P_{\boldsymbol{\theta}}(\mathscr{S}_1,\mathscr{S}_2)}{P_{\boldsymbol{\theta}_0}(\mathscr{S}_1,\mathscr{S}_2)}\right)\right]$$
$$= \log\left[\sum_{A\in\mathscr{A}_1}\sum_{B\in\mathscr{A}_2} h_{\boldsymbol{\theta}_0}(A,B)\frac{P_{\boldsymbol{\theta}}(A,B)}{P_{\boldsymbol{\theta}_0}(A,B)}\right]$$
$$= \log\left[\sum_{A\in\mathscr{A}_1}\sum_{B\in\mathscr{A}_2} \pi(A,B)P_{\boldsymbol{\theta}_0}(A,B)\frac{P_{\boldsymbol{\theta}}(A,B)}{P_{\boldsymbol{\theta}_0}(A,B)}\right]$$
$$= \log\left[\sum_{A\in\mathscr{A}_1}\sum_{B\in\mathscr{A}_2} \pi(A,B)P_{\boldsymbol{\theta}}(A,B)\right]$$
$$= \log\left[\sum_{A\in\mathscr{A}_1}\sum_{B\in\mathscr{A}_2} h_{\boldsymbol{\theta}}(A,B)\right] = 0.$$

□

If the parameter space Θ is finite, then Lemma 2 implies directly that the Maximum Likelihood Estimator (MLE) $\hat{\boldsymbol{\theta}}$ is weakly consistent because it shows that the likelihood is larger at $\boldsymbol{\theta}_0$ than that at any other $\boldsymbol{\theta} \in \Theta$.

Theorem 2 *Under the same assumptions as in Lemma 2, if Θ is finite, then the MLE of $\boldsymbol{\theta}$ exists, is unique with probability tending to one, and is weakly consistent.*

Proof Suppose $\hat{\boldsymbol{\theta}}_n$ maximizes $L_1(\boldsymbol{\theta}|(A_1, B_1), \dots, (A_n, B_n))$ over Θ. Let $\Theta = \{\boldsymbol{\theta}_0, \boldsymbol{\theta}_1, \dots, \boldsymbol{\theta}_m\}$ with $\boldsymbol{\theta}_0$ being the true parameter, and E_{jn}'s be the events that

$$\sum_{i=1}^{n}\log\left[\frac{P_{\boldsymbol{\theta}_j}(A_i,B_i)}{P_{\boldsymbol{\theta}_0}(A_i,B_i)}\right] < 0, \ \text{for } j = 1, 2, \dots, m.$$

By Lemma 2,

$$\lim_{n\to\infty} \boldsymbol{P}_{\boldsymbol{\theta}_0}(E_{jn}) = 1, \ \text{for } j = 1, \dots, m.$$

From Bonferroni's inequality, we have

$$\boldsymbol{P}_{\boldsymbol{\theta}_0}\left(\bigcap_{j=1}^{m} E_{jn}\right) \ge 1 - \sum_{j=1}^{m} \boldsymbol{P}_{\boldsymbol{\theta}_0}(E_{jn}^c),$$

where A^c is the complement of the event A. Therefore,

$$\lim_{n\to\infty} \boldsymbol{P}_{\boldsymbol{\theta}_0}(\bigcap_{j=1}^{m} E_{jn}) = 1, \ \text{for } j = 1, \dots, m,$$

which means that $\hat{\boldsymbol{\theta}}$ is unique and weakly consistent. □

Theorem 3 *Let $(\mathscr{S}_1, \mathscr{S}_2)$ be the bivariate CAR of the random vector (X, Y), and let the model for (X, Y) be $\mathscr{F} = \{f(x, y|\boldsymbol{\theta}) : x, y \in R, \boldsymbol{\theta} \in \Theta \subseteq R^p\}$. Such that*
(i) The joint density of $(\mathscr{S}_1, \mathscr{S}_2)$,

$$h_{\boldsymbol{\theta}}(\mathscr{S}_1 = A, \mathscr{S}_2 = B) = \pi(A, B) P_{\boldsymbol{\theta}}(A, B) = \pi(A, B) \int_B \int_A f(x, y|\boldsymbol{\theta}) dx dy,$$

has same finite support $\mathscr{A}_1 \times \mathscr{A}_2$ for $\boldsymbol{\theta} \in \Theta$;
(ii) The parameter space Θ is an open ball in R^p.
(iii) $h_{\boldsymbol{\theta}}(A, B)$ is continuous in $\boldsymbol{\theta}$ for all $(A, B) \in \mathscr{A}_1 \times \mathscr{A}_2$.
Then there exists a sequence of local maxima, $\{\hat{\boldsymbol{\theta}}_n, n \geq 1\}$, of the likelihood function L_1 given in (16), which converges almost surely (a.s.) to the true parameter $\boldsymbol{\theta}_0 \in \Theta$ as $n \to \infty$.

Proof Let $\varepsilon > 0$, we need to show that if n is sufficiently large, then there exists a local maximum $\hat{\boldsymbol{\theta}}_n$ of $L_1(\boldsymbol{\theta}|(A_1, B_1), \ldots, (A_n, B_n))$ such that

$$\hat{\boldsymbol{\theta}}_n \in B(\boldsymbol{\theta}_0, \varepsilon) \text{ with probability one.}$$

Here, $B(\boldsymbol{\theta}_0, \varepsilon)$ denote the open ball centered at $\boldsymbol{\theta}_0$ with radius ε.
Let

$$L^*(\boldsymbol{\theta}|(A_1, B_1), \ldots, (A_n, B_n)) = \frac{1}{n} \sum_{i=1}^{n} \log \left(\frac{P_{\boldsymbol{\theta}}(A_i, B_i)}{P_{\boldsymbol{\theta}_0}(A_i, B_i)} \right)$$

$$= \frac{1}{n} \log L_1(\boldsymbol{\theta}|(A_1, B_1), \ldots, (A_n, B_n)) - \frac{1}{n} L_1(\boldsymbol{\theta}_0|(A_1, B_1), \ldots, (A_n, B_n)).$$

It suffices to show that $L^*(\boldsymbol{\theta}|(A_1, B_1), \ldots, (A_n, B_n))$ has a sequence of local maxima converging a.s. to $\boldsymbol{\theta}_0$. By the strong law of large numbers, for each $\boldsymbol{\theta} \in \Theta$,

$$L^*(\boldsymbol{\theta}|(A_1, B_1), \ldots, (A_n, B_n)) \to E_{\boldsymbol{\theta}_0} \left[\log \left(\frac{P_{\boldsymbol{\theta}}(\mathscr{S}_1, \mathscr{S}_2)}{P_{\boldsymbol{\theta}_0}(\mathscr{S}_1, \mathscr{S}_2)} \right) \right], \quad \text{a.s.} \quad \boldsymbol{P}_{\boldsymbol{\theta}_0}.$$

When $\boldsymbol{\theta} \neq \boldsymbol{\theta}_0$, Jensen's Inequality applied with the strictly convex function $\log(x)$ yields

$$E_{\boldsymbol{\theta}_0} \left[\log \left(\frac{P_{\boldsymbol{\theta}}(\mathscr{S}_1, \mathscr{S}_2)}{P_{\boldsymbol{\theta}_0}(\mathscr{S}_1, \mathscr{S}_2)} \right) \right] < \log \left[E_{\boldsymbol{\theta}_0} \left(\frac{P_{\boldsymbol{\theta}}(\mathscr{S}_1, \mathscr{S}_2)}{P_{\boldsymbol{\theta}_0}(\mathscr{S}_1, \mathscr{S}_2)} \right) \right] = 0.$$

Let $\Psi = \{\boldsymbol{\theta} \in \Theta | \boldsymbol{\theta} = \boldsymbol{\theta}_0 \pm \frac{1}{k} \boldsymbol{e}_p\}$, where $\boldsymbol{e}_p = (1, \ldots, 1)^T \in R^p$ and k is a positive integer. Then for each $\boldsymbol{\theta} \in \Theta$, there is a set $\mathscr{N}_\theta$ such that $\boldsymbol{P}_{\boldsymbol{\theta}_0}[\mathscr{N}_\theta^c] = 0$ and when $(A_1, B_1), \ldots, (A_n, B_n)) \in \mathscr{N}_{\boldsymbol{\theta}}$,

$$L^*(\boldsymbol{\theta}|(A_1, B_1), \ldots, (A_n, B_n)) \to E_{\boldsymbol{\theta}_0} \left[\log \left(\frac{P_{\boldsymbol{\theta}}(\mathscr{S}_1, \mathscr{S}_2)}{P_{\boldsymbol{\theta}_0}(\mathscr{S}_1, \mathscr{S}_2)} \right) \right] < 0. \tag{17}$$

Since Ψ is countable, we have

$$\boldsymbol{P}_{\boldsymbol{\theta}_0}(\mathcal{N}) = 0, \quad \text{where} \quad \mathcal{N} = \bigcup_{\boldsymbol{\theta} \in \Psi} \mathcal{N}_{\boldsymbol{\theta}}^c.$$

Therefore, when $((A_1, B_1), \ldots, (A_n, B_n)) \notin \mathcal{N}$, (17) holds for any $\boldsymbol{\theta} \in \Psi$. Let $\boldsymbol{\theta}_1 = \boldsymbol{\theta}_0 - \frac{1}{k_1}\boldsymbol{e}_p$ with $\frac{1}{k_1} < \varepsilon$, and $\boldsymbol{\theta}_2 = \boldsymbol{\theta}_0 + \frac{1}{k_1}\boldsymbol{e}_p$ with $\frac{1}{k_2} < \varepsilon$, then

$$\boldsymbol{\theta}_0 - \boldsymbol{e}_p\varepsilon < \boldsymbol{\theta}_1 < \boldsymbol{\theta}_0 < \boldsymbol{\theta}_2\boldsymbol{\theta}_0 + \boldsymbol{e}_p\varepsilon,$$

where $<$ means componentwise less than. Further more, for $((A_1, B_1), \ldots, (A_n, B_n)) \notin \mathcal{N}$ and n large enough, we have

$$L^*(\boldsymbol{\theta}_1|(A_1, B_1), \ldots, (A_n, B_n)) < 0, \qquad L^*(\boldsymbol{\theta}_2|(A_1, B_1), \ldots, (A_n, B_n)) < 0,$$

and $L^*(\boldsymbol{\theta}_0|(A_1, B_1), \ldots, (A_n, B_n)) = 0$.

By the hypothesis, $L^*(\boldsymbol{\theta}|(A_1, B_1), \ldots, (A_n, B_n))$ is a continuous function of $\boldsymbol{\theta}$, this implies $\boldsymbol{\theta}_0 - \boldsymbol{e}_p\varepsilon, < \hat{\boldsymbol{\theta}}_n < \boldsymbol{\theta}_0 + \boldsymbol{e}_p\varepsilon$, which maximizes $L^*(\boldsymbol{\theta}|(A_1, B_1), \ldots, (A_n, B_n))$. □

Example 2 Let (X, Y) be a discrete random vector having binomial marginal distributions with parameter (m_1, θ_1) and (m_2, θ_2), respectively. m_i, $i = 1, 2$, is the number of independent and identical trials in a binomial experiment which is known, and $\theta_i \in (0, 1)$, $i = 1, 2$, is the probability of success in each trial. The marginal densities is

$$f_i(k_i|\theta_i) = C_{k_i}^{m_i}\theta_i^{k_i}(1-\theta_i)^{m_i-k_i}, \qquad i = 1, 2.$$

Suppose $m_1 = 4$ and $m_2 = 3$ and their copula is FGM-copula-$C_\rho(u, v) = uv(1 + \rho(1-u)(1+v))$. First, we can estimate two marginal distributions parameter θ_1 and θ_2. Suppose that the set observations are given as follows,

$$(A_1, B_1) = (\{4\}, \{1, 2\}), (A_2, B_2) = (\{2, 3\}, \{1, 3\}), (A_3, B_3) = (\{3, 4\}, \{2, 3\}),$$
$$(A_4, B_4) = (\{4\}, \{1, 3\}), (A_5, B_5) = (\{2, 4\}, \{2, 3\}), (A_6, B_6) = (\{3\}, \{1, 2\}).$$

We can get the joint density $f(x, y|\rho)$ by copula $C_\rho(u, v)$.

$$f(x, y|\rho) = \begin{cases} C_\rho(F_X(x), F_Y(y)), & \text{if } x = 0, y = 0, \\ C_\rho(F_X(x), F_Y(y)) - C_\rho(F_X(x-1), F_Y(y)), & \text{if } x \neq 0, y = 0, \\ C_\rho(F_X(x), F_Y(y)) - C_\rho(F_X(x), F_Y(y-1)), & \text{if } x = 0, y \neq 0, \\ C_\rho(F_X(x), F_Y(y)) - C_\rho(F_X(x), F_Y(y-1)) & \\ -C_\rho(F_X(x), F_Y(y-1)) + C_\rho(F_X(x-1), F_Y(y-1)), & \text{if } x \neq 0, y \neq 0. \end{cases}$$

Then by maximizing

$$L_1(\rho|(A_1, B_1), \ldots, (A_n, B_n)) = \prod_{i=1}^{6} \boldsymbol{P}_\rho(A_i, B_i),$$

where $\boldsymbol{P}_\rho(A_i, B_i) = \sum_{x \in A_i} \sum_{y \in B_i} f(x, y|\rho)$. After calculations, we get $\hat{\theta}_1 = 0.889, \hat{\theta}_2 = 0.75$, and $\hat{\rho} = 0.216$.

References

1. Dempster, A.P.: Upper and lower probabilities induced by a multivalued mapping. Ann. Math. Stat. **28**, 325–339 (1967)
2. Gill, R.D., Van der Laan, M.J., Robins, J.M.: Coarsening at random: characterizations, conjectures, counter-examples. Springer Lect. Notes Stat. **123**, 149–170 (1997)
3. Gill, R.D., Grnwald, P.D.: An algorithmic and a geometric characterization of coarsening at random. Ann. Stat. **36**(5), 2409–2422 (2008)
4. Grunwald, P., Halpern, J.: Updating probabilities. J. Artif. Intell. Res. **19**, 243–278 (2003)
5. Heitjan, D.F., Rubin, D.B.: Ignorability and coarse data. Ann. Stat. **19**(4), 2244–2253 (1991)
6. Jaeger, M.: Ignorability for categorical data. Ann. Stat. **33**, 1964–1981 (2005)
7. Li, B., Wang, T.: Computational aspects of the coarsening at random model and the Shapley value. Inf. Sci. **177**, 3260–3270 (2007)
8. Nguyen, H.T.: An Introduction to Random Sets. CRC Press, Boca Raton (2001)
9. Schmelzer, B.: Characterizing joint distributions of random sets by multivariate capacities. Int. J. Approx. Reason. **53**, 1228–1247 (2012)
10. Schmelzer, B.: Joint distributions of random sets and their relation to copulas. Dependence and association concepts through copulas. In Modeling Dependence in Econometrics, Springer International Publishing pp. 155–168 (2014)
11. Shafer, G.: A Mathematical Theory of Evidence. Princeton University Press, New Jersey (1976)
12. Tsiatis, A.A.: Semiparametric Theory and Missing Data. Springer, New York (2006)
13. Tsiatis, A.A., Davidian, M., Cao, W.: Improved doubly robust estimation when data are monotonely carsened, with application to longitudinal studies with dropout. Biometrics **67**(2), 536–545 (2011)
14. Wei, Z., Wang, T., Panichkitkosolkul, W.: Dependence and association concepts through copulas. In Modeling Dependence in Econometrics, Springer International Publishing pp. 113–126
15. Wei, Z., Wang, T., Nguyen, P.A.: Multivariate dependence concepts through copulas, Int. J. Approx. Reason. **65**, 24–33 (2015)
16. Wei, Z., Li, B., Wang, T.: The joint distribution of the discrete random set vector and bivariate coarsening at random models (submitted) (2015)

Brief Introduction to Causal Compositional Models

Radim Jiroušek

Abstract When applying probabilistic models to support decision making processes, the users have to strictly distinguish whether the impact of their decision changes the considered situation or not. In the former case it means that they are planing to make an *intervention*, and its respective impact cannot be estimated from a usual stochastic model but one has to use a causal model. The present paper thoroughly explains the difference between conditioning, which can be computed from both usual stochastic model and a causal model, and computing the effect of intervention, which can only be computed from a causal model. In the paper a new type of causal models, so called compositional causal models are introduced. Its great advantage is that both conditioning and the result of intervention are computed in very similar ways in these models. On an example, the paper illustrates that like in Pearl's causal networks, also in the described compositional models one can consider models with hidden variables.

1 Introduction

In this paper, by *causal models* we understand (multidimensional) probability distributions with specified *causal relations* among the variables, for which the distribution is defined. We accept the philosophy of Judea Pearl [16], whose causal networks are Bayesian networks [6, 13] in which orientation of directed edges is interpreted as causal relations. In our models, Bayesian networks are substituted by *compositional models* [7], and the causal relations are encoded in the ordering describing how the multidimensional model is assembled from its low-dimensional parts. In this way, this paper is the natural continuation of paper *Brief Introduction to Causal Compositional Models* [8] presented at the 6th International Conference of the Thailand Econometric Society in 2013.

R. Jiroušek (✉)
Faculty of management, University of Economics, Prague,
Jarošovská 1117/II, 37701 Jindřichův Hradec, Czech Republic
e-mail: radim@utia.cas.cz

V.-N. Huynh et al. (eds.), *Causal Inference in Econometrics*,
Studies in Computational Intelligence 622, DOI 10.1007/978-3-319-27284-9_12

Let us stress at the very beginning the main difference between classical multidimensional probabilistic models (joint distributions without causal relations) and causal models. This idea is aptly expressed by Judea Pearl (page 22 in [16]): *A joint distributions tells us how probable events are and how probabilities would change with subsequent observations, but a causal model also tells us how these probabilities would change as a result of external interventions— such as those encountered in policy analysis, treatment management, or planning everyday activity. Such changes cannot be deduced from a joint distribution, even if fully specified.*

To illustrate this statement consider just two related events: **s**—smoke is in a room; **a**—fire alarm is on. Denoting π a two-dimensional distribution (fourfold table), π can fully describe the stochastic relation between these two events (and their negations). We can see from π that these events are dependent but there is no way to deduce (just from this fourfold table), which event is a cause and which is an impact. In fact, there is not even a way to deduce from this fourfold table whether this relation is causal or not. As Pearl says, we can read from π *how the probabilities change with subsequent observations*. We can read what we quite naturally expect that $\pi(\mathbf{a}|\mathbf{s}) > \pi(\mathbf{a})$, and $\pi(\mathbf{s}|\mathbf{a}) > \pi(\mathbf{s})$. These relations say that if we see a smoke in the room then we expect that the alarm sounds, and also if we hear the alarm we can expect there is a smoke in the room.

However, the situation changes when, instead of *observations*, we consider *interventions*. For this, we use Pearl's notation: $do(\mathbf{s})$ means that we make a smoke in the room, for example, by smoking a cigar, or burning a piece of paper. Analogously, by the intervention $do(\mathbf{a})$ we understand the situation when we set the alarm siren on disregarding whether there is a smoke in the room or not. This can be realized just by pushing an alarm test push-button, which is used to test the functionality of the alarm siren. In this case, we naturally expect that $\pi(\mathbf{a}|do(\mathbf{s})) > \pi(\mathbf{a})$, because we expect that the alarm performs well. On the other hand, if we push the test button, it activates the alarm siren but we cannot expect that it fills the room with smoke. Therefore, obviously, $\pi(\mathbf{s}|\mathbf{a}) = \pi(\mathbf{s})$.

The importance of causal models and their popularization follows, among others, also from the fact that quite often we witness confusing the intervention for observation. As a typical example the rather frequent piece of news can serve: *Researches from the Top World Institute proved that regular drinking of New-Drink averts the development of a Special disease*, which is followed by a recommendation to start/increase drinking *New-Drink*. Such a report is usually based on a screening among several thousand respondents initiated by the producer of *New-Drink*. If the screening questionnaire was properly prepared then, as a rule, at least one from the monitored diseases evinces significantly higher incidence in the group of those who do not drink *New-Drink* in comparison with the group of *New-Drink* consumers. Even if we did not object to such a goal oriented research, we must object to the conclusion that drinking *New-Drink* averts the development of the *Special disease*, because making anybody to start drinking *New-Drink* is an intervention, the impact of which cannot be derived from the observational data. As we will see later in Sect. 6, starting drinking *New-Drink* may have actually a negative impact on the genesis of *Special disease*.

We would be carrying coals to Newcastle if we brought more reasons to highlight the importance of causal models. For this, see, for example, [5, 17]. What is really new on our models is that we do not use (directed) graphs to represent the asymmetric relations of causes and effects. Instead, we use so called *compositional models* to represent multidimensional probability distributions. These models are based on a non-commutative *operator of composition*, and it is this non-commutativity what makes representation of asymmetric relation possible. Therefore, before introducing causal models in Sect. 4, we have to start the exposition by introducing the notation in Sect. 2, and a brief recollection of compositional models in Sect. 3. The advantages of the new approach will be seen in Sect. 5, where we will show how to compute the impact of conditioning and intervention. The power of the described causal compositional models arises from the possibility to eliminate hidden variables, which will be illustrated in Sect. 6 on a simple example.

2 Notation and Basic Concepts

This paper is self-contained, its reading does not require the preliminary knowledge of neither causal nor compositional models. Nevertheless, to facilitate its reading for the reader familiar with [8], we use notation from the preceding paper. We consider a finite *set of variables* $N = \{u_1, u_2, \ldots, u_n\}$, each variable $u \in N$ having a finite (non-empty) *set of values* that is denoted by $\mathbb{X}_u$. The set of all combinations of the considered values (we call these combinations *states*) is denoted $\mathbb{X}_N = \times_{u \in N} \mathbb{X}_u$. Analogously, for a subset of variables $K \subset N$, the set of all states is $\mathbb{X}_K = \times_{u \in K} \mathbb{X}_u$.

Distributions[1] of the considered variables are denoted by Greek letters $\kappa, \lambda, \ldots$ possibly with indices; for $K \subseteq N$, we can consider distribution $\kappa(K)$, which is a $|K|$-dimensional probability distribution, and $\kappa(\mathbf{x})$ denotes the value of distribution κ for state $\mathbf{x} \in \mathbb{X}_K$.

For a probability distribution $\kappa(K)$ and $J \subset K$, we will often consider a *marginal distribution* $\kappa^{\downarrow J}$ of κ, and, analogously, $\mathbf{y}^{\downarrow J}$ denotes the *projection* of $\mathbf{y} \in \mathbb{X}_K$ into $\mathbb{X}_J$, i.e., $y^{\downarrow J}$ is the state from $\mathbb{X}_J$ that is got from y by deleting all the values of variables from $K \setminus J$. Note that we do not exclude situations when $J = \emptyset$, for which we get $\kappa^{\downarrow \emptyset} = 1$.

Like in [8], the most important notion of this paper is the *operator of composition*. Recall, it is a binary operator that constructs from two probability distributions, say $\kappa(K)$ and $\lambda(L)$, one distribution of variables $K \cup L$. Before presenting its definition let us introduce the concept of dominance. Having two distributions defined for the same set of variables $\pi(K)$ and $\kappa(K)$, we say that κ *dominates* π (in symbol $\pi \ll \kappa$) if for all $\mathbf{x} \in \mathbb{X}_K$, for which $\kappa(\mathbf{x}) = 0$ also $\pi(\mathbf{x}) = 0$.

[1] Instead of probability distributions we could speak, equivalently, about probability measures on $\mathbb{X}_N$. From the computational point of view it is important to realize that such a distribution/measure as a set function can be, thanks to the additivity of probability, represented by a point function $\mathbb{X}_N \to [0, 1]$.

Definition 1 For two arbitrary distributions $\kappa(K)$ and $\lambda(L)$, for which $\kappa^{\downarrow K\cap L} \ll \lambda^{\downarrow K\cap L}$, their *composition* is, for each $\mathbf{x} \in \mathbb{X}_{L\cup K}$, given by the following formula[2]

$$(\kappa \rhd \lambda)(\mathbf{x}) = \frac{\kappa(\mathbf{x}^{\downarrow K})\lambda(\mathbf{x}^{\downarrow L})}{\lambda^{\downarrow K\cap L}(\mathbf{x}^{\downarrow K\cap L})}.$$

In case $\kappa^{\downarrow K\cap L} \not\ll \lambda^{\downarrow K\cap L}$, the composition remains undefined.

Let us summarize the most important properties of the composition operator that were proved in [7]

Proposition 1 *Suppose $\kappa(K)$, $\lambda(L)$ and $\mu(M)$ are probability distributions for which $\lambda^{\downarrow K\cap L} \gg \kappa^{\downarrow K\cap L}$. Then the following statements hold:*

1. Domain: *$\kappa \rhd \lambda$ is a distribution for $K \cup L$.*
2. Composition preserves first marginal: $(\kappa \rhd \lambda)^{\downarrow K} = \kappa$.
3. Reduction: *If $L \subseteq K$ then, $\kappa \rhd \lambda = \kappa$.*
4. Non-commutativity: *In general, $\kappa \rhd \lambda \neq \lambda \rhd \kappa$.*
5. Non-associativity: *In general, $(\kappa \rhd \lambda) \rhd \mu \neq \kappa \rhd (\lambda \rhd \mu)$.*
6. Associativity under a special condition I: *If $K \supset (L \cap M)$ then, $(\kappa \rhd \lambda) \rhd \mu = \kappa \rhd (\lambda \rhd \mu)$, if the right hand side formula is defined.*
7. Associativity under a special condition II: *If $L \supset (K \cap M)$ then, $(\kappa \rhd \lambda) \rhd \mu = \kappa \rhd (\lambda \rhd \mu)$, if the right hand side formula is defined.*
8. Stepwise composition: *If M is such that $(K \cap L) \subseteq M \subseteq L$ then, $(\kappa \rhd \lambda^{\downarrow M}) \rhd \lambda = \kappa \rhd \lambda$.*
9. Exchangeability: *If $K \supset (L \cap M)$ then, $(\kappa \rhd \lambda) \rhd \mu = (\kappa \rhd \mu) \rhd \lambda$, if the right hand side formula is defined.*
10. Simple marginalization: *Suppose M is such that $(K \cap L) \subseteq M \subseteq K \cup L$. Then $(\kappa \rhd \lambda)^{\downarrow M} = \kappa^{\downarrow K\cap M} \rhd \lambda^{\downarrow K\cap M}$.*

3 Compositional Models

As said in the preceding section, for two probability distributions $\kappa_1(K_1)$ and $\kappa_2(K_2)$ their composition (if defined) is a probability distribution of variables $K_1 \cup K_2$. For example, if κ_1 is two-dimensional and κ_2 four-dimensional distribution, their composition $\kappa_1 \rhd \kappa_2$ is a probability distribution whose dimensionality equals four, five or six. Generally, for $|K_1|$-dimensional distribution κ_1 and $|K_2|$-dimensional distribution κ_2, the dimensionality of the composed distribution $\kappa_1 \rhd \kappa_2$ equals $|K_1| + |K_2| - |K_1 \cap K_2|$. This trivial consideration leads us to a natural idea that the multiple application of the operator of composition may result in a multidimensional distribution. Such a multidimensional distribution, called a *compositional model*, will be used in this paper to represent causal models. To avoid some technical

[2] Define $\frac{0\cdot 0}{0} = 0$.

problems and the necessity of repeating some assumptions to excess, let us make the following three conventions that were made also in [8].

Whenever we speak about a distribution κ_k, if not explicitly specified otherwise, the distribution κ_k will always be assumed to be a distribution of variables K_k. Thus, for example, $\kappa_2 \triangleright \kappa_1 \triangleright \kappa_4$, if it is defined, will determine the distribution of variables $K_1 \cup K_2 \cup K_4$.

Our second convention pertains to the fact that the operator of composition is not associative. To avoid having to write too many parentheses in the formulae, we will apply the operators from left to right. Thus considering a multidimensional compositional model

$$\pi(K_1 \cup \ldots \cup K_m) = \kappa_1 \triangleright \kappa_2 \triangleright \kappa_3 \triangleright \ldots \triangleright \kappa_m$$

we always mean

$$\kappa_1 \triangleright \kappa_2 \triangleright \kappa_3 \triangleright \ldots \triangleright \kappa_m = (\ldots((\kappa_1 \triangleright \kappa_2) \triangleright \kappa_3) \triangleright \ldots \triangleright \kappa_{m-1}) \triangleright \kappa_m,$$

and the parentheses will be used only when we want to change this default ordering.

The last convention is of a rather technical nature. To avoid repeating the assumption on dominance under which the operator of composition is defined, we will always assume that all the compositions we speak about are defined.

Recall from [8] that a compositional model $\kappa_1 \triangleright \kappa_2 \triangleright \kappa_3 \triangleright \ldots \triangleright \kappa_m$ is said to be *perfect* if all the distributions from this sequence are marginals of the distribution $\kappa_1 \triangleright \kappa_2 \triangleright \ldots \triangleright \kappa_m$, i.e., if for all $i = 1, 2, \ldots, m$

$$(\kappa_1 \triangleright \kappa_2 \triangleright \ldots \triangleright \kappa_m)^{\downarrow K_i} = \kappa_i.$$

It is important to realize that each compositional model can be transformed into a perfect model: for all compositional models $\pi = \kappa_1 \triangleright \kappa_2 \triangleright \ldots \triangleright \kappa_m$ the following equality holds true

$$\kappa_1 \triangleright \kappa_2 \triangleright \ldots \triangleright \kappa_m = \pi^{\downarrow K_1} \triangleright \pi^{\downarrow K_2} \triangleright \ldots \triangleright \pi^{\downarrow K_m}. \tag{1}$$

For the respective proof see Theorem 10.9 in [7].

When considering compositional models just for the efficient representation of multidimensional distributions (like in [8], or [14]), we can take advantage of several properties making possible to rearrange distributions in a sequence without changing the resulting distribution. This is the characteristic property of so called *decomposable* models, in particular. However, as we will see in the next section, for causal models the ordering of distributions is important, because it describes the given causal relations. Therefore, not to spoil these relations, possibilities to change the ordering of the distributions in a sequence are strictly limited. In connection with this, notice that only one of the properties from Proposition 1, the Exchangeability property (Property 9), makes the swap of distributions in a sequence possible. This property will be discussed in more details in the next section.

Let us present, now, how to compute a conditional distribution using the operator of composition. Consider distribution $\kappa(K)$, variable $u \in K$, and its arbitrary value $\mathbf{a} \in \mathbb{X}_u$. Adopting notation from [20], let us introduce a one-dimensional distribution,[3] which fixes the given value of variable u:

$$\delta_{\mathbf{a}}(u) = \begin{cases} 1 & \text{if } u = \mathbf{a}, \\ 0 & \text{otherwise.} \end{cases}$$

So, distribution $\delta_{\mathbf{a}}(u)$ carries the sure information that $u = \mathbf{a}$, which, as already showed in the TES 2013 paper [8], enables us to compute conditional probability distribution (for a formal proof see Theorem 2.3 in [2])

$$\kappa(L|u = \mathbf{a}) = (\delta_{\mathbf{a}}(u) \triangleright \kappa(K))^{\downarrow L}, \tag{2}$$

for any $L \subseteq K \setminus \{u\}$.

4 Causal Models

Let us consider a set of variables $N = \{u_1, u_2, \ldots, u_n\}$, and for each variable $u_i \in N$ let $\mathfrak{C}(u_i) \subset N$ be the set of its causes. In this paper we consider only *Markovian* models [16], i.e., the models in which variables can be ordered (without loss of generality we assume it is the ordering $u_1, u_2, \ldots, u_n$) such that the causes are always before their effects. So, we assume that

$$u_k \in \mathfrak{C}(u_i) \implies k < i,$$

which, as the reader certainly noticed, means that $\mathfrak{C}(u_1) = \emptyset$, and excludes feedback models from our consideration.

To keep with the above notation, denote $K_i = \mathfrak{C}(u_i) \cup \{u_i\}$, and thus $\kappa_i(K_i)$ denote the distribution describing local behavior of u_i. The corresponding causal compositional model is the probability distribution $\pi(N)$ expressed in the following way

$$\pi(N) = \kappa_1(K_1) \triangleright \kappa_2(K_2) \triangleright \ldots \triangleright \kappa_n(K_n). \tag{3}$$

To realize how the causal relations are encoded into this causal model notice that, having a causal model $\pi(N)$ given in a form of Eq. (3), one can ascertain the set of causes for each variable u in a simple way: first find $i = \min\{k : u \in K_k\}$, and then

$$\mathfrak{C}(u) = K_i \cap \{1, \ldots, i-1\} = K_i \setminus \{u\}. \tag{4}$$

---

[3]In [8] this degenerated distribution was denoted by $\pi_{|u;\mathbf{a}}$, which appeared to be slightly misleading.

In other words, if $u, v \in K_i$, and $u \notin \{K_1 \cup \ldots \cup K_{i-1}\}$, and $v \in \{K_1 \cup \ldots \cup K_{i-1}\}$ then, $v \in \mathfrak{C}(u)$. Recalling the properties of perfect models in the preceding section, we realize that a compositional model can be uniquely defined also by a multidimensional distribution π and a sequence $K_1, K_2, \ldots, K_n$, because of Equality (1). However, this is just a theoretical property because the representation of a joint probability distribution π for the number of variables corresponding to practical problems, and the subsequent computations of the necessary marginals $\pi^{\downarrow K_i}$, is usually not possible.

From what has been said above, we see that the ordered sequence of sets $K_1, K_2, \ldots, K_n$ bears the information about the causal relations in the considered causal model. Nevertheless, it is important to realize that though we cannot arbitrarily change the ordering of distributions in model (3), the ordering is usually not unique. Recall that among the properties from Proposition 1 there is one (Property 9) that makes the replacement of distributions possible. Nevertheless, as it can immediately be seen, the condition under which Property 9 holds true, guarantees that the above presented way of causes identification specifies the same sets of causes before and after the application of Exchangeability rule.

Thus, having a causal model represented in a compositional form, we can use it for inference by the application of any of the computational procedures described in [1, 7, 8], decompose it into two submodels [1], read the conditional independence relations incorporated in the model using procedures from [10], and compute conditional distributions as described in the preceding section. Naturally, we have to apply all the procedures with caution, not to make the forbidden change of distribution ordering in the sequence, keeping in mind that the only legal swap is that according to Property 9 of Proposition 1.

5 Intervention

In this section we consider a causal compositional model

$$\pi(N) = \kappa_1(K_1) \triangleright \kappa_2(K_2) \triangleright \ldots \triangleright \kappa_n(K_n),$$

variable $u \in N$, and its value $\mathbf{a} \in \mathbb{X}_u$. From Eq. (2) we see how to compute the respective conditional distribution for any $L \subseteq N \setminus \{u\}$:

$$\pi(L|u = \mathbf{a}) = \left(\delta_{\mathbf{a}}(u) \triangleright \pi(N)\right)^{\downarrow L} = \Big(\delta_{\mathbf{a}}(u) \triangleright \left(\kappa_1(K_1) \triangleright \kappa_2(K_2) \triangleright \ldots \triangleright \kappa_n(K_n)\right)\Big)^{\downarrow L}.$$

To find an analogous formula for computing the result of intervention, we have to follow the respective ideas of Judea Pearl (page 23 in [16]): *Note the difference between the action do*$(u = \mathbf{a})$ *and the observation* $u = \mathbf{a}$. *The effect of the latter is obtained by ordinary Bayesian conditioning, ..., while that of the former by conditioning* in a modified model. The modification *mirrors the difference between seeing and doing*. We are doing the intervention regardless any causes of variable u, which means that in the modified model we have to consider $\mathfrak{C}(u) = \emptyset$. Let us find a general form of this modified model. Denote $i = \min\{k : u \in K_k\}$. Obviously, we must consider a modified causal compositional model

$$\kappa_1(K_1) \triangleright \ldots \triangleright \kappa_{i-1}(K_{i-1}) \triangleright \kappa_i(u) \triangleright \kappa_{i+1}(K_{i+1}) \triangleright \ldots \triangleright \kappa_n(K_n),$$

which is equivalent, due to Property 3 of Proposition 1, to

$$\kappa_1(K_1) \triangleright \ldots \triangleright \kappa_{i-1}(K_{i-1}) \triangleright \kappa_i(u) \triangleright \kappa_i(K_i) \triangleright \kappa_{i+1}(K_{i+1}) \triangleright \ldots \triangleright \kappa_n(K_n). \quad (5)$$

Applying Property 9 of Proposition 1 to Expression (5) $(i - 1)$times we get

$$\begin{aligned} &\kappa_1(K_1) \triangleright \ldots \triangleright \kappa_{i-1}(K_{i-1}) \triangleright \kappa_i(u) \triangleright \kappa_i(K_i) \triangleright \kappa_{i+1}(K_{i+1}) \triangleright \ldots \triangleright \kappa_n(K_n) \\ &= \kappa_1(K_1) \triangleright \ldots \triangleright \kappa_{i-2}(K_{i-2}) \triangleright \kappa_i(u) \triangleright \kappa_{i-1}(K_{i-1}) \triangleright \kappa_i(K_i) \triangleright \ldots \triangleright \kappa_n(K_n) \\ &= \ldots = \kappa_i(u) \triangleright \kappa_1(K_1) \triangleright \ldots \triangleright \kappa_n(K_n) \end{aligned} \quad (6)$$

(notice the last modification is also possible because K_1 is a singleton different from $\{u\}$). The advantage of this formula is that we do not need to know, which set K_i contains u, and the effect of intervention $\pi(L|do(u = \mathbf{a}))$ is computed as a conditioning in the model (6)

$$\pi(L|do(u = \mathbf{a})) = \Big(\delta_{\mathbf{a}}(u) \triangleright \big(\kappa_i(u) \triangleright \kappa_1(K_1) \triangleright \kappa_2(K_2) \triangleright \ldots \triangleright \kappa_n(K_n)\big)\Big)^{\downarrow L}.$$

The right hand side expression from this formula can further be modified applying n times Property 7 of Proposition 1

$$\begin{aligned} &\delta_{\mathbf{a}}(u) \triangleright \Big(\kappa_i(u) \triangleright \kappa_1(K_1) \triangleright \kappa_2(K_2) \triangleright \ldots \triangleright \kappa_n(K_n)\Big) \\ &= \delta_{\mathbf{a}}(u) \triangleright \Big(\kappa_i(u) \triangleright \kappa_1(K_1) \triangleright \kappa_2(K_2) \triangleright \ldots \triangleright \kappa_{n-1}(K_{n-1})\Big) \triangleright \kappa_n(K_n) \\ &= \ldots = \delta_{\mathbf{a}}(u) \triangleright \Big(\kappa_i(u)\Big) \triangleright \kappa_1(K_1) \triangleright \kappa_2(K_2) \triangleright \ldots \triangleright \kappa_{n-1}(K_{n-1}) \triangleright \kappa_n(K_n) \\ &= \delta_{\mathbf{a}}(u) \triangleright \kappa_1(K_1) \triangleright \kappa_2(K_2) \triangleright \ldots \triangleright \kappa_{n-1}(K_{n-1}) \triangleright \kappa_n(K_n), \end{aligned}$$

where the last modification is just the application of Property 3 of Proposition 1. Thus we see that the computations of conditioning, and the effect of intervention in causal compositional models differ just in a pair of parentheses:

$$\pi(L|u = \mathbf{a}) = \Big(\delta_{\mathbf{a}}(u) \triangleright \big(\kappa_1(K_1) \triangleright \kappa_2(K_2) \triangleright \ldots \triangleright \kappa_n(K_n)\big)\Big)^{\downarrow L}, \quad (7)$$

and

$$\pi(L|do(u = \mathbf{a})) = \Big(\delta_{\mathbf{a}}(u) \triangleright \kappa_1(K_1) \triangleright \kappa_2(K_2) \triangleright \ldots \triangleright \kappa_n(K_n)\Big)^{\downarrow L}. \quad (8)$$

Let us illustrate this computations on the simplest example mentioned in Introduction.

Smoke-Alarm Example

Consider two binary variables s and a with values $\{\mathbf{s}^+, \mathbf{s}^-\}$ and $\{\mathbf{a}^+, \mathbf{a}^-\}$, respectively. In accordance with the example in Introduction, we assume that $\mathfrak{C}(s) = \emptyset$ and $\mathfrak{C}(a) = \{s\}$, which means that we consider a causal model

$$\pi(s, a) = \kappa(s) \rhd \kappa(s, a).$$

Computing conditioning and intervention according to Formulae (7) and (8), respectively, one gets (Properties 8, 3 and 10 are used)

$$\begin{aligned}
\pi(a|s = \mathbf{s}^+) &= \Big(\delta_{\mathbf{s}^+}(s) \rhd \big(\kappa(s) \rhd \kappa(s, a)\big)\Big)^{\downarrow\{a\}} = \Big(\delta_{\mathbf{s}^+}(s) \rhd \kappa(s, a)\Big)^{\downarrow\{a\}} \\
&= \kappa(a|s = \mathbf{s}^+), \\
\pi(s|a = \mathbf{a}^+) &= \Big(\delta_{\mathbf{a}^+}(a) \rhd \big(\kappa(s) \rhd \kappa(s, a)\big)\Big)^{\downarrow\{s\}} = \Big(\delta_{\mathbf{a}^+}(a) \rhd \kappa(s, a)\Big)^{\downarrow\{s\}} \\
&= \kappa(s|a = \mathbf{a}^+), \\
\pi(a|do(s = \mathbf{s}^+)) &= \Big(\delta_{\mathbf{s}^+}(s) \rhd \kappa(s) \rhd \kappa(s, a)\Big)^{\downarrow\{a\}} = \Big(\delta_{\mathbf{s}^+}(s) \rhd \kappa(s, a)\Big)^{\downarrow\{a\}} \\
&= \kappa(a|s = \mathbf{s}^+), \\
\pi(s|a = do(\mathbf{a}^+)) &= \Big(\delta_{\mathbf{a}^+}(a) \rhd \kappa(s) \rhd \kappa(s, a)\Big)^{\downarrow\{s\}} = \Big(\delta_{\mathbf{a}^+}(a) \rhd \kappa(s)\Big)^{\downarrow\{s\}} = \kappa(s).
\end{aligned}$$

6 Hidden Variables

Quite often, when solving practical problems we face the fact that some causes are unobservable. For example, going back to the New-Drink example from Introduction, we do not know why some people like this beverage, and some others not. This obliges us to incorporate into the model a hidden variable that influences the taste, and therefore also the behavior of respondents. Such a variable is called *hidden*, or *unobservable*, because we do not have any corresponding data. The goal of this section is to show that though hidden variables do not cause problems when computing conditional distributions, they may be the source of unsurpassable obstacles for the computations of the effect of an intervention.

Consider a causal compositional model

$$\pi(N) = \kappa_1(K_1) \rhd \kappa_2(K_2) \rhd \ldots \rhd \kappa_n(K_n),$$

with (for the sake of simplicity) one hidden variable $w \in N$. It means that we have data at out disposal from which we can estimate only marginals of $\pi(N \setminus \{w\})$.

When computing conditional distribution (for some $u \in (N \setminus \{w\})$, and $\mathbf{a} \in \mathbb{X}_u$) using Formula (7) we get rid of variable w just using Property 10 of Proposition 1

$$\pi(L|u=\mathbf{a}) = \left(\delta_{\mathbf{a}}(u) \triangleright \pi(N)\right)^{\downarrow L} = \left(\delta_{\mathbf{a}}(u) \triangleright \left(\kappa_1(K_1) \triangleright \kappa_2(K_2) \triangleright \ldots \triangleright \kappa_n(K_n)\right)\right)^{\downarrow L}$$
$$= \left(\delta_{\mathbf{a}}(u) \triangleright \left(\kappa_1(K_1) \triangleright \kappa_2(K_2) \triangleright \ldots \triangleright \kappa_n(K_n)\right)^{\downarrow N\setminus\{w\}}\right)^{\downarrow L}. \qquad (9)$$

The computation of the effect of an intervention is analogously simple[4] only when there exists $k < n$ such that $L \cup \{u\} \subseteq M = K_1 \cup \ldots \cup K_k$, and $w \notin M$. In this case

$$\pi(L|do(u=\mathbf{a})) = \left(\delta_{\mathbf{a}}(u) \triangleright \kappa_1(K_1) \triangleright \kappa_2(K_2) \triangleright \ldots \triangleright \kappa_n(K_n)\right)^{\downarrow L}$$
$$= \left(\left(\delta_{\mathbf{a}}(u) \triangleright \kappa_1(K_1) \triangleright \kappa_2(K_2) \triangleright \ldots \triangleright \kappa_n(K_n)\right)^{\downarrow M}\right)^{\downarrow L}$$
$$= \left(\delta_{\mathbf{a}}(u) \triangleright \kappa_1(K_1) \triangleright \kappa_2(K_2) \triangleright \ldots \triangleright \kappa_k(K_k)\right)^{\downarrow L}.$$

In other situations one has to try to modify Formula (8) using the properties from Proposition 1. In case that this effort fails, one should try to modify the causal model into a form that makes the computations of the intervention possible. The description of these possibilities is far beyond the scope of this "brief introduction", so we will illustrate them just on the New-Drink example.

New-Drink Example

Assume there was a statistical survey showing that a disease $\mathbf{d}^+$ has a lower incidence among *New-Drink* consumers than among those who do not drink *New-Drink*. Since not all people like *New-Drink*, we assume that tendency to drink this beverage is influenced by another cause, say *genetic disposition*, which influences also development of disease $\mathbf{d}^+$. So, let us start considering a simplest possible causal model with three variables

b – drinking *New-Drink*	$\mathbb{X}_b = \{\mathbf{b}^+, \mathbf{b}^-\}$	$\mathfrak{C}(b) = \{g\}$,
d – disease	$\mathbb{X}_d = \{\mathbf{d}^+, \mathbf{d}^-\}$	$\mathfrak{C}(d) = \{b, g\}$,
g – genetic disposition	$\mathbb{X}_g$ unknown	$\mathfrak{C}(g) = \emptyset$,

i.e., the causal compositional model

$$\pi(b, d, g) = \kappa_1(g) \triangleright \kappa_2(b, g) \triangleright \kappa_3(b, d, g).$$

In this case, unfortunately, the computations of

$$\pi(d|do(b=\mathbf{b}^+)) = \left(\delta_{\mathbf{b}^+}(b) \triangleright \kappa_1(g) \triangleright \kappa_2(b, g) \triangleright \kappa_3(b, d, g)\right)^{\downarrow \{d\}}$$

---

[4]Let us stress here that we do not speak about computational complexity of the respective procedures, which may be pretty high even for computation of Formula (9). For a solution of computational problems see [15].

is obviously impossible; because we can estimate only $\kappa_2(b)$, and $\kappa_3(b,d)$. The only way, how to overcome this problem, is to introduce an additional observable variable. Since the *New-Drink* producer claims that, say, the positive impact of drinking their beverage is based on the fact that it decreases the level of cholesterol, let us add the result of the respective laboratory test into the model, and consider, now, a new causal model

$$\pi(b,c,d,g) = \kappa_1(g) \triangleright \kappa_2(b,g) \triangleright \kappa_3(b,c) \triangleright \kappa_4(c,d,g),$$

which means that now we are considering variables

b – drinking *New-Drink*	$\mathbb{X}_b = \{\mathbf{b}^+, \mathbf{b}^-\}$	$\mathfrak{C}(b) = \{g\}$,
c – cholesterol	$\mathbb{X}_c = \{\mathbf{c}^{high}, \mathbf{c}^{low}\}$	$\mathfrak{C}(c) = \{b\}$,
d – disease	$\mathbb{X}_d = \{\mathbf{d}^+, \mathbf{d}^-\}$	$\mathfrak{C}(d) = \{c, g\}$,
g – genetic disposition	$\mathbb{X}_g$ unknown	$\mathfrak{C}(g) = \emptyset$.

Now, though not simple, the computation of

$$\pi(d|do(b=\mathbf{b}^+)) = \left(\delta_{\mathbf{b}^+}(b) \triangleright \kappa_1(g) \triangleright \kappa_2(b,g) \triangleright \kappa_3(b,c) \triangleright \kappa_4(c,d,g)\right)^{\downarrow\{d\}}$$

is possible, regardless the fact that from the available data we can estimate neither κ_1, nor κ_2, nor κ_4, but only κ_3. The computations (for the one page computations, which we do not repeat here, the reader is referred either to [9], or to [2]; in the latter source they were performed even in a more general form) take advantage of the fact that the available data allows also the estimation of the three-dimensional distribution of variables c, d, g, which do not appear in the definition of the model. Denoting the estimate of this three-dimensional distribution κ_4 we get

$$\pi(d|do(b=\mathbf{b}^+)) = \left(\delta_{\mathbf{b}^+}(b) \triangleright \kappa_3(b,c) \triangleright \left(\kappa_3(b)\cdot\kappa_3(c) \triangleright \kappa_4(b,c,d)\right)^{\downarrow\{c,d\}}\right)^{\downarrow\{d\}},$$

which is quite different from the conditional distribution that can be, for this example, computed

$$\pi(d|b=\mathbf{b}^+) = \left(\delta_{\mathbf{b}^+}(b) \triangleright \kappa_4(b,c,d)\right)^{\downarrow\{d\}}.$$

So, it may easily happen that $\pi(d|b=\mathbf{b}^+) < \pi(d)$, and simultaneously $\pi(d|do(b=\mathbf{b}^+)) > \pi(d)$. The reader can check these inequalities with the data from Table 1, for which we get

$$\pi(d=\mathbf{d}^+) = 0.033,$$
$$\pi(d=\mathbf{d}^+|b=\mathbf{b}^+) = 0.027,$$
$$\pi(d=\mathbf{d}^+|do(b=\mathbf{b}^+)) = 0.044.$$

Table 1 Frequency table for *New-Drink Example*

$b = \mathbf{b}^+$				$b = \mathbf{b}^-$			
$c = \mathbf{c}^{high}$		$c = \mathbf{c}^{low}$		$c = \mathbf{c}^{high}$		$c = \mathbf{c}^{low}$	
$d = \mathbf{d}^+$	$d = \mathbf{d}^-$	$d = \mathbf{d}^+$	$d = \mathbf{d}^-$	$d = \mathbf{d}^+$	$d = \mathbf{d}^-$	$d = \mathbf{d}^+$	$d = \mathbf{d}^-$
0.010	0.122	0.008	0.520	0.009	0.263	0.006	0.062

From these values one can see that regardless the value of conditional probability $\pi(d = \mathbf{d}^+|b = \mathbf{b}^+) = 0.027$ may seem promising, the impact of intervention $\pi(d = \mathbf{d}^+|do(b = \mathbf{b}^+)) = 0.044$ is in fact negative.

7 Conclusions

The goal of this paper is twofold. For the readers not familiar with causal networks it can serve as a brief introduction to causal models, from which they can learn the difference between the notions of conditioning and intervention. The reader familiar with this type of models can see a rather unusual way of causal model representation without oriented graphs. This algebraic modeling results in elegant formulae enabling us to compute both conditioning and the effect of intervention in a surprisingly similar way. Though the respective formulae differ just in a pair of parentheses, the result may be substantially different, as illustrated on the example in the last section.

Another advantage of compositional models consists in the fact that compositional models where introduced not only within the framework of probability theory, but also in possibility theory [4, 21], theory of belief functions [3, 12, 18], and recently also for Shenoy's Valuation-Based Systems [11, 19]. Thus, causal models can easily be extended into the all above mentioned theoretical frameworks.

Acknowledgments This work was supported in part by the National Science Foundation of the Czech Republic by grant no. GACR 15-00215S.

References

1. Bína, V., Jiroušek, R.: Marginalization in multidimensional compositional models. Kybernetika **42**(4), 405–422 (2006)
2. Bína, V., Jiroušek, R.: On computations with causal compositional models. Kybernetika **51**(3), 525–539 (2015)
3. Dempster, A.P.: Upper and lower probabilities induced by a multivalued mapping. Ann. Math. Stat. **38**(2), 325–339 (1967)
4. Dubois, D., Prade, H.: Possibility Theory: An Approach to Computerized Processing of Uncertainty. Plenum Press, New York (1988)

5. Hagmayer, Y., Sloman, S., Lagnado, D., Waldmann, M.R.: Causal reasoning through intervention. In: Gopnik, A., Schulz, L. (eds.) Causal Learning: Psychology, Philosophy, and Computation, pp. 86–101. Oxford University Press, Oxford (2002)
6. Jensen, F.V.: Bayesian Networks and Decision Graphs. IEEE Computer Society Press, New York (2001)
7. Jiroušek, R.: Foundations of compositional model theory. Int. J. Gen. Syst. **40**(6), 623–678 (2011)
8. Jiroušek, R.: Brief introduction to probabilistic compositional models. Uncertainty analusis in econometrics with applications. In: Huynh, V.N., Kreinovich, V., Sriboonchita, S., Suriya, K. (eds.) AISC 200, pp. 49–60. Springer, Berlin (2013)
9. Jiroušek, R.: On causal compositional models: simple examples. In: Laurent, A. et al. (eds.) Proceedings of the 15th International Conference on Information Processing and Management of Uncertainty in Knowledge-Based Systems. Part I, CCIS 442, Springer International Publishing, Switzerland, pp. 517–526 (2014)
10. Jiroušek, R., Kratochvíl, V.: Foundations of Compositional Models: structural properties. Int. J. Gen. Syst. **44**(1), 2–25 (2015)
11. Jiroušek, R., Shenoy, P.P.: Compositional models in valuation-based systems. Int. J. Approx. Reason. **55**(1), 277–293 (2014)
12. Jiroušek, R., Vejnarová, J., Daniel, M.: Compositional models of belief functions. In: de Cooman, G., Vejnarová, J., Zaffalon, M. (eds.) Proceedings of the Fifth International Symposium on Imprecise Probability: Theories and Applications, Praha, pp. 243–252 (2007)
13. Lauritzen, S.L.: Graphical Models. Oxford University Press, Oxford (1996)
14. Malvestuto, F.M.: Equivalence of compositional expressions and independence relations in compositional models. Kybernetika **50**(3), 322–362 (2014)
15. Malvestuto, F.M.: Marginalization in models generated by compositional expressions. Kybernetika **51**(4), 541–570 (2015)
16. Pearl, J.: Causality: Models, Reasoning, and Inference, Second Edition. Cambridge University Press, Cambridge (2009)
17. Ryall, M., Bramson, A.: Inference and Intervention: Causal Models for Business Analysis. Routledge, New York (2013)
18. Shafer, G.: A Mathematical Theory of Evidence. Princeton University Press, Princeton (1976)
19. Shenoy, P.P.: A valuation-based language for expert systems. Int. J. Approx. Reason. **3**(5), 383–411 (1989)
20. Tucci, R.R.: Introduction to Judea Pearl's Do-Calculus (2013). arXiv:1305.5506v1 [cs.AI]
21. Vejnarová, J.: Composition of possibility measures on finite spaces: preliminary results. In: Bouchon-Meunier, B., Yager, R.R. (eds.) Proceedings of 7th International Conference on Information Processing and Management of Uncertainty in Knowledge-based Systems IPMU'98, Editions E.D.K. Paris, pp. 25–30 (1998)

A New Proposal to Predict Corporate Bankruptcy in Italy During the 2008 Economic Crisis

Francesca di Donato and Luciano Nieddu

Abstract Timely Corporate failure prediction is a major issue in today's economy especially considering the financial crisis that has affected the World Economy in the last decade. Any prediction technique must be reliable (good recognition rate, sensitivity and specificity), robust and able to give predictions with a sufficient time lag to allow for corrective actions. In this paper we have considered the case of Small-Medium Enterprises (SMEs) in Italy during the 2008 crisis, introducing a non-parametric classification algorithm to predict corporate failure based on financial indicators up to 8 years in advance.

1 Introduction

Since the early 60s great interest has been focused on the ability of financial indicators to predict or at least give information on the possible state of insolvency of a firm. Beaver's [5] and Altman's [2] seminal papers have given rise to many studies devoted to exploring the use of accounting information to predict business failures.

The failure of a limited company is related to two strictly connected situations (see, e.g., [2, 7, 19]):

- The inability to pay financial obligations when they come due (i.e. lack of liquidity and very high leverage).
- The inability to generate operating profits (i.e. negative or very low income and profitability).

Such failure in fulfilling its obligations is measured by a systematic worsening in the values of performance ratios (both financial and profitability ratios) [16].

F. di Donato · L. Nieddu (✉)
UNINT Rome University for International Studies,
Via C. Colombo 200, 00145 Rome, Italy
e-mail: l.nieddu@unint.eu

F. di Donato
e-mail: francesca.didonato@unint.eu

V.-N. Huynh et al. (eds.), *Causal Inference in Econometrics*,
Studies in Computational Intelligence 622, DOI 10.1007/978-3-319-27284-9_13

The possibility to predict, with a sufficient time lag, the state of distress of a firm can be useful for a number of reasons, namely the availability of an efficient and timely classification algorithm could be useful for:

- the firm itself, that, if the time span is large enough, could be allowed to take actions in order to try to correct the state of distress and try to avoid bankruptcy,
- the banks or financial institutions that could avoid lending money to firms that are destined to fail or are likely to be in distress in the forthcoming future. The financial investment sector could improve the risk return trade-off from investments by not investing in businesses that are bound to fail.
- the companies willing to try to establish long-term relationships with other firms and therefore eager to get involved with companies that will not likely fail in the future, thus increasing the longevity and viability of their business relationships.
- regulators that could promptly identify business destined to failure, preventing illegal activities, such as avoiding taxes or diluting debt [12].

When dealing with corporate failure one of the main difficulties is a non unambiguous definition of "failure". The state of failure of a firm could be defined as:

- actual filing for bankruptcy or liquidation (see, e.g., [2, 9])
- suffering financial stress or an inability to pay financial obligations [5].

When financial distress leads to bankruptcy usually it is difficult to discern the precise moment that bankruptcy occurs. According to McKee [18] a firm goes through various stages of financial distress before bankruptcy (i.e. inadequate income and liquid asset position, difficulties with paying the invoices) and any of those stages could be considered as failure.

In our work we will use the "filing for bankruptcy" as definition of failure.

In his seminal work, Beaver [5] used only univariate statistics on US market data to predict bankruptcy, finding out a high predictive ability of financial ratios up to 5 years before failure. Altman [2], applied Linear Discriminant Analysis (LDA) [11]. Altman's LDA outperformed Beavers model for one year prediction intervals. He went back up to 5 years prior to failure but the results deteriorated already at 3 years yielding a prediction rate below 50 % which is below the random recognition rate for a two-class problem. It could be argued that a good recognition rate one year prior to failure is not sufficient to enact strategies to save the firm.

Following Altman's model, many studies have been conducted in the USA using accounting data to predict big corporate failures in different industries. Before Storey's [20] work, only few studies dealt with the failure of small-medium firms (SMEs) [10].

Hall [13] studied the factors affecting small companies failure distinguishing failing and surviving firms but only considering the construction sector. Recently di Donato and Nieddu [9] have analyzed the effect of various classes of indicator (financial or profitability) on the prediction of SMEs in the Italy for a period spanning the latest economic crisis using very well established standard classification techniques to assess what type of indicators is best suited to timely predict a possible state of distress for a firm.

The reason to distinguish the set of indicators relies mainly in the peculiar structure of SMEs in Italy which are often under-capitalized and mostly rely on external financial sources provided financial institutions, which then need reliable prediction models to evaluate the risk of a possible failure and then of a possible loan loss.

Another important issue when dealing with failure prediction models is how far in advance the model is able to accurately predict bankruptcy. Many studies have a high accuracy rate one year prior to failure although some argue that such short time frame could not be enough, in general, to allow to repair the financial situation of the firm or for a lending institution to extricate itself without incurring the risk of a significant loan loss. In general, the farther back in time we go the less the accuracy level of the model is and therefore its usefulness. In the literature there are studies going back up to 6 years prior to failure [6].

The main goal of this paper is to present a new approach to predict corporate failures using a non parametric statistical method. The algorithm that is being proposed is a supervised classification algorithm, i.e. it must be trained on a data-set (training set) of already classified firms.

The results concern a retrospective study of the financial statements of a sample of 50 active firms and 50 failed firms randomly selected over a period of 12 years.

To our knowledge, besides the work of Appetiti [3] and di Donato and Nieddu [9], there are no other studies trying to propose a model to predict corporate failure in Italy for SMEs although they represent around 90 % of the overall Italian enterprises.

The layout of the paper is as follows. In Sect. 2 the proposed algorithm will be presented, while in Sect. 3 the data and the experimental setup will be outlined. Finally in Sect. 4 some conclusions will be drawn.

2 The Algorithm

The algorithm presented in this paper is a supervised classification algorithm, i.e. a data set, called training set, of elements with known classes is supposed to be available. For each element in the training set a vector of measurements (pattern vector) is available together with the class the object belongs to.

The performance of the algorithm is assessed via cross-validation [21].

Given a training set of n pattern vectors in $\mathbb{R}^p$, let us assume a partition defined on the data-set, i.e. each pattern vector is assigned to one and only one of K known classes. Let us assume a Euclidean norm defined on the data-set and let ψ be a function from $\mathbb{R}^p$ onto the set $\mathscr{C} = \{1, 2, \ldots, K\}$ which maps each pattern vector $\mathbf{x}_j,\ j = 1, \ldots, n$ into the class $c \in \mathscr{C}$ that it belongs to. The function $\psi(\cdot)$ is what is called classifier. The aim of any classification algorithm is to get an estimate of such a function based on the dataset at hand.

The proposed algorithm works as follows: compute the barycentre of each class and compute the distance of each vector from each barycentre. If each vector is closer to the barycentre of its class the algorithm stops, otherwise there will be a non empty set $\mathscr{M}$ of pattern vectors which belong to a class and are closer to a barycentre of a

---

Step1: **Let**

- $\mathbf{x}_j$, $j = 1, \dots, n$ be the pattern vectors in the training set
- $\mathbf{B}_0$ be the set of K initial barycentres $\mathbf{b}_i$, $i = 1, \dots, K$

Step2:

Compute the distances of each $\mathbf{x}_j$ from all the $\mathbf{b}_i \in B_0$
Let $\mathscr{M}$ be the set of $\mathbf{x}_w$ that are closer to a barycentre of a class different from their own.
$t \leftarrow 0$

Step3: **while** $\mathscr{M} \neq \emptyset$

- **Let** $\mathbf{x}_s, \in \mathscr{M}$ be the vector with the greatest distance from its own barycentre.
- $c \leftarrow \psi(\mathbf{x}_s)$
- **Let** $\mathbf{B}_{t+1} \leftarrow \mathbf{B}_t \cup \mathbf{x}_s$
- for all the elements of class c perform a k-means routine using as starting points the barycentres of $\mathbf{B}_{t+1}$ that belong to class c
- $t \leftarrow t+1$
- **Compute** the distances of each $\mathbf{x}_j$ from all the $\mathbf{b}_i \in \mathbf{B}_t$
- **Let** $\mathscr{M}$ be the set of $\mathbf{x}_w$ that are closer to a barycentre of a class different from their own.

end

Fig. 1 Algorithm in meta-language

different class. These vectors are those that would be misclassified using a minimum distance criterion to assign elements to a class.

In $\mathscr{M}$ select the pattern vector $\mathbf{x}_w$ that is farthest from the barycentre of its class. This pattern vector will be used as a seed for a new barycentre for class $\psi(\mathbf{x}_w)$.

A k-means algorithm [17] will then be performed for all the pattern vectors in class $\psi(\mathbf{x}_w)$ using, as starting points, the set of barycentres for class $\psi(\mathbf{x}_w)$ and the vector $\mathbf{x}_w$. Once the k-means has been performed, the set of barycentres for class $\psi(\mathbf{x}_w)$ will be increased by one more element.

It should be noticed that the barycentres at the new iterations need not be computed for all classes, but only for class $\psi(\mathbf{x}_w)$, since the barycentres for the other classes have remained unchanged.

In the following step, the distance of each pattern vector from all the barycentres is computed anew, and therefore the set $\mathscr{M}$ is updated (see Fig. 1).

If the set $\mathscr{M}$ after computing the distances of all the elements from all the barycentres is not empty then there are still elements in the training set that are misclassified. Therefore the pattern vector in $\mathscr{M}$ which is farthest from a barycentre of its own class is once again selected to serve as a seed for a new barycentre. This procedure iterates until the set $\mathscr{M}$ is empty.

Upon convergence all the elements in the training set will be closer to a barycentre of their own class.

Upon convergence, the sets of barycentres can be used to classify new elements (query points) assigning the new element to the class of the barycentre it is closest to. If elements from the training set are used as query points, then the algorithm always classify them correctly because, once converged, all pattern vectors in the training set are closer to a centroid of their own class.

The algorithm yields a set of barycentres which, in the worst case, are in a number equal to the number of elements in the training set and which has a lower bound in the number of classes.

The aim of this algorithm is to find sub-classes in the data-set which can be used to classify new vectors of unknown class. It is worth noticing that if the partition defined on the data-set is consistent with the features considered, i.e. if the pattern vectors are linearly separable, then the algorithm generates a number of barycentres equal to the number of classes. On the other hand, if the classes in the data-set are not linearly separable, then the algorithm continues splitting the classes until the sub-classes obtained are linearly separable. It is obvious that it can continue splitting until all the sub-classes are composed of only one vector (singleton). It must be stressed that it will not converge if two vectors in the training set belong to different classes and are represented by the same pattern vector. This problem can be easily overcome increasing the dimension of the vector space.

The algorithm can be generalized allowing for impurity in the result, i.e. the recursive partitioning of the feature space can be performed until the percentage of elements that are closer to a barycentre of another class has decreased under a certain threshold which can be set to a value different from zero. This can be helpful when the training set has been classified with error (imperfect supervisor): in this case allowing for impurity in the sub-classes can prevent the algorithm from over-fitting the data.

3 Experimental Results

3.1 The Data

In this study, we will be testing the proposed algorithm on an original sample of 100 non-listed Italian SMEs during the years 2000–2011, 50 that filed for bankruptcy and 50 still operating at the end of 2011, using business sector as stratifying variable, choosing only firms with turnover in the range 250 million euros at the beginning of the analyzed period. The sample was randomly selected from the firms operating in Italy at the year 2000.

The case of Italian SMEs is peculiar because these companies mainly depend on external source of finance, basically provided by banks. In such a bank-oriented industry, any subsequent contagion on the inter-banking market side would be able to jeopardize the principal source of external finance for the firms, because of banks tightening the credit access to their borrowers.

Following the approach of Abdullah et al. [1], financial companies and property industries were not considered in the analysis since their ratios are highly volatile. Besides, the interpretation of the ratios is slightly different since financial companies, for example, have different nature of income and expenses from non-financial companies.

Due to the longitudinal nature of the data, the sample dimension will decrease from 100 companies at 2000 to a plateau of 50 companies at 2011. The data used in this work were collected through CERVED database (www.cerved.com), related to economic and financial data of Italian non-listed companies. Using the available financial statements of each firm, the most common ratios for every year in the period 2000–2011 have been computed.

According to Barnes [4] we selected the ratios throughout the criterion of popularity, meaning their frequency of appearance in the literature [6]. Following there are the ratios that have been used, that could be further grouped into two sub-groups:

- Profitability Ratios, related to the economic dimension of the company: Return on Equity, Capital Turnover, Net Income/Total Assets, Return on Investment, Earning/Sales, Return on Sales, Financial Interests/Ebitda, Financial Interest/Sales;
- Leverage and Liquidity Ratios, related to the financial dimension of the company: Financial Debts/Equity, Short Term Bank Loan/Working Capital, Cash Flow/Total Debt, Structure Ratio 1, Structure Ratio 2, Working Capital/Total Assets, Quick Ratio, Working Capital Cycle, Financial Debt/Working Capital, Current Ratio, Retained Earnings/Total Assets.

All these ratios will be used to test the ability of the proposed algorithm to predict bankruptcy up to 8 years prior to failure.

3.2 *Performance Assessment*

Once a classifier has been trained on the available data-set, its performance must be determined. The performance of the proposed classification method will be evaluated according to correct recognition rate, sensitivity and specificity. Namely, consider a statistical test that allows choosing between two hypotheses (H_0 and H_1). Let H_0 be the firm belongs to the non-failed ones and H_1 be the firm belongs to the failed ones:

- *sensitivity* of a test is the statistical power of the test and is related to the type II error (non rejecting H_0 when it is false). It is the probability of recognizing a failed firm as failed. It can be estimated using the proportion of firms that have failed that are actually recognized by the test. A highly sensible test is good for ruling out the condition under testing. Positive results in a highly sensitive test are not useful to rule in the firm as being failed since sensitivity does not take into account false-positives. Consider for instance a bogus test that would classify all firms as having failed: such a test would have a perfect sensitivity but would not be useful in determining those that actually fail.
- *specificity* of a test is related to the type I error of a statistical test (rejecting the null hypothesis when it is true). It is the probability of recognizing a sound firm as sound. It can be estimated using the proportion of sound firms that have been recognized as "healthy" by the test. Specificity is therefore the ability of a test to exclude a condition correctly. Specificity is not useful for ruling out a hypothesis.

A fake test that would classify all firms as healthy would have a perfect specificity. If a firm tests positive (failed) to a highly specific test than it would have a great probability of being a failed firm.

- *correct recognition rate* of a test is the probability of correctly classifying a new element. It can be estimated using the proportion of correctly classified firms over the total number of firms.

Unbiased estimates of these quantities can be obtained via cross-validation, i.e. part of the sample is selected to train the classifier and part, independent of the previous one, is used to assess the performance of the classifier. Usually a k-fold cross-validation scheme is used. A special version of the k-fold cross validation is the leave one out scheme (LOO), where in turn each element of the sample is singled out to be tested on the classifier trained on the remaining $n-1$ elements. The performance of the classifier is then a synthesis (usually an average) of the outcomes on each unit. LOO is particularly useful when the data-set at hand is not large and therefore splitting the data-set in two could cause too much variability in the results. In such a situation as many elements as possible should be retained in the training set.

The use of holdout sample and cross-validation is not so frequent in the literature related to bankruptcy as it should be. Although in the specialized literature it was suggested the need of an independent sample to test the classifier [15], several works have continued testing the performance of various techniques on different sets of variables using only resubstitution error. The estimates obtained using resubstitution are biased estimates of the real performance, giving an error which is, on average, lower than the actual error. Any algorithm trained on a data-set will perform well on the data-set it has been trained on. To get a glimpse at the actual potential performance of the proposed methodology a cross-validation approach must be used which provides a nearly unbiased estimate [14] of the future error rate. With LOO this small bias is further reduced. Besides, using LOO or k-fold cross-validation assures that all the elements of the data-set will in turn be tested, avoiding any subjectivity in the choice of the hold-out sample.

The performance of the proposed technique will be compared with that of Classification Trees [8]. The reason for such a choice is twofold: first both Classification Trees (CARTS) and the proposed technique are non parametric classification algorithms, i.e. no prior assumption is necessary on the data-set. Second they both partition the pattern space into disjoint regions and therefore use a similar approach to the estimation of the classifier.

Although the data is longitudinal in nature, the study we have carried out is a cross-sectional study: the failed companies have been considered at various years prior to failure, from 1 to 8 years. Similar studies have only considered data up to 6 years prior to failure (see, e.g., [6] for a detailed review). Each distressed company was randomly matched with a healthy company belonging to the same industry sector and had the closest total assets. The criteria were set as control factors to ensure minimum bias in the selection of the control sample used in the development of the failure prediction model.

Due to the randomness of the matching mechanism, the selection mechanism has been repeated 300 times to get an average estimate of the performance of each prediction technique.

3.3 Results

In Table 1 the average correct recognition rates, sensitivities and specificities over 300 trials for the proposed method have been displayed for up to 8 years prior to failure. The corresponding standard deviations for the rates over the 300 trials have been reported as well.

As it was to be expected, the average correct recognition rate decreases as the time lag increases although the performance experiences a great decrease only after 7 years prior to failure. Until 6 years prior to failure the recognition rate is always greater than 80 % with a variability around 2–3%. Also the standard deviation of the performances increases when the time lag is over 6 years. This may be due to the fact that, with such a large time lag, the financial performance ratios were not able to signal a distress situation probably because there was none to signal.

Table 1 details also the average sensitivities for the same set of experiments. Sensitivities are the probabilities of predicting a firm as failed given that the firm has failed. It is also known as test power. The result of a highly sensitive test can be used as an indicator of the health of the firm, since such a test can be used to rule out the condition under testing (failure). From the obtained results, the most sensitive results (i.e. the most powerful) are those that obtained in the time span that goes from 3 to 5 years prior to failure.

Finally, the average specificities and their standard deviations over the 300 trials have been displayed. A test with high specificity is good as a warning signal, since a firm that is classified as failed by a highly specific test is likely to actually be a failed

Table 1 Proposed technique: average recognition rates sensitivities and specificities and corresponding standard deviations over 300 replications

Offset (years)	Correct recogn. rate		Sensitivity		Specificity	
	Average	Std. dev.	Average	Std. dev.	Average	Std. dev.
1	0.8697	0.0288	0.7788	0.0497	0.9605	0.0232
2	0.8326	0.0310	0.7224	0.0499	0.9429	0.0289
3	0.8878	0.0264	0.8259	0.0442	0.9497	0.0271
4	0.8776	0.0256	0.8192	0.0389	0.9349	0.0265
5	0.8891	0.0298	0.8448	0.0483	0.9335	0.0273
6	0.8387	0.0414	0.7883	0.0707	0.8892	0.0501
7	0.7620	0.0518	0.7391	0.0735	0.7848	0.0824
8	0.6731	0.0639	0.5927	0.0991	0.7535	0.0984

Table 2 CARTS: average recognition rates sensitivities and specificities and corresponding standard deviations over 300 replications

Offset (years)	Correct recogn. rate		Sensitivity		Specificity	
	Average	Std. dev.	Average	Std. dev.	Average	Std. dev.
1	0.9159	0.0260	0.9299	0.0239	0.9019	0.0412
2	0.8487	0.0529	0.8341	0.0581	0.8634	0.0640
3	0.8108	0.0528	0.7676	0.0780	0.8539	0.0628
4	0.8101	0.0638	0.8024	0.0878	0.8177	0.0677
5	0.7926	0.0630	0.7393	0.0793	0.8459	0.0761
6	0.7124	0.0869	0.6630	0.1091	0.7618	0.1098
7	0.7339	0.0954	0.6993	0.1160	0.7686	0.1120
8	0.6301	0.1476	0.5878	0.1967	0.6724	0.1504

firm. The inverse is not true, i.e. a company that is classified as healthy by a highly specific test does not mean that it will not fail. The performance with the highest specificities are those obtained up to 6 years prior to failure. Once again the standard deviation of the performance has the same trend it showed for the recognition rate.

In Table 2 the results for correct recognition rate, sensitivity and specificity obtained using CARTs on the same data-sets have been displayed together with the corresponding standard deviations over the 300 trials.

CARTs perform better than the proposed technique w.r.t. correct recognition rate and sensitivity for 1 and 2 years prior to failure. In all the other cases the proposed technique more than holds its own. It always shows a better recognition rate, sensitivity and specificity than CARTs for the time span from 3 to 8 years prior to failure. It shows a better specificity over CARTs. As the time lag increases the variability of the performance of CARTs increases much more than the corresponding one for the proposed technique, suggesting a more robust behavior.

4 Conclusions

We have introduced a new non parametric method to predict the possible failure of a firm with a time lag from 1 to 8 years prior to failure.

We performed a cross sectional study based on a sample of 100 Italian non listed SMEs over the time period from 2000 to 2011, considering 50 firms that have declared bankruptcy during this time period and 50 still active on the market at 2011.

In this paper the performance of the proposed method has been compared to that of classification trees.

The proposed algorithm has a very high recognition rate, resulting in a very high prediction power, sensitivity and specificity and therefore is good at providing a nice trade off between the ability to detect firms that are likely to fail in the short time and to rule out firms that will not.

The recognition rates are quite good, and the predictive ability of the methods has been tested up to 8 years prior to failure. These results are in line with those available in the literature going from 90 % correct recognition rate one year prior to failure to 66 % 8 years prior to failure.

References

1. Abdullah, N.A.H., Ahmad, A.H., et al.: Predicting corporate failure of malaysians listed companies: comparing multiple discriminant analysis, logistic regression and the hazard model. Int. Res. J. Financ. Econ. **15**, 201–217 (2008)
2. Altman, E.I.: Financial ratios, discriminant analysis and prediction of corporate bankruptcy. J. Financ. **23**(4), 589–609 (1968)
3. Appetiti, Sandro: Identifying unsound firms in Italy: an attempt to use trend variables. J. Bank. Financ. **8**(2), 269–279 (1984)
4. Barnes, P.: The analysis and use of financial ratios: a review article. J. Bus. Financ. Account. **14**, 449–461 (1987)
5. Beaver, W.H.: Financial ratios as predictors of failure. JAR **4**, 71–111 (1966)
6. Bellovary, J.L., Giacomino, D.E., Akers, M.D.: A review of bankruptcy prediction studies: 1930 to present. J. Financ. Educ. **33**, 1–42 (2007)
7. Bhimani, A., Gulamhussen, M.A., Da Rocha Lopes, S.: The role of financial, macroeconomic, and non-financial information in bank loan default timing prediction. Eur. Account. Rev. **22**(4), 739–763 (2013)
8. Breiman, L., Friedman, J., Olshen, R., Stone, C.: Classification and Regression Trees. Wadsworth and Brooks, Monterey (1984)
9. di Donato, F., Nieddu, L.: The effects of performance ratios in predicting corporate bankruptcy: the Italian case. In: Delibasic, B., Hernández, J.E., Papathanasiou, J., Dargam, F.C.C., Zaraté, P., Ribeiro, R.A., Liu, S., Linden, I. (eds.) Decision Support Systems V—Big Data Analytics for Decision Making—First International Conference, ICDSST 2015, Belgrade, Serbia, May 27–29, 2015, Proceedings. Lecture Notes in Business Information Processing, vol. 216, pp. 61–72. Springer, Berlin (2015)
10. Edmister, R.O.: An empirical test of financial ratio analysis for small business failure prediction. J. Financ. Quant. Anal. **7**, 1477–1493 (1972)
11. Fisher, R.A.: The use of multiple measurements in taxonomic problems. Ann. Eugen. **7**, 179–188 (1936)
12. Gepp, A., Kumar, K.: Business failure prediction using statistical tecniques: a review. Bond University ePublication (2012)
13. Hall, Graham: Factors distinguishing survivors from failures amongst small firms in the uk construction sector*. J. Manag. Stud. **31**(5), 737–760 (1994)
14. Hastie, T., Tibshirani, R., Friedman, J.H.: The Elements of Statistical Learning: Data Mining, Inference, and Prediction. Springer, New York (2001)
15. Jones, F.L.: Current techniques in bankruptcy prediction. J. Account. Lit. **6**, 131164 (1987)
16. Laitinen, E.K.: Financial ratios and different failure processes. J. Bus. Financ. Account. **18**(5), 649–673 (1991)
17. MacQueen, J.B.: Some methods for classification and analysis of multivariate observations. In: Le Cam, L.M., Neyman, J. (eds.) Proceedings of the Fifth Berkeley Symposium on Mathematical Statistics and Probability, vol. 1, pp. 281–297. University of California Press (1967)
18. McKee, T.E.: Rough sets bankruptcy prediction models versus auditor signaling rates. J. Forecast. **22**, 569–586 (2003)

19. Ohlson, J.A.: Financial ratios and probabilistic prediction of bankrucy. JAR **18**(1), 109–131 (1980)
20. Storey, D.J., Keasey, K., Wynarczyk, P., Watson, R.: he performance of small firms: profits, jobs and failures. University of Illinois at Urbana-Champaign's Academy for Entrepreneurial Leadership Historical Research Reference in Entrepreneurship (1987)
21. Watanabe, S.: Pattern Recognition: Human and Mechanical. Wiley, New York (1985)

Part II
Applications

The Inflation Hedging Ability of Domestic Gold in Malaysia

Hooi Hooi Lean and Geok Peng Yeap

Abstract Among the investment assets, gold is historically been thought as a powerful inflation hedge to many households in Malaysia. This paper examines and compares the hedging properties of gold against both consumer and energy inflation risks in Malaysia. Using the monthly domestic gold price, we test the long-run and short-run relationships between gold return and consumer inflation as well as energy inflation. We find that gold investment in Malaysia is a good hedge against consumer inflation and energy inflation in the long run but not for the short run. We also could not find any evidence of short-run causality between gold return and both consumer and energy inflations.

1 Introduction

According to Worthington and Pahlavani [45], gold is durable, transportable, generally acceptable and easily authenticated. Moreover, gold fulfils the basic functions of money as a reliable store of value and has long been regarded as an effective inflation hedge. Gold is proved to be the most effective portfolio diversifier and outperform other financial assets held by typical US investors [14]. Gold is a counter-cyclical commodity and it is always perceived as a safe haven asset [4–6, 37]. Nevertheless, gold is now getting lesser attention of investors. This is because the holding period for gold is usually longer than other financial assets. Moreover, unlike other type of financial assets, gold does not have periodic payments.[1]

[1] Bonds provide coupon payments and principal upon the maturity date. Stocks provide dividend that could offer inflation protection to the investors.

H.H. Lean (✉) · G.P. Yeap
Economics Program, School of Social Sciences, Universiti Sains Malaysia,
Penang, Malaysia
e-mail: hooilean@usm.my

G.P. Yeap
e-mail: gpyeap@gmail.com

V.-N. Huynh et al. (eds.), *Causal Inference in Econometrics*,
Studies in Computational Intelligence 622, DOI 10.1007/978-3-319-27284-9_14

There are several ways to define inflation hedging of an asset. As defined in Bodie [10], a security is an inflation hedge if and only if its real return is uncorrelated with the rate of inflation. Bekaert and Wang [8] clarify that the minimum requirement for an asset to be considered as a good inflation hedge is to show a positive correlation between its nominal returns and inflation. Ghosh et al. [22] suggest that gold is an effective inflation hedge in the long-run if the nominal price of gold and the general price level are moving together. According to Ghosh et al. [22], gold must be able to maintain its real value over time to be considered as an effective inflation hedge.

Although inflation rate is low in Malaysia, several sudden events such as unanticipated oil price changes and the depreciation of *Ringgit*, will cause unexpected change in the inflation. When oil price hikes, people will change their expectation on the general price level because oil is one of the most important resources for production and transportation. A rise in oil price leads to higher production cost and induces supply side inflation. On the demand side, higher energy price leads to a lower household's disposable income and reduces the household's consumption [43]. Thus, through the income effect, households lose their purchasing power due to the inflation associated with increasing oil price (Kilian [29]).

The expected inflation is basically the market view on the future general price level in the country. When price changes are expected, the expected inflation is apparently incorporated into the current asset prices. It is considered less risky because people are ready and plan their investment portfolio based on the expected inflation. As a result, the existence of inflation risk is mainly the concern about unexpected inflation. The unexpected inflation is the unpredictable component of inflation and may have serious impact on people real wealth. The unexpected inflation is an uncertainty and it has been shown as one of the important causes of inflation [12, 13]. When an unexpected change on general price level occurs, analysts will revise their forecasts on future inflation to account for the unexpected changes. Hence, we also examine whether gold is able to hedge the expected and unexpected inflation.

This study examines the inflation hedging ability of gold in Malaysia. We contribute to the literature in several aspects. First, two types of inflation are considered in this study, the consumer inflation and energy inflation. Previous studies have examined the long-run and short-run relationships between oil price and gold price but few consider energy inflation.[2] Second, there are many studies on the inflation hedging ability of gold in the developed markets but the studies in emerging markets are relatively fewer. We add to the literature by examining the case in Malaysia. Third, we decompose the inflation into expected and unexpected components to analyse the risk involves in gold investment to the unexpected changes in both consumer and energy prices. Lastly, we make the first attempt to examine the causal relationship between gold return and both consumer and energy inflation.

---

[2]Breitenfellner et al. [11] on energy price and house price is the only study found.

The remainder of the paper is organised as follows. The following section provides an overview of gold investment in Malaysia. It follows by literature review on gold and inflation and the relationship between gold and oil markets in Sect. 3. Section 4 presents the data and methodology. Section 5 reports the empirical results and the last section concludes the study.

2 Gold investment in Malaysia

Gold investment in Malaysia is divided into two major trading medium—the physical gold and the paper gold [23]. The physical gold, in the form of gold jewellery, gold bar, gold coin and gold bullion, is the most traditional way of investment. Gold jewellery is easily available at goldsmiths and it is the most common type of gold trading in Malaysia. Besides serving the purpose of investment, gold jewelleries mainly act as luxury jewellery wear especially for women.

Another type of physical gold investment is gold bullion coins. *Kijang Emas* Gold Bullion Coins is the Malaysia's gold bullion coins which is issued by the Bank Negara Malaysia and can be purchased at selected Maybank branches. The *Kijang Emas* is available in three different sizes, 1 Troy ounce, $\frac{1}{2}$ Troy ounces and $\frac{1}{4}$ Troy ounces with fine gold content of 99.99 %. The purchase and selling prices of *Kijang Emas* are quoted daily based on the international gold price [33]. Moreover, the physical gold bars and coins can also be purchased at some banks and jewellery shops in a variety of weights and sizes. These gold bars and coins are traded based on the daily buying and selling prices.

The paper gold is available in the form of gold saving or investment account which is offered by banks. The investors will be given a passbook to record all the transactions of buying and selling. The buying and selling price of gold is quoted by each individual bank in Malaysia *Ringgit* per gram based on 999.9 (24 Karat) gold price. This type of gold investment is gaining popularity today as it is more convenient and investors do not have to worry about the safe storage of holding physical gold [24]. Investors or account holder will not be given dividend or interest based on their investment. Investors will only gain if they sell the gold at a higher price than their initial buying price. It is a long term saving and is believed to provide a good hedge against inflation like holding physical gold. Another type of paper gold is Gold Exchange Traded Funds (Gold ETF). Gold ETF is listed in the stock market and allow investors to buy and sell like shares during stock market hours. The Gold ETF will invest in physical gold or gold related companies and then issue shares to its investors. Investors of Gold ETF gain exposure to gold price where the value of the shares will increase with the increase price of gold bullion [23].

3 Literature Review

The relationship between inflation and gold. Studies that consider the long run inflation hedging ability of gold investment based on consumer inflation are Taylor [41], Ghosh et al. [22], Worthington and Pahlavani [45], Wang et al. [44], Beckmann and Czudaj [7], Shahbaz et al. [39] and Bampinas and Panagiotidis [4]. Among these studies, Worthington and Pahlavani [45], Shahbaz et al. [39] and Bampinas and Panagiotidis [4] account for structural breaks in their analysis and show that gold prices and inflation rate are strongly cointegrated in the long run.

Wang et al. [44] and Beckmann and Czudaj [7] assess the nonlinear relationship between the inflation rate and gold prices. Wang et al. [44] find the that inflation does not affect god return in high momentum regime or low momentum regime in Japan which shows the inability of short run gold investment to hedge against inflation in Japan. The result is opposite in the US. It shows positive bilateral causality between gold return and inflation which means that gold investment in the US could hedge against inflation in the short run. Beckman and Czudaj [7] test using two measures of inflation e.g. consumer price index and producer price index and they find significant results for both inflation measures. They conclude that gold price moves in accordance with inflationary expectation and it is an effective hedge against inflation.

Taylor [41] study based on two sample periods namely Post-War period from 1968M1 to 1996M4 and Pre-War period from 1914M1 to 1937M12. The results show that gold acts as short run hedge during particular periods, specifically in the second OPEC oil crisis in 1979. In contrast, gold was not hedge against inflation around the first oil crisis in 1973/74. Ghosh et al. [22] find a cointegration relationship between gold price, U.S. price index and world price index. Gold return and inflation rate seem to show positive relation in the short run but it may be caused by other influences such as convenience yield, leasing rate and the supply and demand dynamics. Andrangi et al. [1] also find inflation hedging potential for gold in the long run based on positive relationship between CPI and gold prices.

Some other authors study the inflation hedging of gold based on expected and unexpected inflation. Feldstein [18] find that a higher expected inflation causes a rise in the relative price of land and gold. It shows a positive relationship between expected inflation and gold prices. Among other papers that study the impact of expected inflation on gold prices are Jaffe [28], Larsen and McQueen [30], Adrangi et al. [1], Blose [9] and Bampinas and Panagiotidis [4].

Earlier study by Jaffe [28] uses Treasury bill as a proxy for expected inflation. A significant positive relation between gold return and unexpected inflation is found but show negative sign. This result suggest that gold is not a good inflation hedge but the author believes that there is a long run relationship between gold and inflation as gold remarkably preserves its purchasing power over long run. Larsen and McQueen [30] study the hedging ability of gold and real estate and their securitized form, gold stocks and REITs. They use three different measures of expected inflation namely Treasury bill, ARIMA and naive model and find that gold return has significant positive relationship with unexpected inflation at 10 % level. As such, gold is a good hedge against inflation but gold stock is not as it shows insignificant relationship with expected and unexpected inflation.

Moreover, Adrangi et al. [1] attempt the study on real gold return. They find a significant positive relationship between real gold returns and expected inflation but unexpected inflation is found not affecting gold prices. The test on long run relationship show significant positive results and they confirm gold is a long run hedge against inflation. However, Blose [9] does not find gold price changes to significantly react to expected and unexpected changes in consumer price index.

In Malaysia, the inflation hedging ability of domestic gold is examined by Ghazali et al. [20]. Using monthly data from July 2001 to November 2011, they can only find positive relationship between gold return and inflation with insignificant coefficient. Both expected and unexpected inflation are also found to have positive relationship with gold return but the coefficients are not significant as well. They conclude that domestic gold is not a store of value and does not help Malaysian investors to hedge against inflation in the short run. Ibrahim and Baharom [27] have studied the investment role of gold in Malaysia and they find that gold does not perform its role as a strong hedging, diversifying and safe haven asset. The role played by gold as an investment asset may vary depending on different market conditions. Ibrahim [26] further confirm the diversification ability of domestic gold against the Malaysian stock market. Domestic gold also tends to be a good hedging asset during declining stock market conditions. On the other hand, more recent studies by Ghazali et al. [19] and Ghazali et al. [20, 21] show that Malaysian domestic gold acts as a hedge against stock market but not a safe haven during extreme market conditions.

The relationship between oil and gold prices. The literature that examines the direct linkage between oil prices and gold market has received more attention. The earlier study that finds a strong positive response of gold prices to the changes in crude oil prices is Baffes [3] based on the annual data from 1960 to 2005 and a simple econometric model. Baffes [3] opine that the demand for precious metals is expected to increase when crude oil price spikes as people view these metals as more secure ways for storing wealth. However, using exponential generalised autoregressive conditional heteroskedasticity (EGARCH) model, Hammoudeh and Yuan [25] find insignificant effect of oil price shocks on gold returns and its volatility. These authors show that gold is a better investment during crises or high inflation times as it has significant negative leverage effect.

Similar result is found by Soytas et al. [40] in Turkey who examine the long run and short run dynamic relationship between world oil prices, local gold and silver sport prices and the Turkish lira/USD exchange rate. The world oil price does not Granger cause the price of gold in emerging economy in the long run. Domestic gold acts as a safe haven in Turkey during the country's currency devaluation. However, the domestic spot prices of gold seem to have significant positive impacts on oil price in the short run. Sari et al. [38] only find weak long run relationship between oil price return and gold return thereby oil explaining 1.7 % of gold price returns. In their opinion, gold is considered safe haven asset due to their strong response to inflationary expectations. In the long run, investors may benefits from diversifying the risk of price fluctuations into gold since there is only weak evidence found between oil and gold prices.

With the aforementioned empirical results, Narayan et al. [35] contribute a theoretical framework for the understanding of the relationship between oil and gold prices. Their theoretical motivation is explained by the inflation channel where they argue that a rise in oil price will lead to the increase in general price level. The higher price level causes inflation which in turn drives up the price of gold. Hence, an oil price rise causes a rise in gold price. Their results reveal that oil market and gold market are cointegrated and they believe that gold could be used to hedge against inflation.

Zhang and Wei [46] analyze the cointegration relationship and causality between gold and oil markets. Their results indicate that crude oil and gold prices are high positively correlated. They evident a long-term relationship between oil and gold market and only linear Granger causality exist from oil price to gold price. Moreover, Reboredo [37] examines the role of gold as a hedge or safe heaven against oil price movement. His results reveal that gold and oil are dependent significantly which imply that gold is not a hedge against oil prices but he has evidence to show that gold is an effective safe haven during oil market stress.

In conclusion, most of the above studies show that gold is an effective hedge against inflation in the long run. The short run hedging ability against expected and unexpected inflation remains mixed and inconclusive due to different country of study and different sample period. None of the studies on the link between oil price and gold price express the changes in oil price as energy inflation. The idea of the term energy inflation is initiated from Breitenfellner et al. [11] who use this term in the study of housing price.

4 Data and Methodology

Data. The domestic gold price is represented by the selling price of one troy ounce of *Kijang Emas* that collected from Bank Negara Malaysia. Inflation rate is computed from Consumer Price Index (CPI) collected from the International Financial Statistics. We use the benchmark West Texas Intermediate crude oil price and the data is collected from the U.S. Energy Information Administration. The sample period is from July 2011 until April 2015. All data are in monthly frequency and are transformed into natural logarithm series.

Estimating expected and unexpected inflation. The theory of hedging against inflation refers to Fisher [17] hypothesis that nominal interest rate is the total of expected real rate of return and expected rate of inflation. Fama [15] believes there are two sources of inflation uncertainty, i.e. the expected inflation and unexpected inflation; and the uncertainty about future expected inflation would affect the return of a portfolio. As a result, Fama and Schwert [16] propose a model that reflects the unanticipated component of inflation. The model suggests an asset return as a linear function of expected and unexpected inflation in which it has been extensively used to examine the inflation hedging characteristic of an asset.

Expected inflation is the market expectation on general price level in the coming period. There are three popular measures of expected inflation: (1) time series method, (2) Treasury bill method and (3) survey forecast. The time series method utilizes univariate autoregressive integrated moving average (ARIMA) to model the expected inflation rate. The fitted value of inflation rate is taken as expected inflation and the residual is the unexpected inflation. In the gold market studies, several authors have used this method such as Larsen and McQueen [30] and Adrangi et al. [1].

The second method is considered as 'traditional' method in the inflation hedging literature. It follows Fama and Schwert [16] framework where Treasury bill rate is used as a proxy for expected inflation. Expected inflation is calculated as Treasury bill rates minus constant real rate of return. This method is not common in the literature because it assumes constant real rate of return which is less appropriate in the time-varying model. The last approach is a survey-based approach where expected inflation is collected from survey on inflationary expectations. The adequacy of this method remains debatable although some authors regard this method as better forecast of expected inflation than other alternatives [2, 42]. We will not consider this method in this study due to unavailability of survey data of expected inflation in Malaysia.

In this study, we employ the first measure, i.e. an ARIMA model to estimate the expected and unexpected inflation. We choose the best-fitted model with the lowest value of Schwarz Information Criteria (SIC) to estimate the expected and unexpected inflation. The estimated value is used as expected inflation. The residual which is the actual inflation minus expected inflation, will be the unexpected inflation.

Table 1 Descriptive statistics

Variables	Mean	Std. Dev.	Skewness	Kurtosis
GP	7.9391	0.5134	−0.2886	1.6991
CP	4.5558	0.0985	−0.0396	1.6435
EP	4.1191	0.4825	−0.6663	2.3421
GR	0.0088	0.0491	−0.3845	3.8444
CI	0.0019	0.0043	3.6837	35.8635
ECI	0.0021	0.0020	0.4054	10.8011
UCI	−0.0001	0.0038	4.3155	39.3503
EI	0.0045	0.0886	−0.9934	4.9775
EEI	0.0010	0.0467	−1.1628	6.4387
UEI	0.0035	0.0786	−0.2338	2.7918

Note GP gold price; *CP* consumer price index; *EP* energy price (as proxy by WTI crude oil price); *GR* gold return; *CI* consumer inflation; *ECI* expected consumer inflation; *UCI* unexpected consumer inflation; *EI* energy inflation; *EEI* expected energy inflation and *UEI* unexpected energy inflation. Inflation = Expected Inflation + Unexpected Inflation

The summary statistics of the data are presented in Table 1. The level series show that mean gold price is relatively higher than the consumer and energy prices over the sample period. The energy price is five times more volatile than the consumer price. Gold has an average monthly return of 0.88 % which is higher than the average consumer inflation (0.19 %) and energy inflation (0.45 %). Energy inflation shows a much higher standard deviation than consumer inflation over the sample period. It seems that gold investment in Malaysia is exposed to higher energy inflation risk than the risk comes from consumer inflation.

For the expected and unexpected components of inflation, the expected consumer inflation has a mean of 0.21 % which is higher than the expected energy inflation (0.10 %). Conversely, the mean of unexpected energy inflation (0.35 %) is higher than the mean of unexpected consumer inflation (−0.01 %). We note that the unexpected component has a bigger weight in energy inflation but this is not the case for consumer inflation. A possible explanation is that a sudden hike of oil price causes unexpected increase in consumer inflation. In anticipation of the higher consumer inflation, central bank will tighten the monetary policy to stabilize the general price level (Kilian [29]; Mozes and cooks [34]; [11]). So, the unexpected component of consumer inflation is smaller because the unexpected changes in oil price have been reflected in the expected component of consumer inflation.

Methodology. The analysis of this study is divided into two parts. The first part is to examine the long run relationship between gold price and consumer price and energy price respectively. The second part is to test the inflation hedging ability of gold based on the Fama and Schwert [16] model.

Long run relationship. Johansen cointegration test is applied to examine the existence of long-run equilibrium relationship between gold and consumer prices and between gold and energy prices. After the establishment of cointegration, we proceed to estimate the long-run relationship using Fully Modified Ordinary Least Square (FMOLS) method with the following equations:

$$GP_t = \alpha + \beta_1 CP_t + \varepsilon_t \tag{1}$$

$$GP_t = \alpha + \beta_1 EP_t + \varepsilon_t \tag{2}$$

where Eq. (1) shows the relationship between gold price (GP) and consumer price (CP) and Eq. (2) is to test the link between gold price and energy price (EP). To examine the inflation hedging ability of gold in the long run, the coefficients (β_1) of CP and EP must be positive and significant.

Relationship between gold return and inflation. Following Lee and Lee [32] and Lee [31], we adopt the model developed by Fama and Schwert [16] to test the relationship between gold return and two types of inflation. There are two models. The first model tests whether gold is an effective hedge against actual inflation:

$$GR_t = \alpha + \beta CI_t + e_t \tag{3}$$

where GR represents gold return and CI is the actual consumer inflation. Gold is considered as a hedge against actual inflation if β is significantly greater than zero.

The second model is based on Eq. (4):

$$GR_t = \alpha + \beta_1 ECI_t + \beta_2 UCI_t + e_t \tag{4}$$

where ECI and UCI are the expected and unexpected inflation respectively. In Eq. (4), gold is hedged against expected inflation if $\beta_1 > 0$ while it is considered a hedge against unexpected inflation if $\beta_2 > 0$. If these coefficients, β_1 and β_2, are equal to 1, gold is a complete hedge against expected and unexpected inflation respectively. In addition, gold is a partial hedge against inflation if these coefficients are less than 1, but greater than zero. A negative coefficient shows inability of gold to hedge against inflation.

To examine the impact of energy inflation on gold returns, Eqs. (3) and (4) are re-written as:

$$GR_t = \alpha + \beta EI_t + e_t \tag{5}$$

$$GR_t = \alpha + \beta_1 EEI_t + \beta_2 UEI_t + e_t \tag{6}$$

where EI is actual energy inflation, EEI represents expected energy inflation and UEI represent unexpected energy inflation. The parameters β, β_1 and β_2 have similar interpretation as consumer inflation.

Granger causality test. Granger causality test based on the vector error correction model (VECM) is applied to determine the causality direction between gold return and consumer inflation and between gold return and energy inflation. The following VECM models will be investigated.

$$\begin{bmatrix} \Delta GP_t \\ \Delta CP_t \end{bmatrix} = \begin{bmatrix} \alpha_1 \\ \alpha_2 \end{bmatrix} + \begin{bmatrix} A_{11,1} & A_{12,1} \\ A_{21,1} & A_{22,1} \end{bmatrix} \begin{bmatrix} \Delta GP_{t-1} \\ \Delta CP_{t-1} \end{bmatrix} + \dots + \begin{bmatrix} A_{11,k} & A_{12,k} \\ A_{21,k} & A_{22,k} \end{bmatrix} \begin{bmatrix} \Delta GP_{t-k} \\ \Delta CP_{t-k} \end{bmatrix} + \begin{bmatrix} \phi_1 \\ \phi_2 \end{bmatrix} \times [ECT_{t-1}] + \begin{bmatrix} \varepsilon_{1t} \\ \varepsilon_{2t} \end{bmatrix} \tag{7}$$

$$\begin{bmatrix} \Delta GP_t \\ \Delta EI_t \end{bmatrix} = \begin{bmatrix} \alpha_1 \\ \alpha_2 \end{bmatrix} + \begin{bmatrix} A_{11,1} & A_{12,1} \\ A_{21,1} & A_{22,1} \end{bmatrix} \begin{bmatrix} \Delta GP_{t-1} \\ \Delta EP_{t-1} \end{bmatrix} + \dots + \begin{bmatrix} A_{11,k} & A_{12,k} \\ A_{21,k} & A_{22,k} \end{bmatrix} \begin{bmatrix} \Delta GP_{t-k} \\ \Delta EP_{t-k} \end{bmatrix} + \begin{bmatrix} \phi_1 \\ \phi_2 \end{bmatrix} \times [ECT_{t-1}] + \begin{bmatrix} \varepsilon_{1t} \\ \varepsilon_{2t} \end{bmatrix} \tag{8}$$

where ECT_{t-1} is lag error correction term which is estimated as below:

$ECT_{t-1} = GP_{t-1} - a - b_1 CP_{t-1}$ for Equation (7) and $ECT_{t-1} = GP_{t-1} - a - b_1 EP_{t-1}$ for Equation (8).

Long-run causality exists if the coefficient of ECT_{t-1} is negative and statistical significant. The existence of short-run causality is indicated by the significance of the χ^2 test between the variables.

5 Results

We perform the augmented Dickey-Fuller (ADF) and Phillip-Perron (PP) unit root tests to access the stationarity of the series. The results are presented in Table 2. The results clearly show that all level series are not stationary and their first differences are stationary which suggest that all series are I (1).

Next, we test for the long run equilibrium between gold and consumer prices and between gold and energy prices respectively. The results in Table 3 show a long-run equilibrium exists among these variables. Trace statistic and max-eigen statistic show that there is at least one cointegrating equation between gold price and consumer price. The results also show that there are at least two cointegration equations between gold price and energy price. Normalized cointegrating coefficients for both consumer price and energy price are both positive and statistically significant. This infers that a one percent increase in consumer price leads to 7.72 percent increase in Malaysian gold price in the long run and a one percent increase in energy price leads to 0.96 percent increase in the Malaysian gold price in the long run.

Table 2 Unit root test

Level series			Return series		
Variable	ADF	PP	Variable	ADF	PP
GP	−1.7745	−1.9723	GR	−15.5489***	−15.9324***
CP	−0.2609	−0.1434	CI	−9.0360***	−8.9141***
OP	−2.0895	−2.0161	ECI	−5.5183***	−10.7718***
			UCI	−12.7925***	−12.7925***
			EI	−8.9793***	−9.0242***
			EEI	−8.5812***	−4.3958***
			UEI	−12.9670***	−13.4771***

Note *** indicates significant at 1 % level

Table 3 Johansen Cointegration Test

	r = 0		r = 1		Normalized cointegrating coefficients	
	Trace statistic	Max-Eigen statistic	Trace statistic	Max-Eigen statistic	CP	EP
GP and CP	29.1379**	25.6372***	3.5007	3.5007	7.7214***	–
GP and EP	29.7394***	18.4372**	11.3022**	11.3022**	–	0.9606***

Note ** and *** indicate significant at 5 % and 1 % respectively

Table 4 FMOLS estimation results

	$GP_t = \alpha + \beta_1 CP_t + \varepsilon_t$	$GP_t = \alpha + \beta_1 EP_t + \varepsilon_t$
α	−14.7615***	4.0221***
β	4.9828***	0.9524***
R^2	0.9139	0.7139

Note *** indicates significant at 1 % level

Table 4 reports the FMOLS results of long run inflation hedging ability of gold. Consistent with the normalised equations above, the coefficients of CP and EP are positive and statistically significant at 1 % level. This indicates that gold price varies positively with the consumer price and energy price in the long run and suggests that gold hedges both consumer inflation and energy inflation over the long term period of time. The results are also consistent with the findings of Taylor [41], Ghosh et al. [22], Worthington and Pahlavani [45] and Shahbaz et al. [39]. So far, none of the study in Malaysia attempts to test the long-run relationship between gold price and inflation. Our results reveal that Malaysian gold maintains its value against both types of inflation over long investment horizons. Malaysian domestic gold is a good hedge against consumer inflation ($\beta > 1$) and a partial hedge against energy inflation ($0 < \beta < 1$). As Malaysian domestic gold shows its ability to hedge against inflation over a long investment horizon, investors who wish to store the value of wealth can hold gold for a longer period of time. The hedging ability is more useful for consumer inflation than the energy inflation.

VECM is estimated to examine the short-run dynamics between gold return and consumer and energy inflation respectively. The results are reported in Table 5. The lagged error correction term (ECT) for both consumer and energy inflation are negative and significant. None of the independent variable is significantly affect the changes in gold price (or gold return). We also could not find any significant short-run

Table 5 VECM results

Panel A: gold return and consumer inflation		Panel B: gold return and energy inflation	
Variable	Coefficient	Variable	Coefficient
ECT(−1)	−0.0310*	ECT(−1)	−0.0295**
ΔGP(−1)	−0.2025**	ΔGP(−1)	−0.1741**
ΔGP(−2)	−0.0754	ΔGP(−2)	−0.0314
ΔGP(−3)	−0.0274	ΔEP(−1)	−0.0364
ΔCP(−1)	0.3663	ΔEP(−2)	−0.0609
ΔCP(−2)	−0.8822		
ΔCP(−3)	−0.0175		
Granger causality	χ^2	*Granger causality*	χ^2
ΔCP → ΔGP	0.8875	ΔEP → ΔGP	3.0265
ΔGP → ΔCP	2.6918	ΔGP → ΔEP	0.8568

Note * and ** indicate significant at 10 % and 5 % respectively

Table 6 OLS estimation results

Variables	Equation (3)	Equation (4)	Equation (5)	Equation (6)
Constant	0.0100**	0.0084	0.0103***	0.0099**
GR(−1)	−0.1982**	−0.2014**	−0.1996**	−0.2030***
Actual inflation	0.2329	–	0.0407	–
Expected inflation	–	0.7900	–	−0.0444
Unexpected inflation	–	0.1069	–	0.0718
F-statistic	3.3441**	2.3445*	3.7766**	3.1984**

Note *, ** and *** indicate significant at 10 %, 5 % and 1 % respectively

causality between gold return and consumer inflation and energy inflation respectively. Hence, Malaysian gold is not hedging against both consumer and energy inflation in the short run.

The best fitted model for consumer inflation and energy inflation respectively is ARIMA (2,1,4). The inflation hedging ability of gold based on Eqs. (3), (4), (5) and (6) is tested using the Ordinary Least Square (OLS) method. The estimation result shows there is autocorrelation problem in all models. We solve the problem by adding the lagged of gold return into the models. The results are presented in Table 6. Gold return show positive relationship with the actual, expected and unexpected consumer inflation but are not significant. This indicates that domestic gold in Malaysia is not a significant consumer inflation hedge in the short run supporting Ghazali et al. (2012). Conversely, the result contradicts with studies on consumer inflation such as Adrangi et al. [1] and Larsen and McQueen [30] who find significant relation with unexpected inflation in the U.S.

For the energy inflation, gold return is positively related to the actual and unexpected energy inflation but negatively related to the expected energy inflation. However, all coefficients are not significant as well showing that domestic gold in Malaysia is not a good hedge against energy inflation in the short run. The finding is in line with Hammoudeh and Yuan [25] who reported insignificant relationship between oil price shock and gold return.

Comparing the consumer inflation and energy inflation, higher coefficient of actual consumer inflation showing that domestic gold seems to demonstrate better ability to hedge the consumer inflation than the energy inflation. While Malaysian domestic gold investment showing its ability to protect investors against consumer inflation, the expected energy inflation adversely affects the domestic gold return. It is likely to have some important implications to Malaysian gold investors as the country's fuel price is currently under managed float system. Domestic gold investors are exposed to greater volatility of world crude oil price. Investors may not be able to protect their wealth due to the inflation stemming from rising price of energy. As domestic gold only shows its inflation hedging ability in the long run, Malaysian gold investors may lose their purchasing power over inflation in the short run especially against the energy inflation.

Conclusion. This study investigates the relationship between gold and consumer prices and between gold and energy prices in Malaysia. We would like to know whether domestic gold provides a reliable hedge against both consumer and energy inflation in Malaysia. We find that gold investment is neither hedge the short-run inflation risk against the consumer inflation nor energy inflation. Bampinas and Panagiotidis [4] also document that gold is a poor inflation hedge in the short run but as the investment horizon increases they may provide adequate long-run hedging abilities. This argument holds true for the case of Malaysia.

The domestic gold investors should consider both consumer and energy inflation in making their investment decisions. The government's move to abolish fuel subsidy renders Malaysian households and investors to more uncertainty in energy inflation. Moreover, the introduction of Goods and Services Tax (GST) has severely increased the consumer inflation in the country. These two recent policies may cast investors with the worry of greater inflationary pressure. Investors who wish to protect their wealth against the inflation risk can consider gold in their investment portfolio for long term investment horizon. Indeed, the disability of domestic gold in hedging the short-run inflation should not be ignored especially for the case of poor hedge against unexpected energy inflation. Gold investors may need to gain higher returns from other investment alternative that can provide more protection against energy inflation.

Acknowledgments The authors would like to acknowledge the Fundamental Research Grant Scheme 203/PSOSIAL/6711417 by Ministry of Education Malaysia and Universiti Sains Malaysia.

References

1. Adrangi, B., Chatrath, A., Raffiee, K.: Economic activity, inflation and hedging: the case of gold and silver investment. J. Wealth Manag. **6**(2), 60–77 (2003)
2. Ang, A., Bekaert, G., Wei, M.: Do macro variables, asset markets, or surveys forecast inflation better? J. Monet. Econ. **54**, 1163–1212 (2007)
3. Baffes, J.: Oil spills on other commodities. Resour. Policy **32**, 126–134 (2007)
4. Bampinas, G., Panagiotidis, T.: Are gold and silver a hedge against inflation? A two century perspective. Int. Rev. Financ. Anal. forthcoming
5. Baur, D.G., Lucey, B.M.: Is gold a hedge or a safe haven? An analysis of stocks, bonds and gold. Financ. Rev. **45**, 217–229 (2010)
6. Baur, D.G., McDermott, T.K.: Is gold a safe haven? international evidence. J. Bank. Financ. **34**, 1886–1898 (2010)
7. Beckmann, J., Czudaj, R.: Gold as an inflation hedge in a time-varying coefficient framework. N. Am. J. Econ. Financ. **24**, 208–222 (2013)
8. Bekaert, G., Wang, X.: Inflation risk and inflation risk premium. Econ. Policy **25**(64), 755–860 (2010)
9. Blose, L.E.: Gold prices, cost of carry, and expected inflation. J. Econ. Bus. **62**, 35–47 (2010)
10. Bodie, Z.: Common stocks as a hedge against inflation. J. Financ. **31**(2), 459–470 (1976)
11. Breitenfellner, A., Cuaresma, J.C., Mayer, P.: Energy inflation and house price corrections. Energy Econ. **48**, 109–116 (2015)
12. Buth, B., Kakinaka, M., Miyamoto, H.: Inflation and inflation uncertainty: the case of Cambodia, Lao PDR, and Vietnam. J. Asian Econ. **38**, 31–43 (2015)

13. Cukierman, A., Meltzer, A.: A theory of ambiguity, credibility, and inflation under discretion and asymmetric information. Econometrica **54**(5), 1099–1128 (1986)
14. Dempster, N., Artigas, J.C.: Gold: inflation gedge and long-term strategic asset. J. Wealth Manag. **13**(2), 69–75 (2010)
15. Fama, E.F.: Inflation uncertainty and expected returns on treasury bills. J. Polit. Econ. **84**(3), 427–448 (1976)
16. Fama, E.F., Schewert, G.W.: Asset returns and inflation. J. Financ. Econ. **5**, 115–146 (1977)
17. Fisher, I.: The Theory of Interest. McMillan, New York (1930)
18. Feldstein, M.: Inflation, tax rules, and the prices of land and gold. In: Feldstein, M. (ed.) Inflation, Tax Rules, and Capital Formation, pp. 221–228. University of Chicago Press, USA (1983)
19. Ghazali, M.F., Lean, H.H., Bahari, Z.: Is gold a hedge or a safe haven? an empirical evidence of gold and stocks in Malaysia. Int. J. Bus. Soc. **14**(3), 428–443 (2013)
20. Ghazali, M.F., Lean, H.H., Bahari, Z.: Is gold a good hedge against inflation? empirical evidence in Malaysia. J. Malays. Stud. (Kajian Malaysia), **33**(1), 69–84 (2015)
21. Ghazali, M. F., Lean, H. H., Bahari, Z.: Sharia compliant gold investment in Malaysia: hedge or safe haven? Pac.-Basin Financ. J. **34**, 192–204 (2015)
22. Ghosh, D., Levin, E.J., Macmillan, P., Wright, R.E.: Gold as an inflation hedge? Stud. Econ. Financ. **22**(1), 1–25 (2004)
23. GoldAbout Investment. http://goldaboutinvestment.com/ (2015). Retrieved 30 June 2015
24. Hafizi, A.M., Noreha, H., Norzalita, A.A., Hawati, J.: Gold investment in Malaysia: its operation, contemporary applications and shariah issues. In: Conference on Malaysian Islamic Economics and Finance, pp. 5–6 (2012)
25. Hammoudeh, S., Yuan, Y.: Metal volatility in presence of oil and interest rate shocks. Energy Econ. **30**, 606–620 (2008)
26. Ibrahim, M.H.: Financial market risk and gold investment in an emerging market: the case of Malaysia. Int. J. Islam. Middle East. Financ. Manag. **5**(1), 25–34 (2012)
27. Ibrahim, M., Baharom, A.H.: The role of gold in financial investment: a Malaysian perspective. J. Econ. Comput. Econ. Cybern. Stud. **45**, 227–238 (2011)
28. Jaffe, J.F.: Gold and gold stocks as investments for institutional portfolios. Financ. Anal. J. **45**(2), 53–59 (1989)
29. Kilian, L.: The economic effects of energy price shocks. J. Econ. Lit. **46**(4), 871–909 (2008)
30. Larsen, A.B., McQueen, G.R.: REITs, real estate and inflation: lessons from the gold market. J. R. Estate Financ. Econ. **10**, 285–297 (1995)
31. Lee, C.L.: The inflation-hedging characteristics of Malaysian residential property. Int. J. Hous. Mark. Anal. **7**(1), 61–75 (2014)
32. Lee, C.L., Lee, M.L.: Do European real estate stocks hedge inflation? Evidence from developed and emerging markets. Int. J. Strateg. Prop. Manag. **18**(2), 178–197 (2014)
33. Malayan Banking Berhad. http://www.maybank2u.com (2011). Kijang Emas Gold Bullion Coins. Retrieved 13 July, 2015
34. Mozes, H. A., Cooks, S.: The impact of expected and unexpected inflation on local currency and US dollar returns from foreign equities. J. Invest, **20**(2), 15–24 (2011)
35. Narayan, P.K., Narayan, S., Zheng, X.: Gold and oil futures markets: are markets efficient? Appl. Energy **87**, 3299–3303 (2010)
36. Newell, G.: The inflation-hedging characteristics of Australian commercial property. J. Prop. Financ. **7**(1), 6–20 (1996)
37. Reboredo, J.C.: Is gold a hedge or safe haven against oil price movements? Resour. Policy **38**, 130–137 (2013)
38. Sari, R., Hammoudeh, S., Soytas, U.: Dynamics of oil price, precious metal prices, and exchange rate. Energy Econ. **32**, 351–362 (2010)
39. Shahbaz, M., Tahir, M.I., Ali, I., Rehman, I.U.: Is gold investment a hedge against inflation in Pakistan? A co-integration and causality analysis in the presence of structural breaks. N. Am. J. Econ. Financ. **28**, 190–205 (2014)

40. Soytas, U., Sari, R., Hammoudeh, S., Hacihasanoglu, E.: World oil prices, precious metal prices and macroeconomy inTurkey. Energy Policy **37**, 5557–5566 (2009)
41. Taylor, N.J.: Precious metals and inflation. Appl. Financ. Econ. **8**(2), 201–210 (1998)
42. Thomas, L.B.: Survey measures of expected US inflation. J. Econ. Perspect. **13**(4), 125–144 (1999)
43. Tsai, C.L.: How do US stock returns respond differently to oil price shocks pre-crisis, within the financial crisis, and post-crisis? Energy Econ. **50**, 47–62 (2015)
44. Wang, K.M., Lee, Y.M., Nguyen Thi, T.B.: Time and place where gold acts as an inflation hedge: an application of long-run and short-run threshold model. Econ. Model. **28**, 806–819 (2011)
45. Worthington, A.C., Pahlavani, M.: Gold investment as an inflationary hedge: cointegration evidence with allowance for endogenous structural breaks. Appl. Financ. Econ. Lett. **3**(4), 259–262 (2007)
46. Zhang, Y.J., Wei, Y.M.: The crude oil market and the gold market: Evidence for cointegration, causality and price discovery. Resour. Policy **35**, 168–177 (2010)

To Determine the Key Factors for Citizen in Selecting a Clinic/Division in Thailand

Lee Tzong-Ru (Jiun-Shen), Kanchana Chokethaworn and Huang Man-Yu

Abstract This paper presents an integrated methodology to find out key factors that affects people choose for different types of clinic and hospital department. The requirements of the methodology not only consider factors before, during and after treatment, but also identified clinic, dental clinic, aesthetic clinic, dental department in hospital, department of family medicine in hospital, and department of orthopedics. Although there are multiple and contradictory objectives to be considered respectively, grey relational analysis (GRA) can sort out key factors to each clinic/department and be the decision maker.

1 Introduction

Nowadays, people go to hospital not only for illness, but also try to become more attractive. Since most of people can afford medical treatment, the demand to clinic and hospital now not only effectiveness, but also comfortable environment, advanced equipment, high quality service, and kindness of staffs, etc. Patients can be redefined as customers since clinic and hospital focus on promoting medical environment to be more attracted. The key factors present the tendency of customers when they

L. Tzong-Ru (Jiun-Shen) (✉)
Department of Marketing, National Chung Hsing University, No.250, Guoguang Rd., South Dist., Taichung City 40227, Taiwan, ROC
e-mail: trlee@dragon.nchu.edu.tw

K. Chokethaworn
Department of Economics, Chiang Mai University, 239 Huay Kaew Road, Muang District, Chiang Mai 50200, Thailand
e-mail: patarakn41@gmail.com

H. Man-Yu
Department of Philosophy, National Taiwan University, No. 1, Sec. 4, Roosevelt Rd., Zhongzheng Dist., Taipei City 100, Taiwan, ROC
e-mail: lemonde6201@gmail.com

V.-N. Huynh et al. (eds.), *Causal Inference in Econometrics*,
Studies in Computational Intelligence 622, DOI 10.1007/978-3-319-27284-9_15

choose select a clinic/hospital. Therefore, these key factors are significant to medical organizations.

There are ranges of reasons for people to choose a hospital or clinic. Lanes Hospital choice: A summary of the Key Empirical and Hypothetical Findings of the 1980s (1988) is one of the earliest papers that provided complete factors that affect peoples choice. More scholars have noticed topics and studied more factors since then.

In order to provide better service quality, clinics and hospitals have to know what are the key factors for patients when they chose medical organization. Following the trend of clinic/hospital taking enterprise approach, the paper collected the factors that affect customers when they select a clinic/hospital. In the end of the paper, the result of analysis and conclusion is presented.

2 Literature Review

2.1 Decision Customers Make Before Going to Clinic/Hospital

Due to not all the people go to clinic/hospital because of illness, the paper applied the word customer instead of patient. Thanks to competitive medical industries, we take medical service with more choice and affordable price. Nowadays, we do the survey and compare different clinics before we go just like buying a product. Nevertheless, service is much harder to be compared with products since service is intangible and unquantifiable.

Engel (1995) described five stages model of consumer buying process to reveal the process when consumers make decision. To find out the key factors that affect customer to choice a clinic/hospital, the paper redefined EKB consumer buying process into consumer receiving treatment process as follows.

Stages 1. Basic Purchase Decision When a potential customer has a need (e.g. toothache, tooth correction, orthopedics, etc.), he would see how serious the situation is, and decides if he really has to go to the clinic/hospital.

Stages 2. Product Category Decision If the potential customer determines to go to clinic/hospital, he needs to think which levels of the clinic/hospital he should go.

Stages 3. Brand Purchase Decision If the need is not emergency, the potential customer would do a bit of research about the clinic/hospital he might go to. For example, he would go to the clinic/hospital that he trust after determining which levels of clinic/hospital.

Stages 4. Channel Purchase Decision The potential customer would evaluate the place and if he is able to go. For example, he would think about if he should walk or drive, how to drive there, etc.

Stages 5. Payment Decision When every decision mentioned above has been considered, payment decision would become an issue, such as pay by cash or pay by card.

2.2 *Grey Relational Analysis (GRA)*

Grey relational analysis (GRA) is grounded on the concept of Grey System Theory, which was firstly formally introduced by Deng [1]. Grey system theory uses white, grey and black to refer to how well do people know about the system. White means all the information can be fully realized, while black means nothing about the construction, figure can be known. But in most of the situation, at least some incomplete information can be hold.

Grey relational analysis (GRA) conquers the issues that need a large number of data in statistic, and can deal with incomplete information. It is suitable for studies that take limitation time or restriction interviewees (e.g. a large number of doctors or nurses).

Grey relational analysis (GRA) is an impact evaluation model that measures the degree of similarity or difference between two sequences based on the grade of relation [2]. It can also be used to find out key factors even with incompletely information by comparing the similarity and difference between two sequences based on grade of relation [2]. Grey Relational Analysis (GRA) has been successfully used in variety field [3]. It does not need either statistic resume or statistic software to compile. The method is specific and easy to calculate. Once the medical organizations learn the algorithm, they can find out the key factors by themselves through excel. It can be a help to identify different key factors. Also, the organization can build their own data to compare which factors are more important to customers in certain time.

The main process of Grey Relational Analysis (GRA) is as steps as follows. For better understanding, the paper takes clinic data as example. And the rest of the clinic and department in hospital followed the same process.

Step 1. List the result from the response questionnaire. Calculate the difference between full marks and response marks. The numbers under three treatments are 23 ($= 10 + 8 + 5$) questions inquired in the questionnaire. The numbers below to No are quantity of questionnaires. The rest of numbers refers to the difference between full marks and response marks. For example, the answer to question 1 for the first questionnaire is 4, which become 1 ($= 5 - 4$) after calculation (Tables 1, 2 and 3).

Step 2. Calculate the different sequence based on Eq. 1. Find out the max and min difference in all the factors. Then calculate the different sequence based on Eq. 1. In Eq. 1, ζ is the identification coefficient and is equal to 0.5, $X_0 = X_0(1)$, $X_0(2)$, ..., $X_0(k)$ is the comparison sequences. A Likert five-point scale is used to evaluate the criteria for calculating grey relational coefficients of all factors, and the highest score is 5. Moreover, each factor is graded with a score of 5, 4, 3, 2, or 1, $X_i(k)$ is

Table 1 List of calculated difference between factors

No.	Before treatment			During treatment			After treatment		
	1	...	10	1	...	8	1	...	5
1	1	...	0	0	...	2	1	...	1
2	1	...	1	1	...	1	0	...	1
3	2	...	1	1	...	0	0	...	1
...		...			...			...	
180	0	...	0	0	...	2	2	...	2

Table 2 List of grey relational coefficient

No.	Before treatment			During treatment			After treatment		
	1	...	10	1	...	8	1	...	5
1	0.33	...	1	1	...	0.42	0.67	...	0.5
2	0.5	...	0.5	0.5	...	0.6	1	...	0.5
3	0.5	...	0.5	0.5	...	1	1	...	0.5
...		...			...			...	
180	0.5	...	1	1	...	0.43	0.5	...	0.33

Table 3 List of grey relational degree

Rank	Key factor	Question no.	Grey relational grade
1	Doctor's skill	15	0.79
2	Result of treatment	20	0.79
3	Doctor's explanation before treatment	14	0.76
4	Reputation of the clinic	13	0.75
5	Other staff's attitude	11	0.74
6	Doctor's attitude	8	0.74

the score that the kth respondent answers factor i, where $i = 1, 2, \ldots, 17$, $k = 1, 2, \ldots, n$, and n is the number of valid questionnaires.

$$r[X_0(k), X_i(k)] = \gamma_{0i}(k) = \frac{\Delta \min + \zeta \Delta \max}{\Delta X_{0i}(k) + \zeta \Delta \max} \tag{1}$$

Step 3. Calculated the grey relational coefficient according to Eq. (2). The grey relational degree, which is equal to the arithmetic mean of the grey relation coefficients, was calculated. The grey relational degree represents the relationship between sequence and comparison sequence. If the changes in two factors have the same trend, this means that the extent of synchronous change is high, as well as the extent of the correlation. Then, grey relational degree, which is equal to grey relational coefficient under equal weighted index, is calculated.

$$\Gamma(X_0, X_i) = \Gamma_{0i} = \sum_{k=1}^{n} \beta_k \gamma[X_0(k), X_i(k)] \quad (2)$$

Step 4. Sort out key factors by raking the sequences. Finally, Re-order the relational degree by descending. The higher relational degree is, the more important the factor is.

The above data calculated are for clinic. But all the data for clinics and departments in hospital were analyzed in the same process. The line graphs, which refer to the results, can be found in Sect. 4. Analysis Results.

3 Questionnaire Design

People go to dental and aesthetic clinic not all because of illness or scars. Conversely, they go for better looking and gain more confidence in most of the cases. Therefore, the paper used the words customer instead of patient. The paper had collected factors from papers and trend.

Based on medical clinic by types, total amount of clinic and dental clinic are over 50 %, the questionnaire included these two types of clinic. Since esthetic clinic has become a trend these years, esthetic clinic is considered in the paper as well. Also, the paper thought about the different departments in the hospital, so dental department, department of family medicine in hospital and department of orthopedics were discussed in the paper. In short, the paper adopted six types of medical organization, including clinic, dental clinic, aesthetic clinic, dental department in hospital, department of family medicine in hospital and department of orthopedics.

Likert five-point scale is the criteria in the paper. Customers marked their opinions about how important the factors in by choice point from one point Very unimportant to five points Very important.

The factors that affect customers choices are presented in three stages in the questionnaire: before treatment, during treatment, and after treatment. All the factors in three stages are explained as follows.

Factors before treatment

(1) Number of Units in the clinic

Since doctors have different specialty, and customers think over doctors specialties when decide a clinic/hospital [4, 5].

(2) Expert Recommendation

Bayus [6] claimed that whether positive or negative evaluation makes huge influence to customer before making choice with insufficient information. Realizing about other doctors recommendation is one of the most efficient ways to know the credibility of doctors or clinics/hospitals [7, 8].

(3) Social Recommendation on the Website
Gu [9] noticed that customers do more research before doing high involvement treatment, such as dental correction and plastic surgery. So the questionnaire contained social recommendation on the website.

(4) Distance to the clinic
Boscarino and Steiber [4], Malhotra [7], Wolinsky and Kurz [8], Lane and Lindquist [5] viewed distance to the clinic as a significant factor to customers in 20th century since distance to the clinic is another vital cost beside money for customers.

(5) Parking Convenience
People would be more willing to go a shopping mall that provides parking space than the mall that does not. In the same reason, parking convenience could be an influential factor to customers of clinic/hospital. So parking convenience is contained in the questionnaire.

(6) Waiting Time
Longer waiting time refers to less dominant time of customers. Holtmann and Olsen [10] shared similar opinion in their paper. The paper brings waiting time into the questionnaire as well.

(7) Explanation of the Reservation
More and more clinics/hospitals provide reservation service before treatment. Some organizational clinics/hospitals even only accept reservation. Owing to the trend, the paper takes explanation of the reservation as a question item in questionnaire.

(8) Reputation of the Clinic
Trust and reputation systems represent a significant trend in decision support for Internet mediated service provision [11]. An improper evaluation can negatively affects reputation and profitability of a clinic/hospital. Therefore, the paper adopted reputation of the clinic in questionnaire.

(9) Doctors Background
Doctors background is one of the prime information that customers know about a doctor. Professional background somehow implies better skills than normal one. Since dental treatment, orthopedics require especially skilled technique, a doctor can acquire more credit if he has professional or experienced background.

(10) Past Experience at the Clinic
Kuehn [12] noted that customers make decision by their experience accumulated in the past. Unpleasant experience could cause customers look for another clinic/hospital that make them feel more comfortable. Past experience of customers decides whether they would come back to a clinic/hospital again.

Factors during treatment

(11) Doctors Attitude
Doctor is the key person who gets alone with customers in the treatment procedure. Naturally, the paper included doctors attitude as one of the factors.

(12) Nurses' Attitude
Nurses attitude is an important issue that mentioned in many medical related papers

[5, 8, 13]. Otani [14] defined nurses caring as nurses response when customers need assistance. In spite of different title, nurses caring is contained in nurses attitude. So the paper combined the element into nurses attitude in the questionnaire.

(13) Other Staffs Attitude

Other staffs refer to clinic staff except doctors and nurses, i.e. pharmacists, counter staff etc. Although other staffs do not interact with customer as much as doctors and nurses do, their attitude still affect customers choice since they are included in the treatment procedure. In addition, counter staffs present first impression of the clinic or hospital. And pharmacists are usually in the last service station, which cause deep impression to customers. Because of the reasons, the paper contained other staffs attitude in questionnaire.

(14) Doctors Explanation before doing treatment

Boudreaux [13] notified that use anticipatory guidance (i.e. explain to patient what to expect next) influents the most in all the factors that affect satisfaction in his paper. Doctors explain to next phase is highly related to customers satisfaction with clinic (Koichiro et al. 2005). Based on the studies, the paper put doctors explanation before doing treatment into questionnaire as well.

(15) Doctors Skill

Doctors skill is deeply correlated with the expectation result of customers. Customers want to reduce the risk of fall result down to the basement. Thus, the paper takes doctors skill in the questionnaire.

(16) Physical Environment

Physical environment is one of the considerable factors in the paper, and including light, quietness, and sanitation. Noise and dirty environment cause healthy people disease, not mention about customers in the clinic/hospital.

(17) Privacy of Treatment Room

Inhorn discussed about privacy in clinic in 2004, he also said that medical privacy speak about doing no harm to informant, and whether oral and written guarantees of anonymity and confidentiality can assuage informants anxieties and fears. Thus, the paper sees privacy of treatment room as a factor in questionnaire.

(18) Medical Equipment

Earlier medical related papers define medical equipment as if equipment can operate smoothly or is the equipment clean [4, 5, 7, 8]. But people care more about if the equipment is advanced nowadays. No matter which definition, medical equipment is mentioned in the decades, so medical equipment is considered in questionnaire of the paper.

Factors after treatment

(19) Cost of Treatment

Cost of treatment reflects how much service could customers have in treatment procedure. Customers gain more satisfaction if the cost lower than their expectation. Boscarino and Steiber [4], Wolinsky and Kurz [8], and Lane and Lindquist [5] discussed cost of treatment in their papers. For the reasons, the paper included cost of treatment in questionnaire.

(20) Result of Treatment
Since customers spend money, time, and effort on survey, result of treatment is definitely what they concern about. And result of treatment is the direct way that how customers evaluate the treatment. So the questionnaire takes result of treatment in.

(21) Could Pay by Credit Card
The service in dental clinic and aesthetic clinic are usually expensive. Because Credit card allows customers enjoy first, pay latter, customers might more eager to go to a clinic that accept credit card. Thus, could pay by credit card is one of the questions item in questionnaire.

(22) Could Pay by Installment
For the same reason in (21), the expense of service in dental clinic and aesthetic clinic could be a burden to customers. If they are allowed to pay by installment, the services will be more affordable. Therefore, the paper included in the questionnaire could pay by installment as well.

(23) Explanation for Residential Care and Medication Usage
Explanation for residential care and medication usage is an extension of the relation between the clinic and customers. Since pharmacist response to inform customers suggested use of medicine, the following care etc., there are even some papers regard pharmacists attitude as question in the questionnaire (Woilinsky and Kurz 1984), [5]. Through the reasons, the paper takes explanation for residential care and medication usage as well.

The questionnaire turns to doctors and nurses who have worked in medical organization that included in the paper (clinic, dental clinic, aesthetic clinic, dental department in hospital, departmenrt of family medicine in hospital, and department of orthopedics). The questionnaires had been released and collected during November to December in 2014. And total 180 valid questionnaires (30 for) had been received.

The questionnaire is divided into three sections. An introduction of this study, and appreciation are given in the first section. Next, some simple explanations and evaluated questions before, during and after treatment are offered in the second section. The paper adopted Likert five-point scale that interviewees marked from 1 point (very unimportant) to five points (very important) to present their preference to each question. Finally, the basic information of the interviewees inquired in the third section.

4 Analysis Results

The paper collected that mentioned about the influential factors to customers when choose a clinic/department. Also we used grey relational analysis (GRA) to sort out key factors for each type.

Daniel [15] pointed out that there are three to six factors for enterprise to survive in an industry. Identify three to six key factors does not promise success to the enterprise, but the enterprise might never be success without three to six key factors. The paper

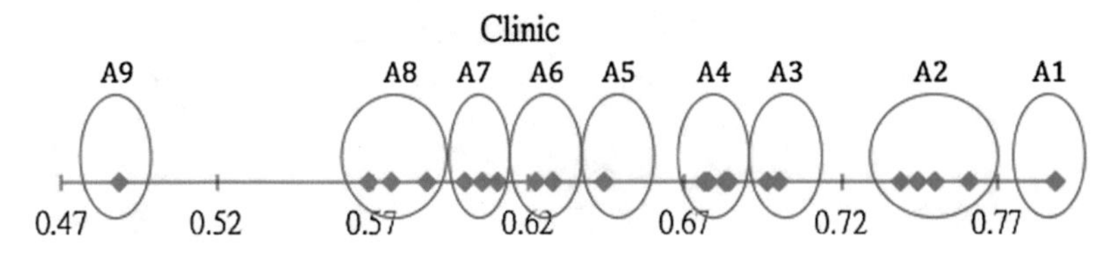

Fig. 1 Grey relational degree of clinic

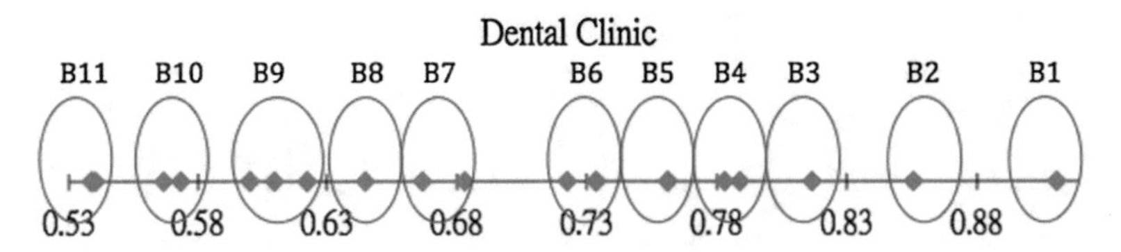

Fig. 2 Grey relational degree of dental clinic

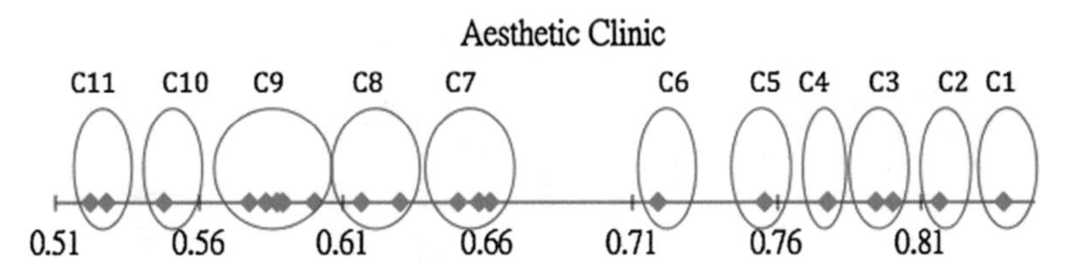

Fig. 3 Grey relational degree of aesthetic clinic

accepts the point, and adopted three to six key factors in 23 factors. By the restriction of length, the paper only discusses key factors of each types of clinic/department in the hospital.

(1) Clinic

After calculated process, the analysis result of clinic is revealed through Fig. 1 Grey relational degree of clinic. The six factors are 'doctor's skill' (0.79), 'result of treatment' (0.79), 'doctor's explanation before doing treatment' (0.76), 'reputation of the clinic' (0.75), 'other staff's attitude' (0.74), and 'doctor's attitude' (0.74).

(2) Dental Clinic

According to Fig. 2 Grey relational degree of dental clinic, group B1–B5 contained six key factors. They are: 'doctor's explanation before doing treatment' (0.91), 'medical equipment' (0.86), 'doctor's skill' (0.81), 'doctor's background' (0.79), 'past experience at this clinic' (0.78), and 'reputation of the clinic' (0.76).

(3) Aesthetic Clinic

By Fig. 3 Grey relational degree of aesthetic clinic, we can find six factors form group C1–C5. The six key factors are: 'expert recommendation' (0.83), 'past experience at this clinic' (0.82), 'doctor's skill' (0.8), 'result of treatment' (0.8), 'doctor's background' (0.79), and 'doctor's explanation before doing treatment' (0.78).

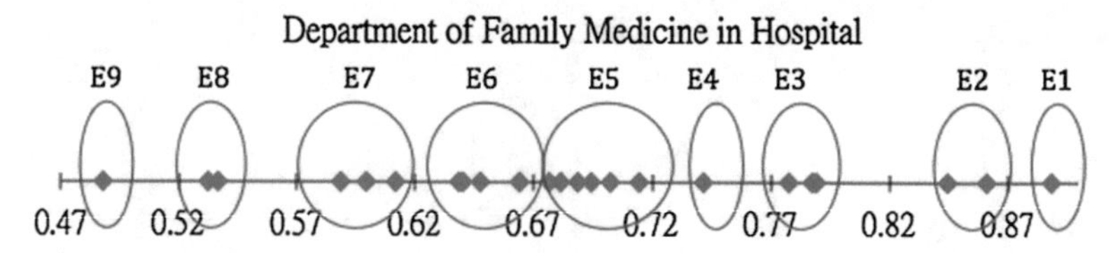

Fig. 4 Grey relational degree of department of family medicine in hospital

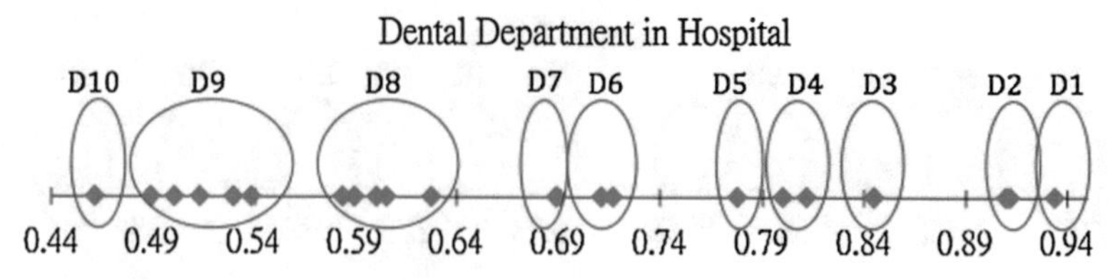

Fig. 5 Dental Department in Hospital

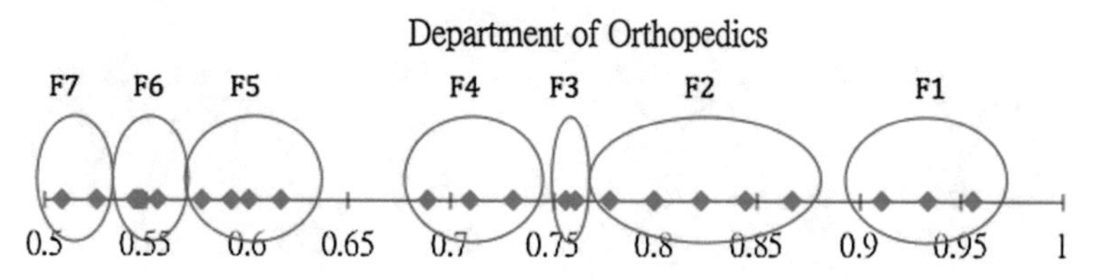

Fig. 6 Grey relational degree of department of orthopedics

(4) Department of Family Medicine in Hospital

Since key factors 'expert recommendation' has the some grey relational degree with key factors nurses attitude, six key factors for department of family medicine in hospital can find in group E1–E3. The six key factors in order are: 'could pay by credit card' (0.89), 'result of treatment' (0.86), 'explanation for residential care and medication usage' (0.84), 'doctor's explanation before doing treatment' (0.79), 'expert recommendation' (0.78), and 'nurses attitude' (0.78) (Fig. 4).

(5) Dental Department in Hospital

According to the calculation result of grey relational degree for dental department in hospital, key factor 'doctor's skill' has the same result as 'could pay by credit card'. Also, key factors 'doctor's explanation before doing treatment' and key factors 'medical equipment' own the same result. We can find Key factors for dental department in group D1–D3. It will be exceed over six key factors if included group D4, so five key factors are adopted here. There are: 'result of treatment' (0.93), 'doctor's skill' (0.91), 'could pay by credit card' (0.91), 'doctor's explanation before doing treatment' (0.84), and 'medical equipment' (0.84) (Fig. 5).

(6) Department of orthopedics

In Fig. 6 Grey relational degree of department of orthopedics, the paper will take more than six factors if contained group F2. So there are key factors for department of orthopedics in group F1: 'doctor's skill' (0.96), 'medical equipment' (0.93), 'past experience at the clinic' (0.91), and 'doctor's explanation before doing treatment' (0.87).

The paper reveals the key factors for each types of clinic or department in hospital in this section. Every type of medical organization has different combination of factors. These key factors are the main points that the clinics or departments for improving service quality. In all the key factors, doctors explanation before doing treatment is the factors that shows in six types medical organization (clinic, dental clinic, aesthetic clinic, dental department in hospital, department of family medicine in hospital, and department of orthopedics). Based on the results, the condition for clinic to survive in the medical market neither necessary to buy expensive equipment to compete with big clinic, nor too much worry about accessible location. In a word, doctors explanation before doing treatment, is the easy but vital to customers.

5 Conclusion

The study sorted out key factors for clinic, dental clinic, aesthetic clinic, dental department in hospital, department of family medicine in hospital, and department of orthopedics. It is appropriate to use methodology Grey Relational Analysis (GRA) since the limitation of quantity of interviewees. Grey Relational Analysis (GRA) prevents the issue that standard statistical method has to conquered, insufficient data. The paper collected total 180 questionnaires, 30 for each kind of medical organization. The key factors are sorted out and available by 30 questionnaires for each medical organization.

The results that we found in this paper are as follows:

1. 'Doctor's explanation before treatment' crosses all kinds of medical organizations.
2. 'Doctor's skill' covers all kinds of medical organizations except department of family medicine in Thailand.
3. 'Result of treatment' is important to all kinds of medical organizations except dental clinic in Thailand.
4. The key factors that affect people when they determine a medical organization are mostly the factors during, and after treatment.
5. 'Expert recommendation', a factor that is before treatment, is an exception, and is important to aesthetic and department of family medicine.

Different medical organizations develop different key factors through different symptom and demand. For example, customers put more attention on expert commend for aesthetic clinic, while they care more about doctors skill for department of orthopedics. Although aesthetic clinic and department of orthopedics operate similar functions, there are different factors between two organizations to notice. Thus, it is noteworthy to know about the key factors indicated. The key factors can be new management focus points to these organizations. The key factors can be the indicators for clinics and hospitals to improve and provide better service quality. If whole medical industries environment raises to higher service standard, it will attract more patients to these organizations. In such a way, the income of these organizations will increase. Hence, it can be an impact to domestic economic.

References

1. Deng, J.L.: Introduction grey system. J. Grey Syst. **1**(1), 1–24 (1989)
2. Zhang, S.F., Liu, S.Y.: A GRA-based intuitionistic fuzzy multi-criteria group decision making method for personnel selection. Expert Syst. Appl. **38**, 11401–11405 (2011)
3. Wei, G.W.: GRA method for multiple attribute decision making the incomplete weight information in intuitionistic fuzzy setting. Knowl.-Based Syst. **23**, 243–247 (2010)
4. Joseph, B., Steiber, S.R.: Hospital shopping and consumer choice. J. Health Care Mark. **2**(2), 23–25 (1982)
5. Lane, P.M., Lindquist, J.D.: Hospital Choice: a summary of the key empirical and hypothetical findings ofthe 1980s. J. Health Care Mark. **8**(4), 5–20 (1988)
6. Bayus, B.L.: Word of mouth: the indirect effects of marketing efforts. J. Advert. Res. **25**(3), 31–39 (1985)
7. Malhotra, N.K.: Stochastic modeling of consumer preferences for health care institutions. J. Health Care Mark. **3(fall)**, 18–36 (1983)
8. Wolinsky, F.D., Kurz, R.S.: How the public chooses and views hospital. Hosp. Health Serv. Adm. **29**, 58–67 (1984)
9. Gu, B., Park, J., Konana, P.: The impact of external word-of-mouth sources on retailer sales of high-involvement products. Inf. Syst. Res. **23**(1), 182–196 (2012)
10. Holtmann, A.G., Olsen Jr, E.O.: The demand for dental care: a study of consumption and household production. J. Hum. Resour. **11**(4), 546–560 (1976)
11. Josang, A., Roslan, I., Coin, B.: A survey of trust and reputation systems for online service provision. Decis. Support Syst. **43**, 618–644 (2007)
12. Kuehn, A.A.: Consumer brand choice as a learning process. J. Adver. Res. **2**, 10–17 (1962)
13. Edwin, B.D., d'Autremont, S., Karen, W., Glenn, J.N.: Predictors of emergency department patient satisfaction: stability over 17 months. ACAD EMERG MED **11**(1), 51–58 (2004)
14. Otani, K., Waterman, B., Faulkner, K.M., Boslaugh, S., Claiborne Dunagan, W.: How patient reactions to hospital care attributes affect the evaluation of overall quality of care, willingness to recommend and willingness to return. J. Health Care Manag. **55**(1), 25–57 (2010)
15. Daniel, R.D.: Management information crisis. Harv. Bus. Rev. **35**(5), 111–121 (1961)
16. Boudreaux, E.D., O'Hea, E.L.: Patient satisfaction in the emergency department: a review of the literature and implications for practice. J. Emerg. Med. **26**(1), 13–26 (2004)
17. Engel, J.F., Blackwell, R.D., Miniard, P.W.: Consumer Behavior, 7th edn. Dryden Press series in marketing, Hinsdale (1993)
18. Marcia, I.C.: Privacy, privatiztion, and the politics of patronage: ethnographic challenges to penetrating the secret world of middle eastern, hospital-based in vitro fertilization. Soc. Sci. Med. **2004**(59), 2095–2108 (2004)

ARIMA Versus Artificial Neural Network for Thailand's Cassava Starch Export Forecasting

Warut Pannakkong, Van-Nam Huynh and Songsak Sriboonchitta

Abstract Thailand is the first rank cassava exporter in the world. The cassava export quantity from Thailand influences cassava trading in international market. Therefore, Thailand's cassava export forecasting is important for stakeholders who make decision based on the future cassava export. There are two main types of cassava export which are cassava starch and cassava chip. This paper focuses on the cassava starch, which is around 60 % of the total cassava export value, including three following products: native starch, modified starch and sago. The cassava starch export time series from January 2001 to December 2013 are used to predict the cassava starch export in 2014. The objectives of this paper are to develop ARIMA models and the artificial neural network (ANN) models for forecasting cassava starch export from Thailand, and to compare accuracy of the ANN models to the ARIMA models as benchmarking models. MSE, MAE and MAPE are used as accuracy measures. After various scenarios of experiments are conducted, the results show that ANN models overcome the ARIMA models for all three cassava starch exports. Hence, the ANN models have capability to forecast the cassava starch exports with high accuracy which is better than well-known statistical forecasting method such as the ARIMA models. Moreover, our finding would give motivation for further study in developing forecasting models with other types of ANN models and hybrid models for the cassava export.

Keywords Cassava · Forecasting · ARIMA · Neural network

W. Pannakkong (✉) · V.-N. Huynh
School of Knowledge Science, Japan Advanced Institute of Science and Technology, Ishikawa, Japan
e-mail: warut@jaist.ac.jp

V.-N. Huynh
e-mail: huynh@jaist.ac.jp

S. Sriboonchitta
School of Economics, Chiang Mai University, Chiang Mai, Thailand
e-mail: songsak@econ.cmu.ac.th

V.-N. Huynh et al. (eds.), *Causal Inference in Econometrics*,
Studies in Computational Intelligence 622, DOI 10.1007/978-3-319-27284-9_16

1 Introduction

Cassava (*Manihot esculenta Crantz*) is one of most important source of calories for the world's population, after rice and maize [4]. According to statistic of FAO in 2012, cassava production quantity is the ninth rank of agricultural production in the world and main producers are Nigeria, Thailand, Indonesia, Brazil and Congo.

Forecasting the international trade of agricultural products is difficult because demand and supply are affected by many unpredictable factors that interact in a complex manner [2]. International trade of cassava in 2012, major exporters are Thailand and Vietnam, and major importer is China. In the cassava products trading, Free on Board (FOB) price—the price of goods that has been placed on the ship at a port of shipment—at Bangkok is used as the reference price and it is driven by supply and demand in the market. Recently, there are factors that influence the supply of the market such as increasing of environmental problems (e.g., drought and pests), cause decreasing of cassava production. For the demand, China needs more cassava for animal food and for producing ethanol in beverage and energy industries. Similarly, domestic demand in Thailand is increasing due to growth of ethanol using as an alternative energy for vehicles. Therefore, the cassava price would be increased and more fluctuated because demand is increased while supply is deceased and the related factors are increased in term of number and uncertainty [12].

Past researches related to other agricultural products (e.g., rice, sugar cane, natural rubber and durian) prove that ANN models can outperform the statistical forecasting techniques [2, 7, 9, 11]. Objectives of this research are to develop autoregressive integrated moving average (ARIMA) models and artificial neural network (ANN) models for forecasting the cassava export from Thailand and to compare accuracy of the ANN models to the ARIMA models as benchmarking models. Results of this research would be useful in decision making for stakeholders such as Thai government, Thai Tapioca Starch Association (TTSA) and traders in cassava future trading market. Moreover, it is a challenge to develop a forecasting model to dominate the ARIMA model which is an effective and well-known time series forecasting tool.

In Thailand, the cassava is an important agricultural product because it is the third rank in production quantity and the seventh rank in export value (Table 1). Around 68 % of total production is exported. There are two main types of the cassava export which are cassava starch and cassava chip (or pallet) [13]. The cassava starch export value is 50,037 million Baht which is around 60 % of total the cassava export value. The remaining 40 % is the cassava chip which has export value 33,817 million Baht. In term of the export value, the cassava starch export has more impact in economic than the cassava chip and pallet. Thus, the cassava starch is focused in this paper.

There are three kinds of the cassava starch export including native starch, modified starch and sago. The native starch is extracted from cassava root. The modified starch is the native starch that has been modified by a chemical or a physical process in order to change its molecular structure to obtain suitable properties for various purpose in

Table 1 Thailand's export value of major agricultural products (in THB)

Item	2009	2010	2011	2012
Total export value	5,194,445	6,176,170	6,707,851	7,082,333
Value of major agricultural products	964,945	1,135,754	1,444,996	1,341,826
Natural rubber	174,984	296,380	440,547	336,304
Rice and products	183,433	180,727	208,253	158,433
Fishes and products	97,566	99,039	112,179	131,369
Cassava and products from cassava	50,581	66,889	77,689	84,322
Sugar and products	68,748	76,327	116,950	132,129
Fruits and products	60,757	63,072	81,334	77,307
Shrimps and products	93,605	101,141	110,665	96,522
Chicken meat and products	48,847	52,223	60,295	67,751
Vegetables and products	19,482	19,238	21,420	21,035
Residues and waste, prepared animal fodder	14,891	18,023	19,583	16,772
Other agricultural products	152,051	162,695	196,081	219,882

industries. The sago is also made from the native starch but there is no change in the properties. In sago manufacturing process, the native starch is formed into globular shape and screened its size by sieving.

2 Literature Review

The forecasting related to several agricultural products, the ARIMA models, the ANN models and other conventional time series forecasting methods are developed and compared in several past researches. Pokterng and Kengpol [11] design and develop the models that are capable to forecast the quantity of fresh durian production in Thailand. The ANN models and four time series models which are moving average model (MA), weight moving average model (WMA), single exponential smoothing model and Holt's linear exponential smoothing model are compared. Correlation analysis is applied for screening the input variables for ANN. The finding is that the ANN has the least value of mean absolute percentage error (MAPE). Moreover, the production quantities optimizing the profit in each region are suggested by the linear programming (LP) model. Additionally, Udomsri et al. [14] construct and compare various five conventional time series models (e.g., moving average, deseasonalised, exponential smoothing, double exponential smoothing and regression), and the ANN models in order to forecast demand of Thailand's durian for export markets. The results reveal that the ANN model is the most accurate model for durian chip with the lowest MAPE. However, deseasonalised model is the most accurate model for fresh durian, frozen durian and durian paste which means that the ANN model is not always the best model. After getting the models, the appropriate quantity of each type of durian for domestic and export markets are determined by the linear programming (LP).

Co and Boosarawongse [2] compare forecasting accuracy of exponential smoothing methods, the ARIMA models and the ANN models in Thailand's rice export forecasting. The results show that the ANN model outperforms the other models in several accuracy measures (e.g., MAE, MAPE, MSE and RMSE) because the ANN model can track the dynamic non-linear trend, seasonality and their interaction better.

Kosanan and Kantanantha [7] construct forecasting models for Thailand's natural rubber by the ARIMA model, the ANN model and support vector machine (SVM). The ANN model is the most accurate model obtaining the lowest MAPE. However, the ARIMA model obtain lower MAPE than the SVM. Moreover, Pattranurakyothin and Kumnungkit [10] determine the suitable ARIMA model for fitting natural rubber's export sale from Thailand and the result is that the ARIMA model gives the lowest MAPE.

Obe and Shangodoyin [9] develop the ANN model to forecast sugar cane production in Nigeria. The ANN based model can get 85.70 % accuracy. Thus, the ANN based model can be applied for forecasting with a high accuracy.

Zou et al. [16] investigate the Chinese food grain price forecasting performance of the ARIMA model, the ANN model and the hybrid model. The hybrid model is a linear combination of the ARIMA model and the ANN model. The outcomes reveal that ANN model overcomes the ARIMA model. The hybrid model is more accurate than ANN model in term of MAE, MSE and MAPE. However, the ANN model is the best model for tracking the turning point.

Until now, there are technical reports of Thai government and domestic researches which uses statistical time series forecasting techniques such as exponential smoothing linear regression analysis and ARIMA to forecast annual cassava export [12]. However, to our best knowledge, there is no past research that applies the ANN model to forecast the cassava starch export from Thailand.

3 Cassava Starch Export Time Series

Historical time series of cassava starch export from Thailand is used for the analyses in this paper. The time series is obtained from Thai Tapioca Starch Association. There are three types of cassava starch export which are native starch, modified starch and sago, and their characteristics are respectively shown in Figs. 1, 2 and 3 that express non-stationary characteristics such as trend and seasonality of the cassava starch export.

The time series data start from January 2001 to December 2014. They are recorded in monthly (168 months). The data from January 2001 to December 2013 are used for models fitting and training. The data in last year, January 2014 to December 2014, are applied for models validation which such data are treated as unseen future data that the forecasting models do not know and do not use as an input.

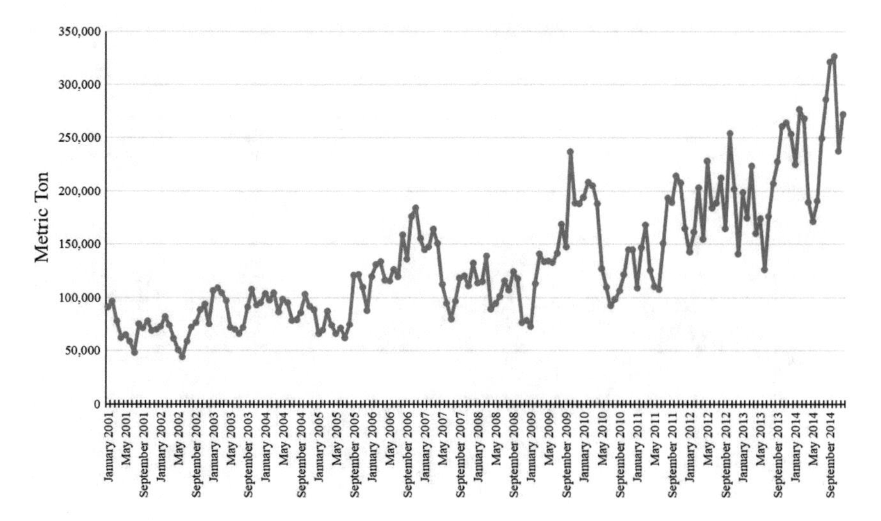

Fig. 1 Native starch export quantity

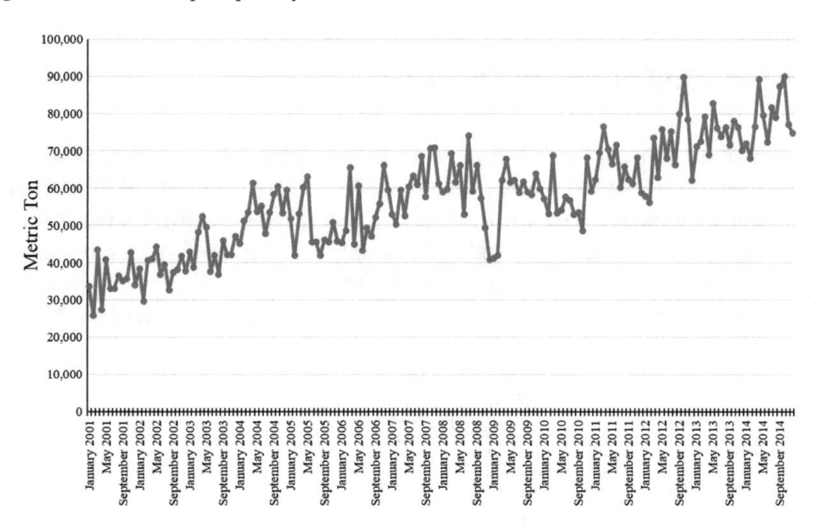

Fig. 2 Modified starch export quantity

In addition, Thai government often has a policy to help cassava farmers every year. The farmers can pawn their cassava to the government. The cassava must be redeemed within three months otherwise farmers' cassava will be seized to the government cassava stock. Normally, the government decides to clear the stock in every September. This policy would cause annual seasonality of the time series.

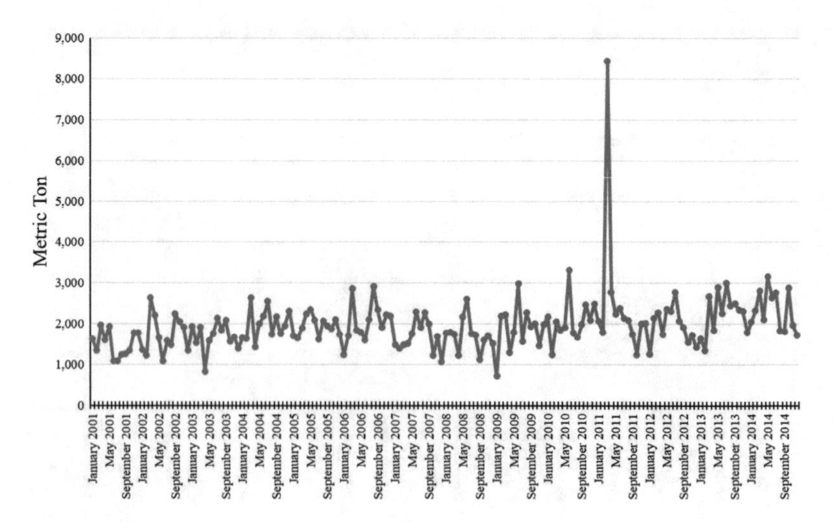

Fig. 3 Sago export quantity

4 Forecasting Accuracy Measures

Accuracy of the forecasting in this paper is measured by mean square error (MSE), mean absolute error (MAE) and mean absolute percentage error (MAPE). Their mathematical formulas are shown in (1)–(3). The parameters involved in the formulas are following: actual value in period t, Z_t; average of forecasted values in period t, $\hat{Z}_t$; and total number of forecasted period, N. To interpret the forecasting accuracy of these three methods, lower value means better accuracy of the forecasting result.

$$\text{MSE} = \frac{1}{N}\sum_{t=1}^{N}(Z_t - \hat{Z}_t)^2 \tag{1}$$

$$\text{MAE} = \frac{1}{N}\sum_{t=1}^{N}|Z_t - \hat{Z}_t| \tag{2}$$

$$\text{MAPE} = \frac{1}{N}\sum_{t=1}^{N}\frac{|Z_t - \hat{Z}_t|}{Z_t} \tag{3}$$

5 ARIMA Models for Cassava Starch Export Forecasting

The ARIMA model is a well-known statistical method time series forecasting invented by Box and Jenkins in 1970 [1]. This method is integration of autoregressive (AR) model and moving average (MA) model, that can deal with non-stationary time series such as the cassava starch export. This section explains details of using the ARIMA model to predict future cassava starch export, and evaluation of its forecasting accuracy.

5.1 ARIMA Models

To understand the ARIMA model, the AR model and the MA model should be presented first. The AR model, (4), expresses a time series value at time t, Z_t, by regressing lagged values $Z_{t-1}, \ldots, Z_{t-p}$ as independent variables, which produces a constant c, and adding a random error term a_t. The amount of the lagged values used in the AR model depends on an order of AR, p. The MA model, (5), represents Z_t as weight sum of previous random errors a_t and time series mean, μ. An order of MA, q, specifies how many previous random errors included in the MA model.

$$Z_t = c + \phi_1 Z_{t-1} + \phi_2 Z_{t-2} + \cdots + \phi_p Z_{t-p} + a_t = c + \sum_{i=1}^{p} \phi_i Z_{t-i} + a_t \tag{4}$$

$$Z_t = \mu + a_t - \theta_1 a_{t-1} - \theta_2 a_{t-2} - \cdots - \theta_q a_{t-q} = \mu + a_t - \sum_{j=1}^{q} \theta_j a_{t-j} \tag{5}$$

A combination of the AR model and the MA model is autoregressive moving average (ARMA) model as shown in (6). We can use the backward shift operator, B, which is defined by $B^i Z_t = z_{t-i}$, to produce (7). After that, the AR term is moved to the left side and (7) is then rearranged to be (8). If an autoregressive operator $\phi_p(B)$ is defined by $\phi_p(B) = 1 - \phi_1 B - \phi_2 B^2 - \cdots - \phi_p B^p$ and a moving average operator $\theta_q(B)$ is defined by $\theta_q(B) = 1 - \theta_1 B - \theta_2 B^2 - \cdots - \theta_q B^q$, AR model can be written economically as (9).

$$Z_t = c + \sum_{i=1}^{p} \phi_i Z_{t-i} + a_t - \sum_{j=1}^{q} \theta_j a_{t-j} \tag{6}$$

$$Z_t = c + \sum_{i=1}^{p} \phi_i Z_t B^i + a_t - \sum_{j=1}^{q} \theta_j a_t B^j \tag{7}$$

$$\left(1 - \sum_{i=1}^{p} \phi_i B^i\right) Z_t = c + \left(1 - \sum_{j=1}^{q} \theta_j B^j\right) a_t \tag{8}$$

$$\phi_p(B) Z_t = c + \theta_q(B) a_t \tag{9}$$

The ARMA model is suitable for stationary time series which has no trend and a constant variance. However, normally, time series has some trend that means the time series usually is non-stationary so that another step called differencing with a degree of differencing, d, is needed to transform the time series to be stationary. The differencing can be applied by replacing Z_t in (9) with $(1 - B)^d Z_t$ to get (10) which is called autoregressive integrated moving average (ARIMA) model or ARIMA(p, d, q).

$$\phi_p(B)(1 - B)^d Z_t = c + \theta_q(B) a_t \tag{10}$$

Moreover, when pattern or event is repeated over a time span, it means that there is seasonality in time series, and the time span of repeating seasonal pattern, s, can be determined. To deal with this situation, seasonal autoregressive operator, $\Phi_P(B^s)$; seasonal moving average operator, $\Theta_Q(B^s)$ which are defined as:

$$\Phi_P(B^s) = 1 - \Phi_1 B^s - \Phi_2 B^{2s} - \cdots - \Phi_P B^{Ps}$$
$$\Theta_Q(B^s) = 1 - \Theta_1 B^s - \Theta_2 B^{2s} - \cdots - \Theta_Q B^{Qs}$$

and seasonal differencing, $(1 - B^s)^D$, are added to (10) and it can then be written as (11) and called seasonal autoregressive integrated moving average (SARIMA) model or ARIMA$(p, d, q) \times (P, D, Q)_s$, where P is an seasonal order of autoregressive; Q is an order of seasonal moving average and D is a degree of seasonal differencing.

$$\phi_p(B)\Phi_P(B^s)(1 - B)^d (1 - B^s)^D Z_t = c + \theta_q(B)\Theta_Q(B^s) a_t \tag{11}$$

In this paper, SARIMA model is chosen because the cassava is a commodity product which has cycles of production, harvesting, export and the government policy which leads seasonality in the time series. Inputs of the SARIMA models are cassava starch export quantities from January 2001 to December 2013 as $Z_1, Z_2, \ldots, Z_{156}$. In addition, time span of repeating seasonal pattern, s, is 12 months. Box et al. [1] shows method to manually select the suitable parameters of the SARIMA model to fit the time series. However, IBM SPSS Statistics software is used to determine appropriate values of the parameters for the best-fit model of each cassava starch export. After that, model validation is conducted. The best-fit models are applied to forecast the future export for next 12 months $Z_{157}, Z_{158}, \ldots, Z_{168}$. Forecasting accuracy is evaluated by comparing the forecasted exports and actual exports in 2014.

5.2 *Forecasting Accuracy of the ARIMA Models*

This part shows the best-fit models of ARIMA for each cassava starch export, which are analyzed by the IBM SPSS Statistics software. The forecasting accuracy measures MSE, MAE and MAPE of validation period are computed to express effectiveness of the ARIMA models in the cassava starch export forecasting. The best-fit model

Table 2 Accuracy measures of the best-fit ARIMA models

Cassava export	ARIMA model	MSE	MAE	MAPE (%)
Native starch	ARIMA(1, 1, 0) $(0, 1, 1)_{12}$	1,717,324,982.39	32,888.64	15.15
Modified starch	ARIMA(1, 1, 0) $(0, 1, 1)_{12}$	48,712,546.51	5,281,85	6.31
Sago	ARIMA(1, 0, 1) $(1, 0, 0)_{12}$	276,612.18	423.72	16.43

for the native starch and the modified starch is ARIMA$(1, 1, 0)(0, 1, 1)_{12}$. For the sago, the best-fit model is ARIMA$(1, 0, 1)(1, 0, 0)_{12}$. The accuracy measures of these ARIMA models are shown in Table 2.

Regarding the results in Table 2, the modified starch can be forecasted with excellent accuracy (i.e., MAPE 6.31 %). However, the forecasting result of the native starch and the sago have good accuracy with lower than 20 % of MAPE which means that the ARIMA models have capability to forecast the cassava starch export time series and can be use as a benchmarking method for this paper.

6 Artificial Neural Network Models for Cassava Starch Export Forecasting

Artificial neural network (ANN) model is a mathematical model that was developed based on concept of human brain neuron working. Advantages of the ANN model over traditional statistical methods are self-learning and not making assumption of characteristic of the data [15].

The structure of the ANN model is identified by number of layers and nodes (neurons). There are three kinds of layer such as input layer, hidden layer and output layer. Normally, the number of the input layer and the output layer is only one but the number of the hidden layer is one or more. However, one hidden layer is enough to fit any continuous function [5].

At each node in the ANN model (Fig. 4), inputs p from previous layer are aggregated as (12) to produce a net input n. Then, the net input n is passed through a transfer function to compute an output a which will be an input of next layer.

$$n = \sum_{i=1}^{R} p_i w_i + b \tag{12}$$

After the output of the ANN model (i.e., forecasted export) is generated from the output node, the output is compared with the target (i.e., actual export) to compute a gap. The ANN learns to minimize the gap via an training algorithm attempting to determine weight w and bias b that fit to relationship between the inputs and the target.

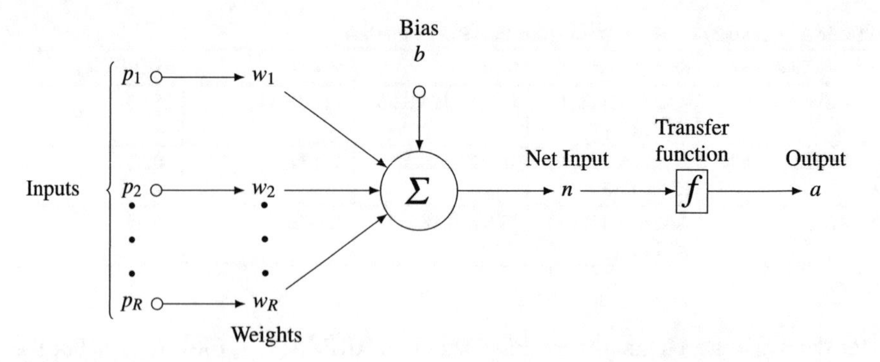

Fig. 4 An artificial neuron

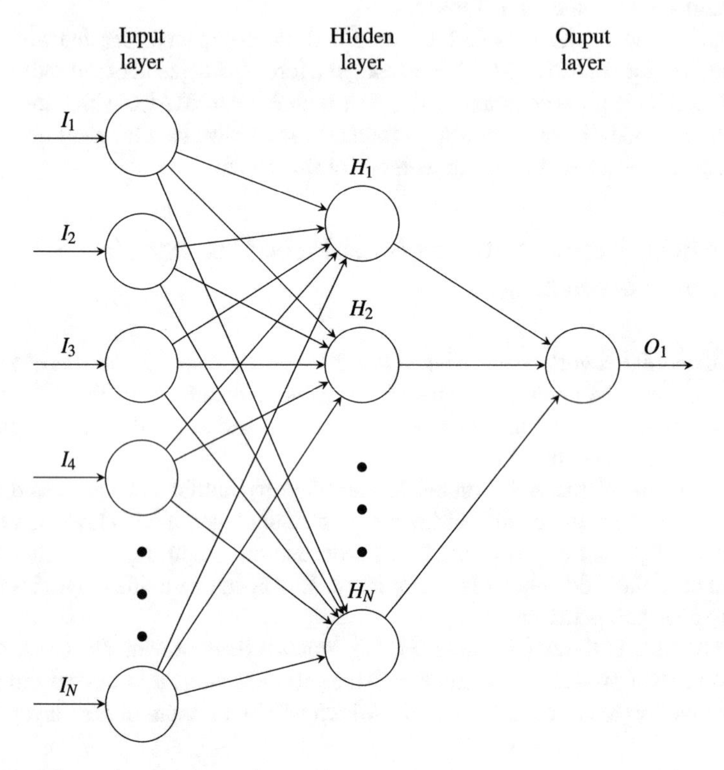

Fig. 5 Feed-forward artificial neural network

The proper number of the node in each layer depends on type of problem and the ANN architect because until now, there is no exact theoretical knowledge to find the best way to identify number of layer and node in the network. Thus, trial and error method is widely used to find the ANN model structure [6, 15]. Feed-forward ANN [3] is used in this research (Fig. 5). The feed-forward ANN model used in this research has three layers (one layer for each layer type). Levenberg-Marquardt algorithm with Bayesian regularization [8] is used as training algorithm for the ANN models in this paper.

This section consists of three main parts in order to identify an appropriate structure of the ANN models. First, input variables are selected and screened to design scenarios of the input layer. Second, output variable of the output layer is defined and a method to get reliable results from the output layer is explained. Third, trial and error experiments with purpose of getting suitable number of the hidden nodes are demonstrated.

6.1 Input Layer

The input layer consists of the input nodes representing set of independent variables related to a dependent variable (i.e., export quantity in period t, Z_t). Theoretically, there are two types of the input variables such as technical variables and fundamental variables [6]. The technical input variables are lagged values including variables calculated from the lagged values as well. The fundamental input variables are economic variables believed that they have effect to the dependent variable. In this paper, nine input variables are considered which consist of three lagged values (i.e., Z_{t-1}, Z_{t-3} and Z_{t-12}), moving averages of Z_t (i.e., MA(3) and MA(12)), an annual seasonal index and three time indices (i.e., sequence, month and quarter). The lagged values, the moving averages and the annual seasonal index are the technical variables. The three time indices are fundamental variables.

Three scenarios of the input nodes are formed to test effect of different input types on the forecasting accuracies. First scenario, the input nodes include only time indices. This scenario is an intention to test how well the ANN models can forecast unseen future while including only simple time indices as the input nodes. Second scenario, all nine input variables are included. Third scenario, the input variables are screened by correlation analysis (Tables 3, 4 and 5). The inputs that have statistically significant correlation to the export quantity Z_t will be chosen as the input nodes.

Table 3 Correlation analysis of ANN inputs for native starch

Type	Input	Correlation	p-value
Fundamental	Sequence	0.786	<0.001*
	Month	0.107	0.183
	Quarter	0.108	0.181
Technical	Z_{t-1}	0.862	<0.001*
	Z_{t-3}	0.717	<0.001*
	Z_{t-12}	0.571	<0.001*
	MA(3)	0.842	<0.001*
	MA(12)	0.767	<0.001*
	Seasonal index	0.534	<0.001*

*Significant at 5 % (p-value < 0.05)

Table 4 Correlation analysis of ANN inputs for modified starch

Type	Input	Correlation	p-value
Fundamental	Sequence	0.827	<0.001*
	Month	0.072	0.369
	Quarter	0.064	0.431
Technical	Z_{t-1}	0.801	<0.001*
	Z_{t-3}	0.715	<0.001*
	Z_{t-12}	0.740	<0.001*
	MA(3)	0.828	<0.001*
	MA(12)	0.829	<0.001*
	Seasonal index	0.484	<0.001*

*Significant at 5 % (p-value < 0.05)

Table 5 Correlation analysis of ANN inputs for sago

Type	Input	Correlation	p-value
Fundamental	Sequence	0.264	0.001*
	Month	−0.018	0.826
	Quarter	−0.071	0.380
Technical	Z_{t-1}	0.165	<0.040*
	Z_{t-3}	0.131	<0.102
	Z_{t-12}	0.181	<0.024*
	MA(3)	0.204	<0.011*
	MA(12)	0.206	<0.010*
	Seasonal index	0.906	<0.001*

*Significant at 5 % (p-value < 0.05)

Table 6 ANN inputs for native starch and modified starch

Scenario 1: fundamental	Scenario 2: fundamental and technical	Scenario 3: correlated
Sequence	Sequence	Sequence
Month	Month	Z_{t-1}
Quarter	Quarter	Z_{t-3}
	Z_{t-1}	Z_{t-12}
	Z_{t-3}	MA(3)
	Z_{t-12}	MA(12)
	MA(3)	Seasonal index
	MA(12)	
	Seasonal index	

Table 7 ANN inputs for sago

Scenario 1: fundamental	Scenario 2: fundamental and technical	Scenario 3: correlated
Sequence	Sequence	Sequence
Month	Month	Z_{t-1}
Quarter	Quarter	Z_{t-12}
	Z_{t-1}	MA(3)
	Z_{t-3}	MA(12)
	Z_{t-12}	Seasonal index
	MA(3)	
	MA(12)	
	Seasonal index	

From the correlation analysis, month and quarter are removed for all cassava starch exports; moreover, for sago, Z_{t-3} is cut out as well. The input nodes of these three scenarios for the cassava starch export are summarized in Tables 6 and 7.

6.2 Output Layer

The output layer is the last layer with only one node which is represented by fitted and forecasted cassava starch export quantities. The output node aggregates the outputs from the hidden layer to compute the net input n as (12). After that, linear transfer function represented by (13), is used to transform the net input n into the output a. Then, the outputs are recorded for every experimental run for further analyses.

$$a = n \tag{13}$$

Generally, each experimental run, the output is not the same on each run because the weights are randomly initialized and the outputs depend on them. To ensure that the average of the output is reliable, the experiments are repeated until the amount of recorded outputs is enough to provide the average with margin of error which is below 5 %. The margin of error can be calculated as shown in (14).

$$\text{Margin of error} = \frac{\text{Half width of 95 \% confidence interval of average}}{\text{Average}} \tag{14}$$

6.3 Hidden Layer

The hidden layer is the layer between the input layer and the output layer. For each hidden node, the input variables of the input layer are aggregated by (12) to

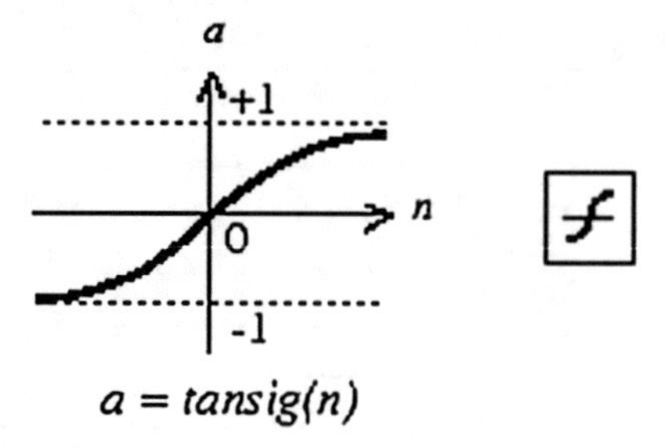

Fig. 6 Tan-sigmoid transfer function

compute the net input n. Then, the net input n is passed through a non-linear transfer function which is tan-sigmoid transfer function presented as (15) and Fig. 6. Finally, the output a is produced and sent to the next layer (the output layer). In the trial and error experiment, the number of hidden nodes is varied from one to ten in order to find the appropriate number which give the lowest MAPE. Ninety ANN models are constructed based on three cassava starch exports, three input scenarios and ten scenarios of the hidden nodes. The number of experimental runs for each ANN model and the proper number of the hidden nodes are presented in following parts.

$$a = \frac{2}{1 + e^{-2n}} - 1 \tag{15}$$

6.3.1 Number of Experimental Runs

To obtain the reliable average of the output which are forecasted quantities in 2014, replication run is required. The experiment is repeated until the average of the output has at most 5 % of margin of error. From the result, the numbers of experimental runs are not the same for each scenario due to difference of the input variables and the number of hidden nodes. Characteristic interpretations of the number of experimental runs, when the input variables and the numbers of hidden nodes are varried, are presented below.

For the native starch (Fig. 7), the ANN models with fundamental inputs require two replications except when the number of hidden nodes is seven, the number of runs is eight replications. The ANN models with fundamental and technical inputs need four replications when the number of hidden nodes is one. After increasing the numbers of hidden nodes to two and three, the numbers of runs are dropped to three and two replications. Then, increasing the numbers of hidden nodes from four to nine, the numbers of runs are increased rapidly from seven to 116 replications. However, when the number of hidden nodes is ten, the number of runs is reduced to 40 replications. The ANN models with significant correlated inputs require two replications when the numbers of hidden nodes are one, two and three. The numbers of runs are fluctuated when the numbers of hidden nodes are four or more. There is

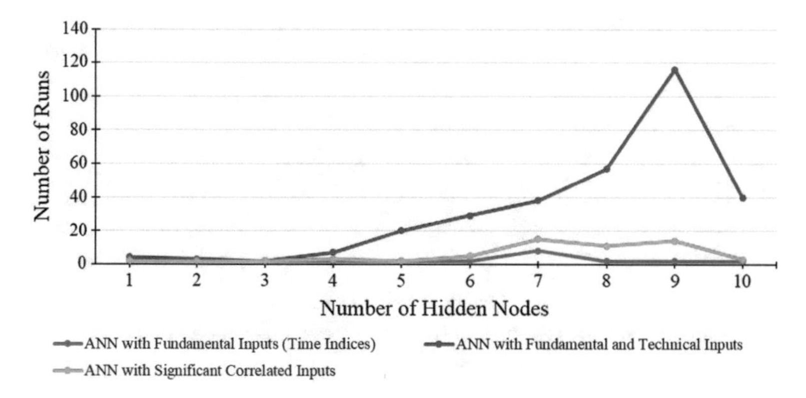

Fig. 7 Number of experimental runs for native starch

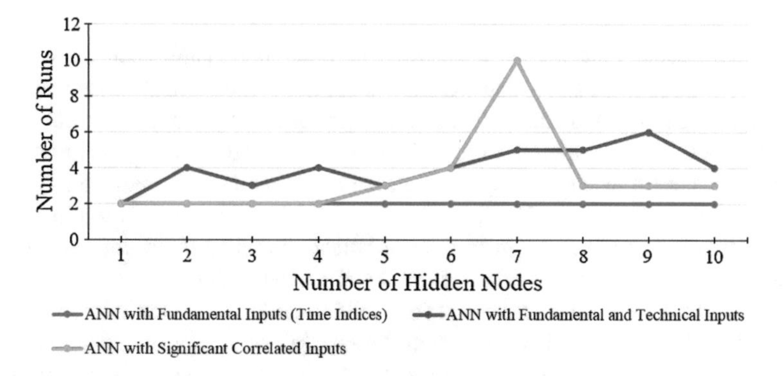

Fig. 8 Number of experimental runs for modified starch

increasing trend of the numbers of runs from two to 15 replications when the numbers of hidden nodes are four to seven. From eight to ten hidden nodes, the numbers of runs show decreasing trend from 11 to three replications.

In case of the modified starch (Fig. 8), the ANN models with fundamental inputs, from one to ten hidden nodes, require only two replications. The numbers of runs of the ANN models with fundamental and technical inputs have increasing trend from two to six replications when the numbers of hidden nodes are one to nine. Nevertheless, the number of runs is dropped down to four replications when the number of hidden nodes is ten. For the ANN models with significant correlated inputs, from one to four hidden nodes, the numbers of runs are two replications. From five to seven hidden nodes, the numbers of runs are increased from three to ten replications. However, eight hidden nodes or more, the numbers of runs are decreased to three replications.

For sago (Fig. 9), for all input scenarios, the numbers of runs are two for one to three hidden nodes. The numbers of runs of the ANN models with fundamental inputs are still two replications until the number of hidden nodes is five. At six hidden nodes, the number of runs is promptly increased to 16 replications. Then,

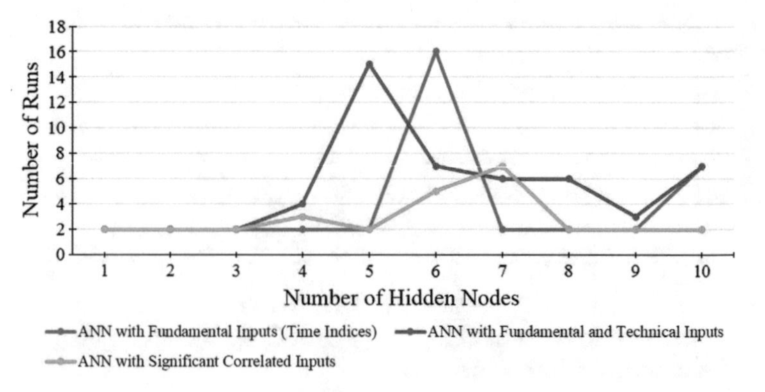

Fig. 9 Number of experimental runs for sago

the numbers of runs are moved back to two replications again when the numbers of hidden nodes are seven to nine. At ten hidden node, the number of runs is raised to seven replications. For the ANN models with fundamental and technical inputs, the numbers of runs are increased from four to 15 replications when the numbers of hidden nodes are four and five. After that, the numbers of runs are decreased from seven to three replications when the numbers of hidden nodes are six to nine. Then, the number of runs is increased to seven replications when the number of hidden nodes is ten. The number of runs of the ANN models with significant correlated inputs is slightly increased to three at four hidden nodes. Then the number of runs is turned back to two replications when the number of hidden nodes is five. The numbers of runs starts increasing from five to seven replications when the numbers of hidden nodes are six and seven. After that, the numbers of runs are decreased to two replications again when the numbers of hidden nodes are eight to ten.

In summary, in order to get the reliable average of outcome, minimum and maximum numbers of the runs are two and 116 replications. In addition, it is difficult to give clearly explanation of relationship between the number of the experimental runs and the number of hidden nodes because there is no consistent pattern and relationship among them. According to the results, however, we may suggest that for the ANN models, using same number of the experimental run for every scenario may not a good approach because it may not give the reliable results or consume more time without necessity.

6.3.2 Number of Hidden Nodes

In this part, after obtaining the reliable average forecasted export quantities for each ANN model from the previous part, they are compared with the actual export quantities, and MSE, MAE and MAPE for all 90 ANN models are computed. However, MAPE is chosen for the interpretation because it expresses the meaning of error in percentage which is a standardized scale and can be understood easily. The suitable

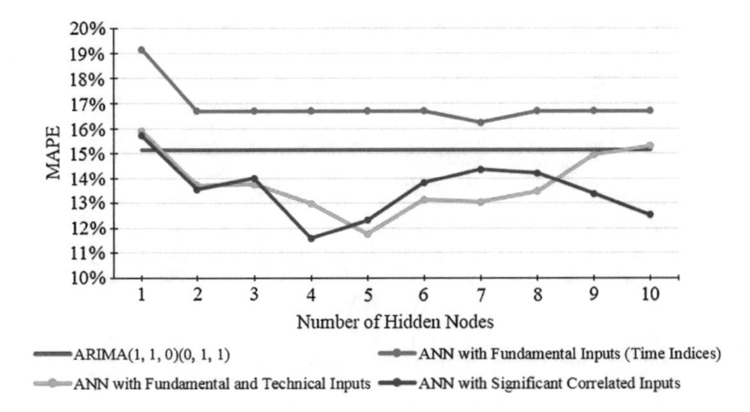

Fig. 10 Forecasting accuracy of ANN models for native starch

number of hidden nodes for each ANN model is the number that gives the lowest MAPE. The MAPEs of each model are plotted with various numbers of hidden nodes and their patterns are interpreted. Moreover, MAPEs of the best-fit ARIMA models are also plotted as a benchmark.

The MAPEs of ANN models for the native starch are presented in Fig. 10. The MAPEs of the ANN models with fundamental inputs are higher than the ARIMA models in all cases. The ANN models with fundamental and technical input and the ANN models with significant correlated inputs seem to have the same pattern of the MAPE changing when the numbers of hidden nodes are varied. When the numbers of hidden nodes are two to nine, their MAPEs are lower than the ARIMA model. However, when the number of hidden nodes is nine, the MAPE of the ANN model with fundamental and technical input is kept increasing until it is over the MAPE of the ARIMA model when the number of hidden nodes is ten. On the other hand, the MAPE of the ANN model with significant correlated inputs is reduced continuously. In addition, one hidden node gives the highest of the MAPE in every the ANN models. The suitable numbers of the hidden nodes for the ANN models with fundamental inputs, fundamental and technical input and significant correlated inputs are seven, five and four respectively. The best model for the native starch is the ANN(7-4-1) model with significant correlated inputs.

In case of the modified starch (Fig. 11), the ARIMA model outperforms almost all the ANN models except the ANN model with fundamental and technical inputs at one hidden node, and the ANN models with significant correlated inputs at one and seven hidden nodes. The MAPEs of the ANN models with fundamental inputs are highest at one hidden node, then they are decreased continuously. Besides, the MAPEs of the ANN models with fundamental and technical inputs are lowest at one hidden node. When the numbers of hidden nodes are increased, the MAPEs are also increased until nine hidden nodes, then the MAPEs are dropped at ten hidden nodes. For the ANN models with significant correlated inputs, their MAPEs seem to be stable on a mean even the numbers of hidden nodes are changed. The suitable numbers of

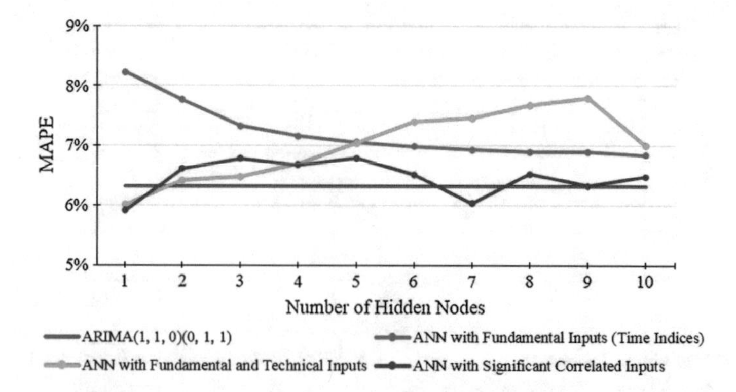

Fig. 11 Forecasting accuracy of ANN models for modified starch

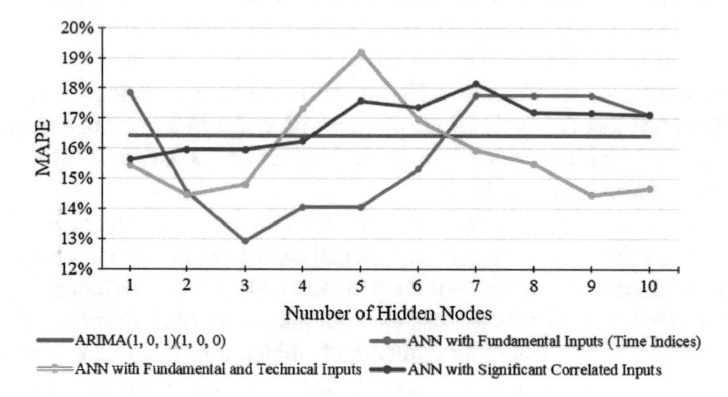

Fig. 12 Forecasting accuracy of ANN models for sago

the hidden nodes for the ANN models with fundamental inputs, fundamental and technical input and significant correlated inputs are ten, one and one respectively. The best model for the modified starch is the ANN(7-1-1) model with significant correlated inputs.

For the sago (Fig. 12), there are around half of the ANN models that can overcome the ARIMA model. The ANN models with fundamental inputs have lower MAPEs than the ARIMA model when the numbers of hidden nodes are two to six. The MAPEs are increased when the numbers of hidden nodes are close to one and seven. The ANN models with fundamental and technical inputs get the highest MAPE at five hidden nodes. Farther from the five hidden nodes gives lower MAPEs. Nevertheless, there is increasing trend of the MAPEs when the numbers of hidden nodes are one and ten. The ANN models with significant correlated inputs with one to four hidden nodes give lower MAPEs than the ARIMA model but there is slightly increasing of the MAPEs when the numbers of hidden nodes reach to four. At five hidden nodes, the MAPE is shifted up continuously above the ARIMA model until seven hidden nodes. Then, the MAPEs are reduced a little bit but they are still above the ARIMA model. The suitable numbers of the hidden nodes for ANN model with fundamental inputs, fundamental and technical input and significant correlated inputs are three, nine and one respectively. Surprisingly, the best model for the sago is not the ANN model with significant correlated inputs as the previous two cassava starch exports but it is the ANN(3-3-1) model with fundamental input which includes only the simple time indices as the input nodes. The screening input variables by the correlation analysis removes month and quarter variables from the input nodes because they are not significant correlated to the export quantity. However, it does not give the best prediction quality. Therefore, this result reveals that using significant correlated inputs does not guarantee better result than non-screening inputs.

In summary, the results show that, for all cassava starch exports, there is at least an ANN model with appropriate number of hidden nodes can outperform ARIMA models. However, there is some ANN model that cannot surpass ARIMA in all numbers of hidden nodes. The effect of changing the number of hidden nodes is varied among different type of the cassava starches and the ANN input scenarios. Moreover, it seems to be strange that in all ANN models, some additional hidden nodes do not improve the accuracy.

7 Comparison of the ANN Models with the ARIMA Models

The forecasting performances are summarized in Table 8 that compares all models of ANN and ARIMA to obtain the best forecasting model for each cassava export which can be determined by the lowest MAPE. The ANN(7-4-1) and the ANN(7-1-1)

Table 8 Summary of forecasting performances of ARIMA versus ANN

Cassava export	Forecasting model	MSE	MAE	MAPE (%)
Native starch	ARIMA(1, 1, 0) $(0, 1, 1)_{12}$	1,717,324,982.39	32,888.64	15.15
	ANN(3-7-1)[a]	3,073,242,729.02	44,625.84	16.23
	ANN(9-5-1)[b]	1,033,191,608.52	27,663.97	11.76
	ANN(7-4-1)[c]	**954,105,575.48**	**27,609.28**	**11.58**
Modified starch	ARIMA(1, 1, 0) $(0, 1, 1)_{12}$	48,712,546.51	5,281.85	6.31
	ANN(3-10-1)[a]	53,068,291.63	5,665.33	6.84
	ANN(9-1-1)[b]	43,992,351.32	5,026.13	6.01
	ANN(7-1-1)[c]	**43,781,080.25**	**4,961.80**	**5.91**
Sago	ARIMA(1, 0, 0) $(1, 0, 0)_{12}$	276,612.18	423.72	16.43
	ANN(3-3-1)[a]	**186,966.78**	**309.26**	**12.91**
	ANN(9-1-1)[b]	188,874.53	363.29	15.64
	ANN(6-1-1)[c]	188,170.14	363.05	15.63

[a] ANN with fundamental inputs (Time Indices)
[b] ANN with technical and fundamental inputs
[c] ANN with significant correlated inputs

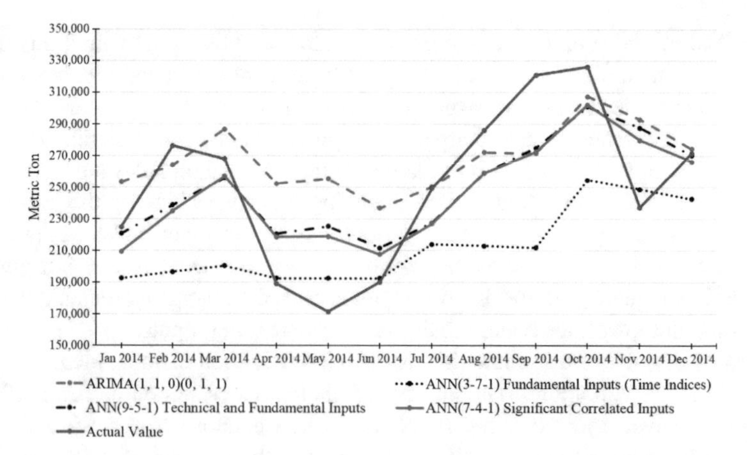

Fig. 13 Actual versus forecasted export quantities of native starch

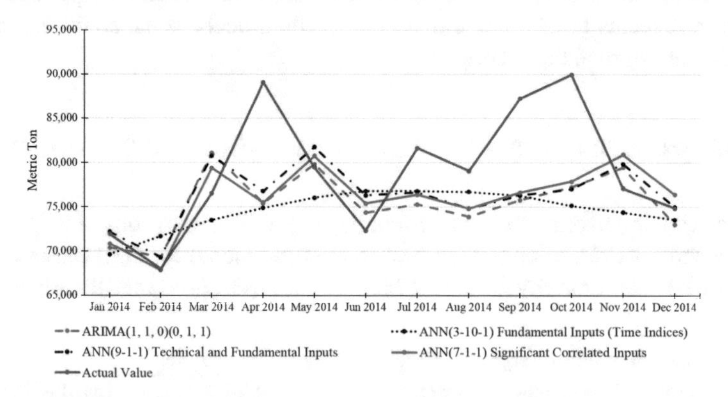

Fig. 14 Actual versus forecasted export quantities of modified starch

models with significant correlated inputs are the most suitable models for the native starch and the modified starch respectively. The ANN(6-1-1) model with fundamental input variables is the most suitable model for the sago.

The forecasted cassava export quantities of the best models of the ANN models and the ARIMA models are compared with the actual cassava export quantities in order to find out reasons why the ANN models can outperform the ARIMA models. The forecasted and actual cassava starch export quantities of the native starch, modified starch and sago are plotted in Figs. 13, 14 and 15 respectively.

From Fig. 13, the best ANN model for the native starch is the ANN(7-4-1) model with significant correlated inputs. The forecasted export quantities from January 2014 to September 2014 of the ANN(7-4-1) model are close to actual export quantities than the ARIMA(1, 1, 0)(0, 1, 1)$_{12}$ model. However, after September 2014, the forecasted export quantities are quite same. For other ANN models, the forecasted export quantities of the ANN(9-5-1) model with technical and fundamental inputs

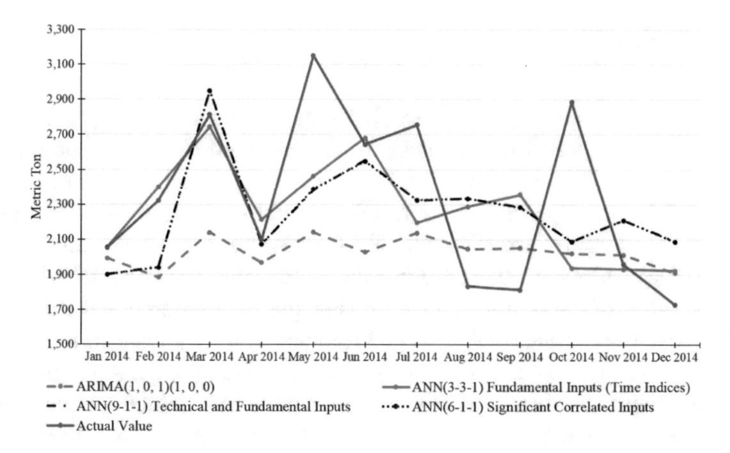

Fig. 15 Actual versus forecasted export quantities of sago

are quite the same as the ANN(7-4-1) model with significant correlated inputs. In case of ANN(3-7-1) model with fundamental inputs, it seems to have low ability to track the actual export quantities.

For the modified starch (Fig. 14), the best model is the ANN(7-1-1) model with significant correlated inputs. The ANN(7-1-1) model and the ARIMA(1, 1, 0)(0, 1, $1)_{12}$ model have almost the same forecasted pattern but from April 2014 to December 2014, the ANN(7-1-1) model give a little bit more accurate forecasted export quantities. In addition, the ANN(9-1-1) model with technical and fundamental inputs produces approximately similar forecasted export quantities as the the ANN(7-1-1). The ANN(3-10-1) with fundamental inputs has low capability to trace the pattern of the actual export quantities.

In case of sago (Fig. 15), the best model for the sago is the ANN(3-3-1) model with fundamental input. The ANN(3-3-1) obviously outperforms the ARIMA(1, 0, 0)(1, 0, $0)_{12}$ model. The ARIMA model has low proficiency to track the pattern of actual export quantities. The ANN(9-1-1) with technical and fundamental inputs gives nearly the same forecasted export quantities as the ANN(6-1-1) with significant correlated inputs.

In summary, the ANN models with significant correlated inputs are the best model and the ANN models with fundamental inputs has low ability to track the pattern of the forecasted cassava export quantities except for the sago. The ARIMA models are quite accurate model for the native starch and the modified starch. However, they have low forecasting capability for the sago. Perhaps, an outlier in the sago time series causes this problem.

8 Conclusion

The ARIMA models and the ANN models are developed to forecast the three types of cassava starch export; the native starch, the modified starch and the sago. The data from January 2001 to December 2013 are used to predict the cassava starch export in 2014. The experiments are run for several replications to obtain reliable results with 5 % margin of error. The models' forecasting accuracies are evaluated and compared. The results show that the ANN models can overcome the ARIMA models in all three types of cassava starch export. For the native starch and the modified starch, the ANN(7-4-1) and the ANN(7-1-1) models with significant correlated inputs respectively give the most accurate forecasting. Surprisingly, the ANN(3-3-1) with fundamental input variables, which uses only simple time indices as the inputs, is the best model for the sago. In case of the sago, it shows that applying the correlation analysis to screen the inputs for the ANN models does not always give the good result. Thus, we should not rely on only scenario using the significant correlated variables as the inputs but the other scenarios, which may include non-significant correlated variables, should also be experimented and compared. There are limitations in the experiments of this paper. In the model validation, one year ahead prediction—export quantities in year 2014—are considered. Additionally, the ANN model structure is feed-forward with one layer of hidden node and the numbers of hidden nodes are varied from 1 to 10.

In conclusion, the feed-forward ANN models show their capability in forecasting the cassava starch export with higher accuracy than the ARIMA models. Therefore, these models are useful for the stakeholders who make a decision based on the future cassava starch export. In future work, two years ahead prediction will be conducted in the model validation. The reason for why some additional hidden nodes in all cases do not improve the accuracy will be investigated. Moreover, it is interesting to continue developing other type of the ANN model (e.g., recurrent ANN model) and the hybrid model which is a combination of ANN model and other models.

References

1. Box, G., Jenkins, G., Reinsel, G.: Time Series Analysis: Forecasting and Control. Wiley Series in Probability and Statistics. Wiley, New York (2008)
2. Co, H.C., Boosarawongse, R.: Forecasting Thailands rice export: statistical techniques versus artificial neural networks. Comput. Ind. Eng. **53**(4), 610–627 (2007)
3. Dayhoff, J.A.: Neural Network Architectures: An Introduction. MIT press, Cambridge (1995)
4. FAO. Why cassava (2008). http://www.fao.org/ag/agp/agpc/gcds/index_en.html. Accessed 07 April 2015
5. Hecht-Nielsen, R.: Theory of the backpropagation neural network. In: International Joint Conference on Neural Networks, IJCNN, pp. 593–605. IEEE (1989)
6. Kaastra, I., Boyd, M.: Designing a neural network for forecasting financial and economic time series. Neurocomputing **10**(3), 215–236 (1996)

7. Kosanan, O., Kantanantha, N.: Thailands para rubber production forecasting comparison. In: Proceedings of the International MultiConference of Engineers and Computer Scientists, vol. 2 (2014)
8. MacKay, D.J.: A practical Bayesian framework for backpropagation networks. Neural Comput. **4**(3), 448–472 (1992)
9. Obe, O.O., Shangodoyin, D.K.: Artificial neural network based model for forecasting sugar cane production. J. Comput. Sci. **6**(4), 439–445 (2010)
10. Pattranurakyothin, T., Kumnungkit, K.: Forecasting model for para rubbers export sales. KMITL Sci. Technol. J. **12**(2), 198–202 (2012)
11. Pokterng, S., Kengpol, A.: The forecasting of durian production quantity for consumption in domestic and international markets. KMUTNB: Int. J. Appl. Sci. Technol. **3**(3), 7–18 (2013)
12. Sunasuan, S., Sombatvichathorn, Y., Thanapongpipat, S., Duangmanee, K., Woradit, S.: The study of agricultural products of cassava. Technical report, The Office of the Agricultural Futures Trading Commission of Thailand (AFTC) (2011)
13. TTSA. Analysis to Thai tapioca market. Technical report, Thai Tapioca Starch Association (TTSA) (2010)
14. Udomsri, N., Kengpol, A., Ishii, K., Shimada, Y.: The design of a forecasting support models on demand of durian for export markets by time series and ANNs. KMUTNB: Int. J. Appl. Sci. Technol. **4**(2), 49–65 (2013)
15. Zhang, G., Patuwo, B.E., Hu, M.Y.: Forecasting with artificial neural networks: the state of the art. Int. J. Forecast. **14**(1), 35–62 (1998)
16. Zou, H., Xia, G., Yang, F., Wang, H.: An investigation and comparison of artificial neural network and time series models for Chinese food grain price forecasting. Neurocomputing **70**(16), 2913–2923 (2007)

Copula Based Volatility Models and Extreme Value Theory for Portfolio Simulation with an Application to Asian Stock Markets

Apiwat Ayusuk and Songsak Sriboonchitta

Abstract Many empirical works used risk modeling under the assumption of Gaussian distribution to investigate the market risk. The Gaussian assumption may not be appropriate for risk estimation techniques in some situations. In this study, we used the extreme value theory (EVT) to examine more precisely the tail distribution of market risk and incorporate high dimensional copulas to explore the dependence between stock markets. We gathered data of stock markets from Asean countries (Thailand, Singapore, Malaysia, Indonesia and the Philippines) to simulate the portfolio analysis during and post subprime crisis. The results found that D-vine copula GARCH-EVT model can simulate the efficient frontier of portfolios greater than other models. Furthermore, we also found the positive dependence for the overall markets.

1 Introduction

For asset allocation models, the risk-return characteristics are the most important issue for investors to consider. The conventional portfolio theory uses standard deviation and linear correlation coefficient to measure portfolio risk under multivariate normal distribution. To construct the optimal portfolio, this theory uses the risk-return framework to allocate assets by minimizing the risk of the portfolio subject to the portfolio return being greater or equal to the risk free rate.

The Value at Risk (VaR) is one of the most important and popular tool to measure the financial risk. It measures the maximum amount of loss that is not exceeded on a given confidence interval. An alternatively risk measure is the Conditional

A. Ayusuk (✉) · S. Sriboonchitta
Faculty of Economics, Chiang Mai University, Chiang Mai 50200, Thailand
e-mail: pai.mr.flute@gmail.com

S. Sriboonchitta
e-mail: songsakecon@gmail.com

A. Ayusuk
Faculty of Liberal Arts and Management Sciences, Department of Business Economics, Prince of Songkla University, Suratthani 84000, Thailand

V.-N. Huynh et al. (eds.), *Causal Inference in Econometrics*,
Studies in Computational Intelligence 622, DOI 10.1007/978-3-319-27284-9_17

VaR (CVaR), which is used to estimate the expected loss from VaR. Rockafellar and Uryasev [30] showed a representation of CVaR based approachs to optimize portfolios. Moreover, Artzner et al. [3] and Rockafellar and Uryasev [31] explained that VaR is not coherence whereas CVaR satisfies the properties of the risk of a diversified portfolio, which are the sub-additive and convex properties. For these reasons, CVaR has the advantages over VaR.

The most widely used econometric approach to volatility modeling is the family of autoregressive conditional heteroscedasticity (ARCH), which is introduced by Engle [10]. It assumes that the conditional variance takes into account the conditional heteroskedasticity inherent in time with the assumption of normally distributed innovations. Bollerslev [6] then improved the ARCH to generalized ARCH (GARCH) model, which can yields VaR and CVaR as well.

In recent years, the EVT has been utilized to analyze financial data. It is a statistical tool to examine the extreme deviations from the median of probability distribution. It is very popular and useful for modeling in rare events. Hence, the EVT can be an alternative for an effective framework to estimate the tail of financial series when there are extreme financial events, such as the Asian financial crisis, Subprime crisis and European debt crisis. Embrechts et al. [9] provided examples for applications of EVT in finance and insurance. Bali [5], Wang et al. [34], Ren and Giles [29] and Jess et al. [16] applied EVT to calculate VaR for risk management.

The EVT based method combines ideas from the GARCH models with the tail of the innovations distribution using EVT to estimate VaR and CVaR. Exemplary works by McNeil and Frey [24] introduced EVT based method (or conditional EVT models) to forecast VaR. Karmakar [18] applied this method to estimate VaR in different percentiles for negative and positive BSE India returns. Furio and Climent [11] found that GARCH-EVT model is more accurate than the GARCH models assuming Gaussian or Student's t distribution innovations for VaR simulation analysis. Meanwhile, Allen et al. [2] used both unconditional and conditional EVT models to forecast VaR. Marimoutou et al. [22] found that this model performs better than other methods without EVT, such as conventional GARCH, historical simulation and filtered historical simulation.

To study the dependence among stock markets using traditional methods, Pearsons correlation has been the most commonly used in empirical works. However, Pearsons correlation used to measure the degree of linear dependence between multivariate normally distributed data. More precisely, Copulas can relax the dependence structure beyond normal distribution. Moreover, the copula is flexible as it can be used to analyze linear, nonlinear or tail dependence. In the context of the copula in financial studies, Embrechts et al. [8] introduced copula in finance to relax the assumption of dependence structures between random returns. Patton [28] explained an overview of copula based models for financial applications. In the study of multivariate copulas, Kole et al. [19] and Wang et al. [35] found that the multivariate t copula is the best measure of the dependence structure between multiple assets because it can capture the dependence both in the center and the tails. Aas et al. [1] introduced the flexible way to set the pair copula construction, namely, D and C-vine copula. The recent study such as Low et al. [21], Hernandez [14], Ayusuk and Sriboonchitta [4], Mensi et al. [25] have applied vine copula with applications to portfolio management.

There are researchers on the effects of the subprime crisis. Hemche et al. [13] found the dynamic linkages between the US and developed stock markets (as France, Mexico, Italy and the UK) with strong comovements in times of financial crisis. The correlation between the US and other markets (as China, Japan, Tunisia, Egypt and Morocco) were weak and thus they suggested that the investors should also invest in some emerging countries. Moralesa and Callaghan [26] and Wang [33] suggested that the US stock markets are less generating effects into the Asian stock markets. In 2015, Asean Economic Community (AEC) is set to be implemented. There will be free trade of goods, services, skilled labor and investment capital following the liberalization and most countries in AEC are still emerging economies. Hence, to take advantage of the portfolio allocation for international stock market, this study focused on VaR and CVaR based on the econometric approaches with the application on the Asean stock markets during and post subprime crisis.

In this paper, the primary objective is to compare the econometric approaches to portfolio simulation. These econometric approaches include the multivariate t copula GARCH-EVT, C-vine copula GARCH-EVT and D-vine copula GARCH-EVT. The secondary objectives to measure the dependence among Asean stock markets.

The remainder of this paper is organized as follows. In Sect. 2, we provide details about the GARCH model, EVT, copulas and the portfolio simulation procedure. In Sect. 3, discuss the data selection, descriptive statistics and the results of the empirical work. In the final section, we present concluding remarks.

2 Methodology

2.1 Marginal Models

Generally, data on market returns present conditional heteroscedasticity. Hence, this study focuses on the marginal returns through the autoregressive conditional heteroskedasticity model. To capture the asymmetry property under the sense that shocks not have the exact same impact on volatility in between negative and positive shocks, we used the GJR GARCH model that was proposed by Glosten et al. [12].

$$r_t = \beta_0 + \beta_1 r_{t-1} + \varepsilon_t = \beta_0 + \beta_1 r_{t-1} + \sigma_t z_t \tag{1}$$

$$\sigma_t^2 = \mu + \alpha\varepsilon_{t-1}^2 + \theta\sigma_{t-1}^2 + \gamma\varepsilon_{t-1}^2 I_{t-1} \tag{2}$$

where r_t is a market return at time t, $I_{t-1} = 1$ if $\varepsilon_{t-1} < 0$, $I_{t-1} = 0$ if otherwise, $\beta_0, \beta_1, \mu, \alpha, \theta, \gamma$ are parameters. For stationarity and positivity, the GJR GARCH model has the following properties: $\alpha > 0, \theta > 0, \gamma > 0, \alpha + \gamma > 0$ and $\alpha + \theta + \gamma/2 < 1, \varepsilon_t = \sigma_t z_t$ is residual return, σ_t is the volatility of the return and z_t

is standardized residual that must satisfy independently and identically distributed. Traditionally, the standardized residuals follow a normal distribution.

2.2 The Distributions of Standardized Residuals

In this study, we focus on EVT, which is an appropriate approach to define the behavior of extreme tail observations. We apply the semi parametric approach to generate the standardized residuals of the GJR GARCH model. To capture the extreme tails, we use the generalized Pareto distribution (GPD) to select the extreme tails that are peaks over the threshold. To capture the interior distribution, we define by using the Gaussian kernel distribution ($\varphi(z)$). The distribution is given by

$$F(z) = \begin{cases} \frac{k_{u^L}}{n}\left(1+\eta^L\left(\frac{u^L - z}{\vartheta^L}\right)\right)^{-(\eta^L)^{-1}} & , z < u^L \\ \varphi(z) & , u^L < z < u^R \\ 1 - \frac{k_{u^R}}{n}\left(1+\eta^R\left(\frac{u^R - z}{\vartheta^R}\right)\right)^{-(\eta^R)^{-1}} & , z > u^R \end{cases} \quad (3)$$

where u^L and u^R are lower (L) and upper (R) thresholds, z is the standardized residuals that excess over the thresholds, k_{u^L} and k_{u^R} are the number of observations that excess over thresholds, n is the number of observation, ϑ^L and ϑ^R are the scale parameters, η^L and η^R are the shape parameters.

2.3 Copula Approach

A copula is a function that connects univariate marginals to construct the multivariate distribution with uniformly distributed marginals $U(0, 1) \longrightarrow [0, 1]$. It also can be used to portray the dependency of random variables in each event. This study used the copula approach for describing the dependence between international markets. Originally, Sklar [32] introduced the important theorem for copula function as follows

Theorem 1 *Let $x_1, \ldots, x_n$ are random variables for $i = 1, \ldots, n$, $F_1(x_1), \ldots, F_n(x_n)$ are the continuous marginal distributions and $F(x_1, \ldots, x_n)$ be a multivariate distribution. Then, n-dimensional copulas $C(\cdot) : [0, 1]^n \longrightarrow [0, 1]$ can be defined by*

$$F(x_1, \ldots, x_n) = C(F_1(x_1), \ldots, F_n(x_n)) \quad (4)$$

Inversely, Eq. (4) can be written as

$$C(u_1, \ldots, u_n) = F(F_1^{-1}(u_1), \ldots, F_n^{-1}(u_n)) \quad (5)$$

where $F_i^{-1}(u_i)$ are the inverse distribution function of the marginals and $u_i \in [1, 0]$.

We can determine the copula density $c(u_1, \ldots, u_n)$ by using n order partial derivative as follows

$$c(u_1, \ldots, u_n) = \frac{\partial^n C(u_1, \ldots, u_n)}{\partial u_1, \ldots, \partial u_n} \tag{6}$$

According to the joint density function $f(x_1, \ldots, x_n)$, it can be defined by n order partial derivative of a multivariate distribution as follows

$$f(x_1, \ldots, x_n) = \frac{\partial^n F(x_1, \ldots, x_n)}{\partial x_1, \ldots, \partial x_n} \tag{7}$$

$$f(x_1, \ldots, x_n) = \prod_{i=1}^{n} f_i(x_i) \frac{\partial^n C(u_1, \ldots, u_n)}{\partial u_1, \ldots, \partial u_n} \tag{8}$$

$$f(x_1, \ldots, x_n) = \prod_{i=1}^{n} f_i(x_i) \cdot c(u_1, \ldots, u_n) \tag{9}$$

Equation (9) shows that the joint density function is the combination between the copula density and the product of marginal densities. In the study of copulas, Mashal and Zeevi [23], Breymann et al. [7], Kole et al. [19] and Wang et al. [35] suggested that t copula is the better measure of the dependency structure for multiple assets. Hence, this study considered t copula for measuring the market dependence. We can define a multivariate t copula for n dimensional as follows

$$C^t(u_1, \ldots, u_n) = t_{\nu,\Sigma}(t_\nu^{-1}(u_1), \ldots, t_\nu^{-1}(u_n)) \tag{10}$$

where $t_{\nu,\Sigma}$ is the distribution function of the multivariate t copula, Σ is a correlation matrix and ν is the degree of freedom. Moreover, This study also applied C and D-vine structures with t copula to determine the market dependence. The two vine copulas were introduced by Aas et al. [1]. In n dimensions, $\frac{n(n-1)}{2}$ is the number of pair copula, $n-1$ is the number of trees in vine copulas and $n!$ is the number of possible tree structures. To select the tree structures, this study determines the appropriate ordering of the tree structures by choosing the maximum of absolute empirical Kendall's tau values for all bivariate copula. C and D-vine density functions can be defined by

$$f(x_1, \ldots, x_n) = \prod_{i=1}^{n} f_i(x_i) \prod_{j=1}^{n-1} \prod_{k=1}^{n-j} c_{j,j+k|1,\ldots,j-1} \left(F(x_j|\mathbf{v_1}), F(x_{j+k}|\mathbf{v_1}) \right) \tag{11}$$

$$f(x_1, \ldots, x_n) = \prod_{i=1}^{n} f_i(x_i) \prod_{j=1}^{n-1} \prod_{k=1}^{n-j} c_{k,k+j|k+1,\ldots,k+j-1} \left(F(x_k|\mathbf{v_2}), F(x_{k+j}|\mathbf{v_2}) \right) \tag{12}$$

where $\mathbf{v_1} = x_1, \ldots, x_{j-1}$, $\mathbf{v_2} = x_{k+1}, \ldots, x_{k+j-1}$, j is the tree in vine copulas, k is the edge in each tree, $c_{j,j+k|1,\ldots,j-1}$ in Eq. (11) and $c_{k,k+j|k+1,\ldots,k+j-1}$ in Eq. (12) are bivariate copula densities. In order to compute the conditional distribution functions $F(x|\mathbf{v})$ in Eqs. (11) and (12) by following Joe [17], as in Eq. (13)

$$F(x|\mathbf{v}) = \frac{\partial C_{xv_j|\mathbf{v_{-j}}}\left(F(x|\mathbf{v_{-j}}), F(v_j|\mathbf{v_{-j}})\right)}{\partial F(v_j|\mathbf{v_{-j}})} \tag{13}$$

where the vector v_{-j} is the vector v_j that excludes the component v_j. $C_{xv_j|\mathbf{v_{-j}}}$ is the bivariate copula distribution between x and v_j that is taken conditional on v_{-j}. The estimated dependence parameters of various copulas are obtained by maximum likelihood (see Aas et al. [1]).

2.4 Portfolio Simulation

We forecast one-day-ahead for VaR, CVaR based on t copula GARCH-EVT at 95 and 99 % confidence level with the procedures as follows:

(1) We estimate the parameters of the GARCH model for each market return series. We obtain the standardized residuals over the threshold follow the generalized Pareto distribution (GPD), because GPD can capture the upper and lower tails. Additionally, we also use the Gaussian kernel estimation for the interior part.
(2) We transform each standardized residuals (z_t) of each univariate distribution to approximate i.i.d. uniform data (u_t) on [0, 1] by using empirical distribution functions and then fit t copula for estimating its parameter.
(3) Given the parameters of copula function, we simulate the uniform series 100,000 dimensional time series and obtain the standardized residuals by using the inverse functions of the estimated marginals.
(4) We converse the standardized residuals from step (3) into the returns at $t + 1$, calculate the empirical one-day-ahead VaR, CVaR at 95 % and 99 % confidence level, and optimize the portfolio based on CVaR minimization problem at 99 % confidence level (or $Min_{w \in W} CVaR$) by following the procedure of Rockafellar and Uryasev [30, 31].

3 Empirical Results

We used the daily data of five main stock market indices in Asean countries from DataStream: The indices composed of SET index (Thailand:TH), Straits Times index (Singapore:SP), KLSE Composite index (Malaysia:MS), JSX Composite index (Indonesia:ID) and PSE Composite (the Philippines:PP). We defined the market returns by $r_t = log(p_t) - log(p_{t-1})$. Following Horta et al. [15] and Lee et al. [20],

Table 1 Descriptive measures for Asean markets

Index	TH	SP	MS	ID	PP
A: crisis period					
Mean	−0.000221	−0.000341	−9.38E–05	0.000210	−0.000220
Max	0.086167	0.102705	0.057165	0.190719	0.083854
Min	−0.085892	−0.129279	−0.102374	−0.257802	−0.136399
S.D.	0.018798	0.020622	0.012513	0.026211	0.019945
Skewness	−0.104720	−0.034376	−0.873542	−1.404863	−0.687729
Kurtosis	6.331662	8.140576	13.30447	27.81636	9.093367
Jarque–Bera	243.3070 [0.0000]	577.0605 [0.0000]	2384.952 [0.0000]	13618.47 [0.0000]	851.9586 [0.0000]
ADF statistics	−20.55669 [0.0000]	−21.93688 [0.0000]	−21.54766 [0.0000]	−22.41699 [0.0000]	−21.35143 [0.0000]
B: after crisis period					
Mean	0.000667	0.000141	0.000304	0.000651	0.000818
Max	0.057515	0.029001	0.047228	0.070136	0.055419
Min	−0.058119	−0.037693	−0.026757	−0.092997	−0.069885
S.D.	0.011965	0.008567	0.006207	0.012800	0.011732
Skewness	−0.367571	−0.423877	0.108370	−0.822461	−0.500209
Kurtosis	6.294566	4.813557	8.424547	10.33282	6.756750
Jarque–Bera	509.4336 [0.0000]	179.1765 [0.0000]	1317.675 [0.0000]	2524.951 [0.0000]	675.7222 [0.0000]
ADF statistics	−30.01431 [0.0000]	−30.38082 [0.0000]	−29.13288 [0.0000]	−22.97933 [0.0000]	−22.97933 [0.0000]

Note In parentheses are standard errors of the coefficient estimates

this study focuses on subprime crises period and then we divide it into sub periods: the subprime crisis period (1 August 2007–29 December 2009) and the post subprime crisis period (4 January 2010–29 December 2014).

Table 1 shows summary statistics. We found that almost all markets of the average yield (mean of market return) are negative during the subprime crisis. SP has the most negative returns. After the subprime crisis, the average yield has a positive sign in every market and the standard deviation (SD) is less than a period of the subprime. The Jarque–Bera rejects the null hypothesis which indicated that returns of the markets are not following the normality assumption. The ADF test approved the stationary property of all markets.

Table 2 shows GJR GARCH parameter estimation. The mean equation is in the simplest form of first autoregressive ($AR(1)$). The Q-statistics confirm that the marginals mostly accept the null hypotheses which suggested that there are no serial correlations and satisfy an i.i.d. assumption for almost all the markets. Then, we transform standardized residuals into the uniform ($U[0, 1]$) by using the empirical distribution functions. The Kolmogorov–Smirnov test (KS-test) is used to test the null hypothesis that the transformed data are uniformly distributed, because all data series

Table 2 Parameter estimates for AR(1)-GJR GARCH-EVT models

Index	TH	SP	MS	ID	PP
A: crisis period					
Mean equation					
β_0	0.000549 [0.000665]	0.000135 [0.000651]	0.000199 [0.000392]	0.000541 [0.000688]	0.000142 [0.000678]
β_1	0.009793 [0.045931]	0.001870 [0.046813]	0.065287 [0.044449]	0.11054 [0.046162]	0.086547 [0.046992]
Variance equation					
μ	1.69e–005 [8.32e–006]	3.22e–006 [3.16e–006]	1.17e–005 [5.24e–006]	0.000137 [3.16e–005]	7.79e–005 [2.59e–005]
α	0.83871 [0.051076]	0.91095 [0.023375]	0.80145 [0.059008]	0.38876 [0.083337]	0.57183 [0.097427]
θ	0.045528 [0.036726]	0.035349 [0.020904]	0.018407 [0.028609]	0.000000 [0.040953]	0.063385 [0.048436]
γ	0.13214 [0.064542]	0.1004 [0.040892]	0.23656 [0.095621]	0.88391 [0.25269]	0.30257 [0.12236]
Q(2)	9.0621 [0.0108]	0.8644 [0.6491]	0.8474 [0.6546]	4.3689 [0.1125]	0.3434 [0.8423]
Q(6)	16.4338 [0.0116]	13.7795 [0.0322]	3.5713 [0.7345]	6.9328 [0.3271]	5.8217 [0.4435]
KS-statistics	0.0027 [0.4771]	0.003 [0.3156]	0.003 [0.3438]	0.0036 [0.1599]	0.0032 [0.2709]
Jarque–Bera	82.0629 [0.0000]	85.3962 [0.0000]	789.3312 [0.0000]	1316.8 [0.0000]	231.8684 [0.0000]
B: after crisis period					
Mean equation					
β_0	0.001186 [0.000282]	0.000277 [0.000210]	0.000306 [0.000146]	0.000988 [0.000289]	0.000791 [0.000293]
β_1	0.027503 [0.032322]	0.010368 [0.03082]	0.083798 [0.028665]	0.004194 [0.032237]	0.085756 [0.033591]
Variance equation					
μ	5.95e–006 [1.68e–006]	8.02e–007 [3.40e–007]	1.97e–006 [6.75e–007]	6.29e–006 [1.99e–006]	8.96e–006 [2.63e–006]
α	0.83591 [0.029077]	0.92912 [0.015984]	0.85629 [0.03229]	0.87223 [0.027457]	0.81495 [0.036924]
θ	0.035536 [0.024681]	0.014864 [0.018464]	0.024771 [0.021801]	0.019357 [0.025768]	0.020671 [0.025987]
γ	0.16857 [0.041524]	0.085624 [0.0243]	0.14827 [0.043269]	0.125 [0.04037]	0.18828 [0.048925]
Q(2)	3.0135 [0.2216]	0.2932 [0.8636]	2.4969 [0.2870]	4.1248 [0.1271]	0.7148 [0.6995]
Q(6)	6.5855 [0.3609]	7.2224 [0.3008]	3.3907 [0.7585]	24.2223 [0.0005]	8.8284 [0.1835]
KS-statistics	0.0161 [0.9339]	0.0164 [0.945]	0.0227 [0.6376]	0.0208 [0.7443]	0.0154 [0.9651]
Jarque–Bera	114.0268 [0.0000]	40.9454 [0.0000]	1319.4 [0.0000]	968.1900 [0.0000]	160.1151 [0.0000]

Note In parentheses are standard errors of the coefficient estimates
In parentheses of Q, KS and JB-statistics are p-value for testing the null hypothesis

Table 3 GPD estimation of each markets residuals

Index	TH	SP	MS	ID	PP
A: crisis period					
u_r	1.1663	1.2225	1.1225	1.1281	1.1864
ϑ_r	0.4929	0.7693	0.5947	0.5580	0.4916
η_r	0.1237	−0.0712	0.1764	0.1827	0.1447
u_l	−1.3113	−1.3290	−1.1794	−1.2537	−1.2979
ϑ_l	0.4835	0.4938	0.3622	0.6470	0.6831
η_l	0.0875	0.0625	0.4554	0.2680	−0.0168
B: after crisis period					
u_r	1.2222	1.2172	1.0998	1.0367	1.1718
ϑ_r	0.5245	0.4665	0.4004	0.4673	0.4097
η_r	0.0178	−0.0380	0.3282	0.0805	0.1942
u_l	−1.3012	−1.2407	−1.1444	−1.1542	−1.2279
ϑ_l	0.6973	0.7277	0.7322	0.6031	0.5931
η_l	−0.0715	−0.1496	0.0092	0.1908	0.0282

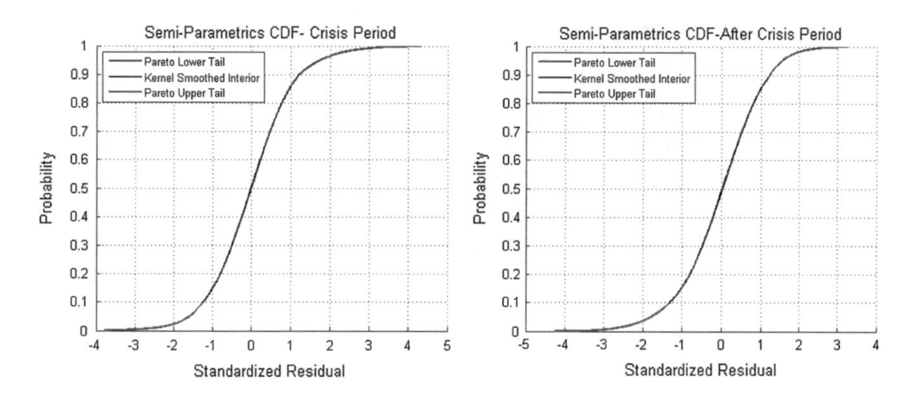

Fig. 1 Semi-parametric CDFs of Singapore residuals

support the null hypothesis and use this results to carry out the copula procedure. Jarque–Bera statistics suggested that the standardized residuals of are non-normality distribution. These findings from statistical testing confirm that the GJR GARCH model can apply EVT to handle on the standardized residuals.

Table 3 shows parameter estimation of extreme value theory, we use the GPD in our study where ϑ, η are the scale parameter and the shape parameter and we fixed the threshold value u at 10 % level of confidence. Figure 1 is a sample of the CDF by using semi-parametric form of Singapore market, in the subprime period, obviously, the valued of upper tail was higher than the after-crisis period.

Table 4 shows the values of Kendall's rank correlation,which were computed by using the parameter of the multivariate t copula function from Eq. (10). The results

Table 4 The matrixes of the Kendall's tau from the multivariate t copula

Index	TH	SP	MS	ID	PP
A: crisis period					
TH	1				
SP	0.4410	1			
MS	0.3524	0.4289	1		
ID	0.3970	0.4866	0.3741	1	
PP	0.2306	0.2808	0.3498	0.2721	1
B: after crisis period					
TH	1				
SP	0.3435	1			
MS	0.2520	0.3291	1		
ID	0.3206	0.3831	0.3550	1	
PP	0.2274	0.2574	0.2662	0.2890	1

Table 5 The matrixes of the Kendall's rank correlation from C and D vine copula

A: crisis period				B: after crisis period			
C-vine copola		D-vine copula		C-vine copula		D-vine copula	
τ_{51}	0.2501	τ_{41}	0.4024	τ_{51}	0.2386	τ_{31}	0.2654
τ_{52}	0.3101	τ_{43}	0.3668	τ_{52}	0.2681	τ_{41}	0.3262
τ_{53}	0.3685	τ_{51}	0.2500	τ_{53}	0.2760	τ_{52}	0.2681
τ_{54}	0.2749	τ_{52}	0.3101	τ_{54}	0.2965	τ_{53}	0.2760
$\tau_{31\|5}$	0.2851	$\tau_{21\|5}$	0.3876	$\tau_{31\|5}$	0.1931	$\tau_{32\|5}$	0.2676
$\tau_{32\|5}$	0.3307	$\tau_{31\|4}$	0.2094	$\tau_{32\|5}$	0.2676	$\tau_{43\|1}$	0.2836
$\tau_{43\|5}$	0.2826	$\tau_{54\|1}$	0.1817	$\tau_{43\|5}$	0.2865	$\tau_{51\|3}$	0.1522
$\tau_{21\|53}$	0.3100	$\tau_{42\|51}$	0.3007	$\tau_{21\|53}$	0.2447	$\tau_{21\|53}$	0.2448
$\tau_{41\|53}$	0.2655	$\tau_{53\|41}$	0.2548	$\tau_{41\|53}$	0.2057	$\tau_{54\|31}$	0.1488
$\tau_{42\|531}$	0.2627	$\tau_{32\|541}$	0.1576	$\tau_{42\|531}$	0.1892	$\tau_{42\|531}$	0.1886

Note 1 = TH, 2 = SP, 3 = MS, 4 = ID, 5 = PP

show that five markets have a monotonic relationship because of the Kendall's tau is more than zero. During the crisis, the highest relationship is SP & ID, SP & TH and SP & MS, respectively. While, PP & ID has the weakest relationship. After the crisis, the strongest relationship is still SP & ID, ID & MS and SP & MS, respectively. While, TH & PP has the weakest relationship.

Table 5 shows the values of Kendall's tau that compute by using the parameter of vine copula function follow Eqs. (11) and (12). In Table 5, we found that five markets have a positive dependence. The highest relationship is ID & TH by D-vine copula in during and after the crisis period.

Table 6 shows the simulation results of one step ahead forecasting in portfolio risk using the multivariate t copula GARCH-EVT, C-vine copula GARCH-EVT and

Table 6 Portfolio risk of the equally weighted market strategy

	The multivariate t copula	C-vine copula	D-vine copula
	GARCH-EVT	GARCH-EVT	GARCH-EVT
A: crisis period			
$VaR_{0.95}$	0.0148	0.0144	0.0147
$VaR_{0.99}$	0.0246	0.0217	0.0224
$CVaR_{0.95}$	0.0212	0.0215	0.0218
$CVaR_{0.99}$	0.0330	0.0325	0.0328
B: after crisis period			
$VaR_{0.95}$	0.0137	0.0137	0.0141
$VaR_{0.99}$	0.0207	0.0208	0.0215
$CVaR_{0.95}$	0.0183	0.0203	0.0205
$CVaR_{0.99}$	0.0263	0.0306	0.0307

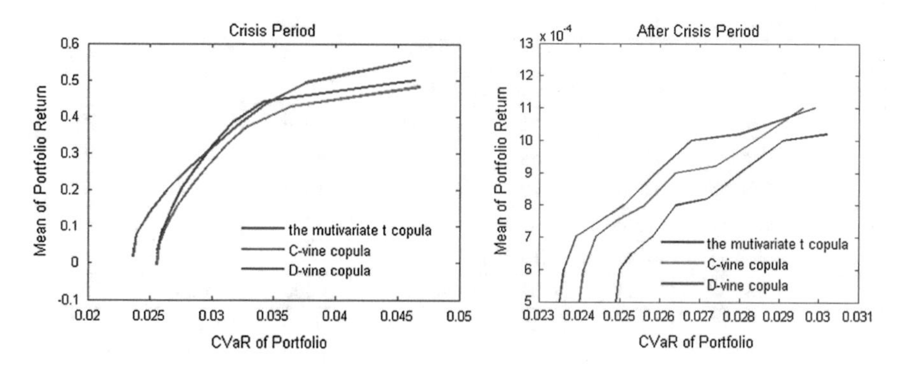

Fig. 2 Efficient frontier from minimizing CVaR at 99 % confidence level

D-vine copula GARCH-EVT under the same strategies for all markets. The simulation results found that VaR and CVaR values during the crisis are higher than after-crisis at 0.95 and 0.99 significant level. In the crisis period, the multivariate t copula GARCH-EVT gives the values of VaR higher than both vine copula GARCH-EVT and D-vine copula GARCH-EVT gives the values of VaR and CVaR higher then C-vine copula GARCH-EVT . Then we also found that the computational of VaR and CVaR using D-vine copula GARCH-EVT gives higher value than the multivariate t copula GARCH-EVT and the C-vine copula GARCH-EVT after the crisis period.

Figure 2 shows the efficient frontier of Asean portfolio by minimizing portfolio risk. A left Fig. 2 is the efficient frontier during the crisis period and a right Fig. 2 is the efficient frontier after the crisis period. From the Fig. 2, we can conclude that at the same level of CVaR, D-vine copula GARCH-EVT generates portfolio return higher than the multivariate t copula GARCH and C-vine copula

Table 7 The optimal portfolio weights in efficient frontier for Asean markets

Portfolios	TH	SP	MS	ID	PP	Return	$CVaR_{0.99}$
A: crisis period by the multivariate t copula GARCH-EVT							
1	0.0483	0.0483	0.6923	0.1032	0.1080	0.0343	0.0257
2	0.1129	0.1129	0.4703	0.1306	0.1734	0.1507	0.0268
3	0.1752	0.1752	0.2400	0.1801	0.2295	0.2671	0.0289
4	0.2646	0.2646	0.0000	0.0000	0.4709	0.4417	0.0343
B: crisis period by C-vine copula GARCH-EVT							
1	0.1506	0.0000	0.5981	0.0000	0.2513	0.0506	0.0258
2	0.1982	0.0589	0.3768	0.0000	0.3661	0.1584	0.0273
3	0.2372	0.1306	0.1612	0.0000	0.4711	0.2661	0.0297
4	0.0000	0.5310	0.0000	0.0000	0.4690	0.4277	0.0365
C: crisis period by D-vine copula GARCH-EVT							
1	0.2902	0.1128	0.0000	0.0000	0.5969	0.0781	0.0240
2	0.2700	0.2532	0.0784	0.0000	0.3984	0.1966	0.0264
3	0.2566	0.3569	0.1833	0.0000	0.2032	0.3152	0.0300
4	0.0083	0.5311	0.4606	0.0000	0.0000	0.5524	0.0376
D: after crisis period by the multivariate t copula GARCH-EVT							
1	0.0000	0.0000	0.5158	0.3826	0.1016	0.0006	0.0250
2	0.0000	0.0000	0.3575	0.5084	0.1341	0.0007	0.0258
3	0.0000	0.0000	0.1997	0.6349	0.1654	0.0008	0.0272
4	0.0000	0.0000	0.0000	0.8604	0.1396	0.0010	0.0302
E: after crisis period by C-vine copula GARCH-EVT							
1	0.3534	0.0642	0.5117	0.0707	0.0000	0.0006	0.0241
2	0.4165	0.1553	0.3626	0.0656	0.0000	0.0007	0.0249
3	0.4930	0.2324	0.2099	0.0648	0.0000	0.0009	0.0264
4	0.6017	0.3682	0.0000	0.0301	0.0000	0.0011	0.0296
F: after crisis period by D-vine copula GARCH-EVT							
1	0.4440	0.0423	0.3284	0.0000	0.1853	0.0006	0.0236
2	0.2887	0.1340	0.3893	0.0000	0.1880	0.0007	0.0244
3	0.1309	0.2273	0.4455	0.0000	0.1964	0.0009	0.0259
4	0.0000	0.4152	0.5826	0.0000	0.0022	0.0011	0.0299

GARCH-EVT during the crisis period. After the crisis period, D-vine copula GARCH-EVT gives portfolio return higher than C-vine copula GARCH-EVT and the multivariate t copula GARCH-EVT at the same level of CVaR. Finally, We calculate the optimal weights of the portfolio at the efficient frontier as Table 7. All three approaches generate the return and CVaR of the portfolio in the crisis period higher than the post crisis period.

4 Conclusions

In this study, we adopt copula based volatility models to measure the dependence between Asean stock markets and then we used a semi-parametric approach from extreme value theory to capture the tail distribution of standardized residuals from the data in the context of the subprime crisis. We examine the portfolio simulation produced by each model and emphasize comparing three models. The models consist of the multivariate t copula GARCH-EVT, C-vine copula GARCH-EVT and D-vine copula GARCH-EVT. Regarding dependence, all copulas provide evidence of positive dependence in every pair. The dependences are mostly strong between Singapore and other markets by the multivariate t copula, which may imply that Singapore market plays an important role in Asean markets. Meanwhile, the risk measure was simulated with equally weighted strategy. This result indicates that D-vine copula GARCH-EVT can be estimates VaR and CVaR greater than C-vine copula GARCH-EVT and the multivariate t copula GARCH-EVT in the post subprime crisis period. The values of VaR and CVaR during the subprime crisis are higher than those after the subprime crisis. Moreover, the results of the portfolio optimization problem using CVaR objective show that D-vine copula GARCH-EVT is a more efficient tool to simulate the portfolio optimization. Finally, the optimal portfolio weights suggest that the international investors should concentrate on the Malaysian market at the high risk and return, and should invest in Thailand market at the low portfolio risk and return after the subprime crisis.

Acknowledgments The authors are thankful to Dr. Supanika Leurcharusmee and Dr. Kittawit Autchariyapanitkul for reviewing the manuscript. First author was supported from Prince of Songkla University-Ph.D. Scholarship.

References

1. Aas, K., Czado, C., Frigessi, A., Bakken, H.: Pair-copula constructions of multiple dependence. Insur.: Math. Econ. **44**(2), 182–198 (2009)
2. Allen, D.E., Singh, A.K., Powell, R.J.: EVT and tail-risk modelling: evidence from market indices and volatility series. North Am. J. Econ. Financ. **26**, 355–369 (2013)
3. Artzner, P., Delbaen, F., Eber, J.M., Heath, D.: Coherent measures of risk. Math. Financ. **9**, 203–228 (1999)
4. Ayusuk, A., Sriboonchitta, S.: Risk analysis in Asian emerging markets using canonical vine copula and extreme value theory. Special Issue on Copula Mathematics and Econometrics. Thai J. Math. 59–72 (2014)
5. Bali, T.G.: An extreme value approach to estimating volatility and value at risk. J. Bus. **76**, 83–108 (2003)
6. Bollerslev, T.: Generalized autoregressive conditional heteroskedasticity. J. Econom. **31**, 307–327 (1986)
7. Breymann, W., Dias, A., Embrechts, P.: Dependence structures for multivariate high-frequency data in finance. Quant. Financ. **3**, 1–14 (2003)

8. Embrechts, P., McNeil, A.: Straumann, correlation and dependence properties in risk management: properties and pitfalls. In: Dempster, M. (ed.) Risk Management: Value at Risk and Beyond. Cambridge University Press, Cambridge (2002)
9. Embrechts, P., Resnick, S., Samorodnitsky, G.: Extreme value theory as a risk management tool. North Am. Actuar. J. **3**, 30–41 (1999)
10. Engle, R.F.: Autoregressive conditional heteroskedasticty with estimates of the variance of U.K. inflation. Econometrica **50**, 987–1008 (1982)
11. Furi, D., Climent, F.J.: Extreme value theory versus traditional GARCH approaches applied to financial data: a comparative evaluation. Quant. Financ. **13**, 45–63 (2013)
12. Glosten, L.R., Jagannathan, R., Runkle, D.E.: On the relation between the expected value and the volatility of the nominal excess return on stock. J. Financ. **48**, 1779–1801 (1993)
13. Hemche, O., Jawadi, F., Maliki, S.B., Cheffou, A.I.: On the study of contagion in the context of the subprime crisis: a dynamic conditional correlation multivariate GARCH approach. Econ. Model (2014). In press
14. Hernandez, J.A.: Are oil and gas stocks from the Australian market riskier than coal and uranium stocks? Dependence risk analysis and portfolio optimization. Energy Econ. **45**, 528–536 (2014)
15. Horta, P., Lagoa, S., Martins, L.: The impact of the 2008 and 2010 financial crises on the Hurst exponents of international stock markets: implications for efficiency and contagion. Int. Rev. Financ. Anal. **35**, 140–153 (2014)
16. Jess, R., Ortiz, E., Cabello, A.: Long run peso/dollar exchange rates and extreme value behavior: value at risk modeling. North Am. J. Econ. Financ. **24**, 139–152 (2013)
17. Joe, H.: Families of m-variate distributions with given margins and $m(m-1)/2$ bivariate dependence parameters. Distributions with Fixed Marginals and Related Topics. Institute of Mathematical Statistics, California (1996)
18. Karmakar, M.: Estimation of tail-related risk measures in the Indian stock market: an extreme value approach. Rev. Financ. Econ. **22**, 7985 (2013)
19. Kole, E., Koedijk, E., Verbeek, M.: Selecting copulas for risk management. J. Bank. Financ. **31**(8), 2405–2423 (2007)
20. Lee, Y.H., Tucker, A.L., Wang, D.K., Pao, H.T.: Global contagion of market sentiment during the US subprime crisis. Global Financ. J. **25**, 1726 (2014)
21. Low, R.K.Y., Alcock, J., Faff, R., Brailsford, T.: Canonical vine copulas in the context of modern portfolio management: are they worth it? J. Bank. Financ. **37**, 3085–3099 (2013)
22. Marimoutou, V., Raggad, B., Trabesi, A.: Extreme value theory and value at risk: application to oil market. Energy Econ. **31**(4), 519–530 (2009)
23. Mashal, R., Zeevi, A.: Beyond correlation: extreme co-movements between financial assets. Unpublished, Columbia University (2002)
24. McNeil, A., Frey, R.: Estimation of tail-related risk measures for heteroscedastic financial time series: an extreme value approach. J. Empir. Financ. **7**(3–4), 271–300 (2000)
25. Mensi, W., Hammoudeh, S., Reboredo, J.C., Nguyen, D.K.: Are Sharia stocks, gold and U.S. treasuries hedges and safe havens for the oil-based GCC markets?. Emerg. Mark. Rev. (2015). In press
26. Moralesa, L., Andreosso-OCallaghan, B.: The current global financial crisis: do Asian stock markets show contagion or interdependence effects? J. Asian Econ. **23**, 616–626 (2012)
27. Naoui, K., Liouane, N., Brahim, S.: A dynamic conditional correlation analysis of financial contagion: the case of the subprime credit crisis. Int. J. Econ. Financ. **2**(3), 85–96 (2010)
28. Patton, A.J.: Copula-based models for financial time series. In: Andersen, T.G., Davis, R.A., Kreiss, J.-P., Mikosch, T. (eds.) Handbook of Financial Time Series. Springer, New York (2009)
29. Ren, F., Giles, D.E.: Extreme value analysis of daily Canadian crude oil prices. Appl. Financ. Econ. **20**(12), 941–954 (2010)
30. Rockafellar, R.T., Uryasev, S.: Optimization of conditional value-at-risk. J. Risk **2**, 21–41 (2000)
31. Rockafellar, R.T., Uryasev, S.: Conditional value-at-risk for general loss distributions. J. Bank. Financ. **26**, 1443–1471 (2002)

32. Sklar, A.: Fonctions de rpartition n dimensions et leurs marges. Publ. Inst. Statist. Univ. Paris **8**, 229–231 (1959)
33. Wang, L.: Who moves East Asian stock markets? The role of the 2007–2009 global financial crisis. J. Int. Financ. Mark. Inst. Money **28**, 182–203 (2014)
34. Wang, Z., Wu, W., Chen, C., Zhou, Y.: The exchange rate risk of Chinese yuan: using VaR and ES based on extreme value theory. J. Appl. Stat. **37**, 265–282 (2010)
35. Wang, Z.R., Chen, Z.R., Jin, Z.R., Zhou, Z.R.: Estimating risk of foreign exchange portfolio: using VaR and CVaR based on GARCHEVT-Copula model. Phys. A **389**, 4918–4928 (2010)

Modeling Dependence of Health Behaviors Using Copula-Based Bivariate Ordered Probit

Kanchit Suknark, Jirakom Sirisrisakulchai and Songsak Sriboonchitta

Abstract This study simultaneously determines the factors affecting each pairing of health behaviors such as alcohol-consumption and physical activity, tobacco-consumption and physical activity, and alcohol-consumption and tobacco-consumption. The measure of dependence between these pairs was quantified using the copula approach. The Copula-based Ordered Probit Model was used to control any common unobserved factors that might affect the random errors related to each pair of health behaviors. The results is more efficient parameter estimates, in terms of lower standard errors, in comparison with separate estimations. Moreover, understanding the dependencies between ordinal choices for each pair of health behaviors gives useful information for designing more efficient health care programs.

Keywords Copula · Bivariate ordered probit · Alcohol consumption · Tobacco consumption · Physical activity

1 Introduction

Thailand is a medium-high income country where morbidity and mortality are primarily related to chronic rather than infectious diseases. Cardiovascular disease is the main cause of death with cancer as the next highest [11]. The risk factors for raising the mortality rate were health behaviors. For example, alcohol consumption, smoking, poor eating habits and diet, urban air pollution, obesity, physical inactivity, and unsafe sex [4]. Health behaviors are particularly important factors for health policy planning.

The explicit burden on society due to health-risk behaviors, particularly alcohol and tobacco consumption, includes health care costs, productivity loss, property damage costs, costs of criminal justice as well as law enforcement. To reduce health-risk behaviors, Thailand should aim to reduce alcohol consumption and prevent initiation of drinking. While Thailand already implements alcohol related policies,

K. Suknark (✉) · J. Sirisrisakulchai · S. Sriboonchitta
Faculty of Economics, Chiang Mai University, Chiang Mai, Thailand
e-mail: kanchitsook@hotmail.com

V.-N. Huynh et al. (eds.), *Causal Inference in Econometrics*,
Studies in Computational Intelligence 622, DOI 10.1007/978-3-319-27284-9_18

such as high alcohol taxation, restricted alcohol sale times, more effective measures at the societal level to control alcohol consumption and alcohol-related harms are still required. The national survey in 2011 reported that about 17.7 million people or 20.8 % of the population aged 15 years and over are alcohol users. Men used alcohol at a higher rate than women [10].

Equally, tobacco consumption control policies have been implemented to reduce tobacco consumption and prevent initiation of smoking, especially in younger people. Current policies include high rates of tobacco taxation, control of tobacco advertising, non-smoking areas and bans on smoking in public places, workplaces, public transport, schools and other areas and facilities, supporting quit-smoking programs and publicity campaigns. These policies have been shown to be successful in decreasing the proportion of smokers in the Thai population (aged 15 years and older) from 32 % in 1991 to 20 % in 2013 [10].

Since 2010, the Thai Health Promotion Foundation has promoted physical activity in the Thai population by sponsoring and supporting several public campaigns nationally on the benefits of physical activity and advising people on the effective levels of frequency, duration and intensity required to achieve physical fitness. Such programs have also been supported at the local and regional level in many areas of the country. Most of the projects are mainly focused on increasing perceptions, attitudes, and practices related to physical activity generally [3]. The national survey in 2011 reported that about 26.1 % of the population played some form of sport or physical exercised, but this is actually a decrease of about 3 % when compared with the 2007 levels [10].

The previous studies on the factors affecting alcohol consumption, tobacco consumption, and physical activity were based on a single equations [3, 6, 7, 9]. In this paper, we simultaneously determined the factors affecting each pair of some important health behaviors including alcohol-consumption and physical activity pair, tobacco-consumption and physical activity pair, and alcohol-consumption and tobacco-consumption pair, and attempted to quantify the dependence measures between these pairs using the copula approach. A bivariate ordered probit model was used to control for the common unobserved factors that might affect the random errors in each pair of health behaviors. If these random errors are ignored, and not correlated, inefficiency in parameter estimation is likely [1]. Moreover, understanding the dependencies between the ordinal choices for each pair of health behaviors will give information useful for designing more efficient health care programs.

2 Data

The data used in this study are from the Thai National Health Examination Survey, No. 4 (NHES IV) from 2009. The data consists of a sample of 20,450 individuals. The ordered dependent variables are alcohol consumption (Y_1), tobacco smoking (Y_2), and physical activity in leisure time (Y_3). The independent variables are sex, age, income, chronic diseases, marital status, education level, and occupation. The

Table 1 Main statistics and description of variables

Variables	Description	N	Mean	SD	Min.	Max.
Y1	Level of alcohol consumption	20450	0.446	0.697	0	3
Y2	Level of Tobacco consumption	20450	0.052	0.339	0	3
Y3	Level of Physical Activity	20450	2.201	0.845	0	3
Sex	1 if individual is male; 0 otherwise	20450	0.524	0.499	0	1
Age	In year	20450	52.917	18.236	14	98
Income	In 1,000 Baht	20450	3.310	5.698	0	32.480
Bachelor	1 if individual graduated from Bachelor degree or higher; 0 otherwise	20450	0.061	0.24	0	1
Agr	1 if individual works in agricultural sector; 0 otherwise	20450	0.176	0.381	0	1
Whi	1 if individual is white-collar worker	20450	0.035	0.184	0	1
Police	1 if individual works as police or soldier; 0 otherwise	20450	0.012	0.108	0	1
Labor	1 if individual is in labor sector; 0 otherwise	20450	0.48	0.499	0	1
Married	Marital status where 1 indicates married; 0 otherwise	20450	0.636	0.481	0	1
pe_bmi25	1 if individual has body mass index more than 25; 0 otherwise	20450	0.348	0.476	0	1
pe_tc200	1 if individual has chloresterol level more than 200; 0 otherwise	20450	0.561	0.496	0	1
qlhealth	Self health quality assessment, where 5 is the highest level	20450	3.708	0.867	0	5
NCD	Number of chronic diseases	20450	0.632	0.959	0	10

alcohol consumption variable (Y_1) was stated as an amount of ethanol consumption on average per day in a year, and was classified into four levels: 0 for non-alcohol consumption; 1 for less than or equal to 40 g of ethanol on average per day (considered to be a responsible level of consumption); 2 for 41–60 g of ethanol on average per day (a harmful level); and 3 for over 61 g of ethanol on average per day (hazardous level). For the tobacco consumption variable (Y_2), measured as an amount of cigarettes per day, it can be classified into four levels: 0 for non-smoking; 1 for up to 10 cigarettes per day; 2 for more than 10 and up to 20 cigarettes per day; 3 for more than 20 cigarettes per day. For the physical activity variable (Y_3), the levels of physical activity or exercise in leisure time were: 0 for non-physical activity; 1 for low level of activity; 2 for moderate level of activity; and 3 for high level of activity. These are obviously indicative levels rather than attempting to quantify physical activity by number of hours or some other more precise measure. Table 1 presents the description of variables and main statistics.

3 Copula-Based Bivariate Ordered Probit Models

A Bivariate Ordered Probit Model is a system of two equations that can be used to model a simultaneous relationship of two ordinal outcome variables. The traditional Bivariate Ordered Probit Model uses the bivariate normal distribution to model the dependence between two equations [1]. In this study, we used a copula distribution function to model the dependence between two ordinal outcome responses. This is more flexible than the bivariate normal distribution. The Copula Function is a joint distribution with uniform margins. Let $U_1, \ldots, U_q$ be the possibly dependent uniform random variables on [0, 1]-interval. Copula can be defined as

$$C_\theta(u_1, \ldots, u_q) = \Pr(U_1 \leq u_1, \ldots, U_q \leq u_q), \tag{1}$$

where $C_\theta(\cdot, \ldots, \cdot)$ is a Copula Function with the dependent parameter θ, and u_m, for $m = 1, \ldots, q$ is a realization of U_m. The Copula Function must be grounded and increasing on the unit hypercube on its domain $[0, 1]^q$ (see, [5] for more details). By Sklar's Theorem (1959), for q marginal distribution functions, $F_1(\cdot), \ldots, F_q(\cdot)$ and $(z_1, \ldots, z_q)$ are arbitrary, we can derive the joint distribution $H(\cdot, \ldots, \cdot)$ for the random variables, $Z_1, \ldots, Z_q$ as follows:

$$C_\theta(F_1(z_1), \ldots, F_q(z_q)) = \Pr(F^{-1}(U_1) \leq z_1, \ldots, F^{-1}(U_q) \leq z_q) \equiv H(z_1, \ldots, z_q), \tag{2}$$

where $Z_m = F_j^{-1}(U_m), m = 1, \ldots, q$. Thus, we can construct a joint distribution function from a set of margins by using the Copula Function to combine them.

Now we can start deriving our copula-based bivariate ordered probit model. Suppose that each individual i selects the level of two dependent ordinal responses based on the following system of two equations:

$$Y_{i1}^* = X_{i1}\beta_1 + \varepsilon_{i1}, \tag{3}$$
$$Y_{i2}^* = X_{i2}\beta_2 + \varepsilon_{i2}, \tag{4}$$

where i indexes individual $i = 1, \ldots, N$, Y_{i1}^* and Y_{i2}^* are latent variables, X_{i1} and X_{i2} are the $K \times N$ matrices of explanatory variables, β_1 and β_2 are conformable vectors of parameters to be estimated, and ε_{i1} and ε_{i1} are random errors.

We can model the observed level of two dependent ordinal responses, Y_{i1}, and Y_{i2} by the following threshold crossing conditions:

$$Y_{ij} = r_j \quad \text{if} \quad \tau_{r_j,j} \leq Y_{ij}^* < \tau_{r_j+1,j}, \quad r_j = 1, \ldots, R_j, \quad j = 1, 2 \tag{5}$$

where R_j are the number of ordinal levels of $Y_i j$ and $\tau_{r_j,j}$ are threshold parameters to be estimated from the model, with $\tau_{1,j} = -\infty$ and $\tau_{R_j,j} = +\infty$. The joint distribution of the individual selected the level of two ordinal response outcomes can be expressed as follows:

$$\begin{aligned}
&\Pr(\tau_{r_1,1} \leq Y_{i1}^* < \tau_{r_1+1,1}, \tau_{r_2,2} \leq Y_{i2}^* < \tau_{r_2+1,2}) \\
&\quad = \Pr(\tau_{r_1,1} \leq X_{i1}\beta_1 + \varepsilon_{i1} < \tau_{r_1+1,1}, \tau_{r_2,2} \leq X_{i2}\beta_2 + \varepsilon_{i2} < \tau_{r_2+1,2}) \\
&\quad = \Pr(\tau_{r_1,1} - X_{i1}\beta_1 \leq \varepsilon_{i1} < \tau_{r_1+1,1} - X_{i1}\beta_1, \tau_{r_2,2} - X_{i2}\beta_2 \leq \varepsilon_{i2} < \tau_{r_2+1,2} - X_{i2}\beta_2) \\
&\quad = C_\theta(F_1(\tau_{r_1+1,1} - X_{i1}\beta_1), F_2(\tau_{r_2+1,2} - X_{i2}\beta_2)) \\
&\qquad - C_\theta(F_1(\tau_{r_1,1} - X_{i1}\beta_1), F_2(\tau_{r_2+1,2} - X_{i2}\beta_2)) \\
&\qquad - C_\theta(F_1(\tau_{r_1+1,1} - X_{i1}\beta_1), F_2(\tau_{r_2,2} - X_{i2}\beta_2)) \\
&\qquad + C_\theta(F_1(\tau_{r_1,1} - X_{i1}\beta_1), F_2(\tau_{r_2,2} - X_{i2}\beta_2)).
\end{aligned}$$

For the traditional Bivariate Ordered Probit Model, the marginal distribution $F_1(\cdot)$ and $F_2(\cdot)$ are specified as the standard normal distribution and the copula function is specified as a Gaussian copula. Therefore, the traditional Bivariate Ordered Probit Model is the special case of copula-based Bivariate Ordered Probit Model. To capture a wider range of dependencies and distributional shapes of random errors, we use different type of copula functions and a mixture of two normal components for random errors.

The most general form of normal mixtures can be expressed as

$$F_j(z) = \pi_j \Phi\left(\frac{z - \mu_{j1}}{\sigma_{j1}}\right) + (1 - \pi_j)\Phi\left(\frac{z - \mu_{j2}}{\sigma_{j2}}\right) \tag{6}$$

where Φ is the standard normal distribution, π_j is the mixing parameter, μ_{j1} and μ_{j2} are location parameters, and σ_{j1} and σ_{j2} are dispersion parameters. The location and dispersion parameters have to be constrained to satisfy the mean and variance normalizations as follows:

$$\pi_j\mu_{j1} + (1 - \pi_j)\mu_{j2} = 0, \quad \pi_j(\sigma_{j1}^2 + \mu_{j1}^2) + (1 - \pi_j)(\sigma_{j2}^2 + \mu_{j2}^2) = 1. \tag{7}$$

This normal mixtures distribution can capture the varieties of skewness or bimodality in the shape of random errors.

The log-likelihood of the copula-based Bivariate Ordered Probit Model is given by

$$\begin{aligned} LL = \sum \log(&C_\theta(F_1(\tau_{r_1+1,1} - X_{i1}\beta_1), F_2(\tau_{r_2+1,2} - X_{i2}\beta_2)) \\ &- C_\theta(F_1(\tau_{r_1,1} - X_{i1}\beta_1), F_2(\tau_{r_2+1,2} - X_{i2}\beta_2)) \\ &- C_\theta(F_1(\tau_{r_1+1,1} - X_{i1}\beta_1), F_2(\tau_{r_2,2} - X_{i2}\beta_2)) \\ &+ C_\theta(F_1(\tau_{r_1,1} - X_{i1}\beta_1), F_2(\tau_{r_2,2} - X_{i2}\beta_2))). \end{aligned}$$

The corresponding vector of parameters β_1, β_2, $\tau_{r_1,1}$, $\tau_{r_2,2}$, parameters of random errors π_1, μ_{11}, μ_{12}, σ_{11}, σ_{12}, π_2, μ_{21}, μ_{22}, σ_{21}, σ_{22}, and dependence parameter θ can be estimated simultaneously using the maximum likelihood estimation. This study uses STATA software [8] and user written command BICOP [2] to estimate all parameters in the models.

4 Results and Discussion

We consider both Frank copula and Gaussian copula that allow for both positive and negative dependence. For the marginal distribution of each residual, we consider three specifications including specifying each marginal as a standard normal distribution, and specifying one of the random errors as a normal-mixture distribution and another as a standard normal distribution, and specifying each marginal as a normal-mixture distribution. For all three pairs, the best fitted model (in terms of Akaike Information Criteria, AIC) is the Frank copula with standard normal distribution for both random errors. In comparison with the two separate univariate ordered probit model (independent copula), we found that the estimated standard errors of bivariate models are lower than those of univariate models (the results are not shown here). However, the differences are very small (five digits after the decimal point) corresponding with the low level of correlation between each random error.

4.1 Factors Affecting Alcohol Consumption and Physical Activity Behaviors

Table 2 presents the model estimation results for the level of alcohol consumption and physical activity behaviors pair. The first dependent variable to be discussed is alcohol consumption level. The explanatory variables included in the model that significant are age, income, high cholesterol, gender, non-communication diseases, occupation, education, and Body Mass Index. The coefficient interpretations are: (1) young individuals, individuals with higher income, individuals with lower cholesterol

Table 2 Parameter estimates for level of alcohol consumption and level of physical activity model

Variables	Y1		Y3	
	Coeff.	Std. err	Coeff.	Std. err
Sex	0.956	0.019	0.143	0.017
Age	−0.012	0.001	−0.005	0.001
Income	1.48E−05	1.71E−06	−2.40E−06	1.60E−06
Bachelor	−0.08	0.042	0.035	0.038
Agr	0.249	0.031	0.523	0.027
Whi	0.179	0.057	0.125	0.052
Police	0.251	0.079	0.194	0.077
Labor	0.168	0.027	0.339	0.023
Married	0.025	0.02	0.097	0.018
pe_bmi25	−0.034	0.02	0.059	0.018
pe_tc200	−0.062	0.019	−0.053	0.017
qlhealth	−0.011	0.011	0.09	0.009
NCD	−0.12	0.011	−0.041	0.009
$\tau_{1,j}$	−1.626	0.072	−2.244	0.068
$\tau_{2,j}$	−0.135	0.071	−0.576	0.064
$\tau_{3,j}$	0.107	0.071	0.205	0.064
θ	0.623	0.060		
LL	−35,863.919			

of 200 mg/dl or a lower number of chronic diseases, individuals who have education lower than bachelor degree, and individuals who are non-obese (BMI $<$ 25) are more likely to alcohol consumption; (2) males are more likely to consume alcohol than females; (3) individuals who work in the agricultural sector and work in risky occupations such as police and soldiers are more likely to consume alcohol than white-collar workers and those from the labor sector.

The second dependent variable is physical activity level. The explanatory variables included in the model that are significant are age, high cholesterol, health quality assessment, gender, non-communicable diseases, occupation, married status, and Body Mass Index. The coefficient interpretations are: (1) young individuals, individuals with higher health quality assessment, individuals with lower cholesterol of 200 mg/dl or lower, number of chronic diseases, individuals who are married, and individuals who are non-obese (BMI $<$ 25) are more likely to undertake physical activities; (2) males are more likely to undertake physical activities than females; (3) individuals who work in the agricultural sector are more likely to undertake physical activities than those from the other sectors.

4.2 Factors Affecting Tobacco Consumption and Physical Activity Behaviors

Table 3 presents the model estimation results of tobacco consumption and physical activity behaviors. For the first dependent variable, namely, the level of tobacco consumption, the explanatory variables included in the model that significant are age, quality of health assessment, gender, non-communication diseases, occupation only agriculture and labor, married, and Body Mass Index. The coefficient interpretations are: (1) Young individuals, individuals who lower health quality assessment or lower number of chronic diseases, individuals who education lower than bachelor degree, and individuals who non-obese ($BMI < 25$) are more likely to tobacco consumption; (2) male are more likely to alcohol consumption than female; (3) individuals who work in agricultural sector are more likely to tobacco consumption than labor sector.

For the second dependent variable, which is physical activity level, the explanatory variables included in the model that significant are age, high cholesterol, health quality assessment, gender, non-communication diseases, occupation, married status, and Body Mass Index. The coefficient interpretations are: (1) Young individuals, individuals who higher health quality assessment, individuals who lower cholesterol 200 mg/dl or lower number of chronic diseases, individuals who married, and

Table 3 Parameter estimates for level of tobacco consumption and level of physical activity model

Variables	Y2		Y3	
	Coeff.	Std. err	Coeff.	Std. err
Sex	1.426	0.088	0.143	0.017
Age	−0.009	0.002	−0.005	0.001
Income	5.70E−06	3.85E−06	−2.40E−06	1.64E−06
Bachelor	−0.533	0.131	0.035	0.038
Aagr	0.259	0.072	0.523	0.027
Whi	0.048	0.159	0.124	0.052
Police	0.135	0.156	0.194	0.077
Labor	0.161	0.069	0.339	0.023
Married	−0.06	0.049	0.097	0.018
pe_bmi25	−0.157	0.05	0.06	0.018
pe_tc200	0.06	0.043	−0.053	0.017
qlhealth	−0.098	0.024	0.09	0.009
NCD	−0.08	0.029	−0.041	0.009
$\tau_{1,j}$	−0.569	0.174	−2.244	0.068
$\tau_{2,j}$	−0.413	0.174	−0.576	0.064
$\tau_{3,j}$	0.051	0.176	0.205	0.064
θ	−0.528	0.170		
LL	−23,904.808			

individuals who non-obese (BMI $<$ 25) are more likely to physical activities; (2) male are more likely to physical activities than female; (3) individuals who work in agricultural sector and labor are more likely to physical activities than the other sector.

4.3 *Factors Affecting Alcohol Consumption and Tobacco Consumption Behaviors*

Table 4 presents the model estimation results of alcohol consumption and tobacco consumption behaviors. The estimated parameters are similar to the previous subsections. More information from Table 4 is just the dependence parameter estimation, which will be discussed in the next subsection. The marginal effects of each dependent variable are shown in Tables 5, 6 and 7.

Table 4 Parameter estimates for level of alcohol consumption and level of tobacco consumption model

Variables	Y1		Y2	
	Coeff.	Std. err	Coeff.	Std. err
Sex	0.956	0.019	1.429	0.088
Age	−0.012	0.001	−0.009	0.002
Income	1.48E−05	1.71E−06	5.00E−06	3.87E−06
Bachelor	−0.081	0.042	−0.529	0.131
Aagr	0.249	0.031	0.263	0.072
Whi	0.182	0.057	0.046	0.158
Police	0.253	0.079	0.154	0.155
Labor	0.169	0.028	0.156	0.069
Married	0.025	0.02	−0.056	0.05
pe_bmi25	−0.034	0.02	−0.15	0.05
pe_tc200	−0.062	0.019	0.066	0.043
qlhealth	−0.01	0.011	−0.1	0.024
NCD	−0.12	0.011	−0.081	0.029
$\tau_{1,j}$	−1.622	0.072	−0.581	0.175
$\tau_{2,j}$	−0.131	0.071	−0.425	0.175
$\tau_{3,j}$	0.111	0.071	0.039	0.176
θ	0.979	0.178		
LL	−17,171.767			

Table 5 Marginal effects for level of alcohol consumption

Variables	Level of alcohol consumption			
	Level 0	Level 1	Level 2	Level 3
Age	0.0037	−0.0026	−0.0003	−0.0008
Income	−4.65e−06	3.28e−06.	4.13e−07	0.000000957
pe_tc200	0.0195	−0.0138	−0.0017	−0.004
qlhealth	0.0033	−0.0023	−0.0003	−0.0007
Sex	−0.3007	0.2119	0.0267	0.0621
NCD	0.0377	−0.0266	−0.0033	−0.0078
Agr	−0.0786	0.0554	0.007	0.0162
Whi	−0.0573	0.0404	0.0051	0.0118
Police	−0.0792	0.0558	0.007	0.0164
Labor	−0.0533	0.0375	0.0047	0.0111
Married	−0.0077	0.0054	0.0007	0.0016
Bachelor	0.0255	−0.018	−0.0022	−0.0053
Pe_bmi25	0.0109	−0.0077	−0.001	−0.0022

Table 6 Marginal effects for level of tobacco consumption

Variables	Level of tobacco consumption			
	Level 0	Level 1	Level 2	Level 3
Age	0.0005	−0.0001	−0.0002	−0.0002
Income	−3.14E−07	7.19E−08	1.44E−07	9.81E−08
pe_tc200	−0.0034	0.0008	0.0015	0.0011
qlhealth	0.0052	−0.0012	−0.0024	−0.0016
Sex	−0.0753	0.0172	0.0344	0.0237
NCD	0.0042	−0.0009	−0.0019	−0.0014
Agr	−0.0136	0.0031	0.0062	0.0043
Whi	−0.0017	0.0004	0.0008	0.0005
Police	−0.0068	0.0016	0.0031	0.0021
Labor	−0.0082	0.0019	0.0037	0.0026
Married	0.0031	−0.0007	−0.0014	−0.001
Bachelor	0.0281	−0.0064	−0.0128	−0.0089
Pe_bmi25	0.0081	−0.0019	−0.0037	−0.0025

4.4 Dependence Measures of Health Behaviors Pairs

The dependence parameters for three different pairs are significant, indicating the need to model these behaviors simultaneously. The dependence parameter estimated from the Frank copula bivariate ordered probit for alcohol consumption and physical activity behaviors is 0.623. This dependence parameter can be interpreted as

Table 7 Marginal effects for level of physical activity

Variables	Level of physical activity			
	Level 0	Level 1	Level 2	Level 3
Age	0.0002	0.0014	0.0004	−0.002
Income	7.00E−08	6.53E−07	1.85E−07	−9.08e−07
pe_tc200	0.0016	0.0147	0.0041	−0.0204
qlhealth	−0.0026	−0.0246	−0.0069	0.0341
Sex	−0.0042	−0.0392	−0.0111	0.0545
NCD	0.0012	0.0113	0.0032	−0.0157
Agr	−0.0154	−0.1433	−0.0405	0.1992
Whi	−0.0036	−0.0339	−0.0096	0.0471
Police	−0.0057	−0.0534	−0.0151	0.0742
Labor	−0.0099	−0.0929	−0.0262	0.129
Married	−0.0028	−0.0265	−0.0075	0.0368
Bachelor	−0.001	−0.0097	−0.0027	0.0134
Pe_bmi25	−0.0017	−0.0162	−0.0046	0.0225

a concordance measure (Kendall's Tau) equal to 0.07. The dependence parameter estimated from the Frank copula bivariate ordered probit for tobacco consumption and physical activity behaviors is −0.528. This dependence parameter can be interpreted as a concordance measure (Kendall's Tau) equal to −0.06. For the parameter estimated from the Frank copula bivariate ordered probit for alcohol consumption and tobacco consumption behaviors, the dependence parameter is 0.979, corresponding with 0.108 as the concordance measure. The concordance measure for all three models are quite small but statistically significant. Thus, we can not ignore these dependencies in model estimation.

5 Concluding Remarks

From the empirical results previously discussed, the followings are the recommended policies designed to reduce health-risk behavior and increase health inducing behavior for Thai citizens:

(a) Campaigns aimed at reducing alcohol consumption should have a greater focus on workers in the agricultural sector and in risky occupations.

(b) The empirical results show that there is a negative correlation between tobacco consumption behavior and physical activity behavior. Thus, anti-smoking policies would have a more positive impact when the policy makers promote physical activity campaign.

(c) Finally, the empirical results confirm that there is some dependence between alcohol and tobacco consumption as discussed in the Alcohol Alert, 2007. This study found that people who smoke are much more likely to drink, and people who drink are much more likely to smoke. Thus, the alcohol consumption reduction policies and anti-smoking policies would have more positive impact when they are more closely associated.

For further study, the copula-based ordered probit model should be generalized to a multivariate model. However, the main concern on this issue is the curse of dimensionality. When the level of ordinal outcomes and the number of outcomes itself increase, it will give more computational burden on model estimation. Practitioners have to consider about the trade-off between computational cost and efficiency gain.

References

1. Greene, W.H., Hensher, D.A.: Modeling Ordered Choices: A Primer. Cambridge University Press, Cambridge (2010)
2. Hernandez-Alava, M., Pudney, S.: BICOP: a Stata command for fitting bivariate ordinal regressions with residual dependence characterised by a copula function and normal mixture marginals, Working Paper Series No. 2015-02 (2015). https://www.iser.essex.ac.uk/research/publications/working-papers/understanding-society/2015-02
3. Katewongsa, P., Sawangdee, Y., Yousomboon, C., Choolert, P.: Physical activity in Thailand: the general situation at national level. J. Sci. Med. Sport **18**, 100–101 (2014)
4. Lopez, A.D., Mathers, C.D., Ezzati, M., Jamison, D.T., Murray, C.J.L.: Global Burden of Disease and Risk Factors. A copublication of Oxford University Press and The World Bank (2006)
5. Nelsen, R.B.: An Introduction to Copulas, 2nd edn. Springer, New York (2006)
6. Praponsin, S.: Factor affecting alcohol consumption of Thai people. In: The 3rd National Alcohol Conference "Alcohol : No Ordinary Commodity", Richmond Hotel. Nonthaburi, pp. 159, 21–22 November 2007
7. Sirirassamee, T., Sirirassamee, B., Jampaklay, A., Borland, R., Fong, G.T.: Risk factors of tobacco use among thai adolescents: finding from international tobacco control policy survey Southeast Asia (ICT-SEA). J. Med. Assoc. Thai **92**, Suppl 3 (2009)
8. StataCorp.: Stata statistical software: Release 13. StataCorp LP, College Station, TX (2013)
9. Suwannashote, P.: Factors affecting alcohol consumption behavior: case study of the air technical students of the Royal Thai Air Force. Mahidol University, Bangkok (2009)
10. The National Statistical Office of Thailand (2011). Executive summary: The 2011 survey on population behavior in playing sport or physical exercising and mental health
11. World Health Organization: Noncommunicable Diseases (NCD) Country profiles. WHO Documentation Production Service, Geneva (2014)

Reinvestigating the Effect of Alcohol Consumption on Hypertension Disease

Kanchit Suknark, Jirakom Sirisrisakulchai and Songsak Sriboonchitta

Abstract The researchers reinvestigate the effect of alcohol consumption on hypertension from observational data, taken from the Thai National Health Examination Survey. In the observed samples, the treatment assignment is not ignorable, thus using treatment as a dummy variable in the statistical model will lead to the bias estimation of treatment effects. Factors affecting self-selection (drink/not drink) may cause the dummy variable of treatment to be correlated with random errors in the outcome model, which leads to the biased parameters estimation. We propose to use copula-based endogenous switching regression for ordinal outcomes as the more appropriate model for treatment effect estimation. The new results should give us more a accurate and reliable treatment effect for causal inference.

Keywords Alcohol consumption · Hypertension disease · Copula · Endogenous switching regression

1 Introduction

Hypertension is a major risk factor for cardiovascular disease. Tobacco consumption and alcohol consumption are the most important avoidable causes of cardiovascular diseases worldwide [14]. In Thailand, cardiovascular disease and hypertension are major health problems. The cardiovascular morbidity rate and Hypertension have been reported as an important cause of morbidity for past several years (1) The cardiovascular disease was ranked fourth largest cause of death in 2012 (32.9 people per 100,000) (2) Hypertension was the third largest cause of death (37.4 people per 100,000). Both causes are increasing in frequency [22]. Table 1 presents the causes of death in Thailand.

Epidemiological and experimental investigations have established a close association between alcohol consumption and hypertension [6]. A number of population studies have almost unanimously shown an empirical link between high levels of

K. Suknark (✉) · J. Sirisrisakulchai · S. Sriboonchitta
Faculty of Economics, Chiang Mai University, Chiang Mai, Thailand
e-mail: kanchitsook@hotmail.com

V.-N. Huynh et al. (eds.), *Causal Inference in Econometrics*,
Studies in Computational Intelligence 622, DOI 10.1007/978-3-319-27284-9_19

Table 1 Causes of death in Thailand (per 100,000 inhabitants)

Cause groups	2008	2009	2010	2011	2012
1. Malignant neoplasm, all forms	87.6	86.3	91.2	95.2	98.5
2. Accident and poisonings	55.1	55.6	51.6	52.8	51.6
3. Hypertension and cerebrovascular disease	24.7	24.7	31.4	–	37.4
4. Disease of the heart	29.8	29.0	28.9	31.4	32.9
5. Pneurmonia and other diseases of lung	23.0	22.9	25.7	26.3	26.1

Source Bureau of Policy and Strategy, Office of the Permanent Secretary, Ministry of Public Health

alcohol consumption and hypertension. In many former studies, risk factors were assessed using multiple regression models; i.e. age, gender, family history, weight and height for hypertension [2, 9, 12, 15, 21, 24, 25].

In 1985, [15] studied the direct effects of alcohol consumption. He received responses from 46 male drinkers (22–55 years) to a questionaire. Similar research has been conducted by Yadav et al. [25] and Taraman et al. [21]. They studied both tobacco and alcohol consumption among people in Southern India [21]. The results suggest that hypertension becomes more frequent among women of advancing age who are alcohol users, especially if they smoke [25]. Yadav studied the prevalence of Hypertension in northern India among subjects of approximately 30 years of age. The results indicate that increasing age, body mass index, obesity and impaired glucose tolerance were significantly associated with Hypertension [25]. They all used regression analysis in their research. Oyunbileg used regressive methods to assess the correlation between alcohol consumption and arterial hypertension among the population of the Gobi region of China [12].

In Thailand, multiple regression modeling research shows an empirical link between high levels of alcohol consumption and hypertension. Leelarassme et al. studied a sample in Bangkok [7]. Pati and Siviroj studied a sample in Northern Thailand [13, 19]. Howteerakul studied people living in the rural areas in Thailand [5]. In these studies the researchers used data collected in a national survey [13, 19].

In this paper, the researchers reinvestigate the effect of alcohol consumption on the levels of hypertension. Moreover, the researchers investigate whether alcohol consumption may be a cause of hypertension by introducing the switching regression model for observational data analysis. This investigation of causality has been done using the Neyman-Rubin counterfactual framework [17]. In this framework, groups were compared to extract the differences that are due to changes in the level of alcohol consumption and the affect on levels of hypertension.

Ideally, we just want to compare two groups that only differ with regard to whether they consumed alcohol or not. Two groups should be identical in all relevant characteristics, so that we can deduce changes due to the effect of alcohol consumption on hypertension level. This idea can be done in practice by randomly assigning individuals to each group by using the same probability of being assigned to either the control or treatment groups. The characteristics of the individuals in each group may be viewed as interchangeable and are the same, at least on average, if we have a large enough sample.

The potential outcomes framework [10] uses a what-if scenario as the baseline for making causal inferences. This baseline allows each individual the same chance for potential outcomes. However, we can only observe the individual in one treatment condition. Under the interchangeable assumption and given that each group only differs with regard to the treatment condition, the causal effect can then be statistically estimated as the difference in the means of an outcome of interest between the control and treatment groups [1].

To consistently estimate the average treatment effect, we have to satisfy the ignorable treatment assignment assumption [16]. This condition ensures that the outcome of interest is independent of the treatment assignment mechanism [10]. This assumption can be satisfied when we assign each individual randomly to the treatment and control groups. In other words, we can only consistently estimate the average treatment effect if there is no selection bias in the treatment assignment.

In our study, the survey participants self-selected themselves into treatment condition (whether to drink alcohol or not), thus it is hard to believe that the independence assumption holds in this situation. In the observed data, the treatment assignment is not ignorable, thus using treatment as a dummy variable in the statistical model will lead to the bias estimation of the treatment effect. Factors affecting the self-selection might cause the dummy variable of treatment condition to be correlated with the random errors in the outcome model, which leads to the biased parameters estimation. By introducing the endogenous switching regression model for ordered outcomes, the new treatment effect estimation should give us more accurate and reliable results for causal inference.

2 Data

This study uses the Thai National Health Examination Survey, No. 4 (NHES IV) data of 2009. The ordered dependent variables are the blood pressure levels of participants. Blood pressure levels may be classified into three groups of individuals as follows (Thailand context). First, those with an average diastolic blood pressure (DBP) and systolic blood pressure (SBP) of <80/120 mmHg as normal blood pressure level; second, those with DBP and SBP of 80–90/120–140 mmHg as pre-hypertension level; and third those with DBP and SBP of greater than 90/140 mmHg as regarded as having hypertension. The independent variables are gender, age, income, chronic diseases (if any), marital status, level of education, and occupation. Table 2 presents a description of the variables and related statistics.

Table 2 Description of variables and statistics

Variables	Description	N	Mean	SD	Min.	Max.
BP	Level of blood pressure	20,450	0.446	0.697	0	2
Gender	1 if individual is male; 0 otherwise	20,450	0.524	0.499	0	1
Age	In Year	20,450	52.917	18.236	14	98
Income	In Baht	20,450	3.31	5.698	0	32.48
Bachelor	1if individual graduated in Bachelor degree or higher; 0 otherwise	20,450	0.061	0.24	0	1
Agr	1 if individual works in agricultural sector; 0 otherwise	20,450	0.176	0.381	0	1
Whi	1 if individual is white collar	20,450	0.035	0.184	0	1
Police	1 if individual works as police or soldier; 0 otherwise	20,450	0.012	0.108	0	1
Labor	1 if individual is classified as labor; 0 otherwise	20,450	0.48	0.499	0	1
Married	Marital status where 1 indicates married; 0 otherwise	20,450	0.636	0.481	0	1
pe_bmi25	1 if individual has body mass index more than 25; 0 otherwise	20,450	0.348	0.476	0	1
pe_tc200	1 if individual has cholesterol level more than 200; 0 otherwise	20,450	0.561	0.496	0	1
qlhealth	Health quality assesment where 5 is the highest level	20,450	3.708	0.867	0	5
NCD	Number of chronic diseases	20,450	0.632	0.959	0	10

3 Switching Regression Model for Level of Hypertension

In the sample used to estimate the effect of alcohol consumption on the level of hypertension, the participants are not randomly drawn from the population from which we wanted to draw inferences, but participants who self-selected themselves into treatment. The approach to self-selection used, is that proposed by Heckman [4]. The assumption is that the self-select mechanism may be modeled by a binary choices model. The switching regression model, was supplemented with copula by using copula to model the correlation between the random errors from a decision model and outcome models [4].

Consider two decisions, $S = 0, 1$, where 1 is 'drink alcohol' and 0 is not. Let $S^* = Z\gamma + \nu$ be the latent variable for the decision mechanism. The decision rule is the following condition

$$S = \begin{cases} 1 & \text{if } s^* > 0 \\ 0 & \text{if } s^* \leq 0 \end{cases}$$

where Z is the matrix of the explanatory variables explaining the self-select mechanism, and γ is the corresponding vector of parameters to be estimated. The individuals are observed either in decision $S = 0$, or in decision $S = 1$, but never in both.

Consider the outcome of interest, the level of hypertension $Y_s = 0, \ldots, J$, can be modeled using the latent variable framework and can be determined by the following condition:

$$Y_s = j \quad \text{iff} \quad \kappa_{s,j-1} < Y_s^* \leq \kappa_{s,j}, \quad s = 0, 1, \quad j = 0, \ldots, J \tag{1}$$

where $\kappa_{s,j}$ are the threshold values, which form a partition of the real line, i.e., $\kappa_{s,0} = -\infty$, $\kappa_{s,J} = \infty$, and $\kappa_{s,j} > \kappa_{s,j-1}$ for all j.

Let $Y_0^* = X\beta_0 + \varepsilon_0$ be the latent variable for the individual decision not to drink $S = 0$, and $Y_1^* = X\beta_1 + \varepsilon_1$ be the latent variable for the individual to consume alcohol $S = 1$, where, X is the vector of all the explanatory variables, β_0 and β_1 are the vector of the parameters to be estimated.

As previously discussed, there might be some unobserved factors affecting both the self-selected mechanism and the response outcome, therefore the probability of observing $Y_s = j$ depends on the self-selected variable S. Given that S and Y_s are not necessarily independent. We have

$$\begin{aligned} Pr(Y_0 = j, S = 0|X, Z) &= Pr(\kappa_{0,j-1} - X\beta_0 \leq \varepsilon_0 < \kappa_{0,j} - X\beta_0, \nu \leq -Z\gamma) \\ &= Pr(\varepsilon_0 < \kappa_{0,j} - X\beta_0, \nu \leq -Z\gamma) \\ &\quad - Pr(\varepsilon_0 \leq \kappa_{0,j-1} - X\beta_0, \nu \leq -Z\gamma) \end{aligned}$$

$$\begin{aligned} Pr(Y_1 = j, S = 1|X, Z) &= Pr(\kappa_{1,j-1} - X\beta_1 \leq \varepsilon_1 < \kappa_{1,j} - X\beta_1, \nu \leq -Z\gamma) \\ &= Pr(\varepsilon_1 < \kappa_{1,j} - X\beta_1) - Pr(\varepsilon_1 < \kappa_{1,j-1} - X\beta_1) \\ &\quad - Pr(\varepsilon_1 \leq \kappa_{1,j} - X\beta_1, \nu \leq -Z\gamma) \\ &\quad + Pr(\varepsilon_1 \leq \kappa_{1,j-1} - X\beta_1, \nu \leq -Z\gamma) \end{aligned}$$

To model the above probability, we have to specify the appropriate joint distribution functions. In this paper, we suggest combining the marginal distributions (ε_s and ν) by using copula.

Copula was introduced in 1959 by Abe Sklar. Sklar's Theorem succinctly states that there is a copula function that connects multivariate distributions to their univariate marginal distributions being uniformly distributed on the unit interval [0,1]. Therefore, copula is the joint distribution of a multivariate uniform random vector. Introduction and standard reference on copula theory can be found in [11].

For a bivariate joint distribution H with marginal distributions F_1 and F_2, the copula $C : [0, 1]^2 \rightarrow [0, 1]$, which combines these two marginal distributions, can be expressed as follows:

$$H(x_1, x_2) = C\left\{F_1(x_1), F_2(x_2)\right\}, (x_1, x_2) \in \mathbf{R}^2 \tag{2}$$

The copula function is uniquely determined for the continuous random vector (F_1, F_2). For a discrete random vector, the copula function is unique only over the Cartesian product of the range of the marginal distribution function [3]. Thus, in discrete cases the mapping from two marginal distributions and copula to a bivariate joint distribution is not one-to-one. However, the region outside the Cartesian product of the range of the marginal distribution function is not of interest [11]. Moreover, [3] demonstrated that parametric modeling of discrete random vector by copula acquires dependence properties in a way that is similar to the continuous case.

For any copula, the marginal distribution implied by bivariate copula are $C(u, v) \leq C(u, 1) = u$ and $C(u, v) \leq C(1, v) = v$, for all $0 \leq u, v < 1$, and so $W(u, v) = max(u + v - 1, 0) \leq C(u, v) \leq min(u, v) = M(u, v)$. The copula $M(u, v)$ and $W(u, v)$ are called the Frechet upper bound and Frechet lower bound, respectively. We can interpret the Frechet lower bound as the copula with the maximum negative dependence and Frechet upper bound as the copula with the maximum positive dependence. In modeling switching regression, it is essential that the copula should allow for both positive and negative dependence, since the direction of the selection bias can be in both directions. We should not restrict the direction of selection bias a priori. The selection pattern should be explained by the data itself.

Copula has had limited use in the endogenous switching regression models. Some, but not all examples, are [18, 23] for modeling endogenous switching regression in count outcomes, [20] for modeling endogenous switching regression of continuous variables and [8] for modeling endogenous switching regression in ordered outcomes.

In this paper, we consider six copula functions, namely, the Normal copula, the FGM (Farlie-Gumbel-Morgenstern) copula, AMH (Ali-Mikhail-Haq) copula, the t copula, and the Frank copula. However, only the Normal copula, t copula, and Frank copula can reach the Frechet lower bound and upper bound, and thus can span the full range of dependence.

For any given copula, the two required joint distribution, $Pr(Y_0 = j, S = 0|X, Z)$ and $Pr(Y_1 = j, S = 1|X, Z)$ are fully determined. Therefore,

$$\begin{aligned} Pr(Y_0 = j, S = 0|X, Z) = {} & C_0(F_1(\kappa_{0,j-1} - X\beta_0), F_2(-Z\gamma); \theta_0) \\ & - C_0(F_1(\kappa_{0,j-1} - X\beta_0), F_2(-Z\gamma); \theta_0) \end{aligned}$$

$$\begin{aligned} Pr(Y_1 = j, S = 1|X, Z) = {} & C_1(F_1(\kappa_{1,j-1} - X\beta_1), 1; \theta_1) - C_1(F(\kappa_{1,j} - X\beta_1), 1; \theta_1) \\ & - C_1(F_1(\kappa_{1,j} - X\beta_1), F_2(-Z\gamma); \theta_1) \\ & + C_1(F_1(\kappa_{1,j-1} - X\beta_1), F_2(-Z\gamma); \theta_1) \end{aligned}$$

where $C_0(u, v)$ and $C_1(u, v)$ are copula functions and F_1 and F_2 are marginal functions which can be either normal or logistic distribution which correspond to the Probit and Logit models, respectively.

4 Results and Discussion

A total of three models were estimated using the Independence copula, the Normal copula, and the Frank copula. We selected the best fitted model based on Akaike Information Criteria (AIC), which is the Frank copula model. Table 4 shows the log-likelihood values for the Independence copula and the Frank copula models. A likelihood ratio test rejects the Independence copula model in the Frank copula model. Therefore, the results provided here are only from the Independence copula and the Frank copula models. In the following subsection, we will discuss the results from the Frank copula model for the policy implications.

4.1 Binary Choice Equation for Alcohol Consumption

Table 3 gives the results of the selection equation. The results of the binary outcome equation of self-selected alcohol consumption provide the effects of the variable on the propensity toward alcohol consumption relative to non-alcohol consumption. All the parameter estimates were statistically significant at the standard level. The coefficient interpretations are: (1) young individuals, individuals income, individuals who are married are more likely to consume alcohol; (2) males are more likely to use alcohol than females; (3) individuals who work in the agricultural sector and work in high risk occupations such as the police and military are more likely to use alcohol than white-collar workers and those in the labor sector.

Table 3 Estimation results of selection equation for alcohol consumption

Variables	Independent		Frank	
	Coeff.	Std.err	Coeff.	Std.err
Selection equation				
Intercept	−0.347	0.050	−0.343	0.050
Age	−0.014	0.001	−0.014	0.001
Income	1.34E−05	1.88E−06	1.36E−05	1.88E−06
Sex	0.946	0.02	0.946	0.020
Agr	0.345	0.032	0.347	0.032
Whi	0.200	0.061	0.203	0.061
Police	0.315	0.088	0.317	0.088
Labor	0.181	0.029	0.184	0.029
Bachelor	−0.044	0.044	−0.042	0.045
Married	0.048	0.022	0.048	0.022

4.2 Factors Affecting Hypertension Level for Non-alcohol Users

Table 4 presents the model estimation results of hypertension levels for non-alcohol users. The significant explanatory variables included in the model are age, income, high cholesterol, gender, non-communicable diseases, occupation, education, and Body Mass Index. The coefficient interpretations are: (1) older individuals, individuals with lower income, individuals who have higher cholesterol 200 mg/dl, or the number of chronic diseases, or obese (BMI $<$ 25) individuals, or individuals with education lower than bachelor degree level are more likely to develop hypertension; (2) females are more likely to have hypertension than males; (3) individuals who work in the agricultural sector are more likely to have hypertension.

4.3 Factors Affecting Hypertension Level for Alcohol Users

Table 4 also presents the model estimation results of hypertension levels for alcohol users. The explanatory variables included in the model that are significant are age, income, high cholesterol, gender, non-communicable diseases, occupation, education, and Body Mass Index (BMI). The coefficient interpretations are: (1) older individuals, individuals who have a higher health quality assessment, individuals who have higher cholesterol 200 mg/dl, or higher numbers of chronic diseases, or individuals who are obese (BMI $<$ 25), individuals whose educational levels are lower than bachelor degree, and individuals who are married are more likely to have

Table 4 Estimation results of blood pressure level equations

Regime	Independent				Frank			
	Alcohol user		Non-alcohol user		Alcohol user		Non-alcohol user	
Variables	Coeff.	Std.err	Coeff.	Std.err	Coeff.	Std.err	Coeff.	Std.err
Intercept	−1.696	0.075	−1.718	0.100	−1.838	0.073	−1.701	0.126
Age	0.026	0.001	0.022	0.001	0.027	0.001	0.022	0.001
Income	−1.68E−06	2.38E−06	4.26E−06	2.44E−06	−4.91E−06	2.28E−06	4.12E−06	2.51E−06
Sex	0.198	0.023	0.538	0.032	−0.138	0.046	0.528	0.051
Agr	−0.010	0.035	−0.040	0.047	−0.083	0.036	−0.044	0.051
Whi	−0.066	0.075	0.033	0.086	−0.102	0.072	0.031	0.087
Police	−0.038	0.126	0.041	0.108	−0.112	0.115	0.038	0.109
Labor	0.080	0.029	0.124	0.044	0.039	0.029	0.122	0.045
Bachelor	−0.176	0.053	−0.195	0.061	−0.156	0.051	−0.195	0.062
Married	−0.032	0.022	−0.085	0.032	−0.042	0.022	−0.085	0.032
qlhealth	0.006	0.012	−0.040	0.017	0.006	0.011	0.040	0.017
NCD	0.149	0.011	0.160	0.019	0.143	0.011	0.160	0.019
pe_bmi25	0.481	0.022	0.511	0.031	0.460	0.022	0.512	0.031
pe_tc200	0.252	0.022	0.250	0.028	0.236	0.021	0.245	0.028
$\kappa_{s,1}$	1.028	0.014	1.045	0.018	0.976	0.022	1.045	0.018
θ					−2.583	0.588	−0.097	0.412
LL	−30334.15				−30324.16			

Table 5 Predicted probabilities of blood pressure level within sample for the Frank copula model

Outcome	Mean of predicted probabilities						
	Sober			Drink			Difference of mean
	Min	Mean	Max	Min	Mean	Max	
Normal	0.035	0.52	0.942	0.009	0.404	0.907	0.116
Moderate	0.053	0.289	0.375	0.084	0.34	0.399	−0.051
High	0.005	0.191	0.797	0.009	0.256	0.904	−0.065

hypertension; (2) males are more likely to hypertension than females; (3) individuals who work in the agricultural sector are more likely to have hypertension.

4.4 Effect of Alcohol Consumption on Blood Pressure Level

From Table 4, the dependence parameter θ tells us about the direction of self-selection biases. The t-test is used for hypothesis testing. The null hypothesis that $\theta = 0$ implies that there is no self-selection bias. If the null is rejected, the quantification of the selection effects can be computed by comparing the outcome distribution of $\Pr(Y_0 = j|S = 1)$ with the counterfactual predicted distribution $\Pr(Y_0 = j|S = 0)$ of an individual who chooses to consume alcohol but is hypothetically allocated to non-alcohol user regime [8].

The parameter θ for alcohol user regime is negative and significant. This indicates that the two random errors (ε_0 and ν) tend to move in the opposite direction. The negative correlation means that the alcohol user counterfactual blood level of those who actually chose not to drink are below than that of an average. For the non-alcohol user regime, the dependence parameter is not significant, indicating that the blood pressure level distribution of those who are non-alcohol users do not differ from the distribution of an arbitrary individual with the same characteristics.

To quantify the effect of alcohol consumption on the probability of each level of blood pressure, we compute the average of predicted probabilities of each level of blood pressure for alcohol user and non-alcohol user regimes. Table 5 shows the results of the mean of predicted probabilities of outcome within the sample. The results indicate that alcohol consumption is more likely to lead to higher level of blood pressure and hypertension disease.

5 Concluding Remarks

This paper applied a copula-based endogenous switching regression for ordinal outcomes to examine the effect of alcohol consumption on levels of hypertension, using the data from National Health Examination Survey in 2009. We present the Frank copula and Independence copula models in this paper. We found statistical evidence for positive self-selection on alcohol users

From the empirical results previously discussed, the following are the recommended policy designs to reduce the levels of hypertension in alcohol users. In non alcohol users, other policies may help reduce levels of hypertension:

(a) The protection and prevention program for hypertension for non-alcohol users, should focus more on the needs of women, individuals from lower income groups and those with lower educational levels.
(b) For alcohol users, the protection and prevention program for hypertension should be focus more on male alcohol users, undergraduates, and manual workers.
(c) The high risk group includes individuals who are older,have higher numbers of chronic diseases, high cholesterol, and who may be obese. These people should receive regular follow-up medical examinations and take appropriate measures, including life-style changes to prevent hypertension.

References

1. Antonakis, J., Lalive, R.: Counterfactuals and causal inference: methods and principles for social research. In: Morgan, S.L., Winship, C. (eds.) Structural Equation Modeling, pp. 152–159. Cambridge University Press, Cambridge (2011)
2. Fuchs, F.D., Chambless, L.E., Whelton, P.K., Nieto, F.J., Heiss, G.: Alcohol consumption and the incidence of hypertension. Hypertension **37**, 1242–1250 (2001)
3. Genest, C., Neslehova, J.: A primer on copulas for count data. Astin Bull. **37**, 475–515 (2007)
4. Heckman, J.: Sample selection bias as a specification error. Econometrica **47**, 153–161 (1979)
5. Howteerakul, N., Suwannapong, N., Sittilerd, R., Rawdaree, P.: Health risk behaviours, awareness, treatment and control of hypertension among rural community people in Thailand. Asia Pac. J. Public Health **18**, 3 (2006)
6. Klatsky, A.L.: Alcohol and hypertension. Health effects of moderate alcohol consumption: a paradigmatic risk factor. Clin. Chim. Acta. **246**(12), 91–105 (1996)
7. Leelarasamee, A., Aswapokee, N., Muangmanee, L., Charoenlarp, K., Jaroonvesama, N.: Prevalence of hypertension in Thai population in Bangkok. J. Med. Assoc. Thail. **61**(5), 301–308 (1968)
8. Luechinger, S., Stutzer, A., Winkelmann, R.: Self-selection models for public and private sector job satisfaction. Res. Labor Econ. **30**, 233–251 (2010)
9. Mlunde, L.: Knowledge, attitude and practices towards risk factors for hypertension in Kinondoni Municipality, Dar es Salaam. DMSJ **14**(2), 59 (2007)
10. Morgan, S.L., Winship, C.: Counterfactuals and Causal Inference: Methods and Principles for Social Research. Cambridge University Press, New York (2007)
11. Nelsen, R.B.: An Introduction to Copulas, 2nd edn. Springer, New York (2006)
12. Oyunbileg, D., Bolorm, I., Tsolmon, U., Chimedsuren, O.: Environmental and occupational risk factors effect to arterial hypertension: correlation between arterial hypertension and lead. J. Geosci. Env. Prot. **3**, 60–65 (2015)
13. Pati, K.: Hypertension in Nan Province. Faculty of Graduate Studies. Mahidol University (2004)
14. Pechacek, T.F., Asma, S., Blair, N., Eriksen, M.P.: Tobacco, global and community solutions. In: Yusuf, S., Cairns, J.A., Camm, A.J., Fallen, E.L., Gersh, B.J. (eds.) Evidence-Based Cardiology, vol. 10313, 2nd edn. BMJ Books, London (2003)

15. Puddey, I., Lawrence, B., Beilin, J., Vandongen, R., Rouse, I.L., Rogers, P.: Evidence for a direct effect of alcohol consumption on blood pressure in normotensive men. A randomized controlled trial. Hypertension **7**(5), 707 (1985)
16. Rosenbaum, P.R., Rubin, D.R.: The central role of the propensity score in observational studies for causal effects. Biometrika **70**(1), 41–55 (1983)
17. Shenyang, G., Fraser, M.W.: Propensity Score Analysis. Statistical Methods and Applications. SAGE Publications, Los Angeles (2010)
18. Sirisrisakulchai, J., Sriboonchitta, S.: Factors affecting hospital stay involving drunk driving and non-drunk driving in Phuket, Thailand, modeling dependence in econometrics. Adv. Intell. Syst. Comput. **251**, 479–489 (2014)
19. Siviroj, P., Pengpid, S., Peltzer, K., Morarit, S.: Alcohol use and associated factors among older adults in northern Thailand. J. Hum. Ecol. **42**(2), 187–193 (2013). (Kamla-Raj)
20. Smith, M.: Using copulas to model switching regimes with an application to child labour. Econ. Rec. **81**, 547–557 (2005)
21. Venkataraman, R., Kumar, S., Kumaraswamy, M., Singh, R., Pandey, M., Tripathi, P., George, V.J., Dahal, P., Vaibhav, P.: Smoking, alcohol, and hypertension. Int. J. Pharm. Pharm. Sci. **5**(4), 28–32 (2013)
22. The Permanent Secretary Ministry of Thailand. Statistical Thailand. Bureau of Policy and Strategy (2013)
23. van Ophem, H.A.: Modeling selectivity in count-data models. J. Bus. Econ. Stat. **18**(4), 503–511 (2000)
24. Xin, X., He, J., Frontini, M.G., Ogden, L.G., Motsamai, O.I., Whelton, P.K.: Effects of alcohol reduction on blood pressure. A meta-analysis of randomized controlled trials. Departments of Epidemiology and Biostatistics, School of Public Health and Tropical Medicine, and Department of Medicine, School of Medicine, Tulane University, New Orleans, La; and Louisiana State Office of Public Health (M.G.F.). New Orleans. Hypertension **38**, 1112–1117 (2001)
25. Yadav, S., Boddula, R., Genitta, G., Bhatia, V., Bansal, B., Kongara, S., Julka, S., Kumar, A., Singh, H.K., Ramesh, V., Bhatia, E.: Prevalence and risk factors of pre-hypertension and hypertension in an affluent north Indian population. Departments of Endocrinology and Pathology. Indian J. Med. Res. **128**, 712–720 (2008)

Optimizing Stock Returns Portfolio Using the Dependence Structure Between Capital Asset Pricing Models: A Vine Copula-Based Approach

Kittawit Autchariyapanitkul, Sutthiporn Piamsuwannakit, Somsak Chanaim and Songsak Sriboonchitta

Abstract We applied the vine copulas, which can measure the dependence structure of uncertainty in portfolio investments. C-vine and D-vine copulas based on capital asset pricing models were used to exhibit portfolio risk structure in the content of asset allocation. With this approach, we employed the Monte Carlo simulation and the empirical results of C-vine and D-vine copulas to determine the expected shortfall of an optimally weighted portfolio. Furthermore, we used the condition Value-at-Risk (CVaR) model with the assumption of C-vine and D-vine joint distribution to gain the maximum returns in portfolios.

Keywords CAPM · Vine-copulas · CVaR · Conditional value at risk

1 Introduction

An important task of financial institutions is evaluating the exposure to market and credit risks. Market risks arise from variations in prices of equities, commodities, exchange rates, and interest rates. Credit risks refer to potential losses that might occur because of a change in the counterparty's credit quality such as a rating migration or a default. The dependence on market or credit risks can be measured by changes in the portfolio value, or gains and losses.

The classical portfolio theory was originally conceived by Markowitz in 1952, the idea that explained the return of the portfolio by mean and variance. Since econometrics concerns quantitative relations in modern economic life, its analysis consists

K. Autchariyapanitkul (✉)
Faculty of Economics, Maejo University, Chiang Mai, Thailand
e-mail: Kittawit_a@mju.ac.th

S. Piamsuwannakit · S. Sriboonchitta
Faculty of Economics, Chiang Mai University, Chiang Mai 52000, Thailand

S. Chanaim
Faculty of Science, Chiang Mai University, Chiang Mai 52000, Thailand

V.-N. Huynh et al. (eds.), *Causal Inference in Econometrics*,
Studies in Computational Intelligence 622, DOI 10.1007/978-3-319-27284-9_20

mainly of determining the impact of a set of variables on some other variable of interest. For example, we wish to determine how return on market X affects return on asset Y in a stock exchange. Now this problem is a regression problem, namely, capital asset pricing model (CAPM). We regress the values of the variable of interest Y, usually called the dependent variable in the explanatory variable X, often called the independent variable. This regression problem is formulated by Sharpe [1] and Lintner [2].

Many pieces of research on the CAPM model is used to explain the diversification of the risk parameter and the performance of portfolios. The investigated issue from Zabarankin et al. [3] purposed drawdown parameter in CAPM model to provide tools for hedging against market drawdowns. Fabozzi and Francis, Levy used CAPM measure risk parameter for a various period. The contributions to the CAMP are the papers of Vassilios [4], Chochola et al. [5], Zhi et al. [6].

A typical risk assessment situation is this. Consider a portfolio consisting of n assets whose possible losses are random variables $X_1, X_2, \ldots, X_n$. We are interested in the overall risk of the portfolio at some given time, i.e., the total loss $Y = X_1 + X_2 + \cdots + X_n$. The value-at-risk (VaR) is a commonly used methodology for estimating of risks. The essence of the VaR computations is an estimation of high quantiles (see, Autchariyapanitkul et al. [7]) in the portfolio return distributions. Usually, these computations are based on the assumption of normality of the financial return distribution. However, financial data often reveal that the underlying distribution is not normal. The standard value-at-risk is $F_Y^{-1}(\alpha)$, the maximum possible total loss at level $\alpha \in [0, 1]$, i.e.,

$$P(Y > F_Y^{-1}(\alpha)) \leq 1 - \alpha$$

In order to obtain the distribution F_Y of Y, we need the joint distribution of $(X_1, X_2, \ldots, X_n)$, since, clearly, we cannot assume that the $X_i's$ are mutually independent. A multivariate normal distribution will not work, since empirical work of Mandelbrot and Fama showing that financial variables are rather heavy-tailed. Not only we need copulas to come up with a realistic multivariate model (i.e., a joint distribution for $(X_1, X_2, \ldots, X_n)$), but we also need copulas to describe quantitatively the dependence among assets.

Vine copulas started with Harry Joe in 1996. He gave a construction of multivariate copulas in terms of bivariate copulas, expressed in terms of distribution functions. Thus, it suffices, besides estimating the marginals, to come up with a high dimensional copula to arrive at a joint distribution for the marginal. In one hand, while lots of parametric bivariate copulas models exist in the literature, there seems not to be the case for higher dimensional copulas. On the contrary, we want a high dimensional copula to capture, say, pairwise dependencies between capital asset pricing models. First, We modeled pairwise dependencies by bivariate copulas and then glue them together to obtain the global high dimensional copula. Zhang et al. [8] used vine copula methods estimate CVAR of the portfolio based on VaR measurement, and showed that D-vine copula model is superior to C-vine and R-vine copulas. Also, to study construct dependence structure, So and Yeung [9] used the time varying vine copulas based GARCH model to show that Kendalls tau and linear correlation of the stock return change over time. Moreover, an enormous number of papers about vine copulas that we can found in a study of Aas et al. [10], Gugan and Maugis [11], Roboredo and Ugolini [12].

In this paper, we intend to use C-vine and D-vine copulas to examine the dependence structure between CAPM models. Then, use the joint distribution that minimize expected shortfall with respect to the expected returns to show the optimal weight of stocks in portfolios. Similarly to the work of Autchariyapanitkul [13] introduced multivariate t-copula to optimize stock returns in portfolio analysis.

This study concentrated on the top 50 largest companies by market capitalization on the Stock Exchange of Thailand ($SET50$). With this method, we used it to measure the risk of a multi-dimensional stock returns in portfolios. Thus, the primary benefaction of this paper can be reviewed as follows: First, we emphasize that the dependence structure is determined by vine copulas and evaluates the complicated nonlinear relations among financial portfolio management. Second, we use the high-dimensional of bull ship stocks show the notable proportion of stocks to the returns of the portfolios. In this studied the selection of the optimal portfolio depends on the underlying assumption on the behavior of the assets under various situations. An unreliable model for dependence structure can cause the damage on portfolios.

The remains of this paper is designed as follows: Sect. 2 provides a short theoretical framework of copulas, covering C-vine and D-vine copulas. Section 3 conducts the empirical results, and final Section gives the conclusion and extension.

2 Copulas and Vine Copulas

Consider the situation where we know the marginal distributions F and G of the random variables X and Y, respectively (or to be more realistic in term of their estimates). We wish to model and quantify, among other things, the correlation between X and Y. So far, It is all about Sklar's theorem that says: If H is the joint distribution of (X, Y), then there is a copula C such that

$$H(x, y) = C(F(x), G(y))$$

for $(x, y) \in \mathbb{R}^2$.

However, everybody only looks as the "nice" case where both F and G are *continuous*. It is a nice case since the Sklar's theorem becomes:

(i) The copula C is *unique*.

(ii) It can be extracted as

$$C(u, v) = H(F^{-1}(u), G^{-1}(v))$$

(iii) C characterizes dependence structures and dependence measures (with desirable properties). For example,

$$C(u, v) = uv \Longleftrightarrow X \perp Y$$

$$C(u, v) = u \wedge v \Longleftrightarrow F(X) = G(Y)$$

$$C(u, v) = (u + v - 1) \vee 0 \Longleftrightarrow F(X) = 1 - G(Y)$$

and dependence measures for (X, Y) can be defined nicely in terms of C (with invariant property).

2.1 Vine Copulas

Suppose, we have a data set on random vector of interest, let say, $X = (X_1, X_2, \ldots, X_d)$. We are focused in making inference about some function of X, e.g., $Y = \varphi(X) = \sum_{i=d}^{d} \alpha_i X_i$ (say, in financial (portfolio) investments), where, e.g., the interest is on deriving the value-at-risk $VaR_\alpha(F_Y)$.

We need the joint distribution H_X of X to determine the distribution F_Y of Y in order to derive

$$VaR_\alpha F(Y) = F_Y^{-1}(\alpha) = inf\{y \in R : F_Y(y) \geq \alpha\} \tag{1}$$

The accurate specification of F_Y is crucial! It comes from the specification of H_X. Now, we have data on X and wish to specify a joint H_X which seems to generate the observed data (a problem of curve fitting). Moreover, since the dependencies among the components X_i, $i = 1, 2, \ldots, d$ are of enormous importance, they should be captured as accurate as possible. Thus, the problem of specifying H_X should take into account, at least, two things in mind: generating the observed data, and modeling pairwise dependencies faithfully.

To accomplish the above program, first recall that, according to Sklar's theorem, we have

$$H_X(x_1, x_2, \ldots, x_d) = C(F_1(x_1), F_2(x_2), \ldots, F_d(x_d)) \tag{2}$$

2.2 Drawable Vine (D-vine)

The decomposition of the joint density in terms of bivariate (pairwise) copulas and marginals is Drawable and hence is called a D-vine. With this drawable vine copula, the joint density is obtained simply by multiplying all (bivariate) copula densities appeared in the tree together with all marginal densities (Figs. 1 and 2).

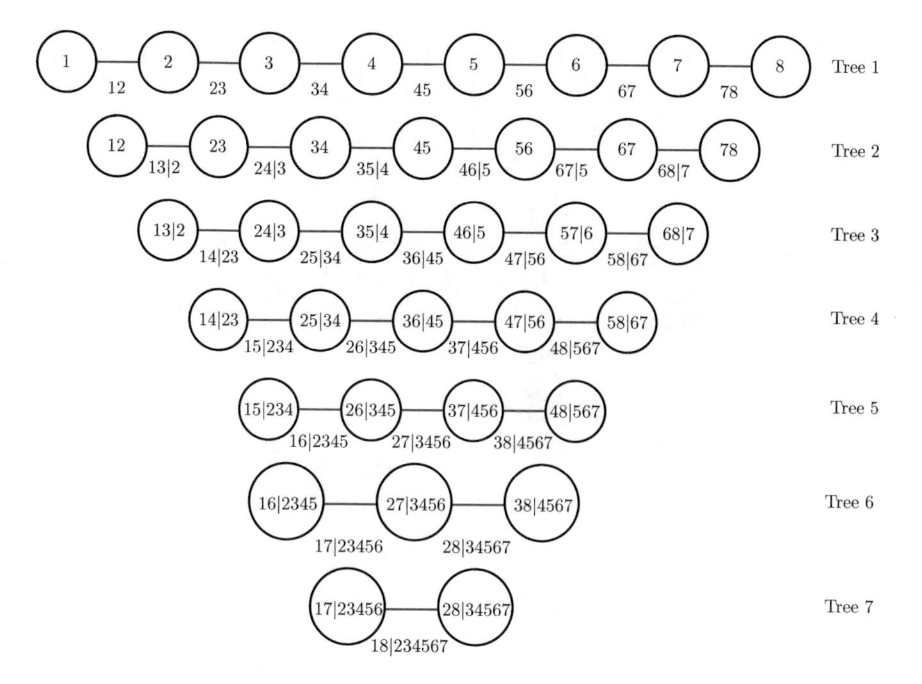

Fig. 1 D-vine

The usefulness of graphical displays is this. When trying to model dependencies in a multivariate model (i.e., we do not know the joint distribution!), we choose a D-vine, according to important pairwise dependencies of interest. We have a "formula" to arrive at the joint distribution, i.e., to come to a model capturing the dependencies of interest. How to use D-vine copulas to build multivariate models? In general, we should figure out that, any d-dimensional copula density can be decomposed in $\frac{d(d-1)}{2}$ different ways. $d = 8, X = (X_1, X_2, X_3, X_4, X_5, X_6, X_7, X_8)$ a possible D-vine is resulting in the multivariate (density) model

$$
\begin{aligned}
f(x_1, x_2, \ldots, x_8) = \prod_{i=1}^{8} f_i(x_i) & \cdot c_{12}c_{23}c_{34}c_{45}c_{56}c_{67}c_{78} \\
& \cdot c_{13|2}c_{24|3}c_{35|4}c_{46|5}c_{57|6}c_{68|7} \\
& \cdot c_{14|23}c_{25|34}c_{36|45}c_{47|56}c_{58|67} \\
& \cdot c_{15|234}c_{26|345}c_{37|456}c_{48|567} \\
& \cdot c_{16|2345}c_{27|3456}c_{38|4567} \\
& \cdot c_{17|23456}c_{28|34567} \\
& \cdot c_{18|234567}
\end{aligned} \tag{3}
$$

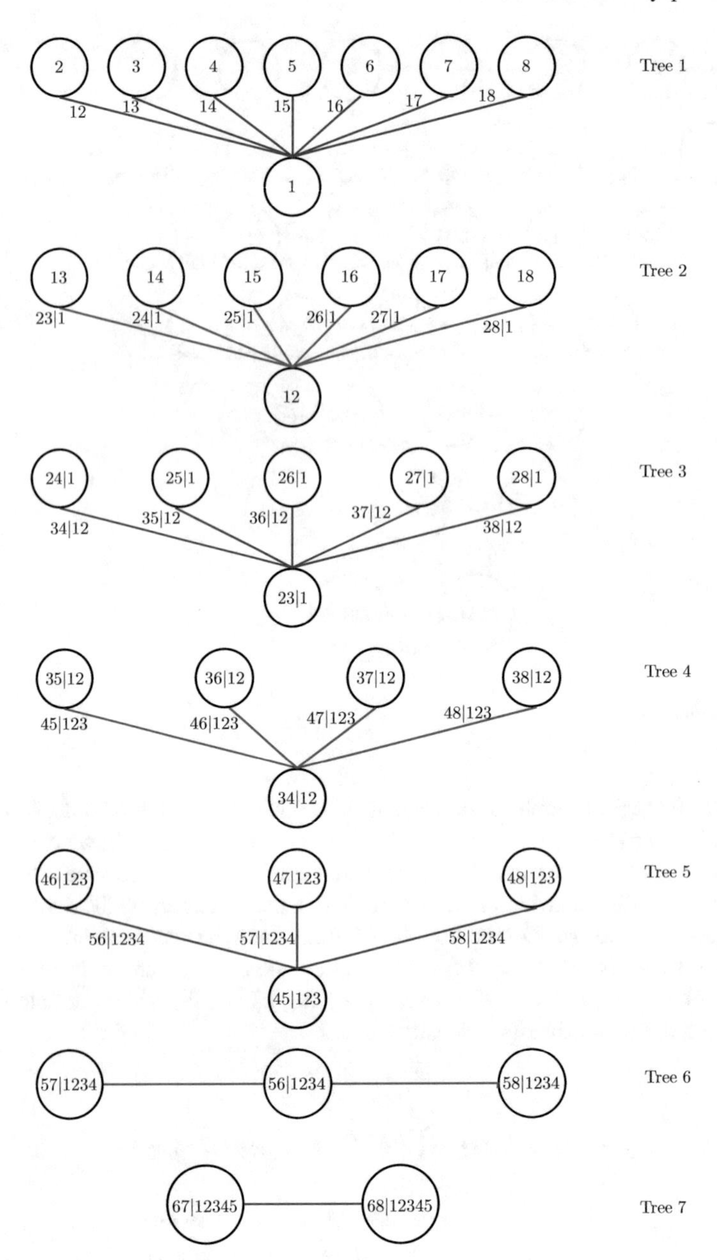

Tree 1
2
3
4
5
6
7
8
12
13
14
15
16
17
18
1
Tree 2
13
14
15
16
17
18
23|1
24|1
25|1
26|1
27|1
28|1
12
Tree 3
24|1
25|1
26|1
27|1
28|1
34|12
35|12
36|12
37|12
38|12
23|1
Tree 4
35|12
36|12
37|12
38|12
45|123
46|123
47|123
48|123
34|12
Tree 5
46|123
47|123
48|123
56|1234
57|1234
58|1234
45|123
Tree 6
57|1234
56|1234
58|1234
Tree 7
67|12345
68|12345

Fig. 2 C-vine

2.3 Canonical Vine (C-vine)

A C-vine is a regular vine such that each tree T_j has a unique node of degree d-j. The node with maximal degree in T_1 is the root, for eight-dimension (d = 8) C-vine copulas can written as

The decomposition of joint densities in terms of C-vines copulas is illustrated as follows

$$
\begin{aligned}
f(x_1, x_2, \ldots, x_8) = \prod_{i=1}^{8} f_i(x_i) & \cdot c_{12}(F_1, F_2) \cdot c_{13}(F_1, F_3) \cdot c_{14}(F_1, F_4) \cdot c_{15}(F_1, F_5) \\
& \cdot c_{16}(F_1, F_6) \cdot c_{17}(F_1, F_7) \cdot c_{18}(F_1, F_8) \cdot c_{23|1}(F_{2|1}, F_{3|1}) \\
& \cdot c_{24|1}(F_{2|1}, F_{4|1}) \cdot c_{25|1}(F_{2|1} \cdot F_{5|1}) \cdot c_{26|1}(F_{2|1} \cdot F_{6|1}) \\
& \cdot c_{27|1}(F_{2|1} \cdot F_{7|7}) \cdot c_{28|1}(F_{2|1} \cdot F_{8|1}) \cdot c_{34|12}(F_{3|12}, F_{4|12}) \\
& \cdot c_{35|12}(F_{3|12}, F_{5|12}) \cdot c_{36|12}(F_{3|12}, F_{6|12}) \cdot c_{37|12}(F_{3|12}, F_{7|12}) \\
& \cdot c_{38|12}(F_{3|12}, F_{8|12}) \cdot c_{45|123}(F_{4|123}, F_{5|123}) \cdot c_{46|123}(F_{4|123}, F_{6|123}) \\
& \cdot c_{47|123}(F_{4|123}, F_{7|123}) \cdot c_{48|123}(F_{4|123}, F_{8|123}) \cdot c_{56|1234}(F_{5|1234}, F_{6|1234}) \\
& \cdot c_{57|1234}(F_{5|1234}, F_{7|1234}) \cdot c_{58|1234}(F_{5|1234}, F_{8|1234}) \\
& \cdot c_{67|12345}(F_{6|12345}, F_{7|12345}) \cdot c_{68|12345}(F_{6|12345}, F_{8|12345}) \\
& \cdot c_{78|123456}(F_7, F_8)
\end{aligned} \tag{4}
$$

3 An Application and Empirical Results

3.1 Capital Asset Pricing Model:CAPM

The Capital Asset Pricing Model (CAPM) was formerly conceived by Sharpe [1] and Lintner [2]. The CAPM is the linear combination of the expected excess return on asset and expected market returns. A linear function of CAPM model can be addressed as follows:

$$E(R_A) - R_F = \beta_0 + \beta_1 E(R_M - R_F), \tag{5}$$

where $E(R_A)$ and R_M describe the expected return on stock and the expected market returns, sequentially, β_0 show the intercept and R_F is the risk-free rate. $E(R_M - R_F)$ is the expected risk premium, and β_1 is the risk parameter. We can calculate the systematic risk of each stock by this mathematical statement

$$\beta_1 = \frac{cov(R_A, R_M)}{\sigma_M^2}, \tag{6}$$

where σ_M^2 is the variance of the expected market returns. Given CAPM equation of each stock returns, we can calculate the joint dependency structure via C-vine and D-vine to carry out the optimization process.

3.2 Optimal Portfolio with Conditional Value at Risk via Vine-Copulas

We start our calculation of VaR and CVaR of an equally weighted portfolio and then, the optimal portfolio can be constructed by minimizing CVAR subject to maximum returns. The procedure of optimization, we refer to the paper from Autchariyapanitkul [13]. The following formula can show as below:

$$\text{Min } CVAR = E[r_p | r \leq r_\alpha], \quad (7a)$$

$$\text{subject to } E(r_p) = w_1 E(r_1) + w_2 E(r_2) + \cdots + w_n E(r_n), \quad (7b)$$

$$w_1 + w_2 + \cdots + w_n = 1, \quad (7c)$$

$$0 \leq w_i \leq 1, \text{ where } i = 1, 2, \ldots, n,$$

where r_α is the lower $\alpha - quantile$, and r_p is the return on individual asset at time t.

We use vine copulas to extract dependence structure between CAPM equations and then use the solutions of C-vine and D-vine copulas parameters to create an efficient portfolio and find the optimal solutions for the expected returns with minimum lost.

Now, we simulate the error terms of each stock form the CAPM equations by using the estimated vine-copulas to generate a set of 1,000,000 samples. Then, we obtained a possible price of each stock under CAPM models and vine-copulas to optimization problem.

3.3 Data

The data contains 260 weekly returns during 2010–2014 are retrieved from DataStream, we calculate the log returns on the tracking stocks. The data consist of the returns from the 8 big capitalization companies such as Banpu Public Company Limited (BANPU), Bank of Ayudhya Public Company Limited (BAY), Bangkok Bank Public Company Limited (BBL), Central Pattana Public Company Limited (CPN), Land and Houses Public Company Limited (LH), Pruksa Real Estate Public Company Limited (PS), Thanachart Capital Public Company Limited (TCAP) and Thai Oil Public Company Limited (TOP). Table 1 supplies a summary of the variables (Table 2).

Table 1 Summary statistics

	SET50	BANPU	BAY	BBL	CPN	LH	PS	TCAP	TOP
Mean	0.0025	−0.0036	0.0028	0.0020	0.0055	0.0015	0.0020	0.0016	−0.0002
Median	0.0041	−0.0051	0	0.0025	0.0048	0.0000	0.0000	0.0000	0.0000
Max.	0.0706	0.1802	0.1341	0.1002	0.1268	0.1638	0.1650	0.1475	0.1377
Min.	−0.0766	−0.1324	−0.1658	−0.1039	−0.1406	−0.1947	−0.1926	−0.1581	−0.2173
SD.	0.0253	0.0422	0.0425	0.0344	0.0426	0.0503	0.0567	0.0382	0.0413
Skew.	−0.3412	0.1823	−0.1674	0.1789	−0.1494	0.1803	−0.3417	−0.2854	−0.3675
Kurt.	3.8347	4.6789	4.2171	3.2542	3.6822	4.2344	3.8459	4.6276	6.2557
J.B.	12.5927	31.9757	17.2625	2.0867	6.0089	17.9156	12.8106	32.2261	120.6802
PROB.	0.0080	0.0010	0.0034	0.3097	0.0454	0.0030	0.0077	0.0010	0.0010

All values are the log return

Table 2 Parameters estimation from CAPM models

	BANPU	BAY	BBL	CPN	LH	PS	TCAP	TOP
β_0	−0.0060	0.0003	−0.0005	0.0030	−0.0014	−0.0009	−0.0008	−0.0028
	(0.0021)	(0.0021)	(0.0014)	(0.0022)	(0.0024)	(0.0029)	(0.0018)	(0.0020)
β_1	0.9653	1.0098	1.0248	0.9695	1.2592	1.2617	0.9556	1.0543
	(0.0846)	(0.0835)	(0.0554)	(0.0855)	(0.0956)	(0.1153)	(0.728)	(0.0773)
σ^2	0.0012	0.0012	0.00005	0.0012	0.0015	0.0022	0.0009	0.0010
R^2	0.3350	0.3620	0.5700	0.3320	0.4020	0.3170	0.4000	0.4190
KS test	0.0811	0.7856	0.4211	0.8055	0.4854	0.6835	0.4326	0.0678

3.4 Experimental Results

Given equations from (5) and (6), we can estimate parameters of CAPM models as the following

Tables 3 and 4 show the estimation results for C-vine and D-vine copulas, respectively.

Given a market return $R_M = 0.01$ and a risk free rate $R_F = 0$, we considered all possible ordered of vine-copulas with the lowest AIC. Note that, this method does not guaranteed the best ordered of vine-copulas but in this paper we only have one set of vine-copulas with the lowest AIC. In general, it is possible to have many set of vine-copulas with the same minimum AIC values. Then, we compare the AIC values of the C-vine and D-vine copulas, we found that the D-vine copula structure gives a better results. We can use values of the D-vine copula to estimate the CVAR and efficient portfolio with the maximum expected return for a minimum loss.

Table 5 shows the expected returns of VaR and CVaR at levels of 1, 5 and 10 % with an equally weighted stock. We notice that the estimated CVaR converges to –1.4289, –1.8687 and –2.7599 at 10, 5 and 1 % levels in period $t + 1$, respectively.

We applied the Monte Carlo simulation to produce a set of 1,000,000 samples. Then, provided a significant level of 5 %, we optimized the portfolio by employing the

Table 3 Estimated Results of C-vine copula

Pairs	Families	Parameter 1	Parameter 2	AIC
*1, 2	Frank	−1.1185 (0.4130)	–	−5.2231
1, 3	Gumbel	1.0630 (0.0370)	–	−9.6632
1, 5	Frank	−1.2610 (0.4220)	–	−6.7918
2, 4|1	Clayton	0.2010 (0.0697)	–	−10.3094
2, 5|1	Clayton	0.2151 (0.0784)	–	−8.1258
3, 4|1, 2	Rotated BB8	−1.2777 (0.2005)	−0.9540 (0.0753)	−4.8876
3, 5|1, 2	Gaussian	−0.1833 (0.0595)	–	−6.7201
3, 7|1, 2	Gaussian	−0.1952 (0.0593)	–	−7.8331
4, 5|1, 2, 3	Rotated Gumbel	1.1671 (0.0527)	–	−19.7762
4, 6|1, 2, 3	Rotated BB8	1.2728 (0.1545)	0.9608 (0.0572)	−4.9537
4, 7|1, 2, 3	Frank	0.8335 (0.4005)	–	−2.3229
7, 8|1, 2, 3, 4, 5, 6	Frank	0.7930 (0.4031)	–	−1.8701

() standard error is in parenthesis, 5 % level of significant. *1 = BANPU, 2 = CPN, 3 = TOP, 4 = PS, 5 = LH, 6 = TCAP, 7 = BBL, 8 = BAY

mean-CVaR model and received the efficient frontier of the portfolio under different expected returns, as displayed in Fig. 3.

Eventually, we also obtained the optimal weight of the portfolios varies to the CVAR. Table 6 exposes some of the results of optimal weight with the expected returns in the frontier.

Table 4 Estimated Results of D-vine copula

Pairs	Families	Parameter 1	Parameter 2	AIC
*1, 2	Gaussian	−0.1928 (0.0586)	–	−7.8570
2,3	Frank	0.8544 (0.4129)	–	−2.2602
4,5	Frank	−1.1185 (0.4133)	–	−5.2231
6,7	Clayton	0.1536 (0.0737)	–	−3.9440
1,3|2	Rotated BB8	−1.3060 (0.2259)	−0.9489 (0.0833)	−5.9783
3,5|4	Clayton	0.2010 (0.0696)	–	−10.3093
5,7|6	Survival BB8	1.4622 (0.2565)	0.9138 (0.0972)	−9.5663
1,4|2,3	Gumbel	1.05627 (0.0349)	–	−9.0128
3,6|4,5	Survival BB8	1.2374 (0.1098)	0.9864 (0.0202)	−7.9990
4,7|5,6	Frank	−0.9626 (0.4092)	–	−3.4873
3,7|4,5,6	Survival Gumbel	1.1833 (0.0530)	–	−20.7914
1,7|2,3,4,5,6	Rotated Clayton	−0.1281 (0.0648)	–	−3.4141
2,8|3,4,5,6,7	Frank	0.8412633 (0.4084)	–	−2.2348

() standard error is in parenthesis, 5 % level of significant. *1 = TOP, 2 = BBL, 3 = PS, 4 = BANPU, 5 = CPN, 6 = TCAP, 7 = LH, 8 = BAY

Table 5 Expected shortfall of equally weighted portfolios

	Expected Returns	VaR	CVaR
10 %	0.9405	–0.7537	–1.4289
5 %	0.9405	–1.2657	–1.8687
1 %	0.9405	–2.2458	–2.7599

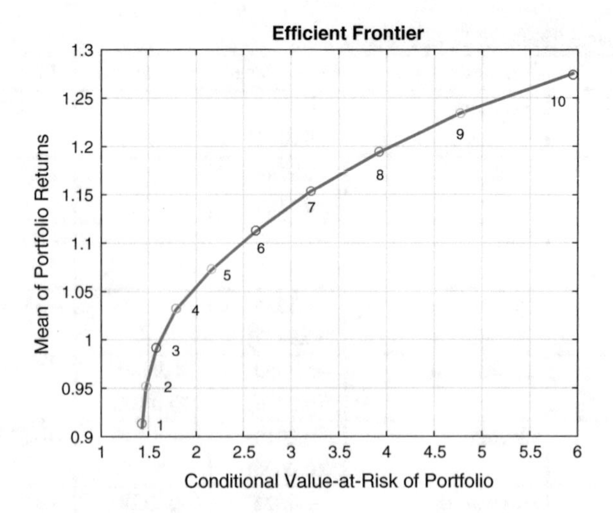

Fig. 3 The efficient frontiers of CVaR under mean

Table 6 Optimal weighted portfolios for CVAR 5 %

Portfolios	$BANPU_{w_1}$	BAY_{w_2}	BBL_{w_3}	CPN_{w_4}	LH_{w_5}	PS_{w_6}	$TCAP_{w_7}$	TOP_{w_8}	Returns
1	0.1086	0.1024	0.2725	0.1179	0.0759	0.0154	0.1229	0.1845	0.9129
2	0.0604	0.1152	0.2866	0.1415	0.0715	0.0246	0.1187	0.1814	0.9522
3	0.0130	0.1259	0.2979	0.1662	0.0695	0.0339	0.1153	0.1782	0.9916
4	0.0000	0.1449	0.2857	0.2270	0.0695	0.0467	0.0873	0.1391	1.0320
5	0.0000	0.1676	0.2652	0.3020	0.0667	0.0618	0.0500	0.0867	1.0727
6	0.0000	0.1872	0.2512	0.3798	0.0626	0.0729	0.0143	0.0320	1.1133
7	0.0000	0.1985	0.1801	0.4810	0.0530	0.0874	0.0000	0.0000	1.1539
8	0.0000	0.1961	0.0530	0.6114	0.0329	0.1065	0.0000	0.0000	1.1942
9	0.0000	0.1164	0.0000	0.7714	0.0000	0.1123	0.0000	0.0000	1.2340
10	0.0000	0.0000	0.0000	1.0000	0.0000	0.0000	0.0000	0.0000	1.2736

4 Concluding Remarks

In this paper, we have determined the risk in portfolio management by employing CVaR and used the mean-CVaR model to optimize portfolios. We used the C-vine and D-vine copula to measured dependence structure between capital asset pricing model (CAPM) affects the returns of portfolios. We carried our analysis in two steps. First, we examined the dependence structure of stock returns obtained from CAPM equations. Second, we investigated how the dependence structure of the asset pricing model influences portfolio optimization. We used an optimization procedure to allocate risk in the portfolios. It is feasible to reason that vine copulas can be explained dependency structure of the asset in the portfolio management.

Acknowledgments The authors thank Prof. Dr. Hung T. Nguyen for his helpful comments and suggestions.

References

1. Sharp, W.F.: Capital asset prices: a theory of market equilibrium under conditions of risk. J. Financ. **19**(3), 425–442 (1964)
2. Lintner, J.: The valuation of risk assets and the selection of risky investments in stock portfolios and capital budgets. Rev. Econ. Stat. **47**(1), 12–37 (1965)
3. Zabarankin, M., Pavlikov, K., Uryasev, S.: Capital asset pricing model (CAPM) with drawdown measure. Eur. J. Oper. Res. **234**(2), 508–517 (2014)
4. Papavassiliou, V.G.: A new method for estimating liquidity risk: Insights from a liquidity-adjusted CAPM framework. J. Int. Financ. Mark. Inst. Money **24**, 184–197 (2013)
5. Chochola, O., Huskova, M., Praskova, Z., Steinebach, J.G.: Robust monitoring of CAPM portfolio betas II. J. Multivar. Anal. **132**, 58–81 (2014)
6. Da, Zhi, Guo, Re-Jin, Jagannathan, Ravi: CAPM for estimating the cost of equity capital: Interpreting the empirical evidence. J. Financ. Econ. **103**(1), 204–220 (2012)
7. Autchariyapanitkul, K., Chainam, S., Sriboonchitta, S.: Quantile regression under asymmetric laplace distribution in capital asset pricing model. Econom. Risk **583**, 219–231 (2015). Springer International Publishing
8. Zhang, B., et al.: Forecasting VaR and CVAR of stock index portfolio: a vine copula method. Physica A: Stat. Mech. Appl. **416**, 112–124 (2014)
9. So, M.K.P., Cherry, Y.T.Y.: Vine-copula GARCH model with dynamic conditional dependence. Comput. Stat. Data Anal. **76**, 655–671 (2014)
10. Aas, K., Czado, C., Frigessi, A., Bakken, H.: Pair-copula constructions of multiple dependence. Insur. Math. Econ. **44**(2), 182–198 (2009)
11. Guegan, Dominique, Maugis, P.A.: An econometric study of vine copulas. Int. J. Econ Financ. **2**(5), 2–14 (2011)
12. Reboredo, Juan C., Ugolini, A.: A vine-copula conditional value-at-risk approach to systemic sovereign debt risk for the financial sector. N. Am. J. Econ. Financ. **32**, 98–123 (2015)
13. Autchariyapanitkul, K., Chainam, S., Sriboonchitta, S.: Portfolio optimization of stock returns in high-dimensions: A copula-based approach. Thai J. Math. (Special Issue on: Copula Mathematics and Econometrics) 11–23, (2014)

Analysis of Transmission and Co-Movement of Rice Export Prices Between Thailand and Vietnam

Duangthip Sirikanchanarak, Jianxu Liu, Songsak Sriboonchitta and Jiachun Xie

Abstract Copulas have become one of the most significant new tools to measure nonlinear dependence structure and tail dependence. Combining time-varying copulas and VAR model with kernel density function, this paper proposes a new method, called the time-varying copula-based VAR model, to analyze the transmission and co-movement of rice export prices between Thailand and Vietnam. The time-varying BB1 and BB7 copulas are proposed to measure asymmetric tail dependences. The main findings of this study reveal that there exists obvious co-movement between rice export prices of Thailand and Vietnam, and the time-varying BB7 copula has a better performance than others. In addition, the price transmission between the two markets is bi-directional, and the Vietnamese price is more suitable as price leader in terms of the results of impulse response functions.

Keywords Price transmission · Causality tests · Time-varying copulas · Rice export

1 Introduction

Thailand and Vietnam are the the world largest and the second largest rice exporters occupying about 28 and 17 % of global rice market share. In the past five years, Thai and Vietnam rice exports have been highly competitive due to rice export prices have gone down during that time. To a large extent, a relatively small amount of rice being traded in Thailand and Vietnam can generate a large influence on the world rice prices. The fluctuations of rice export prices in Thailand and Vietnam are correlated, and the two rice export markets are competing each other as evident by the co-movement of export rice prices of Thailand and Vietnam. The co-movement implies that the price movement in a rice export market will affect that in another one

D. Sirikanchanarak · J. Liu (✉) · S. Sriboonchitta
Faculty of Economics, Chiang Mai University, Chiang Mai, Thailand
e-mail: liujianxu1984@163.com

J. Xie
School of Statistics and Mathematics, Yun Nan University of Finance and Economics, Kun Ming 650221, China

V.-N. Huynh et al. (eds.), *Causal Inference in Econometrics*,
Studies in Computational Intelligence 622, DOI 10.1007/978-3-319-27284-9_21

in the same direction. Price transmission is extremely important for the functioning of competitive markets as these price signals help policy makers, buyers and sellers understand what would be in the future. Then, the issues of price transmission and co-movement between Thailand and Vietnam are of considerable interest to international rice market researchers, policy makers, and investors. Considering the issues above, the objectives of this paper are to examine the co-movement between rice export prices of Thailand and Vietnam, to test whether there do exist price transmissions between Thailand and Vietnam rice export markets, and to judge whether Thailand or Vietnam is actually a price leader on the basis of direction and magnitude of price transmission.

In recent years, there have been few studies with focus on testing the rice export price relations between Thailand and Vietnam. In addition, the Granger causality test, vector auto-regression (VAR) and cointegration regression models have been usually used to explore the relationship between the rice export prices of Thailand and Vietnam. For example, Ghoshray [4] used cointegration regression approach to test the price relations between Thailand and Vietnam rice export markets. Some empirical evidences suggested that the two rice prices have a long-run relationship, and Thailand is a price leader for higher quality grades. John [9] used the Granger causality test and VAR model to examine price transmission among five major rice exporters including Thailand, Vietnam, USA, Pakistan and Argentina. He found that there exist price transmissions across these major rice export markets and Asia acts as a price leader. Vietnam's export price is a more suitable world reference price than Thailand's. John [8] found that the price transmission between Thailand's domestic and export rice markets is bi-directional by using VAR and Granger causality test.

The VAR and cointegration regression models assume that the rice prices follow a multivariate normal or Student-t distribution with linear dependence. However, the actual relationship between the rice export prices of Thailand and Vietnam is possibly non-linear or asymmetrical. Also, the linear correlation may fail to capture the potentially asymmetric dependence between the rice export prices of Thailand and Vietnam. Therefore, we use the time-varying copula-based VAR models to describe the dynamic non-linear correlations and tail dependences between the rice export prices of Thailand and Vietnam. The time-varying copula-based VAR models allow for better flexibility in joint distributions than bivariate normal or Student-t distributions. On the one hand, the VAR model, Granger causality test and impulse response function are used to interpret the causality relationship and test the price transmission between the rice export prices of Thailand and Vietnam. On the other hand, the VAR model is employed to filter the rice export price returns of Thailand and Vietnam, thereby transforming the residuals in VAR model into Uniform (0, 1) by Gaussian kernel cumulative distribution function.

The contribution of this research is fourfold: First, a new approach, the time-varying copula-based VAR with kernel density function model is proposed. Second, we propose the time-varying BBl and BB7 copulas to estimate and forecast the co-movement between the rice export prices of Thailand and Vietnam. BB1 and BB7 copulas not only measure the dynamic non-linear correlation, but also capture the dynamic and asymmetric tail dependences between the rice export prices of

Thailand and Vietnam. Third, a simulation test is invented to judage the accuracy of the prediction of non-linear correlation between the rice export prices of Thailand and Vietnam. Last, our contribution also is to show especially practitioners in the area of agricultural economics how to use time-varying copula-based models so that they can obtain more reliable conclusions from their data.

The remainder of this paper is organized as follows. In Sect. 2, we describe the methodology corresponding to the time-varying copula-based VAR with kernel density function model. In Sect. 3, empirical study is described and results are presented. Section 4 gives us some conclusions.

2 Methodology

2.1 VAR Models

VAR model is appropriate for characterizing the multivariate relationship among the price series. Compared it with structural equation models, VAR model is more simplicity, and it does not need any economic theories to support the relationship between the studied variables. For the empirical purposes of our study, the model involves Thai rice export price (y) and Vietnam rice export price (x). Therefore, the VAR model can be expressed by

$$y_t = \theta_1 + \sum_{i=1}^{k} \psi_{1i}\, y_{t-i} + \sum_{i=1}^{k} \varphi_{1i}\, x_{t-i} + z_{1t} \tag{1}$$

$$x_t = \theta_2 + \sum_{i=1}^{k} \psi_{2i}\, y_{t-i} + \sum_{i=1}^{k} \varphi_{2i}\, x_{t-i} + z_{2t}, \tag{2}$$

where k is the number of lagness, z_t is a vector of random disturbances. We use ordinary least squares to estimate the VAR models. However, the VAR model is generally applied to study the causality relations and assess what impact each of the variables has on one another. So, the Granger causality tests and impulse response functions (IRFs) within the VAR model are used to assess price transmission between the rice export prices of Thailand and Vietnam. From the Granger causality tests, we can infer whether the price transmission exists or not, and what is the direction of the price transmission between the rice export prices of Thailand and Vietnam. While IRFs may help us identity the magnitude of price transmission between the rice export prices of Thailand and Vietnam. There are many papers and books that explain the Granger causality tests and IRFs in details, such as Gujarati [5], Enders [3] and John [9] etc.

2.2 Copulas

Researchers and academics always used some classical families of bivariate distributions such as bivariate normal and Student-t distributions to estimate the pairwise dependence of different variables empirically. It is well known that bivariate normal and Student-t distributions are symmetric with linear correlation parameters. In addition, there exist some limitations in bivariate Student-t distribution, such as each element of bivariate Student-t distribution has a univariate Student-t distribution with the same degrees of freedom parameter. Copula models could circumvent this limitation because their joint distribution can be decomposed into two parts: marginal distribution and dependence structure.

Sklar [14] first gave the definition of a copula as follows: Let $x = (x_1, x_2)$ be a random vector with the joint distribution function H and the marginal distribution F_1, F_2. Then, there exists a function C called copula such that

$$H(x_1, x_2) = H(F_1^{-1}(u_1), F_2^{-1}(u_2)) = C(F_1(x_1), F_2(x_2)), \tag{3}$$

where $u_1 = F_1(x_1)$ and $u_2 = F_2(x_2)$. In this study, we consider that the marginal distributions F_1 and F_2 are Gaussian kernel cumulative distribution functions. The formula for F_1 and F_2 can be expressed as

$$\hat{F}_i(z) = \frac{1}{n \cdot h} \sum_{t=1}^{n} K(\frac{z - z_t}{h}), \tag{4}$$

where the Gaussian kernel K is the standard Gaussian cumulative distribution function, and the optimal choice of bandwidth h is $1.06\widehat{\sigma}n^{-1/5}$ in the Gaussian kernel. The $\widehat{\sigma}$ is the standard deviation of the samples, and n is the number of the samples. The residuals z_t are generated by the VAR model. Because the non-parametric kernel density estimation looses the assumption of marginal distribution, then we use the kernel density estimation to fit the marginal distributions of rice export prices, which should be more efficient than and superior to an empirical distribution or t-GARCH (see Sriboonchitta et al. [15], Huang et al. [6]).

There are many advantages of using copulas. For example, they can be used to measure both rank dependence and tail dependence. One could use Kendall's tau to measure the monotonic dependence between variables:

$$\tau = \frac{c - d}{c + d} = \frac{2(c - d)}{n(n - 1)}, \tag{5}$$

where c is the number of concordant pairs, and d is the number of discordant pairs. It is well known that the copula can be used to measure the Kendall's tau coefficient (Nelsen [10]), as follows:

$$\tau_{x,y} = \tau_k = 4 \iint_{[0,1]^2} C(u_1, u_2)\, dC(u_1, u_2) - 1, \tag{6}$$

where C is the copula, and u_1 and u_2 are the values of the kernel cumulative distribution functions. One advantage of using the copula over the Kendall's tau is that the former can measure the upper tail and lower tail dependences:

$$\lambda_L = \lim_{u \to 0^+} P[Y \leq F_2^{-1}(u) | X \leq F_1^{-1}(u)] = \lim_{u \to 0^+} \frac{C(u, u)}{u} \tag{7}$$

and

$$\lambda_{up} = \lim_{u \to 1^-} P[Y > F_2^{-1}(u) | X > F_1^{-1}(u)] = \lim_{u \to 1^-} \frac{1 - 2u + C(u, u)}{1 - u}, \tag{8}$$

where X and Y are continuous random variables. To capture the co-movement between the rice export prices of Thailand and Vietnam, we follow the framework of Wu et al. [16] to construct the time-varying Gaussian, T, Clayton, Gumbel, BB1 and BB7 copulas. The time-varying Gaussian and T copulas are given as

$$\rho_t^* = \alpha + \beta \rho_{t-1}^* + \gamma (u_{1,t-1} - 0.5)(u_{2,t-1} - 0.5), \tag{9}$$

where we define $\rho_t^* = -ln \frac{(1 - \rho_t)}{(\rho_t + 1)}$ for guaranteeing the dependence parameter with the interval $(-1, 1)$, and ρ_t is the correlation coefficient and $0 \leq \beta < 1$. The time-varying Clayton and Gumbel copulas can be specified as follows:

$$\tau_t^* = \alpha + \beta \tau_{t-1}^* + \gamma (u_{1,t-1} - 0.5)(u_{2,t-1} - 0.5), \tag{10}$$

where $\tau_t^* = -ln \frac{(1 - \tau_t)}{(\tau_t + 1)}$, and τ_t represents the rank correlation Kendall's tau. Both BB1 and BB7 copulas have two parameters θ and δ which control asymmetric tail dependences for variables. The formulas of BB1 and BB7 copulas are followed from Joe and Hu [7].

$$C_{BB1}^{\theta,\delta}(u_1, u_2) = (1 + [(u_1^{-\theta} - 1)^{\delta} + (u_2^{-\theta} - 1)^{\delta}]^{1/\delta})^{-1/\theta}, \tag{11}$$

where $\theta \in (0, +\infty)$ and $\delta \in [1, +\infty)$, $\lambda_{up} = 2 - 2^{1/\delta}$, $\lambda_{low} = 2^{-1/(\delta\theta)}$

$$C_{BB7}^{\theta,\delta}(u_1, u_2) = 1 - (1 - [1 - (1 - u_1)^{\theta})^{-\delta} + (1 - [1 - (1 - u_2)^{\theta})^{-\delta} - 1)^{-1/\delta})^{1/\theta}, \tag{12}$$

where $\theta \in (1, +\infty)$ and $\delta \in [0, +\infty)$, $\lambda_{up} = 2 - 2^{1/\theta}$, $\lambda_{low} = 2^{-1/\delta}$. To measure the dynamic tail dependences of rice export price between Thailand and Vietnam, we use the time-varying BB1 and BB7 copulas to capture dynamic Kendall's tau and tail dependences between the rice export prices of Thailand and Vietnam. The formulas of the time-varying BB1 and BB7 copulas can be expressed as

$$\theta_t = \Lambda(\alpha + \beta \cdot \theta_{t-1} + \gamma (u_{1,t-1} - 0.5) \cdot (u_{2,t-1} - 0.5)) \tag{13}$$

$$\delta_t = \lambda(\alpha + \beta \cdot \delta_{t-1} + \gamma (u_{1,t-1} - 0.5) \cdot (u_{2,t-1} - 0.5)), \tag{14}$$

where the $\Lambda(x)$ and $\lambda(x)$ are used to ensure the parameters θ and δ fall within their intervals, respectively. In the time-varying BB1 copula $\Lambda(x) = exp(x)$ and $\lambda(x) = exp(x) + 1$ are defined, while $\Lambda(x) = exp(x) + 1$ and $\lambda(x) = exp(x)$ for the time-varying BB7 copula. Among all the time-varying copulas, the preferable time-varying copula is selected in terms of AIC and SIC.

2.3 Model Validation

The time-varying copulas are used to estimate and forecast the dynamic dependences between the rice export prices of Thailand and Vietnam. Then, how to evaluate the forecasted abilitiy of the preferable time-varying copula, which is a crucial problem for us and practitioners in the area of applying copulas. Thus, we propose a copula ratio approach to evaluate the forecast function of the preferable time-varying copula. A copula ratio is defined as the preferable copula C_p divided by the empirical copula C_e. The formula is given as

$$CR = \frac{C_p(u_{1,t+1}, u_{2,t+1}; \hat{\Theta}_{t+1})}{C_e(u_{1,t+1}, u_{2,t+1})} \tag{15}$$

where $\widehat{\Theta}_{t+1}$ repesents the estimated parameters of the preferable time-varying copula at t+1 period. The CR should be close to one if the model has a good prediction. A confidence interval can be calculated from Monte Carlo simulation method. We simulate 10,000 values of copula distribution from the preferable time-varying copula with $\widehat{\Theta}_{t+1}$. Therefore, the simulated copula ratio CR_s^i for each time can be expressed as

$$CR_s^i = \frac{C_p(u_{1,i}, u_{2,i}; \hat{\Theta}_{t+1})}{C_e(u_{1,t+1}, u_{2,t+1})} \tag{16}$$

where $i = 1, 2, \ldots, 10,000$, and $u_{1,i}$ and $u_{2,i}$ are from copula simulation. If the CR drops in the credible interval, such as 95 % confidence interval, then our prediction should be regarded as correct.

3 Empirical Results

3.1 The Data

This study uses monthly export rice prices for Thailand and Vietname white rice 5 % broken rice (F.O.B) from January 1996 to December 2014. The data were obtained from the World Bank and both prices are measured in a common currency, namely US dollars per metric ton. From Fig. 1, we can find that Thailand and Vietnam rice export

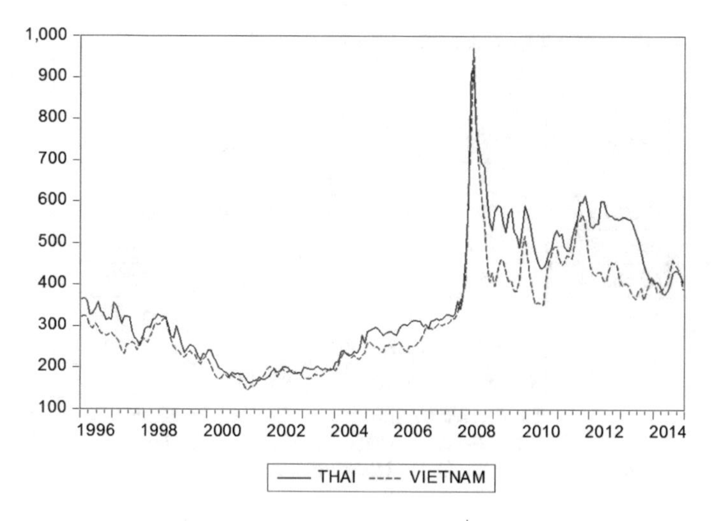

Fig. 1 Rice export price (F.O.B) of Thailand and Vietnam (USD per metricton)

prices have seemed to move with each other closely. It can be clearly seen that both prices increased to record levels in 2008 that was called the period of "Global rice crisis". Hereafter, both prices seemed to have fluctuated much more than previous years. There is a large gap between the rice prices of Thailand and Vietnam from 2011 to 2014. This is because Thai government implemented the paddy pledging program during this period. We use the difference between the logarithmic rice export prices for Thailand and Vietnam to calculate the log returns, and there are 228 observations for each country. This is one way to guarantee that the returns are stationary. We partition the data into two parts: in-sample and out-of-sample. The in-sample data from January 1996 to December 2013, with 216 observations were used for estimating the copula-based VAR model. So, there are 12 months left as out-of-sample data from January 2014 to December 2014. The out-of-sample data were used to predict the Kendall's tau and tail dependences of rice export prices of Thailand and Vietnam.

3.2 *Causality Tests and Impulse Response*

First, we should decide the number of lag order for VAR model. Table 1 reports the results of VAR lag order selection criteria. There are five selection criteria including likelihood ratio (LR), final prediction error (FPE), Akaike information criterion (AIC), Schwarz information criterion (SIC) and Hannan-Quinn information criterion (HQIC). We can find that the lagness should be two in terms of FPE, AIC and HQIC. Therefore, we use a VAR(2) structure to model the price returns of Thailand and Vietnam. Hereafter, Granger causality tests are performed under two lags.

Table 1 The results of VAR lag order selection criteria

Lag	LogL	LR	FPE	AIC	SIC	HQIC
0	642.7011	NA	9.86e−06	−5.8511	−5.8202	−5.8386
1	680.5834	74.7268	7.24e−06	−6.1605	−6.0677*	−6.1230
2	687.4661	13.4510	7.05e−06*	−6.1869*	−6.0321	−6.1244*
3	688.5750	2.1469	7.24e−06	−6.1605	−5.9438	−6.0730
4	693.9888	10.3827*	7.15e−06	−6.1734	−5.8948	−6.0609
5	697.8584	7.3505	7.15e−06	−6.1722	−5.8317	−6.0347
6	700.5020	4.9733	7.24e−06	−6.1598	−5.7574	−5.9973
7	704.1570	6.8092	7.27e−06	−6.1566	−5.6924	−5.9691
8	706.6101	4.5253	7.37e−06	−6.1425	−5.6164	−5.9300

Note that: * indicates lag order selected by the criterion; LR: sequential modified LR test statistic (each test at 5 % level); FPE: Final prediction error; AIC: Akaike information criterion; SIC: Schwarz information criterion; HQIC: Hannan-Quinn information criterion

Table 2 Pairwise Granger Causality Tests

Null Hypothesis:	Obs	F-Statistic	Prob.
Vietnam does not Granger Cause Thailand	213	15.0767	0.0000
Thailand does not Granger Cause Vietnam		10.2185	0.0001

Table 2 reports the results of Granger causality tests. The results of the tests show that both the null hypotheses are rejected at 0.01 significant level, which implies there does exist causality relationships between Thailand and Vietnam. Also, the short run price transmission exists with bi-directional price transmission between Thailand and Vietnam. The analysis is consistent with opinion of the market and the results of past study (John [9]). We would suggest that Thailand and Vietnam rices are almost completely substitute goods. Therefore, we can conclude that the rice export prices of Thailand and Vietnam are important to each other.

Figure 2 displays the results of impulse response function. The IRFs are estimated by using Sim's Cholesky factorization which takes the recursive structure of the error variance-covariance matrix. The IRFs are presented in 10 months's period. The results show that shocks originating in Thailand and Vietnam prices will transmit to each other. First, the impact of shocks in Thailand and Vietnam presents a persistent process that is consistent with macroeconomic theory. The impact of shocks gradually decreases to negative value, and then rice export prices of Thailand and Vietnam rebound to normal level. Second, although there exists bi-directional price transmission between Thailand and Vietnam by Granger causality tests, the magnitudes of the price transmission are different. Thailand's shocks in Vietnam market is slightly less than Vietnam's shocks in Thailand, which implies Vietnam is closer to be a price leader than Thailand. Third, Thailand's shocks in Vietnam market persist for 7 months, and Vietnam's shocks in Thailand probably persist for 7 months as well. In addition, the main finding from the IRFs is that price transmission occurs

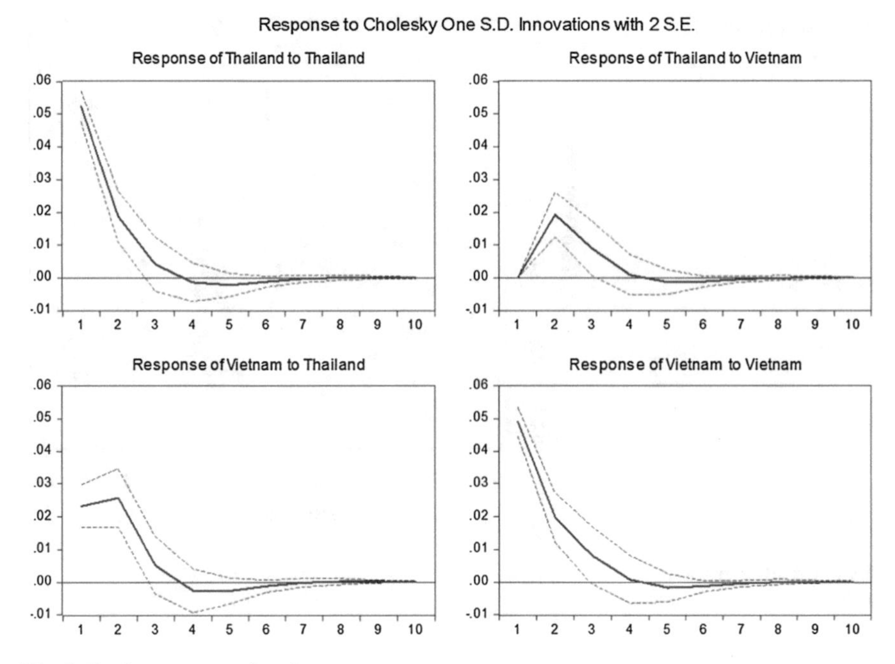

Fig. 2 Impluse response functions

between the two rice markets is so enormous, which can probably be explained by the fact that strong competition prevails between the rice markets of Thailand and Vietnam.

3.3 *Estimate Results of Copulas*

After we estimate VAR(2) model for the returns of rice export prices of Thailand and Vietnam, we generate the residuals from the VAR(2) model. Then, the residuals are transformed into Uniform (0,1) by using cumulative Gaussian kernel distribution. Figures 3 and 4 show the scatter plot of the marginals and the copula density from kernel density estimation (see Duong [2]), respectively. It can be seen that there are obvious tail dependences between the rice export prices of Thailand and Vietnam, and the tail dependences are asymmetric. Seemingly, the lower tail dependence is stronger than the upper tail dependence between the rice export prices of Thailand and Vietnam.

We now turn to estimate the time-varying copulas by using maximum likelihood method. Table 3 presents the estimated results of the time-varying copula models and the values of information criteria. It can be seen that the time-varying BB7 copula has the best performance in terms of AIC and SIC. The second best model

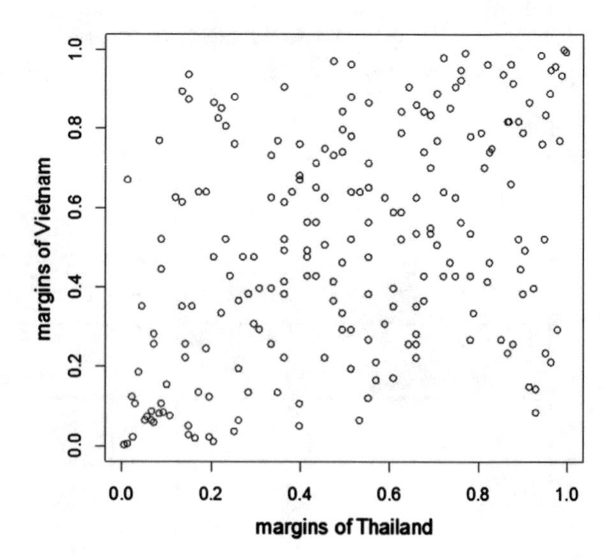

Fig. 3 Scatter plot of Margins

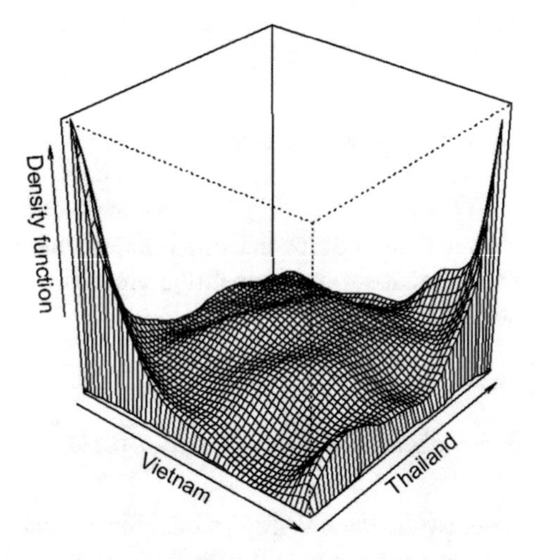

Fig. 4 Copula density function

is the time-varying BB1 copula, while the time-varying T copula is the third best model. These findings can tell us that there exists obviously tail dependences, which is consistent with Figs. 3 and 4. The time-varying Clayton and Gumbel copulas only can capture either lower tail dependence or upper tail dependence, so they cannot properly measure the dependence structure between the rice export prices of Thailand and Vietnam. Also the time-varying Gaussian copula does not fit very well because tail dependences cannot be captured. There are two parameters in BB7 copula. The parameters θ and δ control the magnitudes of lower tail and upper tail dependences, respectively. We can see the autoregressive parameters β in the time-varying BB7 copula equal to 0.976 and -0.856 for the parameters θ and δ, implying a high degree of persistence pertaining to the dependence structure and tail dependences between

Table 3 The results from the time-varying copula models

Copulas		α	β	γ	DoF	LL	AIC	BIC
Gaussian	ρ	0.025	0.995***	−0.535**		30.390	−54.781	−58.701
		(0.022)	(0.018)	(0.265)				
T	ρ	0.021	0.999***	−0.601*	5.418***	34.622	−61.244	−67.165
		(0.029)	(0.028)	(0.312)	(1.122)			
Clayton	θ	−0.098	0.863***	−0.358		30.022	−54.044	−57.964
		(0.167)	(0.192)	(0.609)				
Gumbel	θ	0.009	0.999***	−0.405		26.265	−46.53	−50.450
		(0.019)	(0.038)	(0.366)				
BB1	θ	−1.755**	−0.863***	4.131		37.591	−63.182	−71.023
		(0.775)	(0.070)	(2.599)				
	δ	−3.601***	−0.973***	−1.344				
		(0.841)	(0.026)	(0.963)				
BB7	θ	0.019	0.976***	−1.635*		39.266	−66.531	−74.373
		(0.032)	(0.032)	(0.895)				
	δ	−0.942*	−0.856***	2.758**				
		(0.492)	(0.075)	(1.353)				

Note: $*$, $**$, $***$ denoted significance at 10, 5 and 1 % respectively
Source: computation

the rice export prices of Thailand and Vietnam. The latent parameters γ are also significant and display that the latest return information is a meaningful measure. Specially, γ in the time-varying BB7 copula are much larger than others, which means they have a greater short-run response than other copulas. Therefore, we conclude that the information at the last period has effect on the time-varying dependences between the rice export prices of Thailand and Vietnam.

Figure 5 plots the Kendall's tau and tail dependences estimates and forecasts of the rice export prices of Thailand and Vietnam based on the time-varying BB7 copula model. First, the Kendall's tau, a kind of rank correlation, always has positive values implying that Thailand and Vietnam have competitive relationship. Second, we can find that the the smallest Kendall's tau was about 0.2 in 2011. This phenomenon was caused by the pledge paddy program in minor crop season. The biggest correlation happened in 2005 and 2008, respectively. In 2005, Thailand's government modified some policies to increase agricultural income of farmers, such as pledging domestic rice at a higher price than the market price. The global food and financial crisis in 2008 had a large influence on the dependence between the rice export prices of Thailand and Vietnam. Third, the lower tail dependences between the rice export prices of Thailand and Vietnam have volatility clustering due to the persistence of the dependence structure.While the upper tail dependences change wave upon wave. Last, the forecasts of Kendall's tau and tail dependences can effectively link up with their estimates, and keep the same changeful trend. According to our prediction, the upper tail dependences in 2014 seem to gradually increase with months, and the lower tail dependences have a strong fluctuation.

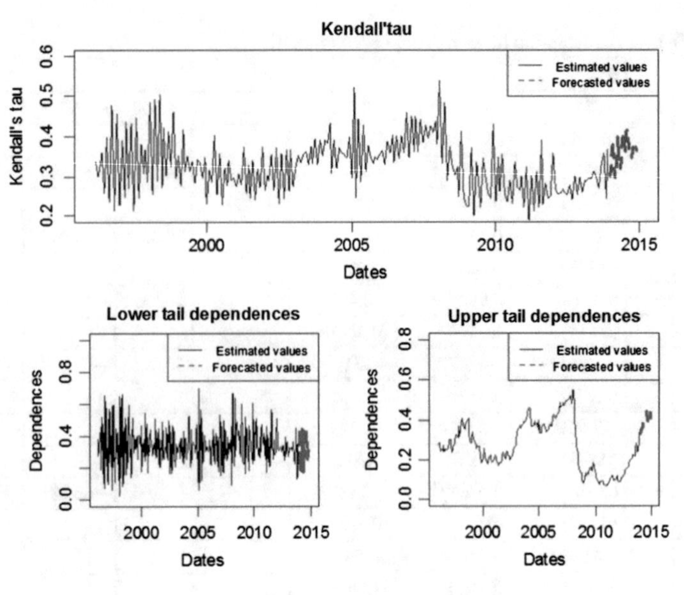

Fig. 5 Results of Kendall's tau estimated and forecasted values

Table 4 The copula ratio test for prediction

Forecast time	Kendall's tau	BB7 copula	Empirical copula	Ratio	Interval of 95 %
Jan-14	0.3001	0.1331	0.1388	0.9587	[0.0642, 6.5595]
Feb-14	0.3869	0.1309	0.1382	0.9472	[0.0931, 6.6095]
Mar-14	0.2951	0.2471	0.2385	1.0358	[0.0314, 3.8491]
Apr-14	0.3966	0.1547	0.1369	1.1297	[0.0953, 6.7420]
May-14	0.3179	0.4458	0.4409	1.0112	[0.0196, 2.0836]
Jun-14	0.4045	0.6558	0.6606	0.9927	[0.0177, 1.4088]
Jul-14	0.3623	0.7412	0.7297	1.0158	[0.0121, 1.2828]
Aug-14	0.415	0.8052	0.7982	1.0088	[0.0139, 1.1604]
Sep-14	0.3551	0.2591	0.2455	1.0555	[0.0384, 3.7714]
Oct-14	0.3457	0.2398	0.2088	1.148	[0.0453, 4.4587]
Nov-14	0.3826	0.1216	0.1194	1.0178	[0.0901, 7.7926]
Dec-14	0.3645	0.0602	0.0837	0.7203	[0.1170, 11.0426]

Source: computation

Table 4 reports the forecasted values of Kendall's tau and the copula ratios. As can be seen in Fig. 5, the Kendall's tau keeps much high values, specially in February, April, June, August and November. This might due to the fact that February, April, June and August are harvest months for second season crops while November is harvest month for main season crops. Harvest season is able to direct the rice export prices to increase or decrease at the same time in Thailand and Vietnam. Moreover, we can see that all copula ratios are very close to 1 and drop in the interval of 95 %, which implies that our prediction is very accurate.

4 Conclusions

The correlation among the price behavior of crops usually is measured by using cointegration regression and VAR model within IRFs and Granger causality test. However, the models are confined to the analysis of the linear correlation, and cannot reflect any time-varying characters. Hence, the time-varying copulas are appropriate for making up for the deficiencies.

This paper describes a model for analyzing the transmission and co-movement of Thai and Vienamese rice prices by the time-varying copula-based VAR model, in which the empirical evidence shows that this method can be quite robust in estimating and forecasting non-linear correlation. The results reveal that there exists price transmission between Thai and Vietnamese rices, and they are causality interrelated. Also, the time-varying BB7 estimates reflect a high degree of persistence pertaining to the dependence structure and tail dependences between the rice export prices of Thailand and Vietnam. Moreover, the predicted Kendall's tau in 2014 describes relatively high correlation, especially in February, April, June, August and November, which coincide with the harvest months. An implication of these findings is that Thailand (Vietnam) has the ability to destabilize the rice market of Vietnam (Thailand). Maybe, both countries also can distort international rice market. According to our findings, we firstly suggest that Thailand and Vietnam should enhance cooperation and communication. Export volume should be managed by them through dialogue and consultation thereby turning competition into cooperation, which is also beneficial to the stability in international rice market. Second, farmers in Thailand and Vietnam need to produce wide variety of rice, which can reduce their competition. Third, Thai and Vietnamese authorities should regularly adjust their minimum export prices together in order to keep their rice exports competitive in world markets. In short, Thailand and Vietnam should enhance communication among nations, such as US, China and India etc., and establish a strategic rice reserve for emergencies and stabilizing markets.

Acknowledgments The authors are very grateful to Professor Vladik Kreinovich and Professor Hung T. Nguyen for their comments. The authors wish to thank the Puey Ungphakorn Centre of Excellence in Econometrics and Bank of Thailand Scholarship for their financial supports.

References

1. Dawe, D.: The changing structure of the World rice market 1950–2000. J. Food Policy **27**, 355–370 (2002)
2. Duong, T.: Package ks. http://cran.r-project.org/web/packages/ks/ks.pdf (2015)
3. Enders, W.: Applied Econometric Time Series. Wiley, New York (1995)
4. Ghoshary, A.: Asymmetric adjustment of rice export prices: the case of Thailand and Vietnam. Int. J. Appl. Econ. **5**(2), 80–91 (2008)
5. Basic Econometrics, Fourth Edition edn. The McGraw-Hill Companies, USA (2004)
6. Huang, J.J., Lee, K.J., Liang, H., Lin, W.: Estimating value at risk of portfolio by conditional copula-GARCH method. Insur.: Math. Econ. **45**, 315–324 (2009)
7. Joe, H., Hu, T.: Multivariate distributions from mixtures of max-infinitely divisible distributions. J. Multivar. Anal. **57**(2), 240–265 (1996)
8. John, A.: Price relations between export and domestic rice markets in Thailand. J. Food Policy **42**, 48–57 (2013)
9. John, A.: Price relations between international rice markets. J. Agric. Econ. (2014)
10. Nelsen, R.: An introduction to Copula. Springer, New York (2006)
11. Poangpongsakorn, N.: The policy economy of Thai rice price and export policies in 2007–2008. In: Dawe, D. (ed.) The Rice Crisis: Markets, Policies and Food Security Food Security. Food and Agriculture Organization of the United Nation, Roome, 191–218 (2010)
12. Siamwalla, A.: A history of rice policies in Thailand (1975). http://ageconsearch.umn.edu/bitstream/135510/2/fris-1975-14-03-169.pdf
13. Siamwalla, A., Heykin, S.: The World rice markets: structure, conduct and performance. Research Report no.39, International Food Policy Research Institute, USA (1983)
14. Sklar, M.: Fonctions de répartition à n dimensions et leurs marges. Publ. Inst. Stat. **8**, 229–231 (1959)
15. Sriboonchitta, S., Nguyen, H.T., Wiboonpongse, A., Liu, J.: Modelling volatility and dependency of agriculture price and production indices of Thailand: static versus time-varying copulars. Int. J. Approx. Reason. **54**, 793–803 (2013)
16. Wu, C.C., Liang, S.S.: The economic value of range-based covariance between stock and bond returns with dynamic copulas'. J. Empir. Financ. **18**, 711–727 (2011)

Modeling Co-Movement and Risk Management of Gold and Silver Spot Prices

Chen Yang, Songsak Sriboonchitta, Jirakom Sirisrisakulchai and Jianxu Liu

Abstract This paper aims to model volatility and correlation dynamics in spot price returns of gold and silver, and examines the corresponding market risk management implications. VaR (value at risk) and ES (expected shortfall) are used to analyze the market risk associated with investments in gold and silver. Many GARCH family models are employed to describe the volatility. This work applied the copula based-GARCH model in the estimation of a portfolio VaR and ES composed of gold and silver spot prices. The empirical results exhibit that the NAGARCH and the TGARCH families performed better than other GARCH family members in describing the volatility of gold and silver returns, respectively. Furthermore, the time-varying T copula has the most appropriate performance in capturing the dependence structure between gold and silver returns. The out-of-sample forecast performance indicates that the time-varying T copula-based GARCH model can measure the VaR and ES with the accurate estimates of gold and silver.

Keywords Value at risk · Expected shortfall · Copulas · Metal prices

1 Introduction

One of the dominant trends we are witnessing right now is the increasing number of individuals and investors globally who seek to own gold and silver in a fashion of securities. Gold and silver are precious metals with uniqueness and non-renewable nature, which can function as commodity and monetary assets. They have played as multifaceted metal down through the centuries, conducting similar features to money in that they act as a medium of exchange, a store of wealth and a unit of value. They also act important roles as precious metals with crucial portfolio diversification properties. As precious metals, they are also largely used in industries, such as jewelry, machine, electronic and medicine as well.

C. Yang · S. Sriboonchitta · J. Sirisrisakulchai · J. Liu (✉)
Faculty of Economics, Chiang Mai University, Chiang Mai 50200, Thailand
e-mail: liujianxu1984@163.com

V.-N. Huynh et al. (eds.), *Causal Inference in Econometrics*,
Studies in Computational Intelligence 622, DOI 10.1007/978-3-319-27284-9_22

That is true—investors can certainly get benefit as long as they invest into the gold and silver markets at an appropriate time. With a solidified upward trend in spite of the falling value of dollar and other currencies, investing in gold and silver can help investors avoid a personal financial crisis. However, in order to be successful, investors must constantly monitor the markets and understand the industry. As a result, model and forecast volatility and dependence structures accurately of gold and silver spot prices seems important for researchers, portfolio managers, investors, and policy makers.

To analyze financial market movement and co-movement is significant for efficient diversity in a portfolio management. Some researchers, like Chang et al. [4] employed multivariate GARCH models to estimate time-varying dependence structures, but this is always grounded on strict restrictions for the sake of guaranteeing a well-defined covariance matrix. Moreover, the VAR (vector auto-regression) and multivariate GARCH models suppose that the asset returns is linear dependence following a multivariate normal or Student-t distribution. This hypothesis has clashed with many empirical researches which display that gold and silver returns are asymmetric, leptokurtic time-varying and fat-tailed, with very different marginal distributions and disparate degrees of freedom parameters. In addition, the actual gold and silver correlation is potentially non-linear, time-varying or asymmetric. Embrechts et al. [5] found that using linear correlation to model the dependence structure revealed many shortcomings. On the one hand, a severe defect of linear correlation is that under non-linear severely increasing transformation it is not invariant; on the other hand, capturing dependence structures of variables is one of the particular difficulties in evaluating VaR (Value at Risk), and linear correlation may fail in capturing the possibly asymmetric dependence between gold and silver.

To overcome these drawbacks mentioned above, we applied copula-based GARCH families to capture the volatility and dependence structures of gold and silver. The copula-based GARCH models show better flexibility of joint distributions comparing with both bivariate normal and Student-t distributions.

Our contribution to the literature is as follow as: first of all, our work proposed GARCH family models to model the volatilities of gold and silver spot prices. The results showed that the NAGARCH model has the most appropriate performance compared with others on measuring the volatility of gold. For sliver, the TGARCH model behaved better than others. Second, we employed copula-based GARCH families to describe the dependence structures of gold and silver spot prices. The copula-based GARCH family models can be applied for capturing the likely leptokurtosis and skewness of gold and silver spot prices. We found that the time-varying copulas outperformed the static copulas with respect to describing dependence structures of gold and silver spot prices. Third, we evaluated VaR and ES through the copula-based GARCH models with Monte Carlo simulation. The results showed that the returns of gold and silver are always existing dependence except for some special periods. For example in 2000–2003, with world economic recession, the dependence between gold and silver was very weak. The tail dependencies of gold and silver also exhibited the same characteristic in that period. VaR and ES backtests across copula-based portfolio showed that the number of violations by the time-varying T copula-GARCH model is in proximity to the expected number of violations. It means our model is correct for gold and silver returns in risk management.

The format for this paper is as follows. Section 2 we introduce the methodologies of many GARCH families, copulas, VaR and ES-based copula-GARCH model in detail. Section 3 presents empirical estimation results and provides the out-of-sample forecasted results of the copula-based GARCH models. Finally, Sect. 4 presents the conclusions.

2 Methodology

2.1 ARMA-GARCH Families

For analysing time series data, GARCH family models have been employed widely, especially when we aim to analyze and forecast volatility. Bollerslve [3] came up with GARCH model to replace the ARCH model in application, then, the GARCH has been extensively used in econometric and economic fields. Following Ling [13], the ARMA (p, q)-GARCH (k, l) model shown as:

$$r_t = c + \sum_{i=1}^{p} \phi_i r_{t-i} + \sum_{i=1}^{q} \psi_i z_{t-i} + z_t, \tag{1}$$

$$z_t = \sigma_t \cdot \varepsilon_t, \tag{2}$$

$$\sigma_t^2 = \omega + \sum_{i=1}^{k} \alpha_i z_{t-i}^2 + \sum_{i=1}^{l} \beta_i \sigma_{t-i}^2, \tag{3}$$

where $\sum_{i=1}^{p} \phi_i < 1, \omega > 0, \alpha_i \geq 0, \beta_i \geq 0$, and $\sum_{i=1}^{k} \alpha_i + \sum_{i=1}^{l} \beta_i < 1$ with σ_t^2 denoting the conditional variance, ω being the intercept, and the GARCH order is defined by (l, k) (GARCH, ARCH). The standardized residual ε_t is supposed to be a Skewed Student-t distribution (SSTD), which can be applied to describe the potentially asymmetric and heavy-tailed features of gold and silver returns. In our study, other GARCH families are also proposed, such as TGARCH, AVGARCH, NAGARCH, GJR-GARCH. The GARCH family models following Alexios [1] are described as follows:

$$\sigma_t^\lambda = \omega + \sum_{i=1}^{k} \alpha_i \sigma_{t-i}^\lambda \left(|z_{t-i} - \eta_{2i}| - \eta_{1i}(z_{t-i} - \eta_{2i})\right)^\delta + \sum_{i=1}^{l} \beta_i \sigma_{t-i}^\lambda. \tag{4}$$

For Eq. (4), when $\lambda = \delta = 2$ and $\eta_{1i} = \eta_{2i} = 0$, it is the sample GARCH model of Bollerslev [3] and when $|\eta_{1i}| \leq 1$, it represents TGARCH (the Threshold GARCH model) of Zakoian [22]; when $\lambda = \delta = 1$ and $|\eta_{1i}| \leq 1$, it exhibits AVGARCH (the Absolute Value GARCH model) of Taylor [20] and Schwert [17]; when $\lambda = \delta = 2$ and $\eta_{1i} = 0$, it displays NAGARCH (the Nonlinear Asymmetric GARCH model) of Engle and Ng [6]; when $\lambda = \delta = 2$ and $\eta_{2i} = 0$, it demonstrates GJR-GARCH (the Glosten-Jagannathan-Runkle GARCH model) of Glosten et al. [8]. Following Fernandez and Steel [7], generated a Skewed Student-t distribution, which exposes both flexibility of tails and possibility of skewness, each entirely controlled by a separate scalar parameter. The formula of Skewed Student-t distribution is shown as

$$P(x_i\,|v,\gamma) = \frac{2}{\gamma + \frac{1}{\gamma}} \left\{ f_v(\tfrac{x_i}{\gamma}) I_{(0,\infty)}(x_i) + f_v(\gamma x_i) I_{(-\infty,0)}(x_i) \right\}, \tag{5}$$

where $f_v(.)$ is unimodal and symmetric around zero, and γ is the skewness parameter that is defined from 0 to ∞; I indicates the indicator function and v is the degree of freedom.

2.2 Copulas

Sklar's theorem (Sklar, [18]), named after Abe Sklar, gives the theoretical foundation for the application of copulas. Sklar's theorem states that every multivariate cumulative distribution function

$$\mathrm{H}(x_1, \ldots, x_d) = P[X_1 \leq x_1, \ldots, X_d \leq x_d] \tag{6}$$

of a random vector $(X_1, X_2, \ldots, X_d)$ with margins $F_i(x) = P[X_i \leq x]$ can be written as

$$\mathrm{H}(x_1, \ldots, x_d) = C(F_1(x_1), \ldots, F_d(x_d)) \tag{7}$$

where C is a copula.

Kendall's tau for measuring a pair (X, Y), distributed following H, can be defined as the disparity between the probabilities of consistency and in consistency with respect to two independent pairs (X_1, Y_1) and (X_2, Y_2) each with distribution H (Jose et al. [9]); namely

$$\tau_{XY} = \Pr\{(X_1 - X_2)(Y_1 - Y_2) > 0\} - \Pr\{(X_1 - X_2)(Y_1 - Y_2) < 0\}. \tag{8}$$

These probabilities can be estimated through integrating over the distribution of (X_2, Y_2). So that, related to copulas, Kendall's tau (τ) turns into

$$\tau_C = 4 \int_0^1 \int_0^1 C(u_1, u_2) dC(u_1, u_2) - 1 \tag{9}$$

where C is the copula associated with (X, Y).

As a mathematical tool, copulas were used in finance to help identify economic risk, market risk, capital adequacy, operational risk and credit risk (Nelsen, [14]). The correlation coefficient is usually used to calculate interdependence of returns of two or more assets. But, correlation only works well with normal distributions, while distributions in financial markets are mostly skewed. The copula, therefore, has been applied to financial areas such as option pricing and portfolio VaR to handle the skewness.

Additionally, copulas can be employed to typify the dependence in tails of the distribution. In terms of tail dependence, two measures known as the upper and the lower tail dependence coefficients are especially helpful in measuring the trend of market crashing or booming together. Following the concept of Aloui et al. [2], the lower and the upper tail dependencies were presented as in the following:

Let X, Y display variables related to marginal function F and G. Then the coefficient of lower tail dependence λ_L and the coefficient of upper tail dependence λ_U are

$$\lambda_L = \lim_{u \to 0+} \Pr[Y \leq G^{-1}(u) \left| X \leq F^{-1} (u)\right.] \tag{10}$$

$$\lambda_U = \lim_{u \to 1-} \Pr[Y > G^{-1}(u) \left| X > F^{-1}(u)\right.]. \tag{11}$$

Our work applied a diverse range of parametric copulas to describe different dependence structures between spot prices of gold and silver. Following Sriboonchitta et al. [19], Gaussian copula, T copula, Frank copula, (rotate) Clayton copula, (rotate) Gumbel copula, (rotate) Joe copula, (rotate) BB1, (rotate) BB6, (rotate BB7) and (rotate) BB8 copula, all are the static copula candidates. This study also employed a series of time-varying copulas [19], such as time-varying T copula, time-varying Gaussian copula, time-varying (rotate) Gumbel copula, and time-varying (rotate) Clayton copula. Both elliptical copula (e.g.: Gaussian copula) and Archimedean's copula (e.g.: Clayton copula) functions are applied to describe diverse dependence structures. One of the advantages of elliptical copulas is specifying the correlation with different levels between the margins; but, these copulas must master radical asymmetry. The property of T copula is symmetric and which can indicate symmetric dependence in the extreme tails. The T copula converges to the Gaussian copula which exhibits dependencies on the two side tails of zero, as long as the degree of freedom increases to infinity. Archimedean copulas are typified by their generator function with many significant properties. They can take upper tail dependence, lower tail dependence, or both; therefore, they can make better description for the reality of the behavior with respect to financial markets.

Gaussian copula has a critical flaw, it cannot capture tail dependence. On the contrary, T copula can capture tail dependence with symmetric structures, which is its biggest advantage compared with Gaussian copula. Both Gaussian and T copulas belong to the elliptical-copula family. The Gumbel copula is an asymmetric copula, weightier in the right tail with higher probability, while it can capture both upper tail and lower tail dependencies. Clayton copula can catch lower tail dependence. The Joe copula can help us capture the upper tail dependence. BB1 and BB7 copulas reflect different tail dependence between upper tail and lower tail, while BB6 and BB8 can capture the upper tail dependence structure.

There are many copulas that cannot exhibit negative tail dependence such as BBX copula and Gumbel, Clayton etc. These copulas will not fit once the bivariate random variable has negative dependence. While they may be rotated and then could be applied again.

2.3 Time-Varying Copulas

Time-varying copula has been considered as the dynamic generalization of a Pearson correlation or Kendall's tau. Pearson's coefficient (ρ) is generally used in the Gaussian copula and the T copula. Moreover, we follow the concept of Wu et al. [21] supposing that the dependence parameters depend on past dependence and historical information $(u_{g,t-1} - 0.5)(u_{s,t-1} - 0.5)$. If both $(u_{g,t-1} - 0.5)$ and $(u_{s,t-1} - 0.5)$ are either bigger or smaller than 0.5, we deduce that the dependence is higher than previously. Let ρ_t^* displays a proper logistic transformation of dependence parameters ρ_t, so that the time-varying parameters ρ_t^* can be shown as:

$$\rho_t^* = \alpha_c + \beta_c \rho_{t-1}^* + \gamma_c (u_{g,t-1} - 0.5)(u_{s,t-1} - 0.5) \tag{12}$$

the proper logistic transformation is employed to guarantee the dependence parameter with the interval $(-1, 1)$, which can be written as $\rho_t^* = -\ln[(1 - \rho_t)/(\rho_t + 1)]$.

2.4 Inference Function for Margins (IFM)

Sklar's theorem was extended by Patton [15] who also introduced the conditional copula function to model time-varying conditional dependence. Let $r_{g,t}$ and $r_{s,t}$ be stochastic variables that indicate gold price and silver price at time t respectively, following marginal conditional cumulative distribution function $u_{g,t} = G_{g,t}(r_{g,t} \mid \Psi_{t-1})$ and $u_{s,t} = G_{s,t}(r_{s,t} \mid \Psi_{t-1})$, where Ψ_{t-1} indicates past information. Then, the conditional copula function $C(u_{g,t}, u_{s,t} \mid \Psi_{t-1})$ can be shown using the two time-varying cumulative distribution functions. Extending Sklar's theorem, the bivariate conditional cumulative distribution functions of stochastic variables $r_{g,t}$ and $r_{s,t}$ can be written as

$$F(r_{g,t}, r_{s,t} \mid \Psi_{t-1}) = C(u_{g,t}, u_{s,t} \mid \Psi_{t-1}). \tag{13}$$

Suppose the cumulative distribution function can be differentiated, and the conditional joint density can be shown as

$$\begin{aligned} f(r_{g,t}, r_{s,t} \mid \Psi_{t-1}) &= \frac{\partial^2 F(r_{g,t}, r_{s,t} \mid \Psi_{t-1})}{\partial r_{g,t} \partial r_{s,t}} \\ &= c(u_{g,t}, u_{s,t} \mid \Psi_{t-1}) \times g_g(r_{g,t} \mid \Psi_{t-1}) \times g_s(r_{s,t} \mid \Psi_{t-1}), \end{aligned} \tag{14}$$

where the conditional copula density function is $c(u_{g,t}, u_{s,t}\,|\Psi_{t-1}) = \partial^2 C(u_{g,t}, u_{s,t}\,|\Psi_{t-1})/\partial u_{g,t}\partial u_{s,t}$ and $g_i(\cdot)$ is the density function in accordance with $G_i(\cdot)$.

From Eq. (14), the likelihood function can be shown as:

$$L_{g,s}(\Theta) = L_g(\Theta_g) + L_s(\Theta_s) + L_c(\Theta_c), \tag{15}$$

where the parameters vectors of marginal distributions of gold and silver are Θ_g and Θ_s respectively and the vector of parameters in the copula function C is Θ_c. When implement the maximum likelihood method over a high dimension case, the problems related to extensive calculation and estimator accuracy will be faced by optimization process. Therefore, we apply the two-stage estimation method, known as inference functions for margins (IFM), to evaluate the parameters of our copula-based GARCH models. Joe [10, 11] demonstrated that this estimate is approximated well and asymptotically effective to the maximum likelihood estimator under some regularity conditions. Consequently, the estimators can be calculated effectively without losing any real information.

2.5 VaR and ES-based Copula-GARCH Model

The VaR and ES are estimated by using copula-based GARCH with Monte Carlo simulation. First, use the estimation results of the preferred copula to generate the random number 10,000. Second, apply the inverse function of the corresponding marginal distribution (Skewed Student-t distribution) of each variable to get the standardized residuals. Third, forecast the value of each variable at the t+1 period by using the GARCH model; thus, 10,000 possible values are generated at the t+1 period for each variable, which can be expressed as

$$r_{m,n,t+1} = \hat{c} + \sum_{i=1}^{p} \hat{\phi}_{m,i} r_{m,t-i+1} + \sum_{i=1}^{q} \hat{\psi}_{m,i} z_{m,t-i+1} + \hat{\sigma}_{m,t+1}\varepsilon_{m,n,t+1}, \tag{16}$$

where $n = 1, 2, \ldots, 10000$, m equals to the number of variables, $\varepsilon_{m,n,t+1} = F_m^{-1}(u_{m,n,t+1})$, and $u_{m,n}$ is from the simulation of the preferable copula. Then, the portfolio return approximately equals to the following:

$$R_{t+1} = w \cdot r_{1,t+1} + (1-w) \cdot r_{2,t+1} \tag{17}$$

where the portfolio weights of asset 1 and asset 2 are w and $1-w$ respectively. Last, the VaR can be defined as:

$$VaR_\alpha(W)_{t+1}\,|\Psi_t = \inf\{R_{t+1} : F(R_{t+1}) \geq \alpha\} \tag{18}$$

where F is empirical cumulative distribution function of a portfolio return R at time t+1.

In risk management, VaR is probably the most popular risk measure. But, it does not satisfy the property of subadditivity. There is an alternative method which is called ES, which satisfies the property of subadditivity and provides a more conservative measure of losses relative to VaR. By following the method discussed by Rockafellar and Uryasev [16], the ES can be formulated as follows:

$$ES_{\beta}(W) = \{VaR_{\beta}(W) + \frac{1}{q(1-\beta)} \sum_{k=1}^{q} [-W^T R_k - VaR_{\beta}(W)]^+\}, \quad (19)$$

where $[t]^+ = \max(t, 0)$, $\sum_{i=1}^{n} w_i = 1$, q represents the number of samples generated by Monte Carlo simulation. $VaR_{\beta}(W)$ is the VaR under the β confidence level and the W portfolio allocations, and R_k is the kth vector of simulated returns.

3 Empirical Results

3.1 The Data

Both gold and silver spot prices time series were taken from the EcoWin database from 4th Jan 2000 to 31th Dec 2014, yielding a total of 3885 observations. These are calculated to logarithm form and the market returns r_t is performed by equation (20):

$$r_t = \log(\frac{p_t}{p_{t-1}}), \quad (20)$$

where r_t is daily returns, p_t is price at time t, p_{t-1} is price at previous time t−1. We divided the data into two groups, sample-in and sample-out data. The sample-in data contain the first 3365 observations, and the leftover 520 observations are sample-out data for the test.

Table 1 exhibits the descriptive statistics for the two assets. Over the sample period, all of the series exhibit approximately zero mean, and negative skewness which indicate both silver and gold returns are skewed to the left. In terms of kurtosis, the values of both gold and silver returns are greater than three, thereby implying the empirical observations of returns have fatter tails in comparison with the normal distribution. In other words, their distributions are leptokurtic. Analogously, since the Jarque–Bera statistics are large and significant, the supposition of the Skewed Student-t distribution should be appropriate in our study.

Table 1 Data description and statistics

	Gold	Silver
Mean	0.0004	0.0003
Median	0.0004	0.0000
Maximum	0.1039	0.1393
Minimum	−0.0888	−0.1698
Std. Dev.	0.0114	0.0196
Skewness	−0.2783	−1.1179
Kurtosis	8.6894	11.6410
Jarque-Bera	5291.270	12899.01
Probability	0.0000***	0.0000***

Note: this table represents the descriptive statistics for daily gold and silver spot prices returns for the sample period from 4th Jan 2000 to 31th Dec 2014. The symbols *, **, and *** present statistical significance at the 10, 5 and 1 % levels respectively

3.2 Volatility Analysis

The ARMA-GARCH families were applied to analyze the volatility of gold and silver returns respectively. Table 2 and 3 report the results of selection criteria of 5 ARMA-GARCH families. We used the Akaike's information criterion (AIC), Bayesian information criterion (BIC), Shibata's information criterion (SIC) and Hannan-Quinn's information criterion (HQIC) to evaluate the models' performance. The model selection criteria AIC, BIC, SIC and HQIC suggest the best fit of data for gold is ARMA (1, 0)-NAGARCH (1, 1) model. NAGARCH model is a generalization of the standard GARCH model which takes the asymmetric through shift into consideration.

Table 2 The results of selection criteria of ARMA-GARCH family models for gold returns

Information criteria	ARMA(1,0)-GARCH(1,1)	ARMA(1,0)-TGARCH(1,1)	ARMA(1,0)-AVGARCH(1,1)	ARMA(1,0)-NAGARCH(1,1)	ARMA(1,0)-GJRGARCH(1,1)
Akaike	−6.3425	−6.3511	−6.3505	−6.3524	−6.3471
Bayes	−6.3297	−6.3365	−6.3341	−6.3379	−6.3326
Shibata	−6.3425	−6.3511	−6.3505	−6.3525	−6.3471
Hannan-Quinn	−6.3379	−6.3459	−6.3446	−6.3472	−6.3419

Table 3 The results of selection of ARMA-GARCH family models for silver returns

Information Criteria	ARMA(0,0)-GARCH(1,1)	ARMA(1,0)-TGARCH(1,1)	ARMA(1,1)-AVGARCH(1,1)	ARMA(1,0)-NAGARCH(1,1)	ARMA(1,0)-GJRGARCH(1,1)
Akaike	−5.4124	−5.4330	−5.4318	−5.4220	−5.4220
Bayes	−5.4033	−5.4203	−5.4155	−5.4093	−5.4083
Shibata	−5.4124	−5.4330	−5.4318	−5.4220	−5.4210
Hannan-Quinn	−5.4091	−5.4284	−5.4260	−5.4174	−5.4165

Table 4 Parameter estimates of TGARCH model of silver and NAGARCH model of gold

Parameter	Silver		Gold	
	ARMA(1,0)-TGARCH(1,1)		ARMA(1,0)-NAGARCH(1,1)	
	Estimate	S. E	Estimate	S. E
c	–	–	0.0006***	0.0001
ϕ_1	−0.0379**	0.0157	−0.0482***	0.0164
ω	0.0000	0.0000	0.0000	0.0000
α_1	0.0714***	0.0089	0.0370***	0.0032
β_1	0.9502***	0.0049	0.9330***	0.0006
η_{11}	−0.3084***	0.011	–	–
η_{21}	–	–	−0.8481***	0.0476
γ	0.8993***	0.0164	0.9738 ***	0.0226
ν	3.5238***	0.2929	5.4182***	0.6945
LLF	9148.02		10696	

Signif.codes:'***'0.001; '**'0.01; '*'0.05
Source computation

It implies there does exist significant leverage and size effects for gold and silver. And the best fit of data for silver is ARMA (1, 0)-TGARCH (1, 1) model, which has the most appropriate performance compared to others in terms of 4 kinds of information criteria. TGARCH model adds an additional term to account for possible asymmetries, and it permits a rotation of news impact, but does not allow a shift.

Table 4 presents the maximum likelihood results of parameter estimates of TGARCH model of silver and NAGARCH model of gold. We fitted ARMA (1, 0)-TGARCH (1, 1) and ARMA (1, 0)-NAGARCH (1, 1) models for the returns series of silver and gold, respectively. All parameters in the TGARCH model for silver and NAGARCH model for gold are significant except ω. Thus we concluded that these models are adequate. The asymmetry parameters, γ, are significant and less than 1 for both silver and gold returns, indicating that both silver and gold returns are skewed to the left. Furthermore, in the variance equations, the parameters α_1 and β_1 are significant and similarly interpreting that silver and gold returns have volatility clustering. The asymmetry rotation parameter η_{11} is significant and negative, which means that negative shocks introduce more volatility than positive shocks of the same size in the next period. And the asymmetry shift parameter η_{21} is negative, which means that the news impact curve is shifted to the left by the distance 0.8481.

3.3 *Estimated Results of Copulas and VaR*

Table 5 presents copula modeling results of the 18 static copula models and the time-varying T copula. The AIC criterion is used for model selection here. In terms of the values of AIC, T copula is superior to other selected static copula models because

Table 5 Results of the static and the time-varying T copulas

Copulas	Parameters	Values	S. E	Kendall's tau	Tail dependence		AIC
					lower	upper	
Gaussian	ρ	0.6695***	0.0079	0.4669	0	0	−1994.939
T	ρ	0.7027***	0.0093	0.4961	0.4078	0.4078	−2337.68
	DoF	3.7228***	0.2829				
Clayton	θ	1.3713***	0.0400	0.4068	0.6032	0	−1819.255
Gumbel	θ	1.8704***	0.0264	0.4653	0	0.5514	−2010.991
Frank	θ	5.9277***	0.1358	0.5104	0	0	−2137.205
Joe	θ	2.0665***	0.0366	0.3695	0	0.6015	−1500.528
BB1	θ	0.5369***	0.0422	0.4861	0.4310	0.4288	−2222.77
	δ	1.5341***	0.0319				
BB6	θ	1.001***	0.3402	0.4653	0	0.5516	−2008.544
	δ	1.8691***	0.4251				
BB7	θ	1.6571***	0.0406	0.4591	0.5154	0.4806	−2145.736
	δ	1.0459***	0.0462				
BB8	θ	6***	0.4535	0.4858	0	0	−2057.008
	δ	0.6492***	0.0308				
R-Clayton	θ	1.2152***	0.0375	0.3780	0	0.5653	−1536.972
R-Gumbel	θ	1.9190***	0.0273	0.4789	0.5649	0	−2171.298
R-Joe	θ	2.2168***	0.0394	0.3996	0.6329	0	−1782.112
R-BB1	θ	0.2666***	0.0370	0.4886	0.5056	0.6691	−2227.907
	δ	1.7253***	0.0353				
R-BB6	θ	1.001***	0.1154	0.4789	0.5651	0	−2169.03
	δ	1.9177***	0.1474				
R-BB7	θ	1.8784***	0.0427	0.4592	0.5537	0.4163	−2140.111
	δ	0.7908***	0.0447				
R-BB8	θ	6***	0.6144	0.4993	0	0	−2136.598
	δ	0.6665***	0.0429				
Time-varying T copula	α_c	0.0007	0.0020				−2774.11
	β_c	0.9815***	0.0031				
	γ_c	0.6346***	0.0973				
	φ	8.4724***	0.3135				

* Indicates statistical significance at the 5 % level.
** Indicates statistical significance at the 1 % level
*** Indicates statistical significance at the 10 % level
Source computation

it has the smallest AIC value. ρ is Pearson's correlation coefficient which is used to describe the dependence structure in Gaussian and T copulas. The estimated coefficient ρ in T copula model is significant, which shows silver and gold has significant linear relationship. The value of Kendall's tau is close to 0.5, which indicates that the rank correlation between silver and gold is not very strong. Moreover, the estimated parameters of the upper and lower tail dependence are equal to 0.4078, which implies that the dependence between silver and gold returns during bull markets and bear markets is the same. Since the T copula reports the best explanatory ability of all selected static copula models, we choose the time-varying T copula to describe the dependence structure of silver and gold returns series. Comparing static T copula with the time-varying T copula, the time-varying T copula has a better performance in terms of AIC. As can be seen in the time-varying T copula equation, the autoregressive parameter β is approximate to 1, indicating a high degree of persistence concerning the dependence structure between silver and gold returns. The latent parameter γ is significant, which exhibits the newest return information is a relevant measure.

Figure 1 plots estimated and forecasted dependencies and tail dependencies of gold and silver based on the time-varying T copula model. For the dependence between gold and silver, we can see an obvious fluctuation during 2000–2002, in this period, the volatility interval is around 0.65—−0.15. Particularly, in the middle and later periods of 2001, their dependence approximated to 0, and even less than 0, which showed a very weak dependence and negative dependence. Following the stock market crash of NASDAQ in April 2000, American economy gradually fell into crisis, which implicated many countries, whose economy slowed down. Meanwhile, the oil price increased. After the 9.11 happened in America in 2001, the whole world economy went into growth recession. Because of special physical and chemical properties, silver was widely used in electricity, medicine, chemistry, optics materials industries. Due to the slowed down industry development, the industry demand for silver deduced, which caused the price of silver to decline. Unlike silver, gold does not have many industrial uses; even the industry development slowed down, the price of gold did not changed like silver. Apart of this, during the economic recession period, central banks were enthusiastic about increasing gold reservation which urged gold price rising. Furthermore, People prefer investing in gold to investing in stock markets which also resulted in increasing price increasing of gold. Eventually, gold

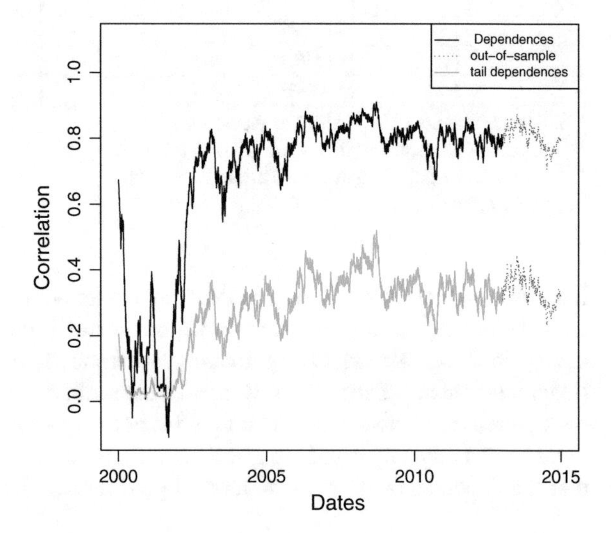

Fig. 1 Estimated and forecasted dependencies and tail dependencies. Source: computation

price increased, while silver price unchanged or even decreased during the world economic recession period, their dependence to be very weak or even negative. After 2002, the dependence of gold and silver returned to a point about 0.8 in 2003, then, their dependence exhibited a stable volatility with the fluctuation around 0.65–0.85 during 2003–2014. As we can see the plots of the forecasted dependence between gold and silver in the figure, it displayed a gentle volatility which is similar to the volatility during 2003–2014, which means that our model for forecasting the dependence of gold and silver shows a good fit. In terms of tail dependence of gold and silver, it consists of two parts during 2000–2014. The tail dependence between gold and silver was decreasing from 2000 to 2002. In this period, their tail dependence was closed to 0, which means the returns of gold changed, the returns of silver didn't change with similar reasons mentioned above. Their returns are independent. After 2002, their tail dependence parameter rose to the point 0.35 in 2003, then, the parameter of tail dependence between gold and silver fluctuated in the interval of 0.2–0.5 during 2003–2014, which indicates that their returns exist as dependence, or their returns simultaneously increased or decreased. The forecasted tail dependence resembles the tail dependence during 2003–2014, which shows a float volatility. These proved that our model matched our study very well.

Figure 2 shows the estimated VaR by using time-varying T copula with GARCH-SSTD model at $\alpha = 0.1$, $\alpha = 0.05$ and $\alpha = 0.01$. As it can be seen, most of the portfolio returns are located above the VaR curves, and the portfolio returns of VaR gradually decreases with the increase of confidence level, both of which are consistent with the reality. The number of violations of the VaR estimation and VaR and ES backtests across copula-based portfolio are shown in Table 6. We can see that the number of violations by the time-varying T copula-GARCH model is approximate to the expected number of violations. Moreover, the Percentage of Failure (PoF) and the Conditional Coverage Likelihood Ratio (CCLR) tests [12] are used to evaluate

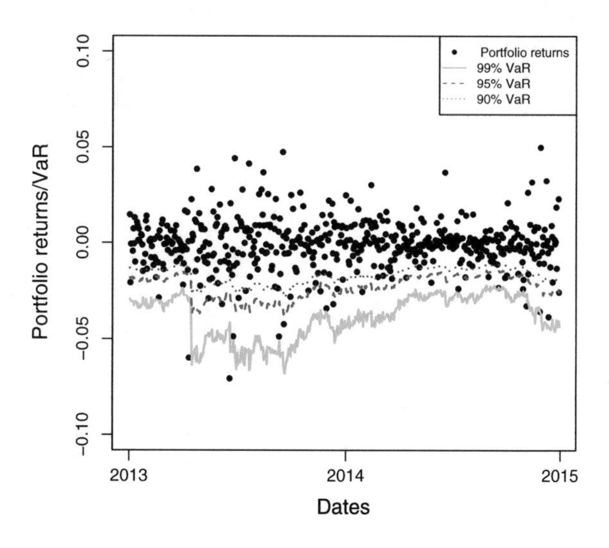

Fig. 2 Estimated VaR using the time-varying T copula with GARCH-SSTD model

Table 6 Backtesting VaR based on the time-varying T copula-GARCH models

	VaR		
	90 %	95 %	99 %
Violations	48	30	6
Expected num.	52	26	5
POF test stats.	0.3499	0.6185	0.1185
CCLR stats.	5.7666	0.5069	1.0904

the performance of the time-varying T copula-GARCH model. Both PoF and CCLR statistics are less than critical values 3.84 and 5.99, respectively. These mean they do not reject the null hypotheses, thus the time-varying T copula-GARCH model is correct for gold and silver returns in risk management. As everyone knows Basel committee replaced VaR with ES in 2013, which means ES is more powerful as complementary tool than VaR. Figure 3 shows the ES plots we estimated using the time-varying T copula with GARCH-SSTD model with a 99 %, 95 % and 90 % levels of confidence. The portfolio return of ES with a 90 % confidence is certainly lower than that with a 95 % confidence; and with a 95 % confidence is certainly lower than that with a 99 % confidence. From the plots, we can see clearly that in the middle and later periods of 2013, all the paths exhibited huge fluctuation. The reasons for such a big volatility are three-fold: first, after the crisis of Cyprus, the Eurozone governments, like Germany came up with harsh additions for giving help to Cyprus. This led the government of Cyprus to declare the selling of gold reserve to raise money to enable debt repayment. Other European countries who sank into the same situation like Cyprus also took the same measure. Large-scale gold underselling

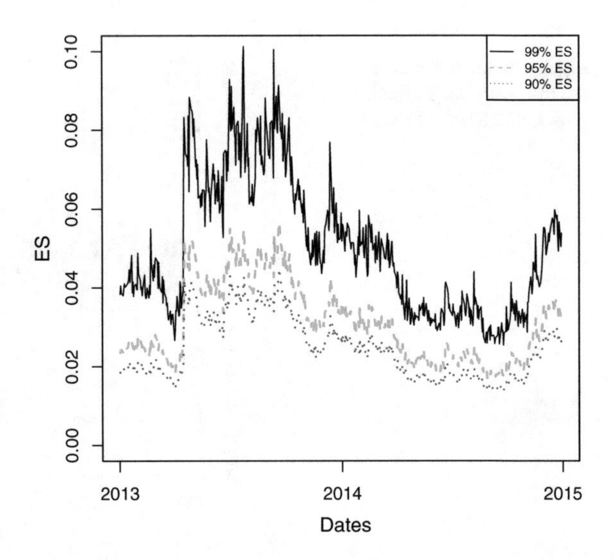

Fig. 3 Estimated expected shortfall using the time-varying T copula with GARCH-SSTD model

urged gold price to drop down. Second, the gold price is in US dollar, meaning that the changing exchange rate against US dollar will somehow influence gold price. After the American economic recovery, people sold gold to get money to invest in stock market. With a hut of gold in the market, gold price went down. Third, a great quantity fund of gold ETF flowed out from the gold market, and this decreased the gold price. The silver price is more sensitive than gold price in terms of precious metal investment. So when faced the situations mentioned above, silver price decreased more significantly than gold price. After 2014, the ES tracks of three different confidence levels display slight volatility, and all tracks present similar fluctuation trajectories, indicating that our model is best suited for calculating ES with respect to a portfolio of gold and silver.

4 Conclusions

Because gold and silver have important and various industrial uses in jewelry, medicine, machine, and electronics, gold and silver investments are always interested by policy makers, portfolio managers and manufacturers. Quantifying the volatility and dependence changes is fundamental in designing risk management strategies for gold and silver investments.

This paper analyzed the VaR and ES predictions of gold and silver with long memory volatility models under the Skewed Student-t distribution. Our empirical results are threefold. First, we applied many GARCH family models to fit our data. According to the model selection criterions, the NAGARCH and TGARCH models with Skewed Student-t distribution were found to have the best fit for matching the conditional variances of gold and silver, respectively, this indicated that the returns of gold and silver have asymmetry and leverage effect. Second, by comparing the performances of selected copula models in our study, we found that the time-varying T copula model is very well suited for our data. For the dependence of gold and silver, we can see an obvious fluctuation during 2000–2003, and their dependence exhibited a stable volatility after 2003. Third, the time-varying T copula-GARCH with Skewed Student-t distribution has been employed to compute for VaR and ES of gold and silver returns. VaR backtesting proved that the time-varying T copula based-GARCH model is correct and accurate to estimate and forecast the VaR and ES of gold and silver returns.

References

1. Alexios, G.: Introduction to the rugarch package (2014). http://cran.r-project.org/web/pakages/rugarch/vignettes/introduction_to_the_rugarch_pakage.pdf
2. Aloui, R., Aïssa, M.S.B., Nguyen, D.K.: Conditional dependence structure between oil prices and exchange rates: a copula-GARCH approach. J. Int. Money Financ. **32**, 719–738 (2013)
3. Bollerslev, T.: Generalized conditional heteroskedasticity. J. Econ. **31**, 307–327 (1986)

4. Chang, C.L., McAleer, M., Tansuchat, R.: Crude oil hedging strategies using dynamic multivariate GARCH. Energy Econ. **33**, 912–923 (2011)
5. Embrechts, P., McNeil, A.J., Straumann, D.: Correlation and dependence in risk management: properties and pitfalls. In: Dempster, M.A.H. (ed.) Risk Management: Value at Risk and Beyond, Cambridge University Press, Cambridge (2002)
6. Engle, R.F., Ng, V.K.: Measuring and testing the impact of news on volatility. J. Financ. **48**, 1749–1778 (1993)
7. Fernandez, C., Steel, M.F.J.: On Bayesian modeling of fat tails and skewness. J. Am. Stat. Assoc. **93**, 359–371 (1998)
8. Glosten, L.R., Jagannathan, R., Runkle, D.E.: On the relation between the expected value and the volatility of the nominal excess return on stocks. J. Financ. **48**, 1779–1801 (1993)
9. Jose, J.Q.M., Jose, A.R.L., Manuel, U.F.: What are copulas? Monografías del Semin. Matem. García de Galdeano. **27**, 499–506 (2003)
10. Joe, H.: Multivariate Models and Dependence Concepts. Chapman & Hall Ltd, London (1997)
11. Joe, H.: Asymptotic efficiency of the two-stage estimation method for copula-based models. J. Multivar. Anal. **94**, 401–419 (2005)
12. Kupiec, P.: Techniques for verifying the accuracy of risk measurement models. J. Deriv. **3**, 73–84 (1995)
13. Ling, S.Q.: Self-weighted and local quasi-maximum likelihood estimators for ARMA-GARCH/IGARCH models. J. Econ. **140**, 849–873 (2007)
14. Nelsen, R.B.: An Introduction to Copulas. Springer, New York (1999). ISBN 0-387-98623-5
15. Patton, A.J.: Applications of Copula Theory in Financial Econometrics. A dissertation submitted in partial satisfaction of the requirements for the degree of Doctor of Philosophy in Economics (2002)
16. Rockafellar, R.T., Uryasev, S.: Conditional value-at-risk for general loss distributions. J. Bank. Financ. **26**, 1443–1471 (2002)
17. Schwert, G.W.: Stock Volatility and the Crash of "87". Rev. Financ. Stud. **3**, 77–102 (1990)
18. Sklar, A.: Fonctions de repartition an dimensions et leurs marges. Publications de l'Institut Statistique de l'Universite de Paris **8**, 229–231 (1959)
19. Sriboonchitta, S., Nguyena, H.T., Wiboonpongse, A., Liu, J.X.: Modeling volatility and dependency of agricultural price and production indices of Thailand: Static versus time-varying copulas. Int. J. Approx. Reason. **54**, 793–808 (2013)
20. Taylor, S.: Modeling Financial Time Series. Wiley, Chichester (1986)
21. Wu, C.C., Chung, H., Chang, Y.H.: The economic value of co-movement between oil price and exchange rate using copula based GARCH models. Energy Econ. **34**, 270–282 (2012)
22. Zakoian, J.M.: Threshold heteroskedastic models. J. Econ. Dyn. Control **18**, 931–955 (1994)

Efficient Frontier of Global Healthcare Portfolios Using High Dimensions of Copula Models

Nantiworn Thianpaen, Somsak Chanaim, Jirakom Sirisrisakulchai and Songsak Sriboonchitta

Abstract This paper aims to find the optimal Global Healthcare Portfolios at different levels of risks and returns to obtain the efficient frontier. The risks are measured by expected shortfall. The dependency of selected stocks in portfolios cannot be ignored. The high-dimension copula-models are used to capture the dependency parameters of the selected stocks. Five largest market capitalization stocks in the global healthcare sector are selected for this analysis. According to the Akaike Information Criterion (AIC), the empirical results show that *t*-copula is better fitted between the *t*- and the Gaussian copulas. Based on the *t*-copula, the result of this study which is the efficient frontier of the global healthcare portfolios is finally shown in Table 4 for related decision makers.

Keywords Copulas · Expected shortfall · Efficient frontier

1 Introduction

This research is devoted to the analysis of the healthcare sector which is a bigger market in the United States than others, globally. The signals from this country will have remarkable effects in healthcare sectors throughout the world. As a result, we selected the five largest market capitalization healthcare equities from the New York Stock Exchange (NYSE). They consist of healthcare stocks which are (1) Abbott Laboratories (abt), (2) GlaxoSmithKline plc (gsk), (3) Johnson & Johnson (jnj), (4) Novartis AG (nvs), and (5) Stryker Corporation (syk). The S&P 500 healthcare index

N. Thianpaen (✉)
Faculty of Management Science, Suratthani Rajabhat University,
Surat Thani 84100, Thailand
e-mail: nantiworn@outlook.com

N. Thianpaen · J. Sirisrisakulchai · S. Sriboonchitta
Faculty of Economics, Chiang Mai University, Chiang Mai 52000, Thailand

S. Chanaim
Faculty of Science, Chiang Mai University, Chiang Mai 52000, Thailand

V.-N. Huynh et al. (eds.), *Causal Inference in Econometrics*,
Studies in Computational Intelligence 622, DOI 10.1007/978-3-319-27284-9_23

(Sphc) is required to be the benchmark of the global healthcare market in this research study.

The objective of this study is to obtain the efficient frontier of those five largest market capitalization healthcare stocks. To estimate the expected return of each stock, we used the Capital Asset Pricing Models (CAPMs) with the S&P 500 healthcare index (Sphc) as the explanatory variable. Since the five stock returns are normally interdependent, to capture the dependency parameter of these stock returns, the Gaussian and the t-copulas are usually employed. Sriboonchitta et al. [4] focused on vine copula-cross entropy evaluation of a dependence structure to reasonably explain more appropriate ordering of the vine copula, for financial risks in agricultural commodities in terms of index return figures. Other studies by Sriboonchitta et al. [5–8] made use of copulas. However, it is better to use the high dimension copula to analyze the efficient frontier since we could obtain all parameters involved in this analysis simultaneously.

From the appropriate copula model selected by AIC, the expected shortfalls (ESs) were estimated. By minimizing the portfolio expected shortfalls with respect to the weight for each stock of the portfolio subject to each level of the expected portfolio return, the efficient frontier of the portfolios is obtained.

This paper is organized as in the following. Section 2 provides the methodology of this study. Section 3 describes the data background of this study. Descriptive statistics of the data in this study are presented in Sect. 4. The empirical results are shown in Sect. 5. Conclusion is drawn in Sect. 6.

2 Methodology

2.1 Capital Asset Pricing Model (CAPM)

The expected excess return for each stock of interest is estimated by

$$r_{it} = \alpha_i + \beta_i Sphc + \varepsilon_{it},\ for\ i = 1, ..., 5, \tag{1}$$

where $r_{1t} = abt,\ r_{2t} = gsk,\ r_{3t} = jnj,\ r_{4t} = nvs,\ and\ r_{5t} = syk$ are stock excess returns. The *Sphc* represents the market excess returns of the S&P 500 healthcare index. The risk free rate used to compute the excess return in this paper is the three-month U.S. Treasury Bill. The market coefficient β is the representative of the risk measure. The smaller value of β, the lower risk of the asset. $\varepsilon_{i,t} \sim N(0, \sigma^2)$ or white noise for each model. The marginal distribution for each stock excess return is investigated and then used to fit the Gaussian and *t*-copulas, after Embrechts et al. [2].

2.2 Copulas

Both the Gaussian and t-copulas are used to fit our data in this study. Following Embrechts et al. [2], the Gaussian copula is the multivariate normal distribution over $\Re^d$ by taking the probability integral transformation. Moreover, the copula reaches to the results of random variables which have continuous distributions and a uniformed distribution [0,1]. The Gaussian copulas can be written in the form of

$$C_R^{Gaussian}(u) = F_R^d(F_1^{-1}(u_1), \ldots\ldots, F_d^{-1}(u_d)), \tag{2}$$

where F_R^d is a joint density function of multivariate normal distribution function for d dimensions.

$$C_R^{Gaussian}(u) = \frac{1}{\sqrt{detR}} exp\left(-\frac{1}{2}\begin{pmatrix} F^{-1}(u_1) \\ \cdot \\ \cdot \\ \cdot \\ F^{-1}(u_d) \end{pmatrix}' \cdot (R^{-1} - I) \cdot \begin{pmatrix} F^{-1}(u_1) \\ \cdot \\ \cdot \\ \cdot \\ F^{-1}(u_d) \end{pmatrix}\right), \tag{3}$$

where Eq. (3) is the density equation for d dimensions of the Gaussian copulas, I is the identity matrix, and R the matrix of correlation.

Compared with Gaussian copula, by Demarta and McNeil [1], the t-copula can detect the symmetric extreme dependence. The t-copula can be evaluated in the form of,

$$C_R^t(u) = F_R^d(F_1^{-1}(u_1), \ldots\ldots, F_d^{-1}(u_d)), \tag{4}$$

where Eq. (4) is the joint density function of t-copulas for d dimensions.

$$C_{\nu,P}^t(u) = \int_{-\infty}^{t_\nu^{-1}(u_1)} \cdots \int_{-\infty}^{t_\nu^{-1}(u_d)} \frac{\Gamma\left(\frac{\nu+d}{2}\right)}{\Gamma\left(\frac{\nu}{2}\right)\sqrt{(\pi\nu)^d|R|}} \cdot \left(1 + \frac{x'R^{-1}x}{\nu}\right)^{-\frac{\nu+d}{2}} dx, \tag{5}$$

where Eq. (5) is the density equation for d dimensions of the t-copulas, ν is the degree of freedom, and R the matrix of correlation.

2.3 Estimation of Efficient Frontier

To estimate the efficient frontier, investors need to minimize portfolio with respect to expected returns. Therefore, the following procedure is utilized.

$$Min \quad ES_\tau(r_i) = E[r_i \mid r_i \leq r_\tau], \tag{6}$$

$$subject \quad to \quad r_i = w_1 r_{(1,t+1)} + w_2 r_{(2,t+1)} + \cdots + w_n r_{(n,t+1)}, \tag{7}$$

where r_i are expected returns on asset i for $i = 1, 2, \ldots, n$, r_τ is the lower τ-*quantile*, $r_{(i,t+1)}$ is the return on each asset at time $t + 1$, and $w_1 + w_2 + \cdots + w_n = 1$, if $n = 5$ then $w_1 + w_2 + w_3 + w_4 + w_5 = 1$.

We prefer to consider the left tail, or negative risk, on "long positions" which are the norm of this literature. Finally, we build efficient frontiers for copula models with weights allocated to ten portfolios of the five assets.

3 Data Background

The research investigates the financial investment of the United States in its healthcare market, where is the most remarkable area in the world. The healthcare data series are extracted from creditable databases such as Thomson Reuters Eikon and Yahoo Finance websites. The series includes potential data sets, which are Abbott Laboratories (abt), Glaxo Smith Kline plc (gsk), Johnson & Johnson (jnj), Novartis AG (nvs), and Stryker Corporation (syk). The interest rate on a three-month maturity for the U.S. Treasury Bill (T-Bill) is substituted for the risk-free rate for CAPMs. The S&P 500 healthcare price index (Sphc) is the representative of the market portfolio for this research. Finally, every data set has a "weekly" set of data from April 4, 2005 to April 17, 2015.

4 Descriptive Statistics of the Data

Table 1 presents the summary statistics of the interest rate on a three-month maturity for the U.S. Treasury Bill (T-Bill), "the market return" (rt-Sphc), and other stock returns (including rt-abt, rt-gsk, rt-jnj, rt-syk, and rt-nvs). Evidently, every data set

Table 1 Typical statistical test of log return data sets

Statistic value	T-Bill	rt-Sphc	rt-abt	rt-gsk	rt-jnj	rt-nvs	rt-syk
Mean return	1.3596	0.1889	0.1889	0.0932	0.1291	0.2077	0.1439
Standard error (se)	0.0837	0.1166	0.1166	0.1296	0.0924	0.1167	0.1505
Median	0.1270	0.1998	0.1998	0.1826	0.2399	0.3061	0.2210
Standard deviation (sd)	1.9105	2.6617	2.6617	2.9589	2.1087	2.6638	3.4350
Skewness	1.0148	−0.4096	−0.4096	−0.7682	−0.8642	−0.8400	−0.9095
Excess kurtosis	−0.7194	4.4579	4.4579	4.6028	9.6758	8.5046	4.2880
Min. value	0.0030	−16.1046	−16.1046	−19.3988	−16.9407	−20.0215	−17.6106
Max. value	5.1890	11.4700	11.4700	10.4391	11.4894	14.9874	11.5681
Observations	521	521	521	521	521	521	521

has 521 observations in the analysis. The market return and five stock returns have a range of value from −20.02 to 14.99 %. Additionally, most mean returns are also close to zero which possibly reflects a normal distribution for each data set. However, the excess kurtosis statistic shows the "skewed Student t distribution" (see Hansen [3]). The statistic kurtosis as in Table 1 tells the shape of an asymmetric probability distribution in residuals. Nonetheless, these log return data sets are prepared to apply for CAPMs.

5 Empirical Applications

We analyze five healthcare stocks which consist of the world's greatest healthcare markets for capitalization, from New York Stock Exchange (NYSE), to exhibit our goal model. Weekly excess returns from five stocks have always relied on the market excess return. The weekly rate of returns on the 3–month U.S. Treasury Bill is also the risk-free proxy rate to cope with CAPMs.

5.1 Estimation Results of CAPMs

Table 2 demonstrates the statistical inference for the capital asset pricing models (CAPMs) of the Sphc or market excess returns and other stock excess returns. Coefficients or $\beta's$ are obtained from Maximum likelihood estimation (MLE) in CAPMs.

Table 2 Statistical inference for capital asset pricing models

CAPMs*	Sphc versus abt	Sphc versus gsk	Sphc versus jnj	Sphc versus nvs	Sphc versus syk
Intercept (α)	0.0005	−0.0005	−1.6323e-05	0.0008	−0.0003
Coefficient (β)	0.8144	0.8054	0.7412	0.7544	1.0509
Standard error (se)	0.0379	0.0459	0.0255	0.0403	0.0489
t-Statistic	21.4700	17.5420	29.1030	18.7030	21.4740
p-Value	1.1894e−73	1.9509e−54	3.8633e−111	4.7302e−60	1.1386e−73
RMSE	0.0194	0.0235	0.0130	0.0206	0.0250
RSQ	0.4700	0.3720	0.6200	0.4030	0.4700
Adjust RSQ	0.4690	0.3710	0.6190	0.4010	0.4690
Kolmogorov–Smirnov (KS) test*	0.1170	0.1004	0.0833	0.2071	0.1690

Note CAPMs* were computed from log returns, RMSE = root mean square error, RSQ = R-squared, and KS test* offers a p-value

The p-values for every CAPM are highly significant at lower than 5 % significance level in every model. This means that market excess returns have a high correlation among stock excess returns of abt, gsk, jnj, nvs, and syk. Moreover, KS statistic values in the table indicate the normal distribution of residuals. Hence, these results mean that we receive all potential CAPMs.

Figure 1 exhibits scatter plots of the market returns (rt-Sphc) versus five stock returns (rt-abt, rt-gsk, rt-jnj, rt-nvs, and rt-syk exclusively). All the plots appear to signify the presence of correlations between rt-Sphc and stock returns of the five items definitively.

In other words, these diagrams provide distinctive evidence of a significant correlation between the market return and each stock return for examining the dependence in the copula modelings. For (1a) rt-Sphc versus rt-abt, (1b) rt-Sphc versus rt-gsk, (1c) rt-Sphc versus rt-jnj, (1d) rt-Sphc versus rt-nvs, and (1e) rt-Sphc versus rt-syk.

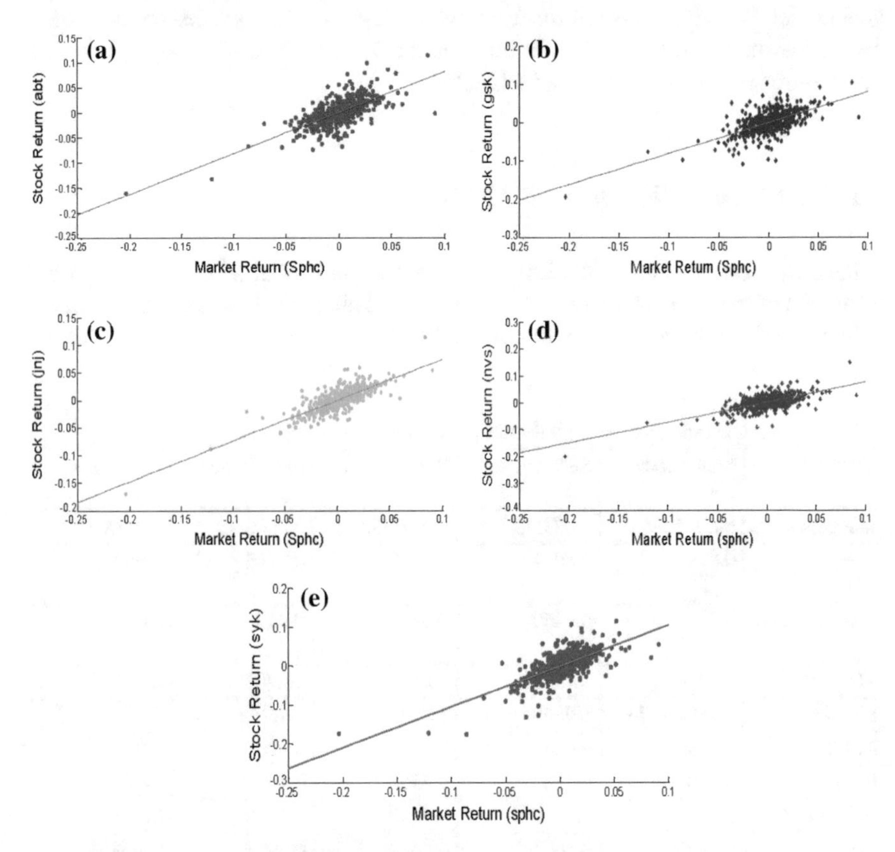

Fig. 1 The scatter plots of the market returns versus five stock returns

5.2 Copula Parameter Estimation Results

Table 3 describes the parameter correlation ($\rho_{i,j}$) in the stock excess returns of the Gaussian copula parameters and the t-copula parameters, respectively.

The AIC value of the Gaussian copula model is 0.572. The AIC value of the t-copula model is −58.50 and degree of freedom (ν) is 7.31. Table 3 is the result of bivariate copula parameters. All of these values will be later employed to find the optimal solutions in regards to the minimized loss of expected returns.

Table 3 Copula parameter values

(a) the Gaussian copula

$\rho_{i,j}$	Parameters	std. err.	z stat.	p-value
1,2	0.0296	0.0449	0.6590	0.5099
1,3	0.0493	0.0446	1.1050	0.2693
1,4	0.0135	0.0448	0.3000	0.7638
1,5	−0.0118	0.0448	−0.2630	0.7922
2,3	−0.1527	0.0433	−3.5260	0.0004***
2,4	0.0012	0.0449	0.0270	0.9783
2,5	−0.0733	0.0445	−1.6470	0.0996
3,4	0.0699	0.0444	1.5750	0.1152
3,5	0.0087	0.0447	0.1940	0.8461
4,5	−0.0315	0.0446	-0.7050	0.4807

(b) the t-copula

$\rho_{i,j}$	Parameters	std. err.	z stat.	p-value
1,2	0.0181	0.0473	0.3820	0.7021
1,3	0.0418	0.0480	0.8700	0.3841
1,4	0.0065	0.0472	0.1370	0.8908
1,5	−0.0068	0.0488	−0.1400	0.8887
2,3	−0.1443	0.0452	−3.1930	0.0014**
2,4	−0.0101	0.0460	−0.2190	0.8270
2,5	−0.0678	0.0472	−1.4350	0.1513
3,4	0.0599	0.0469	1.2770	0.2015
3,5	0.0477	0.0489	0.9760	0.3291
4,5	−0.0025	0.0479	-0.0510	0.9592

Significant codes "***" means significant at 0.001, "**" means significant at 0.01, and "*" means significant at 0.05

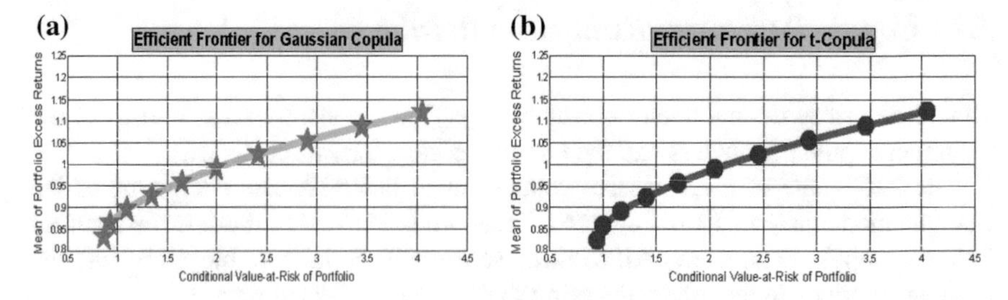

Fig. 2 The alternative efficient frontiers of the mean-CVaR for the market portfolio returns

5.3 *Efficient Frontiers from Copula Models*

In Fig. 2a, b the efficient frontiers of two copula models are at each level of the expected portfolio excess returns presented for two diagrams.

These results are obtained by the minimization for the mean-CVaR of the portfolio. It gives a thorough insight into how the mean of thes portfolio excess returns changes when the mean-CVaR of the portfolio changes. Therefore, the optimization of the portfolio takes the mean-CVaR model by getting the efficient frontier with regard to the expected returns.

The copulas which have been used for building efficient frontiers are (Fig. 2a) for the Gaussian copula and (Fig. 2b) for the *t*-copula. Then, we obtain the efficient frontiers resulting from the multivariate probability of the copula models. Since the *t*-copula is the best fit, it is recommended to use the efficient frontier from *t*-copula (Fig. 2b) for decision making in investment.

5.4 *Optimal Weights of Portfolio in the T-Copula Model*

The weight for each stock of the efficient portfolio is obtained from Eq. (6) and (7) with respect to the *t*-copula model. The portfolio of assets has its own expected return and its corresponding ES or CVaR is calculated at 95 % confidence level.

Table 4 could be read as in the following: for instance, the first row of the table means that it is recommended to allocate 17.38, 15.67, 39.56, 14.56, and 12.83 % for stock abt, gsk, jnj, nvs, and syk, respectively, to get the expected return of 0.8121 percent per day with the expected shortfall of 0.9440 percent at the confidence interval of 95 %. The last row of the table means that it is recommended to invest 100 % in syk stock to obtain 1.0203 % of the daily expected return with the higher expected shortfall of 4.0504 %.

Table 4 Optimal portfolio weights on assets for *t*-copula

Portfolio	abt	gsk	jnj	nvs	syk	Expected return	ES or CVaR
p_1	0.1738	0.1567	0.3956	0.1456	0.1283	0.8121	0.9440
p_2	0.1961	0.1392	0.3061	0.1521	0.2065	0.8370	1.0026
p_3	0.2156	0.1226	0.2212	0.1540	0.2865	0.8617	1.1661
p_4	0.2293	0.1054	0.1388	0.1586	0.3679	0.8864	1.4072
p_5	0.2387	0.0894	0.0592	0.1618	0.4509	0.9111	1.7034
p_6	0.2441	0.0633	0.0000	0.1542	0.5384	0.9351	2.0445
p_7	0.2275	0.0154	0.0000	0.1154	0.6418	0.9582	2.4525
p_8	0.1898	0.0000	0.0000	0.0555	0.7547	0.9795	2.9252
p_9	0.1261	0.0000	0.0000	0.0000	0.8739	1.0005	3.4583
p_{10}	0.0000	0.0000	0.0000	0.0000	1.0000	1.0203	4.0504

6 Conclusion

This paper aims to find the optimal Global Healthcare Portfolios at different levels of risks and returns to obtain the efficient frontier. The interest rate on a three-month U.S. Treasury Bill is also considered to be the representative of the risk free rate in CAPMs. CAPMs help indicative the relationship of all parameter values between the market excess return (Sphc) and stock excess returns (abt, gsk, jnj, nvs, and syk respectively). Although beta parameters apparently do not tell the full efficiency of the systematical risk (a) bit low in sensitivity as regards market movements), the Kolmogorov-Smirnov test readily provides residuals having normality distribution. The *t*-copula is the best model compared to the Gaussian copula because it has the lower AIC. Copulas are useful to compute ES or CVaR issuing novel, and are strong risk-management applications for investors to consider the minimum loss at the given level of expected return of the portfolios. The efficient frontiers defined by curves of optimal risk-return portfolios resulted by practically minimizing ES or CVaR. These efficient frontier curves illustrate the boundary which is the best opportunity of portfolio. They provide the highest expected returns on healthcare stock prices with respect to any defined risk levels. Investors have different utility functions to satisfy their selections with any expected return levels or mean of portfolio levels. Asset's portfolio weights are allocated to ten different portfolios in the copula models that investors have to take into account which side they want to take between the long position and the short-sale. We thus will let them choose to maximize their profit from investments in healthcare equities employing Table 4 optimal weights from the copula model. Consequently, these outcomes can be useful for investors handling risk management in healthcare stocks.

Acknowledgments We are pleased to extend our gratitude to Lhoyd Castillo, the Customer Support Executive, Investment Management, from Thomson Reuters, who provided an underlying track and went searching high and low to acquire the data required to support this contribution. Our thanks go also to Chris Snow, the researcher who participated, and suggested the inclusion of some points that were missing while this research study was being conducted. Finally, we are kindly thankful to the referees to provide beneficial suggestions to this research.

References

1. Demarta, S., McNeil, A.J.: The t Copula and related Copulas. Int. Stat. Rev. **73**, 111–129 (2004)
2. Embrechts, P., Lindskog, F., McNeil, A.: Modelling Dependence with Copulas and Applications to Risk Management. Working Paper, ETH Zurich (1999)
3. Hansen, B.: Autoregressive conditional density estimation. Int. Econ. Rev. **35**, 705–730 (1994)
4. Sriboonchitta, S., Liu, J., Wiboonpongse, A.: Vine copula-cross entropy evaluation of dependence structure and financial risk in agricultural commodity index returns. Adv. Intell. Syst. Comput. **251**, 275–287 (2014)
5. Sriboonchitta, S., Chaiboonsrib, C.: The dynamics Co-movement toward among capital markets in aSEAN exchanges: C-D Vine Copula approach. Procedia Econ. Financ.:Sci. Direct **5**, 696–702 (2013)
6. Sriboonchitta, S., Nguyen, H.T., Wiboonpongse, A., Liu, J.: Modeling volatility and dependency of agricultural price and production indices of Thailand: Static versus time-varying copulas. Int. J. Approx. Reason.:Sci. Direct **54**, 793–808 (2013)
7. Sriboonchitta, S., Liu, J., Kreinovich, V., Nguyen, H.T.: A Vine Copula approach for analyzing financial risk and Co-movement of the Indonesian, Philippine and Thailand stock markets. Adv. Intell. Syst. Comput.:SpringerLink **251**, 245–257 (2014)
8. Sriboonchitta, S., Praprom, C.: Investigation of the dependence structure between imports and manufacturing production index of Thailand using Copulas-Based GARCH Model. Thai J. Math. 73–90 (2014)

Analyzing MSCI Global Healthcare Return and Volatility with Structural Change Based on Residual CUSUM GARCH Approach

Nantiworn Thianpaen and Songsak Sriboonchitta

Abstract This study aims to analyze the Morgan Stanley Capital International (MSCI) world return and volatility of the healthcare price index using daily time series data. Since the data of MSCI healthcare returns cannot be described by linear models, the residual CUSUM GARCH(1,1) model is applied in this paper. The CUSUM test is used to estimate the optimal change point. The findings of this paper are (1) the estimated point is at day 1,201 of the entire daily data set of 4,209 observations; (2) if the change point is not taken into consideration, the estimated parameters of GARCH(1,1) become $\hat{\gamma}_1 + \hat{\beta}_1 \approx 1$, i.e., we encounter the "IGARCH effect", which leads to an infinite variance for a model. The contribution of this paper is the recommendation for the analysis of the change point as the necessary condition, rather than jumping into using the whole data set to estimate all parameters of the model without testing nonlinearity, especially for financial time series data.

Keywords Change point · Residual CUSUM GARCH(1,1) model · Brownian bridge

1 Introduction

The Morgan Stanley Capital International (MSCI) world price index is based on the computed prices of healthcare stocks to represent the performance of the healthcare sector throughout 23 developed markets countries.[1] Furthermore, the index

[1]Developed markets countries are Australia, Austria, Belgium, Canada, Denmark, Finland, France, Germany, Hong Kong, Ireland, Israel, Italy, Japan, Netherlands, New Zealand, Norway, Portugal, Singapore, Spain, Sweden, Switzerland, the UK, and the US.

N. Thianpaen (✉)
Faculty of Management Science, Suratthani Rajabhat University,
Surat Thani 84100, Thailand
e-mail: nantiworn@outlook.com

N. Thianpaen · S. Sriboonchitta
Faculty of Economics, Chiang Mai University, Chiang Mai 52000, Thailand

V.-N. Huynh et al. (eds.), *Causal Inference in Econometrics*,
Studies in Computational Intelligence 622, DOI 10.1007/978-3-319-27284-9_24

relied on the MSCI Global Investable Market Indexes Methodology.[2] This index demonstrated the soaring characteristic trend of financial healthcare assets. Thus, investors might have gained benefits from healthcare capital assets for modern investment strategies. Therefore, this study aims to analyse the behavior of the returns of this healthcare sector for the purpose of investment strategies.

Since the data of MSCI world returns of the healthcare price index using daily time series data cannot be described by linear models, the residual CUSUM GARCH(1,1) model is applied for the analysis in this paper. The CUSUM test is used to estimate the optimal change point. The Quasi Maximum Likelihood (QMLE) has been employed to estimate the data from the beginning up to the change point for the first estimation. The QMLE is applied for the second part of the data after the optimal change point.

Many studies did not consider the nonlinearity of the data and jumped into using the linear time series models with Maximum likelihood estimation (MLE). Those studies include Boonyanuphong and Sriboonchitta [2, 3], Carroll and Chen [4], Chinnakum et al. [6], Do et al. [7], Kiatmanaroch and Sriboonchitta [8, 9], Puarattanaarunkorn and Sriboonchitta [13], Praprom and Sriboonchitta [15, 16], Puarattanaarunkorn and Sriboonchitta [17], Sims et al. [18], Sirisrisakulchai and Sriboonchitta [19], Tang et al. [20], Wichian et al. [21], Wichian and Sriboonchitta [22], and Xiongtoua and Sriboonchitta [23]. However, other contributions which employed the QMLE in an econometric analysis are Lee et al. [10] carried out a study of the parameter changes in time series models related to a CUSUM test by developing a more general conceptual framework. Lee et al. [11] that intended to find out the structural change in the GARCH(1,1) model based on the residual CUSUM test when its statistic values are limited to the sub Brownian bridge. Chevallier [5] detected instability and the presence of outliers in carbon prices based on OLS-Recursive-based CUSUM processes, F-statistics, residual sum of squares and monitoring recursively structural changes. Neto [14] modeled the fully modified least square (FMLS)—based CUSUM statistic which was extended to the traditional OLS-CUSUM to test the null hypothesis of smooth time-varying cointegration in the presence of a structural break. The FMLS residuals, under the null hypotheses, were derived using "Chebyshev time polynomials" specifying the time-varying coefficients.

This paper follows the concepts and notations mostly used by Lee et al. [10–12] and is organized as follows: Methodology is presented in Sect. 2; data are discussed in Sect. 3; Sect. 4 shows the empirical results; Sect. 5 provides policy implication; conclusion is drawn in Sect. 6.

2 Methodology

2.1 GARCH (1,1) Model

The GARCH model which emerged has made the model more flexible in the assumption of the volatility, as it is able to handle the uncertain movement in financial

[2]Visit https://www.msci.com/index-methodology.

security time series data with heteroscedasticity. The model is widely used to fit with dynamic volatility in financial time series. Bollerslev et al. [1] who made use of a GARCH(1,1) model, demonstrated that it would be sufficient for most financial data. Then, the AR(1)-GARCH(1,1) model would be sufficient for most financial data. The AR(1)-GARCH(1,1) may be expressed as

$$x_t = \rho_0 + \rho_1 x_{t-1} + a_t, \tag{1}$$

$$a_t = \zeta_t \sqrt{h_t}, \tag{2}$$

$$h_t = \gamma_0 + \gamma_1 a_{t-1}^2 + \beta_1 h_{t-1}, \tag{3}$$

where $x_t = x_1, \ldots, x_n$ which is *i.i.d.* data, $a_t \sim N(0, h_t)$ or a white noise process, $\zeta_t \sim i.i.d.\ N(0, 1),\ \gamma_1 + \beta_1 < 1,\ \gamma_1 > 0$, and $\beta_1 > 0$.

Since we used AR(1)-GARCH(1,1) model, the null hypothesis and alternative hypothesis could be as specific as

$$H_0 : \theta = (\rho_0, \rho_1, \gamma_0, \gamma_1, \beta_1) \text{ is constant over time}$$
$$H_1 : \theta \text{ changes to be } \theta' = (\rho_0', \rho_1', \gamma_0', \gamma_1', \beta_1')$$

We can reject H_0 when the p-value is theoretically less than the αth-quantile. The residual CUSUM test could be expressed as in Sect. 2.2.

2.2 Residual CUSUM GARCH Model

By Lee et al. [11, 12], assuming x_t satisfies the following equations

$$x_t = \rho_0 + a_t, \tag{4}$$

$$a_t = \zeta_t \sqrt{h_t}, \tag{5}$$

$$h_t = \gamma_0 + \gamma_1 a_{t-1}^2 + \beta_1 h_{t-1}, \tag{6}$$

where we provide $\phi = (\gamma_0, \gamma_1, \beta_1),\ E|a_t|^{4+\delta} < \infty$ and $E|\zeta_t|^{4+\delta} < \infty$ for some $\delta > 0$. The hypotheses to be tested are as follows:

$$H_0 : \phi = (\gamma_0, \gamma_1, \beta_1) \ \text{ is constant}$$
$$H_1 : \phi = (\gamma_0, \gamma_1, \beta_1) \ \text{ is not constant}$$

For the empirical analysis, the following statistic test is

$$\hat{T}_n = \frac{1}{\sqrt{n}\hat{\tau}} \max_{1 \le k \le n} \left| \sum_{t=1}^{k} \hat{\zeta}_t^2 - (\frac{k}{n}) \sum_{t=1}^{n} \hat{\zeta}_t^2 \right|, \tag{7}$$

where $\hat{\tau}^2 = \frac{1}{n}\sum_{t=1}^{n} \hat{\zeta}_t^4 - \left(\frac{1}{n}\sum_{t=1}^{n} \hat{\zeta}_t^2\right)^2$ is employed to find the change point with H_0 and H_1 as follows:

$$H_0 : \hat{T}_n \to \sup_{0 \le u \le 1} \left| B^0(u) \right|, \; n \to \infty$$

$$H_1 : \text{inverse to } H_0$$

where B^0 is the Brownian bridge.

Since T_n is uncorrelated from the GARCH parameters, T_n could detect the structural change in a parameter. Unfortunately, it has no information of the GARCH parameters in Eq. (7). Additionally, $\zeta_t's$ are unobservable so we have to change them by using $\hat{\zeta}_t^2 = \left(x_t - \hat{\rho}_0\right)^2 / \hat{h}_t^2$ so that we obtain the parameters of $\rho_0,\ \gamma_0,\ \gamma_1$, and β_1 in practice. These parameters have the crucial role for capturing the structural change in the GARCH parameters. Additionally, the independently identical distribution of the true residuals still exists for the stationarity. As a consequence, the CUSUM test has more powerful stability, and it is appropriate for anticipating the future change of residuals according to the function of the standard Brownian Bridge.

The new feature form of the residual CUSUM test, according to Lee et al. [11], should be

$$h_t^2 = \varphi + \gamma_1 \sum_{j=0}^{\infty} \beta_1^j a_{t-1-j}^2, \tag{8}$$

defining as,

$$\hat{h}_t^2 = \hat{\varphi} + \hat{\gamma}_1 \sum_{j=0}^{q} \hat{\beta}_1^j \hat{a}_{t-1-j}^2, \tag{9}$$

where $\varphi = \frac{\gamma_0}{1-\beta_1}$, $\hat{a}_t = x_t - \hat{\rho}_0$, when $\hat{\rho}_0,\ \hat{\varphi},\ \hat{\gamma}_1$, and $\hat{\beta}_1$ are estimates of $\rho_0,\ \varphi,\ \gamma_1$, and β_1. q is orderly positve integers.

3 Data Background

This research study determines to use the "MSCI World Health Care Index" which is designed to detect the mid-cap and the large-cap portions through 23 developed-market countries. All the index securities are separated in the health care sector proportional to the Global Industry Classification Standard (GICS), a standardized classification system which is a major group in the world. The MSCI dataset is extracted from dependable sources like the "Thomson Reuters Eikon" database. Lastly, the set of data consists of available full and complete daily data for several years, from January 31, 1996, to January 5, 2015, or 4,209 observations, totally.

3.1 Characteristics of Data

Figure 1 shows the MSCI healthcare price index dynamic data. The time series trend is prone to increase from 1996 to the recent times.

Moreover, this dataset exists in a financial sector, so, first, we intend to work with the volatility factor. Therefore, the dataset is transformed to be log returns of MSCI world healthcare price indexes.

Figure 2 is the version of the MSCI price index returns that elicits the clustering volatility.

This brings about the underlying OLS (Ordinary Least Squares) model which is not workable to cope with those outcomes that keep changing over time or are nonstationary in variance. Instead, the GARCH model will have the major role of analyzing the stochastic problem or uncertainty.

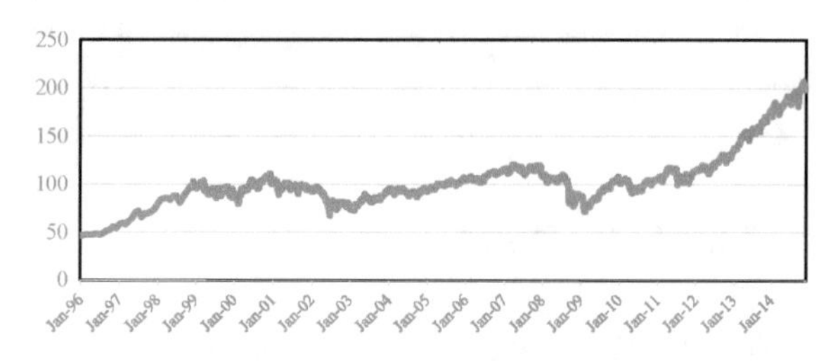

Fig. 1 The MSCI world health care price index data

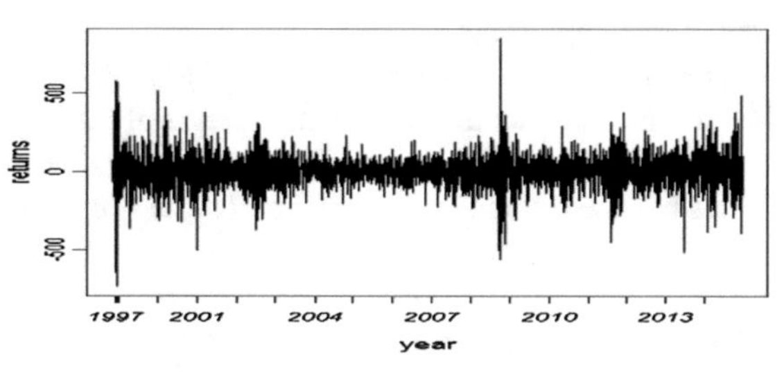

Fig. 2 The MSCI world health care price indexes in the form of log returns

Table 1 Primary statistics for MSCI health care price index returns

Statistics	rt-MSCI
Mean	0.0343
Standard error	0.0161
Median	0.0491
Standard deviation	1.0412
Sample variance	1.0841
Kurtosis	11.3776
Skewness	0.2408
Observations	4,208
Jarque–Bera statistic (p-value)	$<2.2e{-16}$

3.2 Descriptive Statistics

Table 1 shows the fundamental statistic values that will be exploited for subsequent applications. The Jarque–Bera test leads to rejecting the normality of the time series data of the MSCI World healthcare price index returns.

Remarkably, the data are not independent and identically distributed *(i.i.d.)*. Literally not normal distribution, the data contains the leptokurtic kurtosis inasmuch as the excess kurtosis is not equal to zero. These characteristics are critical as affecting the additional analysis in the next step because of the distribution not being a normal type.

4 Empirical Results

Table 2, the ADF test reveals that the log return is stationary regardless whether there was an intercept or not.

The estimated parameters of the AR(1)-GARCH(1,1) model using QMLE (Quasi Maximum Likelihood) for the whole series of data are presented in Table 3.

Upon following Tables 3 and 4, we observe that the total sum of the coefficient values between γ_1 and β_1 is less than 1. The Ljung and Box and LM Arch statistics

Table 2 Augmented Dickey Fuller (ADF) test for returns on MSCI

ADF test	p-value	Degree of freedom
Non-intercept	$< 2.2e{-16}$	−17.6788
Intercept	$< 2.2e{-16}$	−17.7504
Intercept with time trend	$< 2.2e{-16}$	−17.7469

Table 3 Parameters of AR(1)-GARCH(1,1) model

Parameter	Coeff.	Std. error	t-value	p-value
ρ_0	0.033722	0.011252	2.9970	0.00273 **
$AR(1)$	0.078383	0.016820	4.6600	3.16e−6 ***
γ_0	0.017956	0.003766	4.7680	1.87e−6 ***
γ_1	0.107788	0.011741	9.1810	<2e−16 ***
β_1	0.872857	0.012779	68.3050	<2e−16 ***

Note Log likelihood is −5187.785, AIC is 2.468054, and BIC is 2.475592
Significant codes "***" means significant at 0.001, "**" means significant at 0.01, and "*" means significant at 0.05

Table 4 Standardized residuals tests

Test statistics		p-values
Jarque Bera test	Chi^2	0.000000
Shapiro Wilk test	W	0.000000
Ljung Box test R	Q(10)	0.2989930
	Q(15)	0.3561786
	Q(20)	0.4421147
Ljung Box test R^2	Q(10)	0.8485606
	Q(15)	0.9051507
	Q(20)	0.9549539
LM Arch test R	TR^2	0.8534370

reveal that there is no serial correlation problem for residuals resulting from the p-value being more than the significance level 0.05.

However, $\gamma_1 + \beta_1 \approx 1$ which is undesirable. This undesirable result might be due to the nonlinearity of the data. It means that we might have a change point of the data. Thus, we apply the residual based cusum test (see Lee and Lee [12]) and reject $H_0 : \theta = (\rho_0, \rho_1, \gamma_0, \gamma_1, \beta_1)$ which remains the same for the whole series versus H_1 at $\alpha = 0.05$.

Figure 3 plots the entire picture of the definite location where the change point orderly exists at the point of 1,201 from the full dataset of MSCI healthcare price

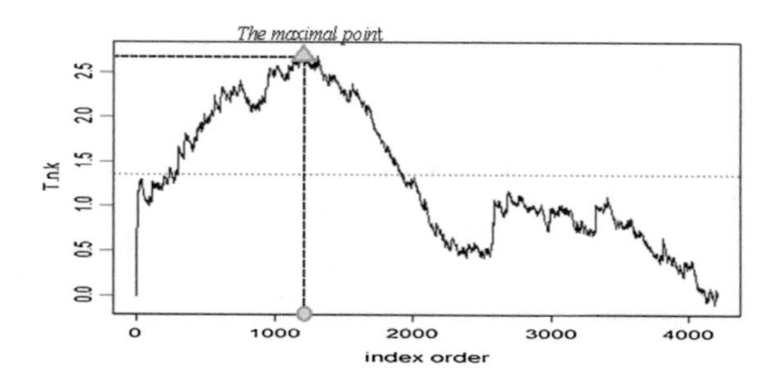

Fig. 3 The change point from the residual CUSUM test for the entire dataset

indices by the CUSUM approach dependent on ζ_t^2. Figure 3 moreover shows the $\hat{T}_{n,k}$ on the vertical axis versus the horizontal axis of the time series. The maximum $\hat{T}_{n,k}$ is at the point of 1,201 from the full data.

Therefore, we apply QMLE (Quasi Maximum Likelihood) for the first period of the daily data from day 1 to day 1,201, and also for the second period from day 1,202 to the end of the series. When the change point is estimated, the dataset is separated into two parts: before the change point up to the change point and after the change point. Then, the two datasets to fit with the AR(1)-GARCH(1,1) model are MSCI1 and MSCI2.

The estimated parameters of interest are shown in Table 5 and Table 7 for the first period and the second period respectively. It is seen that the estimated parameters are different and $\gamma_1 + \beta_1$ in both periods is less than 1, giving desirable results because the GARCH(1,1) is stationary in both periods.

In short, MSCI1 is the representative event of before the change point, and MSCI2 is the representative event for after the change point. It is evident from the data presented from Table 5 to Table 8 that the two AR(1)-GARCH(1,1) models for MSCI1 and MSCI2 do not have the IGARCH effect. Additionally, as far as the coefficients of both the AR(1)-GARCH(1,1) models are concerned, $\gamma_1 + \beta_1$ is less than 1 and less than those from the whole dataset, which is consistent with the condition sufficient for the model.

Table 5 Parameters of AR(1)-GARCH model for MSCI1

Parameter	Coeff.	Std. error	*t*-value	*p*-value
ρ_0	−0.02173	0.02903	−0.7480	0.45422
$AR(1)$	0.13945	0.03107	4.4890	7.16e−06***
γ_0	0.09154	0.02988	3.0630	0.00219**
γ_1	0.13790	0.02466	5.5920	2.25e−08***
β_1	0.80181	0.03628	22.1030	<2e−16***

Note Log likelihood is −1873.910, AIC is 3.134129, and BIC is 3.155352
Significant codes "***" means significant at 0.001, "**" means significant at 0.01, and "*" means significant at 0.05

Table 6 Standardized residuals tests

Test statistics		*p*-values
Jarque Bera Test	Chi^2	6.091866e−08
Shapiro Wilk Test	W	1.872967e−06
Ljung Box test R	Q(10)	0.08377342
	Q(15)	0.0611181
	Q(20)	0.1377031
Ljung Box test R^2	Q(10)	0.6462503
	Q(15)	0.4357432
	Q(20)	0.5892061
LM Arch test R	TR^2	0.2108268

Table 7 Parameters of AR(1)-GARCH Model for MSCI2

Parameter	Coeff.	Std. error	t-value	p-value
ρ_0	0.045779	0.012176	3.7600	0.00017 ***
$AR(1)$	0.047727	0.019752	2.4160	0.01568 *
γ_0	0.018765	0.004089	4.5890	4.46e−06 ***
γ_1	0.094460	0.013516	6.9890	2.77e−12 ***
β_1	0.874602	0.015500	56.4240	<2e−16 ***

Note Log Likelihood is −3282.658 , AIC is 2.186670 , and BIC is 2.196661.
Significant codes: "***"means significant at 0.001, "**"means significant at 0.01, and "*" means significant at 0.05

Table 8 Standardized residuals tests

Test statistics		p-values
Jarque Bera test	Chi^2	0.0000000
Shapiro Wilk test	W	1.865799e−15
Ljung Box test R	Q(10)	0.9846039
	Q(15)	0.7349946
	Q(20)	0.7376519
Ljung Box test R^2	Q(10)	0.07666263
	Q(15)	0.3046282
	Q(20)	0.3769716
LM Arch test R	TR^2	0.1898284

5 Policy Implication

Since the series of data have a change point, we should use the estimated parameters of the second period for policy strategy or some other relevant decisions. For example, the behavior of volatility has changed. Then to forecast one period ahead of the volatility, the estimated parameters from the second period are better recommended in anticipation.

6 Conclusion

Morgan Stanley Capital International (MSCI) world price index of the healthcare sector has demonstrated the soaring characteristic trend of the return. Thus, it is interesting to study the behavior of the MSCI Global Healthcare return and volatility. AR(1)-GARCH(1,1) is the model for this study. QMLE was applied to estimate all parameters of the model using the full data set of 4,209 observations of daily time series. The result reveals that the AR(1)-GARCH(1,1) becomes almost IGARCH, which is undesirable. So, we suspect that the series cannot be described by linear models. This study uses the residual based cusum test to investigate the change point of the data. The result shows that we reject H_0 : all parameters of the AR(1)-GARCH(1,1) are constant for the whole series of this data set versus H_1 : H_0 is not true. From the $\hat{T}_{n,k}$ statistic, we found that the change point is at day 1,201 of the daily time series. We then separated the data into two sets. The first set is from day 1 to day 1,201 while the second from day 1,202 to day 4,209. The QMLE was applied to estimate all parameters of the first and second sets of the data. We found that the estimated parameters of AR(1)-GARCH(1,1) reveal the desirable results: $\gamma_1 + \beta_1$ less than 1. As the series have been changed, the second estimation is recommended to explain the behavior of the series. Therefore, policy makers, investors, and other related agents are recommended to use these estimated parameters to serve their purposes.

Acknowledgments We are thankful to Professor Sangyeol Lee from Seoul National University, Department of Statistics, who inspired us by giving us the great idea of doing this research. Additionally, we then extend our gratitude to Miss Young Mai Lee, Professor Sangyeol Lee's Ph.D. student, who designed the computational part correctly. We greatly appreciate the referees comments to improve our paper substantially. Furthermore, we are indebted to Professor Thierry Denoeux from Université Technologie de Compiègne, who took the time to read this paper. Moreover, we are grateful to Lhoyd Castillo, the Customer Support Executive, Investment Management, from Thomson Reuters, who provided significant data for this work. Many thanks are extended to Puay Ungphakorn Center of Excellence in Econometrics, Faculty of Economics, Chiang Mai University.

References

1. Bollerslev, T., Chou, R.Y., Kroner, K.F.: ARCH modeling in finance: a review of the theory and empirical evidence. J. Econom. **52**, 559 (1992)
2. Boonyanuphong, P., Sriboonchitta, S.: An Analysis of volatility and dependence between rubber spot and futures prices using copula-extreme value theory: implications for risk management. In: Huynh, V.N., Kreinovich, V., Sriboonchitta, S. (eds.) Modeling Dependence in Econometrics in Advances. Intelligent systems and computing, pp. 431–444. Springer, Heidenberg (2014)
3. Boonyanuphong, P., Sriboonchitta, S.: The impact of trading activity on volatility transmission and interdependence among agricultural commodity markets. Thai J. Math. pp. 211–227 (2014)
4. Carroll, R.J., Chen, X.: Mixing and moment properties of various GARCH and stochastic volatility models. Econom. Theory **18**, 17–39 (2000)
5. Chevallier, J.: Detecting instability in the volatility of carbon prices. Energy Econ. **33**, 99110 (2011)
6. Chinnakum, W., Sriboonchitta, S., Pastpipatkul, P.: Factors affecting economic output in developed countries: a copula approach to sample selection with panel data. Int. J. Approx. Reason. **54**, 809–824 (2013)
7. Do, G.Q., Mcaleer, M., Sriboonchitta, S.: Effects of international gold market on stock exchange volatility: evidence from ASEAN emerging stock markets. Econ. Bull. **29**, 599–610 (2009)
8. Kiatmanaroch, T., Sriboonchitta, S.: Dependence structure between world crude oil prices: evidence from NYMEX, ICE, and DME Markets. Thai J. Math. pp. 181–198 (2014)
9. Kiatmanaroch, T., Sriboonchitta, S.: Relationship between exchange rates, palm oil prices, and crude oil prices: a vine copula based GARCH approach: implications for risk management. In: Huynh, V.N., Kreinovich, V., Sriboonchitta, S. (eds.) Modeling Dependence in Econometrics in Advances, Intelligent systems and computing, pp. 399–413. Springer, Heidenberg (2014)

10. Lee, S., Lee, J.: Residual based cusum test for parameter change in AR-GARCH models. In: Huynh, V.N., Kreinovich, V., Sriboonchitta, S. (eds.) Modeling Dependence in Econometrics in Advances, Intelligent systems and computing pp. 101–110. Springer, Heidenberg (2014)
11. Lee, S., Tokutsu, Y., Maekawa, K.: The CUSUM test for parameter change in GARCH (1,1) model. ResearchGate (2000). doi:10.1080/03610920008832494
12. Lee, S., Ha, J., Na, O.: The CUSUM test for parameter change in time series models. Scand. J. Stat. **30**, 781–790 (2003)
13. Neto, D.: The FMLS-based CUSUM statistic for testing the null of smooth time-varying cointegration in the presence of a structural break. Economics Letters **125**, 208211 (2014)
14. Praprom, C., Sriboonchitta, S.: Dependence analysis of exchange rate and international trade of Thailand: application of vine copulas: implications for risk management. In: Huynh, V.N., Kreinovich, V., Sriboonchitta, S. (eds.) Modeling Dependence in Econometrics in Advances, Intelligent systems and computing, pp. 229–243. Springer, Heidenberg (2014)
15. Praprom, C., Sriboonchitta, S.: Extreme value copula analysis of dependence between exchange rates and exports of Thailand: implications for risk management. In: Huynh, V.N., Kreinovich, V., Sriboonchitta, S. (eds.) Modeling Dependence in Econometrics in Advances, Intelligent systems and computing pp. 187–199. Springer, Heidenberg (2014)
16. Puarattanaarunkorn, O., Sriboonchitta, S.: Analysis of volatility and dependence between the tourist arrivals from China to Thailand and Singapore: a copula-based GARCH approach : implications for risk management. In: Huynh, V.N., Kreinovich, V., Sriboonchitta, S. (eds.) Uncertainty Analysis in Econometrics with Applications, Intelligent Systems and Computing, pp. 283–294. Springer, Heidenberg (2014)
17. Puarattanaarunkorn, O., Sriboonchitta, S.: Copula based GARCH dependence model of Chinese and Korean tourist arrivals to Thailand: implications for risk management. In: Huynh, V.N., Kreinovich, V., Sriboonchitta, S. (eds.) Modeling Dependence in Econometrics in Advances, Intelligent systems and computing pp. 343–365. Springer, Heidenberg (2014)
18. Sims, C.A., Waggoner, D.F., Zha, T.: Methods for inference in large multiple-equation Markov-switching models. J. Econom. **146**, 255–274 (2008)
19. Sirisrisakulchai, J., Sriboonchitta, S.: Modeling Dependence of Accident-Related Outcomes Using Pair Copula Constructions for Discrete Data: implications for risk management. In: Huynh, V.N., Kreinovich, V., Sriboonchitta, S. (eds.) Modeling Dependence in Econometrics in Advances, pp. 215–228. Heidenberg, Intelligent Systems and Computing, Springer (2014)
20. Tang, J., Sriboonchitta, S., Yuan, X.: A Mixture of Canonical Vine Copula GARCH Approach for Modeling Dependence of European Electricity Markets. Thai Journal of Mathematics, 165–180 (2014)
21. Wichian, A., Sirisrisakulchai, J., Sriboonchitta, S.: Copula based polychotomous choice selectivity model: application to occupational choice and wage determination of older workers: implications for risk management. In: Huynh, V.N., Kreinovich, V., Sriboonchitta, S., Suriya, K. (eds.) Econometrics of Risk, Intelligent systems and computing pp. 359–375. Springer, Heidenberg (2015)
22. Wichian, A., Sriboonchitta, S.: Econometric analysis of private and public wage determination for older workers using a copula and switching regression. Thai J. Math. pp. 111-128 (2014)
23. Xiongtoua, T., Sriboonchitta, S.: Analysis of volatility of and dependence between exchange Rate and inflation rate in Lao peoples democratic republic using copula-based GARCH approach: implications for risk management. In: Huynh, V.N., Kreinovich, V., Sriboonchitta, S. (eds.) Modeling Dependence in Econometrics in Advances, Intelligent Systems and Computing, pp. 201–214. Springer, Heidenberg (2014)

Indicator Circuits with Incremental Clustering and Its Applications on Classification of Firm's Performance and Detection of High-Yield Stocks in the Medium-Term

Natchanan Kiatrungwilaikun, Komsan Suriya and Narissara Eiamkanitchat

Abstract This paper introduces the indicator circuit with incremental clustering (ICIC) and shows that the ICIC works better than the indicator circuit with reference points (ICRP) for the evaluation of the telecommunications companies' performance presented in Suriya Int. J. Intell. Technol. Appl. Stat. vol 8, pp 103–112 (2015) [4]. Moreover, it also extends the ICIC to detect high-yield stocks in the Stock Exchange of Thailand. It classifies 134 stocks by 6 indicators; E/P ratio (the inverse of P/E ratio), BV/P ratio (the inverse of P/BV ratio), return on equity (ROE), growth of the E/P ratio, dividend growth, and ROE growth with the data at the end of 2013. It justifies the performance of the model by the yield of the stock measured at the peak price of each stock during April 1st, 2014 to March 31st, 2015. The buying date is the first trading day on the second quarter of 2014, when most of the 2013 financial statements have already been announced. Surprisingly, the method detects the low-yield stocks instead of the high-yield ones. Therefore, it acts like a warning signal to investors to avoid the low-yields.

Keywords Indicator circuit · Incremental clustering · Composite index · Investment yield · Stock investment

JEL Classification: C43 · C45 · L96

N. Kiatrungwilaikun (✉) · K. Suriya
Social Research Institute and Faculty of Economics, Chiang Mai University, Chiang Mai 50200, Thailand
e-mail: nkiatrungwilaikun@gmail.com

N. Eiamkanitchat
Social Research Institute and Faculty of Engineering, Chiang Mai University, Chiang Mai 50200, Thailand

V.-N. Huynh et al. (eds.), *Causal Inference in Econometrics*,
Studies in Computational Intelligence 622, DOI 10.1007/978-3-319-27284-9_25

1 Introduction

Indicator circuit (IC) mimics electronic circuit such that electricity flows from the input gate to several nodes inside the circuit and produce the output signal at the end. Each node in the circuit transforms the input signal into an intermediate signal to feed other nodes in the next layer. Then, all the intermediate signals flow into the last layer to produce the final output of the circuit.

Suriya [4] introduces the indicator circuit with reference points (ICRP) and indicator circuit with self-organizing map (ICSOM) that aggregate many indicators into a composite index. In that study, the index measures the efficiency of 1,000 top companies in terms of total revenue in 2003 in Thailand. Then it shows the ranking of the efficiency of 17 companies in telecommunications industry. It also compare the performance between ICRP and ICSOM, and discovers that ICRP is better than ICSOM in terms of the relevance to financial ratios and the distribution of scores of the index.

A problem occurs in the ICRP. The scores of the index are extremely close to one another. It is hard to see which company is better than another one when their scores are quite similar. This little difference of the scores make it almost impossible to separate the companies into the category of high and low efficiency.

In this study, the indicator circuit comes with the incremental clustering. It aims at breaking the scores into more details. The idea is to create more clusters than that appears in the ICRP. The incremental clustering seems to serve this purpose well. Therefore, the study employs the incremental clustering into making an indicator circuit. At the end, it compares the performance of the indicator circuit with incremental clustering (ICIC) and the ICRP.

Apart of the quantification of the efficiency of firms especially in the telecommunications industry. This study also uses the ICIC to detect the high-yield stocks in the Stock Exchange of Thailand. The critical idea behind the model is at the selected indicators; P/E ratio, P/BV ratio, return on equity (ROE), growth of the E/P ratio (the inverse of P/E ratio), dividend growth and ROE growth. These indicators arise by the recommendation from an expert who successfully gains from the stock market, Prof. Songsak Sriboonchitta. Therefore, the selection of these 6 indicators should be called the "Songsak Hypothesis". This study uses the ICIC to test the Songsak Hypothesis. If the hypothesis is true, stocks with the highest scores produced by the indicator circuits should give the highest yields.

2 Indicator Circuit with Incremental Clustering (ICIC)

The construction of ICIC follows these steps. From step 1 to 6, they are similar to the construction of ICRP in the work of Suriya [4]. The difference is at step 7 when the ICRP assign four different points but the ICIC assigns many more points due to increment of the distance to the first point. Another differnce is at step 12–14. The

ICRP adjusts the centroid of each cluster by the mean of the cluster. The number of the adjustment is n times. After the complete adjustment, the ICRP reset its weights and start over again m times. However, the ICIC does not adjust the centroid. It resets the weights m times right after the clustering in each repeat is done.

Step 1: Selection of indicators

This study selects 6 financial indicators which are current ratio, debt-equity ratio, return on asset, return on equity, net profit margin and return on investment.

Step 2: Unit of measurement

ICIC does not normalize the unit of measurement. It separates the units into 5 groups; times, per cent, days, rounds and dollars. The advantage of this method is at the intertemporal comparability. While the normalized unit depends heavily on the maximum value of each indicator of the leading company in each year (normalized to be one), it is hard to compare the composite index over time. The unnormalized unit still keeps the meaning of each financial indicator and does not depend on the leading company in each year, thus the composite index can be compared over time. However, the disadvantage of this method should be noted that the composite index may place heavier weight to an indicator with higher value. Therefore, this trade-off is at the choice of the modeller.

Step 3: Number of layers

There are 3 layers consisting of input layer, latent layer and output layer. Number of outputs is two (Y and Z) whereas number of nodes in the latent layer is 5 that accounts for five different units of measurement assigned in step 2.

Step 4: Initial weights

The initial weights range from zero to one. All the linkages between the input to latent nodes, and between the latent to output nodes are assigned the initial weights.

Step 5: Calculation of the latent variable (L) and the outputs (Y and Z)

The latent variable (L) and the outputs (Y and Z) can be calculated as

$$L_j = \sum_{i=1}^{K} w_{ji} x_i;\ j = 1, 2, \ldots, 5 \quad (1)$$

$$Y = \sum_{j=1}^{5} w_{6j} L_j \quad (2)$$

$$Z = \sum_{j=1}^{5} w_{7j} L_j \quad (3)$$

Step 6: Plot Y and Z on the Euclidean space. The space limits to the area bounded by $(Y, Z) = (0,0)$ until $(Y, Z) = (1,1)$.

Step 7: Set the first point at $(Y_1, Z_1) = (1,1)$. Then set the increment of the distance, d. The second point will be located at $(Y_2, Z_2) = (1\text{-d},1\text{-d})$. The third point will be also located at $(Y_3, Z_3) = (Y_2\text{-}d, Y_2\text{-}d)$. In general, each point $(Y_k, Z_k) = (Y_{k-1}\text{-}d, Z_{k-1}\text{-}d)$.

The number of the points are determined by $k = (1/d) + 1$ to ensure that the points are bounded in the area $Y = [0, 1]$ and $Z = [0, 1]$.

Step 8: Measure the Euclidean distance between a coordinate (Y_j, Z_j) of a firm j and each reference point k by this following formula.

$$d_j = \sqrt{(Y_j - Y_k)^2 + (Z_j - Z_k)^2} \tag{4}$$

where d_i distance between a coordinate (Y_j, Z_j) of a firm j and each reference point k when j = 1, 2,..., n firms and k = 1, 2,..., k groups. Y_j and Z_j is the coordinate of Y and Z for a firm j. Y_k and Z_k is the coordinate of Y and Z at a point k.

Step 9: Compare the Euclidean distance between those calculated in step 8. Choose the point with the shortest distance to represent a group of that firm.

Step 10: Assign a score of 100 to the first cluster (Y_1, Z_1). The second lower cluster will get the score of 100-(10d). The third lower cluster will get the score of 100-(20d). In general, the score of the cluster k is 100-[(k-1)(10d)]. For example, when the increment is set to be 0.01, the last cluster is k=(1/d)+1=101. Then the least score is 90. Finally, measure the score of each firm by the cluster where it belongs. Collect these scores.

Step 11: Adjust the weights (in step 4) with randomized numbers with a randomized sign of positivity or negativity. This is called Δw.

Step 12: Reset the initial weights in step 4. Repeat step 5–10. Iterate this step for m rounds. Collect all the scores of all rounds.

Step 13: Calculate the grand mean of the scores from all the m rounds.

Step 14: Rank the grand mean from the highest to the lowest value.

3 An Application of Indicator Circuit to the Classify the Performance of Top 1,000 Companies

The settings of ICIC model are shown in Table 1.

The production of Y and Z signals by different weights differentiate the firms into many different locations in the Eucledean space. A point in the scatter plot in Fig. 1 represents a firm. Each firm will be assigned to be a member of cluster. It can be imagined that these clusters are located by the diagonal line linking (0,0) and (1,1).

Table 1 The settings and results of the ICIC model

The settings of ICIC model			
Round of clustering	1	Unit of measurement	Unnormalized
Rounds of reweights	3	Number of indicators	6

Source ICIC model

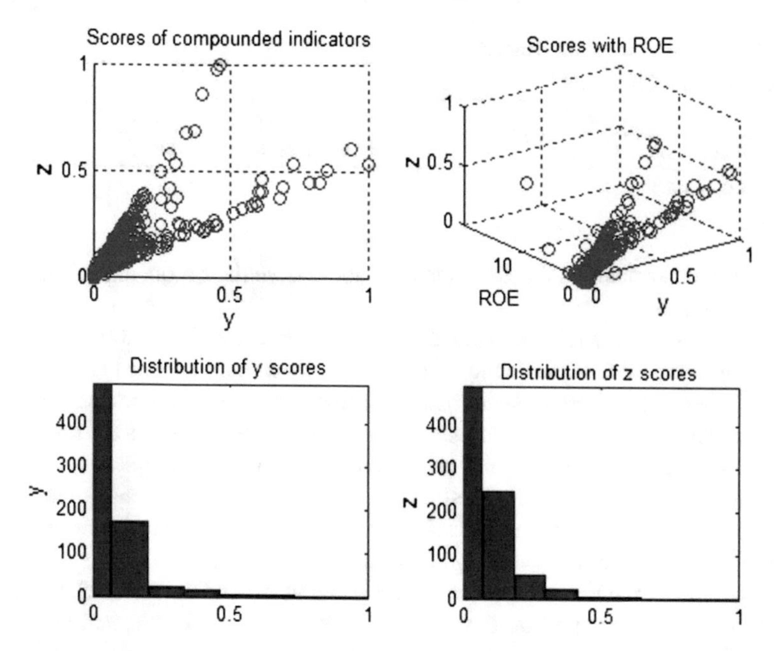

Fig. 1 Scatter plots on the Euclidean space of (Y,Z) and (Y,Z,ROE) with the bar charts showing the distribution of Y and Z

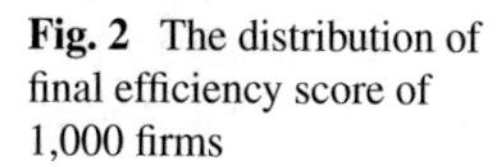

Fig. 2 The distribution of final efficiency score of 1,000 firms

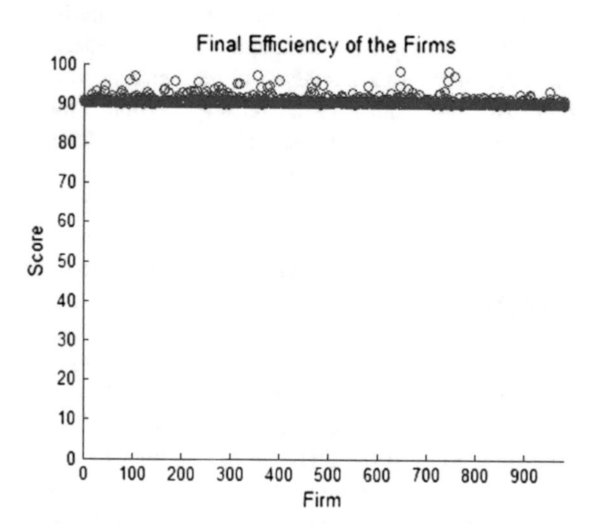

The following figure (Fig. 2) shows the distribution of the final efficiently score of 1,000 firms that top Thailands chart of largest revenue in 2013. The left-hand-side of the figure locates the firm with the largest revenue and vice versa.

It can be seen that the scores are bounded between 90 and 100. Most of the firms stick together at the bottom line near 90. It can be imagined that if the least score is

Table 2 The results of the ICIC model

The results from ICIC model			
Mean score	90.8	Maximum score	98.40
Standard deviation	0.99	Minimum score	90.00

Source ICIC model

set to be 0 rather than 90, these companies should have their score also close to the bottom line of 0.

A reason why the study set the least score at 90 because these 1,000 companies are the top companies of the country. Their efficiencies should not be marked by low scores. When the owners or share holders of the companies look at such the very low score and ask why, the researchers just reply that it only reflects the rank of the efficiently in the relative term, not an absolute term. This answer seems not to be satisfying to them. They need something that shows the high efficiency of the firms as well as reflects the rank among other firms in the country. Therefore, the most compromised range of the score begins from 90.

By this range, the top score shown in Table 2 is 98.40 and the mean of the scores is just 90.80 with the standard deviation of 0.99. These numbers reveal again that most of the firms are located near the bottom line.

To compare the performance between ICIC and ICRP, the study ranks the efficiency of 17 companies in the telecommunications industry. There are several reasons why it selects only these companies. First, the telecommunications companies are at the focus of the further analysis of their efficiencies. This study is a part of a project funded by the regulator of the industry. Second, it may make a long list of companies when the study compares the rank of all the 1,000 companies. If so, it cannot give a clear picture of the ranking results in details.

The results in Table 3 shows some remarkable differences between the performance of ICIC and ICRP. First, the ranks of companies are not the same. Second, the number of companies that share the same rank are different in favour of ICIC (Table 4).

It can be noticed that ICIC places more importance to the current ratio over the profits. Firm G which is in the 5th rank of ICRP moves upward to the 4th place. The ICIC differentiates between firm B and C which are indifferent in ICRP such that now firm B dominates firm C in terms of 5 indicators apart of only the Return on Equity, ROE. It should be noted that the smaller Debt/Equity ratio (D/E ratio) is better than the larger one. This seems to be reasonable.

At the bottom of Table 3, the ICIC also differentiates firm Q and P which share the same rank in ICRP, the 16th place. Now firm Q is at the 13th place while firm P is at the last place. Both firms have 3 indicators that dominates each other. It is hard to judge from the number of better indicators in this case. The largest different is at the D/E ratio. Firm Q has a lower D/E ratio, 3.5579, compared to that of firm P which is 59.5518. It can be seen that the ICIC ranks the 13–17th places mainly by the D/E ratio and also the current ratio.

Table 3 The ranking of the efficiency of telecommunications operators

Rank by ICIC	Rank by ICRP	Firms	Final efficiency score from ICIC model (Out of 100 points)	Current ratio	D/E ratio	ROA	ROE	Net profit margin	ROI
1	1	Firm A	92.83	1.0663	15.0796	0.1848	2.9720	0.0400	38.2953
2	2	Firm B	90.97	2.0143	0.9859	0.4215	0.8371	0.3269	7.3333
2	2	Firm D	90.97	3.5049	0.3992	0.5031	0.7040	0.7280	0.3938
4	5	Firm G	90.87	3.7947	0.3578	−0.2147	−0.2916	−0.2539	−0.4169
5	2	Firm C	90.73	1.0324	30.883	0.1015	3.2370	0.2563	6.1348
6	5	Firm E	90.60	2.5541	0.6435	0.0219	0.0359	0.0500	0.7257
7	5	Firm F	90.50	1.6245	1.6012	0.2153	0.5601	0.3421	1.9732
7	8	Firm K	90.50	2.1985	0.8344	0.0324	0.0595	0.1253	0.1416
9	8	Firm I	90.40	1.7380	1.3550	0.1989	0.4684	0.3457	0.6554
10	8	Firm H	90.33	1.0624	16.0345	0.0735	1.2516	0.1221	2.0515
11	10	Firm J	90.23	1.3671	2.7237	0.0168	0.0625	0.1417	0.0071
12	12	Firm L	90.17	1.2236	4.4723	−0.0011	−0.0061	−0.0010	−0.0008
13	14	Firm N	90.07	1.2489	4.0183	−0.098	−0.4918	−0.0771	−1.5032
13	16	Firm Q	90.07	1.2811	3.5579	−0.3754	−1.7111	−0.6691	−0.265
13	13	Firm M	90.07	1.0438	22.8336	−0.0277	−0.6600	−0.0190	−0.2916
16	14	Firm O	90.03	1.0230	43.5517	−0.0177	−0.7877	−0.0492	−0.5039
17	16	Firm P	90.00	1.0168	59.5518	−0.0455	−2.7538	−0.0552	−0.1912

Source ICIC model from this study and the results of ICRP from Suriya [4]

Table 4 The number of firms that share the same rank

Types of indicator circuit	Total number of firms	Number of firms that share the same rank	Ratio of the number of firms that share the same rank (%)
ICIC	17	7	41.2
ICRP	17	14	82.4

Source ICIC model from this study and the results of ICRP from Suriya [4]

4 An Application of Indicator Circuit to Detect High-Yield Stocks

High-yield stock selection is one of the most important financial topics, as well as the most challenging task for investors. Therefore, the stock selection has received considerable attention from researchers. The stock selection techniques are usually divided in to two disciplines which are fundamental analysis (FA) and technical analysis (TA). Fundamental analysis is considered to be the most appropriate evaluation approach within the medium- to long-term investment, while technical analysis is more appropriate to short-term investment. In both techniques, there are many indicators available. Thus, the selection criteria are varying according to investors' preferences, strategies etc. Several machine learning-based methodologies have been used to distinguished high-yield stock based on fundamental indices.

Quah and Srinivansa [2] use the Artificial Neural Network (ANN) to detect healthy firms based on financial ratios in five categories as inputs: (1) yield factors historical P/E ratio, prospective P/E ratio, and cashflow yield (2) liquidity factors market capitalization (3) risk factors earning per share uncertainty (4) growth factor return on equity (ROE) and (5) momentum factors average of the price appreciation over the quarter with half of its weights on the last month and remaining weights being distributed equally in the remaining two months. They also use the return differences between the stock and the market return (excess returns) as the output of the calculation. In the testing period, the top 25 stocks with the highest output value are selected to form an investment portfolio. The result shows that the ANN is able to beat the market overtime.

Quah [3] introduces three machine learning-based methodologies, including multi-layer perceptrons (MLP), adaptive neuro-fuzzy inference systems (ANFIS) and general growing and pruning radial basis unction (GGAP-RBF) to separate high-performance stock by using fundamental analysis including eleven indices based on Benjamin Grahams common stock selection rules. The eleven indices are P/E ratio, book value per share (BVPS), ROE, dividend payout ratio (DPR), dividend yield (DY), price to book ratio (P/BV), total current assets, Gross debt, weigh average number of shares, current ratio (CR), and earning per share. The results show that the top picked equities have the average appreciation about 40–60 %.

Huang [1] employs support vector regression (SVR) and genetic algorithms (GAs) to detect high-yield stocks. The model used fourteen financial ratios which is separated in six categories: (1) share per rationality—P/E ratio, P/BV ratio, price to sales ratio (P/S), and ROE (2) profitability—ROA, operating profit margin (OPM), net profit margin (NPM) (3) leverage D/E ratio (4) liquidity—current ratio and quick ratio (5) efficiency inventory turnover rate, and receivables turnover rate (6) growth operating income growth rate and net income growth rate. After employing the SVR method, the top-ranked stocks can be selected to form a portfolio. The empirical results show that the cumulative return of selective portfolio can out-perform the benchmark, which is the product of the average yearly returns of the 200 stocks.

Yu et al. [5] adopt support vector machine (SVM) and the principal components analysis (PCA). The study uses seven categories of financial ratios: (1) earning ability EBIT, ROA, and ROE (2) activity ratio turnover of account receivable, turnover of inventory, and turnover of current asset (3) shareholder return earning per share (EPS), P/BV ratio, common stock profitability, P/CF (4) cash ratio EBIT-to-cash ratio, cash-to- asset ratio, and operation ratio (5) growth ratio growth of total asset (6) risk level financial leverage and operation leverage (7) solvency ratio quick ratio, debt-to-asset ratio, EBIT/Interest ratio, and EBIT/Fixed charge ratio. The results show that the annual earning portfolio significantly outperforms those of A-share index of Shanghai Stock Exchange.

4.1 Data

This section attempts to test the ICIC model with fundamental analysis to investigate the high-yield stocks in the Stock Exchange of Thailand during 2013–2014. The six indicators are selected as the inputs according to recommendation from Prof. Songsak Sriboonchitta, who successfully gains from the stock market. The indicator includes P/E ratio, P/BV ratio, ROE, growth of the E/P ratio (the inverse of P/E ratio), dividend growth and ROE growth. After classification, the results will be compared with the highest capital gain yields of each stock during examined period to evaluate the ICIC method.

The study selected stocks listed in the SET100 during 2013–2014 periods and screened only those which have complete financial ratios. After the primary stock screening, the number of stocks used in the ICIC model equals 134 stocks. The selected financial ratios used are those of 2013.

The highest capital gain yields are calculated according to the following equations:

$$capital\ gain\ yield = \frac{the\ selling\ price - the\ buying\ price}{the\ buying\ price} \times 100 \quad (5)$$

where the selling price is the peak price of each stock during April 1st, 2014 to March 31st, 2015, the buying date is the first trading day on the second quarter of 2014, when most of the 2013 financial statements have already been announced.

4.2 The ICIC Results

The settings of ICIC model shown in Table 5 are the same with the previous application.

The distribution of the 134 stocks on the y-z Eucledean space is displayed in Fig. 3. It can be seen that there is one company that tops the table. It is located at the upper-right corner of the graph. Most of the companies gather around the origin, the lower-left corner of the graph. Some better companies are dispersed over the origin but not exceed the half way to the top company.

Figure 4 shows the distribution of final scores of 134-selected stocks based on six indicators mentioned earlier. With the incremental distance of 10 times 0.01 (equals to 0.10) from the highest score, there is one stock that is very distinguished by the score of 100. About fifteen are scored over 90 while the rest are clustered around 90 which is the minimum score by this scoring method.

Table 5 The settings and results of the ICIC model

The settings of ICIC model			
Round of clustering	1	Unit of measurement	Unnormalized
Rounds of reweights	3	Number of indicators	6

Source ICIC model

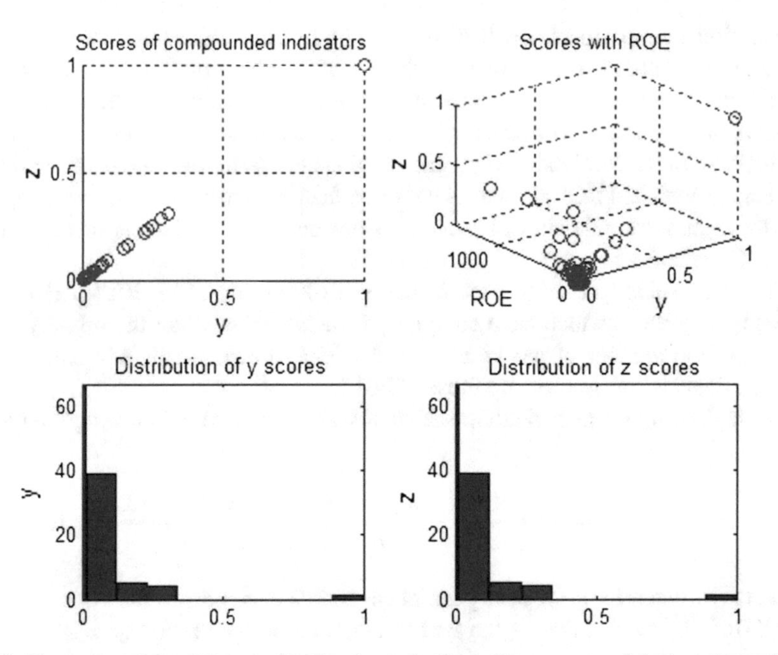

Fig. 3 The scatter plots of the y and z signals on the Eucledian spaces and the bar charts of the distribution of both y signal and z signal

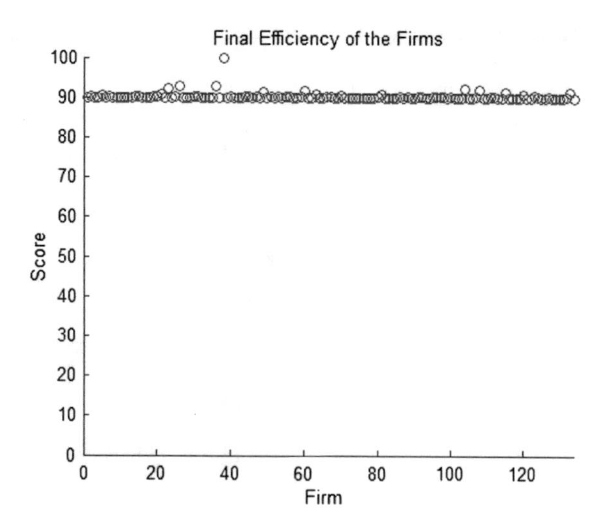

Fig. 4 The distribution of final scores of 134 stocks when the increment is 0.01

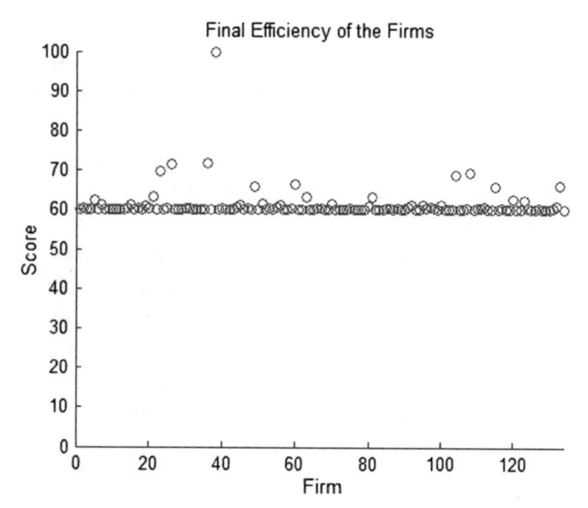

Fig. 5 The distribution of final scores of 134 stocks when the increment is 0.10

Although it can be seen that some variations among the scores when the increment is set at 0.01, it is still hard to classify these stocks into more than two groups. Therefore, the increment is increased to 100 times 0.01 (equals to 1.00) in order to illustrate the clearer result of the clustering.

Figure 5 shows the distribution of final scores of the stocks when the increment is set at 1.00. Therefore, the lowest score is decreased to 60. The result still shows one distinguished stock that scores 100, while there are some stocks scored between 60 and 70. But, most of them are still clustered at the lowest score.

When the increment is set at 200 times 0.01 (equals to 2.00), the lowest result, accordingly, decreases to 26.67. Thus, the distribution of final scores of the stocks is more scattered than the previous results. Still, it can be seen from Fig. 6 that there is

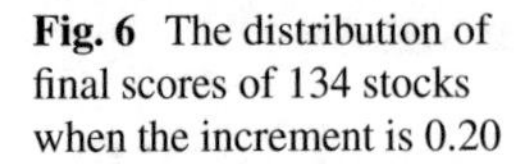

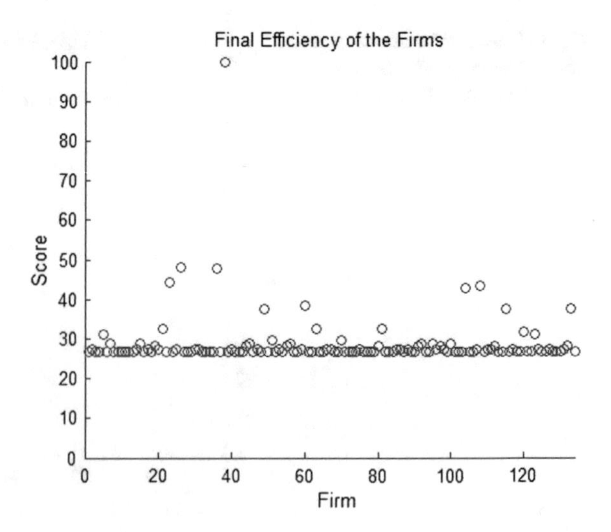

Fig. 6 The distribution of final scores of 134 stocks when the increment is 0.20

one stock with a score of 100. The figure also shows that approximately four stocks are scored between 40 and 50, while around nine of them are scored between 30 and 40. However, the majority of the stocks are still clustered at a bit below 30.

In order to validate the stock classification by the ICIC model, the capital gain yields are used as the performance testers. Companies in the SET usually announce their annual performances within the first quarter of the following year. This study assumes that the investors will carefully examine performances of all the companies in the market before making decision to buy stocks. Therefore, the buying prices are the closed prices of the first trading day in the second quarter, which in this case is April 4th, 2013. For selling prices, investors will seek to earn the highest capital gain yield as possible. Therefore, the ideal selling prices used in the study are the highest closed prices before the next annual performances are announced, which is during April 1st, 2014 to March 31st, 2015.

Table 6 shows the results which divide the 134 stocks into 10 deciles, where the 10th decile is the highest score and the 1st decile is the lowest score. It can be seen in the table that the performance testers average capital gain yield and the results of the ICIC model conformed to each other. The average score of the 10th Decile is 44.28 when the increment is set at 0.20 (200 times of 0.01, thus called 200x), and equals to 69.68 and 92.31 when the increment is set at 0.10 (100 times of 0.01, thus called 100x) and 0.01 (10 times of 0.01, thus called 10x) accordingly. Comparing with the average yield, it can be seen that the average yield of those companies in the 10th Decile is also the highest average yield at 61.31 %. For the 9th Decile, the average scores are 29.53, 61.56, and 90.36, accordingly. The average yield is 57.25 % which also the second highest yield. From the 5th Decile down to the 1st Decile, the average scores equals to one another which are the lowest score when each increment is set. Although the average scores are equal in those deciles, the average yields are different. The 5th Decile has the average yield at 42.71 %. Then the average yield

Table 6 The performance of the ICIC model for the classification of high-yield stocks

Decile	Average score_200x	Average score_100x	Average score_10x	Average capital gain yield	Standard deviation
10th Decile	44.28	69.68	92.31	61.31	54.58
9th Decile	29.52	61.56	90.36	57.25	59.37
8th Decile	27.84	60.64	90.14	35.84	38.85
7th Decile	27.41	60.41	90.08	44.43	35.2
6th Decile	27.08	60.22	90.04	45.28	37.9
5th Decile	26.67	60	90	42.71	35.45
4th Decile	26.67	60	90	39.35	44.88
3rd Decile	26.67	60	90	37.83	57.36
2nd Decile	26.67	60	90	29.79	20.45
1st Decile	26.67	60	90	26.88	38.11

Source Computations

Note The average yield is calculated by simple average method. This is the best when some yields are zero which cannot be calculated by geometric mean and harmonic mean.

decreases at the lower decile. The 1st decile which is given the least score from ICIC also has the lowest score of 26.88 %.

The results in Table 7 using Wilcoxon–Mann–Whitney test indicates that the average capital gain yields between different deciles are statistically different only in the lowest decile. These results confirms that the ICIC can detect the difference between the low-yield stock and the rest. It means that instead of the ability to detect the high-yield stocks, ICIC can rather detect the low-yield stocks.

4.3 The ICIC Results and the Tobit Regression

In order to justify the ICIC result, the Tobit regression is used to predicted capital gain yield of those 134 stocks, and then compared with the ICIC results (Table 8). The comparison shows that Tobit regression overestimates the capital gain yield in the 1st, 2nd and 3rd deciles. These results from Tobit regressions regardless of the inclusion of all variables or only-significant variables are statistically significant different from those of ICIC (Table 9). It confirms that ICIC separates the low-yield stocks from the rest.

Table 7 Results from Wilcoxon–Mann–Whitney (Wilcoxon Rank Sum) Tests

Prob.	9th Decile	8th Decile	7th Decile	6th Decile	5th Decile	4th Decile	3rd Decile	2nd Decile	1st Decile
10th Decile	0.6261	0.1062	0.4118	0.3428	0.3428	0.2282	**0.0612**	0.1062	**0.0280**
9th Decile		0.2282	0.898	0.8576	0.7005	0.2486	0.209	0.2486	**0.0226**
8th Decile			0.3428	0.3173	0.4571	0.8576	0.8175	0.7779	0.3254
7th Decile				0.9387	0.8175	0.4571	0.1303	0.2931	**0.0468**
6th Decile					0.8175	0.4267	0.1303	0.3975	**0.0383**
5th Decile						0.3428	0.3173	0.2702	**0.0311**
4th Decile							0.8175	0.898	0.2954
3rd Decile								0.7005	0.2498
2nd Decile									0.233

Note Ho: Means of two groups are equal. H1: Means of two groups are unequal.

Table 8 Comparison between the results from ICIC and Tobit regression

Decile	ICIC		Tobit with all variables		Tobit with significant variables	
	Average capital gain yield	Standard deviation	Predicted capital gain yield	Standard deviation	Predicted capital gain yield	Standard deviation
10th Decile	61.31	54.58	51.08	9.33	42.31	2.82
9th Decile	57.25	59.37	53.87	46.25	56.39	44.98
8th Decile	35.84	38.85	33.47	6.81	41.16	3.01
7th Decile	44.43	35.2	42	12.38	45.86	11.89
6th Decile	45.28	37.9	36.54	7.57	40.27	1.33
5th Decile	42.71	35.45	39.73	3.53	42.71	2.99
4th Decile	39.35	44.88	38.53	4.49	40.73	1.94
3rd Decile	37.83	57.36	36.01	5.06	41.27	2.33
2nd Decile	29.79	20.45	39.36	3.52	42.61	2.53
1st Decile	26.88	38.11	42.17	9.41	42.07	3.46

Source Computations

Table 9 Results from Wilcoxon sign rank test

Decile	Comparison of the means of yields from ICIC and predicted yields from Tobit	
	ICIC and Tobit (all variables)	ICIC and Tobit (only significant variables)
10th Decile	0.8613	0.4216
9th Decile	0.7532	0.7532
8th Decile	0.4216	0.2489
7th Decile	0.8613	0.7007
6th Decile	0.7532	0.7532
5th Decile	0.6496	0.4216
4th Decile	0.5067	0.4216
3rd Decile	**0.0747**	**0.0277**
2nd Decile	**0.0869**	**0.0330**
1stDecile	**0.0056**	**0.0099**

Note The numbers in the table present the prob-values gained from Wilcoxon sign rank test of matched pairs.
Source Computations

5 Conclusions

This paper introduces the indicator circuit with incremental clustering (ICIC) with its application on the evaluation of telecommunication companiess performance which ICIC shows a better result than the previous version of the indicator circuit with reference points (ICRP) presented in [4]. It also attempts to spot high-yield stocks in the Stock Exchange of Thailand (SET) by adopting the ICIC. It uses 6 financial ratios according to Songsak hypothesis of medium-term stock selection criteria. These indicators are announced at the end of 2013 as indicators in the model. The data include 134 stocks in the SET ranked in the SET100 group during 2013–2014. The capital gain yields of each stock are used to justify the performance of the model. The results show that the scoring generated by the ICIC indicates the low-yield stocks instead of the high-yield ones. This method turns out to be the warning signal for investors to avoid the low-yields. Moreover, the study confirms that the Songsak hypothesis or the criteria for the medium-term stock selection is correct

References

1. Huang, C.: A hybrid stock selection model using genetic algorithms and support vector regression. Appl. Soft Comput. **12**(2), 807–818 (2012)
2. Quah, T., Srinivansa, B.: Improving returns on stock investment through neural network selection. Expert Syst. Appl. **17**(4), 295–301 (1999)
3. Quah, T.: DJIA stock selection assisted by neural network. Expert Syst. Appl. **35**(1), 50–58 (2008)

4. Suriya, K.: Indicator circuit with reference points and the comparison with indicator circuit with self-organizing map. Int. J. Intell. Technol. Appl. Stat. **8**(2), 103–112 (2015)
5. Yu, H., Chen, R., Zhang, G.: A SVM stock selection model within PCA. Proced. Comput. Sci. **31**, 406–412 (2014)

Nonlinear Estimations of Tourist Arrivals to Thailand: Forecasting Tourist Arrivals by Using SETAR Models and STAR Models

Nyo Min, Songsak Sriboonchitta and Vicente Ramos

Abstract The main objective of this study is to evaluate some alternatives to estimate tourism arrivals under the presence of structural changes in the sample size. Several specification of Self-exciting threshold autoregressive (SETAR) model and Smooth transition autoregressive (STAR) model, especially Logistic STAR (LSTAR) are estimated. Once the parameters are estimated, a one period out of sample forecasting is performed to evaluate the forecasting efficiency of the best specifications. The finding from the study is that the STAR model beats SETAR model slightly, and these two groups of models have forecast proficiency at least in the tourism field.

1 Introduction

Accurate forecast of tourism demand is necessary for many agents involved in tourism development at destinations. That is the case for example of governments responsible of public tourism policies governments or for investors around the world when designing business plans in tourism destinations. Predicting long term tourism demand will assist the policy makers in project appraisal and will provide reasonable options for portfolio selection, whereas short term forecast plays major roles in tourism, especially when it has an urgent need for making decisions.

According to the UN World Tourism Organization (UNWTO), tourism is one of the fastest growing economic industries in the world. The UNWTO stated that in 2014 international tourism grew by 4.4 % in 2014 reaching 1.135 billion arrivals and USD 1.5 trillion in export earnings. The trend is expected to remain positive in 2015 with a growth rate between 3 and 4 % in international tourist arrivals [14]. The positive impact of tourism in the economy explains that many countries in the world are paying increasing attention to the development of their potential destinations. As

N. Min (✉) · S. Sriboonchitta
Faculty of Economics, Chiang Mai University, Chiang Mai 50200, Thailand
e-mail: nyo.min@gmail.com

V. Ramos
Department of Applied Economics, University of the Balearic Islands, Palma, Spain
e-mail: vicente.ramos@uib.es

V.-N. Huynh et al. (eds.), *Causal Inference in Econometrics*,
Studies in Computational Intelligence 622, DOI 10.1007/978-3-319-27284-9_26

a result, there is an increase in competition among destinations which demands wise decisions from private investors and policy makers. In addition, tourism industry is extremely sensible to adverse external shocks that can generate dramatic changes in the number of tourism arrivals. The above reasons justify the need to continue the improvement of tourism forecasting and, in particular, to deal with the fact that in many cases, the external shocks are likely to generate a structural change in the behavior of tourism arrivals. Neglecting the presence of structural change(s) is likely to cause incorrect modeling. In other words, taking structural change(s) into account may improve the forecast performance in terms of capturing and exploiting changes in macroeconomic relationship, and detecting and forecasting changes in the long run dynamics correctly [6]. If the time series data are subject to change unexpectedly, considering structural change will lead to correct model selection and provide genuine findings. In classical research environment, many just applied linear forecasting, as a result, the findings provided less accuracy by depending on sample selection. In tourism literature, many researchers proposed more reliable methods to estimate and forecast tourism arrivals. Among them, only some took into account the potential presence of structural change in their datasets. This paper focuses on two families of models that are able to deal with the presence of structural changes in the data, namely, Self-Exciting Threshold Autoregressive (SETAR), and Smooth Transition Autoregressive (STAR). The SETAR model is an extension of the Autoregressive model which allows a higher degree of flexibility in parameters estimation as it possesses a regime switching behavior [13]. The switching is dependent on past values of the time series. It can be estimated using the same or different autoregressive lags in the regimes. STAR models are also an extension of autoregressive models which allow for a higher degree of flexibility by proposing a smooth transition. Instead of using a regime switching approach, the STAR models link two autoregressive parts by using a transition function. The most popular transition functions are exponential function and first and second-order logistic functions. Then the STAR model with exponential function is called Exponential STAR (ESTAR) model and the STAR models with first and second-order logistic functions are called Logistic STAR (LSTAR) models.

This paper uses the case of Thailand tourism which has been affected by several potential structural changes during the last two decades. For instance, Thailand's position in the global rankings dropped from 10th to 14th due to social unrest in 2014 [8]. In particular, arrivals from East Asian countries to Thailand are considered as they correspond this fast growing intraregional market which, led by China, has been growing at a terrific rate. As stated above, the political instability in Thailand in 2014 which decreased tourist arrivals to Thailand by 6.65 %. The number of tourists from some East Asian countries, including Japan, South-Korea, Hong Kong, and most ASEAN countries significantly dropped, whereas the number of tourists from Europe was less affected in 2014 [16]. Therefore the forecasting tourism demands, especially tourist arrivals from East Asian countries, will play an important role in the modern tourism in Thailand.

The paper is organized as follows: Sect. 2 focuses on literature review, concept, and models. Data is presented in Sect. 3 and empirical results with the results appear in Sect. 4. Section 5 presents an out of sample comparison of SETAR and STAR models. The paper will end with the conclusion.

2 Literature Review, Concept, and Models

Although tourism demand can be measured in several ways, tourist arrivals and tourist expenditures are mainly chosen to measure tourism demand for a specific destination [11]. Knowing the tourism demand will provide better understanding of the tourist decision making process. Therefore, many studies tried to measure tourism demand. Generally, there are four groups of measurement criteria, doer criterion: such as the number of tourist arrivals, the number of tourist visits and the visit rate; pecuniary criterion: such as the level of tourist expenditure (receipts) and the share of expenditure (receipts) in income; time-consumed criterion: such as tourist-days, tourist-nights; and distance-travelled criterion: for instance, the distance travelled in miles or kilometers [7]. In classical literature, several researchers just relied on linear models to estimate and predict tourism demand, but some researchers started to use nonlinear models for forecasting tourism demand to handle structural change(s) in data set. Structural change can be defined as the state in which one or more variables are changing over time [9].

The challenges of Modeling and detecting of econometric model's parameter stability have risen since a long time ago. Structural change may be the outcome of factors such as preferences changes, institutional changes and technological progress. Chen [5] said that current estimation and testing methods are formulated without endogeneity and proposed a nonparametric test to check smooth structural changes and abrupt structural breaks with possibly unknown change points in regression models by using potential endogeneity. Andrews and Fair [1] tested for structural change in nonlinear, dynamic, and simultaneous models with limited dependent variables. They tried to extend the classical test for structural change in linear regressions of Chow to nonlinear models. Till now, structural changes issue has not been recognized enough by most of researchers and only few researchers are seeking effective ways to control the structural change(s).

2.1 *Self-exciting Threshold Autoregressive (SETAR) Models*

SETAR models were initiated by Howell Tong in 1977 as an extension of AR models designed to handle changes in the model parameters by the threshold value and delay parameter. Tong and Lim [12] stated that SETAR $(1, K)$ is a linear AR model with order k. Serletis and Shahmoradi [10] said that SETAR model is one of the most popular Threshold Autoregressive (TAR) models. Then, they stated the following model as two regimes SETAR model;

$$x_t = \alpha_0 + \alpha_1 x_{(t-1)} + \ldots + \alpha_p x_{(t-p)} + (\beta_0 + \beta_1 x_{(t-1)} + \ldots + \beta_p x_{(t-p)})\{x_{(t-d)} \le \gamma\} + \varepsilon_t, \tag{1}$$

where $p \ge 1 = autoregressive\ order,\ d = delay\ parameter$ and $\gamma = the\ threshold\ parameter.$

In general, the multi regime SETAR model can be described as follows;

$$\begin{aligned}
®ime\ 1\ \ x_t = \beta_0^{(1)} + \beta_1^{(1)} x_{t-1} + \cdots - \beta_p^{(1)} x_{t-p} + \varepsilon_t \quad if\ x_{t-d} \le r_1,\\
®ime\ 2\ x_t = \beta_0^{(2)} + \beta_1^{(2)} x_{t-1} + \cdots - \beta_p^{(2)} x_{t-p} + \varepsilon_t \quad if\ r_1 \le x_{t-d} \le r_2,\\
&\quad\vdots\\
®ime\ k\ x_t = \beta_0^{(k)} + \beta_1^{(k)} x_{t-1} + \cdots - \beta_p^{(k)} x_{t-p} + \varepsilon_t\ if\ r_{k-1} \le x_{t-d} \le r_k.
\end{aligned} \tag{2}$$

where $p = autoregressive\ order,\ d = delay\ parameter$ and $\gamma_1, \gamma_2, \ldots, \gamma_r = the\ threshold\ parameters.$

The above mentioned model is defined as self-exciting model because the regime r is a function of the past realizations of x_t sequence itself.

2.2 Smooth Transition Autoregressive (STAR) Models

STAR models are also an extension of AR models and introduced by K.S. Chan and H. Tong in their paper [5]. The STAR model can be seen as a two-regime SETAR model, but it has a smooth transition between regimes, or continuum of regimes. The two smooth transition autoregressive [STAR] model for a univariate time series x_t can be described as follows;

$$x_t = (\beta_{(1,0)} + \beta_{(1,1)} x_{(t-1)} + \cdots + \beta_{(1,p)} x_{(t-p)}) \times (1 - G(s_t; \gamma, c)) + (\beta_{(2,0)} + \beta_{(2,1)} x_{(t-1)} + \cdots + \beta_{(2,p)} x_{(t-p)}) G(s_t; \gamma, c) + \varepsilon_t, \tag{3}$$

where $t = 1, \cdots, T$. The transition function $G(s_t; \gamma, c)$ is a continuous function, but bounded between 0 and 1. It can be written as follows;

$$G(s_t; \gamma, c) = (1 + exp\{-\gamma(s_t - c)\})^{-1}, \quad \gamma > 0 \tag{4}$$

and it is called the logistic STAR (LSTAR) model.

If the transition function is written as follows;

$$G(s_t; \gamma, c) = 1 + exp\{-\gamma(s_t - c)^2\}, \quad \gamma > 0 \tag{5}$$

it is called exponential STAR (ESTAR) model. The notations of the equation are $c = threshold\ values\ between\ two\ regimes$, $s_t = transition\ variable$ and $\gamma = parameter\ that\ determines\ the\ smoothness\ of\ the\ change$. If $s_t = x_{(t-d)}$, this model is called a self-exciting TAR [SETAR] model.

The weakness of ESTAR is that if limits $\gamma \to 0$ or $\gamma \to \infty$ are taken, the transition function changes to a constant (equal to 0 and 1, respectively). Then, the model cannot be transformed into a SETAR model. If limits $\gamma \to 0$ is taken in LSTAR model, the logistic function becomes equal to a constant (equal to 0.5) and then the LSTAR model reduces to a linear model [15].

A multi-regime Logistic Smooth Transition Autoregressive (MR-STAR) allows for ESTAR type dynamics and SETAR type dynamics. The ESTAR cannot account for discontinuous case and it can be changed to linear process. This feature generates a sudden shift between regimes. To solve this weakness, the answer is the MR-STAR [2].

The MR-LSTAR or three regimes LSTAR model can be written as follows;

$$\begin{aligned} x_t = &(\beta_{(1,0)} + \beta_{(1,1)}x_{(t-1)} + \cdots + \beta_{(1,p)}x_{(t-p)})G_1 \\ &+ (\beta_{(2,0)} + \beta_{(2,1)}x_{(t-1)} + \cdots + \beta_{(2,p)}x_{(t-p)})G_2 \\ &+ (\beta_{(3,0)} + \beta_{(3,1)}x_{(t-1)} + \cdots + \beta_{(3,p)}x_{(t-p)})G_3 + \varepsilon_t, \end{aligned} \tag{6}$$

where $G_1 = (1 + exp\{\gamma(s_t + c)\})^2, \gamma > 0$, $G_2 = 1 - G_1 - G_3$ and $G_3 = (1 + exp\{-\gamma(s_t - c)\})^{-1}, \gamma > 0$.

The study will not focus on the ESTAR models, but on the LSTAR models to compare with the SETAR models.

3 Data

The dataset used for the model estimation gathers monthly tourist arrivals from East Asia countries including China, Japan, and Koreas to Thailand between January 1997 and October 2014. There are 214 observations in the sample. Visually, the data set projected fluctuation and structural changes. Therefore the study analysis the nature of the dataset and the study found that the dataset has a unit root with ADF statistic −0.318429 and probability 0.9187 according to the Augmented Dickey–Fuller test in EViews (Fig. 1).

The study continued its analysis to detect breakpoints and applied CUSUM test to obtain visual evidence of the structural change. The study found that there were three breaks in the data observed and the break dates were November 1999, August 2003, and December 2010. The study confirmed these breakpoints with the Chow break point test. When the study tested the data with CUSUM test, the following graphic was resulted. This graphic showed that there is a structural change in the sample (Fig. 2).

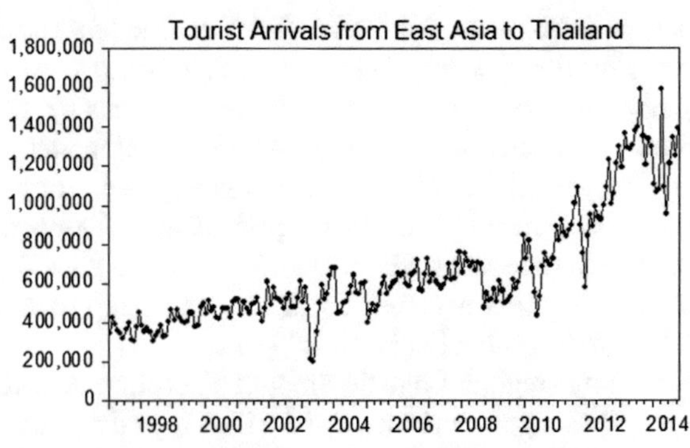

Fig. 1 Tourist arrivals from East Asia to Thailand between January 1997 and October 2014

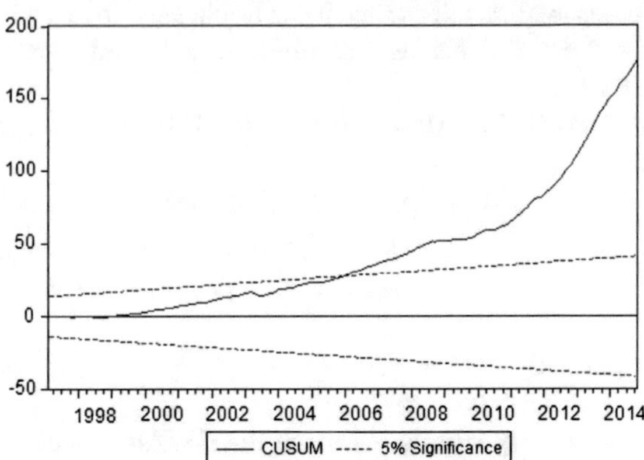

Fig. 2 CUSUM test result on OLS of tourist arrivals from East Asia to Thailand

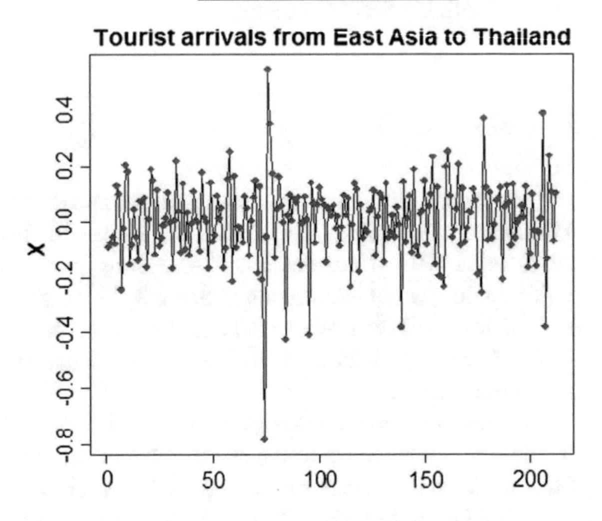

Fig. 3 Log return of tourist arrivals from East Asia to Thailand

The study took log return on the number of tourist arrivals from East Asia to get stationary in the data set. After having a stationary process in the dataset, its statistical properties such as mean, variance, autocorrelation, etc. became constant over time. The dataset taken log return can be seen as follows (Fig. 3).

4 Empirical Results

The study found that Autoregressive (9) is the best $AR(p)$ model for the dataset with the lowest AIC value (−1.148) and BIC value (−0.9853). Although $AR(9)$ and $AR(10)$ are competing, $AR(9)$ slightly performed better than $AR(10)$.

4.1 SETAR Models

The study used embedding dimensions 9 and the threshold value 0 to search for the best SETAR model. The search-process used maximum autoregressive order 9 for low regime, and maximum autoregressive order 9 for high regime. The process searched on 141 possible threshold values within regimes with sufficient (15 %) number of observations and used 11, 421 combinations of thresholds (141), the delay value (1), and lag order (9) (Table 1).

The maximum lag order 5 for low regime and order 9 for high regime, with threshold delay value 0 provided the lowest pooled AIC value. Therefore the selected SETAR model for this study is SETAR (2, 5, 9) with delay 1 and threshold value −0.1234467 (Table 2).

The SETAR (2, 5, 9) can be written as follow;

Table 1 Searching the best SETAR model among Threshold values, lag orders, and pooled AICs

The delay	Lag order for low regime	Lag order for high regime	Threshold values	Pooled AIC
1	5	9	−0.1234467	−253.1171
1	9	9	−0.1234467	−253.1024
1	5	9	−0.1163712	−253.0915
1	5	9	−0.1236115	−252.7790
1	9	9	−0.1236115	−252.6571
1	5	9	−0.1296206	−252.5920
1	9	9	−0.1217031	−252.5855
1	5	9	−0.1217031	−252.4666
1	5	9	−0.1144791	−252.4561
1	9	9	0.1163712	−252.1901

Table 2 SETAR(2,5,9) with d = 1, $\gamma = -0.1234$

Low regime of SETAR (2,5,9)				High regime of SETAR(2,5,9)			
Variable	Coefficient	t-stat	Prob	Variable	Coefficient	t-stat	Prob
C	−0.066399	−1.2623	0.2083	C	0.033300	3.0010	0.0030
AR1	−0.139510	−0.7654	0.4449	AR1	−0.350893	−3.4433	0.0007
AR2	−0.479152	−3.0087	0.0030	AR2	−0.426983	−5.6615	5.277e−07
AR3	0.228376	1.2657	0.2071	AR3	−0.395466	−5.3113	2.939e−07
AR4	0.08752	0.4973	0.6195	AR4	−0.266827	−3.3759	0.0009
AR5	−0.825953	−3.740	0.4881	AR5	−0.191766	−2.5091	0.0129
				AR6	−0.248583	−3.0969	0.0022
				AR7	−0.287709	−3.7382	0.0002
				AR8	−0.035271	−0.4785	0.6328
				AR9	−0.297572	−3.6503	0.0003

$$low\ regime\ x_t = \beta_0^{(1)} + \sum_{i=1}^{5} \beta_i^{(1)} x_{t-i} + \varepsilon_t^{(1)}\ if\ x_{t-d} \leq \gamma,$$

$$high\ regime\ x_t = \beta_0^{(2)} + \sum_{i=1}^{9} \beta_i^{(2)} x_{t-i} + \varepsilon_t^{(2)}\ \ if\ x_{t-d} > \gamma. \tag{7}$$

Or

$$\begin{aligned} low\ regime\ x_t &= -0.066399 - 0.139510x_{t-i} - \cdots - 0.825953x_{t-5} + \varepsilon_t^{(1)} \\ &\quad if\ x_{t-1} \leq -0.1234, \\ high\ regime\ x_t &= 0.0333 - 0.350893x_{t-i} - \cdots - 0.297572x_{t-9} + \varepsilon_t^{(2)} \\ &\quad if\ x_{t-1} > -0.1234. \end{aligned} \tag{8}$$

The SETAR model for the study, has estimated all stable parameters because the necessary condition for stability is $|\beta_j| < 1$, but some AR lags are not significant. Based on the SETAR model selected, the study estimated and forecasted tourist arrivals for one period ahead by using Bayesian estimation for the two regime threshold autoregressive model. This method is called BAYSTAR or Threshold Autoregressive model: Bayesian approach. The BAYSTAR estimated and forecasted on SETAR (2, 5, 9) with Markov Chain Monte Carlo for 10, 000 iterations and 2, 000 burn-in iterations. The simulation results were shown in Table 3.

The deviance information criterion (DIC); a Bayesian method for model comparison is −622.56578 and the highest posterior probability of lag is at 1. Mean Forecast Error for the selected SETAR is 9.85221e−13 and it means the SETAR (2, 5, 9) is likely to make under forecast.

Table 3 The results of BAYSTAR for 10,000 MCMC reiterations

Coefficient	Mean	Median	s.d.	Lower	Upper
$\beta_{1.2}$	−0.3255	−0.3327	0.1279	−0.5582	−0.0471
$\beta 1.5$	−0.3749	−0.3697	0.1442	−0.6749	−0.1062
$\beta_{2.2}$	−0.2533	−0.2536	0.0879	−0.4208	−0.0836
$\beta_{2.3}$	−0.3699	−0.3699	0.0709	−0.5117	−0.2314
$\beta_{2.6}$	−0.2068	−0.2059	0.0832	−0.3700	−0.0453
$\beta_{2.6}$	−0.2440	−0.2442	0.0763	−0.3946	−0.0942
$\beta_{2.9}$	−0.2607	−0.2615	0.0823	−0.4218	−0.0976
σ_1^2	0.0218	0.0213	0.0041	0.0153	0.0313
σ_2^2	0.0140	0.0139	0.0018	0.0110	0.0179
γ	−0.0395	−0.0410	0.0187	−0.0660	−0.0017

Table 4 Potential STAR models for the study

Embedding dimensions	No of regime	AIC	BIC
2	3	−828.2866	−781.6510
3	2	−849.1836	−815.6177
4	2	−842.6261	−802.3471
5	2	−843.4300	−796.4378
6	2	−862.4446	−808.7395
7	2	−846.5283	−786.1001
8	3	−842.7184	−738.6642

4.2 STAR Models

In search of a STAR model for the dataset, the study used embedded dimensions 2 to 8. The search-result showed whether the model is nonlinear with significant p values. The test is called Multiple regimes STAR test (MR-STAR). Table 4 summarized the test results.

Among the STAR models in the Table 4, the study chose the STAR (6) model with embedding dimensions 6, because it had the lowest AIC value (−862.4446), BIC value (−808.7395), and the lowest mean absolute percentage error value (425.6 %). Based on the STAR model selected, the study fitted three LSTAR models. The model selection for the LSTAR is described in Table 5.

The study went on with the LSTAR (6) model with constant as it had the lowest AIC, BIC, and MAPE values (Table 6).

The test used Non-linearity test of full-order LSTAR model against full-order AR model. The result showed that the LSTAR (6) model is a nonlinear model with the F statistic ($F = 3.5828$) and p-value (0.002173). After obtaining the best

Table 5 Three LSTAR models of the study

LSTAR Model	AIC	BIC	MAPE (%)
With constant	−862	−8087393	425.6
Without constant	−853	−806.1383	1133
With controls	−811	−797.3271	236.1

Table 6 LSTAR (6) chosen for the paper

Coefficients and constant	Estimate	Std. Error	t value	Pr(> |z|)
Constant L	0.0125757	0.0105079	1.1968	0.2313879
β_1	−0.4074463	0.0767565	−5.3083	1.107e−07
β_2	−0.4925890	0.0753060	−6.5412	6.104e−11
β_3	−0.4287615	0.0788714	−5.4362	5.443e−08
β_4	−0.3309752	0.0861557	3.8416	0.0001222
β_5	−0.2665477	0.0883131	−3.0182	0.0025427
β_6	−0.3238979	0.0907500	−3.5691	0.0003582
Constant H	−0.0682689	0.0337803	−2.0210	0.0432830
β_1	0.9889003	0.1959962	5.0455	4.523e−07
β_2	0.4323626	0.1786095	2.4207	0.0154900
β_3	0.6037953	0.2042318	2.9564	0.0031123
β_4	0.5600631	0.1931883	2.8991	0.0037429
β_5	0.3589648	0.1615078	2.2226	0.0262438
β_6	0.4452616	0.1840469	2.4193	0.0155511
γ	100.00000	65.127254	1.5355	0.1246720
th	0.5332688	0.006843	77.932	<2.2e−16

LSTAR model for the dataset, 5 periods ahead were predicted sequentially and the results were 0.03975614, −0.0231934, −0.03127030, −0.0357560, and 0.03476227 respectively.

The LSTAR (6) for the study can be written as follows;

$$x_t = \beta_1' y_t + (\beta_2 - \beta_1)' y_t G(s-t; \gamma, c) + \varepsilon_t, \quad (9)$$

where $\beta_1' = (0.0125757, -0.4074463, \ldots, -0.3238979)$,
$\beta_2' = (-0.0682689, 0.9889003, \ldots, 0.4452616)$, and
$y_t = (1, x_{(t-1)}, \ldots, x_{(t-6)})$.

Transition function for the LSTAR (6) can be calculated as follows;

$$G(s_t; \gamma, c) = (1 + exp\{-100(x_{(t-1)} - 0.5333)\})^{-1}. \quad (10)$$

Table 7 Analyzing forecast accuracy between SETAR (2,5,9) and LSTAR (6)

Method	MAE	MFE	MAPE (%)	AIC
SETAR	0.095325	9.85e−13	1289.0	Pooled AIC (−253.1171)
LSTAR	0.092257	−4.85e−13	425.6	AIC(−862.4446)

5 Model Comparing and Discussion

In this section the study compared SETAR (2, 5, 9) and LSTAR (6) and examined which model performed better in forecasting for one period ahead. The forecasted results of SETAR and LSTAR were examined with the real tourist arrivals from East Asia to Thailand in November 2014 (Table 7).

The study used Mean-Absolute-Error (MAE), Mean Forecast error (MFE), and Mean Absolute Percentage Error (MAPE) to recommend the model for forecasting tourist arrivals. In terms of error measurements, the study preferred STAR than SETAR because the LSTAR not only had smaller errors, but also had a smaller AIC value. The Mean Forecast Error (MFE) stated that the SETAR method can underestimate its forecast value and the LSTAR can overestimate its forecasting. In terms of forecasting, the SETAR model used BAYSTAR method to forecast one step ahead with Markov Chain Monte Carlo (MCMC) simulations whereas the LSTAR predicted five steps ahead. To reveal their forecasting efficiency, the study took out one step ahead result forecasted or predicted.

There were forecasted results, 0.03975614 and 0.013681 by LSTAR and SETAR respectively. Afterward examining these nonlinear models SETAR and LSTAR in terms of errors, and AICs, the study went on examining the forecast efficiency of the models with actual tourist arrivals from East Asia in November 2014. There were 1, 472, 427 people coming to Thailand from East Asia in November 2014 whereas the forecasted results of LSTAR and SETAR were 1, 443, 064 and 1, 409, 068 respectively. When the LSTAR model deviated from the actual tourist arrival for 2 %, SETAR only deviated 4.3 %. Although, some AR terms and constant values were not significant, the findings recommended the LSTAR in terms of forecast power. The LSTAR model deviated up to 2 %, whereas the SETAR deviated 4.3 %, but all AR terms in the model were significant. Under the rule of thumb, the SETAR model is more effective to estimate the dataset in a trend, but its power is not sure for immediate change in the dataset that the model is forecasting. Indeed, the forecast powers of these two nonlinear models are very similar and both methods are productive.

6 Conclusion

The study applied nonlinear models, namely SETAR and STAR models and tried to solve bias resulted from structural changes in the data observed. As the study wanted to compare the SETAR and the STAR models, the study used the LSTAR model as a representative of STAR family. Both methods are applicable to forecast tourist arrivals and the result of the study revealed that the LSTAR performs better than the SETAR in forecasting for immediate result. Both models handled structural change in the historical data, but they estimated and predicted the data by assuming there are only two regimes. Indeed, the data projected three break points and multiple regimes in some embedding dimensions m.

The short term forecast is necessary for problematic situations, but medium term and long term forecasts are very important for tourism development. In addition, to reveal the actual forecast power of the nonlinear models SETAR and STAR, similar studies on different sample sizes in terms of observations and time variation are needed and all SETAR family and the STAR family including Exponential STAR (ESTAR) models should be studied. To conclude the study, we recommend that both SETAR and STAR families are applicable to forecast tourist arrivals and both families have powerful estimation and forecasting mechanisms, although the STAR beats SETAR in this study.

References

1. Andrews, D.W.K., Fair, R.C.: Inference in nonlinear econometric models with structural change. Rev. Econ. Stud. **LV**, 615–640 (1988)
2. Bec, F., Salem, M.B., & Carasco, M.:Detecting Mean Reversion in Real Exchange Rates From a Multiple Regime STAR Model, Rochester Center for Economic Research, University Of Rochester, **509**1–35(2004)
3. Chan, K.S., Tong, H.: On estimating thresholds in autoregressive models. J. Timeser. Anal. **7**(3), 178–190 (1986)
4. Chang, C.-L., Sriboonchitta, S., Wiboonpongse, A.: Modelling and forecasting tourism from East Asia to Thailand under temporal and spatial aggregation. Math. Comput. Simul. **79**, 1730–1744 (2007)
5. Chen, B.: Modeling and testing smooth structural changes with endogenous regressors. J. Econom. **185**, 196–215 (2015)
6. D'Agostino, A., Gambetti, L., Giannone, D.: Macroeconomic forecasting. In: Working Paper Series (2010) https://www.ecb.europa.eu/pub/pdf/scpwps/ecbwp1167.pdf. Accessed on 25 June 2015
7. Kim, S.H.:The demand for international travel and tourism to South Korea. Ph.D. thesis, University of Santo Tomas, Manila (1988)
8. Kositchotethana, B.: Thailand drops in global tourism rankings. In: Bangkok Post, http://www.bangkokpost.com/news/general/536583/thailand-drops-in-global-tourism-rankings. Accessed on 25 June 2015, 22 Apr 2015
9. Ngai, L.R., Pissarides, C.A.: Structural change in a multisector model. Am. Econ. Rev. **97**(1), 429–443 (2007)
10. Serletis, A., Shahmoradi, A.: Chaos, self-organized criticality, and SETAR nonlinearity: an analysis of purchasing power parity between Canada and the United State. Chaos Solitons Fractals **33**, 1437–1444 (2007)
11. Song, H., Li, G., Witt, S.F., Fei, B.: Tourism demand modelling and forecasting: how should demand be measured? Tour. Econ. **16**(1), 63–81 (2010)
12. Tong, H., Lim, K.S.: Threshold autoregression, limit cycles and cyclical data. J. R. Stat. Soc. **42**(3), 245–292 (1980)

13. Tong, H.: Non-linear Time Series: A Dynamical System Approach. Oxford University Press, Oxford (1990)
14. UNWTO : Why tourism? In: World Tourism Organization UNWTO, http://www2.unwto.org/content/why-tourism. Accessed on 25 June 2015
15. Van Dijk, D., Tersvirta, T., Franses, P.H.: Smooth transition autoregressive models—a survey of recent developments. Econom. Rev. **21**(1), 1–47 (2002)
16. Vanhaleweyk, G. : Tourist arrivals to Thailand by nationality 2013 and 2014. In: Thai-websites.com (2014) http://www.thaiwebsites.com/tourists-nationalities-Thailand-2014.asp. Accessed on 3 July 2015

Dependence Between Volatility of Stock Price Index Returns and Volatility of Exchange Rate Returns Under QE Programs: Case Studies of Thailand and Singapore

Ornanong Puarattanaarunkorn, Teera Kiatmanaroch and Songsak Sriboonchitta

Abstract This study found the evidences of the dependence between the volatility of stock price index returns and the volatility of exchange rate returns measured against US Dollar and Japanese Yen, and the independence between the volatility of stock price index returns and the volatility of exchange rate returns measured against Euro, in both Thailand and Singapore, under the operation of QE programs. It also found that all bivariate copula of the volatility of stock price index returns—the volatility of Thai Baht/US Dollar exchange rate returns, and the volatility of stock price index returns—the volatility of Thai Baht/Japanese Yen of Thailand, had a degree of dependence greater than that of Singapore. This can be explained that the QE programs can affect capital flows to Thailand and Singapore, and also may have different effects on the volatility of each exchange rate returns and the volatility of stock price index returns, of the individual country. This information can be useful for policy makers and investors so that they can directly focus on avoiding adverse implications from the operation of QE programs, in terms of the risks incurred from the volatility of exchange rate returns and the volatility of stock price index returns.

O. Puarattanaarunkorn
Department of Economics, Faculty of Management Sciences,
Khon Kaen University, Khon Kaen, Thailand
e-mail: pornan@kku.ac.th

T. Kiatmanaroch
Department of Agricultural Economics, Faculty of Agriculture,
Khon Kaen University, Khon Kaen, Thailand
e-mail: teera@kku.ac.th

S. Sriboonchitta (✉)
Faculty of Economics, Chiang Mai University, Chiang Mai, Thailand
e-mail: songsakecon@gmail.com

V.-N. Huynh et al. (eds.), *Causal Inference in Econometrics*,
Studies in Computational Intelligence 622, DOI 10.1007/978-3-319-27284-9_27

1 Introduction

The 2007–2009 financial crisis in the United States, United Kingdom, Japan, and the Euro area, played an important role in the collapse of major businesses. The crisis has caused the decline of economic activities, thus leading to a global recession. Consequently, the Federal Reserve of US (Fed), the Bank of Japan (BOJ), the Bank of England (BOE), and the European Central Bank (ECB) began establishing monetary policies, including the quantitative easing (QE) program, to stimulate an economic growth. The QE policies can temporary increase the monetary base that is used to provide liquidity for short periods. These programs include asset purchases and lending programs that are varied across the Fed, BOJ, BOE, and ECB [1]. The QE can have an effect on the macroeconomy and financial market in particular. For example, Neely [2] showed that the Fed's QE in 2008–2009 had significantly reduced international long-term bond yields and the spot value[1] of the Dollar. Meaning and Zhu [3] showed that the Fed's QE announcement led to depreciations in the nominal effective exchange rates of the US Dollar. Fawley and Neely [1] showed that the purchasing short-term securities by the QE program can have an effect on the exchange rates and stock prices. Likewise, Fratzscher et al. [4] found that the QE pushed up the equity prices worldwide and led to a depreciation on the US Dollar.

The QE programs also had an effect on Asia. Cho and Rhee [5] found that the first round of Fed's QE program significantly contributed to the capital inflows to the 10 large regional economies: China, Hong Kong, India, Indonesia, Japan, the Republic of Korea, the Philippines, Singapore, Taiwan, and Thailand. Moreover, the QE programs have contributed to the appreciation of local currency value and increasing the asset prices. Chen et al. [6] showed that the Fed's QE resulted in lower emerging market Asian bond yields, raised equity price, and caused an appreciation to the local currency. Similarly, the QE caused the capital inflows to Thailand; the Thai Baht against the US Dollar had appreciated, and the stock price index had risen [7].

As it is known, an increase of capital flows creates a demand for and supply of currencies, leading to an interdependence between the stock prices and the exchange rates [8]. Therefore, it is interesting that what is the interdependence between the exchange rates and the stock price index within the ASEAN region, under the operation of QE programs? We consider two countries, namely Thailand and Singapore, and examine the linkages between exchange rates and stock prices index for each country in terms of the dependence between the volatility of exchange rate returns and the volatility of stock price index returns. The reasons for choosing these two countries are because Thailand is a developing country and it has a strategic location as a gateway into Asia, while Singapore is a developed country where its financial markets are fully liberalized.

[1] The current exchange rate or the rate of a foreign-exchange contract for immediate delivery.

Thus, the purpose of this study is to investigate the dependence between the volatility of stock price index returns and the volatility of exchange rate returns that are measured against the US Dollar; Euro; and Japanese Yen for each country; Thailand and Singapore, under the operation of QE programs.

Understanding the dependence between the volatilities of exchange rate returns and stock price index returns, under the operation of QE programs for each country in the ASEAN, may help the ASEAN policy makers and the investors manage the risk, and to avoid adverse implications from the operation of unconventional monetary policies of advanced economies.

To answer the research question of this study, the copula based ARMA-GARCH model was used. As Sriboonchitta et al. [9] argued that to find the dependence by the traditional multivariate analysis imposed some strong assumptions and suggested to use the copula approach that could provide better flexibility for the analysis. Since financial data usually shows the evidence of non-normal distribution with skewness and excess kurtosis that might not be the same margins for each random variable. Thus, using traditional approach that is restricted with normal distribution and linear correlation, is not appropriate. But the copula can cross this restriction, and offers us the flexibility of merging a univariate distribution to get a joint distribution with an appropriate dependence structure. Recently, many studies have used the copula model to find the dependence of financial data; see, e.g. [10–16].

The remainder of this paper is organized as follows: Sect. 2 describes the method which is used for analysis. Section 3 describes the data and the descriptive statistics of the stock price index and the exchange rates. Section 4 presents the empirical results. Lastly, Sect. 5 presents the conclusions and policy implications.

2 Methodology

This study aims to find the dependence between the volatility of exchange rate returns and the stock price index returns for each country, Thailand and Singapore in the ASEAN region. As we have known, the financial data usually shows the evidence of autoregression and volatility clustering. Moreover, they also exhibit evidence of non-normal distribution with skewness and excess kurtosis that might not be the same margins for each random variable [10]. As the traditional multivariate analysis imposed some strong assumptions of normal distribution and linear correlation. One approach of modeling the multivariate dependence is the copula, which provides a better flexibility for the analysis [9].

Therefore, we employed the ARMA-GARCH model and the copula model to analyze the dependence structure of data series. First, the ARMA-GARCH model was used to filter the marginal distribution, $F_i(x_i)$, of each data series. Then the marginal distributions were used to obtain the copula data. After that, the copula model was used to find the dependence between the marginal distributions.

2.1 ARMA-GARCH

The ARMA(p,q)-GARCH(1,1) model for each log-return data series (y_t) can be written in general form as:

$$y_t = a_0 + \sum_{i=1}^{p} a_i y_{t-i} + \sum_{i=1}^{q} b_i \varepsilon_{t-i} + \varepsilon_t \tag{1}$$

$$\varepsilon_t = z_t \sqrt{h_t},\ z_t \sim D \tag{2}$$

$$h_t = \omega_t + \alpha \varepsilon_{t-1}^2 + \beta h_{t-1} \tag{3}$$

Equation (1) presents ARMA(p,q) process where y_{t-i} is an autoregressive term of y_t and ε_t is an error term. Equation (2) then defines this error term as the product between conditional variance h_t and a residual z_t. A residual z_t is assumed to follow an appropriate distribution (D). Equation (3) presents GARCH(1,1) process where $\omega_t > 0$, $\alpha \geq 0$, $\beta \geq 0$ are sufficient to ensure that the conditional variance $h_t > 0$. The $\alpha \varepsilon_{t-1}^2$ represents the ARCH term and α refers to the short run persistence of shocks, while βh_{t-1} represents the GARCH term and β refers to the contribution of shocks to long run persistence $(\alpha + \beta)$.

For the next analysis, the standardized residuals with an appropriate distribution from ARMA(p,q)-GARCH(1,1) model of each data series is transformed to copula data.

2.2 Copula

A copula is the joint distribution of random variables, in which the marginal distribution of each variable is uniform [0,1]. The fundamental theorem of copula is the theorem proposed by Sklar [17] as the Sklar's theorem.

Let H be a joint distribution function with marginal distributions F, G. Then there exists a copula C for all x, y in real line, with the following property:

$$H(x, y) = C(F(x), G(y)) \tag{4}$$

If F, G are continuous, then C is unique. Conversely, if C is a copula and F, G are distribution functions, then the above function $H(x, y)$ in Eq. (4) is a joint distribution function with the marginal distributions F, G.

If H is known, the copula is an Eq. (4) that one can obtain from this expression,

$$C(u, v) = F(F^{-1}(u), G^{-1}(v)) \tag{5}$$

where F^{-1} and G^{-1} are are the inverse distribution functions of the marginals.

2.3 Maximum Likelihood Estimation

The method of maximum pseudo-log likelihood studied by Genest et al. [18] is used for estimation since the marginal distribution functions F and G of random vectors are unknown. Thus, we can construct pseudo copula observations by using the empirical distribution functions to transform the standardized residuals series from the ARMA(p,q)-GARCH(1,1) model to uniform [0,1].

Under the assumption the marginal distributions F and G are continuous, the copula C_θ is a bivariate distribution with density c_θ and pseudo-observations $F_n(X_i)$ and $G_n(Y_i)$, $i = 1, 2, \ldots, n$.

The pseudo-log likelihood function of θ is expressed as

$$L(\theta) = \Sigma_{i=1}^{n} \log[c_\theta(F_n(X_i), G_n(Y_i))]. \tag{6}$$

where $c_\theta = \frac{\partial^2 C_\theta(F_n(x), G_n(y))}{\partial x \partial y}$, $F_n(x) = \frac{1}{n+1}\Sigma_{i=1}^{n} 1(X_i \leq x)$ and $G_n(x) = \frac{1}{n+1}\Sigma_{i=1}^{n} 1$ $(Y_i \leq y)$ are the empirical distributions.

2.4 Copula Families

There are various copula families to measure the dependence of copula. Each copula family has different dependence structure or joint distribution. This study used five copula families. The Gaussian (Normal) copula and the Student's T, which are the Elliptical copulas. The Elliptical copulas are simply the copulas of elliptically contoured distributions and have radical symmetry. The Frank copula, the Rotated Clayton 90°, and the Rotated Gumbel 90°, all are the Archimedean copulas. The Rotated Clayton 90° copula is an asymmetric copula by exhibiting greater dependence in the

Table 1 Copula families

Name	Copula function	Parameter range
Gaussian	$C\,(u_1, u_2; \rho) = \Phi_G(\Phi^{-1}(u_1), \Phi^{-1}(u_2); \rho) = \int_{-\infty}^{\phi^{-1}(u_1)} \int_{-\infty}^{\phi^{-1}(u_2)} \frac{1}{2\Pi\sqrt{(1-\rho^2)}} \times [\frac{-(s^2-2\rho st+t^2)}{2(1-\rho^2)}]dsdt$	$\rho \in (-1, 1)$
Student's T	$C^T\,(u_1, u_2; \rho, \nu) = \int_{-\infty}^{T_\nu^{-1}(u_1)} \int_{-\infty}^{T_\nu^{-1}(u_2)} \frac{1}{2\Pi\sqrt{(1-\rho^2)}} \times [1 + \frac{(s^2-2\rho sT+T^2)}{\nu(1-\rho^2)}]^{-(\frac{\nu+2}{2})}dsdT$	$\rho \in (-1, 1), \nu > 2$
Frank	$C(u_1, u_2; \theta) = -\frac{1}{\theta}\log(1 + \frac{(e^{-\theta u_1}-1)(e^{-\theta u_2}-1)}{e^{-\theta}-1})$	$\theta \in (-\infty, \infty)\backslash\{0\}$
Rotated Clayton 90°	$C(u_1, u_2; \theta) = u_2 - [(1-u_1)^{-\theta} + u_2^{-\theta} - 1]^{-\frac{1}{\theta}}$	$\theta \in (-\infty, 0)$
Rotated Gumbel 90°	$C(u_1, u_2; \theta) = u_2 - exp(-[(-\ln(1-u_1))^{\theta} + (-\ln(u_2))^{\theta}]^{\frac{1}{\theta}})$	$\theta \in (-\infty, -1]$

Source Trivedi and Zimmer [20], Nelson [21], and Fisher [22]

left tail than in the right, and it can capture negative dependence. The Frank copula is a symmetric copula, and it can capture both positive and negative dependences. The Rotated Gumbel 90° copula is an asymmetric copula by exhibiting greater dependence in the right tail than in the left, and it can capture negative dependence. Selecting a family of copulas is based upon information criteria such as Akaike Information Criterion (AIC) by Akaike [19].

The copula function and parameter range of each copula family that are used in this paper are shown in Table 1.

3 Data

The weekly data of the exchange rates and the stock price index of each country from 28/11/2008 to 5/6/2015 (the QE program began in 2008), are used. For Thailand, the stock price index is the Bangkok SET (SET), the exchange rates are Thai Baht/Euro (TEU); Thai Baht/US Dollar (TUS); Thai Baht/ Japanese Yen (TJPY). For Singapore, the stock price index is the Straits Times Index Local Currency (STI), the exchange rates are Singaporean Dollar/Euro (SEU); Singaporean Dollar/US Dollar (SUS); Singaporean Dollar/Japanese Yen (SJPY). The data series are obtained from the Thomson Reuters Datastream database. Each data series are transformed to the log-return $\ln \frac{R_t}{R_{t-1}}$.

Table 2 presents the descriptive statistics of the returns of each data series of Thailand. All data series have a negative skewness except TJPY, and the excess kurtosis. This means that the data series have asymmetric distributions and heavy tail. The null hypotheses of normality of the Jarque-Bera tests are rejected in all data series. The Augmented Dickey-Fuller (ADF) test shows that these data series are stationary at p-value less than 0.01.

Table 2 Data descriptive statistics for returns of stock price index and exchange rates for Thailand

	SET	TEU	TUS	TJPY
Mean	0.0039	−0.0005	−0.0001	−0.0009
Median	0.0072	−0.0006	0.0003	2.51e-05
Maximum	0.0781	0.0452	0.0219	0.0600
Minimum	−0.0775	−0.0504	−0.0251	−0.0407
Std. dev.	0.0246	0.0136	0.0065	0.0136
Skewness	−0.3047	−0.1863	−0.0695	0.0893
Kurtosis	4.0838	3.8604	4.4737	3.7006
Jarque-Bera (p-value)	21.901 (<1.755e-05)	12.456 (<1.82e-07)	31.038 (0.0000)	7.405 (<0.025)
p-value of Dickey-Fuller test	0.01	0.01	0.01	0.01
Number of observations	340	340	340	340

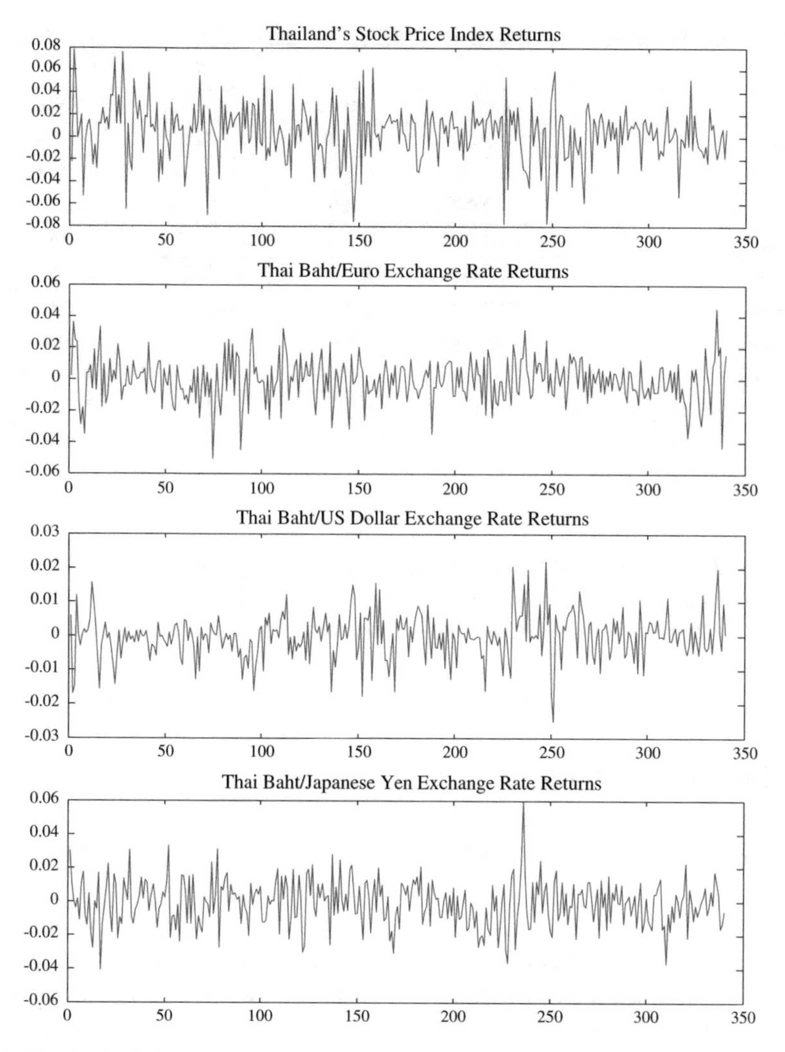

Fig. 1 Stock price index returns and exchange rate returns of Thailand

Figure 1 shows the returns of each data series of Thailand during 28/11/2008–5/6/2015. Each data series has great volatility.

Table 3 presents the descriptive statistics of the returns of each data series of Singapore. All data series have a positive skewness, meaning that the data series have asymmetric distributions and have a long right tail. Moreover, all data series show the excess kurtosis indicating their heavy tails. The null hypotheses of normality of the Jarque-Bera tests are rejected in all data series, except SEU. The Augmented Dickey-Fuller (ADF) test shows that these data series are stationary at p-value less than 0.01.

Table 3 Data descriptive statistics for returns of stock prices index and exchange rates for Singapore

	STI	SEU	SUS	SJPY
Mean	0.0019	−0.0007	−0.0003	−0.0011
Median	0.0023	−0.0008	−0.0002	−0.0014
Maximum	0.1532	0.0329	0.0436	0.0473
Minimum	−0.0671	−0.0338	−0.0261	−0.0450
Std. dev.	0.0223	0.0112	0.0079	0.0143
Skewness	0.8985	0.0262	0.4669	0.3173
Kurtosis	9.7142	3.2047	6.0691	3.8917
Jarque-Bera (p-value)	684.381 (<2.2e-16)	0.632 (<0.729)	145.781 (<2.2e-16)	16.969 (0.0002)
p-value of Dickey-Fuller test	0.01	0.01	0.01	0.01
Number of observations	340	340	340	340

Figure 2 shows the returns of each data series of Singapore during 28/11/2008–5/6/2015. Each data series has great volatility. However, the stock price index returns (STI) of Singapore have a volatility that is less than the stock price index returns (SET) of Thailand.

4 Empirical Results

4.1 Result of ARMA-GARCH Model

Table 4 presents the results of an appropriate ARMA(p,q)-GARCH(1,1) model for each data series of Thailand. The Akaike Information Criterion (AIC) by Akaike (1973) is used to select an appropriate ARMA(p,q)-GARCH(1,1). The results show that ARMA(1,2)-GARCH(1,1) with skewed student T residual is fitted for Thailands stock price index returns (SET) series. For the Thai Baht/Euro exchange rate returns (TEU), ARMA(1,1)-GARCH(1,1) with student T residual is appropriate. The Thai Baht/US Dollar exchange rate returns (TUS), ARMA(0,1)-GARCH(1,1) with skewed normal residual is fitted. The Thai Baht/Japanese Yen exchange rate returns (TJPY), ARMA(1,0)-GARCH(1,1) with skewed normal residual is fitted. Moreover, there exist both short run (α) and long run ($\alpha + \beta$) persistence of volatilities in all data series, except that of the TJPY because a parameter α is insignificant at level 0.05.

Table 5 presents the results of an appropriate ARMA(p,q)-GARCH(1,1) model for each data series of Singapore. The results show that ARMA(2,2)-GARCH(1,1) with student T residual is fitted for Singapore's stock price index returns (STI) series. For the Singaporean Dollar/Euro exchange rate returns (SEU), ARMA(4,4)-

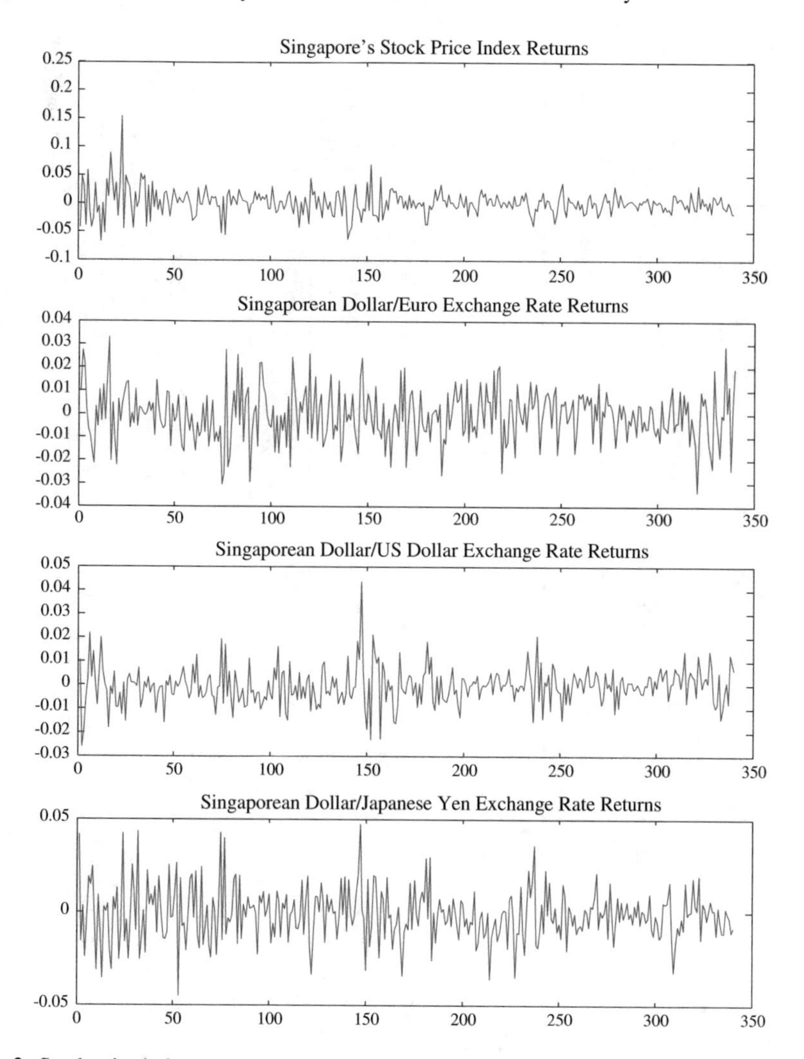

Fig. 2 Stock price index returns and exchange rate returns of Singapore

GARCH(1,1) with normal residual is appropriate. The Singaporean Dollar/US Dollar exchange rate returns (SUS), ARMA(2,2)-GARCH(1,1) with student T residual is fitted. The Singaporean Dollar/Japanese Yen exchange rate returns (SJPY), ARMA(1,1)-GARCH(1,1) with skewed normal residual is fitted. Similar to Thailand data series, there exist both short run (α) and long run persistence ($\alpha + \beta$) of volatilities in all data series.

After that, the standardized residuals of each appropriated ARMA-GARCH model are transformed into uniform [0,1], u_1 and u_2, by using the empirical distribution function. The Kolmogorov-Smirnov (K-S) test and the Box-Ljung test are used to check whether the marginal distributions that we transformed are correctly specified,

Table 4 Results of ARMA(1,2)-GARCH(1,1) with Skewed Student T Residual for SET, ARMA(1,1)-GARCH(1,1) with Student T Residual for TEU, ARMA(0,1)-GARCH(1,1) with Skewed Normal Residual for TUS, and ARMA(1,0)-GARCH(1,1) with Skewed Normal Residual for TJPY

	SET	Std. error (p-value)	TEU	Std. error (p-value)	TUS	Std. error (p-value)	TJPY	Std. error (p-value)
mu	2.143e-03	1.012e-03 (0.034*)	−5.266e-05	6.072e-05 (0.386)	−6.612e-04	3.722e-04 (0.076)	−9.329e-04	7.077e-04 (0.187)
ar1	3.850e-01	1.494e-01 (0.010**)	9.450e-01	5.694e-02 (<2e-16***)	-	-	4.109e-02	5.753e-02 (0.475)
ma1	−4.287e-01	1.576e-01 (0.007**)	−9.324e-01	6.679e-02 (<2e-16***)	1.409e-01	5.688e-02 (0.013*)	–	–
ma2	8.536e-02	5.056e-02 (0.091)	–	–	–	–	–	–
ω	5.013e-05	3.496e-05 (0.152)	8.237e-06	5.859e-06 (0.160)	2.766e-06	1.503e-06 (0.066)	1.672e-05	1.214e-05 (0.169)
α	1.234e-01	5.165e-02 (0.017*)	1.021e-01	4.200e-02 (0.015*)	1.269e-01	5.409e-02 (0.019*)	7.914e-02	4.294e-02 (0.065)
β	8.049e-01	8.204e-02 (<2e-16***)	8.600e-01	5.378e-02 (<2e-16***)	8.124e-01	6.909e-02 (<2e-16***)	8.301e-01	8.940e-02 (<2e-16***)
ν (degree of freedom)	5.202e+00	1.653e+00 (0.002**)	1.000e+01	3.592e+00 (0.005 **)	–	–	–	–
γ (skewness)	8.103e-01	5.702e-02 (<2e-16***)	–	–	9.261e-01	5.630e-02 (<2e-16***)	9.335e-01	7.559e-02 (<2e-16***)
Log likelihood	805.370	–	992.417	–	1,245.117	–	986.6354	–
AIC	−1592.74	–	−1970.83	–	−2478.23	–	−1961.27	–

(continued)

Table 4 (continued)

	SET	Std. error (p-value)	TEU	Std. error (p-value)	TUS	Std. error (p-value)	TJPY	Std. error (p-value)
p-value of K-S test	1	–	1	–	1	–	1	–
p-value of Box-Ljung test	–	–	–	–	–	–	–	–
1st moment	0.948	–	0.690	–	0.120	–	0.929	–
2nd moment	0.868	–	0.126	–	0.265	–	0.328	–
3rd moment	0.683	–	0.801	–	0.101	–	0.946	–
4th moment	0.942	–	0.248	–	0.443	–	0.331	–

Note Significant codes: 0 '***' 0.001 '**' 0.01 '*' 0.05 '.' 0.1 ' ' 1

Table 5 Results of ARMA(2,2)-GARCH(1,1) with Student T Residual for STI, ARMA(4,4)-GARCH(1,1) with Normal Residual for SEU, ARMA(2,2)-GARCH(1,1) with Student T Residual for SUS, and ARMA(1,1)-GARCH(1,1) with Skewed Normal Residual for SJPY

	STI	Std. error (p-value)	SEU	Std. error (p-value)	SUS	Std. error (p-value)	SJPY	Std. error (p-value)
mu	4.925e-03	2.292e-03 (0.032*)	−9.251e-04	5.039e-07 (<2e-16***)	−2.340e-04	2.642e-04 (0.376)	−1.983e-03	1.022e-03 (0.052)
ar1	−8.374e-01	1.291e-01 (8.80e-11***)	−7.426e-01	4.073e-05 (<2e-16***)	8.993e-01	1.241e-01 (4.38e-13***)	−4.794e-01	2.301e-01 (0.037*)
ar2	−7.999e-01	7.397e-02 (<2e-16***)	−6.036e-02	3.933e-05 (<2e-16***)	−6.455e-01	1.678e-01 (0.0001***)	–	–
ar3	–	–	4.659e-01	4.032e-05 (<2e-16***)	–	–	–	–
ar4	–	–	−1.124e-01	4.064e-05 (<2e-16***)	–	–	–	–
ma1	8.986e-01	1.177e-01 (2.26e-14***)	8.222e-01	3.945e-05 (<2e-16***)	−8.471e-01	1.382e-01 (8.78e-10***)	4.595e-01	2.444e-01 (0.060)
ma2	8.474e-01	8.126e-02 (<2e-16***)	1.537e-01	3.959e-05 (<2e-16***)	5.997e-01	1.725e-01 (0.0005***)	–	–
ma3	–	–	−4.793e-01	3.947e-05 (<2e-16***)	–	–	–	–
ma4	–	–	−2.252e-01	3.961e-05 (<2e-16***)	–	–	–	–
ω	2.205e-05	1.206e-05 (0.068)	4.167e-06	3.348e-06 (0.213)	3.242e-06	2.026e-06 (0.110)	7.991e-06	5.637e-06 (0.156)
α	1.944e-01	7.100e-02 (0.006**)	5.708e-02	2.741e-02 (0.037*)	1.357e-01	5.124e-02 (0.008**)	1.173e-01	5.210e-02 (0.024*)
β	7.627e-01	7.314e-02 (<2e-16***)	9.040e-01	4.460e-02 (<2e-16***)	8.147e-01	6.194e-02 (<2e-16***)	8.458e-01	6.100e-02 (<2e-16***)

(continued)

Table 5 (continued)

	STI	Std. error (p-value)	SEU	Std. error (p-value)	SUS	Std. error (p-value)	SJPY	Std. error (p-value)
ν (degree of freedom)	6.252e+00	2.064e+00 (0.003**)	–	–	6.600e+00	2.330e+00 (0.005**)	–	–
γ (skewness)	–	–	–	–	–	–	1.069e+00	7.536e-022 (<2e-16***)
Log likelihood	883.958	–	1084.511	–	1206.148	–	979.82	–
AIC	−1,749.924	–	−2,145.02	–	−2,394.30	–	−1,945.64	–
p-value of K-S test	1	–	1	–	1	–	1	–
p-value of Box-Ljung test	–	–	–	–	–	–	–	
1st moment	0.814	–	0.830	–	0.583	–	0.643	–
2nd moment	0.916	–	0.196	–	0.609	–	0.842	–
3rd moment	0.357	–	0.968	–	0.877	–	0.686	–
4th moment	0.924	–	0.050	–	0.723	–	0.763	–

Note Significant codes: 0 '***' 0.001 '**' 0.01 '*' 0.05 '.' 0.1 ' ' 1

which means that u_1 and u_2 are uniform [0,1] and i.i.d, respectively. The results in Tables 4 and 5 show that the marginal distributions are uniform [0,1] and i.i.d. Therefore, our marginal distributions are not misspecified and can be used for the copula model.

4.2 Result of Copula Model

Copula model is used to find the dependence between the volatility of stock price index returns and the volatility of exchange rate returns measured against the US Dollar; Euro; and Japanese Yen for each country, Thailand and Singapore, under the operation of QE programs. Table 6 presents the results of bivariate copula for Thailand. The results show that the volatility of Thailand's stock price index returns and the volatility of Thai Baht/Euro exchange rate returns (SET–TEU) are independent at a significance level of 0.05. While we found some evidences of dependence between the volatility of Thailand's stock price index returns and the volatility of Thai Baht/US Dollar exchange rate returns (SET–TUS), and also the volatility of Thai Baht/Japanese Yen exchange rate returns (SET–TJPY).

For bivariate copula of SET–TUS, the Student's T copula is selected to describe the dependence structure between SET–TUS by taking into consideration the values of the AIC and a goodness-of-fit test of the Cramér-von Mises (CvM) statistic. The Student's T copula has two copula parameters and symmetric tail dependences. The estimated parameters are, $\theta = -0.478$ and $\nu = 9.235$, the Kendall's tau correlation is -0.317, the lower tail (T^L) and the upper tail (T^U) dependences are 0.0003. These results imply that there exists a relatively small negative dependence between the volatility of Thailand's stock price index returns and the volatility of Thai Baht/US Dollar exchange rate returns, under the operation of QE programs. The changes of Thailand's stock price index are related to the changes of Thai Baht/US Dollar exchange rates in the opposite direction.

From bivariate copula of SET–TJPY, the results also show that there exists a relative small negative dependence, under the operation of QE programs. The Frank copula is chosen to describe the dependence structure with copula parameter value of -2.157 and Kendall's tau correlation value of -0.229, and this bivariate copula shows the evidences of independences in lower and upper tails. The changes of Thailand's stock price index are related to the changes of Thai Baht/Japanese Yen exchange rates in the opposite direction.

From the comparison of Kendall's tau correlations, we found that the volatility of Thailand's stock price index returns was related more to the volatility of Thai Baht/US Dollar exchange rate returns (-0.317) than to the volatility of Thai Baht/Japanese Yen exchange rate returns (-0.229). Therefore, it can be said that the change in inflows—outflows of US Dollars has a greater effect on Thailand's stock price index than the change in inflows—outflows of Yen and Euro.

Table 7 presents the results of bivariate copula for Singapore. Similar to Thailand, the results show the independence between the volatility of Singapore's stock price

Table 6 Results of copula models for Thailand data

Bivariate Copula	Copula	Parameter	Std. error	p-value	Kendall's tau	T^L (lower tail dep.)	T^U (upper tail dep.)	AIC	p-value of CvM
SET–TEU*	Independence	$\theta = 0$	–	–	0	–	–	–	–
SET–TUS	Gaussian	$\theta = -0.467$	0.039	0.000	−0.310	0	0	−81.308	0.65
	Student's T	$\theta = -0.478$ $\nu = 9.235$	0.042 4.580	0.000 0.022	−0.317	0.0003	0.0003	−85.932	0.57
	Frank	$\theta = -3.154$	0.352	0.000	−0.326	0	0	−79.126	0.66
	Rotated Clayton 90°	$\theta = -0.701$	0.096	8.171e-13	−0.259	0	0	−69.772	0.06
	Rotated Gumbel 90°	$\theta = -1.391$	0.058	0.000	−0.281	0	0	−70.438	0
SET–TJPY	Gaussian	$\theta = -0.338$	0.046	5.457e-13	−0.219	0	0	−38.902	0.65
	Student's T	$\theta = -0.338$ $\nu = 200$	0.046 NA	6.681e-13 NA	−0.219	3.927e-50	3.927e-50	−36.741	0.4
	Frank	$\theta = -2.157$	0.336	2.209e-10	−0.229	0	0	−39.640	0.56
	Rotated Clayton 90°	$\theta = -0.380$	0.079	1.098e-06	−0.160	0	0	−27.297	0
	Rotated Gumbel 90°	$\theta = -1.229$	0.049	0.000	−0.187	0	0	−27.592	0.02

Note * The SET–TEU are independent by accepting the null hypothesis of bivariate independence (p-value = 0.75). The result is obtained from independence test for bivariate copula data in the R-package CDVine [23]

Table 7 Results of copula models for Singapore data

Bivariate Copula	Copula	Parameter	Std. error	p-value	Kendall's	T^L (lower tail dep.)	T^U (upper tail dep.)	AIC	p-value of CvM
STI–SEU*	Independence	$\theta = 0$	–	–	0	–	–	–	–
STI–SUS	Gaussian	$\theta = -0.336$	0.046	7.088e-13	−0.218	0	0	−38.553	0.46
	Student's T	$\theta = -0.321$ $\nu = 5.274$	0.055 1.993	7.975e-09 0.004	−0.208	0.012	0.012	−46.726	0.36
	Frank	$\theta = -2.005$	0.342	5.191e-09	−0.214	0	0	−32.234	0.10
	Rotated Clayton 90°	$\theta = -0.427$	0.082	1.753e-07	−0.176	0	0	−35.033	0.20
	Rotated Gumbel 90°	$\theta = -1.261$	0.051	0.000	−0.207	0	0	−38.882	0.04
STI–SJPY	Gaussian	$\theta = -0.391$	0.043	0.000	−0.255	0	0	−53.90	0.11
	Student's T	$\theta = -0.393$ $\nu = 22.032$	0.045 30.122	0.000 0.233	−0.257	2.123e-07	2.123e-07	−52.49	0.13
	Frank	$\theta = -2.478$	0.342	1.421e-12	−0.260	0	0	−50.90	0.16
	Rotated Clayton 90°	$\theta = -0.576$	0.090	2.748e-10	−0.224	0	0	−54.15	0.39
	Rotated Gumbel 90°	$\theta = -1.279$	0.052	0.000	−0.218	0	0	−38.80	0.00

Note * The STI–SEU are independent by accepting the null hypothesis of bivariate independence (p-value = 0.075). The result was obtained from independence test for bivariate copula data in the R-package CDVine [23]

index returns and the volatility of Singaporean Dollar/Euro exchange rate returns (STI–SEU), under the operation of QE programs, at significance level 0.05. While there exist the evidences of dependence between the volatility of Singapore's stock price index returns and the volatility of Singaporean Dollar /US Dollar exchange rate returns (STI–SUS), and also the volatility of Singaporean Dollar/Japanese Yen exchange rate returns (STI–SJPY).

For bivariate copula of STI–SUS, the Student's T copula is selected to describe the dependence structure with estimated parameters of $\theta = -0.321$ and $\nu = 5.274$, the Kendall's tau correlation is -0.208, the lower tail (T^L) and the upper tail (T^U) dependences are 0.012. These results imply that there exists a relatively small negative dependence between the volatility of Singapore's stock price index returns and the volatility of Singaporean Dollar/US Dollar exchange rate returns under the operation of QE programs. The changes of Singapore's stock price index are related to the changes of Singaporean Dollar/US Dollar exchange rates in the opposite direction.

For STI–SJPY, the Rotated Clayton 90° copula is selected to describe the dependence structure with the copula parameter of $\theta = -0.576$, the Kendall's tau correlation at -0.224, and no lower tail (T^L) and upper tail (T^U) dependences. The volatility of Singapore's stock price index returns has a relatively small negative dependence with the volatility of Singaporean Dollar/Japanese Yen exchange rate returns, which implies that the changes of Singapore's stock price index are related to the changes of Singaporean Dollar/US Dollar exchange rates in the opposite direction.

With dissimilar results compared to Thailand, the Kendall's tau correlation of STI–SUS (-0.208) is slightly less than the Kendall's tau correlation of STI–SJPY (-0.224) or the two values are relatively close. Therefore, it implies that the changes of inflows—outflows of US Dollars and also the changes of inflows—outflows of Yen, both have a similar effect on the change of Singapore's stock price index.

4.3 Comparison of Kendall's Tau Correlations Between Thailand and Singapore

Table 8 presents the comparison of the Kendall's tau correlations of each bivariate copula between Thailand and Singapore. Apparently, the Kendall's tau correlations between the volatility of Thailand's stock price index returns and the volatility of exchange rate returns of both; Thai Baht/ US Dollar (SET–TUS) and Thai Baht/ Japanese Yen (SET–TJPY), are greater than the Kendall's tau correlations between the volatility of Singapore's stock price index returns and the exchange rate returns of both; Singaporean Dollar/ US Dollar (STI–SUS) and Singaporean Dollar/ Japanese Yen (STI–SJPY).

This implies that the volatility of capital inflows of both US Dollars and Japanese Yen to Thailand's financial market can have an influence on the volatility of Thailand's stock price index which is greater than Singapore, under the operation of QE programs. This result is consistent to Fig. 3, which presents the capital inflows to each country (% of GDP).

Table 8 Comparison of Kendall's tau correlations

Thailand	Kendall's tau correlation	Singapore	Kendall's tau correlation
SET–TEU	Independence	STI–SEU	Independence
SET–TUS	−0.317	STI–SUS	−0.208
SET–TJPY	−0.229	STI–SJPY	−0.224

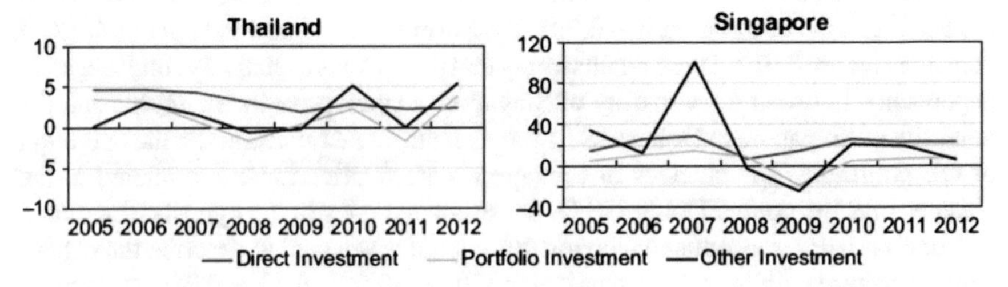

Fig. 3 Capital inflows to Thailand and Singapore (% of GDP) (*Source* Cho and Rhee [5])

Figure 3 shows that the capital inflows to Singapore exceed 10 % of GDP [5], while Thailand is about 5 % of GDP. However, under the operation of QE programs that began from 2008, the capital inflows to Thailand has more volatility than Singapore. This may be due to the fact that Thailand is an emerging financial market and its stock prices have a value that is less than those of Singapore, while Singapore is a regional financial hub. Therefore, the two countries have a different degree of financial market liberalization. This may be the reason why Thailand has more volatility of capital inflows, which in turn causes the volatility in Thailand's exchange rates and stock price index.

5 Conclusions and Policy Implications

This study aims to investigate the dependence between the volatility of stock price index returns and the volatility of exchange rate returns measured against the US Dollar; Euro; and Japanese Yen for each country, Thailand and Singapore, under the operation of QE programs. The empirical results found the evidences of dependence between the volatility of stock price index returns and the volatility of exchange rate returns measured against US Dollar and Japanese Yen for both Thailand and Singapore, but with a different degree of dependence. While it was found the independence between the volatility of stock price index returns and the volatility of exchange rate returns measured against Euro at significance level 0.05 for both Thailand and Singapore.

The difference in degree of dependence can be discussed in two issues. First, there was a different degree of dependence within the individual country. Second, there was a different degree of dependence between the two countries. For Thailand, the results show that the dependence between the volatility of Thailand's stock price index returns and the volatility of Thai Baht/US Dollar exchange rate returns (−0.317) is greater than the dependence between the volatility of Thailand's stock price index returns and the volatility of Thai Baht/Japanese Yen exchange rate returns (−0.229). This is because the Thai Baht/US Dollar exchange rate returns have more volatility than Thai Baht/Japanese Yen exchange rate returns, under the operation of QE programs, as shown in Fig. 1. For Singapore, the dependence between the volatility of Singapore's stock price index returns and the volatility of Singaporean Dollar/US Dollar exchange rate returns (−0.208) is slightly less than the dependence between the volatility of Singapore's stock price index returns and the volatility of Singaporean Dollar/Japanese Yen exchange rate returns (−0.224). This is because Singaporean Dollar/US Dollar exchange rate returns have less volatility than Singaporean Dollar/Japanese Yen exchange rate returns, under the operation of QE programs, as shown in Fig. 2.

When considering the total asset purchase program size of the QE programs, the Fed's asset purchase programs are the largest at $3,152, and are followed by the BOJ at $2,193 and ECB at $432, respectively [1]. Thus, they may have different effects on the capital flows of the US Dollar, Japanese Yen, and Euro to financial markets of Thailand and Singapore, and also may have a different influence on the volatility of each exchange rate returns and the volatility of stock price index returns for the individual country.

Moreover, the comparison results of dependence between Thailand and Singapore show that the dependence between the volatility of stock price index returns and the volatility of exchange rate returns measured against US Dollar and Japanese Yen of Thailand, are greater than those of Singapore; this is shown in Table 8. This may be due to the fact that the two countries have a different response to the QE programs, which causes the sizes of volatility of stock price index returns and the volatility of exchange rate returns to become different in each country.

According to the above findings and interpretation, we can conclude that under the operation of QE programs, there are different degrees of capital flows to each country, thus exerting a different influence on the volatility of stock price index returns and the volatility of exchange rate returns in the individual country. Therefore, the empirical results show the evidences of dependence between the volatility of stock price index returns and the volatility of exchange rate returns in different degree, for Thailand and Singapore. This information can be useful for policy makers that want to directly focus on avoiding adverse implications from the operation of QE programs, such as maintaining the size of reserve and the exchange rates that are in line with the increasing volatility of financial flows. Similarly, the investors should monitor closely the risks incurred from the change of exchange rates that have more volatility under the operation of QE programs, which are also related to the volatility of stock prices.

Acknowledgment This work was supported by Faculty of Management Sciences, Khon Kaen University, Thailand.

References

1. Fawley, B.W., Neely, C.J.: Four stories of quantitative easing. Fed. Reserv. Bank of St. Louis Rev. **95**(1), 51–88 (2013)
2. Neely, C.J.: Unconventional Monetary Policy Had Large International Effects. Federal Reserve Bank of St. Louis Working Papers, 2010
3. Meaning, J., Zhu, F.: The impact of recent central bank asset purchase programmes. BIS Q. Rev. 73–83 (2011)
4. Fratzscher, M., Duca, M.L., Straub, R.:On the international spillovers of US quantitative easing. Working Paper Series No. 1557 June 2013, European Central Bank (2013)
5. Cho, D., Rhee, C.: Effects of quantitative easing on asia: capital flows and financial markets. ADB Economics Working Paper Series No. 350, June, 2013
6. Chen, Q., et al.: International spillovers of central bank balance sheet policies. BIS Papers No. 66. Bank for International Settlements (BIS). Basel
7. Bank of Thailand: Unconventional monetary policy from the beginning to the countdown. Prasiam **1**, 3–9 (2014)
8. Kanas, A.: Volatility spillovers between stock returns and exchange rate changes: International evidence. J. Bus. Financ. Account. **27**(3&4), 447–467 (2000)
9. Sriboonchitta, S., Nguyen, H.T., Wiboonpongse, A., Liu, J.: Modeling volatility and dependency of agricultural price and production indices of Thailand: static versus time-varying copulas. Int. J. Approx. Reason. **54**(6), 793–808 (2013)
10. Puarattanaarunkorn, O., et al.: Dependence structure between TOURISM and TRANS sector indices of the stock exchange of Thailand. Thai J. Math. Spec. Issue: Copula Math. Econ. 199–210 (2014)
11. Kiatmanaroch, T., et al.: Will QE change the dependence between Baht/Dollar exchange rates and price returns of AOT and MINT? Thai J. Math. Spec. Issue: Copula Math. Econ. 129–144 (2014)
12. Autchariyapanitkul, K., et al.: Portfolio optimization of stock returns in high-dimensions: a copula-based approach. Thai J. Math. Spec. Issue: Copula Math. Econ. 11–23 (2014)
13. Sriboonchitta, S., et al.: A vine copula approach for analyzing financial risk and co-movement of the Indonesian, Philippine and Thailand stock markets. In: Huynh, V.N., Kreinovich, V., Sriboonchitta, S. (eds.) Modeling Dependence in Econometrics. AISC, vol. 251, pp. 245–257. Springer, Heidelberg (2014)
14. Boubaker, A., Salma, J.: Greek crisis, stock market volatility and exchange rates in the European monetary union: a var-garch-copula model. Global J. Manag. Bus. Res.: C Financ. **14**(2), 50–60 (2014)
15. Lin, F.: Tail dependence between stock index returns and foreign exchange rate returns a copula approach. In: Vogel, R. (eds.) Proceedings of the New York State Economics Association, vol. 5, pp. 129–139. New York (2012)
16. Ning, C.: Dependence structure between the equity market and the foreign exchange market—a copula approach. J. Int. Money Financ. **29**, 743–759 (2010)
17. Sklar, A.: Fonctions de rpartition n dimensions et leurs marges. Publications de l'Institut de Statistique de L'Universit de Paris **8**, 229–231 (1995)
18. Genest, C., Ghoudi, K., Rivest, L.P.: A semiparametric estimation procedure of dependence parameters in multivariate families of distributions. Biometrika **82**, 543–552 (1995)
19. Akaike, H.: Information theory and an extension of the maximum likelihood principle. In: Petrov, B.N., Csaki, F. (eds.) Proceedings of the Second International Symposium on Information Theory; 1973; Budapest, Akademiai Kiado 267–281 (1973)

20. Trivedi, P.K., Zimmer, D.M.: Copula modeling: an introduction for practitioners. Found. Trends Econ. **1**(1), 1–111 (2005)
21. Nelson, R.B.: An Introduction to Copulas, 2nd edn. Springer, New York (2006)
22. Fisher, M.: Tailoring copula-based multivariate generalized hyperbolic secant distributions to financial return data: an empirical investigation. Discussion Papers, University of Erlangen-Nuremberg, Germany (2003)
23. Schepsmeier, U., Brechmann, E.C.: Statistical inference of C- and D-vine copulas. http://cran.r-project.org/web/packages/CDVine/index.html (2012)

Seemingly Unrelated Regression Based Copula: An Application on Thai Rice Market

Pathairat Pastpipatkul, Paravee Maneejuk, Aree Wiboonpongse and Songsak Sriboonchitta

Abstract This paper introduced the seemingly unrelated regression (SUR) model based on Copula to improve a linear regression system since the conventional SUR model has a strong assumption of normally distributed residuals. The Copula density functions were incorporated into the likelihood to relax the restriction of the marginal distribution. The real dataset of Thai rice was used for an application comparing the conventional SUR model estimated by GLS and the Copula-based SUR model. The result indicated that the Copula-based SUR model performed slightly better than the conventional SUR. In addition, the estimated results showed that Gaussian Copula was the most appropriate function for being the linkage between the marginal distributions. Moreover, the marginal distributions also were tested, and the result showed that a normal distribution and student-t distribution were the best fit for the marginal distributions of demand and supply equations, respectively.

Keywords Seemingly unrelated regression · Copula · Thai rice market

P. Pastpipatkul · P. Maneejuk (✉) · S. Sriboonchitta
Faculty of Economics, Chiang Mai University, Chiang Mai, Thailand
e-mail: mparavee@gmail.com

P. Pastpipatkul
e-mail: ppthairat@hotmail.com

A. Wiboonpongse
Faculty of Economics, Prince of Songkla University, Hat Yai, Thailand

A. Wiboonpongse
Faculty of Agriculture, Department of Agricultural Economics
and Agricultural Extension, Chiang Mai University, Chiang Mai, Thailand

A. Wiboonpongse
Faculty of Economics, Institute for Sufficiency Economy Research
and Promotion, Chiang Mai University, Chiang Mai, Thailand

V.-N. Huynh et al. (eds.), *Causal Inference in Econometrics*,
Studies in Computational Intelligence 622, DOI 10.1007/978-3-319-27284-9_28

1 Introduction

The main interest of this paper is the seemingly unrelated regression (SUR) model, which was first publicized in Arnold Zellner's paper in 1962. At first, the model was used to construct a system of different linear equations where all equations were estimated simultaneously by applying a two-step generalize least squares estimator [16]. Since then, SUR model has been used widely for many applications especially in the estimation of demand -or supply- system; for examples, modeling the demand system for a given commodity when the data sets come from different households and estimating a market system when the model consists of two different equations, mainly demand and supply of any commodity. Looking at its properties, the SUR model could gain more efficiency in the estimation as it can combine information from the different equations by letting errors from the equations be related.

But apart from its advantage, the SUR model comes with a strong assumption that an error term of each equation is assumed to have -or must have- a normal distribution. In spite of its strong assumption, the SUR model still has been used for many studies, e.g., Baltagi [1], Takada et al. [15], Sparks [14], Bilodeau and Duchesne [2], Drton and Richardson [4], and Lar et al. [8]. Although their results turned out to be good, and the models performed well, the fact that the models were estimated under the strong assumption of normal distribution should not be ignored. This gives us a possibility that the SUR model could be improved by getting rid of or relaxing this assumption. For instance, the Gaussian Copula gives the bivariate normal distribution if and only if the margins are normally distributed [17], which is the assumption of the conventional SUR model, otherwise not.

To relax this strong assumption, we take into account an advantage of Copula approach. It is more flexible on account of its ability to link the different marginal distributions of residuals of each equation in the model. In particular it makes the model more realistic and far from the assumption like a normal distribution. Perhaps Copula is a motivation to help us reach the goal relaxing the normal distribution assumption of SUR model. To the best of our knowledge, Copula has often been applied in the regression model for being the joint distribution function of dependent variables, which have different distributions; for examples, Kolev and Paiva [7], Masarotto and Varin [9], Noh et al. [11], and Parsa and Klugman [12]. However, we have never seen Copula applied in SUR model in the analysis of demand and supply of the Thai rice market.

The idea of this paper is that we try to relax the normal distribution assumption of error term of each equation in SUR model by applying Copula approach in the estimation. Then, the equations in SUR model are allowed to have different marginal distributions of residuals and Copula function can link them as a joint distribution. Therefore, it is the purpose of this paper to introduce the Copula-based SUR model as an alternative to the conventional SUR model. The advantage of incorporating Copula in the model will be investigated by making a comparison to the conventional SUR model.

For an overview of this paper, we begin with our intention to improve the SUR model in the introduction. To see an empirical result using this model, the real dataset of Thai rice is employed. Since the market generally works as a system, it has at least demand and supply equations which work together. Moreover, rice is one of the most important agricultural commodities of Thailand, and Thai people have always taken an interest in. According to these reasons, the Thai rice market is concerned to be an application for this paper. The following Sect. 2 is the discussion on methodology used in this paper. Then in Sect. 3, the estimation based Copula will be entirely explained in procedure. The estimated results are presented in Sect. 4.

2 Methodology

2.1 Seemingly Unrelated Regression (SUR) Model

The SUR model introduced by Zellner [16] is a generalization of system of linear regression equations or multivariate regression equations. It consists several regression equations that have their own dependent variables and can also be estimated separately. But the advantage of the SUR model is that it can gain efficiency or improve estimation by combining information from different equations. Consider the structure of SUR model, as it consists several regression equations, let us say M equations, the system of M equations can be shown as in the following.

$$Y_i = X_i \beta_i + \varepsilon_i \quad i = 1, \ldots, M \tag{1}$$

where Y_i is a vector of dependent variables, X_i is a matrix of independent variables or regressors, and β_i is a vector of an unknown parameter called regression coefficients. ε_i is a vector of the error terms. The important assumption of the SUR model is that it assumes that the errors in the different equations are related. Thus we can assume that $E\left[\varepsilon_{ia}\varepsilon_{ib}|X\right] = 0 \;\; ; a \neq b$ whereas $E\left[\varepsilon_{ia}\varepsilon_{ib}|X\right] = \sigma_{ij}$.

That means the SUR model allows non-zero covariance between the error terms of different equations in the model. And the errors are considered to have a normal distribution, that is $\varepsilon_i = \left(\varepsilon_{t,1}, \ldots, \varepsilon_{t,M}\right) \backsim N(0, \Sigma)$ where the sigma matrix, Σ , is a variance-covariance matrix for M equations, such that

$$\Sigma\left(\varepsilon_t \varepsilon_t\right) = \begin{bmatrix} \sigma_{11} & \sigma_{12} & \cdot\,\cdot & \sigma_{1M} \\ \sigma_{21} & \sigma_{22} & & \sigma_{2M} \\ \cdot & & \cdot & \cdot \\ \cdot & & \cdot & \cdot \\ \sigma_{M1} & \cdot & \cdot\,\cdot & \sigma_{MM} \end{bmatrix} \tag{2}$$

The covariance matrix for system is

$$\Gamma = \begin{bmatrix} \sigma_{11}I & \sigma_{12}I & . . & \sigma_{1M}I \\ \sigma_{21}I & \sigma_{22}I & & \sigma_{2M}I \\ . & . & & . \\ . & . & & . \\ \sigma_{M1}I & . & . . & \sigma_{MM}I \end{bmatrix} = \Sigma \otimes I \quad (3)$$

where I is an identity matrix, this system equation can be estimated by

$$\beta_{sure} = (X'\Gamma^{-1}X)^{-1}X'\Gamma^{-1}Y \quad (4)$$

2.2 Copula Approach

The most fundamental theorem, which describes the dependence in Copula, is the Sklar's theorem. Sklar [13] has proposed the linkage between the marginal distributions which is possible to have different distributions with the same correlation, but different dependence structure. The linkage between the marginal distributions is called Copula. Formally, let H be an n-dimensional joint distribution function of the random variables x_n with marginal distribution functions F_n . So, there exist the n-Copula C such that for all X_n.

$$H(x_1, \ldots, x_n) = C(F_1(x_1), \ldots, F_n(x_n)) \quad (5)$$

where C is Copula distribution function of a n-dimensional random variable. If the marginals are continuous, C is unique. Equation 2 defines a multivariate distribution function F. Thus, we can model the marginal distribution and joint dependence separately. If we have a continuous marginal distribution, Copula can be determined by

$$C(u_1, \ldots, u_n) = C(F_1^{-1}(u_1), \ldots, F_n^{-1}(u_n)) \quad (6)$$

where u is uniform [0,1].

2.3 Copula Families

There are two important classes of Copula, namely elliptical Copula and Archimedean Copula. The symmetric Gaussian and t-Copula are the families of Copula in Elliptical Copula while the asymmetric rank, Clayton, Gumbel, Joe, and Ali-Mikhail-Haq (AMH) are five important families in Archimedean Copula.

2.3.1 Elliptical Copula

(1) Gaussian Copula
In the case of n-dimensional Gaussian or Normal Copula, we can rewrite Eq. 3 as

$$C(u_1,\ldots,u_n)=\Phi_n^{\Sigma_n}(\Phi_1^{-1}(u_1),\ldots,\Phi_n^{-1}(u_n)\qquad(7)$$

where $\Phi_d^{\Sigma_d}$ is n-dimensional standard normal cumulative distribution and $\Sigma-n$ variance-covariance matrix. The density of the Gaussian Copula can be calculated from Eq. 4 and has the following form

$$c\left(u_1,\ldots,u_n\right)=\left(\sqrt{det\,\Sigma_n}\right)^{-1}\left(\frac{1}{2}\left(\Phi_1^{-1}\left(u_1\right)\cdots\Phi_n^{-1}\left(u_n\right)\right)\cdot\left(\Sigma_n^{-1}-I\right)\cdot\begin{pmatrix}\Phi_1^{-1}\left(u_1\right)\\ \vdots\\ \Phi_n^{-1}\left(u_n\right)\end{pmatrix}\right)\qquad(8)$$

(2) T-Copula
In the case of n-dimensional student-t or T- Copula, we can rewrite Eq. 3 as

$$c\left(u_1,\ldots,u_n\right)=\int_{-\infty}^{t_v^{-1}(u_1)}\ldots\int_{-\infty}^{t_v^{-1}(u_n)}f_{t_1(v)}\left(x\right)dx\qquad(9)$$

where $f_{t_1(v)}\left(x\right)$ n-dimensional t-density is function with degree of freedom v and t_v^{-1} is the quantile function of a standard univariate t_v distribution. In the estimation, the density of t-Copula is evaluated by

$$c_{v,P}^{t}\left(u_1,\ldots,u_n\right)=\frac{f_{v,P}\left(t_v^{-1}\left(u_t\right)\ldots t_v^{-1}\left(u_t\right)\right)}{\Pi_{i=1}^{n}f_v\left(t_v^{-1}\left(u_t\right)\right)}\qquad(10)$$

where $f_{v,P}$ is the joint density of a $t_d(v,0,P)$-distributed random vector where P is the correlation matrix implied by the dispersion matrix Σ.

2.3.2 Archimedean Copula

According to Genest and MacKay [5], the general form of the Archimedean Copula can be defined as

$$c\left(u_1,\ldots,u_n\right)=\begin{cases}\Phi^{-1}\left(\Phi\left(u_t\right)+\cdots+\Phi\left(u_t\right)\right) & ,if\ \ \Sigma_{t=1}^{n}\Phi\left(u_i\right)\le\Phi\left(0\right)\\ 0 & ,\ \ otherwise\end{cases}\qquad(11)$$

where Φ is a decreasing function and strictly decreasing. The density of this class Copula namely, Frank, Clayton, Gumbel, Joe, and Ali-Mikhail-Haq (AMH), as proposed in Hofert et al. [6], can be written as follows:

(1) Frank Copula

$$c_{\theta}^{F}(u_1,\ldots,u_n)=\left(\frac{\theta}{1-exp(-\theta)}\right)^{d-1}Li_{-(n-1)}\left(h_{\theta}^{F}(u_1,\ldots u_n)\right)\frac{exp\left(-\theta\Sigma_{j=1}^{n}u_j\right)}{h_{\theta}^{F}(u_1,\ldots,u_2)} \tag{12}$$

where $h_{\theta}^{F}(u_1,\ldots,u_n)=\left(1-e^{-\theta}\right)^{1-n}\prod_{j=1}^{n}\left\{1-exp\left(-\theta u_j\right)\right\}$

(2) Clayton Copula

$$c(u_1,\ldots,u_n)=\prod_{k=0}^{n-1}(\theta k+1)\left(\prod_{j=1}^{n}u_i\right)^{-(1+\theta)}\left(1+t_{\theta}(u_1,\ldots u_n)\right)^{-n(n+1/\theta)} \tag{13}$$

(3) Gumbel Copula

$$c_{\theta}^{G}(u_1,\ldots,u_n)=\theta^{n}C_{\theta}(u_1,\ldots,u_n)\frac{\prod_{j=1}^{n}(-logu_i)^{\theta-1}}{t_{\theta}(u_1,\ldots u_n)\prod_{j=1}^{n}u_i} \tag{14}$$

where $P_{n,\alpha}^{G}(x)=\Sigma_{k=1}^{n}\alpha_{nk}^{G}(\alpha)x^{k}$

where $\alpha_{nk}^{G}(\alpha)=(-1)^{n-k}\Sigma_{n=k}^{n}\alpha^{j}s(n,j)S(k,j)=\frac{n!}{k!}\Sigma_{j=1}^{k}\binom{k}{j}\binom{\alpha j}{n}(-1)^{n-j},\ k\in\{1,\ldots,n\}$, s and S are the stirling numbers of the first kind and the second kind, respectively.

(4) Joe Copula

$$c_{\theta}^{J}(u_1,..,u_n)=\theta^{d-1}\frac{\prod_{j=1}^{n}\left(1-u_j\right)^{\theta-1}}{h_{\theta}^{J}(u_1,..,u_n)}\left(1-h_{\theta}^{J}(u)\right)^{\alpha}p_{n,\alpha}^{J}\left(\frac{h_{\theta}^{J}(u_1,..,u_n)}{1-h_{\theta}^{J}(u_1,..,u_n)}\right) \tag{15}$$

where $h_{\theta}^{J}(u_1,\ldots,u_n)=\prod_{j=1}^{n}\left(1-\left(1-u_j\right)^{\theta}\right)$

and $P_{n,\alpha}^{J}(x)=\Sigma_{k=0}^{n-1}\alpha_{nk}^{J}(\alpha)x^{k}$

(5) Ali-Makhail-Haq Copula

$$c_{\theta}^{A}(u_1,\ldots,u_n)=\frac{(1-\theta)^{n+1}}{\theta^{2}}\frac{h_{\theta}^{A}(u_1,\ldots,u_n)}{\prod_{j=1}^{n}u_j^{2}}Li_{-d}\left\{h_{\theta}^{A}(u_1,\ldots,u_n)\right\} \tag{16}$$

where $h_{\theta}^{J}(u_1,\ldots,u_n)=\theta\prod_{j=1}^{n}\frac{u_i}{1-\theta(1-u)}$

3 Estimation of SUR Based Copula

Data

The data used here are dataset related to Thai rice consisting of the demand for rice (Q_t^d), output of grain rice (Q_t^s), export price of Thai rice (P_t^{exp}), Pakistan's exported rice price (P_t^{Paki}), Vietnam's exported rice price (P_t^{Viet}), export price of India's rice (P_t^{Ind}), producer price of rice (P_t^{Farm}), rainfall (R_t), and water storage (W_t). The data sets are monthly frequency data collected from M1/2006 to M9/2014, covering 105 observations. Then we transformed all series data to the log-log before moving to the estimation.

Estimation

In this study, the SUR model with two equations (M = 2) was employed to derive the demand and supply in the rice market. Thus, the bivariate Copula with continuous marginal distribution is conducted in the estimation. First of all, the Augmented Dickey–Fuller test was used to test the stationary of the data series. The estimation procedures SUR based Copula involved four steps. In the first step, the conventional SUR model was estimated using a maximum likelihood technique to obtain the initial values. In the second step, we constructed the SUR Copula likelihood using the chain rule, and we have

$$\begin{aligned}\frac{\partial^2}{\partial u_1 \partial u_2} F(u_1, u_2) &= \frac{\partial^2}{\partial u_1 \partial u_2} C(F_1(u_1), F_1(u_1)) \\ &= f_1(u_1) f_2(u_2)\, c(f_1(u_1) f_2(u_2))\end{aligned} \tag{17}$$

where u_1 and u_2 are the marginal assumption of normal or student-t distribution, $f_1(u_1)$ and $f_2(u_2)$ are normal functions of demand and supply equation and $c(f_1(u_1, f_2(u_2))$ is a density function of Copula function. In this study, Gaussian, T, Frank, Clayton, Gumbel, Joe, and Ali-Mikhail-Haq (AMH) Copula are employed for constructing the joint distribution function of a bivariate random variable with the univariate marginal distribution.

Then, we took a logarithm to transform Eq. 13, we get

$$lnL = \Sigma_{i=1}^{T} (lnl(\theta_1|X_d) + lnl(\theta_2|X_s) + lnf_1(u_1) + lnf_2(u_2) + lnc(F_1(u_1), F_2(u_2))) \tag{18}$$

where $lnl(\theta_1|X_d)$ and $lnl(\theta_2|X_s)$ are the logarithm of the likelihood function of demand and supply equation, respectively, which can be defined as

$$lnL = -\frac{T}{2} ln(2\pi) - \frac{T}{2} ln(\Sigma) \left(\frac{1}{2\Gamma} (Y - X\beta)\prime (Y - X\beta) \right) \tag{19}$$

And $lnc(F_1(u_1), F_2(u_2))$ is a bivariate Copula density assumed for Gaussian, T, Frank, Clayton, Gumbel, Joe,and Ali-Mikhail-Haq Copula (See. Eq. 6–12).

In the third step, we used the maximum likelihood technique again to maximize the SUR Copula likelihood to obtain the final estimated parameters. And the last step involved fitting the SUR Copula likelihood by identifying the appropriate marginal distribution and Copula function. The minimum AIC and BIC were used to find the best-fit the combination between Copula function and marginal distribution.

Finally, we extended the estimated parameters which are obtained from the SUR based Copula to compute the following equations considering the intercept terms of demand and supply $(\hat{\alpha}_1, \hat{\alpha}_2)$ and price elasticity of demand and supply, namely, $\hat{\beta}_1$ and $\hat{\delta}_1$, as follows:

$$\begin{bmatrix} -\hat{\beta}_1 & 1 \\ -\hat{\delta}_1 & 1 \end{bmatrix}^{-1} \begin{bmatrix} \hat{\alpha}_1 \\ \hat{\alpha}_2 \end{bmatrix} = \begin{bmatrix} \kappa_1 \\ \kappa_2 \end{bmatrix}$$

$$CS = \frac{\kappa_1 \times \left(\hat{\beta}_1 - \kappa_2\right)}{2}$$

$$PS = \frac{\kappa_1 \times \left(\kappa_2 - \hat{\delta}_1\right)}{2}$$

Therefore, according to the above equations, in this paper the total social welfare (TS) can be simply measured as $\text{TS} = \text{CS} + \text{PS}$, where CS and PS are the consumer surplus and the producer surplus, respectively.

4 Empirical Results

Checking stationaries of the data used in this paper was done before getting started estimating results. We used Augmented Dickey–Fuller (ADF) unit root test as a tool, and the results are that all variables passed the test at level with probability equals to zero, meaning they are all stationary. Then the Copula-based SUR model were estimated.

4.1 Estimation Results for Demand and Supply

We used Copula in order to improve the SUR model. The estimation of SUR model has been improved by adding Copula density function into the likelihood and we called it SUR Copula likelihood (see Sect. 3). We found very often that the papers working on Copula usually assume any family of Copula for the joint distribution of variables before estimation, e.g., Parsa and Klugman [12], Kolev and Paiva [7], Leon and Wu [3], and Masarotto and Varin [9]. But in this paper we have tried to make things different by testing all possible Copula families and a marginal distribution of standardized residuals, then specifying the family and the marginal distributions which is best fit for the model. Therefore, the results are exhibited in the following table.

The result below is one of our motivations of this paper as we tried to not restrict the type of marginal distribution. Because, sometimes, the error or the marginal distribution of residuals might be non-normal and it might not be the same distributions when the the equations are different. For example, the SUR model which is the linear regression system and consists of several equations whose marginal distributions of residuals may be different. The advantage of Copula is that it can handle this problem, and the empirical results have already been shown as below.

Table 1 shows AIC and BIC criteria for comparison of Copula families as well as the marginal distributions. To choose the best family and marginal distribution for our data, we will look at the minimum AIC and BIC criteria and as we can see, the minimum AIC and BIC are –1480.577 and –1480.301, respectively as underlined. So we should choose Gaussian as the most appropriate Copula for our data, and the most suitable marginal distributions are the normal distribution for demand and student-t distribution for supply.

Since we believe that Copula could improve the SUR model, we tried to compare the results between the Copula-based SUR model, which is our concern, and the conventional SUR model -estimated by a generalized least squares (GLS)- by looking at the sum of squared errors. We link the marginal distributions of demand and supply together using Gaussian Copula function (see Table 1) and the estimated results for demand and supply of the rice market using the Copula-based SUR model are shown in Table 2. The model performed very well, as we can see obviously that the results

Table 1 AIC and BIC criteria for model choice

	Marginal distributions			
Copula	Normal/Normal	Normal/Student-t	Student-t/Student-t	Student-t/Normal
Frank	–146.797	–1395.244	–1394.495	–1397.062
	–146.521	–1394.967	–1394.218	–1396.508
Joe	–66.5821	–1395.244	–1395.059	–1398.008
	–66.3057	–1394.967	–1394.783	–1397.456
Clayton	–98.2111	–1407.17	–1394.008	–1397.031
	–97.9347	–1406.894	–1393.732	–1396.478
AMH	–146.219	–1394.799	–1394.713	–1398.279
	–145.943	–1394.522	–1394.437	–1397.727
Gunbel	–229.379	–1429.266	–1393.800	–1396.391
	–229.103	–1428.99	–1393.523	–1395.837
Student-t	–57.019	–1404.885	–1393.698	–1396.882
	–56.743	–1404.608	–1393.422	–1396.329
Guassian	–985.976	**–1480.577**	–1393.753	–1396.243
	–985.703	**–1480.301**	–1393.477	–1395.691

Source Calculation

Table 2 Estimation of SUR based copula for demand and supply in comparison with SUR estimated by a generalized least squares

	SUR based Copula		SUR (GLS)	
	Coefficient	Std. Error	Coefficient	Std. Error
Demand				
Intercept	–0.00171***	0.000406	0.00195	0.003945
p^{export}	–1.14950***	0.00377	–0.94608***	0.153965
p^{viet}	0.52871***	0.003602	0.44078***	0.113252
p^{paki}	0.05233***	0.015549	–0.14998	0.211111
p^{india}	1.99918***	0.017724	2.24771***	0.231657
Supply				
Intercept	–0.00506***	0.000976	0.00524	0.037353
p^{export}	3.22941***	0.035538	3.22892*	1.359386
Water	1.45118***	0.022175	1.41418	0.847816
Rain	–0.14403***	0.002299	–0.15097	0.08791
p^{farm}	–0.74028***	0.027193	–0.71214	1.039761
Sum of squared errors				
Demand	0.16125		0.16462	
Supply	14.77811		14.77926	

Source Calculation
Note *, **, *** denote rejections of the null hypothesis at the 10, 5, and 1 % significance levels, respectively

of the Copula-based SUR model are significant at level of 0.01 and the coefficient of all variables estimated from this model seem to make sense of the key point of demand and supply theory. Recall, as we know from the theory, the demand will be reduced if the price increases. On the other hand, the supply will do the opposite to the demand. It will be increased if the price increases.

To compare with the conventional SUR model, we estimated the SUR model using one of the well-known estimators for SUR model [10] that is the generalized least squares. The results are shown in Table 2 beside the Copula-based SUR model. We can see that for each variable there is not much difference between the coefficients estimated by the Copula-based SUR and the conventional SUR, but the Copula-based SUR model has a lower value of standard error (SE) for all estimated parameters compared with the result of the conventional SUR. It gives us a possibility that Copula can improve the efficiency of estimation, since SE for all estimated parameters are lower. Because of the advantage of Copula, the covariance estimation is improved. The marginal distributions are not restricted anymore; it can be any distribution and not need to be the same for the different equations -here are demand and supply equations- because Copula can link them together. Unlike the case of SUR model estimated by GLS, the distribution is usually assumed to be normal. If we consider the model comparison and take into account the sum of squared errors (SSE) as a tool, we will see that the Copula-based SUR as a whole is slightly better than the conventional SUR for having lower SSE of both demand and supply equations. It may confirm our belief that Copula can provide an improvement in SUR model.

A comparison of residuals of demand and supply equations between the Copula-based SUR model and the conventional SUR model are illustrated in Fig. 1. These residuals exhibit homogeneity and independence. Figure 1 shows that the residual plots of these two models have a similar structure of residuals. By the assumption of the conventional SUR model, both demand and supply equations are assumed to have a normal distribution for error terms. Figure 1 shows that the structure of residuals seem to correspond well to the assumption. The residual plot of supply equation of the Copula-based SUR model (top right) provides the same structure as the plot of the conventional SUR (bottom right) since both of them have the same distribution of residuals, which is a normal distribution, whereas the residual plot of demand equation of the Copula-based SUR (top left) is slightly different from the plot of the conventional SUR (bottom left) since the distribution of residual of demand equation of the Copula-based SUR is student-t (see Table 1)

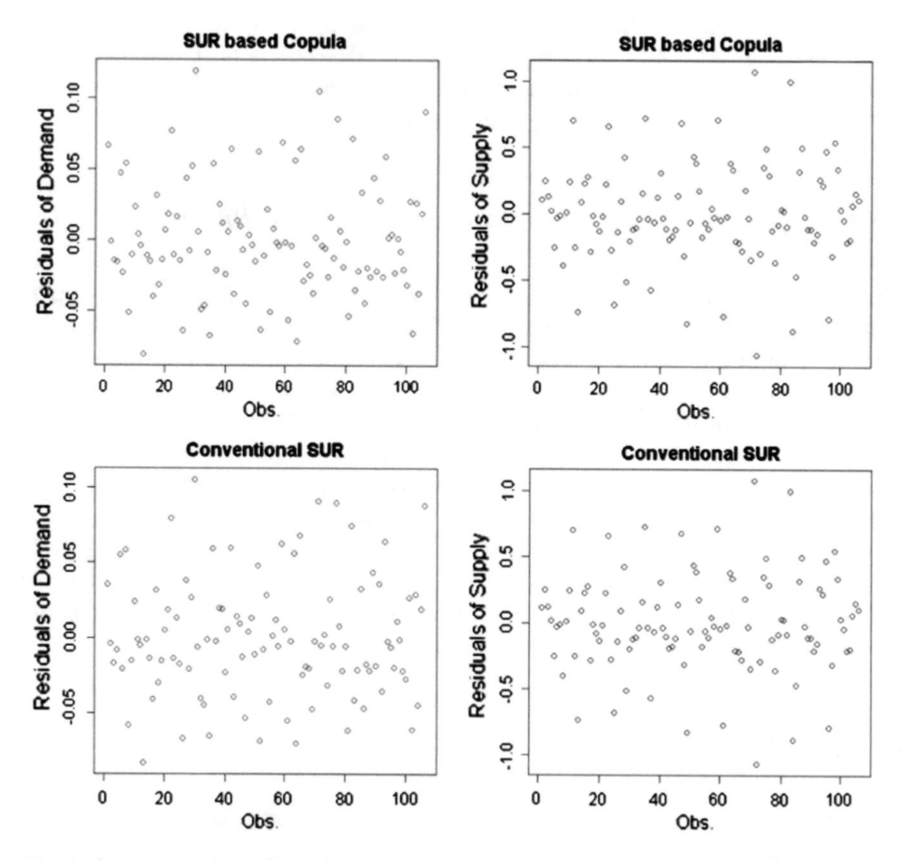

Fig. 1 Residuals of demand (*top left*) and supply (*top right*) equations of the Copula-based SUR and Residuals of demand (*bottom left*) and supply (*bottom right*) of the conventional SUR

Table 3 Welfare measurement

	Amount of welfare (10^{-6})
Demand	1.3682
Supply	4.4595
Total welfare	5.8277

Source Calculation

According to the results, by using this dataset, the Copula-based SUR model provides the similar result to the conventional SUR model (see Table 2 and Fig. 1). It implies that the conventional SUR is acceptable for this dataset. Nevertheless, it would be more desirable to relax the normal assumption of error terms of the SUR model when there is an existence of non-normal distribution in error terms. Even though the estimated results of these two models are not much different, some improvement in SUR model is still be found. Considering the result shown in Table 2, standard errors for all estimated parameters of the Copula-based SUR model are slightly lower than the conventional SUR. As well as the comparison of SSE where the Copula-based SUR model is slightly better than the conventional SUR for having lower SSE. It is possible to conclude that Copula can provide an improvement in SUR model by relaxing that strong assumption.

4.2 Application: Welfare Measurement for Thai Rice Market

According to the results from the previous part, we got the demand and supply equations estimated by the Copula-based SUR model. Then we take into account equilibrium of the market to find the price at the equilibrium point or the market price. Since welfare, which is our concern, can be measured by summing up of consumer surplus (CS) and producer surplus (PS) where both surpluses are related to the equilibrium price level. The consumer surplus is the amount of the willingness to pay above the equilibrium price under the demand curve and the producer surplus is the amount of the willingness to produce or to sell below the market equilibrium price above the supply curve. By the definition, the estimated results for the total welfare of the rice market can be shown in Table 3. The table shows that the amount of overall welfare of the rice market is equal to $5.8277(10^{-6})$ which is the sum of consumer surplus, $1.3682(10^{-6})$ and producer surplus, $4.4595(10^{-6})$.

5 Conclusions

The Copula-based SUR model were introduced in this paper as an alternative to the conventional SUR model, in which requires the assumption of normal distribution of error terms. The Copula density function has been applied into the likelihood to not restrict the marginal distributions, and to relax the normality assumption of errors in the model. The results have shown that Gaussian Copula is the most appropriate function for being the linkage between the marginal distributions where the normal and the student-t distributions are the best fit for the marginal distributions of demand and supply, respectively. The real dataset of Thai rice market is used for making a comparison between the conventional SUR model estimated by GLS and the Copula-based SUR model. The Copula-based SUR model seems to perform slightly better on the ground of having lower standard errors and sum of squared errors. However, it is not much different in the sum of squared errors between both models. This might imply that the conventional SUR model is adequate for this dataset. Nevertheless, our research results might confirm our belief that Copula could improve the SUR model in term of relaxing the normal distribution assumption of error terms.

Acknowledgments The authors are grateful to Puey Ungphakorn Centre of Excellence in Econometrics, Faculty of Economics, Chiang Mai University for the financial support. We also would like to thank Prof. Thierry Denoeux for his insightful comment and encoragement.

References

1. Baltagi, B.H.: The efficiency of OLS in a seemingly unrelated regressions model. Econom. Theory **4**(03), 536–537 (1988)
2. Bilodeau, M., Duchesne, P.: Robust estimation of the SUR model. Can. J. Stat./La Rev. Can. Stat. **73**, 277–288 (2000)
3. de Leon, A.R., Wu, B.: Copula-based regression models for a bivariate mixed discrete and continuous outcome. Stat. Med. **30**(2), 175–185 (2011)
4. Drton, M., Richardson, T.S.: Multimodality of the likelihood in the bivariate seemingly unrelated regressions model. Biometrika **91**(2), 383–392 (2004)
5. Genest, C., MacKay, J.: The joy of copula: Bivariate distributions with uniform marginals. Am. Stat. **40**(4), 280–283 (1986)
6. Hofert, M., Machler, M., McNeil, A.J.: Likelihood inference for Archimedean Copula in high dimensions under known margins. J. Multivar. Anal. **110**, 133–150 (2012)
7. Kolev, N., Paiva, D.: Copula-based regression models: a survey. J. Stat. Plan. Inference **139**(11), 3847–3856 (2009)
8. Lar, N., Calkins, P., Sriboonchitta, S., Leeahtam, P.: Policy-based analysis of the intensity, causes and effects of poverty: the case of Mawlamyine, Myanmar. Can. J. Dev. Stud./Rev. Can. d'tudes dev. **33**(1), 58–76 (2012)
9. Masarotto, G., Varin, C.: Gaussian copula marginal regression. Electron. J. Stat. **6**, 1517–1549 (2012)
10. Moon, H.R., Perron, B.: Seemingly unrelated regressions. New Palgrave Dict. Econ. 1–9 (2006). Retrieved 14 July 2015 from http://www.mapageweb.umontreal.ca/perrob/palgrave.pdf
11. Noh, H., Ghouch, A.E., Bouezmarni, T.: Copula-based regression estimation and inference. J. Am. Stat. Assoc. **108**(502), 676–688 (2013)
12. Parsa, R.A., Klugman, S.A.: Copula regression. Var. Adv. Sci. Risk **5**, 45–54 (2011)
13. Sklar, M.: Fonctions de rpartition n dimensions et leurs marges. Universit Paris **8**, 229–231 (1959)
14. Sparks, R.: SUR models applied to an environmental situation with miss-ing data and censored values. Adv. Decis. Sci. **8**(1), 15–32 (2004)
15. Takada, H., Ullah, A., Chen, Y.M.: Estimation of the seemingly unre-lated regression model when the error covariance matrix is singular. J. Appl. Stat. **22**(4), 517–530 (1995)

16. Zellner, A.: An efficient method of estimating seemingly unrelated regressions and tests for aggregation bias. J. Am. Stat. Assoc. **57**(298), 348–368 (1962)
17. Zivot, E., Wang, J.: Modeling Financial Time Series with S-PLUS vol. 191, pp. 729–730. Springer, Berlin (2007)

Price Transmission Mechanism in the Thai Rice Market

Roengchai Tansuchat, Paravee Maneejuk, Aree Wiboonpongse and Songsak Sriboonchitta

Abstract This study aimed to analyze price transmission in the Thai rice market using the MS-BVECM. We focused on the data set related to Thailands rice price, including Thai white rice price, Thai parboiled rice, Thai paddy price, and World rice price collected from M1/2004 to M3/2014. We estimated the model with two regimes; namely high market price regime and low market price regime. The estimated results showed that there existed some short-run relationships between these rice prices in both regimes. Unlike the long-run, there existed only one long-run relationship (one cointegrating equation) in the high market price regime expressed in the Thai white rice equation. Meanwhile, Thai paddy price has the long-run relationship and short-run adjustment dynamics in the low market price regime. In addition, we found that India's non-basmati rice exports and the paddy price guaranteed at 15,000 THB per ton are two main reasons which caused the switching between these two regimes.

Keywords Price transmission · Markov-switching · Error correction model

R. Tansuchat (✉) · P. Maneejuk · S. Sriboonchitta
Faculty of Economics, Chiang Mai University, Chiang Mai, Thailand
e-mail: roengchaitan@hotmail.com

P. Maneejuk
e-mail: mparavee@gmail.com

A. Wiboonpongse
Faculty of Economics, Prince of Songkla University, Hat Yai, Thailand

A. Wiboonpongse
Faculty of Agriculture, Department of Agricultural Economics
and Agricultural Extension, Chiang Mai University, Chiang Mai, Thailand

A. Wiboonpongse
Faculty of Economics, Institute for Sufficiency Economy Research
and Promotion, Chiang Mai University, Chiang Mai, Thailand

V.-N. Huynh et al. (eds.), *Causal Inference in Econometrics*,
Studies in Computational Intelligence 622, DOI 10.1007/978-3-319-27284-9_29

1 Introduction

Rice has played a vital role in Thailand's economy and creates enormous economic value from exports. It takes around 30 % of primary agricultural exports (Bank of Thailand, 2015), where white rice, and parboiled rice are major exported products of rice since they account for 40 and 27 % of Thailand's export quantity of rice, respectively. Both white rice and parboiled rice actually are the same rice or paddy when they are first harvested—paddy is rice kernels which is harvested directly from rice fields and transported to a processing site. Then, the paddy is put through a step of removing bran and becomes white rice. Some of the white rice is put into the process of parboiling, and then it becomes parboiled rice. In this paper we focus on the parboiled rice that is partially boiled in husk. There are three steps of parboiling including: soaking, steaming, and dehydration. Getting steam underneath pressure can boost nutrients in the grain; the rice can be as nutritious as brown rice. However, these additional steps make the parboiled rice's price high with its enhanced nutritional values.

As we know that both white rice and parboiled rice are milled from the same raw material or paddy but with difference in processes, millers have to decide on allocation of paddy for processing into white rice or parboiled rice or both but in what proportions. Many papers working on Thai rice, e.g., Chulaphan et al. [2], and Wiboonpongse and Chaovanapoonphol [9, 10] investigated that most Thai paddies (approximately 60 %) is used for producing white rice while the rest (approximately 40 %) is used for producing parboiled rice here does not include domestic consumption. There are different markets for both kinds of rice (Global Trade Atlas, 2014), thus export prices of white rice and parboiled rice should not directly affect each other. The price of parboiled rice is normally higher than that of white rice's price as a result of the additional steps of production. But after the Thai government proposed a rice pledging scheme in October 2011, the price of white rice was increased. Then this situation led to higher price for parboiled rice since the input cost—the white rice—was increased. So, there exists a relationship between the prices of white rice and parboiled rice. It seems changing in the white rice price may affect the price of parboiled rice, and conversely.

Moreover, the domestic prices of white rice and parboiled rice normally follow the tendency of world price. Although probably not in the case of white rice any more, Thailand's price may influence the world price or market of parboiled rice as a price leader because Thailand and India are the largest exporters. Their exports share almost 90 % of world's total and other competing countries in the parboiled rice market, i.e., U.S., Uruguay, and Brazil, use their prices as a reference price. So, in this case the world price probably is measured from the domestic rice price—the parboiled rice prices of Thailand and India.

According to these situations as mentioned above, we suspect that first, high price of white rice is transmitted to the parboiled rice price. Furthermore, it is equally important to suspect how the prices of Thai white rice and parboiled rice are transmitted to the Thai paddy price. And the third aspect is related to how the world price is transmitted to the domestic rice price. Therefore, in this paper, the price transmission approach is taken for investigation since we aim to study how the impact of one market in terms of price is transmitted to the others.

Price transmission is the heart of this paper; therefore, an effective tool for measuring the transmission is sought. As many papers, we found that the Error Correction Model (ECM) was having examined used very often, e.g., Sharma [7], Acharya et al. [1], Ifejirika et al. [3], and Rischke [6]. We believed in the efficiency of this model, and employed it as a tool to confirm the equilibrium relations in long-run and cointegration of the data set of Thai rice prices. The Bayesian approach was provided for estimation in order to improve on the estimation. In addition, we also applied an approach of regime switching to this study since we believed that there must be a fluctuation in the prices and at least the market price should have two trends; namely high price and low price. Therefore, we provided the Markov Switching Vector Error Correction Model with the Bayesian estimation (MS-BVECM) to measure the price transmission of Thailand's rice.

For an overview of this paper, we begin with the importance of Thai rice and the price transmission. The following Sect. 2 provides a methodology used in the paper. Then, the estimation will be completely explained in Sect. 3 before eventually showing estimation results in the last Sect. 4.

2 Methodology

2.1 Markov-Switching Vector Error Correction Model

According to MS-VECM of Jochmann and Koop [4], and Sugita [8], the regime dependent of intercept term, the adjustment term, the cointegrating term, the autoregressive term, and the variancecovariance matrix can be written as

$$\Delta y_t = c_{S_t} + \sum_{i=1}^{k} \Gamma_{i,S_t} \Delta y_{t-i} + \Pi_{S_t} y_{t-i} + \varepsilon_t \tag{1}$$

where y_t is n dimension time series data which are transformed to be Δy_t, where $\Delta y_t = y_t - y_{t-1}$ denotes as an integrated order 1 (I(1)), $C_{s(t)}$ is a vector of the state-dependent intercept term, Γ_{i,S_t} stands for the state-dependent $n \times n$ autoregressive parameter matrices of the vector Δy_{t-i}, Π_{S_t} is the state-dependent error correction term, the long-run impact matrices, which are defined by $\Pi_{S_t} = \alpha_{S_t} \beta'_{S_t}$, where α_{S_t} is the state-dependent adjustment term, the state-dependent β_{S_t} is $r \times n$ cointegrating vector, and ε_t is the error term which is assumed to be $N(0, \Sigma_{S_t})$.

In this study, we assume that the two regimes are the high market price regime and the low market price regime. The state variable with the first order Markov process is assumed. Thus, the transition probabilities with 2 regimes can be defined as

$$P = \begin{bmatrix} P_{11} & P_{12} \\ P_{21} & P_{22} \end{bmatrix} \quad (2)$$

2.2 Prior Distributions

In this study, we select flat prior density for the intercept term (c), the autoregressive term (Γ), the adjustment term (α), and the cointegrating vector (β); Inverted Wishart prior for the variancecovariance matrix (Σ), and Dirichlet prior for the transition probabilities (p_{ij}). For a prior for c, Γ, α and β, we assume these parameters as Least Informative Priors, that is, flat priors, where the prior is simply a constant.

$$p(\theta) = k = \frac{1}{b-a}, \quad for \quad a \leq \theta \leq b \quad (3)$$

With a flat prior, the posterior is just a constant times the likelihood,

$$p(\theta, \Sigma, P|x) \alpha Pr(\theta, \Sigma, P) Pr(x|\theta, \Sigma, P) \quad (4)$$

where θ contains c, Γ, α and β which have a uniform distribution from negative infinity to positive infinity thus the posterior distribution will be affected by the own data. Σ is assumed to be Inverse Wishart prior

$$\Sigma_i \sim IW(\Phi_i, h_i), \; i = 1, 2 \quad (5)$$

where $\Phi_i \in R^{nxn}$, as Inverse Wishart prior. For the transition probabilities P, Sugita [8] purposed the likelihood function for the transition probability p_{ij} in the case of two states we obtain

$$\xi(p_{11}, p_{22}||S_T) = p_{11}^{n_{11}}(1-p_{11})^{n_{12}} p_{22}^{n_{22}}(1-p_{22})^{n_{21}} \quad (6)$$

The likelihood function for c, Γ, β, Σ and $\tilde{S}_t$ is given by

$$\xi(B, \beta, \Sigma_1, \ldots, \Sigma_n, \tilde{S}_T|Y)|Y) \infty \left(\prod_{i=1}^{T} \Sigma_i|^{-t_i/2} \right)$$

$$exp\left(-\frac{1}{2} tr \left[\sum_{i=1}^{T} \{ \Sigma_i^{-1} (Y_i - W_i B)'(Y_i - W_i B) \} \right] \right)$$

$$= \xi(B, \beta, \Sigma_1, \ldots, \Sigma_n, \tilde{S}_T|Y) \infty \left(\prod_{i=1}^{T} \Sigma_i|^{-t_i/2} \right)$$

$$exp\left(-\frac{1}{2} tr \left[\sum_{i=1}^{T} \left\{ vec(Y_i - W_i B)'(\Sigma_i \bigotimes I_\tau)(Y_i - W_i B) \right\} \right] \right) \quad (7)$$

where $B = \{c, \Gamma\}$ and $W = (I_1 Z\beta, \ldots, I_{1n} Z\beta)$; $Z = (Y_{p-1}, \ldots, Y_{T-1})'$, and t_i is number of observation. However, these state's probabilities are unknown thus the Hamilton's filter is used to estimate the filter probabilities ($Pr(S_t = j)$) of each state based on the available information set.

Considering Ω_t as the matrix of available information at time t. The Hamilton's filter as provided in Perlin [5] uses the following algorithm.

1. Given initial value of transition probabilities P_{ij} which are the probabilities of switching between regimes.

$$P = \begin{bmatrix} 0.8 & 0.2 \\ 0.2 & 0.8 \end{bmatrix} \tag{8}$$

2. Updating the transition probabilities of each state with the past information including the parameters in the system equation, Ω_{t-1} and P, for calculating the likelihood function in each state Eq. (7) for time t. After that, the probabilities of each state are updated by the following formula

$$Pr(S_t = j|\Omega_t) = \frac{f(y_t|S_{t=j}, \Omega_{t-1})Pr(S_{t=j}|\Omega_{t-1})}{\sum_{j=1}^{2} f(y_t|S_{t=j}, \Omega_{t-1})Pr(S_{t=1}|\Omega_{t-1})} \tag{9}$$

3. Iterating step 1 and 2 for $t = 1, \ldots, T$.

2.3 *Posterior Specifications*

The posterior estimation is computed from the priors and the likelihood functions. Sugita [8] proposed Gibbs Sampling method to sample the initial parameter obtained from the maximum likelihood function Eq. (7). Then, in order to estimate Σ, θ and β, Gibbs sampling can be used to generate sample draws, as follows:

(1) Specify the staring values for p_{ij}^0, θ^0, and Σ^0
(2) Generate $\tilde{S}_\tau = \left\{S_1^j, S_2^j\right\}'$ from $p(\tilde{S}_\tau|\theta^{j-1}, Y)$
(3) Generate the transition probabilities p_{ij}^j from $p(p_{ij}|\tilde{S}_\tau)$
(4) Generate β^j from $p(\beta|s_\tau^j, Y)$
(5) Generate β_j from $p(vec(\beta)|\beta^j, \Sigma^j, S_\tau^j, Y)$
(6) Generate Σ_i^j from $p(\Sigma_i|\theta^j, S_\tau^j, Y)$

Next, repeat step 2–6 in order to generate p_{ik}^{j+1}, β^{j+1}, Σ_i^{j+1} and β^{j+1} and then repeat the previous iteration for N times, $j = 1, \ldots, N$ as a final iterations. Note that, should be large enough to eventually reach a stationary distribution or converge.

Table 1 Descriptive statistics

Price	Mean	SD.	Max	Min
P^5	468.16	163.54	1,047.25	205.95
W^5	450.22	148.78	948.00	215.72
PD	244.31	71.23	409.48	121.02
WP	652.54	288.99	2,349.16	314.16

Source Calculation

3 Data

The data are monthly time series data consisting of Thai parboiled rice 5 %'s price (P^5), Thai White rice 5 %'s price (W^5), Thai paddy price (PD) and World rice price (WP) for the period from January 2004 to March 2015, which were collected from Thomson Reuters DataStream, from Financial Investment Center (FIC), Faculty of Economics, Chiang Mai University. All of them are coded in dollar amounts. We transform these variables into logarithms before estimation. Additionally, descriptive statistics are used to describe the basic features of the data shown in Table 1.

4 Empirical Results

4.1 Unit Root Test

Prior to conducting the Markov-switching with cointegration analysis, it is important to determine the order of integration of the variables and ensure that all the variable series are non-stationary and integrated of the same order. In this study, we employed the ADF test to identify the order of integration of our variables. The results of the ADF test are presented in Table 2, which showed that the logarithm of all variables are I(1). Thus, the first order of integration exists in all the variables.

Table 2 Unit root test with ADF test

Price	Level			First different			Integrated order
	None	c and not t	c and t	None	c and not t	c and t	
P^5	−0.419	−2.226	−2.121	−7.435	−7.413	−7.448	I(1)
W^5	−0.389	−2.76	−2.906	−7.106	−7.087	−7.111	I(1)
PD	−0.157	−2.206	−2.212	−8.447	−8.434	−8.478	I(1)
WP	−0.493	−1.895	−2.79	−12.119	−12.075	−12.098	I(1)

Source Calculation

Table 3 Unit root test with ADF test

Lag	BIC
0	−5.5214
1	−11.9173
2	−11.8081
3	−11.3613
4	−11.0988

Source Calculation
Note t = trend and c = intercept

In this section, we have to specify the lag length for the MS-BVECM model. We employed the vector error correction lag length criteria for the best number of lag lengths. For the VECM lag length criteria based on Bayesian information criteria (BIC), the results are reported in Table 3 which revealed that the BIC values for lag = 1 are lowest in both models. Therefore, in this study, we chose the appropriate lag length p = 1.

4.2 Cointegration Test

To determine the rank or the number of cointegration vectors, trace tests and Maximum-Eigen values are carried out, as shown in Table 3; we tested the rank of the long-run relationship using Johansen's trace test which was obtained from BVECM with a Minnesota prior. In this study, we specified a tightness parameter, a decay parameter, and a parameter for the lags of the variables as 0.10, 0.10, and 0.50, respectively. Based on the results of cointegration test shown in Table 4, the null hypothesis of one or fewer cointegrating vectors is mostly rejected at the 5 % statistically significant level. This indicates that the model has one cointegrating vector; therefore, the study chose r = 1.

Table 4 Cointegration test

	Data trend	None	None	Linear	Linear	Quadratic
	Test type	No I	I	I	I	I
		No T	No T	No T	T	T
Model 1	Trace	1	1	2	1	2
	Max-Eig	1	1	1	1	1

Source Calculation
Note T = trend and I = intercept

4.3 Estimates MS(2)-BVECM(1) of Model 1

The estimated parameters of MS(2)-BVECM(1) model are displayed in Eq. (10) including the estimated variancecovariance matrix (Σ). There are two regime models, which are first regime model for the high market price and second regime model for the low market price. From Eqs. (10) and (11), it is clear that the estimated intercept parameters are statistically significant and differ across regimes. In regime 1, only the parameter of W^5 in the state-dependent error correction term is statistically significant at 10 %, meaning only 5 % white rice price is cointegrated with 5 % parboiled price, paddy price and world rice price, respectively. Thus, 5 % white rice price has the long run relationship and short-run adjustment dynamics, so that the deviation of W^5 from long-run equilibrium is corrected gradually through a series of partial short-run adjustments.

Regime 1

$$\begin{bmatrix} P^5 \\ W^5 \\ PD \\ WP \end{bmatrix} = \begin{bmatrix} 270.26^{***} \\ 263.62^{***} \\ 1.99.55^{***} \\ 242.02^{***} \end{bmatrix} + \begin{bmatrix} 3.591^{**} & -3.362^{**} & 0.926 & 0.344 \\ 2.490^{**} & -2.227^{**} & 0.867 & 0.320 \\ -0.291 & 0.348 & 0.358^{*} & 0.100^{*} \\ 2.836 & -2.874 & 1.175 & 0.526 \end{bmatrix} \cdot \begin{bmatrix} P^5_{t-1} \\ W^5_{t-1} \\ R_{t-1} \\ WP_{t-1} \end{bmatrix} + \begin{bmatrix} -0.502^{*} \\ 0.468^{*} \\ 0.094 \\ -0.339 \end{bmatrix} \cdot \begin{bmatrix} P^5_{t-1} \\ W^5_{t-1} \\ PD_{t-1} \\ WP_{t-1} \end{bmatrix} + \begin{bmatrix} 1.636 & 1.201 & 0.106 & 2.247 \\ 1.201 & 0.933 & 0.067 & 1.657 \\ 0.106 & 0.067 & 0.086 & 0.230 \\ 2.247 & 1.657 & 0.023 & 3.966 \end{bmatrix} \tag{10}$$

In regime 2, because only the parameter of *PD* in state-dependent error correction term is negative and lies between 0 and 1, meaning only *PD* has the long run relationship and short-run adjustment dynamics, and is cointegrated with 5 % parboiled price, 5 % white rice price and world rice price, respectively. In contrast, though the parameter of P^5 and W^5 have negative value, they are greater than 1 in absolute value, meaning they diverge from the long run equilibrium.

Regime 2

$$\begin{bmatrix} P^5 \\ W^5 \\ PD \\ WP \end{bmatrix} = \begin{bmatrix} 385.14^{***} \\ 411.69^{***} \\ 329.93^{***} \\ 117.75^{***} \end{bmatrix} + \begin{bmatrix} 1.051 & -1.464 & 2.542^{*} & 0.002 \\ 0.718 & -1.264 & 2.654^{**} & 0.042 \\ 0.457 & -0.961^{*} & 1.582^{**} & 0.026 \\ -0.482 & 0.085 & 0.234 & 0.424^{**} \end{bmatrix} \cdot \begin{bmatrix} P^5_{t-1} \\ W^5_{t-1} \\ R_{t-1} \\ WP_{t-1} \end{bmatrix} + \begin{bmatrix} -1.062^{**} \\ -1.205^{***} \\ -0.575^{***} \\ 0.371 \end{bmatrix} \cdot \begin{bmatrix} P^5_{t-1} \\ W^5_{t-1} \\ PD_{t-1} \\ WP_{t-1} \end{bmatrix} + \begin{bmatrix} 23.041 & 21.501 & 10.645 & 5.996 \\ 21.501 & 20.283 & 10.088 & 5.479 \\ 10.645 & 10.088 & 5.280 & 2.783 \\ 5.996 & 5.479 & 2.783 & 13.891 \end{bmatrix} \tag{11}$$

Table 5 Matrix of transition probabilities

	Regime 1	Regime 2
Regime 1	0.935	0.065
Regime 2	0.029	0.917

Note $p_{ij} = Pr(s_{t+1} = j|S_t = i)$

Table 6 Regimes and duration

	No. of obs.	Ergodic probability	Duration
Regime 1	47	0.439	15.385
Regime 2	88	0.561	12.048

The matrix of transition probability parameters is presented in Table 5. The estimated probability means the conditional probability based on the information available throughout the whole sample period at future date t. The result shows that the probability of switching from regime 1 (high price) move to regime 2 (low price) is 6.5 %, while remaining in regime 1 is 93.5 %, meaning that in studying period as shown in Table 6, the rice prices are still high and have duration approximately 15 months. Similarly, the probability of switching from regime 2 (low price) move to regime 1 (high price) is 2.9 %, while remaining in regime 2 is 91.7 %. In addition, the duration of regime 2 is approximately 12 months.

Figure 1 displays the result on filtered probabilities and smoothed probabilities for high price regime (regime 1) and low price regime (regime 2). The probabilities lies between 0 and 1. In the case of two regimes, the observations are classified in first regime (high price) if $Pr(s_t = 1|y_t) > 0.5$ and the second regime (low price) if $Pr(s_t = 1|y_t) < 0.5$. The result confirms that during April 2004 November 2007, which is before the period of a temporary banning on export of Indian non-basmati rice, and Thailand rice-pledging scheme, the prices of 5 % white rice are in the low regime (Regime 1).

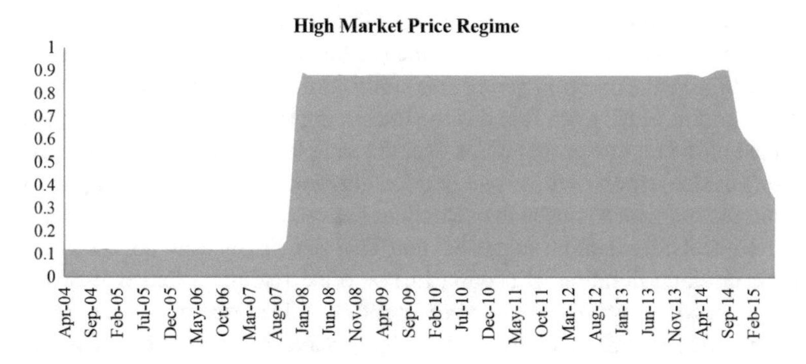

Fig. 1 Estimated probabilities

Suddenly, in Oct 2007, after India banned non-basmati rice exports, the supply of rice in global market fell dramatically, resulting in increased rice prices in the global market. Consequently, all rice prices have switched to high regime from this situation. Later, even though India turned back to exporting rice in September 2011, the rice-pledging scheme for farmers was implemented in Thailand in October 2011. Under this scheme, the paddy price was guaranteed and bought by government at 15,000 Baht per ton, meaning the cost of input for white rice and parboiled rice millers increased dramatically. This scheme also distorted the domestic rice market as well. Therefore, the Thai 5 % white rice price was also high and continued in high price regime with high probability.

Finally, the intervention from Thai government policy in domestic paddy market through the rice-pledging scheme was terminated at the end of February 2014. After that, the probability of high price regime is gradually decreased, because the normal market mechanism has returned to normal operation. In addition, Thai rice exporters have tried to reduce the price of export rice in order to compete with rivals in the global market. Therefore, the Thai 5 % white rice price of Thailand gradually decreased and eventually moved to low price regime after April 2015.

5 Conclusions

This study employed the MS-BVECM to capture the price transmission of Thailand's rice, including Thai white rice price, Thai parboiled rice 5 % price, Thai paddy price, and World rice price collected from January 2004 to March 2014. The results showed that there exist some short-run relationships between these rice prices in both regimes. As for the long-run relationships, we found that there exists only one long-run relationship (one cointegrating equation) expressed in the Thai white rice 5 % equation, which means only 5 % white rice price is cointegrated with 5 % par-boiled price, paddy price and world rice price, respectively. Meanwhile, Thai paddy price has the long run relationship and short-run adjustment dynamics, and it is cointegrated with 5 % parboiled price, 5 % white rice price and world rice price, respectively. The regime probabilities show the persistency in both high and low price market regimes where the probabilities of staying in this regime are larger than 90 %. In addition, the duration period also confirms the long period of time of staying in these regimes. Finally, we found some interesting events that seem to explain the movement of the regime probabilities over the sample period. We found that India's non-basmati rice export and the paddy price guaranteed at 15,000 Baht per ton are two main factors which caused the switching between these two regimes. Based on the above results, it might be suggested that Thai rice prices were not much volatile however, the long duration of staying in low price regime is longer than in high price regime. Moreover, the regime probabilities of Thai rice price seem to switch to the low price regime since the rice-pledging scheme was terminated at the end of February 2014. Consequently, the government should devise other schemes to stabilize and keep the high price of Thai rice to help the Thai farmers.

Acknowledgments The authors are grateful to Agriculture Research Development Agency (Public Organization), and Puey Ungphakorn Centre of Excellence in Econometrics, Faculty of Economics, Chiang Mai University for the financial support.

References

1. Acharya, S.S., Chand, P.R., Birthal, S.K., Negi, D.S.: Market integration and price transmission in India: a case of rice and wheat with special reference to the world food crisis of 2007/08. Food and Agriculture Organization, Rome (2012)
2. Chulaphan, W., Jatuporn, C., Chen, S.E., Jierwiriyapant, P.: Causal impact price transmission of the rice markets in Thailand. Modelling, Identification and Control/770: Advances in Computer Science and Engineering, Phuket, Thailand (2012)
3. Ifejirika, C.A., Arene, C.J., Mkpado, M.: Price transmission and integration of rural and urban rice markets in Nigeria. J. Agric. Sustain. **2**(1), 66–85 (2013)
4. Jochmann, M., Koop, G.: Regime-switching cointegration. Stud. Nonlinear Dyn. Econom. **19**, 35–48 (2011)
5. Perlin, M.: MS regress the MATLAB package for Markov regime switching models (2012). Available at SSRN 1714016
6. Rischke, R.: Transformation of global agri-food systems: trends, driving forces, and implications for developing countries (No. 66). Global Food Discussion Papers. 6. Rischke, R. (2015)
7. Sharma, R.: The transmission of world price signals: the concept, issues, and some evidence from Asian cereal markets. Agricultural Trade and Poverty: Making Policy Analysis Count, pp. 141–160 (2003)
8. Sugita, K.: Bayesian analysis of Markov switching vector error correction model (No. 2006–2013). Graduate School of Economics, Hitotsubashi University (2006)
9. Wiboonpongse, A., Chaovanapoonphol, Y.: Rice marketing system in Thailand. In: International Symposium, Agribusiness Management towards Strengthening Agricultural Development and Trade: III Agribusiness Research on Marketing and Trade (2001)
10. Wiboonpongse, A., Chaovanapoonphol, Y., Battese, G.E.: A quantile regression analysis of price transmission in Thai rice markets. Uncertainty Analysis in Econometrics with Applications, pp. 295–305. Springer, Berlin (2013)

Empirical Relationship Among Money, Output and Prices in Thailand

Popkarn Arwatchanakarn and Akhand Akhtar Hossain

Abstract Using annual data for the period 1953–2013 or a shorter period 1977–2013, this paper investigates the relationship among money, output and prices in Thailand. The empirical results, obtained by three techniques, namely the Engle–Granger cointegration approach, Johansens cointegation approach and the autoregressive distributed lag (ARDL) bounds approach, suggest the presence of a cointegral-causal relation among money, output and prices, especially for the shorter sample period. Predictably, short-run causal relations also exist between money growth and inflation and between money growth and output growth. The empirical results obtained by a structural vector autoregression (SVAR) model are confirmatory, showing that the accumulated impulse responses of output and prices to monetary shocks are positive and significant. The overall results are consistent with classical monetary theory that money matters insofar its impacts on output and prices are concerned.

1 Introduction

The emerging consensus is that the key objective of monetary policy should be price stability, implying low and stable inflation. Price stability plays a key role in maintaining macroeconomic stability and sustaining economic growth [4]. Empirical evidence suggests that low-and stable inflation promotes productive investment and raises economic growth [8, 9]. In contrast, high and volatile inflation affect economic growth and lower social welfare. Despite a sharp decline in inflation across the globe since the 1990s, inflation remains high and volatile in most developing countries, especially in those countries which have weak economic and political institutions [1]. How to control inflation and keep it stable have therefore remained the key issue in the design and conduct of monetary policy. In the contemporary literature, the role

P. Arwatchanakarn (✉) · A.A. Hossain
Newcastle Business School, University of Newcastle, Newcastle, NSW 2308, Australia
e-mail: Popkarn.Arwatchanakarn@uon.edu.au

A.A. Hossain
e-mail: Akhtar.Hossain@newcastle.edu.au

V.-N. Huynh et al. (eds.), *Causal Inference in Econometrics*,
Studies in Computational Intelligence 622, DOI 10.1007/978-3-319-27284-9_30

of money and its growth in inflation generation and its volatility have emerged as a contentious issue [6, 7].

Classical monetary theory suggests a long-run causal linkage between money, output and prices. The classical monetary transmission mechanism takes a stylised form like this; a monetary expansion raises real output initially but only prices in the long run as monetary neutrality condition suggests. As a corollary of this condition, inflation in the long run is considered a monetary phenomenon, in the sense that inflation in the long-run is caused by money growth adjusted for real output growth. Monetarists suggest the presence of a stylised adjustment process of an economy in the event of a one-time increase in the growth rate of the money supply. A rise in the growth rate of the money supply initially raises output growth above its long-run steady equilibrium rate and subsequently output growth falls below its equilibrium level, with attendant disruption to labour markets. Along with the condition of monetary neutrality, monetarists suggest the presence of super-neutrality of money in the long run; that is, the steady-state growth rate of the economy remains independent of the growth rate of the money supply. The initial response of the economy that shows a higher output growth is termed the nominal income effect and the latter response during the period of slower output growth is termed the price effect. These effects represent net additions (both positive and negative) to output growth in relation to what it would have been in the absence of any monetary expansion. The sequencing of such effects reflects the slow increase in prices initially and then they overshoot later such that the rate of inflation equals the money-growth rate over the adjustment period. Super-neutrality of money requires that after all of the dynamics is settled out, the cumulative change in nominal income equals the cumulative change in the price level. This ensures that real output growth remains unaffected by an increase in money growth over the long run [12].

In contrast to such classical views on the role of money in the inflation process, one branch of the monetary literature suggests that under inflation targeting for example, money does not play any role in the inflation process as a central bank can adjust aggregate demand by changing a short-term policy interest rate. While developing such proposition, the role of money in the inflation process is ignored despite the presence of a long-run link between money and prices as classical monetary theory suggests. This has created a paradigm that suggests the phenomenon of monetary policy without money. McCallum and Nelson [13] suggest that the relation between monetary growth and inflation remains valid irrespective of whether the money supply is endogenous or exogenous under a fixed or flexible exchange rate system and whether the money-demand function is stable or not.

The main aim of this paper is to investigate the role of money in the inflation process of Thailand within a cointegration-error-correction modelling framework. For estimation purposes, the paper uses annual data for the period 1953–2013 or a shorter period 1977–2013. The empirical results, obtained by three techniques, namely the Engle–Granger cointegration approach, Johansens cointegration approach and the autoregressive distributed lag (ARDL) bounds approach, suggest the presence of a cointegral-causal relation among money, output and prices, especially over the period 1977–2013. Predictably, short-run causal relations also exist between money growth and inflation and between money growth and output growth. The empirical results obtained by a structural vector autoregression (SVAR) model are confirmatory, showing that the accumulated impulse responses of output and prices to monetary shocks are positive and significant. The overall results are consistent with classical monetary theory that money matters insofar its impacts on output and prices are concerned.

The remainder of the paper is organised as follows. Section 2 reports the data sources and the time-series properties of variables deployed for establishing a relationship among money, output and prices. Section 3 outlines three cointegration approaches to establishing a relationship among money, output and prices, namely the Engle–Granger cointegration approach, Johansens approach and the ARDL bounds approach, and reports empirical results obtained by these techniques. It also reports some confirmatory findings of the relationship among money, output and prices from the structural vector autoregression (SVAR) modelling technique. Section 4 summarises the findings, draws policy implications and makes concluding remarks.

2 The Data and Their Time-Series Properties

The paper uses time-series data for three variables: the consumer price index (CPI), the narrow money stock (NM) and gross domestic product (GDP) at constant prices (RGDP). Data for these variables are obtained from various issues of IMF, *International Financial Statistics*, and World Bank, *World Development Indicator*. For estimation purposes, the data have been transformed into logarithmic forms, such that LCPI is the log of CPI, LNM is the log of NM, and LRGDP is the log of RGDP. The growth rate of the variable is defined as the first-order log-difference of it. For example, inflation is defined as $\Delta LCPI = LCPI - LCPI(-1)$. To determine the time-series properties of each of the variables, two widely-used unit root tests are performed; the augmented Dickey–Fuller (ADF) test and the Kwiatkowski, Phillips, Schmidt and Shin (KPSS) test. The ADF test is based on the null-hypothesis that the series under testing has a unit-root while the KPSS test is based on the hypothesis that the series under consideration is stationary and hence does not have a unit root. The overall results, not reported, suggest that the three variables under consideration have a unit root in a level form but are stationary in the first-order log-difference form. These results suggests that LCPI, LNM and LRGDP are eligible to establish a cointegral relation, as classical monetary theory suggests.

3 Cointegral Relationship Among Money, Output and Prices

3.1 Modelling Approaches and Techniques to Determine a Cointegral Relationship Among Money, Output and Prices

The estimation of a long-run equilibrium relation among money, output and prices hinges on the presence of a cointegral relationship. The null hypothesis that there exists a long-run equilibrium relation among money, output and prices suggests that these variables have common stochastic trends, or are integrated of order one, and are also cointegrated. When the variables are cointegrated, their (permanent) stochastic trends adjust according to equilibrium constraints and the cyclical (transitory) components of the series fit into a dynamic specification in the class of error-correction models [2, 15].

Although there is no ambiguity about the unit-root test results for the three variables (LCPI, LNM and LRGDP) under consideration, caution is taken for modelling purposes. Instead of relying on a single approach to determining the cointegral-causal relation among money, output and prices, this section reports the summary test results obtained from three procedures. The three procedures deployed are: (1) the Engle–Granger two-step residual-based procedure [3]; (2) Johansens system-based reduced-rank regression [10, 11]; and (3) the autoregressive distributed lag (ARDL) bounds-test, developed by Pesaran, Shin and Smith [14]. These tests are complementary and the differences in results obtained from them reflect the procedures sensitivities to uncertainty about the inclusion of constants, trends, and lag terms; especially when the small size is small.

Both the Engle–Granger and the Johansen approach to cointegration require that the variables of focus in a cointegral relationship have a unit root as suggested both by economic theory and by institutional arrangements. The existence or otherwise of such a relationship can be tested for in the presence of intercepts, time-trends and any other stationary or trend-stationary focus-variables. The ARDL bounds-test, however, does not impose restrictions on the time-series properties of regressors in detecting the long-run equilibrium relationship among them, although it is useful to know the time-series properties of those variables to draw inferences.

Establishment of a cointegral relation among variables, which are theoretically related, is an essential step towards detecting the direction of long-run causality, but short-run causality among them is equally important. This can be investigated deploying the Granger-causality test within a vector error correction (VEC) modelling framework, provided that the variables maintain a cointegral relation. It is possible to draw policy inferences if the results obtained from different approaches and techniques are confirmatory.

3.1.1 The Engel–Granger Cointegration

The Engel–Granger cointegration is deployed to determine a long-run relationship among variables that are theoretically correlated and whose time series have the same order of integration [3]. The cointegration testing procedure can be illustrated in generic form. In the first step, the cointegrating coefficient β between Y and X in the form of a regression $Y_t = \alpha + \beta X_t + \mu_t$ is estimated by ordinary least squares (OLS). If the null hypothesis is accepted, that Y and X are integrated of order one and cointegrated with the cointegrating coefficient β, then the expression for the error term $u_t(u_t = Y_t - \alpha - \beta X_t)$ will be stationary; otherwise $u_t = Y_t - \alpha - \beta X_t$ will be nonstationary. Therefore, in the second step, the Augmented Dickey–Fuller (ADF) is used to test for a unit root of the residual [15].

Table 1 reports a summary of cointegration test results as obtained by the dynamic OLS (DOLS) estimation technique. The DOLS estimator is based on a modified version of the cointegrating equation: $Y_t = \alpha + \beta X_t + \mu_t$, in which past, present and future values of the change in X are included. That is, the modification of the estimating equation $Y_t = \alpha + \beta X_t + \mu_t$ takes the form: $Y_t = \alpha + \beta X_t + \sum_{j=k}^{k} \delta_i \Delta X_{t-j} + \mu_t$. Stock and Watson [15] show that when the variables are I(1) and cointegrated, then the DOLS estimator of the cointegration coefficient is efficient and any statistical inference for the estimated cointegration coefficients, based on heteroskedasticity-adjusted standard errors, are valid in large samples. When X is exogenous, the effect on Y_t of a change in X_t can be interpreted as a long-run cumulative multiplier.

Table 1 reports the regression results obtained from the dynamic OLS (DOL) estimation technique. Although the results show only weak evidence of a cointegral relation of money, output and prices for the whole period, they indicate a cointegral relation for a shorter sample period of 1977–2013. The coefficient on income bears a positive sign and is statistically significant. It implies that, in the long-run, a 1 %

Table 1 The results of Engle–Granger cointegration test based on dynamic OLS (DOLS), Dependent variable: *LCPI*[a,b]

Variables	Coefficient (t-ration) [Prob.]	
	Period: 1952–2013	Period: 1977–2013
LNM[b]	0.26 (1.63) [0.10]	0.12 (1.17) [0.25]
LRGDP[b]	0.34 (1.39) [0.16]	0.48 (2.69) [0.01]**
C	−0.03 (−0.03) [0.97]	−0.12 (−0.15) [0.88]
Adjusted R2	0.98	0.98
DW	0.19	0.78
The EG cointegration test statistic by ADF	−1.72 [0.41]	−4.98 [0.00]***

Source Author's estimation and compilation based on IMF, International Financial Statistics
[a]LCPI is regressed on constant, time trend, LNM and LRGDP
[b]LCPI = Logarithm of a consumer price index; LNM = logarithm of narrow money; LRGDP = logarithm of a real GDP
*, ** and *** denote significant level at 90, 95 and 99 % respectively

increase in income raises the price level by 0.48 %. However, the coefficient on money is positive but not highly significant.

3.1.2 The Johansen Cointegration

Johansens cointegration estimation is based on a system method, which can establish multiple cointegral relationships among nonstationary variables given other deterministic and stationary variables. The Johansen procedure is essentially a generalisation of the Dickey–Fuller test for nonstationary of the series. Following Enders [2], the Johansen testing procedure for the presence of a cointegrating vector can be illustrated as follows. Consider a simple first-order autoregressive model of an individual series (y) such that:

$$y_t = \alpha y_{t-1} + \upsilon_t \tag{3.1.1}$$

where υ_t is a randomly distributed error term. For a multivariate case, this model can be expressed as:

$$x_t = A_1 x_{t-1} + \varepsilon_t \tag{3.1.2}$$

Equation 3.1.2 can be re-written as:

$$\Delta x_t = (A_1 - I)\, x_{t-1} + \varepsilon_t \tag{3.1.3}$$

or, in simplified form,

$$\Delta x_t = \pi x_{t-1} + \varepsilon_t \tag{3.1.4}$$

where x_t and ε_t are vectors, I is an $(n \times n)$ identity matrix, A_1 is an matrix of parameters, and $\pi = (A_1 - I)$. The rank of $(A_1 - I)$ equals the number of cointegrating vectors, determined by the number of characteristic roots that are significantly different from zero.

Johansen and Juselius [11] have proposed two likelihood-ratio tests to determine the number of cointegrating vectors: the trace test and the maximal eigenvalue test. These tests are conducted using the following two test statistics:

$$\lambda_{trace}(r) = -T \sum_{i=r+1}^{n} ln\left(1 - \hat{\lambda}_i\right) \tag{3.1.5}$$

$$\lambda_{max}(r, r+1) = -Tln\left(1 - \hat{\lambda}_{r+1}\right) \tag{3.1.6}$$

Table 2 The results of Johansen cointegration test

	Specification with variables: LCPI, LRGDP, LNM 1953–2013		Specification with variables: LCPI, LRGDP, LNM 1977–2013	
Null (alternative)	Maximum eigen-value statistic	Trace statistic	Maximum eigen-value statistic	Trace statistic
Hypothesised number of cointegration vector				
$r = 0$	39.45 (0.02)**	24.30 (0.03)**	55.95 (0.00)***	25.89 (0.00)***
$r = \leq 1$	15.15 (0.22)		30.05 (0.00)***	
$r = 1$		9.99 (0.33)		21.09 (0.01)**
Normalised cointegration vectors				
Variable	Coefficient	t-ratio		
LCPI	1		1	
LNM	−2.32***	−3.08	−0.79***	−2.98
LRGDP	2.89***	2.50	0.88*	1.85
Intercept	−10.97***	−2.44	−6.14***	−2.83
	Coefficient (t-ratio)		Coefficient (t-ratio)	
Error-correction model				
Error-correction model: ecm_{t-1}	−0.001 (−0.17)		−0.07 (−3.66)***	

Note Additional information: restricted intercept and no trend and VAR = 2
Series LCPI, LNM and LRGDP are logarithms of the consumer price index (CPI), the narrow money stock (NM), real GDP (RGDP) respectively
*, ** and *** denote significant level at 90, 95 and 99 % respectively

where $\hat{\lambda}_i$ denotes the estimated values of the characteristic roots (eigenvalues) obtained from the estimated pi matrix, T is the number of usable observations and r is the number of cointegrating vectors.

Table 2 presents the Johansen test results for cointegral relations among money, output and prices for two sample periods: (1) 1953–2013 and (2) 1977–2013. The results suggest a cointegral relationship among money, output and prices. The null hypothesis of no cointegration is rejected by both the trace and maximum eigenvalue tests. In addition, in the shorter 1977–2013, the coefficient on the error correction term (ect) bears a negative sign and is statistically significant. This confirms the existence of a cointegral relation among money, output and prices. The error correction term measures the speed of adjustment toward to the long-run equilibrium which is 7 % within a year.

3.1.3 The ARDL Bounds Test Results

The empirical results obtained by the Engle–Granger and the Johansen methods suggest a weak cointegral relation among money, output and prices. This sub-section reports results obtained by the ARDL bounds-test approach. For a small sample, the ARDL test is superior to Johansens cointegration tests. The Johansen test results, in particular, are sensitive to the inclusion of constants, trends and lag-lengths in the specification. Another important advantage of the ARDL testing approach is that it does not require pre-testing of unit roots for regressors, which enter into a long-run equilibrium relation. However, the critical values needed for determining the presence of a long-run relation have lower and upper bounds. These statistics can be used to determine whether the variables under consideration form a long-run equilibrium relation by taking into account information on whether the variables are stationary, nonstationary or their time-series properties are indeterminate.

The error-correction form of the ARDL model in the variables: *LCPI*, *LNM* and *LRGDP* is specified in the following form:

$$\Delta LCPI_t = \alpha_0 + \alpha_1 Trend + \sum_{i=1}^{n} \beta_i \Delta LCPI_{t-i} + \sum_{i=1}^{n} \gamma_i \Delta LNM_{t-i} + \sum_{i=1}^{n} \varphi_i \Delta LNM_{t-i} + \delta_1 LCPI_{t-1} + \delta_2 LNM_{t-1} + \delta_3 LRGDP_{t-1} + \mu_t \tag{3.1.7}$$

This specification is based on the maintained hypothesis that the time-series properties in the price-level relationship can be approximated by a log-linear *VAR*(*p*) model, augmented with a deterministic intercept and (probably) a trend (*Trend*).

Table 3 reports the estimated F-statistics. The estimated F-statistics obtained for the period 1950–2013 are smaller than the lower bound critical values suggesting that the hypothesis of no cointegration among money, output and prices cannot be rejected at the 95 % level of significance. However, the estimated F-statistics obtained for the period 1977–2013 are found greater than the upper bound critical values, suggesting

Table 3 The F-test statistics for the long-run relationship among money, output and prices

	Specification with variable:LCPI, LRGDP, LNM; F-statistic $F(LCPI \mid LRGDP, LNM)$	Specification with variable:LCPI, LRGDP, LNM; F-statistic $F(LCPI \mid LRGDP, LNM)$
	Annual data, 1950–2013	Annual data, 1977–2013
With intercept and trend		
Lag 1	3.69	6.83**
Lag 2	2.63	4.74
With intercept and no trend		
Lag 1	3.62	6.56**
Lag 2	2.68	5.07**

Note The Critical value bounds of the F-static: K = 2, with an intercept and no trend, are [3.18–4.12] and [3.79–4.12] at 90 and 95 % respectively and those with an intercept and trend is [4.20–5.10] and [4.90–5.87] at 90 and 95 % respectively

* and ** denote significant level at 90 and 95 % respectively

that the hypothesis of no cointegration among these variables is rejected at the 95 % confidence level.

Table 4 reports the estimated long- and short-run coefficients of the relations among money, output and prices as obtained by the ARDL method. The Schwarz Bayesian Criterion (SBC) in used to select the optimal lag length. The results suggest the presence of a long-run relationship among prices, money and output irrespective of the sample period deployed for estimation purposes. In particular, for the period 1977–2013, both the F-statistic and W-statistics were above the upper bound at 95 % confidence level, suggesting the presence of a cointegral relation among money, output and prices. The coefficient of the error-correction term (ect) bears a negative sign and statistically significant. This confirms the existence of a long-run relationship between money, output and prices. The coefficient of the error-correction term implies a 21 % correction within a year toward the long run equilibrium. The coefficient of real GDP is expected to have a negative sign. The coefficient of real GDP bears a positive sign but is not significant. The coefficient of money bears a positive sign and is significant. The results obtained by the ARDL are confirmatory to those obtained by the Engle–Granger and Johansen methods.

3.2 The Granger Causality Test

This section presents results of the presence of a linkage among money, output and prices within the Granger-causality testing framework. Engle–Granger [3] suggest that a cointegral relation can express the presence of long-run causality between nonstationary variables. They also suggest that it is useful to detect the presence of short-run causality among stationary variables using a VEC modelling framework. Table 5 reports the bivariate Granger causality test results among ΔLNM, $\Delta LRGDP$ and $\Delta LCPI$. The test results were generated after selecting the optimal lag length by the Akaike information criterion (AIC). The results indicate that there is short-run causality running from money and output growth to inflation. The short-run impact of monetary expansion on output is suggested. These results are consistent with classic monetary theory which predicts a causal link between money and prices under money neutrality (no long-run relation between money and real output).

3.3 A SVAR Model for Money, Output and Prices

A growing body of literature analyses the monetary policy transmission mechanism by employing the structural vector autoregression (SVAR) model. Hossain [5] has examined the relationship among money, output and prices in nine Muslim-majority countries within a three-variable SVAR model. This study follows Hossain [5]. A SVAR model represents a multivariate system of a set of endogenous variables which maintain feedback relations in a dynamic sense. The variables of the SVAR model are

Table 4 The autoregressive distributed lag (ARDL): long-run relationships among LCPI, LNM and LRGDP (ARDL order selection: SBC)

Regressor	Annual data Specification with variables:LCPI,LNM, LRGDP ARDL(2,2,1) Selected based on SBC (maximum lag = 2) (1953–2013)	Annual data Specification with variables: LCPI, LNM, LRGDP ARDL(2,1,0) Selected based on SBC (maximum lag = 2) (1977–2013)
F-statistic with constant and no trend (CV = critical value)	4.42 CV(95 %):(3.96, 5.04) CV(90 %):(3.29, 4.22)	11.51 CV(95 %):(4.17, 5.36) CV(90 %):(3.41, 4.40)
W-statistic with constant and no trend (CV = critical value)	13.27 CV(95 %):(11.91, 15.12) CV(90 %):(9.87, 12.68)	34.53 CV(95 %):(12.52, 16.06) CV(90 %):(10.25, 13.21)
Regressor	*Coefficient* (*t-ratio*) [*Prob.*]	*Coefficient* (*t-ratio*) [*Prob.*]
LNM	−0.05 (−0.21) [0.83]	0.26 (0.11) [0.02]**
LRGDP	0.81 (2.09) [0.04]**	0.17 (0.21) [0.82]
Intercept	−1.87 (−1.23) [0.22]	1.40 (0.98) [1.41]
Trend		
The Wald test		
H_0: Coefficient on LNM = 0	$\chi^2_{(1)} = 0.05$ [0.83]	$\chi^2_{(1)} = 3.96$ [0.04]**
H_0 : Coefficient on LRGDP = 0	$\chi^2_{(1)} = 4.40$ [0.04]**	$\chi^2_{(1)} = 0.85$[0.36]
Error correction model		
C		
$\Delta LCPI(-1)$	0.46 (4.66) [0.00]***	0.45 (4.07) [0.00]***
ΔLNM	0.08 (1.20) [0.23]	−0.09 (−1.92) [0.06]*
$\Delta LNM(-1)$	0.25 (3.89) [0.00]***	
$\Delta LRGDP$	−0.28 (2.16) [0.03]**	0.03(0.81) [0.42]
ecm_{t-1}	−0.10 (−2.88) [0.00]***	−0.21 (−5.09) [0.00]***
Adjusted R_2	0.46	0.66
DW	1.99	2.23
Diagnostic tests		
LM test for serial correlation	F(1,53) = 0.002 [0.96]	F(1,30) = 1.46 [0.23]
Ramsey's RESET test for misspecification	F(1,53) = 14.12 [0.00]	F(1,30) = 7.54 [0.01]
Normality test	$\chi^2_{(2)} = 4.80$ [0.09]	$\chi^2_{(2)} = 92.82$ [0.00]
Test for heteroskedasticity	F(1,58) = 2.37 [0.13]	F(1,58) = 2.12 [0.15]

Source Author's estimation and compilation based on IMF, International Financial Statistics
Note the numbers inside the parenthesis [] denote the probability value LCPI, LNM and LRGDP are logarithm of a consumer price index, narrow money, real GDP respectively
*, ** and *** denote signficant level at 90, 95 and 99 % respectively

Table 5 The Granger causality between money growth, output growth and inflation based on VEC model: 1977–2013

Null hypothesis: No causality runs from	Optimal lag selection (based on AIC)	Chi-sq. statistic (prob.)	Inference null hypothesis is:
$\Delta LNM \rightarrow \Delta LCPI$	3	17.47 (0.00)	Rejected
$\Delta LNM \rightarrow \Delta LRGDP$	3	9.60 (0.02)	Rejected
$\Delta LCPI \rightarrow \Delta LNM$	3	1.81(0.61)	Not rejected
$\Delta LCPI \rightarrow \Delta LRGDP$	3	0.18 (0.98)	Not rejected
$\Delta LRGDP \rightarrow \Delta LCPI$	3	8.05 (0.04)	Rejected
$\Delta LRGDP \rightarrow \Delta LNM$	3	3.96 (0.26)	Not rejected

Source Author's estimation and compilation based on IMF, International Financial Statistics LCPI, LNM and LRGDP are logarithm of a consumer price index, narrow money, real GDP respectively

then assumed to be contemporaneously and dynamically interdependent. In this case, three variables, namely money (m = log M), output (y = log RGDP) and prices (p = log CPI), were assumed to be contemporaneously and dynamically interdependent. According to classical monetary theory, the monetarists establish a long-run equilibrium relationship with causality running from money to prices. The three variables under consideration can be specified in the following first-order autoregressive form, neglecting intercepts, where t is a time subscript and the β_i and α_i are parameter to be estimated:

$$m_t + \beta_1 y_t + \beta_2 p_t = a_1 m_{t-1} + a_2 y_{t-1} + a_3 p_{t-1} + \varepsilon_t^m \tag{3.3.1}$$

$$\beta_3 m_t + y_t + \beta_4 p_t = a_4 m_{t-1} + a_5 y_{t-1} + a_6 p_{t-1} + \varepsilon_t^y \tag{3.3.2}$$

$$\beta_5 m_t + \beta_6 y_t + p_t = a_7 m_{t-1} + a_8 y_{t-1} + a_9 p_{t-1} + \varepsilon_t^p \tag{3.3.3}$$

The residual ε_t^m, ε_t^y and ε_t^p are assumed to be normally distributed with zero means and constant variances, and they are not serially correlated. The endogenous variables (m, y and p) may have contemporaneous feedback impacts which are captured by the β_i. The structural form of these three equations is specified in the following matrix:

$$\begin{pmatrix} 1 & \beta_1 & \beta_2 \\ \beta_3 & 1 & \beta_4 \\ \beta_5 & \beta_6 & 1 \end{pmatrix} \begin{bmatrix} m_t \\ y_t \\ p_t \end{bmatrix} = \begin{pmatrix} a_1 & a_2 & a_3 \\ a_4 & a_5 & a_6 \\ a_7 & a_8 & a_9 \end{pmatrix} \begin{bmatrix} m_{t-1} \\ y_{t-1} \\ p_{t-1} \end{bmatrix} + \begin{bmatrix} \varepsilon_t^m \\ \varepsilon_t^y \\ \varepsilon_t^p \end{bmatrix} \tag{3.3.4}$$

Equation (3.3.4) can be written as:

$$BY_t = AY_{t-1} + E_t, \tag{3.3.5}$$

where

$$B = \begin{pmatrix} 1 & \beta_1 & \beta_2 \\ \beta_3 & 1 & \beta_4 \\ \beta_5 & \beta_5 & 1 \end{pmatrix}, Y_t = \begin{pmatrix} m_t \\ y_t \\ p_t \end{pmatrix}, A = \begin{pmatrix} \alpha_1 & \alpha_2 & \alpha_3 \\ \alpha_4 & \alpha_5 & \alpha_6 \\ \alpha_7 & \alpha_8 & \alpha_9 \end{pmatrix}, Y_{t-1} = \begin{pmatrix} m_{t-1} \\ y_{t-1} \\ p_{t-1} \end{pmatrix}, E_t = \begin{pmatrix} \varepsilon_t^m \\ \varepsilon_t^y \\ \varepsilon_t^p \end{pmatrix}$$

Pre-multiply Eq. (3.3.5) by the inverse of B, B^{-1} such that:

$$B^{-1}BY_t = B^{-1}AY_{t-1} + B^{-1}E_t, \tag{3.3.6}$$

Matrix B is assumed to be non-singular so that $B^{-1}B = I_{3\times 3}$, Then Eq. (3.3.6) simplifies to:

$$Y_t = B^{-1}AY_{t-1} + B^{-1}E_t \tag{3.3.7}$$

Equation (3.3.7) can be expressed in matrix form as follows:

$$Y_t = \begin{pmatrix} 1 & \beta_1 & \beta_2 \\ \beta_3 & 1 & \beta_4 \\ \beta_5 & \beta_6 & 1 \end{pmatrix}^{-1} \begin{pmatrix} \alpha_1 & \alpha_2 & \alpha_3 \\ \alpha_4 & \alpha_5 & \alpha_6 \\ \alpha_7 & \alpha_8 & \alpha_9 \end{pmatrix} \begin{bmatrix} m_{t-1} \\ y_{t-1} \\ p_{t-1} \end{bmatrix} + \begin{pmatrix} 1 & \beta_1 & \beta_2 \\ \beta_3 & 1 & \beta_4 \\ \beta_5 & \beta_6 & 1 \end{pmatrix}^{-1} \begin{bmatrix} \varepsilon_t^m \\ \varepsilon_t^y \\ \varepsilon_t^p \end{bmatrix} \tag{3.3.8}$$

The reduced-form matrix of Eq. (3.3.8) gives the following simplified form:

$$\begin{bmatrix} m_t \\ y_t \\ p_t \end{bmatrix} = \begin{pmatrix} \delta_1 & \delta_2 & \delta_3 \\ \delta_4 & \delta_5 & \delta_6 \\ \delta_7 & \delta_8 & \delta_9 \end{pmatrix} \begin{bmatrix} m_{t-1} \\ y_{t-1} \\ p_{t-1} \end{bmatrix} + \begin{bmatrix} \upsilon_t^m \\ \upsilon_t^y \\ \upsilon_t^p \end{bmatrix} \tag{3.3.9}$$

In matrix (3.3.9) the δ_i are reduced-form parameters output by the operation $\beta^{-1}A$ and $\left(\upsilon_t^m, \upsilon_t^y, \upsilon_t^p\right)$ are reduced-form errors output by the operation $\beta^{-1}E_t$. In this standard form of the VAR model, the reduced-form error terms $\left(\upsilon_t^m, \upsilon_t^y, \upsilon_t^p\right)$ are composites of three pure shocks to the structural system: $\left(\varepsilon_t^m, \varepsilon_t^y, \varepsilon_t^p\right)$. Since these pure shocks are white noise, the reduced-form shocks $\left(\upsilon_t^m, \upsilon_t^y, \upsilon_t^p\right)$ are assumed to have zero means and constant variances. However, the reduced-form shocks could be correlated as the endogenous variables in the system (m, y, p) may have contemporaneous relations. Consequently, the variance-covariance matrix of the reduced-form error terms can be written as:

$$\Sigma_\vartheta = \begin{pmatrix} \sigma_m^2 & \sigma_{my} & \sigma_{mp} \\ \sigma_{my} & \sigma_y^2 & \sigma_{yp} \\ \sigma_{mp} & \sigma_{yp} & \sigma_p^2 \end{pmatrix} \tag{3.3.10}$$

Only when the endogenous variables (m, y, p) are not contemporaneously related, will the reduced-form shocks be uncorrelated. The variance-covariance matrix can be expressed in the following form:

$$\Sigma_\vartheta = \begin{pmatrix} \sigma_m^2 & 0 & 0 \\ 0 & \sigma_y^2 & 0 \\ 0 & 0 & \sigma_p^2 \end{pmatrix} \quad (3.3.11)$$

3.3.1 Identification and Estimation

The reduced-form equations can be estimated by OLS. These estimates can be given meaningful economic interpretation if the structural systems parameters can be identified by imposing theory-consistent restrictions on contemporaneous relations within the system. There is, however, no consensus on how to impose the restrictions. The Cholesky decomposition technique was deployed to explain the relationship in this study. The Cholesky decomposition technique makes the matrix lower-triangular such that

$$\begin{pmatrix} 1 & 0 & 0 \\ \beta_3 & 1 & 0 \\ \beta_5 & \beta_6 & 1 \end{pmatrix} \begin{bmatrix} m_t \\ y_t \\ p_t \end{bmatrix} = \begin{pmatrix} a_1 & a_2 & a_3 \\ a_4 & a_5 & a_6 \\ a_7 & a_8 & a_9 \end{pmatrix} \begin{bmatrix} m_{t-1} \\ y_{t-1} \\ p_{t-1} \end{bmatrix} + \begin{bmatrix} \varepsilon_t^m \\ \varepsilon_t^y \\ \varepsilon_t^p \end{bmatrix} \quad (3.3.12)$$

The identification procedure adopted here follows classical monetary theory, where prices in the steady-state are determined by the money supply adjusted for potential output. In the specified model, money affects output contemporaneously but causation in the reverse direction is lagged. Both money and output impact prices contemporaneously but price changes impact both money and output with lags. According to classical monetary theory, an increase in money supply raises output initially and its impact on prices increases through time. In the long-run, money is neutral; that is, price equilibrates to a percentage change in the money supply. Following an expansion of money supply, output first rises and then falls while prices rise steadily.

In the present model, money supply shocks are treated as pure shocks. Accordingly, in the first row of the -matrix, the off-diagonal elements are restricted to zero. In the second row, the coefficient of money on output β_3 is non-zero; it should be positive. This follows the standard theory; monetary expansion raises output. However, as prices are thought not to contemporaneously impact on output, the off-diagonal element in the second row is restricted to zero ($\beta_4 = 0$). Money and output have contemporaneous impacts on price. Therefore, in the third row, the coefficients β_5 and β_6 are expected to be non-zero.

In this specification of the transmission mechanism of monetary policy, the ordering of variables in the SVAR model is as follows: m, y and p. The relations between the pure shocks $(\varepsilon_t^m, \varepsilon_t^y, \varepsilon_t^p)$ and the reduced-form shocks $(\upsilon_t^m, \upsilon_t^y, \upsilon_t^p)$ are expressed in the following form:

$$\begin{bmatrix} \varepsilon_t^m \\ \varepsilon_t^y \\ \varepsilon_t^p \end{bmatrix} = \begin{pmatrix} 1 & 0 & 0 \\ \beta_3 & 1 & 0 \\ \beta_5 & \beta_6 & 1 \end{pmatrix} \begin{bmatrix} \vartheta_t^m \\ \vartheta_t^y \\ \vartheta_t^p \end{bmatrix} \quad (3.3.13)$$

In equation form, the pure and structural residuals are expressed in following equations

$$\varepsilon_t^m = \vartheta_t^m \tag{3.3.14}$$

$$\varepsilon_t^y = \beta_3 \vartheta_t^m + \vartheta_t^y \tag{3.3.15}$$

$$\varepsilon_t^p = \beta_5 \vartheta_t^m + \beta_6 \vartheta_t^y + \vartheta_t^p \tag{3.3.16}$$

3.3.2 Impulse Responses and Forecast-Error-Variance Decompositions

This sub-section presents impulse responses of money, output and prices. In order to generate the impulse responses, the restriction is imposed on the specified SVAR. Following the standard proposition, the restriction is that an increase in money supply raises output contemporaneously and raises the price level with lags. The reason behind this proposition is that the presence of short-run rigidities causes the effects of money on prices to be lagged. The ordering of variables is set up as follows: LNM, LRGDP, and LCPI.

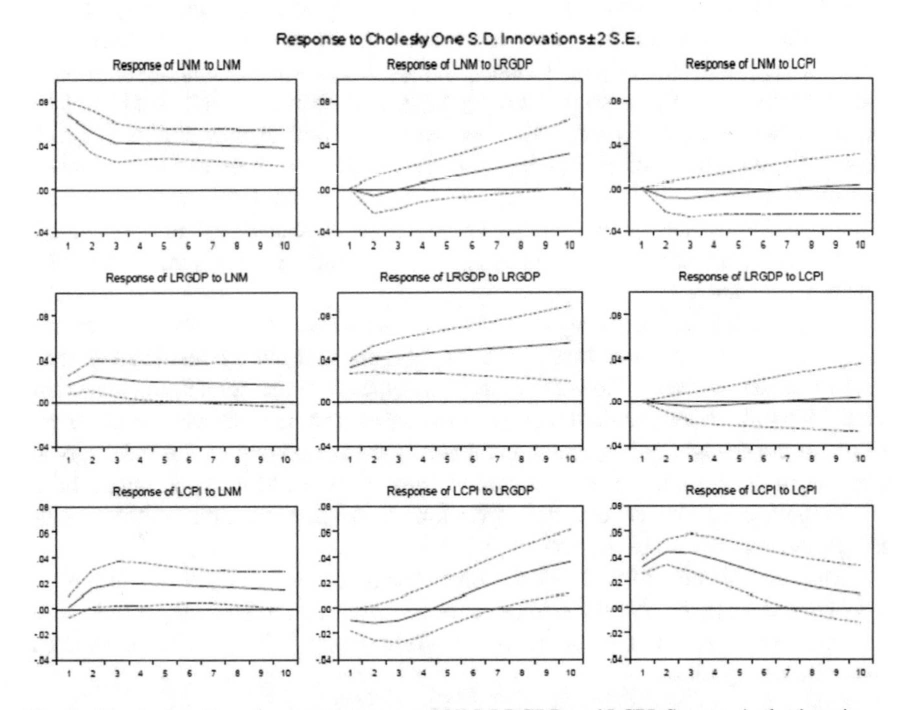

Fig. 1 Accumulated impulse response among LNM, LRGDP, and LCPI. Source: Author's estimation and compilation based on IMF international financial statistics LCPL, LNM and LRGDP are logarithm of a consumer price index, narrow money, real GDP respectively

Figure 1 illustrates the accumulated impulse response of money, output and prices. The impulse responses of output and prices to money are positive as the zero line is below the two standard error confidence bands. These positive responses imply that both output and price level increase when there is a money shock. They also remain steady over a 10-year time-horizon. In particular, these responses are supportive of classical monetary theory predicting that monetary expansion stimulates output in the short-run and that, in the long run, money is neutral. However, there is no significant response to output and prices as the zero line lies between the two-standard-error confidence bands.

Table 6 reports the forecast-error-variance decomposition results for LCPI, LNM and LRGDP. It shows that, for all variables, the dominant source of forecast-error-variance is attributed to own innovations. For real output, they account for more than 80 % over a 10-year forecast horizon. In the case of money, own innovation accounts for 86 % while innovations of output and prices account for approximately 12 and 1 % of the forecast at the end of 10 year horizon. In the case of output, the output innovation is 85 % whereas innovations of money and prices are about 14 and 0 % respectively of the forecast error variance of output after 10 years. In the case of prices, own innovation is 57 % in the forecast error variance after 10 years. Innovations of money and output are accounted for at 16 and 25 % respectively. This implies that output and money play a significant role in determining the forecast error variance of prices. Therefore, policymakers could maintain price stability through money and output.

Table 6 Forecast-error-variance-decomposition: 1952–2013; Cholesky ordering: LNM, LRGDP, and LCPI

Time horizon (years)	Standard error	LNM	LRGDP	LCPI
LNM				
1	0.067157	100.0000	0.000000	0.000000
5	0.113068	96.83791	1.247578	1.914508
10	0.151715	86.43713	12.41367	1.149201
LRGDP				
1	0.036455	20.39326	79.60674	0.000000
5	0.103693	19.98420	79.44785	0.567947
10	0.158675	14.50423	85.13968	0.356091
LCPI				
1	0.033241	0.147250	8.256625	91.59613
5	0.094368	15.09856	4.087373	80.81407
10	0.123818	16.88326	25.79125	57.32548

Source Author's estimation and compilation based on IMF, International Financial Statistics LCPI, LNM and LRGDP are logarithm of a consumer price index, narrow money, real GDP respectively

4 Conclusion and Policy Implications

This paper has used annual data for the period 1953–2013 or a shorter period 1977–2013 to investigate the relationship among money, output and prices in Thailand. The empirical results, obtained by three techniques, namely the Engle–Granger cointegration approach, Johansens cointegation approach and the autoregressive distributed lag (ARDL) bounds approach, suggest the presence of a cointegral-causal relation among money, output and prices, especially for the shorter sample period. Predictably, short-run causal relations also exist between money growth and inflation and between money growth and output growth. The empirical results obtained by a structural vector autoregression (SVAR) model are confirmatory, showing that the accumulated impulse responses of output and prices to monetary shocks are positive and significant. The overall results are consistent with classical monetary theory that money matters insofar its impacts on output and prices are concerned. Therefore, the monetary aggregate could be a supplementary instrument to achieve price stability. Monetary authorities can achieve price stability by implementing a rule-based monetary policy.

References

1. Alesina, A., Stella, A.: The politics of monetary policy. In: Friedman, B.M., Woodford, M. (eds.) Handbook of monetary economics, vol. 3B, pp. 1001–1054. Elsevier, Amsterdam (2011)
2. Enders, W.: Applied Econometric Time Series. Wiley, New Jersey (2010)
3. Engle, R., Granger, C.: Co-integration and error correction: representation, estimation and testing. Econometrica **55**(2), 251–276 (1987)
4. Fischer, S.: Role of macroeconomic factors in growth. J. Monet. Econ. **32**, 485–512 (1993)
5. Hossain, A.: Empirical relationships among money, output and consumer prices in nine Muslim-majority countries. J. Asian Econ. **31–32**, 42–56 (2014)
6. Hosaain, A.: Central banking and monetray polcy in Muslmim-Majority Countries. Edward Elgar, Cheltenham and Northampton (2015)
7. Hosaain, A.: The Evolution of Central Banking and Monetary Policy in the Asia-Pacific. Edward Elgar, Cheltenham and Northampton (2015)
8. Hossain, A., Chowdhury, A.: Monetary and Financial Policies in Developing Countries: Growth and Stabilisation. Routledge, London (1998)
9. Hossain, A., Chowdhury, A.: Open-Economy Macroeconomics for Developing Countries. Edward Elgar, Cheltenham and Northampton (1998)
10. Johansen, S.: Statistical analysis of cointegration vectors. J. Econ. Dyn. Control **12**, 232–254 (1988)
11. Johansen, S., Juselius, K.: Maximum likelihood estimation and inference on cointegration with applications to the demand for money. Oxf. Bulltetin Econ. Stat. **52**(2), 169–210 (1990)
12. Marquis, M.H.: Monetary Theory and Policy. West Publishing Company, New York (1996)
13. McCallum, B.T., Nelson, E.: Money and inflation: some critical issues. In: Friedman, B.M., Woodford, M. (eds.) Handbook of Monetary Economics, vol. 3A, pp. 97–153. Elsevier, Amsterdam (2011)
14. Pesaran, M.H., Shin, Y., Smith, R.J.: Bounds testing approaches to the analysis of level relationships. J. Appl. Econom. **16**(3), 289–326 (2001)
15. Stock, J.H., Watson, M.W.: Introduction to Econometrics. Pearson-Addison Wesley, New York (2012)
16. Hosaain, A.: Central banking and monetary policy in the Asia-Pacific. Edward Elgar, Cheltenham (2009)

Further Readings

17. Kim, S., Roubini, N.: Exchange rate anomalies in the industrial countries: a solution with a structural VAR approach. J. Monet. Econ. **45**(3), 561–586 (2000)
18. Patrawimolpon, P., Rattanalankar, T., Charumilind, C., Ngamchant, P.: A Structural Vector Autoregressive Model of Thailand: A Test for Structural Shifts. Bank of Thailand, Bangkok (2001)
19. Raghavan, M., Silvapulle, P., Athanasopoulos, G.: Structural VAR models for Malaysian monetary policy analysis during the pre- and post-1997 Asian Crisis periods. Appl. Econ. **44**(29), 3841–3856 (2012)
20. Sek, S.K.: Interactions between monetary policy and exchange rate in inflation targeting emerging countries: the case of three East Asian countries. Int. J. Econ. Financ. **1**(2), 27–44 (2009)
21. Vimolsiri, P., Hirunraengchok, A.: Structural Vector Autoregressions: The Case of Thailand. Paper presented at the International symposium on forecasting, Sydney (2004)

The Causal Relationship between Government Opinions and Chinese Stock Market in Social Media Era

Xue Gong and Songsak Sriboonchitta

Abstract China's capital market is still an emerging market. The weaknesses of a transition market are speculative behaviors of investors and heavy government intervention. The recent observed Chinese stock market has experienced an extraordinary rise while the real economics is going down. Behavioral economics tells us that public comments can profoundly affect individual behavior and decision-making. In this information era, the public can be the "collective intelligence" in social media. This study investigates what is the role of Chinese government play in Sina weibo, which is the biggest microblog in China. Are the public posts, which also contain government messages, correlated or a causal relationship of economic indicators? Here we figure out whether measurements of collective public comments derived from large-scale Sina weibo posts are correlated to the value of the Shanghai Composite over time. A Granger causality analysis was finally used to detect the causal relationship between government role in social media and recent Chinese stock market price. The results show that the positive government opinion is the Granger cause of the Chinese stock market.

1 Introduction

The Chinese equity markets have suffered great structural changes and progress in the last twenty years (Gao [1]; Yu and Ashton [2]). After the government established the Shenzhen Stock Exchange and the Shanghai Stock Exchange in December, 1990, the number of listed companies reached 2,683 in the early 2015 from only 10 companies

X. Gong (✉) · S. Sriboonchitta
Faculty of Economics, Chiang Mai University, Chiang Mai 50200, Thailand
e-mail: gongxue.cmu@gmail.com

S. Sriboonchitta
e-mail: songsakecon@gmail.com

V.-N. Huynh et al. (eds.), *Causal Inference in Econometrics*,
Studies in Computational Intelligence 622, DOI 10.1007/978-3-319-27284-9_31

in the early 1990s.[1] With a total market capitalization of One trillion USD (Los and Yu [3]).[2]

The recent state of the China stock market has shocked the world. As CNN reports[3], in 2015 The Shanghai composite index has risen from 3000 point level in January to 6500 point level in June. During six months, the Chinese stock market became the world's most attractive one. Whole nations started to invest in the stock market; not only professorial investors, but also the residents in each household. However, the market has been volatile since the middle of June 2015 and suffer a great drop. One minute shares are up 6 %, the next minute they're down 5 % and jump to a bear market[3].

The internet has fundamentally reshaped the way people's life styles. Also the recent wide usage of social media has changed the way people exchange information and how the government conveys their messages in China. Dating back to 2010, Twitter started to be censored in China. And almost at the same time, other native Chinese microblog services, such as Fanfou and Digu, were being censored. Nowadays, the most popular social media outlet in China is Sina Weibo, which is a twitter-like social media, a mix of Facebook and Twitter. The design follows Twitter, however, the same tweet length limit of 140 for characters can contain more content when it is in Chinese, which makes the weibo, as a microblog have more functions. Moreover, it has long been suggested that Sina has enjoyed a close relationship with the state (Kennedy [4]; Xin [5]).

At the beginning of China's information era in 2009, the government realized it is necessary to use new media to convey the information. In China, this change not only happened with government-related institutions, but also further to government officials. According to the statistics from *People Daily* (the China's biggest newspaper), there are four hundred million registered users (almost 1/3 of China's population). What's more, until March 2015, the verified government weibo users are up to 19,018 users. The public officials account for 72.7 %. The influence of the Sina weibo is not less impactful than the TV or other conventional mediums, and also this media has become one of the most important government mediums. Even the newspaper and TV are cite posts from weibo.

Behavioral economics tells us that public emotions can profoundly affect individual behavior and decision-making (Bollen et al. [2]). How does this affect the Chinese market? The Sina weibo, the biggest social media outlet, is supervised by the government, and at the same time, the stock market is more or less influenced by the government, not by the demand and supply relationship. So what should be expected? Shall we expect that there is a causal relationship between the posts from the government and recent Chinese stock market events?

Granger testing is popular for testing causality in economics using lead or lag relationships across different time series (Granger [6]; Timmermann and Granger [7]). However, Granger testing has not been used directly in text mining in Chinese inter-

[1]Data source: China Securities Regulatory Commission: http://www.csrc.gov.cn/.

[2]Data source: People Daily: http://finance.people.com.cn/n/2015/0616/c66323-27159636.html.

[3]Data source: CNN Money: http://money.cnn.com/2015/07/02/investing/china-stock-markets.

net. In this study, we would like to use this test to understand the causality relationship of the government's posts in weibo and the stock market during the recent bull market.

The rest of the paper is organized as follows: the next section provides the background of this research by critically reviewing existing research on the Chinese stock market, Sina weibo, and Chinese government to provide the rationale for this study. Then, two sets of research questions are formulated to guide the investigation. In the methodology section, the design of the research using a search query mining approach is explained. Findings are then presented in correspondence with each of the research questions. Finally, the conclusion is made.

2 Research Background

With the enormous amount of information potentially available to daily life, the internet, also the social media constitutes an important platform for information exchange (Werthner and Klein [8]; Xiang and Gretzel [9]). Stock market predictions are usually following the random walk theory and the Efficient Market Hypothesis (EMH) (Timmermann and Granger [7]). According to the EMH, the stock market prices are largely driven by news. Research also suggest that news can be predictable by different indicators in online social media (Fama [10]; Schumaker and Chen [11]). This could also work in the case of the China's stock market.

2.1 The China's Stock Market Prediction

China's stock market has grown amazingly rapidly since the establishment of its two exchanges in 1990. However, because the government has much influence on the market (Gao [1]; Los and Yu [3]), the growth of the market has become extremely irregular and has not followed traditional models of stock-market development. In some senses, it is still in the early stages of development. The evidences are as follows: the government kept a control on the issuance of Initial Public Offering (IPOs), and since the large government holdings, many listed companies in China have very low free-float ratios. At the same time, market manipulation and speculation are common because of market structure (Gao [1]).

Stock market predictions are attractive due to the huge profits for investors. But can the stock market really be predicted? The situation in China may be different. The information from the government may dominate the whole market movement, and therefore it could be an important factor to predict the price. To understand the relationship between the news from government and market price is critical.

2.2 Sina Weibo and Government

It is without doubted that the Sina weibo is the most popular microblog in China. However, the Sina weibo has greater than normal state control (both in terms of censorship and close links between state and market). First, it is clear that non-approved microblogs are routinely censored and the unqualified posts are timely deleted. Second, it seems that the weibo is intended to maximize entertainment and to minimize reasoned discussion and debate (Benney [12]). When compared with Twitter and Facebook, weibo has more to do with the entertainment topics. The political topics are rarely being discussed seriously by the public. For instance, a famous Chinese artist, for example, *Zhaowei*, she has the 71,890,245 followers; however, one of the best-known government account, called *People daily* has only 36,911,311 followers.[4] However, the topics about economics are always found to be discussed or be shared freely.

On the other hand, we believe that nowadays the social media is more important than the traditional ones such as the TV, newspaper or other internet channel. Compare with the traditional channels, the way of the government convey information becomes a “collective intelligence”. When the people post and repost, and spread re-repost, the information can go with more people and become the opinions of the mass.

2.3 Social Media and Stock Market

There are several studies about social media mood and stock prices, Bollen et al. [2] used twitter mood to predict the stock market; the results showed that the accuracy of Dow Jones Industrial Average (DJIA) predictions can be significantly improved by the inclusion of specific public mood dimensions, but not others. Kim et al. [13] developed a novel general text mining framework for discovering such causal topics from text. Luo, Zhang, and Duan [14] used vector autoregressive models to show that social media-based metrics (web blogs and consumer ratings) are significant leading indicators of firm equity value.

Since the news from the government in China is important and also the government encourages the government-related units and individuals to open the Sina weibo account, the information could flow from the internet to the stock market via weibo. At the same time, the government also impacted the stock market, the causal relationship between the weibo’s comments from government and stock market may exist.

[4]The data is collected until July 1st 2015.

2.4 Research Rationale

Understanding how the structure and representation of the China governments convey messages via the Sina weibo about the capital market is important. Especially recent stock market event in China, it is a good ground to understand how the government transforms its role in the information era to push the stock market and further to influence economics. Past researches on Chinese social media, to a great extent, only considered interactions between online censorship and post deletion, or the case study of social media behavior under the influence of active censorship. However, our study focuses on the impact of China government behavior in the economic variables by the social media, specifically on the stock market. This is the first time to investigate how the causal relationship between Chinese government power works on the stock market by social media.

One important limitation in the existing literature on social media is that there is a lack of understanding of the role of government messages in social media in terms of stock market, an important aspect of nation's economics. Specifically, currently available information about use of social media was based upon data collected through Sina weibo to analyze the government censorship and control on the China e.g., King et al. [15]. Or, it is the only quanlitative research but without qualitative data to support which gathers the expert's opinion or some experience from the people. Given these limitations, it will be quite interesting to use the big data from social media to see the relationship between the government and the social media, further to the stock market, which publishes every moments by everyone and also government related people or institutes. Since the Chinese stock market is not market driven and cannot be predicted like usual, the government "mood" can be a significantly important factor.

3 Understanding the Causality Among Government Messages in Social Media and Stock Market

This study focuses attention on the causal relationship between social media and stock price, and also the economic role that government plays in social media. To better understand the relationship, the following questions were raised.

Two sets of research questions were formulated, with the first one primarily aiming to understand the extent to how the government is represented through social media:

Q1a. How many verified users in Sina weibo post the information about "stock market" during the study period?

Q1b. What are the verified users consist of? How many of them are related to the government directly?

Q1c. How many of posts have the opinions of "Positive", "Negative" or "Neutral" about the stock market?

In order to further substantiate the causal impact of government messages via social media on stock market, the second set of questions focuses attention on more specific issues related to search, including:

Q2a. Do the topics of "stock market" become 'hot' when value of Shanghai composite become rising?

Q2b. Do the government accounts' positive opinions have causal relation with the stock market price?

Q2c. Do the government accounts' negative opinions have a causal cause on the stock market price?

4 Method and Research Design

4.1 Research Design

In order to answer the above research questions, a data mining exercise was devised. We chose the keywords "stock markets" ("Gu shi" in Chinese) to query Sina weibo's search engine day by day. Content analysis and causality analysis approaches were used to understand the data.

4.2 Coding and Data Analysis

We collected and categorized the weibo accounts included in the search results in a two-step process. First, we counted the everyday weibo posts from Jan 5th, 2015 to July 31st, 2015, totalling 87,094 posts, which were posted by the verified users. It should be noted that the reason we selected the verified users is first to avoid the spams, as the literature suggest, there are a large amount of spams in weibo. And second, we would like to select the influential users, which are the users who have a certain number of followers, say, more than 10,000 followers. Third, there are too many posts per day, it is almost impossible to generate a time series analysis by all users' posts. An example of the verified users can be found in Fig. 1, which has a symbol V besides the account name.

Fig. 1 An snapshot of verified user's profile: *Sina exchange*

The second step involved coding the verified users into different types of users. Due to the large volume of posts, we only analyze the posts from April 1st, 2015 to May 1st, 2015 in details. There are two kinds of authentications for verified users, one is the individual user and another is the institute user. We created four main categories. First, we separate the individual users into two subgroups; they are the individual who works in the government-related institutes, and other individuals. And also the institute groups are divided into government-based institutions and other institutions. Since we focus on the government-based behavior; we select the group of interest, which can represent it. We will focus on two of the sub-groups, they are, the individual who works in government and government-based institutions.

We also separate the positive, negative and neutral opinions toward the stock market. Take the People Daily as an example in Fig. 2, the title is "China Securities Regulatory Commission (CSRC) denied suppressing the stock market", we include it into the positive post. However, if the posts just explained the knowledge of stock market, we will put them in the category of neutral, since there is no "mood" in the post (Fig. 3).

In order to answer the research questions, the data analysis involved two steps.

First, descriptive analyses were conducted to describe government represented in weibo by identifying the unique weibo accounts. The relationship between the weibo post related to the "stock market" and the stock market price and trading volume will be represented.

In the second step, (1) the relationship between total posts and the price of Shanghai composite; and (2) the relationship between the total posts and the volume of the Shanghai exchange.

Finally, a causality analysis was conducted in order to explore the relationship between government "mood" and Shanghai stock market index. This was intended to answer the question whether government "mood" will be a factor to predict the stock market.

Fig. 2 An snapshot of an positive comments example (*Note* the title is "Securities Regulatory Commission (CSRC) denied suppressing the stock market, and claim that the stock market will not slum.")

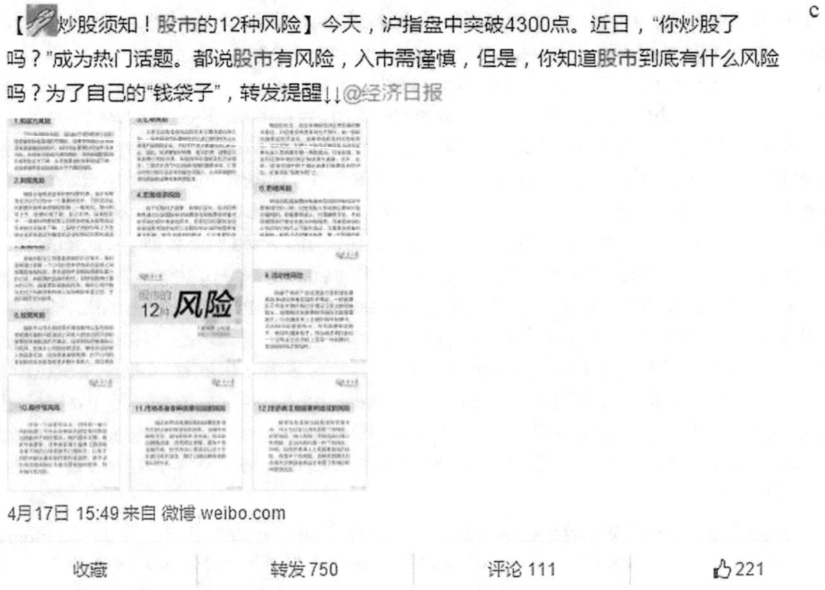

【炒股须知！股市的12种风险】今天，沪指盘中突破4300点。近日，"你炒股了吗？"成为热门话题。都说股市有风险，入市需谨慎，但是，你知道股市到底有什么风险吗？为了自己的"钱袋子"，转发提醒↓↓@经济日报

股市的12种风险

4月17日 15:49 来自 微博 weibo.com

收藏 转发 750 评论 111 221

Fig. 3 An snapshot of an neutral comments example (*Note* the title is "the must-know knowledge of stock market: the twelve kinds of risks in stock market.")

5 Findings

5.1 Social Media Accounts Represented by Government and Non-government Unities

In the following, we took an example of the top ten weibo accounts during our detailed survey week from April 20th, 2015 to April 24th, 2015. We find out that the 10 weibo accounts which posted information about "stock market" had most the highest repost volume (Table 1).

We especially do a case study on the *People Daily* on Shanghai stock market composite in Fig. 4, it seems that the negative moods (comments) followed the drop in stock market price, and the positive moods (comments) followed by rising price.

The total number of posts had some seasonal changes and also showed different patterns. It should be noted that during the weekend, the stock market closes. In January and February, the stock price index is in the level of 3000–3500. During the weekend, the amount of posts are more than in the weekdays. However, after March, when the stock price indexes are rising rapidly (until 6000), the amount of posts are close in two periods, or say, the seasonality disappears. We only searched the verified units and people, since for the all users, it will be too much posts and also some users are spam. In our opinion, the verified users can represent the real opinions of people, and also the voice from the government media. The relationship between

Table 1 Top 10 weibo accounts

Institute name	No. of followers	Total reposts
1. Global market broadcast	323	1016
2. Sina securities	170	1010
3. New fortune	43	1006
4. Sina finance	1323	826
5. CCTV finance	1376	696
6. Global times	348	694
7. Lang's financial talk show	17	651
8. Wall street journal in Chinese	1119	609
9. Financial net	1641	363
10. Caixin.net	186	337

Note The number of the followers is divided by 10,000. We select the period of one week from April 20th, 2015 to April 24th, 2015, this is the period that the price of the stock market go to a peak. The No. 1, 2 and 4 are from the Sina Corp, No. 5 are state-owned channel and the No. 6 (Global times) is closely related to the state

the posts in Sina weibo and the stock price and volume are obvious in Figs. 4 and 5. The comovement can by represented by the Pearson correlation: they are 0.451 and 0.645 (Fig. 6).

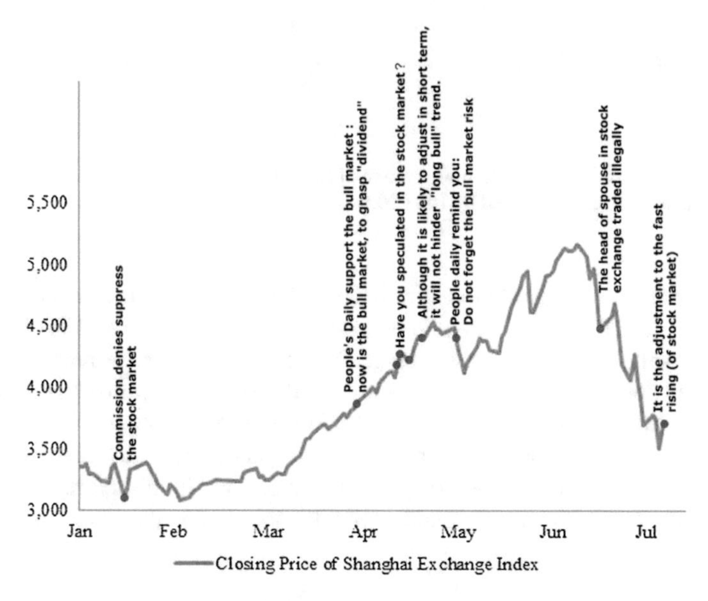

Fig. 4 The closing value of Shanghai composite and positive and negative posts based on *People Daily*

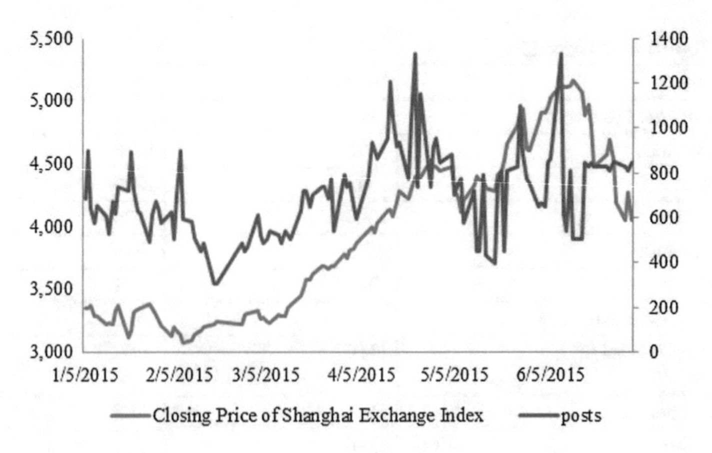

Fig. 5 The relationship between closing value of Shanghai exchange composite and total posts

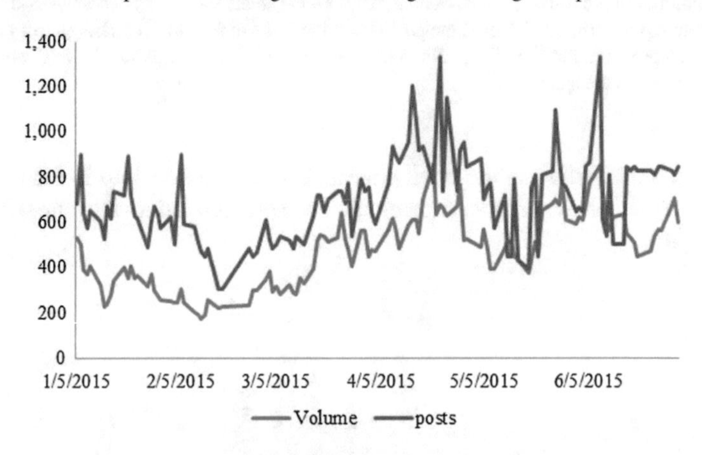

Fig. 6 The relationship between trading volume of Shanghai exchange composite and total posts (*Note* the trading volume is divided by 100,000.)

During our detailed survey week from April 20th 2015 to April 24th 2015, among the total of 1,383 search results, there were 115 (approximately 11 %) identified as search results representing government. This suggests that government institute-based weibo posts, indeed, represent a substantial part of the weibo domain and play an important role. Figure 7 shows the distribution of different subgroups among these social media search results. They are usual individual users, individuals who work in government, government and non-government sectors. In this figure, 53 % of the search results come from the individual who did not work in the government units.

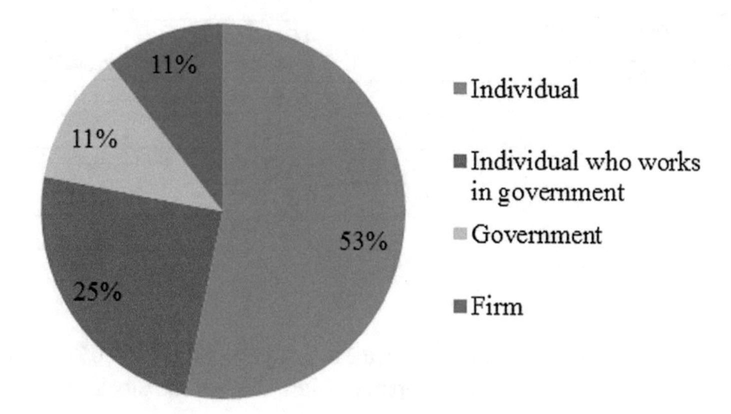

Fig. 7 Composition of search results in Sina weibo

5.2 The Relationship between Government Opinions and Stock Market Price—Granger Causal Test Results

The Granger causality test is a test for determining whether one time series is useful in forecasting another. Granger argued that causality in economics could be tested by measuring the ability to predict the future values of a time series using prior values of another time series (Timmermann and Granger [7]).

In our opinion, there are little factors that can be called "real" causality; econometricians believed that the Granger test finds only "predictive causality". Recent evidence shows that Granger tests can be used in an opinion mining context: predicting stock price movements with a sentiment curve (Bollen et al. [2]).

Granger tests implied that if in time 1 time series A correlate with in time 2 or 3 or more later observations in B, A is said to "cause" B. Let y_t and x_t be two time series. To see if x_t "Granger causes" y_t with maximum p time lag, we run the following equation:

$$y_t = a_0 + a_1 y_{t-1} + \cdots + a_p y_{t-p} + b_1 x_{t-1} + \cdots + b_p x_{t-p} \tag{1}$$

Then, we used F-tests to evaluate the significance of the lagged x terms. We provided the p value of Granger-causality correlation between the government mood and the stock market price in the period of April 1st, 2015 to May 1st, 2015 in Table 2.

Our Shanghai stock market composite time series, denoted S_t, is defined to reflect daily changes in stock market value. That is, its values are the difference between day t and day $t-1$: $S_t = SHC_t - SHC_{t-1}$. We would like to test whether the government mood predicts changes in Chinese stock market values.

Our Granger causality results are shown in Table 2. We can reject the null hypothesis that the government mood time series do not predict SHC values. However, we see that government positive mood has the highest Granger causality relation with

Table 2 P-values of Granger causality tests in the period of April 1st, 2015–May 1st, 2015

Lag	Government positive	Government negative
1 day	0.94	0.421
2 days	0.053*	0.250
3 days	0.022**	0.430
4 days	0.013**	0.500
5 days	0.043**	0.611

Shanghai Composites for lags ranging from 2 to 6 days (p-values < 0.05). The results show that the government moods or opinion did influence the stock market in the period we study.

6 Conclusions and Limitations

China's capital market is still an emerging market. Undoubtedly, the weakness of a transition market is full of speculative behaviors and heavy government intervention. The above characteristics make the basic principles of economics fail, such as the recent observed Chinese stock market has experienced an extraordinary rise, the Shanghai Composite doubled during a year, and later suffered a sharp drop in the last two months. Behavioral economics tells us that public comments can profoundly affect individual behavior and decision-making. People will often change their attitudes to come into line with what they profess publicly. In this information era, the public could be represented by the large number of posts in social media. Recently Chinese government also play a role in it. Are the public posts, which also contain government comments correlated or a causal relationship of economic indicators? Here we investigate whether measurements of collective public comments derived from large-scale Sina weibo posts are correlated to the value of the Shanghai Composite over time. A Granger causality analysis was finally adopted to detect the causal relationship between social media and recent Chinese stock market price.

The limitation of this study is obvious. Since the time span is until July, the study focus on the rising of the stock market but not the falling of the stock market. The future work should focus on the whole period of stock market turmoil.

References

1. Gao, S.: China Stock Market in a Global Perspective, pp. 1–48. Dow Jones Indexes, New York (2002)
2. Bollen, J., Mao, H., Zeng, X.: Twitter mood predicts the stock market. J. Comput. Sci. **2**(1), 1–8 (2011)

3. Los, C.A., Yu, B.: Persistence characteristics of the Chinese stock markets. Int. Rev. Financ. Anal. **17**(1), 64–82 (2008)
4. Kennedy, S.: The stone group: state client or market pathbreaker? China Quart. **152**, 746–777 (1997)
5. Xin, H.: The surfer-in-chief and the would-be kings of content. Media in China: Consumption, Content, and Crisis, pp. 192–199. Taylor and Francis, Hoboken (2002)
6. Granger, C.W.: Investigating causal relations by econometric models and cross-spectral methods. Econom.: J. Econom. Soc. **37**, 424–438 (1969)
7. Timmermann, A., Granger, C.W.: Efficient market hypothesis and forecasting. Int. J. Forecast. **20**(1), 15–27 (2004)
8. Werthner, H., Klein, S.: ICT and the changing landscape of global tourism distribution. Electron. Mark. **9**(4), 256–262 (1999)
9. Xiang, Z., Gretzel, U.: Role of social media in online travel information search. Tour. Manag. **31**(2), 179–188 (2010)
10. Fama, E.F.: Efficient capital markets: a review of theory and empirical work. J. Financ. **25**(2), 383–417 (1970)
11. Schumaker, R.P., Chen, H.: Textual analysis of stock market prediction using breaking financial news: the AZFin text system. ACM Trans. Inf. Syst. (TOIS) **27**(2), 12 (2009)
12. Benney, J.: The aesthetics of microblogging: how the Chinese state controls Weibo. Tilburg paper in culture studies (2013)
13. Kim, H.D., Castellanos, M., Hsu, M., Zhai, C., Rietz, T., Diermeier, D.: Mining causal topics in text data: iterative topic modeling with time series feedback. In: Proceedings of the 22nd ACM International Conference on Conference on Information and Knowledge Management, pp. 885–890. ACM
14. Luo, X., Zhang, J., Duan, W.: Social media and firm equity value. Inf. Syst. Res. **24**(1), 146–163 (2013)
15. King, G., Pan, J., Roberts, M.E.: How censorship in China allows government criticism but silences collective expression. Am. Polit. Sci. Rev. **107**(02), 326–343 (2013)

Firm Efficiency in Thailand's Telecommunication Industry: Application of the Stochastic Frontier Model with Dependence in Time and Error Components

Supanika Leurcharusmee, Jirakom Sirisrisakulchai and Sumate Pruekruedee

Abstract In this study, we measure the ability to produce and provide services of telecommunication firms in Thailand. We propose a methodology to estimate firm technical efficiency using the Stochastic Frontier Analysis (SFA) with copulas to capture both the dependence of the stochastic and inefficiency error components and the dependence of the errors across time. Allowing the dependence between the stochastic error and the aggregate error prevents the estimation of the technical efficiency from being biased. Moreover, allowing the time dependence for the aggregate error improves the efficiency of the model. The results show that the average technical efficiency is 0.54, which indicates a room for improvement in the industry.

Keywords Technical efficiency · Stochastic frontier · Copula · Telecommunication

1 Introduction

The telecommunication industry provides important infrastructure that drives the economy. Due to the limited supply of resources and the economies of scale and scope, the level of competition of the telecommunication industry has been a concern in the literature [18]. Because competition plays its role in forcing firms to produce efficiently, firms in a less competitive industry tend to have lower incentive to improve their production efficiency. Therefore, firm efficiency evaluation and incentive regulations to improve efficiency are important issues for regulators in the telecommunication industry [16, 19].

This study measures and analyzes the ability to produce and provide services of telecommunication firms in Thailand. Specifically, we measure each firm's efficiency using Debreu–Farrell measure of technical efficiency, which is the ability to produce the maximum feasible level of output given a set of inputs. [9, 10].

S. Leurcharusmee (✉) · J. Sirisrisakulchai · S. Pruekruedee
Faculty of Economics, Chiang Mai University, Chiang Mai, Thailand
e-mail: supanika.econ.cmu@gmail.com

V.-N. Huynh et al. (eds.), *Causal Inference in Econometrics*,
Studies in Computational Intelligence 622, DOI 10.1007/978-3-319-27284-9_32

To estimate the technical efficiency, we need to first estimate the production frontier. There are two main methods to estimate the production frontier. The first is the Data Envelopment Analysis (DEA) [12, 13, 19]. The second is the Stochastic Frontier Analysis (SFA) [12, 23]. The two methods differ fundamentally in two aspects. The first aspect is that DEA is a non-parametric model and the SFA is a parametric model. The second aspect is that DEA is a deterministic model and SFA is a stochastic model. In the past decade, the country has been through a series of political unrests resulting in policy unsteadiness. Unobserved shocks should have impacts on the estimation. Moreover, for the purpose of policy suggestions, parametric models can provide more insight information on the factors determining the inefficiency. Therefore, this study applies the SFA model to estimate the technical efficiency of telecommunication firms in Thailand.

In this study, we introduce the SFA model with copula to capture both the dependence of the stochastic and inefficiency error components and the dependence of the errors across time to estimate firms' technical efficiency. The traditional SFA model has an assumption that the stochastic and the inefficiency error components are independent. This is rather a strong assumption as exogenous shocks, which are the sources of the stochastic error, are likely to affect the frontier and firms' efficiency differently. To relax this independence assumption, [17] uses copula to capture the dependence between the stochastic and inefficiency error components. To emphasize the importance of the modification, [22] showed numerically that a positive dependence between the error and the inefficiency components led to an underestimate of the technical efficiency, and vice versa.

To extend [17]'s SFA model for panel data analysis, we generalized two copula-based time dependence SFA approaches developed in [3], by allowing the error components to be correlated and dependent over time within each firm. The first approach allows the inefficiency component to be dependent across time. The second approach allows the aggregate errors, which is the difference between the stochastic and the inefficiency error components, to be dependent. In adopting [3]'s approaches, we use the Pair Copula Construction (PCC) to construct the multivariate distributions. This improves the flexibility of the model as it allows different dependence structures among each pair of the variables. As discussed in [3], the ability to capture the time dependence would improve the efficiency of the estimation.

The results show that the SFA model with dependent error components and time dependence in the aggregate error provides better estimate for the production frontier comparing to the traditional and [17]'s SFA models as it yields a higher log likelihood value. Allowing the dependence in the error components, the estimates for technical efficiency are unbiased. In addition, allowing the time dependence, the estimates are more efficient. Therefore, we can expect lower standard errors for the estimates.

The contribution of this study to develop the SFA with dependent error components and time dependence in the aggregate error is the more accurate and more efficient estimate of firm efficiency. This contribution is particularly important for efficiency studies in the telecommunication industry. This is because the telecommunication industry in many countries are under price-cap regulation and SFA is one of the commonly used methods to estimate price-cap. The inability to capture the dependence between the error and inefficiency components can cause biases in the estimates of technical efficiency and price cap.

2 Firm Efficiency

Since there are many concepts of efficiency in the literature, this section discusses and defines specifically the efficiency estimated in this study. The task of firms is to transform a set of inputs $X \in R_K^+$ into a set of output $Y \in R_M^+$. The transformation process is called production. In the context of production, [8] explains three types of efficiency, which are technical, allocative and economic efficiency. Technical efficiency is the ability to produce the maximum feasible level of output given a set of inputs. For allocative efficiency, a firm is allocatively efficient if it uses the optimal input and output combinations. Economic efficiency requires a firm to both technically and allocatively efficient. To evaluate firm performance in the telecommunication industry, literature focus on technical efficiency more than allocative efficiency. Vogelsang [20] mentions that allocative efficiency is less tangible and may conflict with fairness or distributional concern. This is particularly important for telecommunication industry because key players in this industry are obligated to provide services to all regions in the country. Therefore, some firms are forced to expand the production beyond the optimal size.

To measure technical efficiency, this study follows [11] and uses the concept of output-based Debreu–Farrell style measure of technical efficiency. Let us first consider the single output case. In this case, the relationship between the maximum feasible outputs given a set of inputs is called production function, $Y = g(X)$. An output-based Debreu–Farrell style measure of technical efficiency is

$$TE_i = \frac{Y_i}{g(X_i)}, \tag{1}$$

where Y_i is the level of output of firm i, X_i is the vector of the level of input of firm i and $g(\cdot)$ is the production function. Firm i's technical efficiency TE_i is basically the ratio of firm i's level of output relative to the maximum feasible level of output providing the level of input X_i.

3 Data

The data used in this study are from the balance sheets and income statements of telecommunication firms listed in the website of Thailand's Department of Business Development, Ministry of Commerce. The data include 42 firms from 2010 to 2013.

The price indices are from the Bank of Thailand. The output variable is *real net sales* of each firm calculated using net sales divided by the consumer price index in the telecommunication and transportation sector. The inputs variables are (1) *real fixed asset* and (2) *real operating cost*. The input variables are adjusted using the producer price index.

4 Modeling Time Dependence in Stochastic Frontier Models

We consider a stochastic frontier model in panel data setting of the following form:

$$y_{it} = X_{it}\beta + \varepsilon_{it}, \quad \varepsilon_{it} = v_{it} - u_{it}, \quad u_{it} \geq 0, \tag{2}$$

where $i = 1, 2, \ldots, N$ and $t = 1, 2, \ldots, T$ index firms and time, respectively, y is log of output, X is a vector of inputs, v_{it} are distributed as independent and identically normal distribution, iid $N(0, \sigma_v^2)$, and u_{it} have a half-normal distribution, $N^+(0, \sigma_u^2)$. Some common functional forms for production functions are listed and discussed in [7]. In this paper, we assume the Cobb–Douglas production function.

The non-negative random errors u_{it} represent firm i's technical inefficiency at time t. In panel data setting, we will give basically three approaches, currently available in the literature, in modeling the dependence of inefficiency over time. First and simplest one is to assume the independence of u_{it} over time t and estimate the model using quasi-maximum likelihood estimation [4]. The variation of the first approach can be found in [21], where it is assumed that u_{it} are independent conditional on some explanatory variables. Second, many researchers assume that u_{it} has the multiplicative form as follows:

$$u_{it} = h(z_{it}; \delta)u_i, \tag{3}$$

where u_i is a random variable, $h(z_{it}; \delta)$ is a known function with the vector of unknown parameters δ to be estimated, and z_{it} is the vector of explanatory variables affecting the technical inefficiency. From this representative form, the variation over time is governed by the observable z_{it}. A survey for this approach can be found in [2]. Finally, [3] proposes to model the time dependence in u_{it} using copula. However, the error components v_{it} and u_{it} are assumed independent.

In this paper, we assume that u_{it} is independent over i, but dependent over t. We model time dependence in u_{it} using copula as [3] proposed. However, we relax the assumption of independence in the error components. Following [17], the dependence in the error components can also be modeled using copula approach.

5 Copula

Copula can be briefly defined as a joint distribution function of standard uniform marginals. Let $C(w_1, \ldots, w_T; \rho)$ be the copula distribution and $c(w_1, \ldots, w_T; \rho)$ be the copula density functions with the vector of dependence parameters ρ. Each marginal w_t is a standard uniform marginal (more details of copula can be found in [14]). Copulas can capture the wide range of dependence structures. An important feature of copulas is the ability to model the marginal distributions separately from the dependence structure. This results in the flexible functional forms of the joint distributions. By Sklar's 1959 Theorem, given the marginal distributions, F_t, $t = 1, 2, \ldots, T$ and a joint distribution function $H(\xi_1, \ldots, \xi_T)$, there exists a copula such that

$$H(\xi_1, \ldots, \xi_T) = C(F_1(\xi_1), \ldots, F_T(\xi_T)). \tag{4}$$

The joint density function can be expressed as follows:

$$h(\xi_1, \ldots, \xi_T) = c(F_1(\xi_1), \ldots, F_T(\xi_T)) f_1(\xi_1) \ldots f_T(\xi_T). \tag{5}$$

A higher dimension of multivariate joint distribution can be approximated from conditional and marginal density functions and using the Pair Copula Construction (PCC) technique [1] as follows:

$$h(\xi_1, \ldots, \xi_T) = f_{1|2,\ldots,T}(\xi_1 | \xi_2, \ldots, \xi_T) f_{2|3,\ldots,T}(\xi_2 | \xi_3, \ldots, \xi_T)\ ldots f_T(\xi_T) \tag{6}$$

By Sklar's theorem, it can be shown that the conditional density function on the right hand side of Eq. (6) can be decomposed into the product of a bivariate copula density and a univariate conditional density. This can be done recursively to each of the terms on the right hand side of Eq. (6) until $h(\xi_1, \ldots, \xi_T)$ is decomposed into the product of T(T-1)/2 bivariate copulas [15]. Note that there are many ways for the decomposition in this manner, which are called Vine structures in the literature. The readers are referred to [5] and [6] for more details. For example, the Canonical–Vine structure density function can be expressed as follows:

$$\begin{aligned} h(\xi_1, \ldots, \xi_T) &= \prod_{k=1}^{T} f_k(\xi_k) \\ &\times \prod_{i=1}^{T-1} \prod_{j=1}^{T-i} c_{i,i+j|1,\ldots,i-1}(F(\xi_i|\xi_1, \ldots, \xi_{i-1}), F(\xi_{i+j}|\xi_1, \ldots, \xi_{i-1})|\rho_{i,i+j|1,\ldots,i-1}), \end{aligned} \tag{7}$$

where $c_{i,i+j|1,\ldots,i-1}$ denote bivariate copula densities with parameters $\rho_{i,i+j|1,\ldots,i-1}$.

A major advantage of the PCC approach is the greater flexibility to model a large range of dependence structures, which greatly reduces the computational cost of evaluating the higher dimension of joint distribution.

6 Copula-Based Time Dependence Stochastic Frontier Model

Let the bold letters $\boldsymbol{\varepsilon}$, $\mathbf{u}$, and $\mathbf{v}$ be T-dimensional vectors of the error terms, i.e. $\mathbf{u}_i = (u_{i1}, \ldots, u_{iT})$ and $\mathbf{v}_i = (v_{i1}, \ldots, v_{iT})$. We will explain two approaches in modeling copula-based time dependence stochastic frontier models in this section.

For the first approach, we construct the joint distribution function of $(u_{i1}, \ldots, u_{iT})$ by using pair copula construction approach described in Sect. 4. Let $h_u(\mathbf{u}; \theta, \boldsymbol{\rho})$ be the joint density of the one-sided error vector, $\mathbf{u}_i$, constructed by pair copula construction approach in Eq. (7), and $f_u(\mathbf{u}; \theta)$ denote the marginal density for each individual one sided-error. We assume that u_{it} are independent for each firm i, but within each firm i is dependent over time t. Thus, the joint density of $(u_{i1}, \ldots, u_{iT})$, for each firm i, is

$$\begin{aligned} h_u(\mathbf{u}; \theta, \boldsymbol{\rho}) &= \prod_{k=1}^{T} f_u(u_k) \\ &\times \prod_{l=1}^{T-1} \prod_{j=1}^{T-l} c_{l,l+j|1,\ldots,l-1}(F_u(u_l|u_1, \ldots, u_{l-1}), F_u(u_{l+j}|u_1, \ldots, u_{l-1})|\rho_{l,l+j|1,\ldots,l-1}), \end{aligned} \tag{8}$$

where $F_u(u; \theta) = \int_0^u f_u(s; \theta) ds$ is the cumulative distribution function of the half-normal error terms with parameters vector θ.

Following [17], the joint density of the error components (ε_{it}) can be obtained through the joint density of $f(\mathbf{u}, \boldsymbol{\varepsilon})$ as follows:

$$f(\mathbf{u}, \boldsymbol{\varepsilon}) = f(\mathbf{u}, \mathbf{u} + \boldsymbol{\varepsilon}) = h_u(\mathbf{u}; \theta, \rho) f_v(\mathbf{u} + \boldsymbol{\varepsilon}) c_\gamma(H_u(\mathbf{u}), F_v(\mathbf{u} + \boldsymbol{\varepsilon})), \tag{9}$$

where $H_u(\mathbf{u})$ is the joint distribution of $\mathbf{u}$, and c_γ is the copula density with dependence parameter γ.

Marginalizing out $\mathbf{u}$ gives us

$$f_\Theta(\boldsymbol{\varepsilon}) = \int_0^{+\infty} f(\mathbf{u}, \boldsymbol{\varepsilon}) d\mathbf{u}, \tag{10}$$

or, equivalently,

$$f_\Theta(\boldsymbol{\varepsilon}) = \mathbb{E}_{\mathbf{u}}[f_v(\mathbf{u} + \boldsymbol{\varepsilon}) c_\gamma(H_u(\mathbf{u}), F_v(\mathbf{u} + \boldsymbol{\varepsilon}))], \tag{11}$$

where $\mathbb{E}_{\mathbf{u}}$ represents the expectation with respect to $\mathbf{u}$ and Θ is the vector of all parameters, $\Theta = (\beta, \theta, \boldsymbol{\rho}, \gamma)$, in the model. This expectation can be approximated by

$$\hat{f}_\Theta(\boldsymbol{\varepsilon}) = \frac{1}{M} \sum_{r=1}^{M} f_v(\sigma_u \mathbf{u}_{0,r} + \boldsymbol{\varepsilon}) c_\gamma(H_u(\mathbf{u}_{0,r}), F_v(\sigma_u \mathbf{u}_{0,r} + \boldsymbol{\varepsilon})), \tag{12}$$

where $\mathbf{u}_{0,r}$, $r = 1, \ldots, M$, is a sequence of M random draws from the joint distribution function of $(u_{i1}, \ldots, u_{iT})$ constructed from pair copula construction with half-normal margins.

Finally, the technical inefficiency for each firm can be estimated by the conditional expectations given $\boldsymbol{\varepsilon}$,

$$TE_{\Theta} = \mathbb{E}[\exp(-\mathbf{u})|\boldsymbol{\varepsilon}] \tag{13}$$

$$= \frac{1}{f_{\Theta}(\boldsymbol{\varepsilon})} \int_{0}^{+\infty} \exp(-\mathbf{u}) f(\mathbf{u}, \boldsymbol{\varepsilon}) d\mathbf{u} \tag{14}$$

$$= \frac{\mathbb{E}_{\mathbf{u}}[\exp(-\mathbf{u}) f_v(\mathbf{u}+\boldsymbol{\varepsilon}) c_{\gamma}(H_u(\mathbf{u}), F_v(\mathbf{u}+\boldsymbol{\varepsilon}))]}{\mathbb{E}_{\mathbf{u}}[f_v(\mathbf{u}+\boldsymbol{\varepsilon}) c_{\gamma}(H_u(\mathbf{u}), F_v(\mathbf{u}+\boldsymbol{\varepsilon}))]} \tag{15}$$

The expectations in the nominator and denominator can be approximated by Monte Carlo simulation as usual.

Second approach in modeling time dependence of stochastic frontier model is to assume that the error components, $\boldsymbol{\varepsilon}$, correlated over time within each firm. First, $\boldsymbol{\varepsilon}$ can be obtained as in [17]. Then, the joint density of the error components with time dependence can be formed using copula. The joint density of $\boldsymbol{\varepsilon}$ can be written as

$$h_{\varepsilon}(\boldsymbol{\varepsilon}; \theta, \boldsymbol{\rho}) = \prod_{k=1}^{T} f_{\varepsilon}(\varepsilon_k) \times \prod_{l=1}^{T-1} \prod_{j=1}^{T-l} c_{l,l+j|1,\ldots,l-1}(F_{\varepsilon}(\varepsilon_l|\varepsilon_1, \ldots, \varepsilon_{l-1}), F_{\varepsilon}(\varepsilon_{l+j}|\varepsilon_1, \ldots, \varepsilon_{l-1})|\rho_{l,l+j|1,\ldots,l-1}), \tag{16}$$

This joint density function can be used in the maximum simulated likelihood estimations. Finally, one can estimate technical inefficiency by replacing this joint density function from Eq. (16) into Eq. (13).

7 Empirical Study for Firm Efficiency in Thailands Telecommunication Industry

To estimate the technical efficiency of each firm in Thailand's telecommunication industry, we examine four SFA models including:

Model 1: The SFA model with independent error components and no time dependence.

Model 2: The SFA model with dependent error components but no time dependence.

Model 3: The SFA model with dependent error components and time dependence in the technical inefficiency.

Model 4: The SFA model with dependent error components and time dependence in the aggregate error.

Model 1 is the traditional SFA model. Model 2 is [17]'s SFA model, which uses copula to capture the dependence between the stochastic and inefficiency error components, $\boldsymbol{v}$ and $\boldsymbol{u}$. To estimate Model 2 in this study, we considered three copula functions that allow for both positive and negative dependence including Gaussian, t, and Frank copulas. The results show that the Frank copula provides the highest value of log likelihood. Therefore, the results of Model 2 shown in Table 1 is the results of [17]'s SFA model with Frank copula.

To extend Model 2 for panel data analysis, we applied the copula-based time dependence SFA techniques developed by [3] and proposed Model 3 and 4. For Model 3, we not only allows the inefficiency component $\boldsymbol{u}$ to be dependent with the stochastic error $\boldsymbol{v}$, but also allows the inefficiency component $\boldsymbol{u}$ to be dependent across time. For the dependence between $\boldsymbol{v}$ and $\boldsymbol{u}$, we used Gaussian, t, and Frank copulas to capture the dependence and the results show that Frank copula is the best fit. For the time dependence of $\boldsymbol{u}$, we applied the PCC method to construct the joint distribution of u_{it}. We leave the issues of selecting the optimal ordering of the PCC structures and the combination of different functions of copula in each pair for further research. In this study, we used the Frank copulas to model the joint distribution of u_{it} across time.

For Model 4, we allows the inefficiency component $\boldsymbol{u}$ to be dependent with the stochastic error $\boldsymbol{v}$ and the aggregate error $\boldsymbol{\varepsilon}$ to be dependent across time. By allowing $\boldsymbol{\varepsilon}$ to be dependent across time, the model allows both $\boldsymbol{v}$ and $\boldsymbol{u}$ to be dependent across time. Therefore, this specification is the most flexible among the four models. Similar to Model 2 and 3, Frank copula was chosen to model the dependence between $\boldsymbol{v}$ and $\boldsymbol{u}$. Similar to Model 3, the PCC method was adopted to model the dependence of ε_{it} across time.

7.1 Results and Discussion

The results from the four models discussed in the previous section is shown in Table 1. It can be seen that the estimates for the coefficients of *log(fixed asset)* and *log (op cost)* are robust to the four model specifications. However, as shown in Fig. 1, the estimates for firm technical inefficiency are different.

For the model selection, we compare the log likelihood value *LL* from each of the models. Model 4, which has the most flexible specification, provides the best estimate for the production frontier as it presents the highest log likelihood value, *LL*. In Model 4, the aggregate error ε_{it} are significantly dependent across time. As

Table 1 Estimated parameters and standard errors

Variables	Model 1		Model 2		Model 3		Model 4	
	Coeff.	std.err	Coeff.	std.err	Coeff.	std.err	Coeff.	std.err
Constant	1.376	0.479	1.806	0.537	1.890	0.469	3.498	0.695
Log(fixed asset)	0.140	0.039	0.135	0.040	0.133	0.040	0.162	0.002
Log(op cost)	0.880	0.050	0.885	0.050	0.886	0.050	0.731	0.003
σ_u	0.462	0.300	1.017	0.411	1.108	0.294	0.981	0.169
σ_v	0.657	0.083	0.841	0.165	0.871	0.167	0.934	0.142
γ			4.621	2.435	5.214	2.315	3.586	1.680
ρ_{c12}					−0.418	10.586	18.615	2.945
ρ_{c13}					−2.080	19.761	10.947	2.118
ρ_{c14}					2.188	17.586	9.105	1.832
$\rho_{c23\|1}$					5.468	14.511	5.620	1.361
$\rho_{c24\|1}$					−16.056	63.681	4.273	1.143
$\rho_{c34\|12}$					−6.223	66.406	4.145	1.256
LL	−181.916		−179.007		−178.229		−76.273	

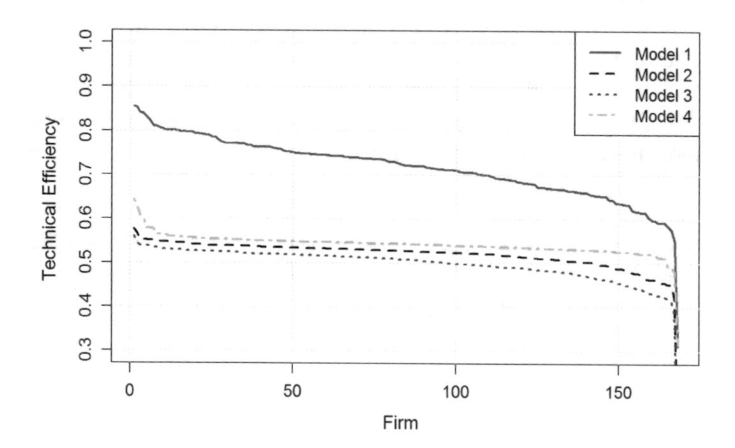

Fig. 1 Technical efficiency plots

discussed in [3], the failure to capture the dependence would cause the frontier estimation to be inefficient.

Figure 1 shows the estimates of technical efficiency from the four models. As discussed in [22], the failure to take into account the dependence between the error components $\boldsymbol{v}$ and $\boldsymbol{u}$ in the model will cause the estimate of the technical efficiency to be biased. Specifically, SFA models that cannot capture a positive dependence between the stochastic and the inefficiency error components will overestimate the technical efficiency and vice versa. The results from this study confirm [22]'s results. As illustrated in Table 1, $\boldsymbol{v}$ and $\boldsymbol{u}$ are positive dependence in this study. As a result,

Model 1 gives higher estimates for technical efficiency than Model 2, 3 and 4, which allow for the dependence between $\boldsymbol{v}$ and $\boldsymbol{u}$. The results show that the average technical efficiency is only 0.54 and the majority of the firms operate inside the efficient frontier.

8 Concluding Remarks

In this study, we developed a copula-based SFA model to estimate the technical efficiency of each firm in Thailand's telecommunication industry. The model allows the inefficiency component $\boldsymbol{u}$ to be dependent with the stochastic error $\boldsymbol{v}$ and the aggregate error $\boldsymbol{\varepsilon}$ to be dependent across time. The results show that our copula-based SFA model yielded higher log likelihood and performed better than existing SFA models. Allowing the dependence between the stochastic error $\boldsymbol{v}$ and the aggregate error $\boldsymbol{\varepsilon}$ prevents the estimation of the technical efficiency from being biased. Moreover, allowing the time dependence for the aggregate error $\boldsymbol{\varepsilon}$ improved the efficiency of the model.

In addition to the advantage from allowing the dependence in time and error components, the SFA model developed in this study allow all the marginals and copulas capturing the dependences to be in different forms. Although the data for this study does not require this advantage because the Frank copula is shown to be superior in all cases, this flexibility can be useful for further studies with different data sets. Finally, for the telecommunication industry in Thailand, the results from this study indicate that there are rooms for technical efficiency improvement in this industry.

Acknowledgments We are highly appreciated and would like to acknowledge the financial and information support from the Telecommunications Economics Research Center at the Faculty of Economics, Chiang Mai University.

References

1. Aas, K., Czado, C., Frigessi, A., Bakken, H.: Pair-copula constructions of multiple dependence. Insur. Math. Econ. **44**, 182–198 (2009)
2. Alvarez, A., Amsler, C., Orea, L., Schmidt, P.: Interpreting and testing the scaling property in models where inefficiency depends on firm characteristics. J. Product. Anal. **25**(3), 201–212 (2006)
3. Amsler, C., Prokhorov, A., Schmidt, P.: Using copula to model time dependence in stochastic frontier models. Econom. Rev. **33**(5—6), 497–522 (2014)
4. Battese, G.E., Coelli, T.J.: A model for technical inefficiency effects in a stochastic frontier production function for panel data. Empir. Econ. **20**, 325–332 (1995)
5. Bedford, T., Cooke, R.M.: Probability density decomposition for conditionally dependent random variables modelled by vines. Ann. Math. Artif. Intell. **32**, 245–268 (2001)
6. Bedford, T., Cooke, R.M.: Vines a new graphical model for dependent random variables. Ann. Stat. **30**(4), 1031–1068 (2002)

7. Coelli, T.J., Rao, D.S.P., Battese, G.E.: An Introduction to Efficiency and Productivity Analysis. Kluwer, Boston (1998)
8. Coelli, T.J., Rao, D.S.P., O'Donnell, C.J., Battese, G.E.: An Introduction to Efficiency and Productivity Analysis. Springer Science & Business Media, Berlin (2005)
9. Debreu, G.: The coefficient of resource utilization. Econom.: J. Econom. Soc. pp. 273–292 (1951)
10. Farrell, M.J.: The Measurement of Productive Efficiency. J. R. Stat. Soc. Ser. A (Gen.) pp. 253–290 (1957)
11. Greene, W.: Fixed and random effects in stochastic frontier models. J. Product. Anal. **23**(1), 7–32 (2005)
12. Li, Y.: A firm-level panel-data approach to efficiency, Total factor productivity, catch-up and innovation and mobile telecommunications reform (1995–2007). ESRC Centre for Competition Policy Working Paper (2009) 09-6
13. Liao, C.H., Lien, C.Y.: Measuring the technology gap of APEC integrated telecommunications operators. Telecommun. Policy **36**(10), 989–996 (2012)
14. Nelsen, R.B.: An Introduction to Copulas, 2nd edn. Springer, New York (2006)
15. Panagiotelis, A., Czado, C., Joe, H.: pair copula constructions for multivariate discrete data. J. Am. Stat. Assoc. **107**(499), 1063–1072 (2012)
16. Sappington, D.E., Weisman, D.L.: Price cap regulation: what have we learned from 25 years of experience in the telecommunications industry? J. Regul. Econ. **38**(3), 227–257 (2010)
17. Smith, M.D.: Stochastic frontier models with dependent error components. Econom. J. **11**(1), 172–192 (2008)
18. Laffont, J.J., Tirole, J.: Competition in Telecommunications. MIT press, Cambridge (2001)
19. Uri, N.D.: Technical efficiency, allocative efficiency, and the implementation of a price cap plan in telecommunications in the United States. J. Appl. Econ. **4**(1), 163–186 (2001)
20. Vogelsang, I.: Incentive regulation and competition in public utility markets: a 20-year perspective. J. Regul. Econ. **22**(1), 5–27 (2002)
21. Wang, W.S., Amsler, C., Schmidt, P.: Goodness of fit tests in stochastic frontier models. J. Product. Anal. **35**, 95–118 (2011)
22. Wiboonpongse, A., Liu, J., Sriboonchitta, S., Denoeux, T.: Modeling dependence between error components of the stochastic frontier model using copula: application to intercrop coffee production in Northern Thailand. Int. J. Approx. Reason (In press) (2015)
23. Yang, A., Lee, D., Hwang, J., Shin, J.: The influence of regulations on the efficiency of telecommunications operators: a meta-frontier analysis. Telecommun. Policy **37**(11), 1071–1082 (2013)

Macroeconomic Factors Affecting the Growth Rate of FDI of AEC Member Countries Using Panel Quantile Regression

Tanaporn Tungtrakul, Kunsuda Nimanussornkul and Songsak Sriboonchitta

Abstract Macroeconomic factors affecting the growth rate of foreign direct investment (FDI) in AEC member countries were investigated using panel quantile regression to investigate the effects of foreign direct investment (FDI) of ASEAN Economic Community (AEC) member countries. Yearly data covering the period of 2001 to 2012 for nine countries except Myanmar were used for the estimation. As the data of Myanmar are limited, this study covered only Singapore, Philippines, Brunei, Cambodia, Indonesia, Laos, Malaysia, Vietnam and Thailand. The independent variables include the growth rate of gross domestic product, the growth rate of exchange rate ratio and inflation. The findings of this study show that the growth rate of gross domestic product affects the growth rate of foreign direct investment (FDI) positively and statistically significant at all levels of quantile except the quantile at 0.75.

Keywords FDI · Economic growth · AEC · Panel quantile regression

1 Introduction

ASEAN Economic Community (AEC) was developed by the Association of Southeast Asian Nations (ASEAN) which consists of 10 countries, namely Singapore, Philippines, Brunei, Cambodia, Indonesia, Laos, Malaysia, Thailand, Vietnam and Myanmar. The goal of the AEC is to increase bargaining power and competitiveness in the international market, to promote marketing and production base with the free movement of goods, services, investment, skilled labor and capital and to reduce the gap between AEC members. Especially, regulations that impede trade and the flow of capital have to be terminated. The results of the implementation of the AEC will contribute directly to foreign direct Investment (FDI). By cancellation of FDI-impeding regulations, the AEC members would become more liberal enhancing the attractiveness of the region in the views of foreign investors [3].

T. Tungtrakul (✉) · K. Nimanussornkul · S. Sriboonchitta
Faculty of Economics, Chiang Mai University, Chiang Mai, Thailand
e-mail: Tanapornecon@gmail.com

V.-N. Huynh et al. (eds.), *Causal Inference in Econometrics*,
Studies in Computational Intelligence 622, DOI 10.1007/978-3-319-27284-9_33

Table 1 shows the values of foreign direct investment inflows of ASEAN economic Community member countries during 2008–2012 in million US dollars, which are on the increasing trend as a result of the global economic recovery and the economic growth of the region. There are also the factors that contribute to steady increasing in investment, employment and funds inflows. Therefore, the AEC has the power to create incentives for foreign investors and can become a major economic power in the world. This will help the global economy recover from the economic downturn. As evident from Table 1, the value of foreign direct investment inflows is unstable. Many factors can affect the growth of foreign direct investment such as the economic and financial situations, gross domestic product (GDP), inflation and exchange rate etc. at varying degree in different countries.

Factors affecting foreign direct investment have been a subject of interest of many scholars. Nonnemberg and Mendonca [9] studied determinants of foreign direct investment in developing countries using panel data to show that the inflation rate, gross domestic product growth and risk are positively correlated with foreign direct investment. Kimino et al. [5] studied case of Japan using panel data. The results of fixed effect show that gross domestic product, foreign exchange rate and wage are positively correlated with the foreign direct investment. However, the value of export and loan interest rate are negatively correlated with the foreign direct investment in Japan. Ang [2] applied GARCH and Error-Correction Model for Malaysia, the study found that the growth rate of gross domestic product is positively correlated with foreign direct investment, while, tax rate and the instability of the exchange rate are discouraging factors.

This paper, therefore, focuses on the macroeconomic factors that affect the rate of growth of foreign direct investment (FDI) in AEC member countries using a panel quantile regression. The quantile regression will provide a particular set of parameters of interest for each level of quantile providing more information for

Table 1 The value of foreign direct investment inflows of ASEAN Economic Community member countries 2008–2012 (Unit: million US dollars)

Country	2008	2009	2010	2010	2012
Brunei	330	371	626	1,208	850
Cambodia	815	539	783	902	1,557
Indonesia	9,318	4,877	13,771	19,241	19,853
Laos	228	190	279	301	294
Malaysia	7,172	1,453	9,060	12,198	10,074
Myanmar	863	973	1,285	2,200	2,243
Philippines	1,544	1,963	1,298	1,816	2,797
Singapore	12,200	24,939	53,623	55,923	59,651
Thailand	8,455	4,854	9,147	7,779	8,607
Vietnam	9,579	7,600	8,000	7,430	8,368
Total	50,504	47,759	97,872	108,998	114,294

decision and policy makers. Since the numbers of years and countries are relatively limited, the panel data are recommended to be employed for this study [8]. So, we can identify the macroeconomic factors that affect the growth rate of foreign direct investment, according to the size or the distribution of the growth rate of foreign direct investment in AEC member countries so as to design appropriate policy. This study is particularly important because Thailand is a member of ASEAN Economic Community (AEC), which has a shortage of funds thus limiting the opportunity of entrepreneurs to access funds [11]. The inflow of foreign direct investment is very important for helping Thailand's economy and is one of the major factors in economic growth. The result of study will be useful for the government for its consideration regarding policy and strategy for encouraging the inflow of capital, the liberalization of investment and other policies.

This paper is organized as the following. Section 2 describes the scope of the data used in this study. Section 3 provides the methodology of this study. Section 4 a discusses the empirical results. Conclusion of this study is drawn in Sect. 5.

2 Data

As Myanmar has limited information, so, this study covers only 9 member countries, Vietnam, Thailand, Singapore, Philippines, Malaysia, Laos, Brunei, Indonesia and Cambodia. The data used in this study are the annual data for 2001–2012 from the World Bank indicators which provide the information on the growth rate of foreign direct investment, the growth rate of gross domestic product, the growth rate of the exchange rate and the inflation rate.

3 Methodology

This section provides details of this study procedure. The details are described as the following.

First step: Conducting the panel unit root test before using these data for estimation to avoid spurious regression. This study employs two methods of unit root test. The first is the first generation of panel unit root test according to Levin et al. [8] which is based on the cross-sectional independence hypothesis. Meanwhile, the second method is the second generation of panel unit root test following Pesaran [4] which is based on the cross-sectional dependency hypothesis.

Second step: Modeling macroeconomic factors affecting the growth rate of foreign direct investment (FDI) in AEC member countries from reviewing literature. The model can be written as an Eq. (1)

$$FDIG_{it} = f(GDPG_{it}, EXG_{it}, INF_{it}) \tag{1}$$

FDIG$_{it}$: The growth rate of foreign direct investments in the country i at time t (unit: percentage)
GDPG$_{it}$: The growth rate of the gross domestic product of country i at time t (unit: percentage)
EXG$_{it}$: The growth rate of the currency exchange rate ratio of the country i at time t the US dollar (in percentage)
INF$_{it}$: Inflation in the country i at the time t (unit: percentage)

where t is the time since the year 2001–2012 and i is a country i with $i = 1, \ldots, 9$, indicating Vietnam, Thailand, Singapore, Philippines, Malaysia, Laos, Indonesia, Cambodia and Brunei.

Third step: The estimation of model (1) using panel quantile regression with penalized fixed effects model to estimate the macroeconomic factors affecting the growth rate of foreign direct investment in AEC member countries.

Koenker and Bassett [6] discussed analysis of the data by quantile regression. The quantile regression is extension of the estimation by ordinary least squares (OLS) that will get the coefficients according to each level of quantile. The quantile regression has more coverage and is effective in the sense of the different conditions of the distribution of variables.

Consider the following equation in which we utilize the concepts and notions by [1] and [6] the others as cited.

$$Y_{it} = X_{it}\beta\alpha_{it} + \mu_i \tag{2}$$

where the Y_{it} and X_{it} are the observed dependent and independent variables, respectively. α_{it} is an unobservable with the assumption of α_{it} is U[0,1] conditional on X_{it} and μ_i is the error term. In addition, α_{it} could be arbitrarily related to all random variables in the Eq. (2) Koenker and Bassett [6]. On the basis of quantile regression, α_{it} is the parameters in the form of fixed effects model which make the estimated β become more efficient. Following Rosen [10], given $\theta(\tau) = (\beta_i, \alpha_i), e_{it}(\tau) = X_{it}[\theta(\alpha_{it}) - \theta(\tau)]$ and α_t is assumed to be independent between the cross section data, $Q_{e_{ij}}(\tau|X_{ij}) = 0$, are not enough to identify $\theta(\tau)$.

Thus, Abrevaya and Dahl [1] proposed the way to estimate $\theta(\tau)$ by using the correlated random effects model and view α_{it} as

$$\alpha_{it}(\tau, X_i, \varepsilon_i) = X_i\Lambda_T(\tau) + \varepsilon_i \tag{3}$$

Therefore, we can rewrite the Eq. (2) as follows

$$Y_{it} = X_{it}\theta(\tau) + \mu_{it} + e_{it} \tag{4}$$

The conditional behavior of Eq. (4) depends on the joint distribution of the μ_{it} and ε_i, however this correlated random effects model might not be appropriate in the estimation. Hence, Koenker [7] treat the α_{it} as parameters to be jointly estimated with $\theta(\tau)$ for any value of q (quantile) and proposed the penalized estimator as

$$(\widehat{\alpha}(\tau), \widehat{\beta}(\tau)) = argmin_{(\alpha,\beta)} \sum_{i} \sum_{j} \rho_{\tau}(y_{ij} - x_{ij}^{\tau}\beta - \alpha_i) + \lambda \sum_{i=1}^{n} |\alpha_i| \quad (5)$$

where, $\rho_t(\mu) = \mu[\tau - I(\mu < 0)]$ and I denotes as the indicator function and λ is a penalization parameter where its value is greater than 0.

This method is essential for this data because they might have relationships between each other. Moreover, we can divide the coefficient into three levels according to the quantile at 0.25, 0.50 and 0.75 that follows

$$min_{(\alpha,\beta)} \sum_{k=1}^{q} \sum_{j=1}^{n} \sum_{i=1}^{m_i} w_k \rho_{\tau k}(y_{ij} - x_{ij}^{\tau}\beta(\tau_k) - \alpha_i) + \lambda \sum_{i=1}^{n} |\alpha_i| \quad (6)$$

4 Empirical Results

The results can be divided into two parts. The first part is panel unit root test and the second part is macroeconomic factors affecting the growth rate of foreign direct investment in AEC member countries using panel quantile regression that follows.

4.1 *The Results of Panel Unit Root Test*

Tables 2 and 3 show the panel unit root test in two methods: The first is the first generation of panel unit root test according to Levin et al. [8] which is based on the cross sectional independency hypothesis and the second generation of panel unit root test following Pesaran [4] which is based on the cross-sectional dependency hypothesis. The dependent variable is the growth rate of foreign direct investments (FDIG). The independent variables are the growth rate of the gross domestic product (GDPG), the growth rate of the currency exchange rate ratio (EXG) and Inflation

Table 2 First generation of panel unit root test

Variables	First generation of panel unit root test					
	None		Intercept		Intercept and trend	
	Statistic	Prob.	Statistic	Prob.	Statistic	Prob.
FDIG	−4.98811***	0.0000	−3.76145***	0.0001	−3.24556***	0.0006
GDPG	−0.94220	0.1730	−6.36247***	0.0000	−6.05404***	0.0000
EXG	−6.15535**	0.0000	−7.94338***	0.0000	−1.6198	0.0526
INF	−0.49497	0.3103	−4.63488***	0.0000	−6.39811***	0.0000

Note ***, ** Show that there is a statistically significant level at 0.01 and 0.05 respectively

Table 3 Second generation of panel unit root test

Variable	Second generation of panel unit			
	Without trend		Trend	
	Statistic	Prob.	Statistic	Prob.
FDIG	−4.836***	0.0000	−3.488***	0.0000
GDPG	−3.679***	0.000	−2.178**	0.0015
EXG	−2.770***	0.003	−0.986	0.162
INF	−1.350*	0.088	−0.536	0.0000

Note ***, **, * Show that there is a statistically significant level at 0.01, 0.05 and 0.1 respectively

(INF) of AEC member countries. It was found that all variables of all countries are stationary or with order of integration equal to zero I(0) from both of the first and second generation panel unit root tests methods.

4.2 The Results of Macroeconomic Factors Affecting the Growth Rate of Foreign Direct Investment in AEC Member Countries Using Panel Quantile Regression

Table 4 shows that the growth rate of gross domestic product is positively correlated with the growth rate of foreign direct investment that means if the growth rate of gross domestic product is increased 1 % then the growth of foreign direct investment will be increased 3.692 and 4.845 % at the quantile 0.25 and 0.5, respectively with 0.10 level of significance. The effects will be greater when the growth rate of foreign direct investment increases, except quantile at 0.75, the growth rate of gross domestic product doesn't affect the growth rate of foreign direct investment. Moreover, the growth rate of the exchange ratio and inflation dont affect the growth rate of foreign direct investment in all levels of quantile.

Table 4 The results of macroeconomics using panel quantile regression

Countries	Quantile	Variable							
		Constant		GDPG		EXG		INF	
		Coeff.	Prob.	Coeff.	Prob.	Coeff.	Prob.	Coeff.	Prob.
AEC	0.25	−51.257	0.047	3.692*	0.067	3.335	0.17	−2.781	0.135
	0.5	−36.675	0.329	4.845*	0.055	2.481	0.399	−2.881	0.252
	0.75	71.091	0.191	0.788	0.853	−0.301	0.958	−2.141	0.67

Note * Show that there are statistically significant at the 90 % confidence level

5 Conclusion

Macroeconomic factors affecting the growth rate of foreign direct investment (FDI) in AEC member countries using a panel quantile regression with limited information for Myanmar were examined The study covered only 9 member countries, Vietnam, Thailand, Singapore, Philippines, Malaysia, Laos, Brunei, Indonesia and Cambodia with secondary data annually since 2001–2012 including 12 years of data. The independent variables used are the growth rate of gross domestic, the growth rate of the foreign exchange rate ratio and inflation of each. The study shows that the growth rate of the gross domestic product (GDPG), the growth rate of the foreign exchange rates ratio (EXG) and inflation (INF) of AEC member countries are stationary or have an order of integration equal to zero I(0) from both of the first and second generation panel unit root tests with LLC and Pesaran methods. The results of panel quantile regression show that the growth rate of gross domestic product are positively correlated with the growth rate of foreign direct investment which means that if the growth rate of gross domestic product is increased 1 % then the growth of foreign direct investment will be increased 3.692 and 4.845 % at the quantile 0.25 and 0.5 respectively with 0.10 level of significance. The effect will be greater when the growth rate of foreign direct investment increases. Except at quantile 0.75 the growth rate of gross domestic product does not affect the growth rate of foreign direct investment. Moreover, the growth rate of the foreign exchange rate ratio and inflation do not affect the growth rate of foreign direct investment in all levels of quantile. Therefore, all AEC member countries should focus on increasing the growth rate of gross domestic product to gain foreign investors confidence in each country's economy.

Acknowledgments We are greatful for financial support from Puay Ungpakoyn Centre of Excellence in Econometrics, Faculty of Economics, Chiangmai University.

References

1. Abrevaya, J., Dahl, C.M.: The effects of birth inputs on birthweight: evidence from quantile estimation on panel data. J. Bus. Econ. Stat. **26**, 379–397 (2008)
2. Ang, J.B.: Determinants of foreign direct investment in Malaysia. J. Policy Model. **30**(1), 185–189 (2008)
3. Asean learning center. Whats the AEC (2012). http://asean.stareducate.net/content/982
4. Im, K.S., Pesaran, M.H., Shin, Y.: Testing for unit roots in heterogeneous panels. J. Econ. **115**(1), 53–74 (2003)
5. Kimino, S., Saal, D.S., Driffield, N.: Macro determinants of FDI inflows to Japan: an analysis of source country characteristics. World Econ. **30**(3), 446–469 (2007)
6. Koenker, R., Bassett, G.: Regression quantiles. Econometrica **46**(1), 33–50 (1978)
7. Koenker, R.: Quantile regression for longitudinal data. J. Multivar. Anal. **91**, 74–89 (2004)
8. Levin, A., Lin, C.F., Chu, C.S.J.: Unit root tests in panel data: asymptotic and finite-sample properties. J. Econ. **108**, 124 (2002)
9. Nonnemberg, M. B., de Mendona, M. J. C.: The determinants of foreign direct investment in developing countries. In Anais do XXXII Encontro Nacional de Economia [Proceedings of

the 32th Brazilian Economics Meeting] (No. 061). ANPECAssociao Nacional dos Centros de Psgraduao em Economia [Brazilian Association of Graduate Programs in Economics] (2004)
10. Rosen, A.: Set identication via quantile restrictions in short panels. Working paper, University College, London (2009)
11. SMEs Thailand. Access to debt financing of SMEs (2013). http://www.smethailandclub.com/knowledges-view.php?id=174

Does Economic Growth Help Reducing Poverty? A Case of Thailand in Recent Data

Wannaphong Durongkaveroj

Abstract Thailand's performance on poverty reduction is obviously impressive because it has already achieved the first goal about halving poverty rate of the Millennium Development Goals promoted by the United Nations Development Programme. The main reason for this success is an outstanding economic growth which is widely accepted to be an efficient tool in eliminating of poverty in many developing countries. This paper attempts to quantitatively estimate the relationship between per capita income and poverty rate in Thailand at national and provincial level using the panel data between 2006 and 2013, and then to suggest the proper policies to accelerate the progress on poverty.

The result reveals that poverty in Thailand is highly elastic to the level of per capita income. For the whole country, a 1 % increase in per capita income will lead to 1.0595 % decrease in poverty. For provincial analysis, it is found that there is a huge difference in the responsiveness of growth on poverty among all provinces. Additionally, it is stated that there are 39 provinces of 75 provinces which have high growth elasticity of poverty while there are 9 provinces having the reversed correlation between per capita income and poverty rate. Lastly, this study also suggests that the government or related organizations should concern more about the different policies in each province in getting the poor out of destitute and low living standard.

1 Introduction

From the first and most important goal of the Millennium Development Goals (MDGs) issued by the United Nations Development Programme (UNDP) to Thailand's National Economic and Social Development Plans issued by the National Economic and Social Development Board (NESDB) of Thailand, poverty reduction is the priority, in term of policy, among all development targets. Since the 5th NESDB's plan (1982–1986), poverty has dramatically declined; meanwhile, Thailand has also

W. Durongkaveroj (✉)
Department of Development Economics, Faculty of Economics,
Ramkhamhaeng University, 10240 Bangkok, Thailand
e-mail: Wannaphongd@gmail.com

V.-N. Huynh et al. (eds.), *Causal Inference in Econometrics*,
Studies in Computational Intelligence 622, DOI 10.1007/978-3-319-27284-9_34

Table 1 Thailand's poverty by regions (Percent of total population in each region)

Regions	2008	2009	2010	2011	2012	2013	Change 2008–2013
Central	12.83	11.18	10.77	10.36	6.94	5.40	−57.90
Southern	16.77	17.03	14.24	10.12	13.32	10.96	−34.63
Northern	29.05	23.38	22.33	16.09	17.40	16.76	−42.30
Northeastern	31.19	27.71	25.26	18.11	19.79	17.37	−44.30
Whole Kingdom	20.43	17.88	16.37	13.22	12.64	10.94	−46.46

Source NESDB (2015)

achieved spectacular economic growth and expanding investment in many sectors which are considered to be the great engine of development. As a result of all positive factors, Thailand is now high human development country [34].

For measuring poverty, the main tool is poverty rate (ratio) which is the percentage of population living below the poverty line. There are two types of the poverty line including the international poverty line issued by the World Bank and the national poverty line normally issued by the national statistic office in each country. According to the World Bank [35], the latest poverty rate at $1.25 a day purchasing-power parity (PPP), as an international poverty line, in Thailand in 2010 is very low where only 0.31 % of its total population is the poor. It is relatively low to other countries in the same region, for example, Indonesia, Vietnam, and China. However, the picture of poverty is different when it is measured by another criteria. Using the national poverty line issued by NESDB [24], poverty rate in Thailand in 2010 is 16.37 %. The reason for higher poverty rate measured by the national poverty line is that the national poverty line is normally derived from an average income of population which is certainly higher than $1.25 PPP per day. Thus, it is clear that the result of poverty depends on the threshold.

Thailand has an impressive outcome in poverty reduction because poverty ratio for the whole country measured by the national poverty line has declined from 32.44 % in 2002 to only 10.94 % in 2013. Nevertheless, the more important issue is the performance in each province. As revealed, there is a huge difference across all regions in an elimination of poverty ([6], Jitsuchon and Richter 2006). It is stated that progress of poverty reduction has been slowest in the Northeast, followed by the North and the South. Interestingly, there is no significant change in the rank of performance over time. The recent data of poverty rate in Thailand measured by the national poverty line in each region are shown in Table 1.

According to Table 1, poverty rate for the whole country declines from 20.43 % in 2008 to 10.94 % in 2013 which means that Thailand is able to halve its poverty rate within five years. In regional performance, the Northeastern part has highest poverty rate, followed by the North and the South. However, the interesting issue is a change of poverty rate between 2008–2013. The central part is the most successful region because its poverty rate is declined by 57.90 %, followed by the Northeastern and the Northern part.

Table 2 Thailand's Per capita income by regions (Thousand Thai Baht)

Regions	2008	2009	2010	2011	2012	2013	Change 2008–2013
Bangkok and Vicinities	321	307	326	332	362	376	17.42
Central	212	204	222	211	229	239	12.51
Eastern	366	339	380	386	424	431	17.59
Western	102	106	111	120	126	127	25.07
Northern	64	65	74	80	93	98	52.44
Northeastern	43	49	56	61	69	75	74.67
Southern	104	100	121	133	126	123	18.44
Whole Kingdom	149	147	164	171	186	193	29.84

Source NESDB (2015)
Notes Central region is separated into four parts including Bangkok and vicinities, Central, Eastern, and Western part

As poverty rate is directly related to the level of income [23, 32], a change in the level of income, thus, will certainly lead to a change in the standard of livings among citizens. The recent data of per capita income of Thailand in each region are shown in Table 2.

According to Table 2, per capita income of Thailand in 2013 is 193 thousand Thai Baht which is increased by 29.84 % from 2008. However, this table displays a situation of income inequality in Thailand as well because there is a difference of per capita income across regions. Per capita income is highest in the East, followed by Bangkok and vicinities, and the central part. However, every region enjoys growth in this indicator, especially the Northeastern part—the poorest region in Thailand—which has the highest growth of per capita income. As a result, it is a good sign throughout the country for an improved living standard.

For the solution of poverty, many literatures highlight the role of economic growth which is referred to an increase in Gross Domestic Product (GDP) or per capita income. The relationship between economic growth and poverty reduction is realized as Growth Elasticity of Poverty (GEP) which can be derived through log-linear regression model and direct calculation. It presents the responsiveness of a change in income to poverty reduction [1, 3, 4, 7, 9, 10, 13, 18, 25, 27, 29, 30]. From the study of GEP in Thailand, poverty is highly elastic to economic growth [8, 15, 17, 20, 26]. Those papers have suggest a single value of GEP which is calculated at national level. However, each province has the different condition in economic development. Thus, it is necessary to consider the value of GEP in regional or provincial level as well. There are few papers considering the regional GEP [2, 22, 28] but there is still no study about regional/provincial GEP for Thailand whose poverty reduction is outstanding as revealed in Table 2.

2 Objectives

The purposes of this study are to quantitatively estimate the economic growth elasticity of poverty at both national and provincial level, and to suggest the proper policies targeting poverty reduction.

3 Methods

Per capita income and poverty rate measured by the national poverty line are collected from the NESDB. The balanced panel dataset covers 1,200 (2006 to 2013) yearly observations for each of the 75 provinces of Thailand (Except Phuket and Bueng Kan due to the missing data). Many empirical tools are employed in the study. As the panel data includes time-series data for each different entity, many stationary tests are conducted in order to avoid the problem of spurious results including the LLC [21], the HT [12], the Breitung [5], as well as the IPS [14] test.

After implementing a series of panel unit root tests, the relationship between per capita income and poverty rate is tested through two techniques including (1) an econometric method through fixed-effects model (FEM), random-effects model (REM), and an ordinary least square (OLS) method for estimating GEP of the whole country, and (2) the direct and simple calculation which is applied the concept of price elasticity of demand for estimating GEP in each province.

For model specification, the basic concept of GEP is obtained from the following expression

$$POV = f(PCI) \tag{1}$$

where POV is the poverty rate and PCI is the per capita income.

This relationship is the stylized idea of development economics which can be further applied to many models, for example, income inequality, education, and health.

Then, the distributions of poverty rate and per capita income are assumed to be lognormal which the growth elasticity of poverty is defined as the percentage change in the poverty rate for one percent increase in per capita income [4]. The equation for FEM can be written as

$$ln(POV)_{it} = a_i + b_1 ln(PCI)_{it} + U_{it} \tag{2}$$

where POV_{it} is the poverty rate in province i during the period t, a_i (i=1,2,3,...,n) is an unknown intercept for each province, PCI_{it} is the per capita income, b_1 is the growth elasticity of poverty, and U_{it} is the error term.

For FEM, an important assumption is that the unobserved/omitted variables do not vary (invariant) over time; in other words, their effects are imposed to be fixed within the entity but their characteristics can be different across the entities (provinces). Also, every change to the dependent variable is stemmed from only specified/known variables. The slope of an independent variable (PCI) is the same for all provinces while the specific intercept varies across provinces but not varies over time [11, 19, 31].

The GEP in random-effects model (REM) is given by the following formula

$$ln(POV)_{it} = a_i + b_1 ln(PCI)_{it} + U_{it} + e_{it} \quad (3)$$

where U_{it} is the error term between provinces and e_{it} is the error term within province. The main assumption of this model is that the error term for each province is not correlated with the predictors [33]. For the process of selecting model between FEM and REM, Hausman test is implemented with the null hypothesis of no correlation between errors and the independent variables. Additionally, Breusch-Pagan Lagrange Multiplier is tested the significant difference across provinces aimed to select the proper model between REM and OLS.

For the sign of the coefficient, poverty rate is expected to be negatively related to per capita income indicating that an increase in per capita income will lead to a reduction in poverty and a decrease in per capita income can cause poverty increase. However, the positive coefficient for per capita income is also possible. There are many studies attempting to explain this result. Positive GEP can occur when country has high income inequality which the benefit of economic growth is not properly trickled down to citizen, especially the poor. It can also take place in the period of recession as an average output of economy decreases but the poor are hurt less than the non-poor [17]. Also, the positive GEP can be referred to the declining benefit of growth as a result of high inflation and financial crisis [22]. In addition, the coefficient can be positive in the case that an increased mean income is still lower than poverty line. Although people have higher income but they can be still the poor if their income is lower than the threshold [10]. Thus, it should be careful for interpretation of this perverse result.

Besides the estimation of GEP through log-linear model, the direct calculation of price elasticity of demand is applied to estimate the provincial GEP. It can be written as

$$GEP_i = \frac{\%(POV)_i}{\%(PCI)_i} \quad (4)$$

where GEP_i is the growth elasticity of poverty (GEP) in province i, $\%(POV)_i$ is the annual change rate in poverty rate in province i during the period 2006–2013, and $\%(PCI)_i$ is the annual change rate in per capita income in province i during the period 2006–2013. An annual change rate is calculated by using Compound Annual Growth Rate (CAGR) method. Thus, this elasticity is expected to be negative because poverty rate normally declines over time.

However, the main disadvantage of this direct calculation is its assumption because it is assumed that a change in income is the only factor affecting poverty. Also, GEP from this method can be positive in two cases. Firstly, poverty rate increases while per capita income increases. This can happen when people spend an additional income to pay debt, instead of consuming to fulfill the basic needs of life. This requires the deeper study towards the characteristics of household income and expenditure. Second, per capita income decreases while poverty rate decreases. This can happen when people are able to live in a better condition without having an increase in income, especially the poor lived in the rural area. The quality of life can be promoted through deriving more public services or in-kind subsidies from government. This requires the deeper study towards specific government policies in each province. Thus, a positive GEP needs a case-by-case explanation because the assumption is strict in explaining poverty with an income level only.

4 Results

The results of the various panel unit root tests are presented in Table 3.

According to Table 3, both variables in this study are stationary with trend under all tests at the 5 % level of significance. Then, FEM, REM, and OLS are employed to derive the relationship between economic growth (Per capita income) and poverty (Poverty ratio). The results are shown in Table 4.

According to Table 4, economic growth helps reducing poverty in Thailand at national level. The result statistically confirms the relationship between per capita income and poverty rate at the 0.05 level of significanc. For FEM, poverty rate

Table 3 Panel unit root tests on poverty rate and per capita income of 75 provinces

Variable	LLC	HT	Breitung	IPS
ln(POV)	−14.1809*** (0.0000)	−0.0636*** (0.0000)	−2.0250** (0.0214)	−4.3047*** (0.0000)
ln(PIC)	−31.0398*** (0.0000)	0.0775*** (0.0002)	−2.3057** (0.0106)	−4.2578*** (0.0000)

Source Author's calculation
Notes (1) Probability is in parenthesis (2) *** and ** indicate the rejection of the null hypothesis of having unit root at least on the 0.01 and 0.05 level of significance, respectively

Table 4 Economic growth elasticity of poverty for the whole country

Statistic Value	FEM	REM	OLS
GEP	−1.1714***	−1.0595***	−0.9645
Constant	16.0064***	14.7294***	13.6459***

Source Author's calculation
Notes (1) FEM is the fixed-effects model, REM is the random-effects model, and OLS is the ordinary least square model (2) *** indicates the statistical significance at the 0.05 level

decreases (increases) by 1.1714 % when there is a 1 % growth (decrease) in per capita income of the society. The equation for FEM can be obtained as

$$ln(POV)_{it} = 16.0064 - 1.1714ln(PCI)_{it} \tag{5}$$

For REM, every 1 % increase (decrease) in per capita income leads to a reduction (growth) of poverty by 1.0595 %. REM can be written as

$$ln(POV)_{it} = 14.7294 - 1.0595ln(PCI)_{it} \tag{6}$$

For OLS model, a 1 % increase (decrease) in per capita income is associated with the 0.9645 % decrease (increase) in poverty head count index. The results from OLS model can be written as

$$ln(POV)_{it} = 13.6459 - 0.9645ln(PCI)_{it} \tag{7}$$

Additionally, the probability of test (F) for all models is less than 0.05 which means that all coefficients in the models are different from zero. Then, to select the most proper model requires additional tests including Hausman test and Breusch-Pagan Lagrange Multiplier test. For the first test, the Prob>Chi2 is greater than 0.05. It fails to reject the null hypothesis which means that errors and regressor are not correlated. In this case, REM is more proper than FEM. Then, the latter test is employed to decide between REM and simple OLS model. The result shows that Prob>Chi2 is less than 0.05. It is able to reject the null hypothesis of zero value of variance across provinces indicating that variance is significantly different across provinces. In this case, REM is more proper than simple OLS mode. Therefore, GEP for this study is −1.0595. It is concluded that poverty in Thailand at national level is highly sensitive to per capita income changes.

For provincial GEP, it is directly calculated from the Eq. 4. The selected results, highest and lowest GEP, are displayed in Table 5.

According to Table 5, there is a huge difference in the responsiveness of poverty to a change in income level. On average, the national GEP measured by this method is −1.633 which is correspondent to the value of GEP measured by the log-linear model. For provincial GEP, poverty is highly elastic to economic growth in many provinces. A 1 % increase in its per capita income in Chon Buri—the highest GEP province—leads to 14.7344 % decrease in poverty. The second highest GEP province is Satun whose value of GEP is −8.7037, followed by Saumut Prakan and Songkhla, respectively. However, GEP is found to be positive in nine provinces including Pathum Thani, Samut Sakhon, Ranong, Phangnga, Trang, Chanthaburi, Mukdahan, Chiang Mai, and Phattalung. For Pathum Thani and Samut Sakhon, poverty rates decline over time but both per capita income are in a downturn over the period 2006–2013. The reason to this perverse result requires an additional information on the distribution of income and expenditure in both provinces. However, it is a special case because Pathum Thani and Samut Sakhon are the only two provinces experiencing

Table 5 Economic growth elasticity of poverty for selected provinces

Provinces	Highest GEP	Provinces	Lowest GEP
Chon Buri	−14.7344	Pathum Thani	140.3150
Satun	−8.7037	Samut Sakhon	123.1867
Samut Prakan	−7.8919	Ranong	10.8290
Songkhla	−7.8434	Phangnga	0.9133
Surat Thani	−7.3268	Trang	0.6412
Samut Songkhram	−6.5594	Chanthaburi	0.4678
Nonthaburi	−5.4163	Mukdahan	0.4669
Nakhon Pathom	−4.7163	Chiang Mai	0.3929
Chumphon	−4.2470	Phattalung	0.1226
Narathiwat	−4.0409	Uttaradit	−0.1194

Source Author's calculation

the negative change of per capita income. For other seven provinces having a positive GEP, both poverty rate and per capita income increase in the selected period.

In conclusion, there are 42 provinces of 75 provinces which their poverty is highly elastic to economic growth because the value of GEP is greater than 1 (39 provinces for negative GEP and 3 provinces for positive GEP). Additionally, there are 33 provinces which their poverty is low elastic to economic growth (27 provinces for negative GEP and 6 provinces for positive GEP). As a result, the poverty responsiveness to economic growth is entirely different in each of the 75 provinces throughout Thailand.

5 Conclusion

To reduce poverty requires not only domestic policies but also international policies. Thailand is one of the developing countries which has made a great progress in fighting poverty as there is only 0.31 % of Thai lived below about $1.25 a day (PPP) in 2010. Additionally, the recent data indicates that Thailand has already achieved the first goal of the MDGs.

This study focuses the role of economic growth in reducing poverty in Thailand by estimating how poverty rate—measured by the national poverty line—responses to a change in per capita income over the period 2006–2013 at national and provincial level. The result reveals that poverty is highly elastic to economic growth. The poverty elasticity is calculated at −1.0595, meaning that for every one percent increase in average per capita income, poverty will decline by 1.0595 %. The value of GEP derived from this study is correspondent to Deolalikar [6], Jitsuchon [15], and Durongkaveroj [8] who estimates GEP equal to −2.2, −1.206, and −1.4445, respectively. However, Ram [26] finds that GEP of Thailand is −7.92 using direct calculation method. Therefore, at national prospect, Thai government should concern more about the quality and distribution of economic growth. The real challenge is how much is growth good for the poor. Moreover, it should be stressed that an economic recession is likely to hit the citizen's well-being because the destitute is highly sensitive to the income level.

For provincial analysis, there is a huge difference in the performance of economic growth in reducing poverty. There are 39 provinces which economic growth is a great tool in tackling poverty. However, this method holds the strong assumption about the relationship between economic growth and poverty which can result in an overestimated GEP. For policy implementation, the great efforts have to be made in some provinces experiencing a negative growth of per capita income and a positive growth of poverty. Also, besides only an acceleration of economic growth, public services and in-cash/in-kind subsidies should be provided to the poor aimed at raising their living standards. Education and health services have to be fully served in the rural areas. Thus, each province around Thailand requires the different magnitude, speed, and direction of efforts to lessen its own poverty.

For further study, more specific policies towards poverty reduction in each province should be examined because they will help us understand deeper about causes and consequences of poverty. Moreover, non-economic variables, e.g. culture and institutions, should be explored in order to elaborately figure out other influential factors affecting the standard of living among citizens.

References

1. Adam, R.: Economic growth, inequality, and poverty: Findings from a new data set. Policy Research Working Papers 2972. The World Bank, Washington (2003)
2. Balisacan, A.M., Pernai, E.M.: Probing beneath cross-national averages: poverty, inequality, and growth in the Philippines. EDR Working Paper Series 7. Asian Development Bank, Manila (2002)
3. Berardi, N., Marzo, F.: The elasticity of poverty with respect to sectoral growth in Africa. Banque de France Working Paper 538 (2015)
4. Bourguignon, F.: The growth elasticity of poverty reduction: explaining heterogeneity across countries and time periods. DELTA Working Papers 2002/03 (2002)
5. Breitung, J., Das, S.: Panel unit root tests under cross-sectional dependence. Stat. Neerl. **59**, 413–433 (2005)
6. Deolalikar, A.B.: Poverty, growth, and inequality in Thailand. ERD Working Paper 8. Asian Development Bank, Manila (2002)
7. Dollar, D., Kraay, A.: Growth is good for the poor. Policy Research Working Paper 2587. The World Bank, Washington (2001)
8. Durongkaveroj, W.: The role of human development in poverty reduction. J. Econ. Ramkhamhaeng Univ. **1**(1) (2015)
9. Durongkaveroj, W., Osathanunkul, R.: Regional multipliers of social accounting matrix and the effective eradication of poverty. Empi. Econom. Quant. Econ. Lett. **2**(4), 39–52 (2013)
10. Fosu, A.K.: Growth, inequality and poverty reduction in developing countries: recent global evidence. UNU-WIDER Working Paper 2011/01 (2011)
11. Greene, W.H.: Econometric Analysis, 6th edn. Prentice Hall, New Jersey (2008)
12. Harris, R.D.F., Tzavalis, E.: Inference for unit roots in dynamic panels where the time dimension is fixed. J. Econom. **91**, 201–206 (1999)

13. Helberg, R.: The growth elasticity of poverty. UNU-WIDER Discussion Paper 2002/21 (2002)
14. Im, K.S., Pesaran, M.H., Shin, Y.: Testing for unit roots in heterogeneous panels. J. Econom. **115**, 53–74 (2003)
15. Jitsuchon, S.: Sources and pro-poorness of Thailand's economic growth. Thammasat Econ. J. **24**(3), 68–106 (2006)
16. Jitsuchon, S., Richter, K.: Thailand's poverty maps: From construction to application. In: Bedi, T., Coudouel, A., Simler, K. (eds.) More than a pretty picture: using poverty maps to design better policies and interventions. The World Bank, Washington (2007)
17. Kakwani, N., Son, H.H.: Pro-poor growth: concepts and measurement with country case studies. Pak. Dev. Rev. **42**(4), 417–444 (2003)
18. Klasen, S., Misselhorn, M.: Determinants of growth semi-elasticity of poverty reduction. The World Bank, Washington (2005)
19. Kohler, U., Kreuter, F.: Data Analysis Using Stata, 3rd edn. Stata Press, Texas (2012)
20. Krongkaew, M., Chamnivickorn, S., Nitithanprapas, I.: Eonomic growth, employment, and poverty reduction linkages: the case of Thailand. Issues in Employment and Poverty Discussion Paper 20 (2006)
21. Lavin, A., Lin, C.-F., Chu, C.-S.J.: Unit root tests in panel data: asymtotic and finite-sample properties. J. Econom. **108**, 1–24 (2002)
22. Le, M.S., Nguyen, D.T., Singh, T.: Economic growth and poverty in Vietnam: evidence from elasticity approach. Griffith Business School Discussion Paper Economics 2014-01 (2014)
23. Nafziger, E.W.: Economic Development, 4th edn. Cambridge University Press, New York (2006)
24. National Economic and Social Development Board Per capita income of population by region and province: 1995–2013 (2015). Available via DIALOG. http://ww.nesdb.go.th. Cited20June2015
25. Ram, R.: Growth elasticity of poverty: alternative estimates and a note of caution. Kyklos **59**, 601–610 (2006)
26. Ram, R.: Income elasticity of poverty: estimates for the ASEAN countries. Empir. Econom. Quant. Econ. Lett. **4**(2), 9–16 (2015)
27. Ravallion, M., Chen, S.: What can new survey data tell us about recent changes in distribution and poverty. World Bank Econ. Rev. **11**(2), 357–382 (1997)
28. Riyana, M., Resosudarmo, B.: Understanding regional poverty in Indonesia: is poverty worse in the east than in the west. Aust. J. Reg. Stud. **11**(2), 141–154 (2005)
29. Roemer, M., Gugerty, M.K.: Does economic growth reduce poverty?. CARE II Discussion Paper 4 (1997)
30. Squire, L.: Fighting poverty. Am. Econ. Rev. **83**(2), 377–382 (1993)
31. Stock, J.H., Watson, M.W.: Introduction to Econometrics, 3rd edn. Pearson, New York (2011)
32. Todaro, M., Smith, S.C.: Economic Development, 10th edn. Pearson, New York (2009)
33. Torres-Reyna, O.: Panel data analysis: fixed and random effects using Stata (2007). Available via DIALOG. http://dss.princeton.edu/training.Cited25June2015
34. United Nations Development Programme (2014) Human development report 2014. Available via DIALOG. http://www.hdr.undp.org/en/global-reports.Cited26June2015
35. World Bank, World development indicators 2015. The World Bank, Washington (2015)

Effect of Quantitative Easing on ASEAN-5 Financial Markets

Pathairat Pastpipatkul, Woraphon Yamaka and Songsak Sriboonchitta

Abstract After the economic crisis in 2007, the United States enter to the economic recession. Thus the central banks (Fed) purposed an unconventional policy and launch various programs in order to restore the weak economic. However, it also generated a spillover effects toward Emerging countries through capital flow. Therefore, the paper aims to provide a new empirical finding by examining the effect of quantitative easing (QE) policy of the United States on Thailand, Indonesian, and the Philippine, Singapore, and Malaysian financial markets (ASEAN-5). In this study, the ASEAN-5 financial markets, comprising the exchange rate market, the stock market, and the bond market are considered. To measure the effect of QE on those markets, we employed the Markov-switching VAR model to study the transmission mechanisms of QE shocks between periods of expansion in the QE program and QE tapering. Moreover, we restrict the structure of the model in order to identify the determinant of the structural change. This paper finds that ASEAN-5 financial markets receive the effect form QE. The treasury securities purchase program seems to generate a larger effect to the ASEAN-5 financial market than other programs. Moreover, the test of best MS-VAR specification, provide the result that MSH(2)-VAR(1) is the best specification model for the exchange rate market and the stock market, while MSIH(2)-VAR(1) is the best specification model for bond markets. This indicates that QE was not the factor leading the ASEAN-5 financial markets switch from one regime to another regime.

Keywords Markov switching model · Structural change · Quantitative easing · ASEAN-5 financial markets

P. Pastpipatkul (✉) · W. Yamaka · S. Sriboonchitta
Faculty of Economics, Chiang Mai University, Chiang Mai, Thailand
e-mail: ppthairat@hotmail.com

W. Yamaka
e-mail: woraphon.econ@gmail.com

V.-N. Huynh et al. (eds.), *Causal Inference in Econometrics*,
Studies in Computational Intelligence 622, DOI 10.1007/978-3-319-27284-9_35

1 Introduction

After the financial crisis in 2007, the U.S. economy was stuck in a liquidity trap and the policy rate was dropped to be low (0.25 %). Consequently, the U.S. Federal Reserve Bank (Fed) has been employed an unconventional policy, which involved substantial expansion of the central bank mortgage-backed securities and the U.S. treasury notes. This policy has come to be known as the quantitative easing (QE) policy. It was designed for the purpose of boosting the weak economy, as well as stimulating real activity. At present, the Fed has announced three rounds. The first round of quantitative easing, QE1, was announced to buy bank debts, mortgage-backed securities (MBS), and treasury securities from financial institutes in late November 2008, which was then extended to October 30, 2009. The asset purchases amounted to approximately 1.7 trillion dollars. In November 2010, the Fed announced a second round of quantitative easing, buying 600 billion dollars of the treasury securities. On September 13, 2012, the Fed announced a third round of quantitative easing, deciding to boost a 40, 000 million dollars per month, open-ended bond purchasing program of agency mortgage-backed securities. Additionally, on December 13, 2012, the Fed announced the expansion of the quantitative easing program, and the financial amount was expanded in order to buy a treasury bill of 45, 000 million dollars per month; the expansion of the QE3 was continued. So, in this round, the Fed injected an amount of 85,000 million dollars per month. In addition, compared to that at the beginning of the Feds programs, its balance sheet had more than 7 trillion U.S. dollars by the end of 2013. However, in the beginning of 2014, the Fed began to decrease its balance sheet (QE tapering).

In general, QE seems to have benefited the U.S. economy, but it also had unintended consequences, especially toward emerging economies (EMEs), through its global liquidity and capital flow. The level of capital outflow to the Asia-Pacific region post the advent of QE is different from the level prior to the advent of QE. The region was found to experience high growth rates and a stable political-economic environment after the advent of QE. The data reveal that the gross capital inflow to the Asia-Pacific region grew more than triple, to 168,483 million U.S. dollars, between 2008 and 2013. The region, however, expressed its concern that the expansion of the Fed balance sheet will cause an inundation of the dollar to the EMEs, thereby causing significant fluctuation in their currencies. In addition, they were also concerned that the QE taper will cause capital outflow from the region, resulting in further volatility in the financial market. In the recent, integration in the global economy has been a key driver of development in the Asian financials, such as the emergence of new capital markets, reduction in capital inflow barrier and foreign exchange restriction, and adoption of flexible exchange rates, in order to increase the opportunities for investment. Thus, the introduction of QE programs by the Fed has impacted Asia in various ways. There is significant evidence that QE programs do have an impact on emerging markets in Asia. For instance, Morgan [17], Chen et al. [4] and Cho and Rhee [5] found a significant impact of the U.S. quantitative easing, in that it lowered the emerging Asian economy bond yields, increased stock prices, and appreciated the domestic currency against the U.S. dollar. Evidently, there is a massive amount of literature that examines the effect of QE in the various countries such as Zammit [19], Giradin and Moussa [10], Bowman et al. [3], Fawley and Neely [11] and Glick and Leduc [9] Albua et al. [1]. However, there is very little research literature on the international spillover of QE, especially its impact on ASEAN-5. They are the newly industrialized countries in Southeast Asia which is largely investment-dependent and which receives the effect of U.S. capital inflow because it is an emerging market which has abolished all capital inflow barriers and foreign exchange restriction. In recent years, ASEAN-5 countries have shown considerable interest in U.S. investors who are in expectation of high returns. The capital inflow to these five countries had increased by approximately 94 % since 2008. International Business Times (2013) had also reported that they received 128.4 billion dollars in foreign investment in 2013.

Thus, our goal is to focus on the effect of QE on ASEAN-5, which includes Thailand, Indonesia, Malaysia, Singapore, and the Philippines, rather than on other countries. In addition, there is substantial empirical finding regarding the effects of QE, using the linear model, such as the results obtained by Krisnamurthy and Vissin-Jorgensen [12], Bowman et al. [3], Glick and Leduc [9], who employed regression analysis, while Chua et al. [7], Moore et al. [16], Christensen and Gillan [6], and Zhu and Yang [20] employed the VAR approach. These models are not appropriate to capture the impact of QE on the real economy which present a different behavior over time thus it is reasonable to assume that there are regime shifts in the macro-econometric system [21]. To the best of our knowledge, there exists a small number of papers that have employed the nonlinear model, as follows: Giradin and Moussa [10] employed MS-FAVAR in order to examine the effectiveness of Japanese monetary policy, while Ledenyov and Levenyov [15] employed the nonlinear DSGE model to estimate the effect of QE in the United States. Our paper takes after these papers, which employed the nonlinear model, but our explicit goal is to capture the effects of QE on the financial markets of ASEAN-5, rather than on other EMEs. Furthermore, we restrict the structure of the model in order to identify the determinant of the regime shift in the model.

In accordance with the above issue, this research paper will be studying the QE effect on the ASEAN-5 markets by using the Markov-switching Vector Autoregressive (MS-VAR) model. MS-VAR is an innovative tool for dating the effect of volatility in the different periods as well as for identifying the factors that lead the ASEAN-5 markets from one state to another state, in addition to signaling a turbulent regime, if any, as an early warning system. Moreover, this model also provides an impulse response with respect to any innovations or shocks in financial markets.

This papers proceeds as follows: the next section describes the MS-VAR modeling methodology, estimation and data. Empirical results are presented in Sect. 3. Finally, Sect. 4 provides the conclusion with a summary of the important results.

2 Methodology

2.1 Markov-Switching VAR Model

The main model for the analysis of the QE effect, which will used here, namely MS-VAR, represents an unobserved state driven by the ergodic Markov process. This method provides a synthesis of the dynamic factor structures and nonlinear approach for the modeling of fluctuations of financial markets. The estimation technique implemented for MS-VAR models is called the Expectation Maximum (EM) algorithm, which consists of two steps. The first step is to compute the expected likelihood and the second step is maximize the expected likelihood to obtain the estimated parameters. Krozig [13] provides a simple MS-VAR process of order p with $S_t = 1, \ldots, k$ regimes where the general form can be written as follows:

$$y_t = v(S_t) + \sum_{i=1}^{p} A_t(S_t) y_{t-i} + u_t. \tag{1}$$

where y_t is the vector of endogenous variables, e_t is the error term with normal distribution ($\varepsilon_t \sim N(0, \Sigma(S_t))$), $v(S_t)$ is regime-dependent vector of intercept term, $A_t(S_t)$ is a regime-dependent matrix of autoregressive coefficients. For k regimes, we can extend Eq. (1) as follows:

$$\begin{bmatrix} y_{t,1} \\ \vdots \\ y_{t,k} \end{bmatrix} = \begin{bmatrix} v_t + A_{11} y_{t-1} + \cdots + A_{1p} y_{t-p} + u_{t,1} \\ \vdots \\ v_t + A_{k1} y_{t-1} + \cdots + A_{kp} y_{t-p} + u_{t,k} \end{bmatrix} \tag{2}$$

where $u_t \sim N(0, \Sigma(S_t))$.

However, for empirical applications, it might be more helpful to use a model where only some parameters are conditioned on the state of the Markov chain, while the other parameters are regime-invariant. In the general MS-VAR specifications, all the parameters are regime-dependent. However, in the empirical applications, it is preferable to consider only some parameters as dependent on the state as this may offer a lot of alternatives for the MS-VAR specification. Krolzig [13] established a common notation to provide simplicity in expressing those models in which the various parameters are subject to shifts with the varying state. The alternative specifications are MS models which allow shifts in mean, intercept, coefficient, and variance across regimes. In order to establish a unique notation for each model, Krolzig [13] specified the following regime-dependent parameters with the general MS(M) term:

M Markov-switching mean
I Markov-switching intercept term
A Markov-switching autoregressive parameters
H Markov-switching heteroskedasticity.

Table 1 Types of MS-VAR models

		MSM		MSI	
		u varying	u invariant	v varying	v invariant
A_j invariant	Σ invariant	MSM-VAR	MVAR	MSI-VAR	VAR
	Σ varying	MSMH-VAR	MSH-MVAR	MSIH-VAR	MSH-VAR
A_j varying	Σ invariant	MSMA-VAR	MSA-MVAR	MSIA-VAR	MSA-VAR
	Σ varying	MSMAH-VAR	MSAH-MVAR	MSIAH-VAR	MSAH-VAR

Source [13]

Common to all MS-VAR specifications, the unobservable regime S_t is governed by first order Markov process, which is defined by the transition probability P (Table 1) in Eq. (3).

$$P_i j = Pr(S_{t+1} = j \,|S_t = i)\,; \sum_{j=1}^{k} p_{ij} = 1, \quad i, j = 1, \ldots, k \tag{3}$$

2.2 *Estimation Procedure*

As in the time series studies, the stationary of the series should be checked prior to the estimation using the MS-VAR model. Thus, we first employ the Augmented DickeyFuller (ADF) test to check the stationary of the 17 observed variables. Second, the best specification from among the various MS-VAR models has to be identified. In this study, we proposed eight MS-VAR specifications with two regimes, as follows:

(1) Markov-switching Intercept Autoregression and Heteroscedasticity: MSIAH (2)-VAR(p)

$$y_t = v(S_t) + \sum_{i=1}^{p} A_t(S_t) y_{t-i} + u_t, \qquad u_t \sim N(0, \Sigma(S_t)); \quad S_t = 1, 2$$

(2) Markov-switching Intercept and Autoregression: MSIA(2)-VAR(p)

$$y_t = v(S_t) + \sum_{i=1}^{p} A_t(S_t) y_{t-i} + u_t, \qquad u_t \sim N(0, \Sigma); \quad S_t = 1, 2$$

(3) Markov-switching Intercept and Heteroscedasticity: MSIH(2)-VAR(p)

$$y_t = v(S_t) + \sum_{i=1}^{p} A_t y_{t-i} + u_t, \qquad u_t \sim N(0, \Sigma(S_t); \quad S_t = 1, 2$$

(4) Markov-switching Intercept: MSI(2)-VAR(p)

$$y_t = v(S_t) + \sum_{i=1}^{p} A_t y_{t-i} + u_t, \qquad u_t \sim N(0, \Sigma; \quad S_t = 1, 2$$

(5) Markov-switching Autoregression: MSA(2)-VAR(p)

$$y_t = v(S_t) + \sum_{i=1}^{p} A_t(S_t) y_{t-i} + u_t, \qquad u_t \sim N(0, \Sigma; \quad S_t = 1, 2$$

(6) Markov-switching Heteroscedasticity: MSH(2)-VAR(p)

$$y_t = v(S_t) + \sum_{i=1}^{p} A_t y_{t-i} + u_t, \qquad u_t \sim N(0, \Sigma(S_t)); \quad S_t = 1, 2$$

(7) Markov-switching Autoregression and Heteroscedasticity: MSAH(2)-VAR(p)

$$y_t = v + \sum_{i=1}^{p} A_t(S_t) y_{t-i} + u_t, \qquad u_t \sim N(0, \Sigma(S_t)); \quad S_t = 1, 2$$

(8) Vector Autoregression: VAR(p)

$$y_t = v + \sum_{i=1}^{p} A_t y_{t-i} + u_t, \qquad u_t \sim N(0, \Sigma)$$

In comparing these models, the model with the lowest value of AIC is preferred. Additionally, AIC has also been used to determine the lag length in the MS-VAR model.

In this study, only two states will be assumed where state 1 is a low growth regime, whereas state 2 is a high growth regime. Thus, we compare the various types of MS-VAR with two regimes and choose the best specification MS-VAR model in each sub-group of the financial markets to estimate the parameters, the transition matrix, and the smoothed regime probabilities. Finally, the impulse response function is estimated in order to demonstrate how the ASEAN-5 financial markets react to a shock caused by the QE policy in each regime.

Following Ehrmann et al. [8], the impulse response function for each regime is defined as follows:

$$\frac{\partial E_t X_{t+h}}{\partial u_{k,t}} | S_t = S_{t+h} = \theta_{k,h}$$

The estimation of the response of the h variable is derived from combining the matrix of the autoregressive parameter (A_t). Thereafter, the initial disturbance vector (u_0) is multiplied with the matrix of the autoregressive parameter (A_t) to obtain the impact responses at $l = 0$ time.

$$\hat{\theta}_{ki,0} = A_i u_0$$

Subsequently, the remaining response vector can be estimated by solving forward for the endogenous variable in Eq. (1), as follows:

$$\hat{\theta}_{ki,0} = \sum_{j=1}^{min(l,p)} A_{ji}^{2-j+1} u_0$$

Data

To find out and understand the impact of the U.S. quantitative easing on the currencies, stock markets, and bond markets of Thailand, Indonesia, the Philippines, Singapore, and Malaysia, the raw weekly data (Wednesday to Wednesday), the mortgage-backed securities purchase program (MBS) of the U.S., the treasury securities purchase program (TS) of the U.S., the Stock Exchange of Thailand Index (SET), the Jakarta Composite Index (JKSE), the Philippine Stock Exchange Composite Index (PSEi), the Singapore Stock Exchange (SGX), the Kuala Lumpur Stock Exchange (KLSE), the Thai Baht/USD (Exth), the Indonesian Rupiah/USD (Exind), the Philippine Peso/USD (Exphp), the Singapore Dollar/USD (Exsin), the Malaysian Ringgit/USD (Exmal), the Thai government bond yields (THY), the Indonesian government bond yields (INDY), the Philippine government bond yields (PHY), the Singapore government bond yields (SY), and the Malaysian government bond yields (MY) were collected from the Thomson Reuters DataStream and from the Financial Investment Center (FIC), Faculty of Economics, Chiang Mai University, and www.federalreserve.gov for the period from December 18, 2009, to June 30, 2014. All of these observations, however, have been transformed to the first difference form in order to make them stationary because the stationary time series is able to avoid the spurious regression problem that is encountered when non-stationary time series is used. Before estimating the parameters, we separated the parameters into three sub-groups of financial markets, as follows:

1. Stock group consisting of RMBS, RTS, RSET, RJKSE, RPSEi, RSGX, and RKLSE
2. Exchange rate group consisting of RMBS, RTS, RExth, RExind, RExphp, RExsin, and RExmal
3. Bond group consisting of RTS, RMBS, RTHY, RINDY, RPHY, RSY, and RMY.

3 Estimation Results

3.1 Unit Root Test

To avoid the phenomenon of false regression caused by the regression analysis of non-stationary time series, the variables should be taken as ADF unit root test before estimating the MS-VAR model. The results of the ADF test, as given in Table 2, suggest that the null hypothesis of the unit root test can be rejected and that all the variables are stationary at the 1, 5, and 10 % confidence levels. This means that we can use these variables as endogenous variables in the MS-VAR model for the purpose of estimating the impact of the QE policy.

3.2 Model Selection

In this study, we preferred MSI-VAR models (with switching intercept) rather than MSM-VAR (with switching mean) models because these models adjust more smoothly after the regime shift. Moreover, the estimation procedure of the MSM-VAR model may have to face the problem of nonlinear optimization if the y_t does not depend on the unobserved variable (S_t) [14].

The various specifications of the MSI-VAR models, as reported in Table 3, were compared for their Akaike information criterion (AIC) value. Among the MS specification values of individual financial markets, the MSH-VAR(1) (Markov-switching and heteroscedastic model with lag 1) appeared to be the best fit in the stock group and the exchange rate group, while MSIH-VAR(1) (Markov-switching intercept and heteroscedastic model with lag 1) appeared to be the best fit in the bond group.

Table 2 Unit root tests results

Variable	Prob.	Variable	Prob.	Variable	Prob.
RMBS	0.000	RExth	0.000	RTHY	0.000
RTS	0.000	Ris EXind	0.000	RINDY	0.000
RSET	0.000	RExphp	0.000	RPHY	0.000
RJKSE	0.000	RExsin	0.000	RSY	0.000
RPHEi	0.000	RExmal	0.000	RMY	0.000
RSGX	0.000		0.000		0.000
RKLSE	0.000		0.000		0.000

Source Calculation

Table 3 MS specification among various MSI models

AIC		Stock market group		Exchange market group		Bond market group	
		v varying	v invariant	v varying	v invariant	v varying	v invariant
A_j invariant	Σ invariant	9473.3	29.497	12325	39.414	8402.8	23.849
	Σ varying	10012	10017	12715	12725	8491.7	8487.2
A_j varying	Σ invariant	9467.2	9464.8	12321	12242	8013.8	8324.6
	Σ varying	9954.2	9957.8	12620	12632	7765	7706.9

Source Calculation

3.3 Estimation Result by Individual Financial Markets

Table 4 presents the final MSH(2)-VAR(1) model for ASEAN-5 stock markets. The variance term ($\sigma(S_t)$) of the MSH(2)-VAR(1) model for each of the two regimes seems to have an economic interpretation. Following the approach proposed in VARGAS III [18], it can be inferred that the value of the variance term in the first regime is less than that in the second regime. Thus, this indicates that regime 1 is in the high

Table 4 Estimate MSH(2)-VAR(1) model of ASEAN-5 stock markets

	RSET	RJKSE	RPHEi	RSGX	RKLSE
Regime-independent intercepts					
Regime 1	0.004***	0.004***	0.005***	0.002***	0.002***
Regime 2	0.004***	0.004***	0.004***	0.004***	0.004***
Regime-independent, Autoregressive parameters at lag 1					
RTS	0.0363	0.089*	−0.026	0.551	0.027
RMBS	−0.001	−0.001	0.002	−0.001	0.001
RSET	−0.0098	0.087	0.049	0.0538	0.026
RJKSE	0.111	−0.195	0.091	−0.091	0.011
RPSEi	0.087	0.118	−0.055	0.075	−0.039
RSGX	0.064	−0.203*	0.165*	−0.008	0.0067
RKLSE	−0.239*	−0.203*	0.03	−0.04	−0.04
Log-likelihood = 5073.3	AIC (10^5) −10.016		BIC (10^5) −9.652		
Variance ($sigma$)	$S_t = 1\ sigma = 0.0136$		$S_t = 2\ sigma = 0.0699$		
	p_{1t}	p_{2t}	Duration	Observation	
Regime 1	0.94	0.36	54.87	100	
Regime 2	0.06	0.64	103.73	185	

Note *, **, and *** denote rejections of the null hypothesis at the 10 %, 5 %, and 1 % significance levels, respectively

growth regime, or the ordinary period, while regime 2 is in the low growth regime, or the turbulent period. In addition, the structure of the model also has an interpretation. The movement of the variables between regimes is governed by the variance of the model.

Furthermore, considering all the stock market equations, the autoregressive parameter, we can see that these do not cause the changes in the economic condition between the two regimes. Thus these variables are regime-independent and their values are not different between the two regimes. Taking into account the QE effectiveness, we can see that the growth rate of the Jakarta Composite Index seems to be significantly driven by the treasury securities purchase program of the U.S., whereas the reaction of the Jakarta Composite Index is negative in the first lag period of the treasury securities purchase program. However, the reactions of other stock markets are not found to be significantly driven by the QE program. These results indicate that the impact of the U.S. QE program is not big on the ASEAN stock markets, except in the case of the Jakarta Composite Index. According to the MundellFleming approach, an increase in money supply will reduce the U.S. interest rate lower than the Indonesian interest rate, so the hot money flows out to take advantage of the higher interest rate in Indonesian financial markets, thereby increasing the Jakarta Composite Index. However, an expansion of the U.S. treasury securities will not cause the Jakarta Composite Index to switch the regime. The regime-switching of the ASEAN-5 stock markets is governed by shocks in ASEAN-5 stock markets and QE programs such as the financial crises and monetary shocks. In other words, any shock in the ASEAN-5 stock markets and the QE program will lead to the probability of regime-varying over time.

Table 5 presents the final MSH(2)-VAR(1) model for ASEAN-5 exchange markets. It is similar to the ASEAN-5 stock markets model. We interpret the regime by considering the variance in each regime. The first regime indicates that the value of variance in the first regime is less than that in the second regime. Thus, this indicates that regime 1 is in the high growth regime, or an expansion of QE and currency appreciation, while regime 2 is in the low growth regime, or a contraction of QE and currency depreciation. The result significantly reveals the existence of some relationship between the QE program and the Singapore exchange market. As is evident from Table 5, it also displays the estimated coefficient in the first lag term. We can see that a change in the autoregressive parameters does not have any influence on the changes in the economic condition between the two regimes. As far as QE effectiveness is concerned, we can see that the Singapore Dollar returns seem to be significantly driven by the U.S. treasury securities purchase program, whereas the reaction of the Singapore Dollar is negative in the first lag period of the U.S. treasury securities purchase program. However, the reactions of other currencies are not found to be significantly driven by the QE program. This means that the expansion of the QE program will cause the capital to flow out to Singapore. Consequently, this will put significant upward pressure on the Singapore Dollar. By contrast, in the case of the Thai Baht, the Philippine Peso, the Indonesian Rupiah, and the Malaysian Ringgit, their capital inflow is not sufficient to cause variations in their currencies because their central banks stand ready to intervene in the foreign exchange market

Table 5 Estimate MSH-VAR(1) model of ASEAN-5 exchange markets

	REXth	REXind	REXphp	REXsin	REXmal
Regime-independent intercepts					
Regime 1	−0.0001***	0.0004***	−0.0002***	0.0002***	−0.0005***
Regime 2	−0.0001***	0.0004***	−0.0002***	0.0002***	−0.0005***
Regime-independent, Autoregressive parameters at lag 1					
RTS	−0.008	−0.0053	0.035	−0.0001**	0.004
RMBS	0.001	−0.0017	0.001	−0.0002	−0.0001
REXth	−0.004	−0.0498	0.074	−0.124	−0.081
REXind	0.035	0.137***	−0.030	−0.023	−0.027
REXphp	−0.025	0.111	−0.121	0.0005	0.005
REXsin	0.004	−0.167*	−0.027	−0.115	0.082
REXmal	0.201***	0.324***	0.207	0.396***	0.020***
Log-likelihood = 6427.38	AIC (10^3) −12.724		BIC (10^3) −12.361		
Variance (*sigma*)	$S_t = 1$ $sigma = 0.0063$		S_t=2 $sigma$=0.0690		
	p_{1t}	p_{2t}	Duration	Observation	
Regime 1	0.96	0.33	25.95	253	
Regime 2	0.04	0.67	3.06	32	

Note *, **, and *** denote rejections of the null hypothesis at the 10 %, 5 %, and 1 % significance levels, respectively

when the volatility of the exchange rate is at a level worse than the level expected by the economy. These results are consistent with the suggestion of Cho and Rhee [5] that most of Asia's exchange rates do not respond to the variations in the QE program, except for the currency of Singapore where the financial market follows the policy of full openness. Singapore is the 4th major financial market in the world thus Singapore will have high potential in attracting substantial capital inflow. In addition, Bank Negara Malaysia (2013) reported that the U.S. QE could increase the Singapore Dollar exchange rate by 21.1.

Furthermore, the structure of the model also has an interpretation. The movement of the variables between regimes is governed by the variance of the model. ASEAN-5 exchange markets are governed by shocks in the ASEAN-5 exchange markets and the QE program, such as financial crises and monetary shocks. In other words, any shock in the ASEAN-5 exchange markets will lead to regime-switching.

Table 6 provides the final MSIH(2)-VAR(1) model for ASEAN-5 bond markets. We interpret the regime by considering the variance in each regime. The first regime demonstrates that the value of variance in the first regime is less than that in the second regime. Accordingly, this indicates that regime 1 is in the high growth regime, or an expansion of the QE program and ASEAN bond markets (price rise causes yield fall), while regime 2 is in the low growth regime, or that the QE program remains in the

Table 6 Estimate MSIH(2)-VAR(1) model of ASEAN-5 bond markets

	RTHY	RINDY	RPHY	RSY	RMY
Regime-independent intercepts					
Regime 1	−0.00002***	0.0004***	−0.0034***	0.0035***	−0.0002***
Regime 2	0.0005***	−0.0002***	0.0018***	−0.001***	−0.0001***
Regime-independent, Autoregressive parameters at lag 1					
RTS	0.006	0.084	0.1	0.169	0.236***
RMBS	−0.001	−0.003	0.001	0.001	−0.008***
RTHY	0.107*	0.092	0.231***	0.274	0.035
RINDY	0.071	−0.07	0.0136	−0.031	−0.058
RPHY	0.015	−0.068	−0.137**	−0.022	0.070**
RSY	−0.0421	0.012	−0.020	−0.157**	0.043
RMY	−0.064	0.223**	−0.035	−0.204	0.0579
Log-likelihood = 4324.85	AIC (10^3) −8.491		BIC (10^3) −8.050		
Variance ($sigma$)	$S_t = 1\ sigma = 0.0173$		$S_t = 2\ sigma = 0.0869$		
	p_{1t}	p_{2t}	Duration	Observation	
Regime 1	0.94	0.58	17.31	269	
Regime 2	0.06	0.42	1.72	17	

Note *, **, and *** denote rejections of the null hypothesis at the 10 %, 5 %, and 1 % significance levels, respectively

same level and a contraction of ASEAN bond markets (price fall causes yield rise). These findings provide that the U.S. mortgage-backed securities purchase program and the U.S. treasury securities purchase program have statistical effect on Malaysian bond yield returns. However, there are heterogeneous effects, as far as the QE program is concerned. The results indicate that the U.S. treasury securities purchase program has a positive effect on Malaysian bond yield returns, while the U.S. mortgage-backed securities purchase program has a negative effect on Malaysian bond yield returns. It is quite surprising that the expansion of the U.S. treasury securities purchase program leads to an increase in the Malaysian bond yield returns. However, this effect may be dependent on the exchange rate regime, degree of financial openness, financial linkage, and size of bond market [16]. The percentage of foreign bond holders in the Malaysian bond market is only 31.2 %. Thus, evidently, the Malaysian bond market is affected more by domestic monetary policy than by foreign policy. In addition, the structure of the model also has an interpretation. The movement of the variables between regimes is governed by the variance and the intercept of the model. So, this indicates that a shock in the ASEAN-5 bond markets and the other factor will lead to regime-switching.

3.4 Regime Probabilities of Individual Financial Markets

The time path of the smoothed probabilities of stock markets, exchange markets, and bond markets are presented in Figs. 1, 2, and 3, respectively. The smoothed probabilities of regime 1 are presented using the solid line, while the smoothed probabilities of regime 2 are presented using the dashed line. The smoothed probabilities stand for the optimal inference of the regime conditional on full observation [2].

Figures 1, 2, and 3 provide the regime probabilities of three financial markets which seem to have similar results. The results of the regime probabilities in the three financial markets are observed to be corresponding to each other. The models are able to capture the periods of both high growth regime and low growth regime. The high growth regime and the low growth regime represent different financial outcomes. Regime 1 of the ASEAN-5 financial markets model is plotted in the top panel of each figure. These regimes are mainly observed in the period of QE implementation, from

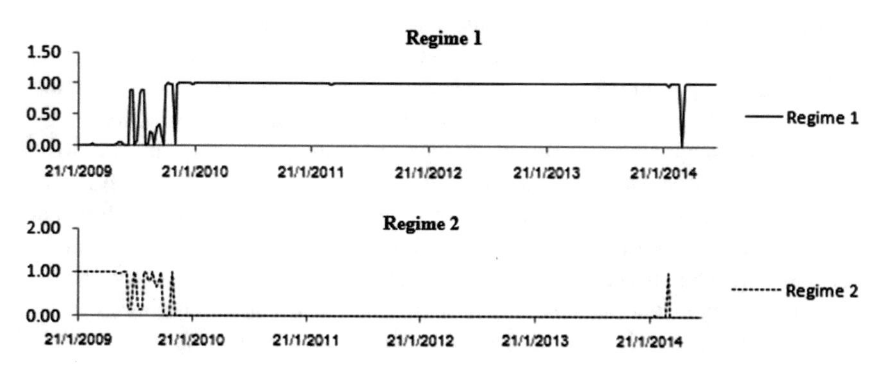

Fig. 1 The regime probabilities of the ASEAN-5 stock markets: regime 1 (*solid line*); regime 2 (*dashed line*)

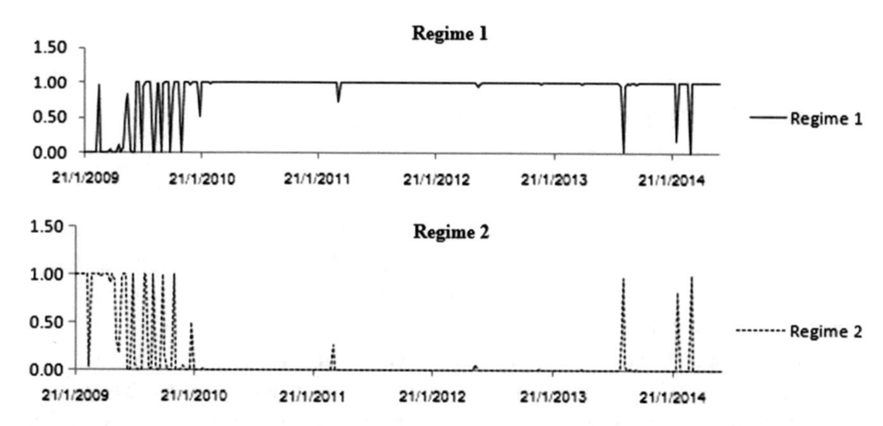

Fig. 2 The regime probabilities of the ASEAN-5 exchange markets: regime 1 (*solid line*); regime 2 (*dashed line*)

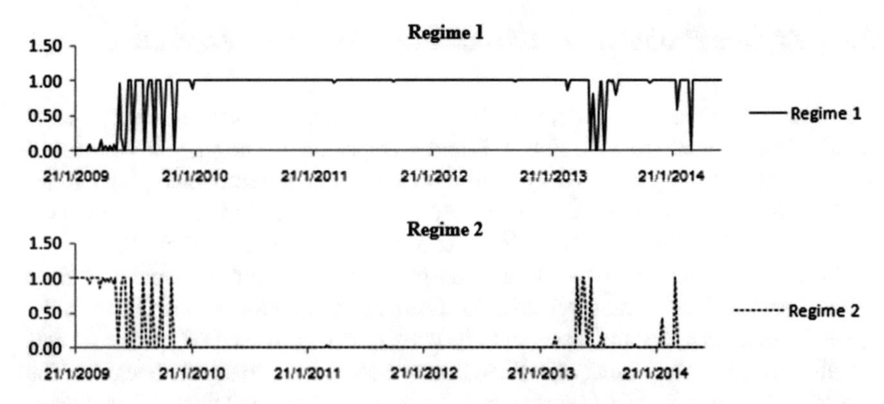

Fig. 3 The regime probabilities of the ASEAN-5 bond markets: regime 1 (*solid line*); regime 2 (*dashed line*)

late 2009 to 2013, which consists of the statement from the Federal Open Market Committee (FOMC), signaling the expansion of the QE program, as mentioned by the chairman, Bernanke, and the Fed announcement of the purchasing bond. The bottom panel of each figure contains regime 2 which seems to observe the period of QE tapering in the beginning of 2014. In addition, the results also show that ASEAN-5 financial markets seem to stay in regime 1 rather than in regime 2. Thus, the ASEAN-5 financial markets have the probability of staying in the high growth regime than in the low growth regime. Moreover, the estimation of the transition matrix for the ASEAN-5 financial markets, as presented in Tables 4, 5, and 6, also confirms that for the ASEAN-5 financial markets, regime 1 is more persistent than regime 2. These results indicate that a shock from the QE program can switch ASEAN-5 financial markets to change between these two regimes. Furthermore, it is evident that the duration of stay in the high growth regime is longer than that in the low growth regime.

3.5 Impulse Response Functions of Individual Financial Markets

The impulse response functions are presented for the best specification of the Markov-switching regime of three ASEAN-5 financial markets. The impulse response functions provide the response of the three financial markets to a shock of the QE program in each regime, as illustrated in Figs. 4, 5, and 6 in this chapter.

In this section, we focus only a shock of those variables which have significance, by following the results presented in Tables 3, 4, and 5. Figure 4 provides the feedback of the QE program in two different regimes. In the high growth regime, the ASEAN-5 stock markets, a shock in RTS has a great and persistent positive effect on RJKSE, which reaches the maximum after about 2 weeks and then falls sharply and reaches

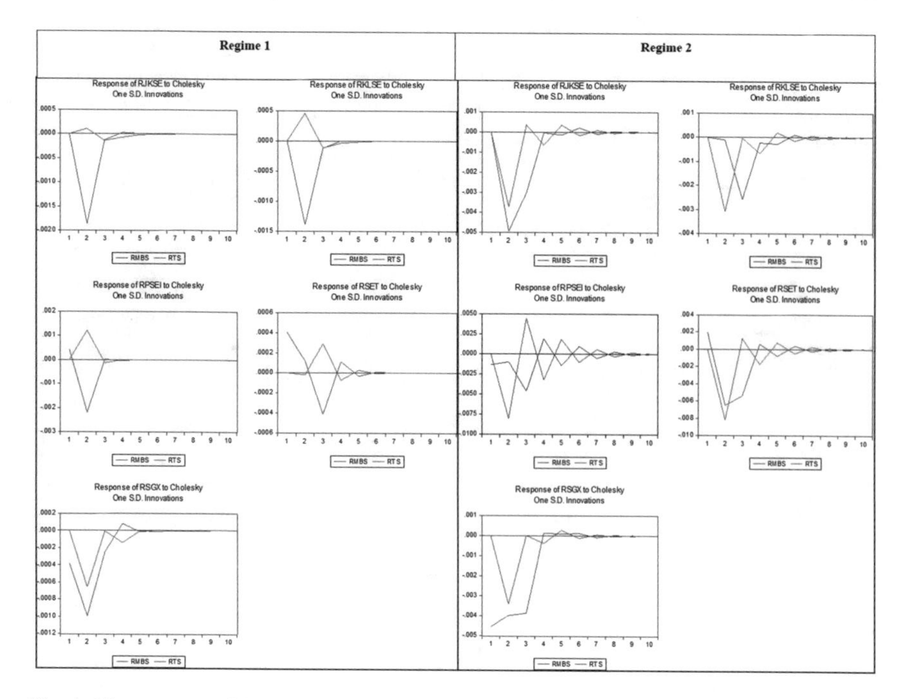

Fig. 4 The response of the ASEAN-5 stock markets to a shock of the U.S. QE program

the steady state within 4 weeks. In the case of the ASEAN-5 exchange markets, as presented in Fig. 5, the result shows that RTS created an initial negative response in RExsin and then moved to equilibrium within 6 weeks. In the case of the ASEAN-5 bond markets, RMY is observed to have risen due to shocks in RTS and RMBS.

In the low growth regime, the ASEAN-5 stock markets, a shock in RTS is found to have a negative effect on RJKSE, which reaches the minimum after 2 weeks and then rises to the steady state within 8 weeks. In the case of the ASEAN-5 exchange markets, a shock in RTS is observed to have a negative effect on RExsin. It leads RExsin to decrease (appreciation) for about 3 weeks, following which it begins to increase (depreciation) to reach equilibrium within 10 weeks. Finally, in the case of the ASEAN-5 bond markets, a shock in RTS creates a positive sharp response in RMY and it then falls to equilibrium in 10 weeks. By contrast, a shock in RMBS creates an impact in the opposite direction from the response to RMY. It causes RMY to initially decrease and then move to equilibrium within 10 weeks.

4 Concluding Remarks

We learned that when the Fed started the quantitative easing (QE) program, various effects were brought on upon the ASEAN-5 financial markets. However, the effects

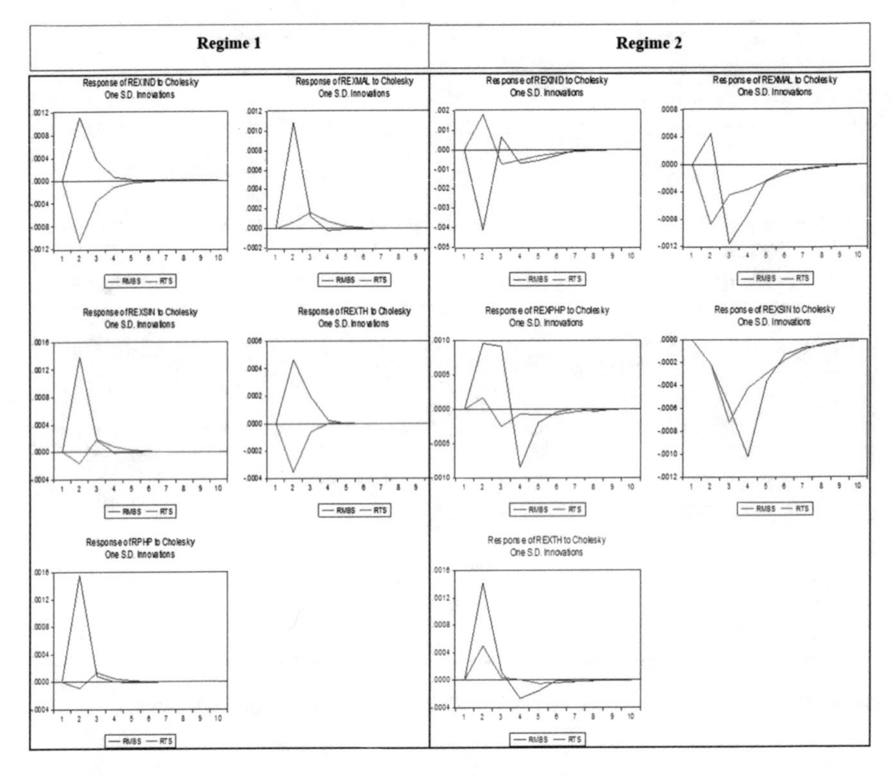

Fig. 5 The response of the ASEAN-5 exchange markets to a shock of the U.S. QE program

of the QE program are likely to occur in some countries. The estimated results show that the treasury securities purchase program appears to be the main impact which drives the ASEAN-5 financial markets. Moreover, the best-fitting MSH(2)-VAR(1) for stock and exchange markets and MSIH(2)-VAR(1) for bond markets seems to have an economic interpretation. The regime shifts in the stock markets as well as the exchange markets are governed by the shock of the QE program and their own shocks, while the regime shifts in bond markets are governed by other factors and their own shocks. This indicates that regime-switching in the ASEAN-5 financial markets is not affected by the QE movement. Furthermore, these models produce regime probabilities in each of the regimes. Similar results are obtained from individual models. Positive effects of the QE program on the ASEAN-5 financial markets, including the announcement of the QE program, the expansion of the purchasing programs, and a positive signal from FOMC, are all placed within the high growth regime. On the other hand, severe events, such as a flooding disaster, speculative shock in financial markets, and QE tapering, are placed in the low growth regime. The results obtained regarding regime probabilities present the ASEAN-5 financial markets as having the possibility to stay in the high growth regime than in the low growth regime. Finally, the impulse response results present the response of the

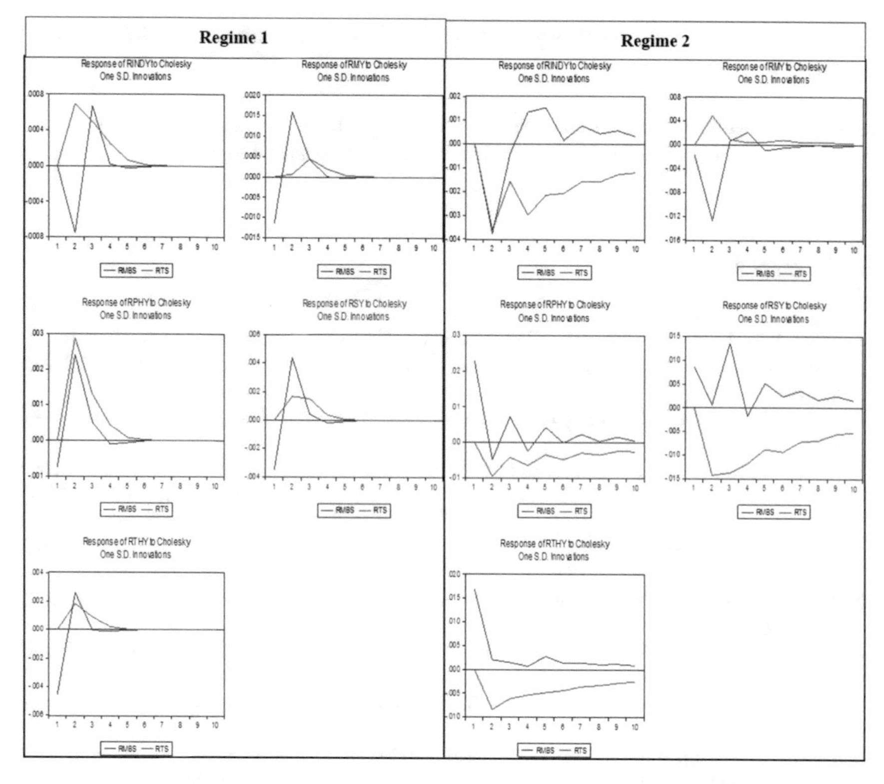

Fig. 6 The response of the ASEAN-5 bond markets to a shock of the U.S. QE program

ASEAN-5 financial markets, especially the Jakarta Composite Index, the Singapore Dollar, and the Malaysian government bond yield, to shocks of the QE program. The impulse response indicates that a shock of QE can affect these markets and linger for a long period, in particular the Singapore Dollar and the Malaysian government bond. Based on the results, we suggest that QE have a direct substantial effect on the ASEAN-5 financials markets. Therefore, if the Fed withdraws the QE program, the move might have an effect on the TIP financial markets. In particular, the treasury securities purchase program is more likely to have an effect on the ASEAN-5 financial markets than any other programs; therefore, the ASEAN-5 governments, the ASEAN-5 central banks, and investors should lay emphasis on the treasury securities purchase program and particularly consider matters with extreme caution when a shock occurs because a shock of the QE program can lead the ASEAN-5 financial markets to switch from one regime to another regime. This work should be regarded as a primary attempt in quantifying the potential of quantitative easing and quantitative tapering on the ASEAN-5 financial markets with regime-switching. Future research in this field can expand the scope of analysis of the impact of QE on other countries with factors such as oil price, interest rates, and gold price, as well as perform counterfactuals based on actual realizations of QE tapering by following its implications.

Acknowledgments The authors are grateful to Puay Ungphakorn Centre of Excellence in Econometrics, Faculty of Economics, Chiang Mai University for the financial support.

References

1. Albu, L.L., Lupu, R., Clin, A.C., Popovici, O.C.: The effect of ECB's quantitative easing on credit default swap instruments in central and Eastern Europe. Proc. Econ. Financ. **8**, 122–128 (2014)
2. Artis, M., Krolzig, H.M., Toro, J.: The European business cycle. Oxf. Econ. Pap. **56**(1), 1–44 (2003)
3. Bowman, D., Cai, F., Davies, S., Kamin, S.: Quantitative easing and bank lending: evidence from Japan (Board of Governors of the Federal Reserve System Working Papers 1018) (2011)
4. Chen, Q., Filardo, A., He, D., Zhu, F.: International spillovers of central bank balance sheet policies (BIS Working Papers 66). Bank for International Settlements, Basel, Switzerland (2012)
5. Cho, D., Rhee, C.: Effects of quantitative easing on Asia: capital flows and financial markets (ADB Working Papers 350) (2013)
6. Christensen, J.H., Gillan, J.M.: Does quantitative easing affect market liquidity? (FRBSF Working Papers 2013-26). Federal ReseeveBank of San Francisco, San Francisco (2013)
7. Chua, W.S., Endut, N., Khadri, N., Sim, W.H.: Bank Negara Malaysia Working Paper Series WP3/2013 (2013)
8. Ehrmann, M., Ellison, M., Valla, N.: Regime-dependent impulse response functions in a Markov-switching vector autoregression model. Econ. Lett. **78**(3), 295–299 (2003)
9. Glick, R., Leduc, S.: The effects of unconventional and conventional US monetary policy on the dollar. Federal Reserve Bank of San Francisco (2013)
10. Girardin, E., Moussa, Z.: The effectiveness of quantitative easing in Japan: new evidence from a structural factor-augmented VAR (2008)
11. Fawley, B.W., Neely, C.J.: Four stories of quantitative easing. Federal Reserve Bank of St. Louis Review, 95 (2013)
12. Krishnamurthy, A., Vissing-Jorgensen, A.: The effects of quantitative easing on interest rates: channels and implications for policy (NBER Working Papers 17555). National Bureau of Economic Research, Cambridge (2011)
13. Krolzig, H.M.: Markov switching vector autoregressions: modelling, statistical inference and application to business cycle analysis. Springer, Berlin (1997)
14. Krolzig, H. M.: Constructing turning point chronologies with Markov-switching vector autoregressive models: the euro-zone business cycle. In: Eurostat Colloquium on Modern Tools for Business Cycle Analysis (2003)
15. Ledenyov, D.O., Ledenyov, V.O.: To the problem of turbulence in quantitative easing transmission channels and transactions network channels at quantitative easing policy implementation by central banks. (Working Paper 5656). University of Cornell (2013)
16. Moore, J., Nam, S., Suh, M., Tepper, A.: Estimating the impacts of US LSAP's on emerging market economies' local currency bond markets (No. 595). Staff Report, Federal Reserve Bank of New York (2013)
17. Morgan, P. J.: Impact of US quantitative easing policy on emerging Asia (ADBI working paper 321). Asian Development Bank Institute, Tokyo (2011)
18. Vargas, G.A., I., II.: Markov switching var model of speculative pressure: an application to the asian financial crisis. (Order No. 1483224, Singapore Management University (Singapore)). ProQuest Dissertations and Theses, 72 (2009)

19. Zammit, R.: Japanese quantitative easing: the effects and constraints of anti deflationary monetary expansions (2006)
20. Zhu, L., Yang, X.: The Study of American Quantitative Easing Monetary Policy's Spillover Effects on China's Inflation. Atlantis Press, Paris (2013)
21. Zivot, E., Wang, J.: Modeling Financial Time Series with S-PLUSv (Vol. 191). n.p. Springer (2007)

Dependence Structure of and Co-Movement Between Thai Currency and International Currencies After Introduction of Quantitative Easing

Pathairat Pastpipatkul, Woraphon Yamaka and Songsak Sriboonchitta

Abstract We analyze the dependence relationship between the Thai currency and international currencies after the introduction of quantitative easing (QE). The daily currency exchange rates of Thailand, European countries, Great Britain, Japan, Indonesia, the Philippines, Singapore, and Malaysia during 2009–2014 are applied in this study. We proposed a Markov-switching dynamic copula approach to test the co-movement between the exchange rates and the Thai Baht. The results show that there is a dependence relationship between the Thai Baht and the other currencies except in the case of the Great British Pound. Additionally, we also found that a high dependence regime has higher volatility than a low dependence regime.

Keywords Markov switching · Dynamic copula · Thai currency · International currencies

1 Introduction

Thailand is a newly industrialized country in Southeast Asia which has a largely investment-dependent economy, with investment accounting for more than 30 % of its gross domestic product (GDP). Thailand has been having ties with other emerging and developed economies, including the United States, Europe, Japan, and ASEAN countries through trade, foreign direct investment, and foreign portfolio investment. After the 2008 financial crisis in the United States, the U.S. Federal Reserve (Fed) proposed an unconventional policy known as quantitative easing (QE). A substantial volume of U.S. capital began inflow into the rest of the world, in particular to Europe and Asia, thereby increasing the fluctuation in those currencies.

Because of the above-mentioned reasons, the movement of the Thai-Baht currency seem to be effected by the movement of other currencies. Thus, in this study, we aim

P. Pastpipatkul (✉) · W. Yamaka · S. Sriboonchitta
Faculty of Economics, Chiang Mai University, Chiang Mai, Thailand
e-mail: ppthairat@hotmail.com

W. Yamaka
e-mail: woraphon.econ@gmail.com

V.-N. Huynh et al. (eds.), *Causal Inference in Econometrics*,
Studies in Computational Intelligence 622, DOI 10.1007/978-3-319-27284-9_36

to study the co-movement between international currencies vis–vis the Thai currency and assess the degree of the exchange rate co-movement between the Thai baht and the currencies of three developed countries, comprising the British Pound Sterling, the Euro, and the Japanese Yen, in addition to the co-movement between the Thai Baht and the currencies of four emerging ASEAN economies, comprising the Malaysian Ringgit, the Indonesian Rupee, the Singapore Dollar, and the Philippines Peso, after the financial crisis in the U.S. in 2008.

As of today, there have been many studies that attempted to assess the relationship between the exchange rates in various countries, using different methods. For instance, Choudhry [5] used the cointegration method and found a strong relationship between the Thai Baht, the Malaysian Ringgit, the Indonesian Rupee, the Korean Won, and the Japanese Yen; using the GARCH approach, Kuhl [9] found out about the dependence volatility between the Euro and the British Pound, while Han Kim, Kim, and Ghi Min [8] found a close link between the Japanese Yen and the currencies of emerging Asian economies. By contrast, there were some empirical studies which could not find a relationship between exchange rates. For instance, using the cointegration and error correction model, Woo [20] found that there was no significant relationship between the Deutsche Mark and the currencies of other major European countries. Bohdalova and Gregus [3] did not find any relationship between the Polish Zloty and the Hungarian Forint. Thus, we can see that different results were obtained in different fields of study and that some could not capture the relationship between the exchange rates. Zhao and Lin [21] suggested that the conventional techniques of evaluating relationships have limitations as regards estimation due to their assumption of linear relationship and normal distribution. Moreover, these methods do not distinguish between the negative and the positive elements in the data and cannot capture the tail dependence. Thus, the results obtained using these conventional models might not capture the correlation of some time series data.

Since the introduction of copulas, and Sklars theorem, copulas have been widely developed and have become a popular innovation tool for measuring the dependency of financial variables including tail dependence. The main feature of this model is distinguished to be the upper tail and the lower tail of the joint distribution among a set of random variables. In addition, it captures the degree and the structure of dependence [12]. Additionally, Patton [13] extended Sklars theorem to define conditional copulas and also proposed copulas with the time-varying factor. Recently, there have been many studies carried out that employed the copula model to measure the dependence between exchange rates, for example, the studies conducted by Patton [12], Antonakis [1], Li [10], Jowaheer and Ameerudden [7], Michelis and Ning [11] and Azam [2]. The results of their studies demonstrated the existence of tail dependence and described a non-linear dependence between exchange rates. This means that copula models are more suitable for capturing the dependence between random variables than conventional models.

As mentioned above, it is evident that copula models has a potential to capture the complicate dependence. Nevertheless, many studies provide a different degree of dependence. This indicates that there are two regimes that may exist high dependence regime and low dependence regime in the dependence between exchange rates. Thus, in this study, we combine a time-varying copula with a regime-switching model, resulting in Markov-switching a dynamic copula which captures asymmetry in the two regimes of dependence, namely high dependence regime and low dependence regime.

Recently, we have observed the emergence of works combining the regime-switching copula models and copulas. Samitas, Kenourgios, and Paltaloidis [14], using multivariate time-varying copulas with Markov-switching, captured the dependence between Brazil, Russia, India, and China (BRICs) stock markets and the United States and United Kingdom stock markets. As far as bivariate models are concerned, Silva Filho and Ziegelmann [17], used dynamic copulas with Markov-switching for the pair of BOVESPA (Brazil and S&P 500 [USA] and FTSE100 [UK], respectively), whereas Fei, Fuertes, and Kalotychon [6] modeled the dependence structure between the iTraxx Europe CDS market and the stock market. However, few studies have investigated the co-movement between the Thai Baht and the currencies of other developed and emerging economies. One of the exceptions is the study carried out by Chokethawon, Chaitip, Sriwichailamphan, and Chaiboosri [4] which, again, employed the Markov-switching dynamic copula approach in order to study the co-movement between the Thai Baht and the Malaysian Ringgit. However, this study has extended their work by capturing the dependence of other currencies. The remainder of the paper is as follows. In the next section, we briefly discuss the basic concepts of copulas and dependence measure, and present the copulas with Markov-switching. In Sect. 3, we describe the data-and-estimation procedure. The estimated results and the investigation of the dynamic dependence structure of each regime are presented in Sect. 4. Section 5 provides the conclusion of this work.

2 Methodology

2.1 Basic Concepts of Copula

According to Sklars theorem [18], any joint distribution $H(x_1, \dots, x_n)$ with marginal distribution $F_1(x_1), \dots, F_n(x_n)$, can be present as follows:

$$H(x_1, \dots, x_n) = C(F_1(x_1), \dots, F_n(x_n)) \quad (1)$$

where $H(x_1, \dots, x_n)$ are the random variable, $F_1(x_1), \dots, F_n(x_n)$ are uniform in the [0,1] and C is copula distribution function of n a marginal distribution. For example, in the bivariate case, the copula density c can be written as

$$c(F_1(x_1), F_2(x_2)) = \frac{h(F_1^{-1}(u\_1), F_2^{-1}(u\_2))}{\prod_{i=1}^{2} f\_i(F_1^{-1}(u\_i))} \quad (2)$$

where

h the density function associated with
f_i the density function of each marginal distribution
c the copula density
u_i random variable which marginally uniformly distribution

In this work, we aim to measure the dynamic dependence (i.e., allowed time-varying) in the copula. Patton [13] extended Sklars theorem to allow for the conditioning of the random variables and suggested allowing the dependence parameter (θ) to vary over time in the ARMA (p,m) process, as follows:

$$\theta_{ct} = \Lambda(\omega + \beta_{ct}\theta_{ct-p} + \alpha_{ct}\Gamma_t) \quad (3)$$

where $\Lambda(\bullet)$ is the logistic transformation for each copula function to keep the parameter in its value all the time, *omega* is intercept term $beta_c t$ at time t is coefficient of dependence parameter at time $t - p$ (θ_{ct}) and is the forcing variable at time t which is defined as

$$\Gamma_t = \begin{matrix} \frac{1}{m}\sum_{j=1}^{m} F_1^{-1}(u_{1,t-j})F_2^{-1}(u_{2,t-j}) & \textit{elliptical} \\ \frac{1}{m}\sum_{j=1}^{m} \left|(u_{1,t-j} - u_{2,t-j})\right| & \textit{Archimedean} \end{matrix} \quad (4)$$

In the Elliptical case, the Gaussian (normal) and student-t copulas, is a quadratic form in copula function which has a symmetrical dependence. The dependence structure related to these copula families is the Pearsons correlation which has the value of its parameter in the interval [18].

In the Archimedean case, for example, the Gumbel, Joe, Clayton, and symmetrized Joe Clayton copulas, they are a copula function form which is mostly used in modeling the dependence structure of financial time series and has an asymmetrical dependence.

There are many functional forms that can be used as copulas. However, in this study, we propose five copula functions with bivariate data to analyze the structure dependence between Thai exchange rate and other exchange rates, as follows.

1. Normal copula

Normal copula is a symmetric function and has no tail dependence. It is defined as follows:

$$C_N(u_1, u_2|p)$$
$$= \int_{-\infty}^{\Phi^{-1}(u_t)} \int_{-\infty}^{\Phi^{-1}(u_2)} \frac{1}{2\pi\sqrt{1-p^2}} exp\left\{-(\frac{r^2 - 2\rho rs + s^2}{2(1-\rho^2}\right\} drds, \rho \subseteq (-1, 1)$$

where ρ is a dependence parameter which is the linear correlation coefficient. Thus, its dynamic equation can be written as

$$\rho_t = \Lambda(\omega_N + \beta_N\rho_{t-1} + \alpha_N \cdot \frac{1}{m}\sum_{j=1}^{m} \Phi^{-1}(u_{1,t-j})\Phi^{-1}(u_{2,t-j})$$

2. student-t copula
Similar to normal copula, student-t Copula has a linear correlation coefficient and has symmetrical tail dependence. However, it can capture some tail dependence. In bivariate cases, it has the following function form:

$$C_T(u_1, u_2|p, v) = \int_{-\infty}^{t_v^{-1}(u_t)} \int_{-\infty}^{t_v^{-1}(u_t)} \frac{1}{2\pi\sqrt{1-p^2}} \left\{1 + (\frac{r^2 - 2\rho rs + s^2}{v(1-\rho^2}\right\}^{\frac{v+2}{2}} drds$$

where ρ and v are linear dependence parameter and degree of freedom, respectively. In addition, its dynamic equation can be written as

$$\rho_t = \Lambda(\omega_{1T} + \beta_{1T}\rho_{t-1} + \alpha_{1T} \cdot \frac{1}{m}\sum_{j=1}^{m} T_v^{-1}(u_{1,t-j})T_v^{-1}(u_{2,t-j})$$

$$v_t = \Lambda(\omega_{2T} + \beta_{2T}\rho_{t-1} + \alpha_{2T} \cdot \frac{1}{m}\sum_{j=1}^{m} T_v^{-1}(u_{1,t-j})T_v^{-1}(u_{2,t-j})$$

3. Gumbel copula
Gumbel copula is an asymmetrical tail dependence copula and has only upper tail dependence. Its functional form can be written as follows:

$$C_G(u_1, u_2|\theta) = exp(-((-logu_1)^\theta + (-logu_2)^\theta)^{\frac{1}{\theta}}), \quad \theta \subseteq [1, \infty]$$

and the dynamic equation is given as follows:

$$\theta_t = \Lambda(\omega_G + \beta_G\theta_{t-1} + \alpha_G \cdot \frac{1}{m}\sum_{j=1}^{m} |u_{1,t-j} - u_{2,t-j}|)$$

4. Clayton copula
Similar to the Gumbel copula, Clayton copula has asymmetrical tail dependence. However, it shows only lower tail dependence. Its functional form can be written as follows:

$$C_C(u_1, u_2|\delta) = (u_1^{-\delta} + u_2^{-\delta} - 1)^{\frac{-1}{\delta}}, \delta \subseteq [0, \infty]$$

and the dynamic equation is given as follows:

$$\delta_t = \Lambda(\omega_{Cl} + \beta_{Cl}\delta_{t-1} + \alpha_{Cl} \cdot \frac{1}{m}\sum_{j=1}^{m} |u_{1,t-j} - u_{2,t-j}|)$$

5. Symmetrized JoeClayton copula (SJC)
The symmetrized JoeClayton (SJC) copula function form can be written as follows:

$$C_{SJC}(u_1, u_2|\lambda_U, \lambda_L) =$$
$$0.5 \cdot C_JC(u_1, u_2|\lambda_U, \lambda_L) + C_JC(1 - (1 - u_t, 1 - u_2|\lambda_U, \lambda_L) + u_1 + u_2 - 1)$$

where C_JC is the JoeClayton copula, known as the BB7 copula.

The SJC copula has both upper tail and lower tail dependence, and its dependence parameters are λ_U and λ_L which are the measures of dependence on the upper tail and lower tail, respectively. In addition, λ_U and λ_L range freely and are not dependent each on other. Its dynamic function form can be written as

$$\lambda_{Ut} = \Lambda(\omega_U + \beta_U \lambda_{Ut-1} + \alpha_{SJCU} \cdot \frac{1}{m} \sum_{j=1}^{m} |u_{1,t-j} - u_{2,t-j}|))$$

$$\lambda_{Lt} = \Lambda(\omega_L + \beta_L \lambda_{Lt-1} + \alpha_{SJCL} \cdot \frac{1}{m} \sum_{j=1}^{m} |u_{1,t-j} - u_{2,t-j}|))$$

While comparing these copula functions with specific dependence structures, it is impossible to compare different functional forms of the copulas. However, Silva Filho and Ziegelmann [15] and Filho, Ziegelmann, and Dueker [16] suggested considering tail dependence measurement, which is defined as follows.

2.2 Copula-GARCH Model

In the bivariate case of the copula-GARCH model, a two-dimensional time series vector, we can write the density function as follows:

$$H(x_t|u_t, h_t) = C_{\theta ct}(F_1(x_{1t}|u_1, h_{1t}), F_2(x_{2t}|u_2, h_{2t}) \tag{5}$$

where $C_{\theta ct}$ is the copula function with the time-varying dependence parameter θ_{ct}. $F_1(x_{1t}|u_1, h_{1t})$ and $F_2(x_{2t}|u_2, h_{2t})$ are the univariate marginal distributions for and which are specified by the GARCH process.

The decomposition in Eq. (5) suggests that we have to specify a marginal distribution via the GARCH process before estimating the copula. By following Tofoli, Ziegelmann, and Silva Filho [19], a univariate ARMA(p,m)-GARCH(q,n) specification is employed to model the marginal distribution. It can be written as follows:

$$y_t = \phi_0 + \sum_{i=1}^{p} \phi_i y_{t-i} + \sum_{j=1}^{m} \theta_j \mathbb{E}_{t-j} + \mathbb{E}_t \tag{6}$$

$$\mathbb{E}_t = h\eta_t \tag{7}$$

$$h_t^2 = \alpha_0 + \sum_{i=1}^{q} \alpha_i \mathbb{E}_{t-i} + \sum_{j=1}^{m} \beta_j h_{t-j}^2 \tag{8}$$

where Eqs. (6) and (8) are the mean and variance equations, respectively. $\mathbb{E}_t$ is a residual, and η_t is innovation process. Then, the best-fit marginal distribution provides a standardized ARMA-GARCH residual which has, then, been transformed into uniform [0,1].

2.3 Regime-Switching Copulas

In the financial time series, the dependence structure of the variable is influenced by a hidden Markov chain with two states [21]. Thus, we propose a Markov-switching copula with time-varying where the dependence parameter θ_{ct} is governed by an unobserved variable (S_t).

In the Markov-switching copula framework, let S_t be a state variable which is assumed to have two states (k = 2), which are the high dependence regime and the low dependence regime. Thus, the joint distribution of x_{1t} and x_{2t} conditional on S_t can be defined as

$$(x_{1t}, x_{2t}|x_{1,t-1}, x_{2,t-1}; S_i = i) - c_t^{S_t}(u_{1t}, u_{2t}|u_{1,t-1}, u_{2,t-1}; \theta_t^{S_t}), \qquad i = 1, 2$$

where $i = 1$ denotes the low dependence regime and $i = 1$ denotes the high dependence regime.

Additionally, the dependence parameter follows the ARMA[1,10] process as proposed in Silva Filho and Ziegelmann [15], with the intercept term switching according to the first order Markov chain with two regimes (k = 2); thus, Eq. (3) can be rewritten as follows:

$$\theta_{ct} = \Lambda((\omega_0(1 - S_t) + \omega_1 S_t) + \beta\theta_{ct-1} + \alpha\Gamma_t \tag{9}$$

The transition probabilities (P)

$$P_{ij} = Pr(S_{t+1} = j|S_i = i) \;\; and \;\; \sum_{1}^{k=2} P_{ij} = 1 \quad i, j = 1, 2 \tag{10}$$

where P_{ij} is the probability that regime i is followed by regime j, and it is convenient to collect the transition probabilities in the transition matrix Q

$$Q = \begin{bmatrix} P_{11} & P_{12} \\ P_{21} & P_{22} \end{bmatrix} \tag{11}$$

Let sum above, the general form of the Markov-switching copula model is specified as follows:

$$C_{S,t}(u_{1t}, u_{2t}|S_t, \theta_{cS,t}) = \sum_{k=1}^{2} 1_{S_{\{t=k\}}} \cdot C_{kt}(u_{1t}, u_{2t}|\theta_{ckt}) \tag{12}$$

with $\theta_{ckt} = \Lambda(\omega_k + \beta_{ckt}\theta_{ckt-1} + \alpha_{ckt} \cdot \frac{1}{m}\sum_{j=1}^{m} |u_{1,t-j} - u_{2,t-j}|)$, where $C_{kt}(\cdot)$ is the copula function with time-varying dependence parameters (θ_{ckt-p})at regime k.

3 Estimation

In this study, we focus our interest on the bivariate dependence between two exchange rates. The data set comprising the Thai Baht (THB), the British Pound Sterling (GBP), the European Euro (EURO), the Japanese Yen (YPY), the Malaysian Ringgit (MYR), the Indonesian Rupee (IDR), the Singapore Dollar (SGD), and the Philippines Peso (PHP), which were pitted against the U.S. Dollar (USD). The data set was collected on a daily frequency during the period from January 3, 2004, to October 1, 2014, covering 1490 observations. All the series were transformed to the difference of the logarithm.

We begin with the identification of marginal distribution by univariate ARMA (p,m)-GARCH (q,n) specification, as described in Eqs. (6), (7), and (8). Then we, transform the marginal into uniform [0,1] using the empirical cumulative distribution function (ecdf). Then we estimate the copula parameter through Maximum Likelihood Estimator (MLE). However, the state variable (S_t) is not observable; therefore, we cannot directly estimate the copula parameter. Hence, we apply the method of Kims filter, as proposed by Silva Filho and Ziegelmann [15], and constructed the log likelihood function of the Markov-switching copula model as

$$l_c(\theta|x_t) = l_{f1}(\theta_1) + l_{f2}(\theta_2) + l_c(\theta_{ct,S_t}) \tag{13}$$

where $l_{f1}(\theta_1)$ and $l_{f2}(\theta_2)$ are the log-likelihood functions related to the marginal distribution which are assumed to fix across the regime. $l_c(\theta_{ct,S_t})$ is the regime dependent log likelihood of the copula function which has the following form:

$$l_c(\theta_{ct,S_t}) = \sum_{t=1}^{T} log(\sum_{S_{t=0}}^{1} c_t(u_1, u_2 \mathrm{d}|S_t, w_{t-1}) Pr[S_t|w_{t-1}]) \tag{14}$$

where w_t is all information set. To evaluate the log-likelihood in Eq. (13), we need to calculate the weight$Pr[S_t|w_{t-1}])$ for $S_t = 1, 2$ because the estimation of the Markov-switching copula needs inferences on the probabilities of S_t, that is, filtered probabilities. The Kims filter algorithms can be explained as the following:
1. Given an initial guess of transition probabilities thus the transition probabilities in the transition matrix Q^0 is given as,

$$Q^0 = \begin{bmatrix} P_{11} & P_{12} \\ P_{21} & P_{22} \end{bmatrix} \tag{15}$$

2. Updating the transition probabilities of each state with the past information including the parameters in the system equation,$theta_{ct-1}$,w_{t-1},and P_{ij}. After that, the probability of each state at time t ($Pr(S_t = k)$)) is to be updated by the following formula

$$Pr(S_t = k)|w_t) = \frac{c_t(u_1, u_2|S_t = k, w_{t-1}) Pr[S_t = k|w_{t-1}]}{\sum_{k=1}^{2} c_t(u_1, u_2|S_t = k, w_{t-1}) Pr[S_t = k|w_{t-1}]} \tag{16}$$

3. Iterating step 1 and 2 for $t = 2, \ldots, T$. Thus, the obtained $S_{t=1}, \ldots, S_{t=T}$ for 2 regimes are called filtered probabilities.

To estimate the smoothed probabilities can be calculated recursively from the filtered probabilities.

After, we obtain the filtered smoothed probabilities, we can proceed to maximize Eq. (13) in order to get the estimated parameter values (θ). In this paper, we used the normal copula, the Clayton copula, the Gumbel copula, the student-t copula, and the symmetrized JoeClayton copula to analyze the co-movement in each pair of the Markov-switching copula model. In addition, we applied the Akaike information criterion (AIC) and the Bayesian information criteria (BIC) to identify the best-fit pair copula family for these models.

4 Empirical Results

The results regarding the descriptive statistics of our data are shown in Table 1. We can see that the variables have different characteristics. MYR, IDR, THB, and EURO can be seen to exhibit excess kurtosis and small negative skewness, while SGD, PHP, GBP, and JPY are also observed to show excess kurtosis, but positive skewness. This means that the marginal distributions of MYR, IDR, THB, and EURO have heavy tails to the left, while the marginal distributions of SGD, PHP, GBP, and JPY have heavy tails to the right. Furthermore, the normality of these marginal distributions are strongly rejected by the JarqueBera test, with prob. = 0. Thus, these findings indicate that normal distribution might not be appropriate for our data. Consequently, we assume that student-t distribution is the appropriate marginal distribution for our data.

Table 1 Summary statistics

Variable	MYR	SGD	IDR	PHP	THB	GBP	JPY	EURO
Mean	2.84E−05	5.13E−05	0.000	1.82E−05	2.30E−05	0.000	0.000	0.000
Median	0.000	6.79E−05	0.000	0.000	0.000	0.000	0.000	0.000
Maximum	0.009	0.012	0.014	0.006	0.006	0.019	0.017	0.009
Minimum	0.01113	0.00941	0.00974	0.00535	0.00984	0.01223	0.01374	0.02005
Std. dev	0.002	0.002	0.002	0.002	0.001	0.002	0.003	0.003
Skewness	0.28983	0.369	0.2652	0.303	0.29111	0.215	0.419	0.3123
Kurtosis	5.923	8.126	10.229	3.830	7.272	6.199	7.015	5.808
Jarque-Bera	551.121	1664.827	3262.221	65.488	1154.013	646.878	1044.598	513.729
Probability	0.000	0.000	0.000	0.000	0.000	0.000	0.000	0.000

Source Calculation

4.1 Marginal Distributions

Table 2 shows the results of the marginal distribution of the student-t distribution with the ARMA(p,m)-GARCH(1,1) model for seven pairs of currencies. The results demonstrate that ARMA(2,2)-GARCH(1,1) with student-t distribution is the best specification for the pairs of THBJPY, THBEURO, THBSGD, THBMYR, THBIDR, while ARMA(5,5)-GARCH(1,1) with student-t distribution and ARMA(12,12)-GARCH(1,1) with student-t distribution are the best specifications for the pairs of THBGBP and THBPHP, respectively. In addition, the KolmogorovSmirnov (KS) test is used as the uniform test for the transformed marginal of these residuals. The result shows that none of the KS test results rejects the null hypothesis. Therefore, it is clear that all the marginal distributions are uniform. Additionally, LjungBox test, which is used as an autocorrelation test on standardized residuals, confirms that there is no autocorrelation in any series.

4.2 Model Fit

In this section, we present the comparison of the various types of families with the following specifications. For the currencies of each of the pairs, the Gaussian copula, the student-t copula, the Clayton copula, the Gumbel copula, and the symmetrized JoeClayton copula analyses have been conducted. The comparison of the families is based on AIC and BIC which is the measure of the family fit. We learned that among the trial runs of several alternative families for the currencies of each of the pairs, the results, Table 3, provide evidence that the Gaussian copula, which has the lowest AIC and BIC values, has the best fit for the pairs THBJPY, THBEURO, THBIND, THBSIN, and THBPHP, while the Clayton copula and the Gumbel copula have the best fit, respectively, for the pairs THBUK and THBMAL.

4.3 Markov-Switching Estimation Results

In this section, we report the results of the dependence structure for the relationships between the currencies of each of the pairs. First, we provide the estimated parameter of each pair copula; next, we graphically present smoothed probabilities concerning the high dependence for the currencies of each of the pairs. Table 4 presents the MS-copula estimation results for the currencies of our seven pairs.

Table 2 Marginal distribution results for seven pairs of currencies

THB-JPY			THB-GBP			THB-Euro		
Mean equation								
Constant	0.0001	0.0001	Constant	−0.0003	0.0003	Constant	0.0001	0.0001***
AR(2)	0.9106***	0.9323***	AR(5)	0.7330***	0.9106***	AR(2)	0.9106***	0.8658***
MA(2)	0.9188***	0.9230***	MA(5)	0.7047***	0.9188***	MA(2)	0.9188***	0.8863
Variance equation								
Constant	0.0001***	0.0001***	Constant	0.0001***	0.0001***	Constant	0.0001***	0.0001***
A(1)	0.1228***	0.0560***	A(1)	0.1199***	0.1228***	A(1)	0.1228***	0.0253***
G(1)	0.8702***	0.9420***	G(1)	0.8734***	0.8702***	G(1)	0.8702***	0.9743***
v	5.125***	4.784***	v	4.578***	7.367***	v	4.578***	6.541***
KS test (prob.)	0.986	0.857	KS test (prob.)	0.579	0.891	KS test (prob)	0.967	0.875
LM test (prob.)	0.624	0.7245	LM test (prob.)	0.6047	0.624	LM test (prob.)	0.624	0.5941
THBSGD			THBMYR			THBIND		
Mean equation								
Constant	0.0003	0.0001	Constant	0.0003	0.0001	Constant	0.0003	0.0001
AR(2)	0.9106***	0.7476***	AR(2)	0.9106***	0.6874***	AR(2)	0.9106***	0.9884***
MA(2)	0.9188***	0.7844***	MA(2)	0.9188***	0.6870***	MA(2)	0.9188***	0.9910***
Variance equation								
Constant	0.0001***	0.0001***	Constant	0.0001***	0.0001***	Constant	0.0001***	0.0001***
A(1)	0.1228***	0.0582***	A(1)	0.1228***	0.1024***	A(1)	0.1228***	0.3471***
G(1)	0.8702***	0.9339***	G(1)	0.8702***	0.8681***	G(1)	0.8702***	0.7947***
v	5.125	8.514	v	5.125	6.114	v	5.125	7.985
KStest (prob.)	0.986	0.9765	KS test (prob.)	0.986	0.9763	KS test (prob.)	0.986	0.9755
LM test (prob.)	0.624	0.2202	LM test (prob.)	0.624	0.1279	LM test (prob)	0.624	0.2903
THBPHP								

(continued)

Table 2 (continued)

THB-JPY			THB-GBP			THB-Euro		
Mean equation								
Constant	0.0001	0.0001						
AR(12)	0.9629***	0.7643***						
MA(12)	0.9809***	0.7507***						
Variance equation								
Constant	0.0001**	0.0001***						
A(1)	0.0708***	0.0944***						
G(1)	0.8818***	0.8764***						
KStest (prob.)	0.987	0.965						
LMtest (prob.)	0.1531	0.8852						

Source Calculation, Note: A = ARCH, G = GARCH

Table 3 Family selection of each pair copula

AIC	THBJPY	THBEURO	THBUK	THBIND	THBSIN	THBMAL	THBPHP
Gaussian	18.4	151.4415	10.5474	174.3308	114.1325	304.4864	291.7933
Clayton	57.1928	191.0311	48.0179	265.5203	193.5323	301.2571	425.4369
Gumbel	44.8389	190.9852	142.0041	268.9001	198.6945	226.0064	427.9861
SJC	47.2748	156.8494	10.55	222.8107	167.3836	260.3542	375.4322
student-t	54.6235	203.5144	157.3989	272.0823	196.8564	315.7836	433.9789
BIC							
Gaussian	50.1989	177.793	42.3462	206.1296	145.9312	336.2852	323.5921
Clayton	110.2581	244.0964	4.9801	318.5856	246.5976	354.255	478.5022
Gumbel	76.6781	222.8244	173.8029	299.9001	230.4933	257.8052	459.7849
SJC	79.0736	183.2009	42.3526	254.6094	199.1824	292.153	407.231
student-t	96.9676	245.8585	199.743	314.4265	239.2006	358.1278	476.3231

Source Calculation

Table 4 Markov-switching dynamic copula results for seven pairs of currencies

	Coefficient	THBYEN	THBEURO	THBIDR	THBSGD	THBPHP
Normal	ω_c^1	0.4777	0.9371***	0.1178***	1.6043***	0.5313***
	ω_c^2	5.455***	1.9580***	2.3391***	0.6978***	2.1709***
	β_c	1.960***	1.1848***	2.0045***	0.1282***	0.5609***
	α_c	0.4919***	0.8133***	0.0607***	0.6923***	0.2089***
	P	0.9999	0.9877	0.8256	0.9703	0.8286
	q	0.9999	0.9913	0.5105	0.9483	0.7266
Log-likelihood		16.4194	89.4925	128.0506	93.3472	207.9930
	Coefficient	THBGBP		Coefficient	THBMYR	
Clayton	ω_c^1	0.00007	Gumbel	ω_c^1	1.1308***	
	ω_c^2	4.1951***		ω_c^2	4.9532***	
	β_c	0.0603***		β_c	0.4020***	
	α_c	0.00004		α_c	0.0548***	
	P	0.9999		P	0.9819	
	q	0.9999		q	0.9938	
Log-likelihood		0.7262			124.1771	

4.3.1 Thai BahtJapanese Yen

The results show that the intercept coefficient in regime 1 (ω_c^1) is lower than the regime intercept coefficient in regime 2 (ω_c^2). Thus, we can interpret regime 1 as the low dependence regime and regime 2 as the high dependence regime. Moreover, we can see that the β_c coefficient, related to the autoregressive parameter component in the dynamic equation, has a negative sign, indicating that THBJPY might not be persistent over time. The α_c, representing the distance from the perfect correlation in the dependence dynamics co-movement of THBJPY, is positive, indicating that greater distance from the perfect correlation can increase the THBJPY dependence. In addition, upon considering the transition probabilities p and q, it can be inferred that the result indicates that the high dependence and the low dependence are persistent.

Upon considering the dependence dynamics for the THBJPY relation plotted in the first panel of Fig. 2, we can see positive dependence in the low dependence regime and negative dependence in the high dependence regime.

4.3.2 Thai BahtEuro

The results shows that the intercept coefficient in regime 1 (ω_c^1) is lower than the regime intercept coefficient in regime 2 (ω_c^2). Thus, we can interpret regime 1 to be the low dependence regime, while regime 2 can be taken as the high dependence regime. In addition, the β_c coefficient has a negative sign, indicating that the relation of THBEURO might not be persistent over time. The α_c has a negative sign, indicating that greater distance from the perfect correlation can reduce the THBEURO dependence. In addition, upon considering the transition probabilities p and q, it can be concluded that the result indicates that the high dependence and the low dependence are persistent.

Upon considering the dependence dynamics for the THBEURO relation plotted in the second panel of Fig. 2, we can see positive dependence in both the high and the low dependence regimes. The values of the degree of dependence parameter are around 0–1 and around 0.4–0.6, respectively, for the high dependence regime and the low dependence regime, along the sample period. This means that the fluctuation in the high dependence regime is larger than that in the low dependence regime.

4.3.3 Thai BahtIndonesian Rupiah

The results show that the intercept coefficient in regime 1 (ω_c^1) is lower than the regime intercept coefficient in regime 2 (ω_c^2). Thus, we can interpret regime 1 to be the low dependence regime while regime 2 can be considered as the high dependence regime. In addition, the β_c coefficient has a negative sign, indicating that the relation of THBIDR might not be persistent through time. The has a negative sign, indicating that greater distance from the perfect correlation can reduce the THBIDR dependence. In addition, the result of the transition probabilities p and q show that the low dependence regime (q) is persistent, while the high dependence regime (p) is not persistent.

Furthermore, we can see that the dynamic dependence value for THBIDR is presented in the third panel of Fig. 2. The result shows that the value of the dependence parameter in the high dependence regime is around 0.4–0.6, while the value of the dependence parameter in the low dependence regime is around 0.1–0.1.

4.3.4 Thai BahtSingapore Dollar

The results show that the intercept coefficient in regime 1 (ω_c^1) is higher than the regime intercept coefficient in regime 2 (ω_c^2). Thus, we can interpret regime 1 to be the high dependence regime while regime 2 can be interpreted as the low dependence regime. Also, both the β_c coefficient and the α_c coefficient have a negative sign, indicating that the relation of THBSGD might not be persistent over time; besides, greater distance from the perfect correlation can reduce the THBSGD dependence. In addition, upon considering the transition probabilities p and q, it can be concluded that the result indicates that the high dependence and the low dependence are persistent.

By observing panel 4 in Fig. 2, it can be seen that the result obtained for the dependence dynamics parameter of the THBSGD is similar to that of the THBEURO. Thus, it is evident that there is higher fluctuation in the high dependence regime than in the low dependence regime.

4.3.5 Thai BahtPhilippines Peso

The results show that the intercept coefficient in regime 1 (ω_c^1) is higher than the regime intercept coefficient in regime 2 (ω_c^2). Thus, we can interpret regime 1 to be the high dependence regime while regime 2 can be considered as the low dependence regime. In addition, we can see that the β_c coefficient, related to the autoregressive parameter component in the dynamic equation, has a negative sign, indicating that THBPHP might not be persistent over time. The α_c, representing the distance from the 100 % correlation in the dependence dynamics co-movement of THB-PHP, is positive, indicating that greater distance from the 100 % correlation can increase the THB-PHP dependence. In addition, upon considering the transition probabilities p and q, it can be inferred that the result indicates that the high dependence and the low dependence are persistent.

Upon taking into consideration the fifth panel of Fig. 2, we can observe that the difference between the dependence parameters in each of the regimes is around 0.6, indicating that the dependence between THBPHP increases substantially when staying in the high dependence regime.

4.3.6 Thai BahtGreat British Pound

Table 4 also presents the estimated parameters for the dynamic structure of lower tail dependence from Clayton copulas for THBGBP. The result shows that the intercept coefficient in regime 1 (ω_c^1) is lower than the regime intercept coefficient in regime 2 (ω_c^2). Thus, we can interpret regime 1 to be the low dependence regime while regime 2 can be interpreted as the high dependence regime. Furthermore, the coefficient has a significant negative sign, indicating that the lower tail dependence of THBGBP might be persistent over time. However, the α_c coefficient has an insignificant negative sign, indicating that there exists no relationship between distance from the perfect correlation and the lower tail dependence of THBGBP. In addition, while considering the transition probabilities p and q, it is evident that the result indicates that the high dependence and the low dependence are persistent. Poor estimated lower tail dependence dynamics parameters were obtained, however, due to the MS-copula seeming not able to capture the dependence dynamic in each regime.

4.3.7 Thai BahtMalaysian Ringgit

Table 4 presents the estimated parameters for the dynamic of upper tail dependence from the Gumbel copulas for THBMYR. The result shows that the intercept coefficient in regime 1 (ω_c^1) is lower than the regime intercept coefficient in regime 2 (ω_c^2). Thus, we can interpret regime 1 to be the low dependence regime while regime 2 can be interpreted as the high dependence regime. Furthermore, the β_c and α_c coefficients have a negative sign, indicating that the upper tail dependence of THBMYR might not be persistent over time; also, greater distance from the perfect correlation can reduce the lower tail dependence THBGBP. In addition, upon considering the transition probabilities p and q, it can be concluded that the results indicate that the high dependence and the low dependence are persistent. Upon considering the dependence dynamics for THBMYR relation plotted in the sixth panel of Fig. 2, we can see that there exists a positive dependence in both the high and the low dependence regimes. The values of the degree of dependence parameter are around 0–0.4 and around 0–0.2, respectively, for the high and the low dependence regimes. This indicates a small dependence relation between THBMYR in both the regimes.

4.4 *Regime Probabilities*

The estimated MS-copula model produces probabilities in each regime for the period 2009–2014, and the high dependence regime probabilities are plotted for the seven pairs of currencies, as illustrated in Fig. 1. We observed heterogeneous results, with regard to the regime dependence among the seven pairs of currencies. The results in Fig. 1 demonstrate that the smoothed probabilities concerning the high dependence regime for the pairs THBYEN and THBGBT, which are plotted in the first and sixth panels, respectively, are likely to have same results. We found that the low dependence regimes for the both pairs have not likely been detected by the MS-copula in this long period. The time series plot of the smoothed probability estimates for regime 2 reveals that it is typically close to 0. On the other hand, the smoothed probability estimates for the high dependence regime are highly erratic. This indicates that there was high dependence for the pairs THBYEN and THBGBT after the introduction of QE. In the case of the smoothed probabilities concerning the high dependence regime for the pairs THBIDR, THBPHP, and THB-MYR, which are plotted in third, fifth, and seventh panels, respectively, we found high volatility along the period of time because the duration of each regime constitutes a short period of

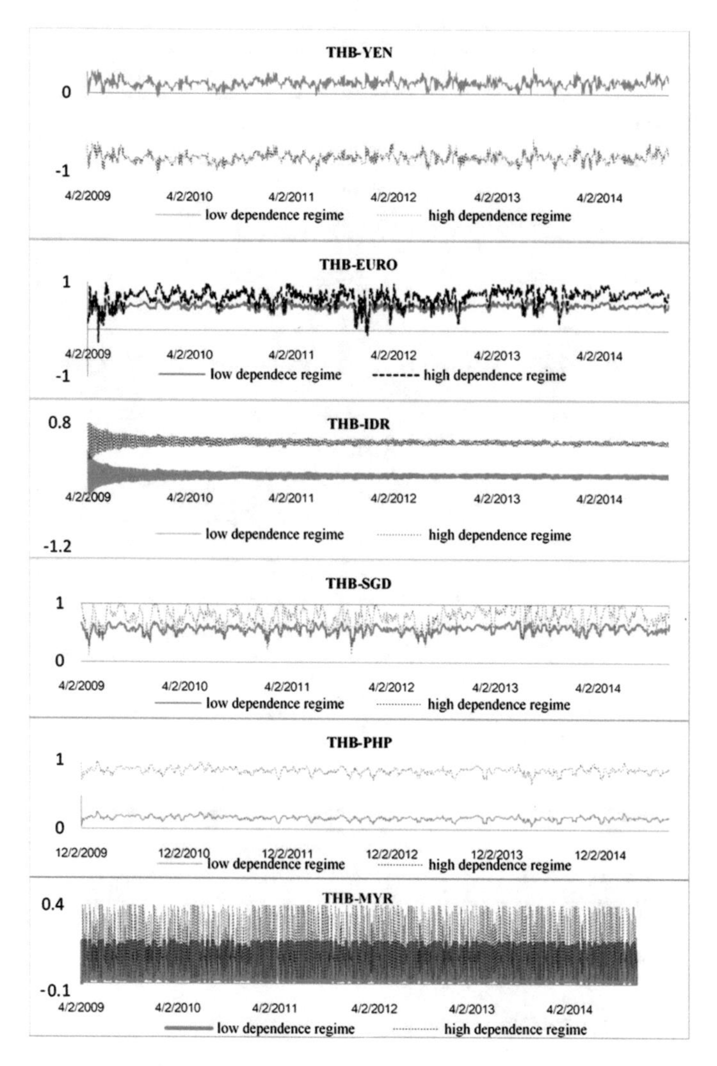

Fig. 1 The smoothed probabilities for the high dependence regime

time. Therefore, this indicates that the dependence for the pairs THBIDR, THBPHP, and THBMYR underwent fluctuation after the introduction of QE. Finally, in the case of THBEURO and THBSGD, plotted in second and fourth panels, respectively, we found that the currencies of both these pairs presented similar results, which indicates that THBEURO and THBSGD are less volatile when compared to the THBIDR, THBPHP, and THB-MYR pairs. However, for the pair THB-EURO, the probability of staying in the high dependence regime is less than the probability of

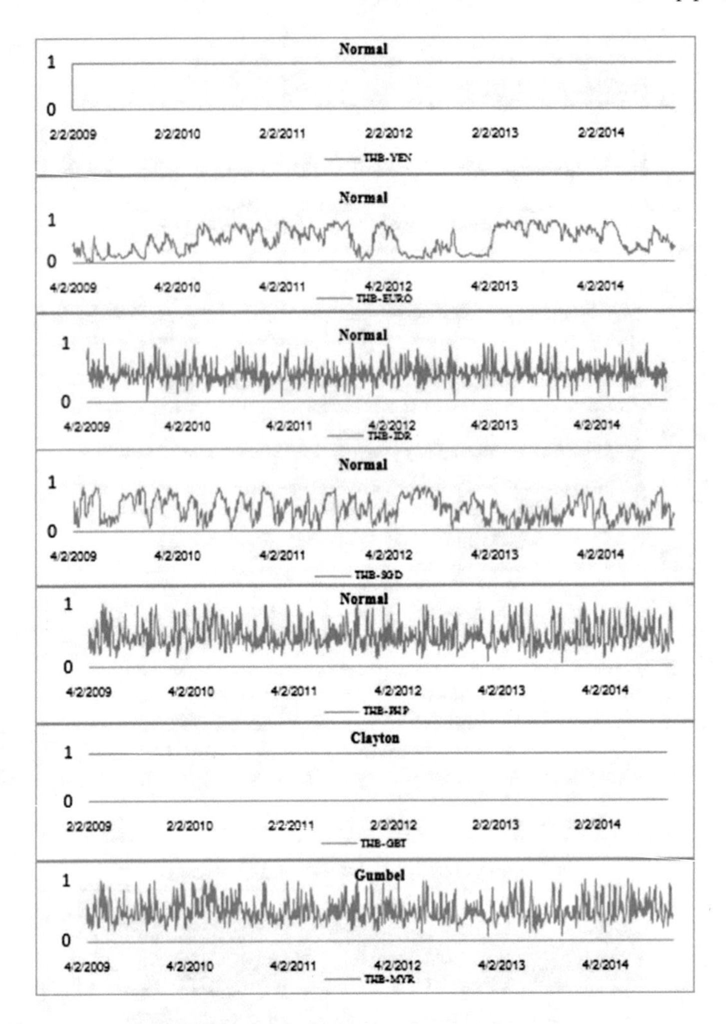

Fig. 2 The smoothed probabilities for the high dependence regime

staying in the low dependence regime, whereas for the pair THB-SGD, the probability of staying in the high dependence regime is more than the probability of staying in the low dependence regime.

5 Conclusion

We analyze the dependence relationship between the Thai currency and international currencies, comprising the Japanese Yen, the Euro, the Great British Pound, the Singapore Dollar, the Philippines Peso, the Indonesian Rupiah, and the Malaysian

Ringgit after the period of QE introduction. We found that there was heterogeneous dependence between the pairs of currency dependence after the introduction of QE. The pair Thai BahtJapanese Yen and the pair Thai BahtGreat British Pound seem to have only positive high dependence over time; however, the dependence dynamics for the pair Thai BahtJapanese Yen is small. This means that there is low positive correlation between the Thai currency and the Japanese Yen after the introduction of QE. In the case of Thai BahtEuropean Euro and Thai BahtSingapore Dollar, there are similar results observed as regards the smoothed probabilities and the dependence dynamics parameter. Their dependence parameters are more likely to fluctuate under the high dependence regime than under the low dependence regime. Finally, in the case of the currency pairs Thai BahtMalaysian Ringgit, Thai BahtIndonesian Rupiah, and Thai BahtPhilippines Peso, similar behaviors are observed regarding the smoothed probabilities. The smoothed probabilities are observed to have fluctuated after the introduction of QE. However, the dependence dynamic parameters provide a different result. The currency pair of Thai BahtPhilippines Peso seems to have the highest correlation in the high dependence regime when compared to the pairs Thai BahtMalaysian Ringgit and Thai BahtIndonesian Rupiah, while the currency pair Thai BahtIndonesian Rupiah seems to have the highest fluctuation in the low dependence regime when compared to the pairs Thai-Baht Malaysian Ringgit and Thai BahtPhilippines Peso. These findings also found the dependence between the Thai Baht and other currencies except for the Great British Pound. We found the trading volume between Thailand and Great Britain to be less than 600 million dollars while the same between Thailand and other countries was found to be more than 1,000 million dollars in September 2014. Additionally, the result also found high volatility in some currencies, especially in the high dependence regime. Based on the results, we suggest that the dependence between the Thai Baht and the other currencies is not constant but varying after the QE introduction. Thus, Thai policy makers and investors should lay emphasis on the co-movement between the Thai currency and the other currencies, especially in the high dependence regime, in order to well prepare for and efficiently manage the high volatility.

Acknowledgments We are greatful for financial support from Puay Ungpakoyn Centre of Excellence in Econometrics, Faculty of Economics, Chiangmai University.

References

1. Antonakakis, N.: Exchange return co-movements and volatility spillovers before and after the introduction of euro. J. Int. Financ. Mark. Inst. Money **22**(5), 1091–1109 (2012)
2. Azam, K.: Copula methods in econometrics (Doctoral dissertation, University of Warwick) (2013)
3. Bohdalov, M., Gregu, M.: Cointegration analysis of the foreign exchange rate pairs. CBU Int. Conf. Proc. **2**, 147 (2014)
4. Chokethaworn, K., Chaitip, P., Sriwichailamphan, T., Chaiboonsri, C.: The dependence structure and co-movement toward between Thai's currency and Malaysian's currency: Markov

switching model in dynamic copula approach (MSDC). Procedia Econ. Financ. **5**, 152–161 (2013)
5. Choudhry, T.: Asian currency crisis and the generalized PPP: evidence from the far east. Asian Econ. J. **19**(2), 137–157 (2005)
6. Fei, F., Fuertes, A. M., Kalotychou, E.: Modeling dependence in cds and equity markets: dynamic copula with markov-switching (2013)
7. Jowaheer, V., Ameerudden, N.Z.: Modelling the dependence structure of MUR/USD and MUR/INR ex-change rates using copula. Int. J. Econ. Financ. Issues **2**(1), 27–32 (2011)
8. Kim, B.H., Kim, H., Min, H.G.: Reassessing the link between the Japanese yen and emerging Asian currencies. J. Int. Money Financ. **33**, 306–326 (2013)
9. Khl, M.: Excess co-movements between the Euro/US dollar and British pound/US dollar exchange rates. US Dollar and British Pound/US Dollar Exchange Rates (2009)
10. Li, X.M.: How do exchange rates co-movement? A study on the currencies of five inflation-targeting countries. J. Bank. Financ. **35**(2), 418–429 (2011)
11. Michelis, L., Ning, C.: The dependence structure between the Canadian stock market and the USD/CAD ex-change rate: a copula approach. Can. J. Econ./Rev. Can. d'conomique **43**(3), 1016–1039 (2010)
12. Patton, A.J.: Applications of copula theory in financial econometrics(Doctoral dissertation. University of California, San Diego) (2002)
13. Patton, A.J.: Modelling asymmetric exchange rate dependence*. Int. Econ. Rev. **47**(2), 527–556 (2006)
14. Samitas, A., Kenourgios, D., Paltalidis, N.: Financial crises and stock market dependence. Working Paper, University of the Aegean (2007)
15. Silva Filho, O. C. D., Ziegelmann, F. A.: Modelling the dependence dynamics through copulas with regime switching (2010)
16. Silva Filho, O.C.D., Ziegelmann, F.A., Dueker, M.J.: Modeling dependence dynamics through copulas with regime switching. Insur. Math. Econ. **50**(3), 346–356 (2012)
17. Silva Filho, O.C., Ziegelmann, F.A.: Assessing some stylized facts about financial market indexes: a Markov copula approach. J. Econ. Stud. **41**(2), 253–271 (2014)
18. Sklar, M.: Fonctions de repartitionan dimensions et leurs marges. Publ. Inst. Statist. Univ. Paris **8**, 229–231 (1959)
19. Tofoli, P. V., Ziegelmann F. A., Silva Filho, O, C.: A comparison study of copula models for European fi-nancial index returns (2013)
20. Woo, K.Y.: Cointegration analysis of the intensity of the ERM currencies under the European monetary system. J. Int. Financ. Mark. Inst. Money **9**(4), 393–405 (1999)
21. Zhao, N., Lin, W.T.: A copula entropy approach to correlation measurement at the country level. Appl. Math. Comput. **218**(2), 628–642 (2011)

Analyzing Financial Risk and Co-Movement of Gold Market, and Indonesian, Philippine, and Thailand Stock Markets: Dynamic Copula with Markov-Switching

Pathairat Pastpipatkul, Woraphon Yamaka and Songsak Sriboonchitta

Abstract In this paper, we analyze the dependency between the Thailand, Indonesia, and the Philippine (TIP) stock markets and gold markets using dynamic copula with the Markov-switching model with 2 regimes, namely high dependence and low dependence regimes, and extend the obtained correlation to measure the market risk. We are particularly interested in examining whether or not gold serves as a hedge in the TIP stock markets. Using daily data from January 2008 to November 2014, we find that the Gaussian copula identifies a long period of high dependence of TIPGOLD returns (market downturn) which coincides with the European debt crisis. However, if we do not take gold into account, the dependence between the TIP returns is lower in both regimes, thereby leading to a higher value at risk (VaR) and expected shortfall (ES). Therefore, gold can serve as a hedging, or a safe haven, for TIP stock markets during market downturns and upturns. Additionally, the Kupiec unconditional coverage and the Christoffersen conditional coverage test are conducted for VaR and ES backtesting. The results reveal that the Gaussian Markov-switching dynamic copula is the appropriate model to estimate a dynamic VaR and ES.

Keywords Markov switching model · Dynamic copula · Gold market · Stock market · Time-varying value at risk and expected shortfall

1 Introduction

Financial and commodity markets have been highly volatile under the free capital mobility in recent years. The stock markets of Thailand, the Philippines, and Indonesia (TIP) are no exception to this phenomenon. After the financial crisis of the United States in 2007, a large capital flowed to the TIP emerging markets. These markets have experienced great growth and contributed a large gain to investors. In

P. Pastpipatkul (✉) · W. Yamaka · S. Sriboonchitta
Faculty of Economics, Chiang Mai University, Chiang Mai, Thailand
e-mail: ppthairat@hotmail.com

W. Yamaka
e-mail: woraphon.econ@gmail.com

V.-N. Huynh et al. (eds.), *Causal Inference in Econometrics*,
Studies in Computational Intelligence 622, DOI 10.1007/978-3-319-27284-9_37

November 2014, the Thailand SET index, the Indonesia JKSE index, and the Philippines PSE index rose 20.81, 18.14, and 22.7 %, respectively, higher than the 2.68 % of Global Dow, 7.7 % of Japan Nikkei, 6.84 % of Hong Kong Hang Seng, 0.19 % of EURO STOXX 50, 2.39 % of Brazil BOVESPA, and 9.33 % Africa Johannesburg All Share. This indicates that the TIP stock markets displayed a strong growth and high returns in the present day. Nevertheless, investment in stocks brings a large share of risk, as well. Thus, the investor needs to diversify the investing to safety places such as the gold market.

Gold market is a market popular to investors. It has played an important role as a safe haven for investment during times of turbulent economic periods and equity crashes. Ang and Chen [2] examined the relationship between volatility and market performance. They suggest that volatility tends to become low as the stock market rises and high as the stock market falls. Therefore, when investors are faced with the high risk, as stock markets fall, they could reduce their high risk by allocating their investment to the gold market. This suggestion seems to correspond with the study of Baur and Lucey [3] and Arouri et al. [4] who found that gold plays a considerable role as a hedge against other stocks.

Recently, many studies have documented the relationship between the stock market and the gold market using different methods Bhunia and Mukhuti [5] confirm that there is no existence of causality between stock and gold price. By contrast, Gilmor et al. [16] and Miyazaki and Hamori [27] used cointegration to explore the association between gold and stock. Baur and Lucey [4] Coudert and Raymond [10], and Arouri et al. [3] used the GARCH approach to investigate both return and volatility spillovers between gold price and stock and found the existence of the volatility spillover effect between gold and stock. Shen et al. [38] used the copula approach and found dependency between gold price, stock price index of gold mining companies, and Shanghai stock index. Of course, in the present day, there exists some literature which is relevant to this study, such Phoong et al. [35] employed a non-linear model, the Markov-switching model, and also found significant effect of gold price on the Malaysia, Singapore, Thailand, and Indonesia stock markets. Meanwhile, Do and Sriboonchitta [11] did not observe any association between gold markets and the ASEAN stock markets as well as any effect from the ASEAN stock markets on the gold markets.

Our empirical analysis shows that there may apparently be a relationship between the stock and the gold markets. Nevertheless, to make the strategic asset allocation, it is necessary to measure a risk in these markets. By now, the most preferred approach for measuring market risk is value at risk (VaR). VaR is widely employed to measure the market risk and to assess the portfolio and market risk associated with financial assets and commodity prices during a particular period of time. However, there is no consensus on the VaR computation, and it is not easily estimated. Thus, there might be present inaccurate estimation of the VaR in financial markets which affect the accuracy of the risk measure [17].

Nowadays, the computation of VaR has been generated by either the parametric approach or the non-parametric approach. For the parametric approach, there already exist many studies on VaR of portfolio, such as the studies conducted by Pafka and Kondor [31] who used the Risk Metrics approach to measure VaR in financial markets. Fss et al. [15], and McAlee and Da Veiga [25] used the GARCH approach to compute VaR. Additionally, the Markov-switching approach was also used to forecast VaR by Billio and Pelizzon [7] and Leon and Lin [21]. As for the non-parametric approach, the copula approach has been widely used in financial applications and has been successful in computing VaR. It has been employed by many studies, for instance, Palaro and Hotta [32], Miller and Liu [26], and Ozun and Cifter [30].

In this study, we adopt the non-parametric models to compute VaR. Caillault and Guegan (2009) who discussed various methods of non-parametric model have suggested that the dynamic copula model seems to be the best for risk measurement. In addition, Mandal and Lagerkvistn [24] also confirm that the copula-based VaR is better than other traditional methods, such as historical simulation, Monte Carlo based simulation, and variance-covariance approach. Therefore, the copula approach was considered as the ideal estimation method for our paper.

Copula was first introduced in Sklar's theorem. It can describe the dependencies between variables and is widely used to measure joint distribution. This method is flexible to use because it eliminates the assumption on joint distribution. Copula was developed by Joe [18], and employed in finance by Embrechts et al. [12]. To model the time-varying dependence, Patton [33] constructed the time-varying copula and attracted the attention of economists to analyses in many topics, especially the measurement of VaR. Recently, there have been many studies that employed the copula model to measure the dependence between stock markets and gold markets, such as Najafabadi et al. [28], and Ostrup and Ellegaard [29].

However, many studies provide different degrees of dependence between upper tail and lower tail. Moreover, many studies found different correlations between downtrend and uptrend of financial time series. This confirms that it may appear to be of two regimes, comprising high dependence regime and low dependence regime; thus, VaR based on this dependency may also have two different regimes. In addition, Lin and Wu [23] suggested a tail dependence and non-linear relationship between financial markets; therefore, we employ a non-linear approach using the Markov-switching dynamic copula to estimate the dependency between these markets, which allows us to calculate the VaR of the portfolio in different regimes. As per our knowledge, the computation of VaR based on Markov-switching dynamic copula in the area of TIP stock markets and gold markets has not been explored yet. Thus, this is the first attempt to compute a dynamic VaR in different regimes in those areas. Nevertheless, VaR is based on some unrealistic assumptions: especially, it does not verify the sub-additivity property of a coherent risk measure; therefore, the expected shortfall (ES) measure is preferable in this study, as well.

Finally, in this study, we aim to compute VaR and ES to measure the portfolio risk comprising TIP stock markets and gold markets, and aim to improve the accuracy of portfolio VaR in different portfolio dependencies. Moreover, we aim to analyze whether gold could reduce the risk in stock markets. Thus, we employ a time-varying copula with regime switching to capture the non-linear relationship between these markets, and extend the model to compute the dynamic VaR and ES.

The remainder of the paper is as follows. In the next section, we briefly explain the basic concept of copulas and dependence measure, and present copulas with Markov-switching. In Sect. 3, we describe a data and estimation procedure. The estimated results and the investigation of the dynamic dependence structure in each regime are discussed in Sect. 4. Section 5 provides the conclusion of this work.

2 Methodology

2.1 Basic Concepts of Copula

According to Sklar's theorem, the dimensional joint distribution copula is such that $F(X_1, \ldots, X_n) = C(F_1(x_1), \ldots., F_n(x_n))$, where the marginal distributions (u) are uniform in the [0,1] interval. Additionally, Schweizer and Skalar [36] present a basic concept for any joint distribution $F(X_1, \ldots, X_n)$ with marginal $F_1(x_1), \ldots., F_n(x_n)$, as follows:

$$F(X_1, \ldots, X_n) = C(F_1(x_1), \ldots., F_n(x_n) \tag{1}$$

The copula density c is obtained by differentiating Eq. (1); thus, we get

$$C(F_1(x_1), \ldots., F_n(x_n) = \frac{h(F_1(x_1), \ldots., F_n(x_n)}{\Pi_{i=1}^{n} \int_i (F_i(x_i))} \tag{2}$$

Thus,

$$c(u_1, \ldots, u_n) \quad = \quad \frac{\partial^n c(u_1, \ldots, u_n)}{\partial u_1 \cdot \ldots . \cdot \partial u_n} \tag{3}$$

In the dynamic case, Patton [34] constructed the time-varying copula by applying Sklar's theorem to conditional distributions. This theorem allows decomposing a conditional joint distribution into marginal distribution. Suppose $X_t = (X_{1t}, X_{2t})$; then, the conditional joint distribution (F_1) and the condition marginal distributions (F_{1t}, F_{2t}) can be written as

$$F_t = C_t(F_{1t}, F_{2t}) \tag{4}$$

where C_t is the conditional copula of X_t. This conditional copula contains a dependence between X_{1t} and X_{2t}. In this dynamic case, Patton [34] suggested allowing the dependence parameter (θ) to vary over time in the ARMA (p,m) process, as follows:

$$\theta_{ct} = \Lambda(\omega + \beta_{ct}\theta_{ct-p} + \alpha \Gamma_t), \tag{5}$$

where $\Lambda(\omega + \beta_{ct}\theta_{ct-p} + \alpha \Gamma_t)$ is the logistic transformation to keep the parameter value in its interval, ω is the intercept term, and is the forcing variable which is defined as

$$\Gamma_t = \begin{cases} \frac{1}{m}\Sigma_{j=1}^{m} F_1^{-1}(u_{1,t-j}) F_2^{-1}(u_{2,t-j}) & elliptical \\ \frac{1}{m}\Sigma_{j=1}^{m} |u_{1,t-j} - u_{2,t-j}| & Archimedean \end{cases} \quad (6)$$

In the elliptical case, it is a quadratic form of the copula function which has symmetrical tail dependence. The dependence structure, related to this function, is the Pearson's correlation which has the value of its parameter in the [−1,1] interval. The most commonly used among these distributions are the Gaussian and the Student-t copulas. As for the Archimedean case, it is an alternative class of copulas which deals with asymmetric tail dependence. This form is different from the elliptical copulas because it cannot derive from the multivariate distribution function and has closed form expressions. Moreover, these copulas allow only positive dependence structures (upper tail), which keeps the parameter in the [1,∞] interval.

There are many functional forms that can be used as copulas. However, in this study, we propose the elliptical copulas to analyze the structure dependence between stock markets and gold markets because those copulas are easily derived from the multivariate distribution function.

1. Gaussian copula

The Gaussian, or Normal, copula is a linear correlation matrix with symmetric function because the upper and the lower tail dependences are equal, and so it has no tail dependence in this function. In the bivariate case, the Gaussian copula expression is defined as follows:

$$C_N(u_1, u_2|p) = \int_{-\infty}^{\Phi^{-1}(u_1)} \int_{-\infty}^{\Phi^{-1}(u_2)} \frac{1}{2¶\sqrt{1-p^2}} exp\left\{-\left(\frac{r^2 - 2prs + s^2}{2(1-p^2)}\right)\right\} drds, \rho \in (-1, 1),$$

And its dynamic dependence (ρ) equation can be written as

$$\rho_{Nt} = \Lambda(\omega_N + \beta_N \rho_{t-1} + \alpha_N \cdot \frac{1}{m}\Sigma_{j=1}^{m} \Phi^{-1}(u_{1,t-j} \Phi^{-1}(u_{2,t-j}).$$

2. Student-t copula

Similar to the Normal copula, the Student-t copula has a linear correlation coefficient and has symmetrical tail dependence. However, it can capture some tail dependence. For example, in the bivariate case, it has the following form of function:

$$C_N(u_1, u_2|p, v) = \int_{-\infty}^{t_v^{-1}(u_1)} \int_{-\infty}^{t_v^{-1}(u_2)} \frac{1}{2¶\sqrt{1-p^2}} exp\left\{1 + \frac{r^2 - 2prs + s^2}{v(1-p^2)}\right\}^{-\frac{v+2}{2}} drds,$$

where ρ is the linear dependence parameter and is the degree of freedom. In addition, its dynamic equation can be written as

$$\rho_{Tt} = \Lambda(\omega_{1T} + \beta_{1T} \rho_{t-1} + \alpha_{1T} \cdot \frac{1}{m}\Sigma_{j=1}^{m} T_v^{-1}(u_{1,t-j} T_v^{-1}(u_{2,t-j}).$$

2.2 Value at Risk with Time-Varying Copula

The value at risk (VaR) can be specified by using the multivariate distribution of the copula functions. Consider the return random variable X_i at time t where $t = 1, \ldots, T$, $R_{i,t}$. VaR is calculated as the quantile α, $0 \leq \alpha \leq 1$ from the inverse function F_l, that is,

$$VaR_{\alpha}^{i} = F_{i.t}^{-1}(\alpha),$$

where $F_{i,t}^{-1}$ is the inverse distribution function of the return distribution, and the quantile α can be determined by

$$Pr(R_{i,j} \leq Var_{\alpha,t}^{i})$$

In the dynamic case, a dynamic copula is proposed into conditional value at risk (CoVaR) by Adrian and Brunnermeier [1]. They presented $CoVaR_{\alpha,\beta,t}^{=}$ as the β quantile of the entire financial system, $R_{s,t}$, conditional on $R_{i,t} = VaR_{\alpha,t}^{i}$. For the Gaussian copula, the explicit form of CoVaR can be defined as

$$CoVaR_{\alpha,\beta,t}^{=} = F_{s,t}^{-1}(\phi(\rho\phi^{-}1(v_t) + \sqrt{1 - p^2\phi^{-1}(\beta)}$$

where ϕ is N(0,1) and $F_{s,t}^{-1}$ is the inverse distribution function of $F_{s,t}$. Then, by using the Gaussian copula density for the conditional quartile u is given by

$$\int_0^{u_t}\int_0^{u_2} \frac{1}{\sqrt{1-p^2}} exp\left(-\frac{\rho^2(s^2+t^2) - 2pst}{2(1-p^2)}\right) dsdt = \alpha^2$$

Similar to the Gaussian copula, the definition for the Student-t copula can be given in explicit form as follows:

$$CoVaR_{\alpha,\beta,t}^{=} = F_{s,t}^{-1}\left(t_v\left(pt_v^{-1}(v_t) + \sqrt{1-\rho^2(v+1)^{-1}\left(v + t_v^{-1}(v_t)^2\right)t_{v+1}^{-1}(\beta)}\right)\right)$$

Then, by using this, the Student-t copula density for the conditional quartile can be given by

$$\int_0^{u_t}\int_0^{u_2} \frac{1}{\sqrt{1-\rho^2}} \frac{\Gamma\left(\frac{v+2}{2}\right)\Gamma\left(\frac{v}{2}\right)}{\Gamma\left(\frac{v+2}{2}\right)^2} \frac{1 + \frac{s^2 - 2st\rho + t^2}{v(1-\rho^2)}}{\left(1+\frac{s^2}{v}\right)^{-\frac{v+1}{2}}\left(1+\frac{t^2}{v}\right)^{-\frac{v+1}{2}}} dsdt = \alpha \cdot \beta$$

Thereafter, we extend this approach to estimate the conditional expected shortfall (COES). According to Karimali and Nomikos [19], let COES be denoted as $CoES_{\alpha,\beta,t}^{=}$ which is the expected shortfall of the return distribution $R_{s,t}$, conditional on $R_{i,t} = VaR^{i}_{\alpha,t}$; then, the explicit form can be written as follows:

$$CoES^{=}_{\alpha,\beta,t} = \frac{1}{\beta}\int_{0}^{\beta} CoVaR^{=}_{\alpha,q,t}dq$$

where $CoES^{=}_{\alpha,q,t} = Pr(R_{s,t} \leq F^{-1}_{s,t}(q)|R_{i,t} = VaR^{i}_{\alpha,t}$.

2.3 Regime-Switching Copula

The financial time series tends to exist a different dependencies over time; for that reason, the dependence structure of the variables may be determined by a hidden Markov chain with two states [40]. Hence, we employ the Markov-switching copula with time-varying which is proposed in Filho and Ziegelmann [13]. This model allows for the dependence parameter (θct) to be governed by an unobserved variable at time t (S_t). Thus,

$$\theta_{ct} = \theta_1(1 - S_1) + \theta_2 S_t \tag{7}$$

Let S_t be the state variable which is assumed to have two states (k=2), which are the high dependence regime and the low dependence regime. Thus, in bivariate case, the joint distribution of x_{1t} and x_{2t}, conditional on S_t, is defined as

$$(x_{1t}, x_{2t}|x_{1,t-1}, x_{2,t-1}; S_t = i) = c^{S_t}_t(u_{1t}, u_{2t}|u_{1,t-1}, u_{2,t-1}; \theta^{S_t}_t), i = 1, 2 \tag{8}$$

where $i = 1$ denotes the low dependence regime and $i = 2$ denotes the high dependence regime. Additionally, with the dependence parameter following the ARMA process, and the intercept term switching according to the first order Markov chain with two regimes, we can extend Eq. (7) to get

$$\theta_{ct} = \Lambda_0(1 - S_t + \omega_t S_t) + \beta\theta_{ct-1} + \alpha\Gamma_t) \tag{9}$$

The unobservable regime (S_t) is governed by the first order Markov chain, which is characterized by the following transition probabilities (P):

$$P_{ij} = Pr(S_{t+1} = j|s_t = i) \quad and \quad \Sigma^{k=2}_{j=1}\rho_{ij} = 1 \qquad i = 1, 2 \tag{10}$$

where $\rho_i j$ is the probability of switching from regime i to j regime, and these transition probabilities can be formed in a transition matrix P, as follows:

$$P = \begin{bmatrix} P_{11} & P_{12} = 1 - P_{11} \\ P_{21} = 1 - P_{22} & P_{22} \end{bmatrix} \tag{11}$$

In summary, the Markov-switching copula model can be specified as follows:

$$C_{S_t,t}\left(u_{1t}, u_{2t}|S_t, \theta_{cS_t,t}\right) = \sum_{k=1}^{2} 1_{\{S_t=k\}} \cdot C_{kt}\left(u_{1t}, u_{2t}|\theta_{ckt}\right), \tag{12}$$

with $\theta_{ckt} = \Lambda(\omega_k + \beta_{ckt}\theta_{ckt-p} + \alpha_k \Gamma_t)$, where $C_{S,t}(.)$ is the copula function with the time-varying dependence parameter at regime $S_t = (1, 2)$. In this study, we employ only the Gaussian and the Student-t functions to estimate the dependency between the variables, where the dependence parameter, ρ, in the dynamic Markov-switching models is calculated as follows:

The dependence parameter, $\rho_{Nt}(S_t)$, for the Gaussian function can be written as

$$\rho_{Nt}\left(S_t\right) = \Lambda\left(\left[\omega_{N1}\left(1 - S_t\right) + \omega_{N1}\left(S_t\right)\right] + \beta_N \rho_{Nt-1} + \alpha_N \cdot \Gamma_t\right)$$

and the dependence parameter, $\rho_{Tt}(S_t)$, for the Student-t copula can be written as

$$\rho_{Tt}\left(S_t\right) = \Lambda\left(\left[\omega_{T1}\left(1 - S_t\right) + \omega_{T2}\left(S_t\right)\right] + \beta_T \rho_{Tt-1} + \alpha_T \cdot \Gamma_t\right).$$

In addition, the degree of freedom for the evolution equation is given by

$$\upsilon_{Tt}\left(S_t\right) = \tilde{\Lambda}(\left[\omega_{\upsilon T2}\left(1 - S_t\right) + \omega_{\upsilon T2}\left(S_t\right)\right] + \beta_{\upsilon T2}\upsilon_{Tt-1} + \alpha_{\upsilon T2} \cdot \Gamma_t).$$

3 Data and Estimation

In this study, we use the data set comprising of the Stock Exchange of Thailand index (SET), Indonesia Stock Exchange index (IDX), the Philippine Stock Exchange (PSE), and the United States spot gold price (GOLD). The data set consists of daily frequency collected from the period of January 3, 2004, to October 1, 2014, covering 1490 observations. All the series have been transformed to the difference of the logarithm.

In order to estimate these four dimensional copula functions, in the first step, we estimate the bivariate copula functions among the six pairs of the gold and stock prices consist of SET-IDX, SET-PSE. SET-GOLD, IDX-PSE, IDX-GOLD, and PSE-GOLD. To compute the bivariate copula function for each pair, first of all, we create the marginal distributions, which are specified using univariate ARMA(p,q)-GARCH(m,n) which can be written as follows:

$$x_t = \alpha_0 + \sum_{i=1}^{p} \alpha_i x_{t-1} + \sum_{j=1}^{q} \beta_j \mu_{t-j} + \mu \tag{13}$$

$$\alpha_t = h_t^{1/2} \varepsilon_t \tag{14}$$

$$h_t^2 = \gamma_0 + \sum_{i=1}^{m} \gamma_i \varepsilon_{t-i}^2 + \delta_0 - \sum_{j=1}^{n} \delta_j h_{t-j}^2, \tag{15}$$

where Eqs. (13) and (15) are mean and variance equation, ε_t respectively. is an standardized residual of a chosen innovation, and, in this paper, we propose three types of distribution for, comprising normal distribution (ε_t, $Normal(0, 1)$) and Student-t distribution (ε_t, $student - t(v)$), as the assumption for the ARMA-GARCH model. Then, the Ljung-Box test of autocorrelation is used to test the autocorrelation in the standardized residuals of the ARMA-GARCH specifications. Additionally, a standardized residual from the best-fit marginal distribution has been transformed into uniform [0,1]. To test whether the standardized residuals have a uniform distribution, we employ the Berkowitz test of goodness of fit.

The second step involves modeling the estimate of the parameter of dependence between the six pairs of variables; in this study, we propose only the elliptical copulas, namely the Gaussian copula and the Student-t copula, because the multivariate extension of the Archimedean copulas suffers from lack of free parameter [12]. Moreover, simulations from elliptical copulas are easy for use to compute the rank correlation and the tail dependence coefficients. In this step, we employ the Markov-switching copula model to estimate the dynamic dependence parameter in the two regimes and to make an initial selection between the two copula functions, the lowest sum of AIC and BIC values is preferred.

To compute the parameters ($\theta_{cSt,t}$) in the Markov-switching dynamic copula model, we have to maximize the copula log-likelihood which has the form as follows:

$$l_c\left(\theta_{cSt,t}\right) = \sum_{t=1}^{T} log c_{St,t}\left(u_{1t}, u_{2t}|\hat{\theta}_1, \hat{\theta}_2; S_t, \theta_{cSt,t}\right) \tag{16}$$

However, the log-likelihoods in Eq. (16) cannot be directly maximized. Thus, we have to decompose the copula density of u_{1t}, u_{2t} and $S_t = (1, 2)$ into the product of the marginal densities and the conditional, we have

$$c_t\left(u_{1t}, u_{2t}|\gamma_{t-1}\right) = \sum_{S_t}^{1} c_t(u_{1t}, u_{2t}|S_t, \gamma_{t-1}) \cdot Pr\left(S_t|\gamma_{t-1}\right)$$

where γ_{t-1} is all the information available up to time $t - 1$. $Pr(S_t|\gamma_{t-1}$ is the possibility of S_t or filtered probabilities for each state. To compute the these filter probabilities, we use Kim's filter and after that we can maximized copula log-likelihood in order to get the estimated parameter. Additionally, the regime probability of multiple pair copulas can be computed by multiplying together the probabilities of each pair.

Then, the obtained parameter of dependence between the six pair copulas will be extended to compute the conditional VaR and the ES of these portfolios in two different regimes, using the following method. First, we simulate the joint-dependent distribution uniform from the fitted copula model via the copularnd function in MATLAB.

In this estimation, we simulate 10,000 replications of the portfolio returns and, subsequently, we multiply the inverse of the marginal distribution with the random variable to obtain ε_{it}^{k} . To find the return of each variable ($x_{it}^{(k)}$), we perform the estimation using the following formula:

$$x_{it}^{(k)} = \hat{u}_{it} + sqrt\,(h_{it}) \cdot \varepsilon_{it}^{(k)}$$

To compute the portfolio return in each period t, we specify an equally weighted portfolio return, that is, $X_{pt} = 0.25x_{1t} + 0.25x_{2t} + 0.25x_{3t} + 0.25x_{4t}$, where $x_{1t}x_{2t}$ $x_{3t}x_{4t}$ are the return of SET, return of IDX, return of PSE, and return of gold, respectively. In addition, we repeat this step to compute the portfolio return at time $t = t + 1, \ldots, t + T$ in order to construct the VaR and the ES with time-varying (CoVaR and CoES). In this computation, we compute all the risk measures at the 1, 5, and 10 % levels.

Additionally, the unconditional coverage of Kupiec [20] and the conditional coverage of Christoffersen [9] are both employed to examine the adequacy and the effectiveness of the CoVaR and CoEs estimates. The Kupiec test is given as follows:

$$LR_{uc} = 2ln\left(\left(\frac{p^{\alpha}}{N}\right)^{p^{\alpha}} \cdot \left(1 - \frac{p^{\alpha}}{N}\right)^{N-p^{\alpha}}\right) - 2ln(\alpha^{p^{\alpha}}\,(1-\alpha)^{N-p^{\alpha}}) \cdot \chi^2_{(1)}$$

The null hypothesis is

$$H_0 : \hat{p} = \tau^*$$

where $\hat{p}$ is the number of violations of N, $NBinomial(T, p)$ divided by the sample size of T,N/T and τ^* is the VaR confidence level.

The Christoffersen test extends the Kupiec test, proposing a VaR measure that exhibits both the independence of violations and the unconditional coverage and becoming the conditional coverage Christoffersen test. The test statistic can be given by

$$LR_{cc} = LR_{uc} + LR_{ind} + \chi^2_{(2)}$$

$$LR_{cc} = LR_{uc} + 2ln\frac{[(1-\pi_0)^{\lambda_{00}}\pi_0^{\lambda_{01}}(1-\pi_1)^{\lambda_{10}}\pi_1^{\lambda_{11}}]}{ln[(1-\pi_0)^{\lambda_{00}+\lambda_{10}}(\pi_1)^{\lambda_{01}+\lambda_{11}}] + \chi^2_{(2)}}$$

The null hypothesis is

$$H_0 : P_{01} = P_{11} = P,$$

where λ is the number of returns in state i while there have been stayed in state j. Note that state 1 denotes that the VaR estimate is violated and state 2 denotes that the VaR estimate is not violated. π is the probability of having an exception that is conditional on state "i" previously.

4 Empirical Results

The descriptive statistics of our data are shown in Table 1. We can see that the variables have different characteristics. SET, IDX, PHE, and GOLD exhibit high kurtosis excess and small positive skewness. This means that the marginal distributions of SET, IDX, PHE, and GOLD have a heavy tail to the right. Furthermore, the normality of these marginal distributions is strongly rejected by the Jarque-Bera test, with probability = 0.000. Thus, these results indicate that normal distribution might not be appropriate for the residual. Consequently, we assume that the student-t distribution is the appropriate marginal distribution for our residual. The result of the augmented Dickey-Fuller test in Table 2 suggests that the null hypothesis of the unit root test can be rejected and that all the variables are stationary at level under the 1 % confidence level.

4.1 Modeling Marginal Distributions

In this section, we create the marginal distributions of the returns, based on the Student-t distribution. The results presented in Table 2 demonstrate that ARMA(2,2)-GARCH(1,1) with student-t distribution is the best specification for SET, IDX, PSE, and GOLD. The results regarding marginal model fitting are shown in Table 2. We can observe that the degrees of freedom coefficient(υ) of all variables indicate a heavy tails. Besides, the high value for the estimated (δ) indicates a high volatility persistence. Therefore, there is a clear suggestion that the use of Student-t distributions become a good choice to describe the joint distribution of our returns.

Table 1 Summary statistics

	SET	IDX	PSE	GOLD
Mean	−0.0002	−0.00034	−0.00039	−0.00034
Median	0.0000	−0.00041	0.0000	−0.00037
Maximum	0.09596	0.10954	0.13089	0.1109
Minimum	−0.06841	−0.07623	−0.07056	−0.07549
Std. dev.	0.013022	0.014589	0.012995	0.013489
Skewness	0.412828	0.662984	1.026522	0.72253
Kurtosis	8.130544	10.9816	12.82065	10.62218
JarqueBera	2018.563	4893.439	7524.348	4498.896
Probability	0.0000	0.0000	0.0000	0.0000
ADF-test(prob.)	0.0000	0.0000	0.0000	0.0000

Source Calculation

Table 2 Univariate ARMA-GARCH models

	SET	IDX	PSE	GOLD
Mean equation				
ω_i	−0.00001***	−0.0009***	−0.0007***	−0.0007***
AR(2)	0.6020***	0.8488***	0.5299***	0.8401***
MA(2)	−0.6074***	−0.8485***	−0.5790***	−0.8714***
Variance equation				
γ_1	0.000001***	0.000002***	0.000004***	0.000004***
γ_2	0.038132***	0.109550***	0.110058***	0.132080***
δ	0.953421***	0.885183***	0.875990***	0.843404***
v_i	5.491568***	4.3389***	5.5767***	4.3113***
BERK-test (prob)	−0.9698	−0.9528	−0.9218	−0.8467
$Q_2(10)(prob)$	−0.6454	−0.1274	−0.6908	−0.7312
AIC (student-t)	−6.121834	−6.182507	−6.074986	−6.219103

Source Calculation
Notes *, **, and ***, respectively, denote significance at the 1, 5, and 10 % levels

Besides, the autocorrelation test (Ljung-Box test) and the goodness-of-fit test (Berkowitz test [BERK-test]) are also shown in this table. The p-value of the BERK-test suggests that the probabilities of the integral transform of the standardized residual are uniform in the [0,1] integral. Additionally, the p-value of the Ljung-Box-test of autocorrelation on standardized residual with 10 lags, $Q^2(10)$, confirms that we cannot reject the 5 % significance level; thus, there is no autocorrelation in any of the series.

4.2 Model Fit

After estimating the marginal distribution, in this section, we need to make an initial choice as regards the best copula function that has the best fit to estimate the structure dependence between each pair return. Therefore, we compared the two types of copula functions with the following specification. For each pair return, the Gaussian copula and the Student-t copula, copula analyses were conducted. The comparison of the functions, which is based on the sum of the AIC and the BIC values which are used to measure the function of fit, is shown in Table 3 reported in appendix. We learned that from the trial runs of these two alternative functions for each pair return, there is enough evidence provided by the results that Gaussian copula, which has the lowest sum of the AIC and the BIC values, is the better one because it has higher ability to explain the dependence structure than the Student-t copula.

Table 3 Family selection of each pair copula

AIC	SET-IDX	SET-PHP	SET-GOLD	IDX-PHE	IDX-GOLD	PHP-GOLD	SUM
Gaussian	564.8006	309.9402	82.5455	438.1732	75.789	14.4392	1485.6877
Student- t	600.9605	342.5209	81.7127	467.5137	97.4548	67.8912	1658.0538
BIC							
Gaussian	642.707	342.8934	115.4988	471.1264	108.7423	47.3924	1728.3603
Student- t	644.8981	386.4585	125.6503	511.4513	141.3924	111.8289	1921.6795

Table 4 Estimated parameters from Markov-switching dynamic copula

	$c_{set,php}$	$c_{set,idx}$	$c_{set,gold}$	$c_{idx,php}$	$c_{idx,gold}$	$c_{php,gold}$
ω_1	0.134***	0.023***	0.011***	0.170***	−1.277***	0.114***
ω_2	1.061***	2.083***	0.111***	3.693***	1.443***	0.114***
β_c	1.610***	−0.090***	2.001***	−1.986***	1.264***	1.860***
α_c	−0.040***	−0.245***	0.043***	0.061***	0.075***	0.081***
$\rho_{Nt}(S_1)$	0.26	−0.031	−0.147	0.0424	0.382	0.211
$\rho_{Nt}(S_2)$	0.827	0.744	0.5164	0.7919	0.425	0.597
Log-likelihood		−706.843		p1	q2	Duration
		Regime 1	p1	0.602	0.397	2.51
		Regime 2	p2	0.308	0.692	3.25

Source Calculation
Notes *, **, and ***, respectively, denote significance at the 1, 5, and 10 % levels

4.3 Results of Estimated Parameters

Table 4 reports the estimated parameters of the Markov-switching dynamic copula, by following Tofoli et al. [40]. We select the Gaussian copula as the best fit model for all pairs of returns since the normal copula has the lowest sum value of the AIC and the BIC values. The results show that the value of the intercept coefficient in regime 1(ω_1) is lower than the value of the regime intercept coefficient in regime 2 (ω_2). Thus, we can interpret regime 1 to be the low dependence regime, while regime 2 is the high dependence regime. Moreover, many recent studies, such as the studies by Tofoli et al. [40] and Karimalis and Nimokis [19], have found that the conditional correlation during market upturns is less than that during market downturns. Thus, this confirms the high dependence regime as the market downturn regime and the low dependence regime as the market upturn regime. Next, we take into consideration the estimated coefficient, β_c, which is related to the autoregressive parameter component in the dynamic equation. Different results have been obtained from these coefficients. We found that the autoregressive parameter components of $c_{set,idx}$, $c_{idx,php}$, $c_{idx,gold}$ *and* $c_{php,gold}$ have a negative sign, indicating that those pair relations might not be persistent over time, while the autoregressive parameter components of $c_{set,php}$ and $c_{set,gold}$ have a positive sign, indicating that those pair relations might not be persistent over time. As for the distance from the perfect correlation in the dependence dynamics co-movement (α_c), the results also provide a different sign for each pair return. We found that the α_c of $c_{set,php}$, $c_{set,idx}$ and $c_{idx,gold}$ has a negative sign, indicating that the greater distance from the perfect correlation can decrease their dependence, while the α_c of $c_{set,gold}$, $c_{idx,php}$ and $c_{php,gold}$ has a positive sign, indicating that the greater distance from the perfect correlation can increase their dependence. For the estimated average Pearson's correlation $\rho_{Nt}(S)$, for each pair return, we obtain a heterogeneous dependence. For $\rho_{set,idx}$, $\rho_{set,gold}$, $\rho_{idx,gold}$ and $\rho_{php,gold}$ we observe negative correlation in regime 1 and positive correlation in regime 2. Notice that gold returns seem to have economic interpretation from these results since gold markets have a negative correlation with TIP stock markets at the time of market upturns but a positive correlation during market downturns. These results could be explained by the safe haven approach: this approach explains that safe haven is defined as the asset that has a negative correlation with another asset in the turbulence market [4]. Therefore, according to the results, we can suggest that gold is a safe haven for TIP stock markets during market downturns. Thus, when the TIP stock market experiences a great decline, investors become less interested in stock. As a result, according to the demand and supply theory, gold becomes more in demand and this leads to its price increase. As for market upturns, such a result provides a positive correlation between the gold market and the TIP stock markets; therefore, we suggest that gold is not a safe haven for TIP stock markets during market upturns. As far as $\rho_{set,php}$, $\rho_{idx,php}$ and $\rho_{idx,php}$ are concerned, we observe the presence of a positive correlation between these pairs in both the regimes. A number of studies, such as the studies conducted by Lim [22] and Sriboonchitta et al. [37] also report the positive dependence and correlation between TIP stock markets. Consequently, we can conclude that TIP stock markets are moving together and that the scope for the diversification of the TIP stock markets to reduce risk is more limited.

In addition, the transition probabilities p and q of all return dependences are also reported in Table 4. We denote the probabilities $Pr(S_t = 1|S_{t-1} = 1)$ by p (low dependence regime) and $Pr(S_t = 2|S_{t-1} = 2)$ by p (high dependence regime). We can notice that neither of the regimes is persistent because of the low values obtained for the probabilities p and q. Moreover, the duration of stay in both the regimes is short, with the duration equaling 2.51 days for the low dependence regime and 3.25 days for the high dependence regime. This result, apparently, indicates that the dependence between these returns has high fluctuation.

4.4 Regime Probabilities

The estimated Markov-switching copula model produces the probabilities of two regimes for the period from 2008 to 2014. In this section, we plotted only the high dependence regime probabilities for all the returns presented in Fig. 1. We try to compare the smoothed probabilities between the dependency of the TIP stock returns (TIP) and the dependency between the TIP-GOLD returns (TIP-GOLD). Note that

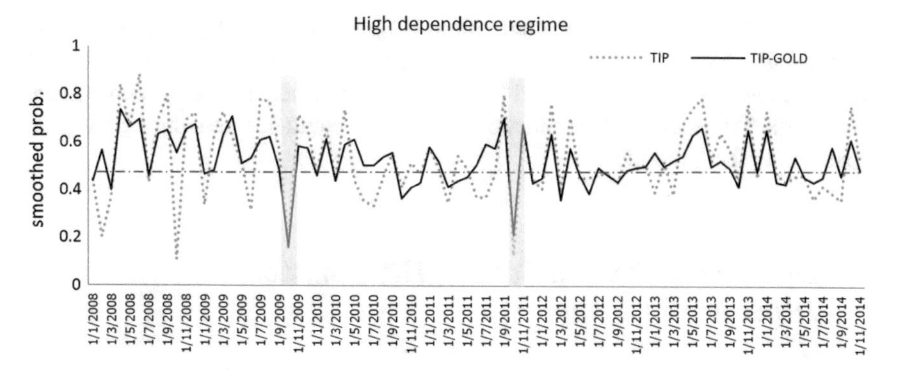

Fig. 1 The smoothed probabilities for the high dependence regime

low dependence regime is indicated as market upturns, while high dependence regime is indicated as market downturns. Based on the results, it is evident that the dependence between the TIP stock returns and the gold return (TIP-GOLD) has mostly taken place in market downturns, except for the period from August 2009 to October 2009 and the period from September 2011 to November 2011 (the gray-shaded line). It is surprising to note that the dependence of these returns stays mostly during market downturns despite the fact that TIP markets have experienced great growth after the introduction of quantitative easing (QE) in 2008. The study of Cho and Rhee [8] found that QE has led to a sharp rebound of capital inflow to the rest of the world, especially to emerging stock markets. Therefore, the dependence of TIP-GOLD (thick line) should stay in the market uptrend rather than the market downtrend. However, in those periods, we also found that the European debt crisis had taken place in many eurozone countries. The investors were concerned on account of the contagion effects on the eurozone and concerned about the uncertainty in the future design of the European monetary union (EMU) with respect to the crisis regarding the exit of Greece. Therefore, TIP stock markets might suffer from this European debt crisis. In addition, this finding is also confirmed by Billio et al. [6]: they found some evidence that sovereign debt problems, in particular Italy and Spain, lead the eurozone economy to enter recession. Thus, we can conclude that the European debt crisis exerted the high negative pressure on the TIP-GOLD markets, leading the market toward entering the turbulence. However, without taking gold into account, if we consider only the case of the dependency of the TIP stock returns (TIP) (dotted line), we find that the probability of staying in the high dependence regime of this portfolio mostly decreases over time. Thus, it is possible to conclude that gold might be the asset that can reduce the dependence of the TIP-GOLD returns.

In addition, let us analyze the two main subperiods of time which tend to stay in the low dependence regime or the market uptrend period, which are from August 2009 to October 2009 and from September 2011 to November 2011. The first period coincided with the expansion of the quantitative easing of the United States (U.S.), and the second period reflected the announcement of the Federal Reserve (FED) that they had decided to launch a new 40 billion USD a month bond purchase program of the agency Mortgage Back Securities (MBS) and also to continue the policy of extremely low rates until at least mid-2015. Therefore, we expect that the dependence between the TIP stock markets and the gold markets could decrease during the great expansion of the QE policy.

4.5 Dynamic CoVaR and CoES in High and Low Dependence Regimes

In this section, we analyze the evolution of the CoVaR and CoES which are computed based on the Gaussian Markov-switching copula and the smoothed probabilities of two dependence regimes, for TIP-GOLD returns. In this computation, we estimate the CoVaR and CoES based on the probabilities which were obtained in the previous section. The results of the dynamic CoVaR and CoES are shown in Fig. 2. Recall that in the previous section, the dependence of TIP-GOLD has mostly taken place in market downturns, in the high dependency regime, except for the period from

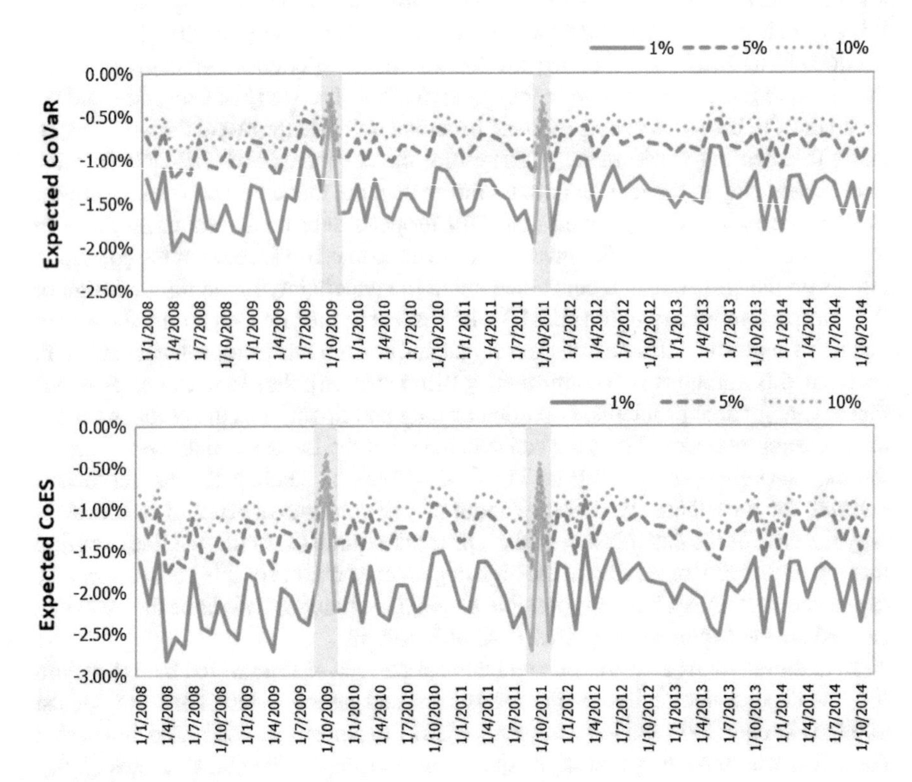

Fig. 2 The dynamic CoVaR and CoES

August 2009 to October 2009 and for the period from September 2011 to November 2011. Considering those two subperiods, we find that the value of CoVaR and CoES decrease sharply and reach the peak loss around 0.5 %, before returning to their previous loss level within 2–3 months. The average gap between the high and the low dependence regimes is around 2 %. In addition, we could see that the average loss in the low dependence regime is less than the average loss in the high dependence regime. Therefore, this indicates that higher values of dependency are associated with higher values of average 1, 5, and 10 % CoVaR and CoES. The results thus confirm our hypothesis that high dependence regime can be considered as indicative of market downturns, while low dependence regime can be taken as indicative of market upturns. This finding also corresponds to the study of Karimalis and Nimokis [19]: they also found that VaR is higher during market downturns than during market upturns.

The CoVaR and CoES estimations with 1, 5, and 10 % levels are reported in Table 5. The estimations reveal the calculated average of 1, 5, and 10 % CoVaR and CoES on an equally weighted portfolio base on the probabilities $Pr(S_t = 1|S_{t-1} = 1)$ for the low dependence regime, and $Pr(S_t = 2|S_{t-1} = 2) = 1$, for the high dependence regime. At each level of risk, we can see that the average loss based on CoVaR and CoES in the low dependence regime is less than the average loss in the high dependence regime. Moreover, we compare the average loss between the high and the low dependence regimes and find that the difference between those two regimes is around 12 %.

A further estimation of the expected CoVaR and CoES are reported in Table 6. We calculated the average expected values of 1, 5, and 10 % CoVaR and CoES on an equally weighted portfolio base on the Gaussian Markov-switching copula and the smoothed probabilities of two dependence regimes (see Fig. 1). For the TIP-GOLD portfolios case, at 1 %, 5 %, and 10 % levels, the estimated average CoVaR values are −1.40, −0.84, and −0.62 %, respectively, while the estimated average CoES values are, respectively, −2.01, −1.27, and −1.01 %. This indicates that for the 1, 5, and 10 % confidence levels of CoVaR, the expected average loss values on an equally weighted portfolio over the next day do exceed 2.01, 1.27, and 1.01 %. For the TIP

Table 5 Average loss computed based on 1, 5, and 10 % daily CoVaR and CoES

Portfolio		Low dependence regime			High dependence regime		
		1 %	5 %	10 %	1 %	5 %	10 %
TIPGOLD	CoVaR	−1.38	−0.87	−0.65	−2.77	−1.67	−1.23
	CoES	−1.85	−1.21	−0.99	−3.83	−2.43	−1.93
TIP	CoVaR	−1.95	−1.24	−0.92	−3.04	−1.79	−1.27
	CoES	−2.66	−1.75	−1.41	−4.23	−2.63	−2.07

Source Calculation

Notes We specify an equally weighted portfolio return, that is, $X_{pt} = 1/3X_{1t} + 1/3X_{2t} + 1/3X_{3t}$, where x_{1t} x_{2t} x_{3t} are the returns of SET, returns of IDX, and returns of PSE, respectively, of the TIP portfolios

Table 6 Average expected loss computed based on 1, 5, and 10 % daily CoVaR and CoES

Portfolio		Expected value		
		1 %	5 %	10 %
TIPGOLD	CoVaR	−1.40	−0.84	−0.62
	CoES	−2.01	−1.27	−1.01
TIP	CoVaR	−1.48	−0.90	−0.65
	CoES	−2.04	−1.30	−1.03

Source Calculation

portfolios case, at 1, 5, and 10 % levels, the estimated average CoVaR values are −1.48, −0.90, and −0.65 %, respectively, while the estimated average CoES values are, respectively, −2.04, −1.30, and −1.03 %. Similarly, in the case of the TIPGOLD portfolio, this result also indicates that for the 1, 5, and 10 % confidence levels of CoVaR, the expected average loss values on an equally weighted portfolio over the next day do exceed −1.48, −0.90, and −0.65 %. We can observe the average loss between the TIP-GOLD and the TIP portfolios and find that gold could reduce the average loss in the TIP stock markets by around 0.02−0.08 %.

From the above results, we find three interesting observations. First, the euro debt crisis had a sizeable negative effect on TIP stock markets along our analysis period through contagion effect. Stracca [39] analyzed the global effects of the euro debt crisis and found a transmission of the euro crisis to countries outside the eurozone. Forbes [14] provided four main channels of contagion, which are trade channels, bank channels, portfolio investors' channels, and wake-up call channels. However, in this study, we focus on the trade channel and the portfolio investor channel because these might play important roles in the TIP stock market. Considering TIP's capital account as a portfolio investor channel, it is obvious that eurozone has been an important investor in TIP in terms of portfolio investment. The value of portfolio investment in Thailand, Indonesia, and the Philippine stock markets are 14.6 billion euros, 24.8 billion euros, and 7 billion euros, respectively, in 2012. Thus, during the market downturn, the euro investors in the TIP stock markets rebalance their portfolio by reducing the proportion of equity, resulting in a higher proportion of safe haven assets such as government bonds and gold, which, in turn, gives rise to capital outflow from the TIP stock markets. Then, considering the current account of TIP as a trade channel, we find that the trade value between TIP and eurozone was 1.9 % shares of the total EU trade, according to the European Commission Directorate-General for Trade (2013). Thus, when eurozone enters a period of crisis, the demand for TIP goods decreases. Consequently, heavy exporters suffer from lower revenue and lower earnings, leading to lower stock prices.

Second, we find that gold acts as a safe haven in the TIP stock markets because the risk is lower when we add gold to the TIP portfolios. According to the Markowitz theory, the chance to decrease the portfolio risk comes from the lack of correlation across assets. We know that TIP stock markets generally move in the same direction; therefore, the scope for diversification within the TIP stock portfolio is limited. However, we find that gold serves as a hedge and safe haven for TIP stock markets during

Table 7 Average expected loss computed based on 1, 5, and 10 % daily CoVaR and CoES

TIPGOLD					TIP	
Copula		α	Kupiec	Christoffersen	Kupiec	Christoffersen
Gaussian	CoVaR	1 %	0.8694	0.9622	1	0.9946
		5 %	1	0.1179	1	0.9352
		10 %	1	0.2324	1	0.8976
	CoES	1 %	0.9999	0.3693	1	0.9997
		5 %	1	0.529	1	0.9997
		10 %	1	0.68	1	0.9985

Source Calculation
Note Kupiec and Christoffersen correspond to the p-values of the respective test

market downturns and upturns. We also find that gold has a negative correlation with TIP stock markets (Table 4). Hence, adding gold can decrease the risk of portfolio; also, gold serves as a safe haven for TIP stock markets.

Third, we observe two main subperiods of time which coincided with the expansion of quantitative easing of the United States (U.S.); it is possible to explain that QE may have had a direct effect on TIP stock markets. This finding does correspond to the findings of Cho and Rhee [8]. They found significant evidence that QE programs have a considerable impact on emerging Asian stock markets.

In order to assess the accuracy of the CoVaR and the CoES estimates, we employ the backtest at 99, 95, and 90 % confidence levels. Backtesting is a process to check the accuracy and the robustness of the VaR model. In this paper, we evaluate the CoVaR and the CoES forecasts by using the likelihood ratio tests proposed by Kupiec [20] and Christoffersen [9]. Table 7 reports the results of the Kupiec and the Christoffersen tests for the daily CoVaR and CoES forecasts. We can observe that both the TIPGOLD and the TIP portfolios, at 1, 5, and 10 % levels, are larger than 0.10. Thus, it is not possible to reject the null hypothesis that the expected proportion of violation (p) is equal to the VaR confidence level (τ). Therefore, the Gaussian Markov-switching copula was concluded as the appropriate model to estimate the VaR and the ES.

5 Conclusions

In recent years, the financial market has been highly volatile under free capital mobility. The stock markets of Thailand, the Philippines, and Indonesia (TIP) are no exception. We found that the TIP stock markets displayed strong growth and high returns; however, these developments bring with them a large proportion of risk, as well. Thus, the investor needs to diversify the investing to safer places such as the gold market. To make the asset allocation, it is necessary to measure the risk in these markets. By now, the most preferred approach for measuring market risk is by assessing the value at risk (VaR) and the expected shortfall (ES). Furthermore, understanding the co-movement between the markets is important for the policy makers and the fund managers to understand and prepare the appropriate policy for investors.

In this study, we employ the Markov-switching dynamic copula, which was proposed in Tofoli et al. [40], to estimate the dependency between these markets over the period from January 2008 to November 2014. We, then, extend the obtained dependency to calculate the VaR and the ES of the portfolio in different regimes. As demonstrated by the results, the Gaussian Markov-switching copula has lower AIC and BIC values when compared with the Student-t Markov-switching copula. The model confirms the significant positive co-movement of TIP stock markets and, significantly, the negative co-movement between the TIP stock returns and gold returns. Moreover, the model also shows that TIP-GOLD portfolios take place mostly during market downturns since the model identifies a long period of high dependence regime as regards our samples. This period coincides with the European debt crisis. However, the model also captures significant decrease in the dependence in two subperiods, that is, from August 2009 to October 2009 and from September 2011 to November 2011, which may be explained by the expansion of quantitative easing of the United States (U.S.). The estimated CoVaR and CoES based on this model show that CoVaR and CoES are higher in high dependence regime or market downturn. When we compare the average loss between TIP-GOLD and TIP portfolios, we find an interesting main issue, which is that gold could reduce the loss in TIP stock markets by around 0.57–0.81 % in market upturns, and 0.27–0.40 % in market downturns.(see Table 5, CoVaR and CoES at 1 % confidence level). This indicates that gold can reduce the risk in market downturn more than in market upturns.

Based on the results, we suggest that gold can be an investment asset for TIP investors. It tends to possess a role as a hedging investment in market downturns or in turbulent periods. Thus, gold can be a safe haven asset which provides diversification benefits in TIP financial markets during periods of market turbulence. Additionally, the investor has to monitor any possible European debt crisis since that would generate tremendous negative pressure on TIP-GOLD markets and would lead the market toward turbulent times.

Future research in this field can expand the scope of other assets such as oil, bond, and currency; tests can be carried out to find out whether or not those assets can serve as a safe haven for TIP stock markets.

Acknowledgments The authors are grateful to Puey Ungphakorn Centre of Excellence in Econometrics, Faculty of Economics, Chiang Mai University for the financial support.

References

1. Adrian, T., Brunnermeier, M.K.: CoVaR (No. w17454). National Bureau of Economic Research. (2011)
2. Ang, A., Chen, J.: Asymmetric correlations of equity portfolios. J. Financ. Econ. **63**(3), 443–494 (2002)
3. Arouri, M.E.H., Lahiani, A., Nguyen, D.K.: World Gold Prices and Stock Returns in China: Insights for Hedging and Diversification Strategies (2013)

4. Baur, D.G., Lucey, B.M.: Is gold a hedge or a safe haven? An analysis of stocks, bonds and gold. Financ. Rev. **45**(2), 217–229 (2010)
5. Bhunia, A., Mukhuti, S.: The impact of domestic gold price on stock price indices an empirical study of Indian stock exchanges. Univers. J. Market. Bus. Res. **2**(2), 35–43 (2013)
6. Billio, M., Casarin, R., Ravazzolo, F., Van Dijk, H.K.: Interactions between eurozone and US booms and busts: a Bayesian panel Markov-switching VAR model. University Ca'Foscari of Venice, Department of Economics Research Paper Series No, 17 (2013)
7. Billio, M., Pelizzon, L.: Value-at-risk: a multivariate switching regime approach. J. Empir. Finan. **7**(5), 531–554 (2000)
8. Cho, D., Rhee, C.: Effects of quantitative easing on Asia: capital flows and financial markets. Singap. Econ. Rev. **59**(03), 1450018 (2014)
9. Christoffersen, P.F.: Evaluating interval forecasts. Int Econ Rev 841–862
10. Coudert, V., Raymond, H.: (2011) Gold and financial assets: are there any safe havens in bear markets? Econ. Bull. **31**(2), 1613–1622 (1998)
11. Do, G.Q., Sriboonchitta, S.: Cointegration and causality among international gold and ASEAN emerging stock markets. Working paper Chiangmai University, pp. 1–11 (2009)
12. Embrechts, P., Lindskog, F., McNeil, A.: Modelling dependence with copulas and applications to risk management. Handb. Heavy Tailed Distrib. Financ. **8**(1), 329–384 (2003)
13. Filho, O.S., Ziegelmann, F.A.: Assessing some stylized facts about finacial market indexes: a Markov copula approach. J. Econ. Stud. **41**(2), 253–271 (2014)
14. Forbes, K.: The "Big C": identifying and mitigating contagion. NBER Working Papers 18465. National Bureau of Economic Research, Cambridge (2012)
15. Fuss, R., Kaiser, D.G., Adams, Z.: Value at risk, GARCH modelling and the forecasting of hedge fund return volatility. J. Deriv. Hedge Funds **13**(1), 2–25 (2007)
16. Gilmore, C.G., McManus, G.M., Sharma, R., Tezel, A.: The dynamics of gold prices, gold mining stock prices and stock market prices comovements. Res. Appl. Econ. **1**(1), 1–19 (2009)
17. Hammoudeh, S., Malik, F., McAleer, M.: Risk management of precious metals. Q. Rev. Econ. Financ. **51**(4), 435–441 (2011)
18. Joe, H.: Multivariate Models and Multivariate Dependence Concepts. CRC Press, Boca Raton (1997)
19. Karimalis, E.N., Nomikos, N.: Measuring systemic risk in the European banking sector: a Copula CoVaR approach (2014)
20. Kupiec, P.H.: Techniques for verifying the accuracy of risk measurement models. J. Deriv. 3(2), 73–84 (1995)
21. Li, M.Y.L., Lin, H.W.W.: Estimating value-at-risk via Markov switching ARCH models an empirical study on stock index returns. Appl. Econ. Lett. **11**(11), 679–691 (2004)
22. Lim, L.K.: Linkages between ASEAN stock markets: a cointegration approach. In: MODSIM 2007 International Congress on Modelling and Simulation. Modelling and Simulation Society of Australia and New Zealand, pp. 1818–1824 (2007)
23. Lin, J., Wu, X.: Smooth Tests of Copula Specification under General Censorship (2013)
24. Mandal, M., Lagerkvist, C.J.: A Comparison of traditional and copula based VaR with agricultural portfolio. In: 2012 Annual Meeting, August 12–14, 2012, Seattle, Washington (No. 124387). Agricultural and Applied Economics Association (2012)
25. McAleer, M., Da Veiga, B.: Forecasting value-at-risk with a parsimonious portfolio spillover GARCH (PSGARCH) model. J. Forecast. **27**(1), 1–19 (2008)
26. Miller, D.J., Liu, W.H.: Improved estimation of portfolio value-at-risk under copula models with mixed marginals. J. Futures Mark. **26**(10), 997–1018 (2006)
27. Miyazaki, T., Hamori, S.: Cointegration with regime shift between gold and financial variables. Int. J. Financ. Res. **5**(4), 90 (2014)
28. Najafabadi, A.T.P., Qazvini, M., Ofoghi, R.: The impact of oil and gold prices shock on Tehran stock exchange: a copula approach. Iran. J. Econ. Stud. **1**(2), 23–47 (2012)
29. Ostrup Christensen, M., Ellegaard, J.: Gold Investment (2014)
30. Ozun, A., Cifter, A.: Portfolio value-at-risk with time-varying copula: evidence from the Americas. J. Appl. Sci. **7**(14) 1916–1923 (2007)

31. Pafka, S., Kondor, I.: Evaluating the riskmetrics methodology in measuring volatility and value-at-risk in financial markets. Phys. A: Stat. Mech. Appl. **299**(1), 305–310 (2001)
32. Palaro, H.P., Hotta, L.K.: Using conditional copula to estimate value at risk. J. Data Sci. **4**, 93–115 (2006)
33. Patton, A.J.: Applications of copula theory in financial econometrics. Doctoral Dissertation, University of California, San Diego (2002)
34. Patton, A.J.: Modelling asymmetric exchange rate dependence. Int. Econ. Rev. **47**(2), 527–556 (2006)
35. Phoong, S.W., Ismail, M.T., Sek, S.K.: A Markov switching vector error correction model on oil price and gold price effect on stock market returns. Inf. Manag. Bus. Rev. **5**(7), 158–169 (2013)
36. Schweizer, B., Sklar, A.: Probabilistic Metric Spaces (1983)
37. Sriboonchitta, S., Liu, J., Kreinovich, V., Nguyen, H.T.: A vine copula approach for analyzing financial risk and co-movement of the Indonesian, Philippine and Thailand stock markets. Modeling Dependence in Econometrics, pp. 245–257. Springer International Publishing, Berlin (2014)
38. Shen, X., Chokethaworn, K., Chaiboonsri, C.: The dependence structure analysis among gold price, stock price index of gold mining companies and Shanghai composite index. Empir. Econ. Quant. Econ. Lett. **2**(4), 53–64 (2013)
39. Stracca, L.: The global effects of the Euro debt crisis (2013)
40. Tofoli, P.V., Ziegelmann, F.A., Silva Filho, O.C.: A comparison Study of Copula Models for European Financial Index Return (2012)

Factors Affecting Consumers' Willingness to Purchase Telehomecare Products

Jakkreeporn Sannork and Wan-Tran Huang

Abstract This paper investigated the influence of various factors-product knowledge and perceived risk, benefit, and value-on consumers' willingness to purchase telehomecare products (e.g., the RFID smart care watch). Of the 500 returned questionnaires, 388 were valid (77.60 %). Structural equation modeling (SEM) was used as the main research method. The results showed that the perceived risk of telehomecare products by consumers had a significant impact on perceived value, but it had no significant impact on willingness to purchase. In contrast, product knowledge significantly influenced the perceived benefit of the product and consumers' willingness to buy.

Keyword Senior citizens · Telehomecare products · Perceived risk · Willingness to purchase · Structural equation modeling (sem)

1 Introduction

The implementation of the National Health Insurance program had changed Taiwanese people's behavior towards medical treatments. Advances in medical technology and improvements in the standard of living have increased the average lifespan, resulting in an aging society. The World Health Organization has defined an aging society as one where 7 % or more of its total population are 65 years or older. According to this definition, revealed by statistics from the Ministry of the Interior (MOI), Taiwan officially became an aging society in 1993. The Council for Economic Planning and Development (CEPD) predicted that the population of senior citizens in Taiwan will have been approximately 3.45 million (14 % of the total population) by

J. Sannork
Agricultural Systems Management Program, Faculty of Agriculture, Chiang Mai University, Chiang Mai, Thailand
e-mail: nn_cmu@hotmail.com

W.-T. Huang (✉)
Department of Business Administration, Asia University, Taiwan, People's Republic of China
e-mail: wthuang@asia.edu.tw

V.-N. Huynh et al. (eds.), *Causal Inference in Econometrics*,
Studies in Computational Intelligence 622, DOI 10.1007/978-3-319-27284-9_38

2018. By 2025, senior citizens will account for 20 % of the total population, making Taiwan a super-aged society. The proportion is expected to continue increasing, and is estimated to reach 39 % by 2060.

A rapidly growing aging population means an increase in the prevalence of chronic diseases, illnesses, and impaired functioning associated with old age. Data from the CEPD [5] indicated that the number of people with dementia or disabilities in Taiwan in 2008 was approximately 390,009, which was expected to reach 900,000 by 2031. The rapidly growing aging population highlights the urgent issue of elderly care. With the increasing need for medical and long-term homecare services, the Taiwanese government must begin to provide such services to achieve their desired goals. One direction for future development is telehealth services. However, using information and communications technology and related equipment, which patients can wear or have them installed in their homes, will provide many benefits to monitoring and preventive homecare services (Barlow et al. [1]). Patients with chronic illnesses will be able to receive immediate attention, have their physiological conditions monitored, and engage in two-way interaction with professionals within the comfort of their own homes. At the same time, the workload and transportation costs of both patients and medical professionals will be greatly reduced. Remaining at home will also increase the patients' freedom of mobility and allow them to manage their illnesses themselves, improve the quality of medical care by eliminating unnecessary medical visits, and provide continuous care.

There are various types of telehomecare products, such as the Radio Frequency Identification (RFID) smart care watch for senior citizens. Although these products are expected to offer great benefits, many may be unapproachable for senior citizens due to unfamiliarity with innovative technologies. A senior citizen who has not used a computer may be concerned about learning how to use an ICT product and worried about potential costs associated with adapting to the new technology. Although, as noted by Park and Lessig [17], product knowledge may ease some of these concerns, and peoples perception of risk may influence their willingness to purchase a new product.

Sweeney and Soutar [21] pointed out that consumers' perceived risk is very important in generating perceived value and quality, which affect willingness to purchase. Product knowledge can influence perceived risk and is an important factor influencing consumers' willingness to purchase (Klerck and Sweeney [12]). In their study, Teas and Agarwal [22] pointed out that perceived risk is a mediating factor between consumers' perception of quality and their perception of value.

This study investigated the relationship between perceived risk and value on consumers' willingness to purchase. We constructed an integrated model from the consumers' point of view based on the literature and investigated the influence of the various factors-product knowledge and perceived risk, benefit, and value-on consumers' willingness to purchase telehomecare products. Through this study, we hope to provide companies with information for the development of marketing policies and strategies.

2 Methodology

2.1 Research Framework

This study comprehensively analyzed relevant theories and literature before proposing a conceptual framework (Fig. 1). Perceived risk, product knowledge, and perceived benefit are antecedent variables for and directly and indirectly affect perceived value and the willingness to purchase.

Hypotheses

From this conceptual framework, we derived our research hypotheses as outlined below.

Relationship between perceived risk/value and the willingness to purchase.

Snoj et al. [20] pointed out that there is a negative relationship between consumers' perceived risk and value. Monroe and Krishnan [15] believed that perceived value has a positive effect on perceived quality, and anticipated cost has a negative effect on perceived value, but perceived value positively affects willingness to purchase. The results of a study by Zeithaml [25] indicated that consumers' willingness to purchase is affected by objective price, perceived quality, perceived value, and product attributes. The willingness to purchase occurs after objective price, perceived quality, perceived value, and product attributes have been generated.

Bruwer et al. [3] believed that the risks consumers perceive during the decision-making process directly affect their willingness to purchase. Lim [13] pointed out that greater perceived risk results in less likelihood of purchase. The results of a study by Park et al. [18] indicated that there is a negative relationship between perceived risk and the willingness to purchase. Grewal et al. [7] found a positive relationship between the perceived value of a product and consumers' willingness to purchase it. With this background, we proposed the following hypotheses.

H1: Perceived risk negatively affects perceived value.

H2: Perceived risk negatively affects the willingness to purchase.

H3: Perceived value positively affects the willingness to purchase.

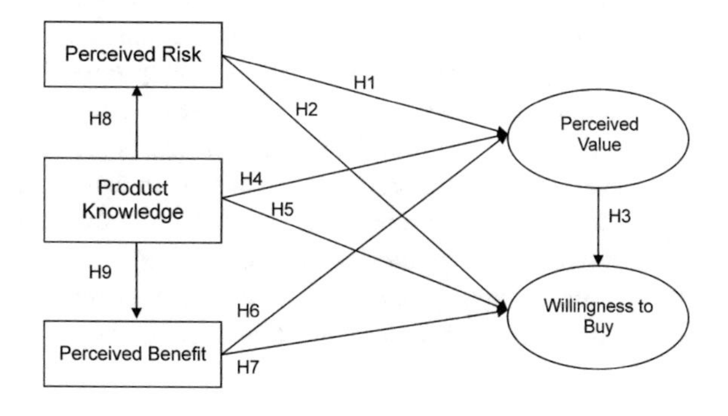

Fig. 1 Conceptual framework

Relationship between product knowledge/perceived value and the willingness to purchase.

Product knowledge can be divided into three major categories: (1) subjective knowledge or perceived knowledge; (2) objective knowledge; and (3) experience-based knowledge (Brucks [2]; Moorman et al. [16]). A study by Rao and Monroe [19] indicated that when consumers select a product, having more product knowledge made them more likely to research factors such as attributes, functions, and price. The fairness of the retail price will be assessed according to the product information and further affects their willingness to purchase. Brucks [2] pointed out that consumers' subjective knowledge affects their choice of search strategy, which affects their final purchasing decision. Horn and Salvendy [9] pointed out that receiving more details about innovative products has a positive impact on consumers' willingness to purchase. Analysis of these theories led to the following hypotheses:
H4: Product knowledge positively affects perceived value.
H5: Product knowledge positively affects the willingness to purchase.

Relationship between perceived benefits/value and the willingness to purchase.

Lovelock [14] believed that when comparing perceived benefits and costs, perceived benefits will positively affect perceived value while perceived costs will negatively affect it. Jen and Hu [11] pointed out that perceived benefits and costs can be used to balance perceived value. They also confirmed that perceived benefits positively affect perceived value. Notably, Zeithaml [25] showed that perceived value is the net value after deducting the cost from the benefits that the consumer gains from purchasing the item. Perceived value has a positive relationship with willingness to purchase. The higher the perceived value is, the higher the willingness to purchase. Chaudhuri [4] also believed that perceived quality is an important factor affecting consumer satisfaction. Higher perceived quality is associated with a greater willingness to purchase. These results lead us to form the following hypotheses:
H6: Perceived benefits have a positive effect on perceived value.
H7: Perceived benefits have a positive effect on the willingness to purchase.

Relationship between product knowledge/perceived risk and perceived benefits.

Klerck and Sweeney [12] pointed out that subjective and objective knowledge of a product directly affects consumers' psychological risk, as well as the functional and physical risks arising from unknown elements. Tuu and Olsen [23] suggested that consumers search for information to help lower perceived risk. More product information is related to a more predictable outcome, which would in turn lower the perceived risk. In addition, Hoyer and Maclnnis [10] believed that consumers compare newly acquired product information with existing knowledge. Product knowledge also affects consumers' subjective view of the product's benefits. With this as a background, we hypothesized the following:
H8: Product knowledge has positive effect on perceived risk.
H9: Product knowledge has a positive effect on perceived benefits

2.2 Pilot Survey

The questionnaire for this study had two main sections. The first section consisted of five variables (perceived risk, product knowledge, perceived benefits, perceived value, and willingness to purchase) and 26 assessment items. These items were designed upon knowledge from the literature, and they were tested and found to conform to acceptable standards. A 5-point likert scale was used to measure the participants' response to the items and their understanding of the relationships between variables. The second section of the questionnaire collected demographic information.

To ensure that the questionnaire was easy to understand and to assess internal consistency, a pilot survey was conducted before the formal survey. The 86 participants were patients of a hospital in central Taiwan and residents of a nursing home. The respondents were asked to provide suggestions to improve the phrasing of the questions. Comments and suggestions received were considered and the questionnaire revised accordingly.

After the data were compiled, factor analysis and Cronbach's coefficient were used to identify questions that should be removed or revised. Of the returned questionnaires, 72 were valid. Through the factor analysis, the loadings for each item exceeded 0.6 % and the explained variance was at least 64.71 %. Thus, all five main variables and 26 items were supported by the factor analysis.

The Cronbach's value for each variable was greater than 0.76, which is in the acceptable range (0.76–0.92). Thus, the measurement tools used for this study were valid and reliable. The results of the factor analysis and reliability tests are shown in Table 1.

2.3 Data Analysis

In this study, structural equation modeling (SEM) was used for data analysis. Confirmatory factor analysis and reliability analysis were also performed for each variable to check the construct validity and reliability, respectively. Next, we analyzed how well the proposed model fit the hypotheses. The fit between the survey data and the research model was evaluated using procedures proposed by Hair et al. [8]. Since perceived value was used as the mediating factor, we also used path analysis to determine whether there was a mediating effect. The strength of a mediating variable's indirect effect is equal to the product of all the path coefficients (Wu [24]). The existence of a mediating effect is proven when two potential indirect effects are greater than the direct effects.

Table 1 Results of factor analysis and reliability tests in the pilot survey

Variable	Assessment item	Factor loading	Cumulative explained variance (%)	Cronbach's value
Perceived risk	PR1	0.822		
	PR2	0.785		
	PR3	0.920	77.04	0.76
	PR4	0.910		
	PR5	0.856		
	PR6	0.861		
Product knowledge	PK1	0.837		
	PK2	0.846		
	PK3	0.796	68.86	0.89
	PK4	0.879		
	PK5	0.788		
Perceived benefit	PB1	0.874		
	PB2	0.870		
	PB3	0.850	70.73	0.90
	PB4	0.835		
	PB5	0.744		
Perceived value	PV1	0.882		
	PV2	0.898		
	PV3	0.863	64.71	0.84
	PV4	0.886		
	PV5	0.648		
Willingness to purchase	BI1	0.883		
	BI2	0.911		
	BI3	0.919	76.35	0.92
	BI4	0.887		
	BI5	0.759		

3 Research Results and Analyses

3.1 Basic Data Analysis

Five interviewers conducted the survey over one month, visiting hospitals and nursing homes. Questionnaires were handed out to patients and residents after we obtained their informed consent to participate. Of the 500 questionnaires issued and returned, 112 were deemed invalid due to incomplete answers. Thus, the effective return rate was 77.6 % (388).

The respondents' basic demographic characteristics were as follows: the participants were 43.6 male and 56.4 % female; most were between 41–50 years old; the majority (64.5 %) had high school or college education; most (38.2 %) had a monthly income in the range of NT$40,001–50,000, followed by NT$30001–40000 (28.6 %); 68.9 % were married and lived with their partners or children. Their main concern was interpersonal relationships among people of the same age group. Several potential telehomecare system users/purchasers were identified and they were found to have relatively stable incomes and a certain amount of knowledge regarding healthcare products and systems.

Most of the respondents (64.6 %) considered themselves their own primary caregiver and source of income. Most of them (91.3 %) listed their children as their emergency contacts. In terms of health, 44.5 % considered their health average, 36.1 % good, 8.7 % poor, 8.4 % excellent, and 2.3 % very poor. When asked if they had fallen down in the previous year, 61.6 % said, "no." Falls were most common in bathrooms/washrooms (30.4 %) and stairs (2.38 %). Of those who fell, 30.7 % sought help by shouting loudly (58.1 %), 29.0 % by phoning a relative or friend, 9.7 % by calling 119 (i.e., the emergency services number in Taiwan), and 3.2 % by ringing a bell for assistance. In most instances, injuries were mild enough for the respondents to attend to themselves or seek help.

3.2 *Descriptive Statistics*

The participants responded to each statement using a 5-point likert scale. The mean and standard deviation for each variable are stated below.

Perceived risk (PR). This section included six statements. Item 6, "I am worried that if the RFID smart care watch I want to purchase malfunctions, it will cost me more time to repair or replace it" had the highest average score of 3.87, which indicated the respondents were most concerned about the inconvenience and time loss arising from product malfunction. Item 4, "I am worried about potential emotional stress or pressure when using the RFID smart care watch that I want to purchase" had the lowest average score of 2.91. The standard deviations were between 0.815 and 1.040, which were greater for items 3 and 4 than for item 1. The standard deviations for the respondents were greater for their understanding of physical and emotional risks compared with the other items.

Product knowledge (PK). This section included five items. Item 2, "I believe that the RFID smart care watch can satisfy my needs for a fast, convenient, healthy, safe, and rich digital lifestyle" had the highest average score at 4.03. This shows that most of the respondents had generally positive objective knowledge about the product and believed in its value. The lowest average score of 2.36 was for item 5, "I can determine the relationship between the price and quality of the RFID smart care watch," which measured the difficulty that the respondents had when matching their knowledge of the quality of a product with its price. The standard deviations for the items were between 0.744 and 0.972, which showed that the respondents had relatively similar amounts of product knowledge.

Perceived benefits (PB). This section consisted of five questions. The highest average score, at 3.65, was for item 1, "I believe that using the RFID smart care watch will keep time during homecare activities and in general," indicating that the respondents were aware of the functional benefits of the product. The lowest average score, at 2.17, was for item 5, "I believe that I have to spend a lot of energy learning how to use the RFID smart care watch." The standard deviations were between 0.716 and 1.052.

Perceived value (PV). Of the five questions in this section, item 2, "I believe that the functions of the RFID smart care watch will bring convenience and efficiency to my life" had the highest average score of 3.94, which indicated that the respondents were aware of the functional value of the product. The lowest average score of 3.17 was for item 5, "When I no longer use the product, I will give it away as a gift or auction it online," reflecting a lack of understanding of the residual value of the product. The standard deviations were between 0.753 and 1.040, with that for item 5 being greater than 1, indicating a relatively large difference among the respondents in terms of the perceived residual value of the product.

Willingness to purchase (BI). This section included five questions. Item 2, "I am willing to buy the RFID smart care watch because it is available on the market" had the highest average score of 4.04. The lowest average score of 3.29 was for item 3, "I will purchase other homecare products similar to the RFID smart care watch." Therefore, the general willingness of the respondents to purchase the product was not entirely generated by impulse. The standard deviations were between 0.736 and 0.903, indicating similarity among the respondents in terms of their willingness to purchase the product.

3.3 Confirmatory Factor Analysis and Reliability Analysis

A confirmatory factor analysis was conducted for each variable to determine its composition further. According to the suggestion by Hair et al. [8], statements with factor loadings of less than 0.5 were removed. This prevented the conceptual model from being affected by statements with a lack of explanatory power. The results of the analysis showed that there were seven items with factor loadings of less than 0.5: PK, PK5, PB3, PB5, PV1, PV5, and BI5.

After these questions were removed, the model was recalculated and the reliability analysis conducted again. The results of the analysis are shown in Table 2. This time, the factor loadings on all of the assessment items for each variable were significant and were greater than 0.6. The average variance extracted (AVE) volume was greater than 0.5 and the composite reliability was greater than 0.75. Therefore, by removing the eight questions with factor loadings of less than 0.5, we were able to achieve good construct validity and internal consistency.

Table 2 Confirmatory factor analysis for each variable ($n = 388$)

Variable	Assessment item	Factor loading	Index reliability R^2	Composite reliability	AVE volume
Perceived risk	1. I am worried that I do not understand the functions of the product	0.776	0.631		
	2. I am worried that the price of the product is different from what I previously thought	0.792	0.673		
	3. I am worried that the product will cause physical discomfort.	0.827	0.761		
	4. I am worried about the potential emotional stress or pressure when using the product.	0.690	0.582	0.777	0.542
	5. I am not worried that my family members or friends have negative perceptions of the product	0.630	0.397		
	6. I am worried that if the product malfunctions, it will cost me more time to repair or replace it	0.858	0.792		
Product knowledge	1. I will take the initiative to search out product information.	0.763	0.582		
	2. I believe the product can satisfy my needs for a fast, convenient, healthy, safe, and rich digital lifestyle	0.819	0.671	0.839	0.636
	3. I understand that the product is the result of the fusion of the Internet and digital technology.	0.828	0.686		
Perceived benefit	1. I believe using the product will save homecare time and time in general	0.604	0.365		
	2. I believe using the product will enhance my image	0.833	0.694	0.797	0.572
	3. I believe using the product will give me access to more homecare functions.	0.810	0.656		
Perceived value	1. I believe using the product will improve my relationships with family members and friends	0.824	0.679		
	2. I believe that the functions of the product will bring convenience and efficiency to my life	0.884	0.781	0.885	0.719
	3. I believe that the money spent on purchasing the product is worth it	0.834	0.696		

(continued)

Table 2 (continued)

Variable	Assessment item	Factor loading	Index reliability R^2	Composite reliability	AVE volume
Willingness to purchase	1. I will recommend this product to my relatives and friends	0.881	0.776		
	2. I am willing to buy the product because it is available in the market	0.969	0.939		
	3. I will purchase other homecare products similar to this product	0.642	0.571	0.941	0.841
	4. I believe that the price of the product is reasonable	0.736	0.659		

Note AVE = average variance extracted, and factor loadings represent how much a factor explains a variable in factor analysis

Table 3 Correlation coefficients and test values of the variables

Variable	Perceived risk	Product knowledge	Perceived benefit	Perceived value	Willingness to purchase	AVE
Perceived risk	1.00	–	–	–	–	0.542
Product knowledge	0.61**	1.00	–	–	–	0.636
Perceived benefit	0.44**	0.42**	1.00	–	–	0.572
Perceived value	0.60**	0.74**	0.45**	1.00	–	0.719
Willingness to purchase	0.62**	0.77**	0.49**	0.74**	1.00	0.733

Note AVE = average variance extracted
**$p < 0.01$

To further distinguish between the variables, we also conducted a discriminant validity analysis in accordance with the method proposed by Fonell and Larcker [6]. If the AVE of a variable is larger than the square of its correlation coefficient with another variable, then there is discriminate validity. The results of the analysis are shown in Table 3. All of the AVE volumes of the variables conformed to the above stated rule, indicating that the assessment items in this model had discriminate validity.

3.4 SEM Analysis of the Conceptual Model

The AMOS 8.0 statistical software was used to conduct an SEM analysis to confirm whether the proposed conceptual model (including the structure of factors) was a good fit. To confirm the various hypotheses, we also used another confirmatory factor analysis to test whether the path variables were significant. The validation results and final conceptual model are shown in Table 4.

The analysis of the degree of conformity of the proposed model is shown in Table 5. The test results showed that the chi-square degree of freedom ratio (df) = 2.197; root mean square error of approximation (RMSEA) = 0.047; root mean square residual (RMR) = 0.049; goodness of fit index (GFI) = 0.941, adjusted goodness of fit index (AGFI) = 0.916; normed fit index (NFI) = 0.939; relative fit index (RFI) = 0.920; incremental fit index (IFI) = 0.971; comparative fit index (CFI) = 0.968; and non-normed fit index (NNFI)/Tucker-Lewis index (TLI) = 0.962. The above stated indexes reached the levels required by the theory, indicating that the proposed model is acceptable.

Table 4 Structural equation modeling analysis and hypothesis validation

Hypothesis	Potential fact path	Path coefficient	Results
H1	Perceived risk –> Perceived value	0.320***	Supported
H2	Perceived risk –> Willingness to purchase	0.091	Not supported
H3	Perceived value –> Willingness to purchase	0.471***	Supported
H4	Product knowledge –> Perceived value	0.523***	Supported
H5	Product knowledge –> Willingness to purchase	0.461***	Supported
H6	Perceived benefit –> Perceived value	0.616***	Supported
H7	Perceived benefit –> Willingness to purchase	0.440***	Supported
H8	Product knowledge –> Perceived risk	0.692***	Supported
H9	Product knowledge –> Perceived benefits	0.845***	Supported

***$p < 0.001$

Table 5 Overall conformity testing of the proposed model

Indices	Criteria	Research finding	Results
df	≤3	2.197 (df = 137)	Fits
RMSEA	<0.05	0.047	Fits
RMR	<0.05	0.049	Fits
GFI	>0.9	0.941	Fits
AGFI	>0.9	0.916	Fits
NFI	>0.9	0.939	Fits
RFI	>0.9	0.920	Fits
IFI	>0.9	0.971	Fits
CFI	>0.9	0.968	Fits
NNFI/TLI	>0.9	0.962	Fits

Note The criteria were based on those proposed by Hair et al. [8]

4 Conclusions

The research hypotheses regarding the RFID smart care watch in this study were created from a review of the relevant literature. The findings revealed a relationship between the variables; the second hypothesis was the only one not supported. Details of the findings are elaborated upon below.

Relationship between perceived risk/benefits and willingness to purchase. The empirical results conformed to and supported the hypothesis "Perceived risk has a negative effect on perceived value." In addition, Snoj et al.'s [20] proposal that there is a negative relationship between consumers' perceived risk and value was also supported.

The empirical results did not support the hypothesis "Perceived risk has a negative effect on willingness to buy." We found that consumers did not view technological products as unapproachable, hence the reduced perceived risk attached to these products. As product knowledge increases, consumers gain a better understanding of the perceived risk and can make a meaningful evaluation process before a purchase. This illustrates the direct effect that product knowledge has on perceived risk, as well as the lack of any significant impact of perceived risk on the willingness to purchase.

In addition, the findings supported the hypothesis, "Perceived value has a positive effect on willingness to purchase." The findings also supported the proposals made by Monroe and Krishnan [15] and Grewal et al. [7], who all believed that there was a positive relationship between the perceived value of a product and consumers' willingness to purchase it. Our results are also consistent with those of Zeithaml [25], which indicated that consumers willingness to purchase is affected by the objective price, perceived quality and value, and product attributes. Another result of this study was that willingness to purchase is generated only after consumers have perceived the value of the product.

The RFID smart care watch possesses the distinguishing characteristic of having multiple functions all at the touch of a button. Different cross functions can be executed flexibly without the need to buy multiple devices. The product's functionality greatly enhanced its value and consumer groups tend to hold it in high regard. Our results demonstrate that consumers' perceived value of the product positively affects their willingness to purchase it.

Relationship between product knowledge/perceived value and willingness to purchase. The results supported the two hypotheses "Product knowledge has a positive effect on perceived value" and "Product knowledge has a positive effect on willingness to purchase." Our findings also support Brucks [2], who believed that consumers' subjective knowledge influenced their search process and affected their final purchasing decision; Rao and Monroe [19], who believed that consumers' product knowledge affected their assessment of a product, which in turn affected their purchasing decision and willingness to purchase; and Horn and Salvendy [9], who believed that consumers' knowledge of more details about an innovative product would have a positive effect on their willingness to purchase it.

The RFID smart care watch is innovative and integrates information technology with the Internet. Consumers became interested and searched for information on the product because they wanted to enjoy more convenience and a better quality of life. After gaining knowledge about the product, they would further assess and test it to confirm whether it is able to satisfy their needs. If it did, their willingness to purchase would be generated, followed by the actual purchase.

Relationship between perceived benefits/value and willingness to purchase. The results of the present study supported the two hypotheses, "Perceived benefits have a positive effect on perceived value" and "Perceived benefits have a positive effect on willingness to purchase." This supported the findings of Jen and Hu [11], who found that consumers willingness to purchase was generated only after they considered the perceived benefits and obtainable value.

As the makers of the RFID smart care watch continue to improve the product and introduce more functions and options in accordance with consumers' needs, consumers will also continue to search for related information, thus enhancing the perceived value of the product. This will include actual money spent and non-monetary expenditure (time and energy invested in the search for information and test of the product), which will have an overall positive effect on perceived value.

Relationship between product knowledge/perceived risk and perceived benefits. The findings of the study supported the hypotheses, "Product knowledge has a positive effect on perceived risk" and "Product knowledge has a positive effect on perceived benefits." The proposal by Rao and Monroe [19] is also supported, that is, more product knowledge is associated with a higher likelihood of attending to product information and problems.

Currently, the RFID smart care watch is still in the first stage of the product's life cycle; therefore, most consumers have only recently started using it, and thus they are still becoming accustomed to its features. Hence, there is a great demand for product knowledge and information, both subjective and objective. Although consumers may have a high degree of perceived risk, this can be tempered by product knowledge. In addition, greater consumer knowledge of the RFID smart care watch is related to the assessment of value and benefits in other words, product knowledge has a positive effect on perceived benefits.

References

1. Barlow, J., Bayer S., Curry, R.: Implementing complex innovations in fluid multi-stakeholder environments: experiences of "Telecare." Technovation **26**(3), 396–406 (2006)
2. Brucks, M.: The effects of product class knowledge on information search behavior. J. Consum. Res. **112**(2), 1–16 (1985)
3. Bruwer, J., Fong, M., Saliba, A.: Perceived risk, risk-reduction strategies (RRS) and consumption occasions. Asia Pac. J. Market. Logist. **25**(3), 369–390 (2013)
4. Chaudhuri, A.: Chronic fatigue syndrome and myalgic encephalomyelitis. Lancet **359**(9318), 1698–1699 (2002)
5. Council for Economic Planning and Development http://www.cepd.gov.tw (2012). Accessed 16 Jan 2013
6. Fonell, C., Larcker, D.F.: Evaluating structural equation models with unobservable variables and measurement error. J. Market. Res. **18**(1), 39–50 (1981)
7. Grewal, D., Monroe, K.B., Krishnan, R.: The effect of price-comparison advertising on buyer's perception of acquisition value, transaction value, and behavioral intentions. J. Market. **62**(2), 46–59 (1998)
8. Hair Jr, J.F., Black, W.C., Babin, B.J., Anderson, R.E.: Multivariate Data Analysis, 7th edn. Prentice Hall, Upper Saddle River (2010)
9. Horn, D., Salvendy, G.: Consumer-based assessment of product creativity: a review and reappraisal. Hum. Factors Ergon. Manuf. **16**(2), 155–175 (2006)
10. Hoyer, W.D., Maclnnis, D.J.: Consumer Behavior, 4th edn. South-Western College Publisher, Ohio (2006)
11. Jen, W., Hu, K.C.: Application of perceived value model to identify factors affecting passengers' repurchase intentions on city bus: a case of the Taipei metropolitan area. Transportation **30**(3), 307–327 (2003)
12. Klerck, D., Sweeney, J.C.: The effect of knowledge types on consumer-perceived risk and adoption of genetically modified foods. Psychol. Market. **24**(2), 171–193 (2007)

13. Lim, N.: Consumer' perceived risk: sources versus consequences. Electron. Commer. Res. Appl. **2**(3), 216–228 (2003)
14. Lovelock, C.H.: Service Marketing: People, Technology, Strategy. 4th edn., pp. 54–71. Prentice-Hall International Edition, New York (2001)
15. Monroe, K.B., Krishnan, R.: The effect of price on subjective product evaluation. In: Jacoby, J., Olson, J.C. (eds.) Perceived Quality: How Consumers View Stores and Merchandise, pp. 209–232. Lexington Books, Lexington (1985)
16. Moorman, C., Diehl, K., Brinberg, D., Kidwell, B.: Subjective knowledge, search locations, and consumer choice. J. Consum. Res. **31**, 673–680 (2004)
17. Park, C.W., Lessig, V.P.: Familiarity and its impact on consumer decision biases and heuristics. J. Consum. Res. **8**(2), 223–230 (1981)
18. Park, J., Lennon, S.J., Stoel, L.: On-line product presentation: effect on mood, perceived risks, and purchase intention. Psychol. Market. **22**(9), 695–719 (2005)
19. Rao, A.R., Monroe, K.B.: The moderating effect of prior knowledge on cue utilization in product evaluations. J. Consum. Res. **15**(2), 253–264 (1988)
20. Snoj, B., Korda, A.P., Mumel, D.: The relationships among perceived quality, perceived risk and perceived product value. J. Prod. Brand Manag. **13**(3), 161–163 (2004)
21. Sweeney, C.J., Soutar, G.N.: Consumer perceived value: the development of multiple item scale. J. Retail. **77**(2), 203–221 (2001)
22. Teas, R.K., Agarwal, S.: The effect of extrinsic product cues on consumers perceptions of quality, sacrifice, and value. J. Acad. Market. Sci. **28**(2), 278–290 (2000)
23. Tuu, H.H., Olsen, S.O.: Certainty, risk and knowledge in the satisfaction-purchase intention relationship in a new product experiment. Asia Pac.J. Market. Logist. **24**(1), 78–101 (2012)
24. Wu, M.L.: Structural Equation Modeling: The Operation and Application of AMOS. Wunan Books Inc., Taipei (2009)
25. Zeithaml, V.A.: consumer perceptions of price, quality, and value: a means-end model and synthesis of evidence. J. Market. **52**(3), 2–22 (1988)

Productivity Convergence in Vietnamese Manufacturing Industry: Evidence Using a Spatial Durbin Model

Phuong Anh Nguyen, Tat Hien Phan and Michel Simioni

Abstract This paper applies the β-convergence regression model in order to assess convergence of total factor productivity among Vietnamese provinces for manufacturing industries. Specifically, we express this model in the form of a Spatial Durbin Model (SDM), which allows us to take into account the presence of omitted variables that can be spatially correlated and correlated with the initial level of productivity. We calculate the annual total factor productivity (TFP) of 63 Vietnamese provinces and 6 manufacturing industries, using the results of the structural estimation of a value-added production function from firm data over the period from 2000 to 2012. The regression of growth rates of TFP over this period on the initial levels of productivity using SDM shows that there is convergence in most industries, i.e. the gap between lower-productivity and higher-productivity provinces decreases. These results also show the importance of modeling the indirect effect of the initial level of productivity of a province on its TFP growth rate, through its effect on neighboring provinces. The inclusion of these indirect effects is made possible by SDM and increases the speed of convergence for most considered manufacturing industries, except for metal and machinery, and transportation and telecommunication.

1 Introduction

After more than 25 years of economic development initiated by the Doi Moi reforms, Vietnam has achieved a number of noteworthy achievements (see [7]). The average growth in real GDP over the period from 2000 to 2012 was approximately 6.5 % per

P.A. Nguyen (✉)
Department of Finance and Banking, School of Business, International University - Vietnam National University of Ho Chi Minh City, Ho Chi Minh City, Vietnam
e-mail: npanh@hcmiu.edu.vn

T.H. Phan
Department of Applied Mathematics, Saigon University, Ho Chi Minh City, Vietnam
e-mail: pthien@sgu.edu.vn

M. Simioni
Toulouse School of Economics, INRA-GREMAQ, Toulouse, France
e-mail: michel.simioni@tse-fr.eu

V.-N. Huynh et al. (eds.), *Causal Inference in Econometrics*,
Studies in Computational Intelligence 622, DOI 10.1007/978-3-319-27284-9_39

year. The poverty headcount in Vietnam fell from nearly 60 % in the early 1990 s to 20.7 % in 2010, according to the World Bank. However, although all provinces in Vietnam have enjoyed growth, the benefits have not been equally spread. Some studies deal with the issue of convergence among Vietnamese provinces. Kokko and Thang [13], and, more recently, [18] focus on income disparities among these provinces. They do not find any sign of a convergence pattern between the richer and poorer provinces. Convergence of productivity has been studied in only one paper. Nguyen et al. [20] employ a spatial econometric framework to investigate labor productivity convergence in the industrial manufacturing sector among Vietnamese provinces over the period 1998–2011. In this article, the authors aim to assess how important the impact of factors such that input mobility, trade relationships, and knowledge spillovers between provinces on the speed of convergence of labor productivity among provinces is. They show theoretical evidence that, in a model with provinces defined as open economies, or, equivalently, a model with spillover effects of technology among provinces, the speed of convergence would be higher than in the case of closed economies. But they do not find empirical evidence supporting this theoretical result. Their empirical results show labor productivity convergence among Vietnamese provinces in the manufacturing industry over the period from 2000 to 2011, i.e. less favored provinces in the early 2000 s experienced labor productivity growth rates that allowed them to catch up with the performance of the most favored provinces. But they obtain the unexpected result that the speed of convergence is slower when considering possible spillover effects between these provinces. Our paper aims to study the convergence of productivity among the provinces of Vietnam for various manufacturing industries, assessing the impact of technological spillovers emphasized by Nguyen et al. [20], but using different methodological choices.

The commonly used approach to quantify convergence is based on the concept of β-regression model (see [1] for an extensive meta-analysis of the existing literature). This approach involves the regression of the average growth rate of productivity over a given period, on the value of productivity at the beginning of this period. A statistically significant negative value of the parameter associated with the initial value of productivity, usually denoted by β, is then the sign of a convergence process. Literature distinguishes then between absolute convergence, i.e. the case where no other explanatory variables than the initial value of productivity are taken into account in the regression model, and conditional convergence when such variables are included in the regression model.

The estimator of the β parameter in the case of absolute convergence quantification can suffer from omitted variable bias. It may happen that, in the specification of the β-regression model, we did not take into account other variables that are related to the level of initial productivity and also play a role in catching up between provinces. When these omitted variables are spatially correlated, a more general regression model, the Spatial Durbin Model (SDM), has been recently proposed to be used. This regression model includes a spatial lag of the dependent and the explanatory variable in addition to the initial productivity rate. Lesage and Pace [16] point out that the SDM is the only model in the family of spatial regression models that will produce unbiased estimates when the underlying data generating process involves omitted factors that are spatially correlated and correlated with the regressors. Moreover the SDM includes the Spatial Autoregressive model (SAR) and the Spatial Error Model (SEM), i.e. models involving spatial dependence either in the dependent variable or in the error term, as special cases. These models have been recently used to introduce the spatial dimension in the analysis of regional β-convergence (see [15] among others). Nguyen et al. [20] estimate a SAR after performing the specific-to-general specification search initially proposed by Anselin [3] and based on Lagrange-multipliers tests developed by Anselin et al. [5]. This procedure reflects the state-of-the-art in spatial econometrics up to 2007 [12] and, unfortunately, does not allow for the choice of a general specification such as the SDM.

In this paper, we use SDM as a suitable specification of the β-convergence regression model, to quantify the productivity convergence among Vietnamese provinces for various manufacturing industries. First, this choice allows us to introduce the spatial dimension of our data and to address the issue of spatial dependence in the analysis of productivity convergence between provinces. Second, it allows us take into account the problem of omitted variables in model estimation. Indeed, we only have annual data on total factor productivity (TFP) at the provincial level over the period from 2000 to 2012. We have no other potential regressors than the initial measure of TFP in 2000 in the β-convergence regression model. Last, the SDM gives the direct effect of the initial level of productivity on the productivity growth rate like the usual non spatial β-convergence regression model. In addition, it also measures the indirect effect of the initial level of productivity of a province on the growth rate of productivity, due to its impact on the performance of neighboring provinces. Taking into account these indirect effects will allow us to assess whether speeds of convergence are faster when spatial effects are taken into account in the β-convergence regression model.

TFP data are recovered from Vietnam's yearly enterprise census which is conducted by Vietnam's General Statistical Office (GSO). We have production data at the firm-level. We build six unbalanced panel data sets of firms for the six following manufacturing industries: (1) food processing industry, (2) textile, leather and wood, (3) gas and chemicals, (4) metal and machinery, (5) construction, and (6) transportation and telecommunications. TFP at the firm level is recovered as the by-product of the structural estimation of the parameters of a value-added production function for each manufacturing industry using control function to account for unobserved productivity shocks as proposed by Levinsohn and Petrin [17]. Firm levels of TFP are aggregated annually at each Vietnamese province level by weighting each firm by its share in total employment in the province.

This paper provides evidence of β-convergence in five Vietnamese manufacturing industries over the period from 2000 to 2012, with the noticeable exception of metal and machinery. Taking into account the indirect effects in the SDM increases the estimated values of speeds of convergence in most manufacturing industries, supporting the theoretical economic reasoning set forth by Nguyen et al. [20]. Speeds of convergence computed using SDM estimation results range from 1.34 % for transportation and telecommunication, to 2.08 % for textile, leather and wood, while they range only from 1.21 % for food processing industry to 1.50 % for transportation and telecommunication when using non spatial β-convergence regression model estimation results.

The rest of the paper is organized as follows. Section 2 presents the main features of the modeling strategy adopted in the paper. Section 3 deals with the issue of TFP measurement using firm-level data. Section 4 presents the main results which are obtained from the estimation of SDM for the six manufacturing industries. Section 5 concludes.

2 The Spatial Durbin Model

2.1 Non-spatial and Spatial β-convergence Models

Let TFP_{kt} denote a measure of total factor productivity in Vietnamese province k at time t. This measure will be defined later. Such measures are available on T years, $t = t_0, \ldots, t_f$, for K provinces, $k = 1, \ldots, K$. Let $RATE_k = \log\left(TFP_{kt_f}/TFP_{kt_0}\right)/T$ be the average growth rate of TFP in province k between over the considered period, and $INITIAL_k = \log\left(TFP_{kt_0}\right)$ be the initial level of TFP in province k at time t_0. Convergence of productivity among provinces is usually assessed by estimating the value of parameter β in the regression model (see, among others, [1])

$$RATE_k = \alpha + \beta \; INITIAL_k + u_k, \;\; k = 1, \ldots, K \tag{1}$$

where u_k is the usual error term. Results on convergence are then summarized using

- the speed of convergence, interpreted as the annual rate of convergence, which is measured by $-\log\left(1 + T\widehat{\beta}\right)/T$
- the "half-life", defined as the time necessary for provinces to fill half the gap from their steady state, or $-\log(2)/\log\left(1 + \widehat{\beta}\right)$

where $\widehat{\beta}$ is the estimated value of β (see, among others, [19]).

As emphasized by Lesage and Pace [16], data collected from regions or provinces are often not spatially independent. This spatial dependence requires special attention when estimating convergence because neglecting this structure would result in biased estimates. Among all models that take into account the spatial dependence between observations in estimating Eq. (1), we have chosen to use a spatial Durbin model (SDM). The need for the use of a SDM arises from the likely presence of omitted variables that may affect TFP growth rate in this equation, that may be spatially autocorrelated, and that may be correlated with the initial TFP level. Lesage and Pace [16] show that the treatment of such omitted variables implies the use of a spatial model that includes both a spatial lag of the dependent variable and of the explanatory variable, namely a SDM (see Appendix 1). It can also be shown that the SDM subsumes not only the SEM and SAR, but also the spatial lag of X model (SLX) and the usual non spatial linear regression model [16].

In our case, SDM can be expressed as

$$RATE_k = \alpha + \rho \sum_{j=1}^{K} w_{kj}\ RATE_j + \beta\ INITIAL_k + \gamma \sum_{j=1}^{K} w_{kj}\ INITIAL_j + v_k \quad (2)$$

The spatial lag $\sum_{j=1}^{K} w_{kj}\ RATE_j$ is the weighted average of the TFP growth rates of the neighboring provinces of province k. Similarly, $\sum_{j=1}^{K} w_{kj}\ INITIAL_j$ is the weighted average of the initial TFPs of the neighboring provinces of province k. If two provinces i and j are spatially close, the weight $w_{ij} \neq 0$, otherwise $w_{ij} = 0$. By convention, a province cannot be a neighbor to itself, so $w_{ii} = 0$. Let W denote the resulting spatial weight matrix of dimension $K \times K$. To simplify the interpretation, this weight matrix is usually row-standardized, so that each row sums to one. The spatial lag operator then corresponds to the weighted average of neighboring observations. Finally, v_k is the usual i.i.d. random error term.

The conventional non spatial β-convergence model (1) focuses on initial level of TFP in the province k ($INITIAL_k$) that exerts an important influence on TFP growth rate in this province. In contrast to the non-spatial β-convergence model, the chosen model (2) also focuses on the impact of characteristics of the neighbouring provinces by including spatial lags of the dependent and independent variables ($\sum_{j=1}^{K} w_{kj}\ RATE_j$ and $\sum_{j=1}^{K} w_{kj}\ INITIAL_j$).

2.2 Direct, Indirect and Total Effects

In models containing spatial lags of the explanatory variables or dependent variables, the interpretation of parameters becomes richer but also more complicated. In the case of SDM, the marginal effect of an explanatory variable on the dependent variable can be decomposed into direct and indirect effects. Lesage and Pace [16] demonstrate how these effects can be computed in the case of SDMs. To illustrate this point, consider the spatial Durbin model defined in (2), which can be rewritten in matrix form as

$$RATE = i_K\alpha + \rho\ W\ RATE + \beta\ INITIAL + \gamma\ W\ INITIAL + v \quad (3)$$

where $RATE$ is the $K \times 1$ vector of TFP growth rates, i_K is a $K \times 1$ vector of ones, and thus, $i_K\alpha$ is a $K \times 1$ vector of constant terms, $INITIAL$ the $K \times 1$ vector of initial TFP rates, and v the $K \times 1$ vector or error terms. Solving for $RATE$, we obtain

$$RATE = (I_K - \rho W)^{-1}(\beta\ INITIAL + \gamma\ W\ INITIAL) + (I_K - \rho W)^{-1} v \quad (4)$$

where I_K denotes the identity matrix of dimension $K \times K$. Then,

$$RATE = S(W)\ INITIAL + V(W)v \tag{5}$$

where $S(W) = V(W)\ (\beta I_K + \gamma W)$ and $V(W) = (I_K - \rho W)^{-1}$. For province k, Eq. (5) becomes

$$RATE_k = S_k(W)\ INITIAL + V_k(W)\ v \tag{6}$$

where $S_k(W)$ and $V_k(W)$ denote the kth rows of the $S(W)$ and $V(W)$ matrices, respectively.

Consider now the impact on $RATE_k$ due to a change in initial growth rate of province j, or

$$\frac{\partial RATE_k}{\partial INITIAL_j} = S_{kj}(W) \tag{7}$$

where $S_{kj}(W)$ denotes the jth element of the kth row of matrix $S(W)$. When no spatial effects are modeled like in Eq. (1), $S(W) = \beta I_K$. The marginal effect in Eq. (7) reduces to β if $j = k$, and to 0 if $j \neq k$. Then, a change in the value of initial TFP of a neighboring province j has only an impact on TFP growth rate of province k when $j = k$. SDM allows to take into account indirect effects when $k \neq j$ through the jth element of the kth row of matrix $S(W)$ whose value is potentially non-zero. This implies that a change in the initial TFP for a single province can potentially affect the TFP growth rate in all other provinces, which represents the fundamental idea of spatial econometrics.

The matrix $S(W)$ can be used to calculate the above mentioned direct and indirect effects, as proposed in Lesage and Pace [16]. Since the impact of change in the explanatory variable differs over the observations, they propose the following scalar measures of these varying impacts. The average direct effect is calculated as the average of the diagonal of the matrix $S(W)$, or,

$$\overline{M}_{\text{direct}} = \frac{1}{K}\text{tr}\,(S(W)) \tag{8}$$

The average total impact is calculated as the average of all derivatives of $RATE_k$ with respect to $INITIAL_j$ for any k and j, i.e.

$$\overline{M}_{\text{total}} = \frac{1}{K} i_K' S(W) i_K \tag{9}$$

Finally, the average indirect effect is calculated as the difference between the average total effect and the average direct effect, i.e.

$$\overline{M}_{\text{indirect}} = \overline{M}_{\text{total}} - \overline{M}_{\text{direct}} \tag{10}$$

In such a way, we are able to calculate the direct, the indirect, and the total average impact effects of the initial TFP level. The average direct effect is the one coming from the same province k. The average indirect effect is the one coming from the other provinces $j \neq k$. The average total effect is the sum of the direct and the indirect effects.

3 Data and Total Factor Productivity Estimation

3.1 Data

The data set is extracted from annual business surveys undertaken by Vietnam's General Statistical Office (GSO) from 2000 to 2012. In this data set, all firms belong to the following six manufacturing industries: (1) food processing industry, (2) textile, leather, and wood products, (3) gas and chemicals, (4) metal products and machinery, (5) construction and (6) transportation and telecommunications.[1]

The data collected for each firm include the type of firm, number of workers (the average number over the year), assets, capital allowance, fixed assets, labor's earnings, salaries, bonuses and social security contribution, financial obligations and profits. These data allow us to compute labor input as the number of workers, capital input as fixed assets, and intermediate input as total revenue minus value added. Value added is used as the output in a value-added production function (see below). Inputs and output are corrected for inflation using various price deflators.

3.2 Total Factor Productivity Estimation

For estimating TFP at the firm level in a given manufacturing industry, we consider a value-added production function of the form

$$Y_{it} = \omega_{it} F(K_{it}, L_{it}) \tag{11}$$

where Y_{it}, K_{it}, and L_{it} denote the value-added of firm i at time t, its stock of capital at the same time, and the labor it makes use of at the same time, respectively. $F(.,.)$ is a function summarizing the technology the firm uses. By construction, firm i's TFP level at time t is given by $\omega_{it} > 0$. Following the choice made recently by Nishida et al. [21], we approximate this value added production function by a Cobb-Douglas form, i.e.

$$\log Y_{it} = \beta_K \log K_{it} + \beta_L \log L_{it} + \log \omega_{it} \tag{12}$$

[1] Kokko and Thang [13] use a similar industrial classification when investigating the impact of foreign direct investment on the survival of domestic private firms in Vietnam.

with β_K and β_L denoting the (value added) output elasticity with respect to the primary inputs: capital and labor. We estimate production function parameters in Eq. (12) using the control function method from Wooldridge [24] that modifies [17] to address the simultaneous determination of inputs and productivity.[2] The estimate of firm-level TFP can be obtained as

$$\widehat{\omega}_{it} = \exp\left(\log Y_{it} - \left(\widehat{\beta}_K \log K_{it} + \widehat{\beta}_L \log L_{it}\right)\right) \quad (13)$$

where $\widehat{\beta}_K$ and $\widehat{\beta}_L$ denote the estimated value of elasticities for the considered industry.[3] TFP in the manufacturing industry j in province k for a given year t is computed as

$$\widehat{TFP}_{jkt} = \sum_{\substack{i \in \text{ manufacturing industry } j \\ i \in \text{ province } k \text{ at time } t}} S_{ijkt}\,\widehat{\omega}_{it} \quad (14)$$

where we use the share of firm i in the total labor cost of sector j, in province k, at time t as weight S_{ijkt}.[4]

4 Results

The estimation of the SDM (2) involves the specification of a spatial weight matrix W. Finding the spatial weight matrix that best reflects the spatial dependence is a key element of spatial econometric analysis. There are several methods to create a spatial weight matrix. The spatial weights could be based on geographical, technological, economic, demographical or political distance. Here, we have the geographic coordinates of the centroids of polygons that define the contours of the provinces on a map (63 spatial units). We can therefore calculate distances between centroids, and use them to define the neighbors of each province. We perform a selection procedure of the spatial weight matrix W using this information. For this, we will take a grid of values between 110 and 200 km.[5] Each value defines the radius of a circle around the centroid of a province and another province will be considered to be its neighbor if the distance between their centroids is less than the selected radius, in which case, its weight is equal to one, otherwise, it is zero. Weights are row-standardized as explained before. We estimate the SDM for each spatial weights matrix thus

[2]The estimation procedure is described in Appendix 2.

[3] Estimation of the parameters in value added production function were carried out in *R* using various functions including *ivreg2* which allows the implementation of the instrumental variables regression proposed by Wooldridge [24].

[4]In a recent paper, [23] choose employment shares when aggregating foreign direct investment at the provincial level in Vietnam. We obtain similar results when using the two types of weights.

[5]110 km is the smallest possible value of radius such that each province has at least one neighbor.

Fig. 1 Province's neighbors for radius = 120 km

defined.[6] We obtain a value of the AIC criterium ([11]) for each estimated SDM, i.e. for each radius, and we select the SDM for which the AIC value is the smallest. For instance, Fig. 1 represents the neighborhood of each province when the chosen radius is 120 km.

Table 1 reports the estimated values of the parameters involved in the SDM (2), and their associated standard errors, for each manufacturing industry. The last column of this table provides the selected value of the radius used to define the spatial weights matrix for each SDM. These distances vary from 110 km for transportation and telecommunication to 160 Km for gas and chemicals. The parameters associated with the initial value of the productivity, or INITIAL, are significantly different from zero no matter what the manufacturing industry is, while those associated with its spatial lagged value, or $W\times$ INITIAL, are less often statistically significant: only

[6]SDM estimations were carried out using the *spdep* package of *R*. See [8, 9].

Table 1 SDMs estimates

Sector	Intercept	INITIAL	$W\times$ INITIAL	ρ	Distance (in km)
Food processing industry	0.0361***	−0.0103***	−0.0058*	−0.0123	120
	(0.0061)	(0.0022)	(0.0034)	(0.1645)	
Textile, leather, wood products	0.0193***	−0.0123***	−0.0049	0.0323	120
	(0.0042)	(0.0029)	(0.0053)	(0.1739)	
Gas, chemicals	0.0149***	−0.0129***	−0.0107**	−0.3763	160
	(0.0029)	(0.0010)	(0.0051)	(0.2597)	
Metal, machinery	0.0049*	−0.0115***	0.0095**	−0.0127	120
	(0.0026)	(0.0018)	(0.0039)	(0.1770)	
Construction	0.0246***	−0.0117***	−0.0015	0.0563	120
	(0.0035)	(0.0017)	(0.0021)	(0.1375)	
Transportation, telecommunication	0.0410***	−0.0137***	0.0004	−0.0816	110
	(0.0066)	(0.0014)	(0.0019)	(0.1414)	

***Parameter significantly different from zero at the 1 % significance level
**Parameter significantly different from zero at the 5 % significance level
*Parameter significantly different from zero at the 10 % significance level

for food processing industry, gas and chemicals, and metal and machinery. The signs and the significance of parameters associated with the initial TFP level seem to indicate convergence in all the manufacturing industries. Last, the parameter ρ, which measures the spatial autocorrelation between TFP growth rates, is never significantly different from zero.

But, all these parameters cannot be interpreted in a usual way when assessing the impact of the initial TFP level on TFP growth rate as shown in Sect. 2. This impact can be assessed using the direct, indirect, and total average effects as explained. These average effects can be calculated directly from the estimated values of the various parameters given in Table 1, but the determination of their statistical distributions is very complicated. Instead, we use a Bayesian Markov Chain Monte Carlo (MCMC) method that provides the empirical distributions of these effects using 10,000 draws for each Vietnamese manufacturing industry. We can then calculate 99 or 95 % confidence intervals from these empirical distributions and evaluate the statistical significance of different effects. Lower and upper bounds of these confidence intervals and the mean and median values of the different effects for each manufacturing industry are reported in Table 2. On the one hand, in line with the results about the statistical significance of the parameters associated with the initial TFP variable in Table 1, the mean value of the direct effect is always significantly different from zero since zero never belongs to either the 95 % confidence interval or the 99 % confidence interval, for all manufacturing industries. On the other hand, the means of the indirect effects are not significantly different from zero, with no exceptions. But the sums of these direct and indirect effects are always significantly

Table 2 SDM effects estimates

Sector	Effect	Lower 99 %	Lower 95 %	Mean	Median	Upper 95 %	Upper 99 %
Food processing industry	Direct	−0.0161	−0.0147	−0.0104	−0.0103	−0.0060	−0.0048
	Indirect	−0.0186	−0.0148	−0.0060	−0.0057	0.0014	0.0034
	Total	−0.0308	−0.0261	−0.0164	−0.0159	−0.0088	−0.0066
Textile, leather, wood products	Direct	−0.0198	−0.0182	−0.0123	−0.0123	−0.0066	−0.0049
	Indirect	−0.0238	−0.0186	−0.0058	−0.0056	0.0051	0.0084
	Total	−0.0397	−0.0327	−0.0182	−0.0178	−0.0061	−0.0015
Gas, chemicals	Direct	−0.0154	−0.0147	−0.0126	−0.0127	−0.0105	−0.0093
	Indirect	−0.0119	−0.0098	−0.0035	−0.0045	0.0012	0.0292
	Total	−0.0249	−0.0226	−0.0161	−0.0171	−0.0114	0.0174
Metal, machinery	Direct	−0.0164	−0.0151	−0.0115	−0.0116	−0.0079	−0.0068
	Indirect	−0.0001	0.0021	0.0098	0.0097	0.0178	0.0218
	Total	−0.0144	−0.0110	−0.0017	−0.0019	0.0081	0.0130
Construction	Direct	−0.0163	−0.0152	−0.0118	−0.0118	−0.0084	−0.0074
	Indirect	−0.0133	−0.0095	−0.0025	−0.0023	0.0028	0.0042
	Total	−0.0279	−0.0230	−0.0143	−0.0140	−0.0075	−0.0054
Transportation, telecommunication	Direct	−0.0175	−0.0166	−0.0137	−0.0137	−0.0110	−0.0101
	Indirect	−0.0043	−0.0028	0.0014	0.0015	0.0050	0.0061
	Total	−0.0192	−0.0173	−0.0123	−0.0122	−0.0082	−0.0069

different from zero with the noticeable exception of metal and machinery. Moreover, for most manufacturing industries, the sum of the direct and indirect effects causes a total effect whose value is higher, in absolute value, than that of the direct effect.

This result becomes clearer when speeds of convergence and half-lifes associated with these total effects are computed and compared to those obtained from the usual β-convergence regression model. All these values are reported in Table 3. These results provide evidence of β-convergence in five Vietnamese manufacturing industries over the period from 2000 to 2012, with the noticeable exception of metal and machinery. Taking into account the indirect effects in the SDM increases the estimated values of speeds of convergence in most manufacturing industries, compared to the values estimated using the β-convergence regression model without spatial effects. Speeds of convergence computed using SDM estimation results range from 1.34 % for transportation and telecommunication, to 2.08 % for textile, leather and wood, while they range only from 1.21 % for food processing industry to 1.50 % for transportation and telecommunication when using non-spatial β-convergence regression model estimation results. The indirect effects lead to a reduction of speed of convergence in the case of transport and telecommunications, from 1.50 to 1.34 %. They even cancel any convergence in the case of metal and machinery. Similar but reverse conclusion can be drawn for half-lifes. Half-lifes decrease for almost all the manufacturing industries when spatial effects are taken into account. They range from 37.74 years for textile, leather and wood products to 56.01 years for transportation and telecommunications. More specifically, a speed of convergence of 2.08 % for

Table 3 Sectorial rates of convergence and half-lifes

Sector	SDM Results		OLS results	
	Speed of convergence (%)	Half-life (years)	Speed of convergence (%)	Half-life (years)
Food processing industry	1.84	41.92	1.21	61.54
Textile, leather, wood	2.08	37.74	1.34	56.01
Gas, chemicals	1.81	42.70	1.38	54.23
Metal, machinery	No convergence		1.41	53.38
Construction	1.58	48.12	1.23	60.45
Transportation, telecommunication	1.34	56.01	1.50	50.25

textile, leather, and wood products, means that if the average growth rate recorded over the period from 2000 to 2012 are sustained in the future, the Vietnamese provinces will need approximatively 37.74 years to narrow half of the distance to their common hypothetical steady state. Note that now the half-life values are larger: from 50.25 in the case of transportation and telecommunication, to 61.54 years in the case of food processing industry, when no spatial effects are incorporated in the β-convergence regression model.

These results are comparable to those obtained in other empirical works addressing the issue of productivity convergence among regions within the same economic entity: country, European Union, etc. For instance, [15] obtained a speed of convergence of 1.51 % (resp. 1.76 %) and a half-life of 54 years (resp. 48 years) using β-convergence regression model (resp. spatial version of this model) when studying labor productivity convergence among European regions. A speed of convergence of 1.44 % with an associated half-life of 67 years were obtained by Arbia et al. [6] when dealing with β-convergence of productivity among Italian provinces over the period form 1951 to 2000. Surprisingly, the convergence rates we get are different from those obtained by Nguyen et al. [20] which are much larger (6 % in the case of the spatial autoregressive model, and 6.3 % in the case without spatial lag). The differences may come from the fact that [20] performs their analysis at the overall manufacturing sector level while we perform ours at a more disaggregated level, and also from the fact that they use of a too restrictive SAR specification compared to SDM.

5 Conclusion

In this paper we have investigated the convergence of productivity in six manufacturing sectors among the sixty three Vietnamese provinces over the period from 2000 to 2012. We employ the spatial econometric approach, but, unlike most papers adopting the specific-to-general specification search approach initiated by Anselin [3] (see, for instance, the recent work of Nguyen et al. [20]), we estimate spatial Durbin models. Thus we want to take into account that we cannot observe certain variables that affect productivity, which can be endogenous, and therefore are omitted in the convergence equation. The results give evidence of convergence among the Vietnamese provinces. They show that in general, convergence speeds obtained from the spatial Durbin model estimates are larger than those obtained from simple estimates by least squares. These results are consistent with the expected positive effect of factor mobility, trade relation, and knowledge spillover at the provincial level in Vietnam emphasized by Nguyen et al. [20], but with two noticeable exceptions: metal and machinery, and transportation and telecommunication sectors where further analysis is needed.

Acknowledgments This research was partially funded by Vietnam National Foundation for Science and Technology Development, grant number II 2.2-2012-18.

Appendix 1: Derivation of the Spatial Durbin Model

Consider a set of n spatial units whose neighborhood can be described by a proximity matrix W. The matrix W is an $n \times n$ deterministic, non-negative spatial weight matrix. The elements of W are used to specify the spatial dependence structure among the observations. If observation unit i is related to observation j, then $w_{ij} > 0$. Otherwise, $w_{ij} = 0$, and the diagonal elements w_{ii} are set to zero as a normalization of the model. The matrix is also standardized to have row-sums of unity. We are interested in estimating the relationship between two variables and we assume that their data generating process can be written as

$$
\begin{aligned}
y &= \beta x + \varepsilon \\
\varepsilon &= \rho W \varepsilon + v \\
v &= \delta x + u
\end{aligned}
$$

where y is an $n \times 1$ vector of observations on the continuous dependent variable, x is an $n \times k$ matrix of observations on the explanatory variable, ε is an $n \times 1$ vector of a spatially correlated omitted variable following a spatial autoregressive process with autoregressive coefficient ρ, and v and u are $n \times 1$ vectors of i.i.d. random error terms. The last equation shows that the omitted variable which is correlated with v, is also correlated with the explanatory variables when $\delta \neq 0$. Using this set of equations, [16] show that the resulting model is a spatial Durbin model (see also [10]). Indeed, the first equation can be written as

$$
y = \beta x + (I - \rho W)^{-1} v
$$

by replacing ε with $(I - \rho W)^{-1} v$ from the second equation. After rearranging terms, we get

$$y = \rho W y + \beta x - \rho W \beta x + v$$

which can be expressed as

$$y = \rho W y + \gamma_1 x + \gamma_2 W x + u$$

using the third equation and defining $\gamma_1 = \beta + \delta$ and $\gamma_2 = \rho\beta$. This last expression is known as the spatial Durbin model as Anselin [4] proposed the adaptation of the Durbin model in time series to the spatial context.

Appendix 2: Estimation Procedure of the Value Added Production Function

This appendix draws from lectures given by Jeffrey Wooldridge at the University of Wisconsin, Madison, in 2008.

Consider a value added production function for firm i in period t, i.e.

$$y_{it} = \alpha + \beta l_{it} + \gamma k_{it} + \omega_{it} + \varepsilon_{it}$$

where y_{it} denotes the firm output, here measured as value added, l_{it} is the labor input, and k_{it} the observed state variable, i.e. capital input, all in logarithmic form. The sequence $\{\omega_{it}, t = 1, \ldots, T\}$ is unobserved productivity, and $\{\varepsilon_{it}, t = 1, \ldots, T\}$ is a sequence of shocks. The key difference between ω_{it} and ε_{it} is that the former is a state variable and, hence, impacts on the firm's decisions. It is not observed by the econometrician, and it can impact on the choice of inputs, leading to the well-known simultaneity problem in production function estimation (see [2]).

The demand of intermediate input as a function of the state variable and unobserved level of productivity of a firm,[7] or

$$m_{it} = m(k_{it}, \omega_{it})$$

where m_{it} denotes the intermediate input. For simplicity, [17] assume that the function $m(.,.)$ is time invariant and show that making mild assumptions on the firm's production function, this demand function is monotonically increasing in ω_{it}. This allows inversion of the intermediate input demand function and allows to write unobserved level of productivity of a firm as a function of the capital and intermediate

[7] Olley and Pakes [22] use the function relating investment to productivity and capital, but only for strictly positive investment. Using intermediate input as the proxy variable instead of investment avoids the problem of zero values.

inputs, i.e.

$$\omega_{it} = g(k_{it}, m_{it})$$

Under the assumption

$$\mathrm{E}\,(\varepsilon_{it}|l_{it}, k_{it}, m_{it}) = 0$$

we get the following regression function

$$\begin{aligned}\mathrm{E}\,(y_{it}|l_{it}, k_{it}, m_{it}) &= \alpha + \beta l_{it} + \gamma k_{it} + g(k_{it}, m_{it}) \\ &= \beta l_{it} + h(k_{it}, m_{it})\end{aligned}$$

where $h(k_{it}, m_{it}) = \alpha + \gamma k_{it} + g(k_{it}, m_{it})$. Since $g(.,.)$ is a general function, γ and α are not identified from the previous regression model.

As we are interested not only by the estimation of β but also of γ, we can follow the direct estimation procedure proposed by Wooldridge [24].[8] In order to identify γ along with β, [24] follows [17, 22] by assuming that

$$\mathrm{E}\,(\varepsilon_{it}|l_{it}, k_{it}, m_{it}, l_{it-1}, k_{it-1}, m_{it-1}, \ldots, l_{i1}, k_{i1}, m_{i1}) = 0$$

The dynamic in the productivity process is also restricted such that

$$\begin{aligned}\mathrm{E}\,(\omega_{it}|k_{it}, l_{it-1}, k_{it-1}, m_{it-1}, \ldots, l_{i1}, k_{i1}, m_{i1}) &= \mathrm{E}\,(\omega_{it}|\omega_{it-1}) \\ &= f(\omega_{it-1}) = f\,(g(k_{it-1}, m_{it-1}))\end{aligned}$$

where the latter equivalence holds for some function $f(.)$ because $\omega_{it-1} = g(k_{it-1}, m_{it-1})$. A consequence of this restriction is that the variable labor input l_{it} is allowed to be correlated with the innovation a_{it} in $\omega_{it} = f(\omega_{it-1}) + a_{it}$ while k_{it}, past values of $l(_{it}, k_{it}, m_{it})$, and functions of these are uncorrelated with a_{it}.

Plugging $\omega_{it} = f\,(g(k_{it-1}, m_{it-1})) + a_{it}$ into the expression of the value added production function gives

$$y_{it} = \alpha + \beta l_{it} + \gamma k_{it} + f\,(g(k_{it-1}, m_{it-1})) + u_{it}$$

where $u_{it} = a_{it} + \varepsilon_{it}$. This equation allows to identify β andγ provided we have the orthogonality condition

$$\mathrm{E}\,(u_{it}|k_{it}, l_{it-1}, k_{it-1}, m_{it-1}, \ldots, l_{i1}, k_{i1}, m_{i1}) = 0$$

[8]Olley and Pakes [22] propose a multi step estimation procedure, also used by Levinsohn and Petrin [17]. The direct estimation procedure proposed by Wooldridge [24] simplifies inference, provides more insights about identification and gives more efficient estimates.

Effectively, k_{it}, k_{it-1} and m_{it-1} act as their own instruments and l_{it-1} as an instrument for l_{it}. The functions $g(.,.)$ and $f(.)$ can be approximated by low-order polynomials and then parameters involved in the previous equation can be estimated using instrumental variable technique.

References

1. Abreu, M., de Groot, H.L.F., Florax, R.J.G.M.: A meta-analysis of β-convergence: the legendary 2%. J. Econ. Surv. **19**, 389–420 (2005)
2. Ackerberg, D., Benkard, C.L., Berry, S., Pakes, A.: Econometric tools for analyzing market outcomes. In: Heckman, J., Leamer, E. (eds.) Handbook of Econometrics, Part A, vol. 6. Elsevier, Amsterdam (2007)
3. Anselin, L.: Spatial Econometrics: Methods and Models. Kluwer, Dordrecht (1988)
4. Anselin, L.: Estimation Methods for Spatial Autoregressive Structures. Cornell University. Regional Science Dissertation and Monograph Series, vol. 8. Ithaca Cornell University, New York (1980)
5. Anselin, L., Bera, A.K., Florax, R.J.G.M., Yoon, M.J.: Simple diagnostic tests for spatial dependence. Reg. Sci. Urban Econ. **26**, 77–104 (1996)
6. Arbia, G., Basile, R., Piras, G.: Using spatial panel data in modelling regional growth and convergence. Working Paper No. 55, Istituto di Studi e Analisa Economica, Roma, Italy (2005)
7. Beresford, M.: Doi moi in review: the challenges of building market socialism in Vietnam. J. Contemp. Asia **38**, 221–243 (2008)
8. Bivand, R.S., Piras, G.: Comparing implementations of estimation methods for spatial econometrics. J. Stat. Softw. **63**, 1–36 (2015)
9. Bivand, R.S., Hauke, J., Kossowski, T.: Computing the jacobian in gaussian spatial autoregressive models: an illustrated comparison of available methods. Geogr. Anal. **45**, 150–179 (2013)
10. Brown, J.P., Florax, R.J.G.M., McNamara, K.T.: Determinants of investment flows in U.S. manufacturing. Rev. Reg. Stud. **39**, 269–286 (2009)
11. Burnham, K.P., Anderson, D.R.: Multimodel inference: understanding AIC and BIC in model selection. Sociol. Method Res. **33**, 261–304 (2004)
12. Elhorst, J.P.: Applied spatial econometrics: raising the bar. Spat. Econ. Anal. **5**(9), 28 (2010)
13. Kokko, A., Thang, T.T.: Foreign direct investment and the survival of domestic private firms in Vietnam. Asian. Dev. Rev. **31**, 53–91 (2014)
14. Kokko, A., Tingvall, T.G.: Regional development and government support to SMEs in Vietnam. Country Economic Report 2005:5, Swedish International Development Cooperation Agency, Stockholm, Sweden (2005)
15. Le Gallo, J., Dall'erba, S.: Spatial and sectorial productivity convergence between European regions, 1975–2000. Pap. Reg. Sci. **87**, 505–526 (2008)
16. LeSage, J.P., Pace, R.K.: Introduction to Spatial Econometrics. CRC Press, Boca Raton (2009)
17. Levinsohn, J., Petrin, A.: Estimating production functions using inputs to control for unobservables. Rev. Econ. Stud. **70**(317), 341 (2003)
18. Nguyen, H.H.: Regional welfare disparities and regional economic growth in Vietnam. Ph.D. thesis, Wageningen University, Holland (2009)
19. Nguyen, K.M., Pham, V.K.: Forecasting the convergence state of per capita income in Vietnam. Am. J. Oper. Res. **3**, 487–496 (2013)
20. Nguyen, K.M., Pham, V.K., Nguyen, V.H.: Using the spatial econometric approach to analyze convergence of labor productivity at the provincial level in Vietnam. J. Econ. Dev. **17**, 5–19 (2015)
21. Nishida, M., Petrin, A., Polanec, S.: Exploring reallocation's apparent weak contribution to growth. J. Prod. Anal. **42**, 187–210 (2014)

22. Olley, S., Pakes, A.: The dynamics of productivity in the telecommunication equipment industry. Econometrica **64**, 1263–1298 (1996)
23. Tran, T.T., Pham, T.S.H.: Spatial spillovers of foreign direct investment: the case of Vietnam. DEPOCEN Working Paper Series No. 2013/12, Hanoi, Vietnam (2013)
24. Wooldridge, J.M.: On estimating firm-level production functions using proxy variables to control for unobservables. Econ. Lett. **104**, 112–114 (2009)

Rural Bank Development and Poverty Reduction in Indonesia: Evidence from Panel Co-Integration and Causality Tests

Laksmi Y. Devi

Abstract The purpose of this study is to identify the possible causal links between rural bank development and poverty in provincial area of Indonesia. This study uses panel data for 27 provinces in Indonesia over the period of 2000–2013. The panel co-integration and Granger causality tests are applied to investigate the relationship between rural bank assets and regional poverty. The results of co-integration tests show that there is no long run relationship between rural bank assets and provincial poverty. Moreover, the Granger causality tests show that there is no evidence of causality relationship between provincial poverty and rural bank assets. This means rural banks still have no significant contribution to reducing provincial poverty as intended by the government.

1 Introduction

Many economists believe that well-functioning financial institutions are essential for economic growth. As early as 1912, Schumpeter argues that financial intermediaries are needed for economic development [26]. This issue has also been explored extensively with empirical evidence that financial depth influences growth [13, 23, 26], as well as evidence of reverse causality [2, 15].

Some studies attempt to narrow the scope by identifying the importance of specific financial institutions, such as rural banks or community banks [12, 14, 27, 30]. The idea is that rural or community banks act differently from large commercial banks. They usually have the advantage of access to local information, better relationships with their customers, and a greater commitment to develop the local community. Hence, they are better placed to monitor and assess the risk of local enterprises [30]. These kinds of banks, mostly found in developing countries, are intentionally

L.Y. Devi (✉)
Department of Applied Economics, Vocational School,
Universitas Gadjah Mada, Yogyakarta, Indonesia
e-mail: lydevi@yahoo.com

L.Y. Devi
Auckland University of Technology, Auckland, New Zealand

V.-N. Huynh et al. (eds.), *Causal Inference in Econometrics*,
Studies in Computational Intelligence 622, DOI 10.1007/978-3-319-27284-9_40

designed to provide financing opportunities to small and medium enterprises (SMEs). Rural banks cover an important gap in the market, caused by the reluctance of commercial banks to finance SMEs. This is because the loans are usually small (less than US$1,000) but carry higher administrative costs than large ones [34]. They are also riskier since the borrowers cannot offer much collateral [3].

The financial system in Indonesia is a bank-based system. Indonesia has a well-functioning stock market, but only the largest corporations are listed in the countrys stock exchange. Hence, it can be said that funding for the majority of businesses in the country is sourced primarily from banks and not through stock markets.[1] According to Fry [20], the dominant role of banks in the financial system is a specific characteristic of a developing country. The banking sector of Indonesia accounted for about 75.2 % share of the total financial sector assets in the first semester of 2012 [18].

Indonesian banking institutions are classified into commercial banks and rural banks. The business activities of rural banks are mainly aimed at SMEs and village communities. Because the banks operate at local level,[2] they are considered to have important roles in local economic development. The number of rural banks in Indonesia in 2012 was 1,653 units, more than 10 times the 120 commercial banks [8]. However, in term of asset, rural banks assets are only 2 % of commercial banks. In the end of 2012, the asset of commercial banks was IDR 4,262,587 billion, whereas the asset of rural banks was IDR 67,397 billion [8]. Despite its relatively small size on Indonesian banking sector, the central bank considers the importance of rural banks particularly in supporting the programme of financial inclusion. This role of rural banks had been stressed by the Governor of Bank Indonesia in his 2008 annual speech. He stated that "*the role of rural banks should be enhanced and directed to provide service to the SME and local economy*".[3] Rural banks existence in Indonesia is intended to alleviate financing constraints. Financial development can support economic growth and thereby reduce poverty. Thus, the objective of this study is to analyse whether rural banks contribute significantly in reducing the provincial poverty rate.

2 Brief Literature Review

2.1 Previous Studies on Financial Institution and Poverty

There are two opposite theories on the role of financial institutions to reduce poverty in developing countries [28]. Some believe that more developed financial institutions

---

[1]Indonesias stock market capitalization (% of GDP) in 2012 was 45.2 %. In the same year, the similar percentage capitalization for Singapore, Malaysia, Philippines, and Thailand was 149.8, 156.9, 105.6, and 99.2 % respectively (ADB, 2013).

[2]Any rural banks may establish a branch office only in the same province as its head office (Bank Indonesia Regulation Number 8/26/PBI/2006 concerning Rural Banks.).

[3]Governor of Bank Indonesias speech on Annual Banking Meeting, 2008. http://www.bi.go.id/web/en/Investor+Relations+Unit/Presentation+and+Speeches/Speeches/bankerdinner2008.htm.

will produce credit constraints for the poor for they cannot provide collaterals needed. Thus, only the rich will benefit from the development. The opposite theory states that better functioning financial intermediaries can serve financial services to larger segments of population. More credit means more entrepreneurship, more firm formation, and economic growth. Many economists argue that growth is the best way to reduce poverty significantly [1]. The other way financial development can reduce poverty is by providing the financially disadvantaged families with low cost loans. Families can use the loans to invest in education and health of their children, an investment to get out of poverty [33]. Holden and Prokopenko [22] conclude that financial development can reduce poverty in indirect or direct ways. The indirect way means that financial development has positive impact on economic growth and growth will reduce poverty. The direct way results from the availability of accessible financial instruments, services, and institutions for poor households.

Burgess and Pande [12] find that opening bank branches in rural locations in India reduces rural poverty. Exploring specifically on microfinance, Inoue and Hamori [24] find that financial intermediaries have significant effects in decreasing the poverty ratio in 90 developing countries. A more recent study reveals that financial deepening in India significantly decreased poverty [21]. Most of the variables that are included in the model of relationship between financial institution and economic growth are also used in the financial institution-poverty model. However, the interaction of these variables with growth may differ from those with poverty [25]. Some studies use Gini ratio (as a proxy of income inequality) instead of poverty rate (for example see [11, 16]). Specifically investigating the causal link between poverty and financial development in developing countries, Perez-Moreno [32] finds that financial development, measured by liquid assets of the financial system as a share of GDP or by money and quasi money[4] as a percentage of GDP, leads to the reduction of moderate poverty in the period of 1970s–1980s, but not in the period of 1980s–1990s. In addition, when Perez-Moreno uses ratio of the value of credits granted by financial intermediaries to the private sector to GDP, similar results do not appear. The results of Perez-Morenos study also show evidence of Granger causality from poverty to financial development.

2.2 *Regional Differences in Indonesia*

The numerous islands of Indonesia can be divided into five major regions: Sumatra, Java and Bali, Kalimantan, Sulawesi, and Eastern Provinces (Maluku, Nusa Tenggara, and Papua). A province is the highest tier of local government in Indonesia. Currently, Indonesia consists of 34 provinces. The provinces have a regional autonomy

[4]Quasi money or near money is *a highly liquid financial asset that do not function directly or fully as a medium of exchange but can be readily converted into currency or checkable deposit. It could be in the form of savings deposits, small-denominated (less than US$ 100,000) time deposits, or money market mutual funds held by individuals* [29].

which means they have rights, authorities, and obligations to manage government affairs and public interest in accordance with applicable laws and regulations. The provinces have the authority to manage their regional revenue and expenditure budget.

There are significant differences in economic development and banking activities among provinces in Indonesia. For example, the economy of DKI Jakarta province, the capital city of Indonesia, grew 6.7 % in 2011, while the economy of Papua province had negative growth of −5.6 % in the same year [9]. Of the main regions, the economies of Java and Bali are the most dominant and accounted for about 58.87 % of Indonesian GDP in 2012. Sumatra is a distant second, accounting for 23.77 % of GDP. Kalimantan, Maluku, Nusa Tenggara, Papua, and Sulawesi, despite their rich natural resources, together account for less than Sumatras share of GDP and less than one third that of Bali and Java. However, because its natural resources, the province of Kalimantan has the smallest number of population living in poverty. In March 2012, the poverty percentage in Kalimantan is 6.61 %. Java and Sumatra is in the second and third place with 9.8 % and 11.53 % population living in poverty. The poverty in Sulawesi and Eastern Provinces exceeds the national average of 11.96 %, accounting for 13 %, and 21.51 % respectively [9]. It is widely believed that the higher the distance of a region from Java, the region will be less prosperous.

In every province of Indonesia, we can find different kinds of banks. For example, in the province of Central Java in 2012, there are 67 commercial banks and 286 rural banks [10]. Commercial banks are in the form of state banks, regional development banks, and private banks. In Indonesia, there are 26 regional development banks. There are three provinces with no regional development banks, namely Gorontalo, North Maluku, and West Papua. Some provinces have joint regional banks, which are the regional development banks of Riau and Riau Islands, the regional development banks of South Sumatra and Bangka-Belitung, the regional development bank of Banten and West Java, and the regional development bank of West Sulawesi and South Sulawesi. In addition to having similar activities of commercial banks, these banks have a specific assignment which is to maintain local government funds. Thus, the third party funds mobilized by the banks are mostly local government funds. In this context, these banks are different from rural banks even though both banks are aimed to promote local economic development.

The commercial banks in Indonesia, particularly large and foreign banks, mostly reside in wealthy provinces. The region of Java and Bali has the largest number of rural banks, about 72.93 % of total rural banks in Indonesia [8]. The following Table 1 describes the total number of rural banks (head offices), percentage of rural banks in each province to total rural banks, deposit at rural banks, and percentage of deposits in each province to total rural banks deposit in 2012.

Table 1 Distribution of rural banks in the provinces of Indonesia

Province	Number of rural banks (unit)	Percentage of rural banks in each province to total rural banks (%)	Deposits at rural banks (billion Rupiah)	Percentage of deposits in each province to total rural banks deposit (%)
West Java	307	18.57	7,746	17.26
Banten	69	4.17	816	1.82
DKI Jakarta	26	1.57	1,063	2.37
Central Java	260	15.73	10,585	23.59
East Java	331	20.02	4,893	10.90
Bengkulu	3	0.18	25	0.06
Jambi	16	0.97	482	1.07
Aceh	5	0.30	61	0.14
North Sumatra	53	3.21	612	1.36
West Sumatra	98	5.93	972	2.17
Riau	32	1.94	678	1.51
Riau Islands	40	2.42	2,735	6.10
South Sumatra	19	1.15	625	1.39
Bangka Belitung	1	0.06	48	0.11
Lampung	26	1.57	3,249	7.24
South Kalimantan	24	1.45	222	0.49
West Kalimantan	19	1.15	783	1.74
East Kalimantan	14	0.85	192	0.43
Central Kalimantan	4	0.24	105	0.23
Central Sulawesi	9	0.54	284	0.63
South Sulawesi	23	1.39	476	1.06
North Sulawesi	17	1.03	588	1.31
Southeast Sulawesi	12	0.73	61	0.14
West Sulawesi	1	0.06	3	0.01
Gorontalo	4	0.24	20	0.04
West Nusa Tenggara	29	1.75	521	1.16
Bali	137	8.29	4,054	9.03
East Nusa Tenggara	9	0.54	186	0.41
Maluku	2	0.12	271	0.60
Papua	6	0.36	203	0.45
North Maluku	2	0.12	8	0.02
West Papua	1	0.06	111	0.25

Source [8]

3 Methodology

3.1 Data

This study uses specific data of Indonesian provinces. There are two variables in this study, namely *lasset* and *lpov*. *lasset* is the logarithm of provincial rural bank assets (in billion Indonesian Rupiah). Data of rural bank assets are from the Banking Statistics Indonesia (*Statistik Perbankan Indonesia*/SPI), which is published by the Central Bank of Indonesia (Bank Indonesia). The data are originally given in nominal values, so that they are converted into real values first before employed in the estimations.

lpov is the logarithm of *pov*. *pov* is the provincial rate of poverty (in percentage). In measuring poverty, we use an indicator by Statistics Indonesia (*Badan Pusat Statistik*/BPS). BPS, in measuring absolute poverty, adopts a basic needs approach. An individual is considered to be poor if and only if, his/her income level is below a defined poverty line. Poverty line is basically a minimum standard expenditure required by an individual to fulfill his/her basic necessity for both food and non-food items. Poverty line is an addition of food poverty line (FPL) and non-food poverty line (NFPL). FPL is the expenditure value of food minimum requirements or is equivalent of 2100 kilocalories per capita per day. NFPL is minimum needs for housing, clothing, education, health, and other basic individual needs. The poverty measurement in Indonesia relies on consumption data taken from the National Socio Economic Survey (SUSENAS), which is collected triennially. BPS releases three poverty measurements: (1) Head-Count Index (P_0) which measures the percentage of the population that is counted as poor; (2) Poverty Gap Index (P_1) which measures the extent to which individuals fall below the poverty line (the poverty gaps) as a proportion of the poverty line; and (3) Poverty Severity Index (P_2) which describes inequality among the poor. In this study, we use the Head-Count Index as a poverty measurement, which is in accordance with the United Nations Development Programme Human Development Reports definition of absolute poverty. The data is the percentage average of rural and urban percentage of poor people in the province, taken from BPS.

The data form a panel data. The period under investigation is from 2000 to 2013. The cross-sections are 27 provinces in Indonesia. The 27 provinces are the number of provinces in Indonesia in 2000. Some new provinces emerge during the study time period, making the total number of provinces in Indonesia at the end of 2012 is 34 provinces. Therefore, data from the new provinces are treated as part of the former provinces before they split. Data for Riau combine Riau, Riau Islands, and Bangka Belitung. Data for West Java combine West Java and Banten. Data for North Sulawesi combine North Sulawesi and Gorontalo. Data for South Sulawesi combine South Sulawesi and West Sulawesi. Data for Papua combine Papua and West Papua. The panel are unbalanced panel data because of several non-available data for two provinces (Southeast Sulawesi, and North Maluku) in the early time series of the study.

3.2 *Model*

3.2.1 Unit Root Test

The first stage in testing for co-integration between variables is to determine the degree of integration of individual time series. We test for the existence of a unit root in the level and first difference of each of the variables (*pov* and *asset*) based on the procedure of Levin, Lin, and Chu (LLC); Hadri; Im, Pesaran, and Shin (IPS); Fisher-Augmented Dickey-Fuller (Fisher-ADF); and Fisher-Phillips Perron (Fisher-PP). All the tests following Autoregressive/AR(1) process for panel data:

$$y_{it} = \rho_i y_{it-1} + z_{it}\gamma_i + \varepsilon_{it} \tag{1}$$

where $i = 1, 2, \ldots, N$ represent the number of provinces, that are observed over periods t = 1, 2, …, T; $y_i t$ stand for each variable under consideration in the respective model (that is *pov* and *asset*); $z_i t$ can represent panel-specific means, panel-specific means and a time trend, or nothing, depending on our options. If $z_i t = 1$, the term $z_i' t \gamma_i$ represents fixed effects. If $z_i' t = (1, t)$, $z_i' t \gamma_i$ represents panel-specific means and linear time trends. ρ_i are the autoregressive coefficients; and $\varepsilon_i t$ are error terms. If $|\rho_i| = 1$ then y_i has a unit root, and if $|\rho_i| < 1$, we could say that y_i is stationary.

There are two assumptions about ρ_i which classify panel unit root tests into two categories. First is the assumption that ρ_i is constant across provinces ($\rho_i = \rho$). LLC and Hadri tests are based on the first assumption. LLC test employs a null hypothesis of a unit root, while Hadri test has a null hypothesis of no unit root. LLC test follows a basic ADF specification:

$$\Delta y_{it} = \alpha y_{it-1} + z_{it}'\gamma_i + \sum_{j=1}^{p_i} \theta_{ij} \Delta y_{it-j} + \mu_{it} \tag{2}$$

where the assumption is that $\alpha = \rho - 1$ and the lag order, ρ_i, varies across cross-sections. $\Delta y_t = y_t - y_{t-1}$. $\mu_i t$ are error terms. The null and alternative hypotheses for the tests are: $H_0 : \alpha = 0$ and $H_1 : \alpha < 0$. Under the null hypothesis, there is a unit root, while under the alternative hypothesis, there is no unit root.

The second assumption is that ρ_i varies across provinces, which is underlying IPS, Fisher-ADF, and Fisher-PP tests. To get a panel specific result, the tests combine individual unit root tests. The first step of IPS test is specifying a separate ADF regression for each province:

$$\Delta y_{it} = \alpha y_{it-1} + z_{it}'\gamma_i + \sum_{j=1}^{p_i} \theta_{ij} \Delta y_{it-j} + \mu_{it}$$

with null hypothesis of $H_0 : \alpha_i = 0$ for all i, while the alternative hypothesis is $H_1 : \alpha_i = 0$ for $i = 1, 2, \ldots, N$ and $H_1 : \alpha_i < 0$ for $i = N + 1, N + 2, \ldots, N$. After estimating the separate ADF regressions, the average of t-statistics for α_i from the individual ADF regressions is calculated as follows:

$$\bar{t} = \frac{1}{N} \sum_{i=1}^{N} t_{p_i}$$

where t_{p_i} is the individual t-statistic.

3.2.2 Co-Integration Test

Once the order of stationarity has been defined, the next step is to apply panel co-integration methodology. We use Pedronis co-integration tests as used by Apergis et al. [2], Dinda and Coondoo [17], Bangake and Eggoh [6]. The first step to conduct Pedroni test is to compute the regression residuals from the hypothesized co-integrating regression. Generally, it may take the form:

$$y_{it} = \alpha_i + \delta_{it} + \beta_{1i}x_{1it} + \beta_{2i}x_{2it} + \cdots + \beta_{Mi}x_{Mit} + \varepsilon_{it}$$
$$\text{for } t = 1, \ldots, T; i = 1, \ldots, N; m = 1, \ldots, M \quad (3)$$

where y_{it} and x_{it} stand for each variable under consideration in the respective model (that is *pov* and *asset*); T refers to the number of observations over time; N refers to the number of individual numbers in the panel; M refers to the number of regression variables; $\beta_{1i}, \beta_{2i}, \ldots, \beta_{Mi}$ are the slope coefficients; α_i is the member specific intercept or fixed effects parameter; and δ_{it} is the deterministic time trend which is specific to individual members of the panel. Pedronis approach includes a number of different statistics for the test of the null hypothesis of no co-integration in heterogeneous panels [31]. The first group of test is termed within dimension (panel co-integration statistics) which includes panel-v, panel rho, panel non-parametric, and panel parametric statistics. The last two statistics are similar to the single equation ADF test. The null hypothesis is $H_0 : \beta_i = 1$ or no co-integration and the alternative hypothesis is $H_1 : \beta_i = \beta < 1$. The other group of tests is called between dimensions (group mean panel co-integration statistics) which includes three tests: group-rho, group-pp, and group-adf statistics. The null hypothesis for this group is $H_0 : \beta_i = 1$ or no co-integration and the alternative hypothesis is $H_1 : \beta_i < 1$.

3.2.3 Causality Test

Pedronis co-integration methodology can identify the presence of long-run relationships. However, it cannot provide estimation by error correction model. To estimate a long-run relationship between variables in a panel framework in the presence of co-integration, I propose to use dynamic OLS (DOLS). Considering a panel model with fixed effect:

$$y_{it} = \alpha_i + x'_{it} + \mu_{it}, i = 1, \ldots, N, t = 1, \ldots, T$$

where y_{it} is a matrix (1, 1), β is a vector of slopes $(k, 1)$ dimension, α_i is an individual effect, μ_{it} is an error term. It is assumed that $x_{it}(k, 1)$ vector is an auto regressive process of the first order difference:

$$x_{it} = x_{it-1} + \varepsilon_{it}$$

The DOLS estimator is derived from the following equation:

$$y_{it} = \alpha_i + x'_{it}\beta + \sum_{j=-q_1}^{j=q_2} c_j \Delta x_{it+j} + v_{it}$$

where c_j is the coefficient of a lead or lag of first differenced explanatory variables. The DOLS model used in this study is:

$$lpov_{it} = \alpha_i + \beta_1 lasset_{it} + \sum_{j=-1}^{1} \beta_j \Delta lasset_{i,t+j} + \mu_{it} \tag{4}$$

The final step is exploring the direction of the panel data causal links among the rural banks assets and regional economic growth. The previous procedures described above are only able to indicate whether or not the variables are co-integrated and if a long run relationship exists between them. To identify the direction of causality, we estimate a panel-based VECM and use it to conduct Granger causality tests. The test is based on Engle–Granger procedure that requires two steps. First, we estimate the long run model specified in Eq. (3) to obtain the estimated residuals. Subsequently, we estimate a Granger causality model with a dynamic error correction. The empirical model is represented by two equations VECM as follows:

$$\Delta lpov_{it} = \theta_{1j} + \lambda_{1i} EC_{i,t-1} + \sum_{k=1}^{m} \theta_{11ik} \Delta lpov_{i,t-k} + \sum_{k=1}^{m} \theta_{12ik} \Delta lasset_{i,t-k} + \mu_{1it} \tag{5}$$

$$\Delta lasset_{it} = \theta_{2j} + \lambda_{2i} EC_{i,t-1} + \sum_{k=1}^{m} \theta_{21ik} \Delta lasset_{i,t-k} + \sum_{k=1}^{m} \theta_{22ik} \Delta lpov_{i,t-k} + \mu_{2it} \tag{6}$$

where *lpov* is the change of provincial poverty, *lasset* is total asset of provincial rural banks, EC is error correction term comes from the DOLS estimation and m, the lag length, is chosen optimally for each region using the Schwarz Bayesian Criterion. The sources of causation can be identified by testing for the significance of the coefficients of the dependent variables in Eqs. (5) and (6). For short run causality, the null hypothesis $H_0 : \theta_{12ik} = 0$ for all i and k in Eq. (5), or $H_0 : \theta_{22ik} = 0$ for all i and k in Eq. (6). The presence of long run causality can be established by testing $H_0 : \lambda_{1i} = 0$ for all i in Eq. (5) or $H_0 : \lambda_{2i} = 0$ for all i in Eq. (6).

However, if there is no co-integration between variables, the model specification is as follows:

$$\Delta lpov_{it} = \theta_{1j} + \sum_{k=1}^{m} \theta_{11ik} \Delta lpov_{i,t-k} + \sum_{k=1}^{m} \theta_{12ik} \Delta lasset_{i,t-k} + \mu_{1it} \quad (7)$$

$$\Delta lasset_{it} = \theta_{2j} + \sum_{k=1}^{m} \theta_{21ik} \Delta lasset_{i,t-k} + \sum_{k=1}^{m} \theta_{22ik} \Delta lpov_{i,t-k} + \mu_{2it} \quad (8)$$

Rejection of the hypothesis of $H_0 : \theta_{12ik} = 0$ indicates that rural bank development leads to the change of regional poverty rate. Rejection of the hypothesis of $H_0 : \theta_{22ik} = 0$ indicates that the change of regional poverty rate leads to rural bank development.

4 Result and Analysis

4.1 Unit Root Test

Unit root tests are conducted for the two variables (log of rural bank assets (*lasset*) and log of provincial rate of poverty (*lpov*)). Table 2 presents the results of unit root tests, using the panel unit root tests of LLC, IPS, Fisher-ADF, Fisher-PP, and Hadri. The null hypothesis of each test is that the variable is non-stationary while the alternative is that the variable is stationary, except for Hadri test which the null hypothesis is the variable is stationary. Firstly, the variables are tested in levels and subsequently on the first difference. The lag length is automatically selected by Eviews using Schwarz Info Criterion (SIC).

Table 2 Panel unit root tests

Series	LLC	IPS	ADF	PP	Hadri
lasset	−2.88434 [0.0020]***	0.54036 [0.7055]	44.7052 [0.8123]	67.9188 [0.0965]*	5.25672 [0.0000]
lpov	−0.64040 [0.2610]	1.61156 [0.9465]	41.8219 [0.8865]	40.7773 [0.9080]	8.46054 [0.0000]***
dlasset	−31.3262 [0.0000]***	−13.8999 [0.0000]***	192.037 [0.0000]***	192.717 [0.0000]***	1.61043 [0.0537]*
dlpov	−26.2091 [0.0000]***	−12.8626 [0.0000]***	191.378 [0.0000]***	194.126 [0.0000]***	1.27225 [0.1016]

Notes *** indicates that the results are significant at the 1 % level, ** indicates that the results are significant at the 5 % level, * indicates that the results are significant at the 10 % level. All tests include fixed effects (individual intercept). Values in [ ] are p-values

Table 2 shows that the tests provide fairly mixed results for *lasset*. LLC and PP tests reject the null hypothesis of non-stationarity, while IPS and ADF do not reject the null hypothesis. In the Hadri test, the null hypothesis is rejected which means that the variable is non-stationary. Based on the results, it is reasonable to conclude that *lasset* is non-stationary in levels. For *lpov*, all LLC, IPS, ADF, and PP tests do not reject the null hypothesis of non-stationary. Hadri test also rejects the null hypothesis. Therefore, *lpov* is non-stationary at level. When unit root tests carried out on the first differences of *lasset*, all tests, with the exception of Hadri test, show that the variable is stationary. For the first difference of *lpov*, all tests indicate that the variable is stationary. Thus, it can be concluded that *lasset* and *lpov* are stationary in first differences and are integrated of order one, $I(1)$.

4.2 Panel Co-Integration

Because the variables are $I(1)$, co-integration tests to examine the presence of a long-run stable relationship between rural bank assets and regional poverty rate can be carried out. Table 3 presents the results of Pedroni panel co-integration tests where the null hypothesis of Pedroni co-integration test is that there is no co-integration between *lasset* and *lpov*, while the alternative hypothesis is that *lpov* and *lasset* is co-integrated.

The results of the panel co-integration tests show that the null hypothesis cannot be rejected. Thus, the co-integration tests do not support the existence of a long run relationship between rural bank assets and regional poverty rate. Because the variables are not co-integrated, panel dynamic OLS (DOLS) cannot be proceeded.

Table 3 Results of Pedroni panel co-integration test

Statistics	Value
Within-dimension	
Panel v-statistics	−1.052998[0.8538]
Panel ρ-statistics	0.176551[0.5701]
Panel PP-statistics	−0.797218[0.2127]
Panel ADF-statistics	−1.015408[0.1550]
Between-dimension	
Group ρ-statistics	1.540375[0.9383]
Group PP-statistics	−0.681687[0.2477]
Group ADF-statistics	−2.072503[0.0191]
Decision	Do not reject H_0

4.3 Granger Causality Test

The Granger causality test is conducted using vector autoregression (VAR) model because the variables are not co-integrated. First, the VAR model is estimated using fixed effect and random effect, so that it can be decided which model is the correct one. The VAR model is specified as follows:

$$\Delta lpov_{it} = \theta_{1j} + \sum_{k=1}^{m} \theta_{11ik} \Delta lpov_{i,t-k} + \sum_{k=1}^{m} \theta_{12ik} \Delta lasset_{i,t-k} + \mu_{1it} \quad (9)$$

$$\Delta lasset_{it} = \theta_{2j} + \sum_{k=1}^{m} \theta_{21ik} \Delta lasset_{i,t-k} + \sum_{k=1}^{m} \theta_{22ik} \Delta lpov_{i,t-k} + \mu_{2it} \quad (10)$$

The Hausman test shows that fixed effect is the correct model, both for (9) and (10). Therefore, the VAR estimation is carried out using the fixed effect model. Table 4 presents results from OLS estimation of the VAR(2). In Eq. (9), the second lag of the dependent variable is significant. That is, *dlpov*(−2) significantly affects *dlpov* in Eq. (9). In Eq. (10), past values of *dlasset* do not significantly affect *dlasset*. In the equation with *dlpov* as the dependent variable, rural bank assets do not Granger leads to change of regional poverty rate. In other words, past values of rural bank assets do not have explanatory power for current change of regional poverty rate. This causality also happens in the equation with *dlasset* as the dependent variable. Past values of regional poverty rate also do not have explanatory power for current rural bank assets.

The next step is to test the null hypothesis $H_0 : \theta_{12ik} = 0$ for all i in Eq. (9) or $H_0 : \theta_{22ik} = 0$ for all i in Eq. (10). The hypothesis testing is conducted with the Wald test after obtaining θ_{12ik} and θ_{22ik} from estimating Eqs. (9) and (10). Table 5 shows that, based on the result of Wald test, past values of *dlasset* does not cause *dlpov*. Conversely, past values of *dlpov* also does not cause *dlasset*.

The results of this study imply that rural banks have not fully functioned as expected by the government, which is to support the development of SMEs, particularly in provincial areas. The development of SMEs was believed to reduce provincial poverty rate. This ineffective role of rural banks could be because rural banks hold

Table 4 Panel VAR

Independent variable	Dependent variable	
	dlpov	*dlasset*
dlpov(−1)	−0.032189[0.6591]	−0.159373[0.6531]
dlpov(−2)	−0.185853[0.0108]*	0.113067[0.7483]
dlasset(−1)	0.001032[0.9442]	−0.070280[0.3280]
dlasset(−2)	−0.022242[0.1255]	−0.049213[0.4853]

Notes * indicates that the results are significant at the 10 % level. Values in [ ] are p-values

Table 5 Results of Wald test

Statistics	Value
$H_0 : \theta_{12ik} = 0$	
F-statistics	1.190567[0.3057]
Chi-square	2.381134[0.3040]
$H_0 : \theta_{22ik} = 0$	
F-statistics	0.158190[0.8538]
Chi-square	0.316380[0.8537]

small share of the total loan disbursement in Indonesia, compared to commercial banks. Particularly for SME loans, rural banks disburse only about 5 % of the total loans. Despite its small share, the government still support the existence of rural banks. The regulation on rural banks, enacted by the central bank in 2006, states that rural banks are expected to be able to provide financial services in particular area with financing focus of SMEs and rural communities. The central bank also encourages an individual and/or a legal entity with a vision of local economic development to be the owner of rural banks. Local owners are expected to have better understanding in economic activities of their community so that they can help the community to grow.

In December 2012, the central bank enacted a regulation that obliges commercial banks to distribute credit for the development of SMEs. Gradually to 2018, the banks have to distribute SME loans of at least 20 % of their total loans. This regulation could hinder the development of rural banks because SMEs are their main customers. Recognizing the potential problem, the central bank initiated APEX program in 2011. In the program, a commercial bank and a rural bank will be involved in a financial and technical cooperation. The (large) commercial bank will act as an APEX (a protector) for the (small) rural bank. Both banks could benefit from the cooperation because it will: (1) broaden services to SMEs and support the development of local economic growth; (2) provide security for rural banks because the large commercial banks could act as a lender of the first resort if the rural banks experiencing liquidity shortage; (3) enhance the roles and contributions of commercial banks financing SMEs; and optimise the use of rural bank funds [7]. By implementing the program, there is a hope that both banks could contribute significantly to reducing Indonesian provincial poverty rate.

5 Conclusion

This study intends to explain the link between rural bank assets and poverty rate at the sub-national (provincial) level in Indonesia. The contribution of this research is to take into account the varieties of each province in explaining the link. This study uses panel data for 27 provinces in Indonesia over the period of 2000–2013. The co-integration tests show that there is no long run relationship between rural bank assets and provincial poverty rate. Results of causality tests also show that there is no causality between rural bank assets and provincial poverty rate in the short run. Rural banks exist because the government want the banks to serve SMEs in provincial area. This means, rural banks are still not effectively functioned as intended. Realizing the strength and weakness of rural banks, the central bank initiated a cooperation program between commercial banks and rural banks. The main objective of the program is that both banks could contribute significantly to financing the SMEs.

Acknowledgments The author wants to thank Professor Tim Maloney, Associate Professor Gail Pacheco, and Dr. Saten Kumar at the Auckland University of Technology for their useful comments and feedbacks for this paper.

References

1. Aghion, P., Howitt, P.W.: The Economics of Growth. MIT Press, Cambridge (2009)
2. Apergis, N., Filippidis, I., Economidou, C.: Financial deepening and economic growth linkages: a panel data analysis. Rev. World Econ. **143**(1), 179–198 (2007)
3. Armendariz de Aghion, B., Morduch, J.: The Economics of Microfinance. MIT Press, Cambridge (2005)
4. Asian Development Bank: Countries/Economies Data. Asian Regional Integration Center (2013). http://aric.adb.org
5. Arestis, P., Demetriades, P.: Financial development and economic growth: assessing the evidence. Econ. J. **107**(442), 783–799 (1997)
6. Bangake, C., Eggoh, J.C.: Further evidence on finance-growth causality: a panel data analysis. Econ. Syst. **35**(2), 176–188 (2011)
7. Bank Indonesia: Generic Model APEX BPR. Bank Indonesia, Jakarta (2011)
8. Bank Indonesia: Bank industries operation. Banking Statistics. Bank Indonesia, Jakarta (2012)
9. BPS: Gross regional domestic product at current market prices by provinces. Statistics Indonesia. BPS, Jakarta (2013)
10. BPS Jawa Tengah: Jawa Tengah in Figures. BPS, Semarang (2013)
11. Bittencourt, M.: Financial development and inequality: Brazil 1985–1994. Econ. Change Restruct. **43**, 113–130 (2010)
12. Burgess, R., Pande, R.: Do rural banks matter? Evidence from the Indian social banking experiment. Am. Econ. Rev. **95**(3), 780–795 (2005)
13. Christopoulos, D.K., Tsionas, E.G.: Financial development and economic growth: evidence from panel unit root and cointegration tests. J. Dev. Econ. **73**(1), 55–74 (2004)
14. Collender, R.N., Shaffer, S.: Local bank office ownership, deposit control, market structure, and economic growth. J. Bank. Financ. **27**, 27–57 (2003)
15. Demetriades, P.O., Huseein, K.A.: Does financial development cause economic growth? Time-series evidence from 16 countries. J. Dev. Econ. **51**(2), 387–411 (1996)
16. Deng, H., Su, J.: Influence of financial development on the income distribution in China. Soc. Sci. Lett. **1**(1), 73–79 (2012)
17. Dinda, S., Coondoo, D.: Income and emission: a panel data-based cointegration analysis. Ecol. Econ. **57**(2), 167–181 (2006)
18. Dominasi Bank di Industri Keuangan Tergerus. Indonesia Finance Today. http://www.indonesiafinancetoday.com/read/36345/Dominasi-Bank-di-Industri-Keuangan-Tergerus. 14 Nov 2012
19. Fowowe, B.: The finance-growth nexus in Sub-Saharan Africa: panel cointegration and causality tests. J. Int. Dev. **23**(2), 220–239 (2011)

20. Fry, M.J.: In favour of financial liberalisation. Econ. J. **107**(442), 754–770 (1997)
21. Hamori, S., Inoue, T.: How has financial deepening affected poverty reduction in India?: empirical analysis using state-level panel data. Appl. Financ. Econ. **22**(4/6), 395–408 (2012)
22. Holden, P., Prokopenko, V.: Financial development and poverty alleviation: issues and policy implications for developing and transition countries. IMF Working Paper WP/01/160 (2001)
23. Honohan, P.: Financial development, growth, and poverty: how close are the links? In: Goodhart, C.A.E. (ed.) Financial Development and Economic Growth Explaining the Links. Palgrave MacMillan, New York (2004)
24. Inoue, T., Hamori, S.: Financial permeation as a role of microfinance: has microfinance actually been helpful to the poor? IDE Discussion Paper, 299 (2011)
25. Jalilian, H., Kirkpatrick, C.: Financial development and poverty reduction in developing countries. Int. J. Financ. Econ. **7**(2), 97–108 (2002)
26. King, R.G., Levine, R.: Finance and growth: Schumpeter might be right. Q. J. Econ. **108**(3), 717–737 (1993)
27. Kendall, J.: Local financial development and growth. Policy Research Working Paper, 4838 (2009)
28. Levine, R.: Finance and growth: theory and evidence. NBER Working Paper, 10766 (2004)
29. McConnell, C.R., Brue, S.L., Flynn, S.M.: Macroeconomics: Principles, Problems, and Policies. McGraw-Hill/Irwin, New York (2012)
30. Meslier-Crouzille, C., Nys, E., Sauviat, A.: Contribution of rural banks to regional economic development: evidence from the Philippines. Reg. Stud. **46**(6), 775–791 (2012)
31. Moudatsou, A., Kyrkilis, D.: FDI and economic growth: causality for the EU and ASEAN. J. Econ. Integr. **26**(3), 554–577 (2011)
32. Perez-Moreno, S.: Financial development and poverty in developing countries: a causal analysis. Empir. Econ. **41**(1), 57–80 (2011)
33. Tiwari, A.K., Shahbaz, M., Islam, F.: Does financial development increase rural-urban income inequality?: cointegration analysis in the case of Indian economy. Int. J. Soc. Econ. **40**(2), 151 (2013)
34. Todaro, M.P., Smith, S.C.: Economic Development. Addison-Wesley, Boston (2012)

Author Index

V.-N. Huynh et al. (eds.), *Causal Inference in Econometrics*,
Studies in Computational Intelligence 622, DOI 10.1007/978-3-319-27284-9